# Marine Geochemistry

Springer
Berlin
Heidelberg
New York
Barcelona
Hong Kong
London
Milan
Paris
Singapore
Tokyo

Horst D. Schulz · Matthias Zabel (Eds.)

# Marine Geochemistry

With 241 Figures and 23 Tables

 Springer

EDITORS
Professor Dr. Horst D. Schulz
University of Bremen
Department of Geosciences
28359 Bremen
Germany
E-mail: hdschulz@geochemie.uni-bremen.de

Dr. Matthias Zabel
University of Bremen
Department of Geosciences
28359 Bremen
Germany
E-mail: mzabel@geochemie.uni-bremen.de

ISBN 3-540-66453-X Springer-Verlag Berlin Heidelberg New York

Library of Congress Cataloging-in-Publication Data applied for
Die Deutsche Bibliothek – CIP-Einheitsaufnahme
Schulz, Horst D.: Marine geochemistry: with 23 tables/Horst D. Schulz; Matthias Zabel. – Berlin; Heidelberg; New York; Barcelona; Hong Kong; London; Milan; Paris; Singapore; Tokyo: Springer, 2000
ISBN 3-540-66453-X

Springer-Verlag Berlin Heidelberg New York
a member of BertelsmannSpringer Science+Business Media GmbH

© Springer-Verlag Berlin · Heidelberg 2000
Printed in Germany

Cover-Design: Erich Kirchner, Heidelberg
Typesetting and Layout: Thomas Wilke
Graphic layout: Matthias Zabel

SPIN 10879956      32/3111-5  4  3  2  1 – Printed on acid-free paper

# PREFACE

Today, most branches of science have been extensively described. As to their objectives and interrelationships they are also well distinguished from the adjoining science disciplines. In this regard, marine geochemistry makes an exception, as does geochemistry in general, for - depending on the scientist's educational and professional career - this particular field of research can be understood more or less in terms of geology, chemistry, biology, even mineralogy or oceanography. Despite some occasional objection, we conceive our interdisciplinary approach to marine geochemistry rather as an opportunity - even if our own origins most certainly lie in the geosciences. R. Chester (1990) emphasized the chemistry of the water column and the relations to continental inputs in his book on "Marine Geochemistry". For us, however, the investigation of the marine surface sediments and the (bio)geochemical processes taking place therein will be of major concern. We therefore see our book as a continuation of what R.A. Berner (1980) initiated with his classical work "Early Diagenesis", with which he had a determining influence in pointing the way ahead. The concept and the contents we present here is addressed to graduated students of earth science who specialize in marine geochemistry.

Before the background of a continually expanding field of research, it appears impossible for a textbook on marine geochemistry to cope with the task of achieving completeness. Some parts of marine geochemistry have been described in more detail elsewhere and with an emphasis placed on a different context. These parts were permitted to be treated in brief. The classical subjects of "Marine Geology" are therefore to be found in J.P. Kennet (1982), or with a different perspective in R.N. Andersen (1986). No book on "Aquatic Chemistry" can hardly be better than the one written by W. Stumm and J.J. Morgan (1996). There is furthermore the textbook written by K. Grasshoff, K. Kremling, and M. Ehrhardt (1999) which is concerned with the analytical measurements in seawater. "Tracers in the Sea" by W.S. Broecker and T.-H. Peng (1982) still remains an essential work of standard, albeit a new edition is dearly awaited for. The important field of "Isotope Geochemistry" is exquisitely represented by the books written by G. Faure (1986), Clark and Fritz (1997) and J. Hoefs (1997), as much as "Organic Geochemistry" is represented by the book published by M.H. Engel and S.A. Macko (1993). "Diagenetic Models and Their Implementation" are described from the perspective of a mathematician by B.P. Boudreau (1997), owing to which we were able to confine ourselves to the geochemist's point of view.

Marine geochemistry is generally integrated into the broad conceptual framework of oceanography which encompasses the study of the oceanic currents, their interactions with the atmosphere, weather and climate; it leads from the substances dissolved in water, to the marine flora and fauna, the processes of plate tectonics, the sediments at the bottom of the oceans, and thus to marine geology. Our notion of marine geochemistry is that it is a part of marine geology, wherefore we began our book with a chapter on the solid phase of marine sediments concerning its composition, development and distribution. The first chapter written by Dieter K. Fütterer is therefore a brief summarizing introduction into marine geology which describes all biochemically relevant aspects related to the subject. Monika Breitzke and Ulrich Bleil are concerned in the following chapter with the physical properties of sediments and sedimentary magnetism in a marine-geophysical context, which we deem an important contribution to our understanding of geochemical processes in the sediment.

In the third chapter, Horst D. Schulz demonstrates that the method to quantify biogeochemical processes and material fluxes in recent sediments affords the analysis of the pore-water fraction. Jürgen Rullkötter subsequently gives an overview of organic material contained in sediments which ultimately provides the energy for powering almost all (bio)geochemical reactions in that compartment. The fifth chapter by Bo. B. Jørgensen surveys the world of microorganisms and their actions in marine sediments. The chapters six, seven, and eight are placed in the order of the oxidative agents that are involved in the oxidation of sedimentary organic matter: oxygen and nitrate (Christian Hensen and Matthias Zabel), iron (Ralf R. Haese) and sulfate (Sabine Kasten and Bo. B. Jørgensen). They close the circle of primary reactions that occur in the early diagenesis of oceanic sediments.

In the ninth chapter, marine carbonates are dealt with as a part of the global carbon cycle which essentially contributes to diagenetic processes (Ralf R. Schneider, Horst D. Schulz and Christian Hensen). Ratios of stable isotopes are repeatedly used as proxy-parametes for reconstructing the paleoclimate and paleoceanography. Hence, in the succeeding chapter, Torsten Bickert discusses the stable isotopes in the marine sediment as well as the processes which bear influence on them. Geoffrey P. Glasby has dedicated a considerable part of his scientific work to marine manganese. He visited Bremen several times as a guest scientist and therefore it goes without saying that we seized the opportunity to appoint him to be the author of the chapter dealing with nodules and crusts of manganese. Looking at the benthic fluxes of dissolved and solid/particulate substance across the sediment/water interface, the processes of early diagenesis contribute primarily to the material budgets in the world's oceans and thus to the global material cycles. In chapter 12, Matthias Zabel, Christian Hensen and Michael Schlüter have ventured to make initial methodological observations on global interactions and balances, a subject which is presently in a state of flux. A summarizing view on hot vents and cold seeps is represented by a chapter which is complete in itself . We are indebted to Peter M. Herzig and Mark D. Hannington for having taken responsibility in writing this chapter, without which a textbook on marine geochemistry would always have remained incomplete. In the final chapter, conceptual models and their realization into computer models are discussed. Here, Horst D. Schulz is more concerned with the biogeochemical processes, their proper comprehension, and with aspects of practical application rather than the demonstration of ultimate mathematical elegance.

This book could only be written because many contributors have given their support. First of all, we have to mention the Deutsche Forschungsgemeinschaft in this regard, which has generously funded our research work in the Southern Atlantic for over ten years. This special research project [Sonderforschungsbereich, SFB 261] entitled *The South Atlantic in the Late Quaternary: Reconstruction of Material Budget and Current Systems*", covered the joint activities of the Department of Geosciences at the University Bremen, the Alfred-Wegener-Institute for Polar and Marine Research in Bremerhaven, and the Max-Planck-Institute for Marine Microbiology in Bremen.

The marine-geochemical studies are closely related to the scientific publications submitted by our colleagues from the various fields of biology, marine chemistry, geology, geophysics, mineralogy and paleontology. We would like to thank all our colleagues for the long talks and discussions we had together and for the patient understanding they have showed us as geochemists who were not at all times familiar with the numerous particulars of the neighboring sciences.

We were fortunate to launch numerous expeditions within the course of our studies in which several research vessels were employed, especially the RV METEOR. We owe gratitude to the captains and crews of these ships for their commitment and services even at times when duty at sea was rough. All chapters of this book were subject to an international process of review. Although all colleagues involved have been mentioned elsewhere, we would like to express our gratitude to them once more at this point. Owing to their great commitment they have made influential contributions as to contents and character of this book.

Last but not least, we wish to thank our wives Helga and Christine. Although they never had the opportunity to be on board with us on any of the numerous and long expeditions, they still have always understood our enthusiasm for marine geochemistry and have always given us their full support.

Horst D. Schulz and Matthias Zabel                                              Bremen, May 1999

# References

Anderson, R.N., 1986. Marine geology - A planet Earth perspective. Wiley & Sons, NY, 328 pp.

Berner, R.A., 1980. Early diagenesis: A theoretical approach. Princton Univ. Press, Princton, NY, 241pp.

Boudreau, B.P., 1997. Diagenetic models and their impletation: modelling transport and reactions in aquatic sediments. Springer-Verlag, Berlin, Heidelberg, NY, 414 pp.

Broecker, W.S. and Peng, T.-H., 1982. Tracer in the Sea. Lamont-Doherty Geol. Observation Publ., 690 pp.

Chester, R., 1990. Marine Geochemistry. Chapman & Hall, London, 698 pp.

Clark, I. and Fritz, P., 1997. Environmental isotopes in hydrogeology. Lewis Publ., NY, 328 pp.

Engel, M.H. and Macko, S.A., 1993. Organic Geochemistry. Plenum Press, 861 pp.

Faure, G., 1986. Principles of Isotope Geology. Wiley & Sons, NY, 589 pp.

Grasshoff, K., Kremling K. and Ehrhardt, M., 1999. Methods of Seawater Analysis. Wiley-VCH, Weinheim, NY, 600 pp.

Hoefs, J., 1997. Stable Isotop Geochemistry. Springer, Berlin Heidelberg NY, 201 pp.

Kennett, J.P., 1982. Marine Geology. Prentice Hall, New Jersey, 813 pp.

Stumm, W. and Morgan, J.J., 1996. Aquatic Chemistry. Wiley & Sons, 1022 pp.

# Acknowledgements

This book would not exist without the help of all the colleagues listed below. With great personal commitment they invested their time to thoroughly review the manuscripts. Along with the authors we are deeply indebted to these reviewers, whose many constructive and helpful comments have considerably improved the contents of this book.

| Reviewer | Institution | review Chapter |
| --- | --- | --- |
| David E. Archer | University of Chicago, USA | 9 |
| Wolfgang H. Berger | Scripps Institution of Oceanography, La Jolla, USA | 10 |
| Ray Binns | CSIRO Exploration & Mining, North Ryde, Australia | 13 |
| Walter S. Borowski | Exxon Exploration Company, Housten, USA | 8 |
| Timothy G. Ferdelman | Max-Planck-Institut für Marine Mikrobiologie, Bremen, Germany | 7 |
| Henrik Fossing | National Environmental Research Institute, Silkebord, Denmark | 8 |
| John M. Hayes | Woods Hole Oceanographic Institution, USA | 10 |
| Bo B. Jørgensen | Max-Planck-Institut für Marine Mikrobiologie, Bremen, Germany | 4 |
| David E. Gunn | Southampton Oceanography Centre, United Kingdom | 2.1 |
| Richard A. Jahnke | Skidaway Institute of Oceanography, Savannah, USA | 12 |
| Karin Lochte | Institut für Ostseeforschung Warnemünde, Germany | 5 |
| Philip A. Meyers | University of Michigan, USA | 4 |
| Jack Middelburg | Netherland Inst. of Ecology, Yerske, Netherlands | 6 |
| Nikolai Petersen | Universität München, Germany | 2.2 |
| Christophe Rabouille | Unité Mixte de Recherche CNRS-CEA, Gif-sur-Yvette, France | 14 |
| Jürgen Rullkötter | Universität Oldenburg, Germany | 5 |
| Graham Shimmield | Dunstaffnage Marine Laboratory Oban, United Kingdom | 3 |
| Doris Stüben | Universität Karlsruhe, Germany | 11 |
| Bo Thamdrup | Odense Universitet, Denmark | 7 |
| Cornelis H. van der Weijden | Utrecht University, Netherlands | 3 |
| John K. Volkman | CSIRO, Hobart, Tasmania, Australia | 4 |

Furthermore we would like to acknowledge the assistance of Bernard Oelkers for translating four chapters and for proof-reading most of the other chapters of the book.

We would also like to express our gratitude to numerous unnamed support staff at the University of Bremen.

# Table of Contents

## 1 The Solid Phase of Marine Sediments

Dieter K. Fütterer

## 2 Geophysical Perspectives in Marine Sediments

## 2.1 Physical Properties of Marine Sediments

Monika Breitzke

## 2.2 Sedimentary Magnetism

ULRICH BLEIL

## 3 Quantification of Early Diagenesis: Dissolved Constituents in Marine Pore Water

HORST D. SCHULZ

## 4 Organic Matter: The Driving Force for Early Diagenesis

JÜRGEN RULLKÖTTER

# 5 Bacteria and Marine Biogeochemistry

BO BARKER JØRGENSEN

# 6 Early Diagenesis at the Benthic Boundary Layer: Oxygen and Nitrate in Marine Sediments

CHRISTIAN HENSEN AND MATTHIAS ZABEL

# 7 The Reactivity of Iron

RALF R. HAESE

# 8 Sulfate Reduction in Marine Sediments

Sabine Kasten and Bo Barker Jørgensen

# 9 Marine Carbonates: Their Formation and Destruction

Ralph R. Schneider, Horst D. Schulz and Christian Hensen

# 10 Influences of Geochemical Processes on Stable Isotope Distribution in Marine Sediments

Torsten Bickert

# 11    Manganese: Predominant Role of Nodules and Crusts

GEOFFREY P. GLASBY

# 12    Back to the Ocean Cycles: Benthic Fluxes and Their Distribution Patterns

MATTHIAS ZABEL, CHRISTIAN HENSEN AND MICHAEL SCHLÜTER

# 13   Input from the Deep: Hot Vents and Cold Seeps

PETER M. HERZIG AND MARK D. HANNINGTON

# 14   Conceptual Models and Computer Models

HORST D. SCHULZ

# Authors

Torsten Bickert — Universität Bremen, Fachbereich Geowissenschaften, 28359 Bremen, Germany, bickert@allgeo.uni-bremen.de

Ulrich Bleil — Universität Bremen, Fachbereich Geowissenschaften, 28359 Bremen, Germany, bleil@zfn.uni-bremen.de

Monika Breitzke — Universität Bremen, Fachbereich Geowissenschaften, 28359 Bremen, Germany, a13f@zfn.uni-bremen.de

Dieter K. Fütterer — Alfred-Wegener-Institut für Polar- und Meeresforschung, 27515 Bremerhaven, Germany, dfuetterer@awi-bremerhaven.de

Geoffrey P. Glasby — Geological Survey of Japan, Marine Geology Department, Japan, glasby@gsj.go.jp

Ralf R. Haese — University of Utrecht, Institute of Earth Sciences, 3584 Utrecht, Netherlands, rhaese@earth.ruu.nl

Mark. D. Hannington — Geological Survey of Canada, Ottawa, Canada, markh@gsc.NRCan.gc.ca

Christian Hensen — Universität Bremen, Fachbereich Geowissenschaften, 28359 Bremen, Germany, hensen@uni-bremen.de

Peter M. Herzig — TU Bergakademie Freiberg, Institut für Mineralogie, 09596 Freiberg, Germany, herzig@mineral.tu-freiberg.de

Bo Barker Jørgensen — Max-Planck-Institut für marine Mikrobiologie, 28359 Bremen, Germany, bbj@postgate.mpi-mm.uni-bremen.de

Sabine Kasten — Universität Bremen, Fachbereich Geowissenschaften, 28359 Bremen, Germany, skasten@uni-bremen.de

Jürgen Rullkötter — Universität Oldenburg, Institut für Chemie und Biologie des Meeres, 26111 Oldenburg, Germany, j.rullkoetter@ogc.icbm.uni-oldenburg.de

Michael Schlüter — GEOMAR - Forschungszentrum für marine Geowissenschaften, 24148 Kiel, Germany, mschlueter@geomar.de

Ralph R. Schneider — Universität Bremen, Fachbereich Geowissenschaften, 28359 Bremen, Germany, rschneid@allgeo.uni-bremen.de

Horst D. Schulz — Universität Bremen, Fachbereich Geowissenschaften, 28359 Bremen, Germany, hdschulz@uni-bremen.de

Matthias Zabel — Universität Bremen, Fachbereich Geowissenschaften, 28359 Bremen, Germany, mzabel@uni-bremen.de

# 1 The Solid Phase of Marine Sediments

Dieter K. Fütterer

## 1.1 Introduction

The oceans of the world represent a natural depository for the dissolved and particulate products of continental weathering. After its input, the dissolved material consolidates by means of biological and geochemical processes and is deposited on the ocean floor along with the particulate matter from weathered rock. The ocean floor deposits therefore embody the history of the continents, the oceans and their pertaining water masses. They therefore provide the key for understanding Earth's history, especially valuable for the reconstruction of past environmental conditions of continents and oceans. In particular, the qualitative and quantitative composition of the sedimentary components reflect the conditions of their own formation. This situation may be more or less clear depending on preservation of primary sediment composition, but the processes of early diagenesis do alter the original sediment composition, and hence they alter or even wipe out the primary environmental signal. Hence, only an entire understanding of nature and sequence of processes in the course of sediment formation and its diagenetic alteration will enable us to infer the initial environmental signal from the altered composition of the sediments.

Looking at the sea-floor sediments from a geochemical point of view, the function of particles, or rather the sediment body as a whole, i.e. the solid phase, can be quite differently conceived and will vary with the perspective of the investigator. The "classical" approach – simply applying studies conducted on the continents to the oceans – usually commences with a geological-sedimentological investigation, whereafter the mineral composition is recorded in detail. Both methods lead to a more or less overall geochemical description of the entire system. Another, more modern approach conceives the ocean sediments as part of a global system in which the sediments themselves represent a variable component between original rock source and deposition. In such a rather process-related and globalized concept of the ocean as a system, sediments attain special importance. First, they constitute the environment, a solid framework for the geochemical reactions during early diagenesis that occur in the pore space between the particles in the water-sediment boundary layer. Next to the aqueous phase, however, they are simultaneously starting material and reaction product, and procure, together with the porous interspaces, a more or less passive environment in which reactions take place during sediment formation.

## 1.2 Sources and Components of Marine Sediments

Ocean sediments are heterogeneous with regard to their composition and also display a considerable degree of geographical variation. Due to the origin and formation of the components various sediment types can be distinguished: Lithogenous sediments which are transported and dispersed into the ocean as detrital particles, either as terrigenous particles – which is most frequently the case – or as volcanogenic particles having only local importance; biogenous sediments which are directly produced by organisms or are formed by accumulation of skeletal fragments; hydrogenous or authigenic sediments which precipitate directly out of solution as new formations, or are formed de novo when the particles come into contact with the solution; finally, cosmogenic sediments which are only of secondary importance and will therefore not be considered in the following.

1

### 1.2.1   Lithogenous Sediments

The main sources of lithogenous sediments are ultimately continental rocks which have been broken up, crushed and dissolved by means of physical and chemical weathering, exposure to frost and heat, the effects of water and ice, and biological activity. The nature of the parent rock and the prevalent climatic conditions determine the intensity at which weathering takes place. Information about these processes can be stored within the remnant particulate weathered material, the terrigenous detritus, which is transported by various routes to the oceans, such as rivers, glaciers and icebergs, or wind. Volcanic activity also contributes to lithogenous sediment formation, however, to a lesser extent; volcanism is especially effective on the active boundaries of the lithospheric plates, the mid-ocean spreading ridges and the subduction zones.

The major proportion of weathered material is transported from the continents into the oceans by rivers as dissolved or suspension load, i.e. in the form of solid particulate material. Depending on the intensity of turbulent flow suspension load generally consists of particles smaller than 30 microns, finer grained than coarse silt. As the mineral composition depends on the type of parent rock and the weathering conditions of the catchment area, it will accordingly vary with each river system under study. Furthermore, the mineral composition is strongly determined by the grain-size distribution of the suspension load. This can be seen, for example, very clearly in the suspension load transported by the Amazon River (Fig. 1.1) which silt fraction (>4-63 μm) predominantely consists of quartz and feldspars, whereas mica, kaolinite, and smectite predominate in the clay fraction.

It is not easy to quantify the amount of suspension load and traction load annually discharged by rivers into the oceans on a worldwide scale. In a conservative estimative approach which included 20 of the probably largest rivers, Milliman and Meade (1983) extrapolated this amount to comprise approximately $13 \cdot 10^9$ tons. Recent estimations (Milliman and Syvitski 1992) which included smaller rivers flowing directly into the ocean hold that an annual discharge of approximately $20 \cdot 10^9$ tons might even exist.

Under the certainly not very realistic assumption of an even distribution over a surface area of $362 \ km^2$ which covers the global ocean floor, this amount is equivalent to an accumulation rate of $55.2$ tons $km^{-2}yr^{-1}$, or the deposition of an approximately 35 mm-thick sediment layer every 1000 years.

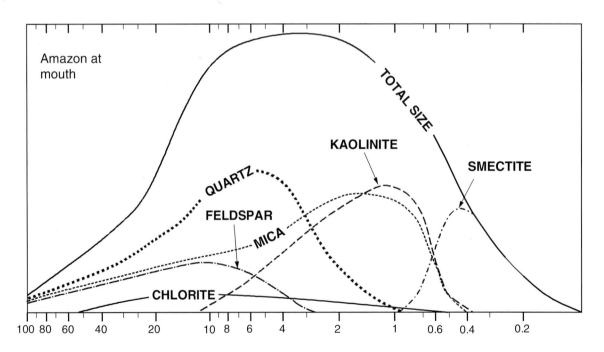

**Fig. 1.1**   Grain-size distribution of mineral phases transported by the Amazon River (after Gibbs 1977).

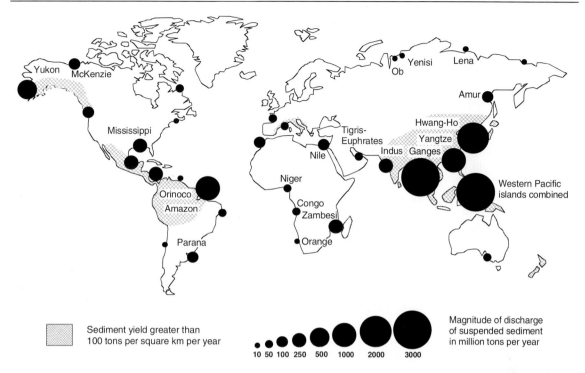

**Fig. 1.2** Magnitude of annual particulate sediment discharge of the world's major rivers. The huge amount of sediment discharge in southeast Asia and the western Pacific islands is due to high relief, catchment, precipitation and human activity (Hillier 1995).

Most of the sediment transported to the coastline by the rivers today is deposited on protected coastal zones, in large estuaries, and on the shelves; only a rather small proportion of the sediment is transported beyond the shelf edge and reaches the bottom of the deep sea. The geographical distribution of the particulate discharge varies greatly worldwide, depending on the geographical distribution of the respective rivers, amount and concentration of the suspended material. According to Milliman and Syvitski (1992), the amount of suspension load is essentially a function of the surface area and the relief of the catchment region, and only secondarily does it depend on the climate and the water mass of the rivers. Apart from these influences, others like human activity, climate, and geological conditions are the essential factors for river systems in southeast Asia.

The south east Asian rivers of China, Bangla Desh, India, and Pakistan that drain the high mountain region of the Himalayan, and the rivers of the western Pacific islands (Fig. 1.2), transport just about one half of the global suspension load discharged to the ocean annually. This must naturally also exert an effect on the sedimentation

rates in the adjoining oceanic region of the Indo-Pacific.

Sediment transport by icebergs which calve from glaciers and inland ice into the ocean at polar and subpolar latitudes is an important process for the discharge and dispersal of weathered coarse grained terrigenous material over vast distances. Due to the prevailing frost weathering in nival climate regions, the sedimentary material which is entrained by and transported by the ice is hardly altered chemically. Owing to the passive transport via glaciers the particles are hardly rounded and hardly sorted in fractions, instead, they comprise the whole spectrum of possible grain sizes, from meter thick boulders down to the clay-size fraction.

As they drift with the oceanic currents, melting icebergs are able to disperse weathered terrigenous material over the oceans. In the southern hemisphere, icebergs drift from Antarctica north to 40°S. In the Arctic, the iceberg-mediated transport is limited to the Atlantic Ocean; here, icebergs drift southwards to 45°N, which is about the latitude of Newfoundland. Coarse components released in the process of disintegration and melting leave behind "ice rafted detritus" (IRD), or

3

"drop stones", which represent characteristic signals in the sediments and are of extremely high importance in paleoclimate reconstructions.

In certain regions, the transport and the distribution carried out by sea ice are important processes. This is especially true for the Arctic Ocean where specific processes in the shallow coastal areas of the Eurasian shelf induce the ice, in the course of its formation, to incorporate sediment material from the ocean floor and the water column. The Transpolar Drift distributes the sediment material across the Arctic Ocean all the way to the North Atlantic. Glacio-marine sedimentation covers one-fifth of present day's ocean floor (Lisitzin 1996).

Terrigenous material can be carried from the continents to the oceans in the form of mineral dust over great distances measuring hundreds to up to thousands of kilometers. This is accomplished by eolian transport. Wind, in contrast to ice and water, only carries particles of finer grain size, such as the silt and clay fraction. A grain size of approximately 80 μm is assumed to mark the highest degree of coarseness transportable by wind. Along wind trajectories, coarser grains such as fine sand and particles which as to their sizes are characteristic of continental loess soil (20-50 μm) usually fall out in the coastal areas,

whereas the finer grains come to settle much farther away. The relevant sources for *eolian dust transport* are the semi-arid and arid regions, like the Sahel zone and the Sahara desert, the Central Asian deserts and the Chinese loess regions (Pye 1987). According to recent estimations (Prospero 1996), a total rate of approximately $1\text{-}2\cdot10^9$ tons $yr^{-1}$ dust is introduced into the atmosphere, of which about $0.91\cdot10^9$ tons $yr^{-1}$ is deposited into the oceans. This amount is, relative to the entire terrigenous amount of weathered material, not very significant; yet, it contributes considerably to sediment formation because the eolian transport of dust concentrates on few specific regions (Fig. 1.3). Dust from the Sahara contributes to sedimentation on the Antilles island Barbados at a rate of 0.6 mm $yr^{-1}$, confirming that it is not at all justified to consider its contribution in building up deep-sea sediments in the tropical and subtropical zones of the North Atlantic as negligible. Similar conditions are to be found in the northwestern Pacific and the Indian Ocean where great amounts of dust are introduced into the ocean from the Central Asian deserts and the Arabic desert.

According to rough estimates made by Lisitzin (1996), about 84% of terrigenous sediment input into the ocean is effected by fluvial transport,

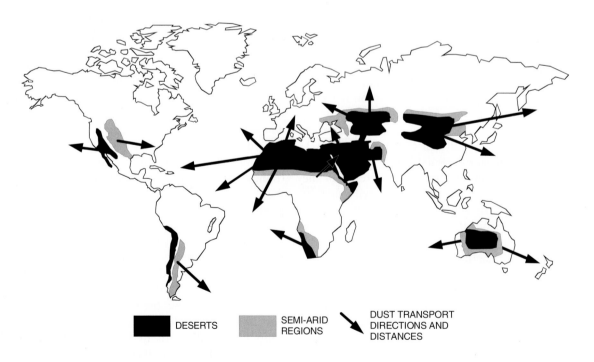

|  | DESERTS |  | SEMI-ARID REGIONS |  | DUST TRANSPORT DIRECTIONS AND DISTANCES |

**Fig. 1.3** The world's major desert areas and semi-arid regions and potential long-distance eolian dust trajectories and oceanic depocenters (Hillier 1995).

somewhat more than 7% by eolian transport, and less than 7% is due to the activity of icebergs.

A distinctly less, but still not insignificant proportion of lithogenous sediment is formed by volcanic activity which is quite often coupled with processes of active subduction at the continental plate boundaries. A large proportion of pyroclastic fragments become wind-dispersed over large areas, whereafter they are usually retraced in oceanic sediments as finely distributed volcanic glass. Yet, the formation of single, distinct, cm-thick tephra layers might also occur in the deep sea where they represent genuine isochronous markers which can be used for correlation purposes and the time calibration of stratigraphic units. Layers of ash deposits in the eastern Mediterranean are prominent examples indicative of the eruption of the volcanic island Ischia in prehistoric times of more than 25,000 years ago, and Santorin about 3500 years ago.

Locally, there may be a frequent occurrence of tephra layers and significant concentrations of finely dispersed volcaniclastic material in deep-sea sediments especially in the proximal zones of volcanic activity, like in marginal zones of the modern Pacific Ocean. In a recent evaluation of tephra input into the Pacific Ocean sediments based on DSDP and ODP data Straub and Schmincke (1998) estimate that the minimum proportion of volcanic tephra corresponds to 23 vol.% of the existing Pacific oceanic sediments.

Lithogenous detrital components of marine sediments, despite all regional variability, include only few basic minerals (Table 1.1). With the exception of quartz, complete weathering, particularly the chemical weathering of metamorphic and igneous rock, leads to the formation of clay minerals. Consequently, this group represents, apart from the remaining quartz, the most important mineral constituent in sediments; clay minerals make up nearly 50% of the entire terrigenous sediment. To a lesser degree, terrigenous detritus contains unweathered minerals, like feldspars. Furthermore, there are mica, non-biogenous calcite, dolomite in low quantities, as well as accessory heavy minerals, for instance, amphibole, pyroxene, apatite, disthene, garnet, rutile, anatase, zirconium, tourmaline, but they altogether seldom comprise more than 1% of the sediment. Basically, each mineral found in continental rock - apart from their usually extreme low concentrations - may also be found in the oceanic sediments. The percentage in which the various minerals are present in a sediment markedly depends on the grain-size distribution.

The clay minerals are of special importance inasmuch as they not only constitute the largest proportion of fine-grained and non-biogenous sediment, but they also have the special geochemical property of absorbing and easily giving off ions, a property which affords more detailed observation. Clay minerals result foremost from the weathering of primary, rock forming aluminous silicates, like feldspar, hornblende and pyroxene, or even volcanic glass. Kaolinite, chlorite, illite, and smectite which represent the four most important groups of clay minerals are formed partly under very different conditions of weathering. Consequently, the analysis of their qualitative and quantitative distribution will enable us to draw essential conclusions on origin and transport, weathering and hydrolysis, and therefore on climate conditions of the rock's source region (Biscay 1965, Chamley 1989). The extremely fine-granular structure of clay minerals, which is likely to produce an active surface of 30 $m^2g^{-1}$ sediment, as well as their ability to absorb ions internally within the crystal structure, or bind them superficially by means of reversible adsorption, as well as their capacity to temporarily bind larger amounts of water, all these properties are fundamental for us to consider clay sediments as a very active and effectively working "geochemical factory".

Clay minerals constitute a large part of the family of phyllosilicates. Their crystal structure is characterized by alternation of flat, parallel sheets, or layers of extreme thinness. For this reason clay minerals are called layer silicates. Two basic types of layers, or sheets make up any given clay mineral. One type of layer consists of tetrahedral sheets in which one silicon atom is surrounded by four oxygen atoms in tetrahedral configuration. The second type of layer is composed of octahedron sheets in which aluminum or magnesium is surrounded by hydroxyl groups and oxygen in a 6-fold coordinated arrangement (Fig. 1.4). Depending on the clay mineral under study, there is still enough space for other cations possessing a larger ionic radius, like potassium, sodium, calcium, or iron to fit in the gaps between the octahedrons and tetrahedrons. Some clay minerals - the so-called expanding or swelling clays - have a special property which allows them to incorporate hydrated cations into their structure. This process is reversible; the water changes the

**Table 1.1**   Mineralogy and relative importance of main lithogeneous sediment components.

| | relative importance | idealized composition |
|---|---|---|
| Quarz | +++ | $SiO_2$ |
| Calcite | + | $CaCO_3$ |
| Dolomite | + | $(Ca,Mg)CO_3$ |
| Feldspars | | |
|   Plagioclase | ++ | $(Na,Ca)[Al(Si,Al)Si_2O_8]$ |
|   Orthoclase | ++ | $K[AlSi_3O_8]$ |
| Muscovite | ++ | $KAl_2[(AlSi_3)O_{10}](OH)_2$ |
| **C l a y   m i n e r a l s** | | |
| Kaolinite | +++ | $Al_2Si_2O_5(OH)_4$ |
| Mica Group | | |
|   e.g. Illite | +++ | $K_{0.8-0.9}(Al,Fe,Mg)_2(Si,Al)_4O_{10}(OH)_2$ |
| Chlorite Groupe | | |
|   e.g. Chlorite s.s. | +++ | $(Mg_{3-y}Al_1Fe_y)Mg_3(Si_{4-x}Al)O_{10}(OH)_8$ |
| Smektite Groupe | | |
|   e.g. Montmorillonite | +++ | $Na_{0.33}(Al_{1.67}Mg_{0.33})Si_4O_{10}(OH)_2 \cdot nH_2O$ |
| **H e a v y   m i n e r a l s, e. g.** | | |
| Amphiboles | | |
|   e.g. Hornblende | + | $Ca_2(Mg,Fe)_4Al[Si_7,Al_{22}](OH)_2$ |
| Pyroxene | | |
|   e.g. Augite | + | $(Ca,Na)(Mg,Fe,Al)[(Si,Al)_2O_6]$ |
| Magnetite | - | $Fe_3O_4$ |
| Ilmenite | - | $FeTiO_3$ |
| Rutile | - | $TiO_2$ |
| Zircon | - | $ZrO_2$ |
| Tourmaline | - | $(Na,Ca)(Mg,Fe,Al,Li)_3Al_6(BO_3)_3Si_6O_{18}(OH)_4$ |
| Garnet | | |
|   e.g. Grossular | - | $Ca_3Al_2(SiO_4)_3$ |

volume of the clay particles significantly as it goes into or out of the clay structure. All in all, hydration can vary the volume of a clay particle by 95%.

Kaolinite is the most important clay mineral of the two-layer group, also referred to as the 7-Ångstrom clay minerals (Fig. 1.4), which consist of interlinked tetrahedron-octahedron units. Illite and the smectites which have the capacity to bind water by swelling belong to the group of three-layered minerals, also referred to as 10-Ångstrom clay minerals. They are made of a combination of two tetrahedral and one octahedral coordinated sheets. Four-layer clay minerals, also known as 14-Ångstrom clay minerals, arise whenever a further autonomous octahedral layer emerges between the three-layered assemblies. This group comprises the chlorites and an array of various composites.

Apart from these types of clay minerals, there is a relatively large number of clay minerals that possess a mixed-layered structure made up of a composite of different basic structures. The result is a sheet by sheet chemical mixture on the scale of the crystallite. The most frequent mixed-layer structure consists of a substitution of illite and smectite layers.

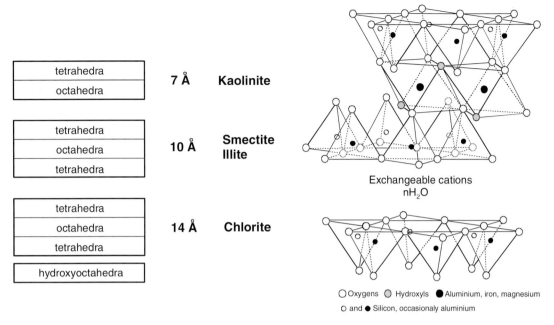

| | | |
|---|---|---|
| tetrahedra | | |
| octahedra | 7 Å | **Kaolinite** |

| | | |
|---|---|---|
| tetrahedra | | |
| octahedra | 10 Å | **Smectite** |
| tetrahedra | | **Illite** |

| | | |
|---|---|---|
| tetrahedra | | |
| octahedra | 14 Å | **Chlorite** |
| tetrahedra | | |
| hydroxyoctahedra | | |

Exchangeable cations
nH$_2$O

○ Oxygens  ◎ Hydroxyls  ● Aluminium, iron, magnesium
○ and ● Silicon, occasionaly aluminium

**Fig. 1.4** Schematic diagram of clay mineral types: Left: According to the combination of tetrahedral- and octahedral- coordi-nated sheets; Right: Diagrammatic sketch of the structure of smectite (after Hillier 1995 and Grim 1968).

Kaolinite is a regularly structured di-octahe-dral two-layer mineral and preferentially devel-ops under warm and humid conditions, by chemi-cal weathering of feldspars in tropical soil. Good drainage is essential to assure the removal of cat-ions released during hydrolysis. Abundance and distribution of kaolinite reflects soil-forming pro-cesses in the area of its origin which is optimal in lateritic weathering in the tropics. Owing to the fact that the occurrence of this mineral in ocean sediments is distinctly latitude-dependent, it is of-ten referred to as the "mineral of low latitudes".

Illite is a three-layered mineral of the mica group and not really a specified mineral, instead the term illite refers to a group of mica-like min-erals in the clay fraction; as such, it belongs to the most frequently encountered type of clay min-erals. Illites are formed as detrital clay minerals by fragmentation in physical weathering. Chemi-cal weathering (soil formation) in which potas-sium is released from muscovite also leads to the formation of illites. For this reason, illite is often referred to as incomplete mica or hydromica. The distribution of illites clearly reflects its terrestrial and detrital origin which is also corroborated by K/Ar-age determinations made on illites obtained from recent sediments. There are as yet no indica-tions as to *in-situ* formations of illites in marine environments.

Similar to the illites are the cation-rich, ex-pandable, three-layered minerals of the smectite group. The smectites, a product of weathering and pedogenic formation in temperate and sub-arid zones, hold an intermediate position with regard to their global distribution. However, smectites are often considered as indicative of volcanic en-vironments, in fact smectite formation due to low-temperature chemical alteration of volcanic rocks is even a quite typical finding. Similarly, finely dispersed particles of volcanic glass may transform into smectite after a sufficiently long exposure to seawater. Smectites may also arise from muscovite after the release of potassium and its substitution by other cations. There is conse-quently no distinct pattern of smectite distribution discernible in the oceans.

The generally higher occurrence of smectite concentrations in the southern hemisphere can be explained with the relatively higher input of vol-canic detritus. This is especially the case in the Southern Pacific.

Chlorite predominately displays a triocta-hedral structure and is composed of a series of three layers resembling mica with an interlayered sheet of brucite (hydroxide interlayer). Chlorite is mainly released from altered magmatic rocks and from metamorphic rocks of the green schist facies as a result of physical weathering. It therefore characteristically depends on the type of parent rock. On account of its iron content, chlorite is prone to chemical weathering. Chlorite distinctly displays a distribution pattern of latitudinal zonation and due to its abundance in polar regions it is considered as the "mineral of high latitudes" (Griffin et al. 1968). The grain size of chlorite minerals is - similar to illite - not limited to the clay fraction (<4 µm), but in addition encompasses the entire silt fraction (4-63 µm) as well.

## 1.2.2 Biogenous Sediments

Biogenous sediments generally refer to bioclastic sediments, hence sediments which are built of remnants and fragments of shells and tests produced by organisms - calcareous, siliceous or phosphatic particles. In a broader sense, biogenous sediments comprise all solid material formed in the biosphere, i.e. all the hard parts inclusive of the organic substance, the caustobioliths. The organic substance will be treated more comprehensively in Chapter 4, therefore they will not be discussed here.

The amount of carbonate which is deposited in the oceans today is almost exclusively of biogenous origin. The long controversy whether chemical precipitation of lime occurs directly in the shallow waters of the tropical seas, such as the banks of the Bahamas and in the Persian Gulf (Fig. 1.5), during the formation of calcareous ooids and oozes of acicular aragonite, has been settled in preference of the concept of biomineralization (Fabricius 1977). The tiny aragonite

**Fig. 1.5** SEM photographs of calcareous sediments composed of aragonite needles which are probably of biogenic origin. *Upper left:* slightly etched section of a calcareous ooid from the Bahamas showing subconcentric laminae of primary ooid coatings; *upper right:* close-up showing ooid laminae formed by small acicular aragonite needles. *Lower left:* silt-sized particle of aragonite mud from the Persian Gulf; *lower right:* close-up showing details of acicular aragonite needles measuring up to 10 µm in length, scalebar 5 µm.

needles (few micrometers) within the more or less concentric layers of calcareous ooids have been considered for a long time as primary precipitates from seawater. However, further investigations including the distribution of stable isotopes distinctly evidenced their biological origin as products of calcification by unicellular algae. Yet, one part of the acicular aragonite ooze might still originate from the mechanical disruption of shells and skeletal elements.

Although marine plants and animals are numerous and diverse, only relatively few groups produce hard parts capable of contributing to the formation of sediments, and only very few groups occur in an abundance relevant for sediment formation (Table 1.2). Relevant for sediment formation are only carbonate minerals in the form of aragonite, Mg-calcite and calcite, as well as bio-

genic opal in the form of amorphous $SiO_2 \cdot nH_2O$. The sulfates of strontium and barium as well as various compounds of iron, manganese, and aluminum are of secondary importance, yet they are of geochemical interest, e.g. as tracers for the reconstruction of past environmental conditions. For example, the phosphatic particles formed by various organisms, such as teeth, bone, and shells of crustaceans, are major components of phosphorite rocks which permit us to draw conclusions about nutrient cycles in the ocean.

Large amounts of carbonate sediment accumulate on the relatively small surface of the shallow shelf seas - as compared to entire oceans surface - of the tropical and subtropical warm water regions, primarily by few lime-secreting benthic macrofossil groups. Scleritic corals, living in symbiosis with algae, and encrusting red algae

**Table 1.2** Major groups of marine organisms contributing to biogenic sediment formation and mineralogy of skeletal hard parts. Foraminifera and diatoms are important groups of both plankton and benthos. x = common, (x) = rare (mainly after Flügel 1978 and Milliman 1974).

| | Aragonite | Aragonite + Calcite | Mg-Calcite | Calcite | Calcite + Mg-Calcite | Opal | divers |
|---|---|---|---|---|---|---|---|
| **Plankton** | | | | | | | |
| Pteropods | x | | | | | | |
| Radiolarians | | | | | | x | celestite |
| Foraminifera | (x) | | x | x | x | | |
| Coccolithophores | | | | x | | | |
| Dinoflagellates | | | x | | | | organic |
| Silicoflagellates | | | | | | x | |
| Diatoms | | | | | | x | |
| **Benthos** | | | | | | | |
| Chlorophyta | x | | | | | | |
| Rhodophyta | x | | x | | (x) | | |
| Phaeophyta | x | | | | | | |
| Sponges | x | | x | | | x | celestite |
| Scleratinian corals | x | | | | | | |
| Octocorals | x | | x | | | | |
| Bryozoens | x | x | | | | | |
| Brachiopods | | | | x | | | phosphate |
| Gastropods | x | x | | | | | |
| Pelecypods | x | x | | x | | | |
| Decapods | | | x | | | | phosphate |
| Ostracods | (x) | | x | | | | |
| Barnacles | (x) | | x | | | | |
| Annelid worms | x | x | x | | | (x) | phosphate |
| Echinoderms | | | x | | | | phosphate |
| Ascidians | x | | | | | | |

constitute the major proportion of the massive structures of coral reefs. Together with calcifying green algae, foraminifera, and mollusks, these organisms participate in a highly productive ecosystem. Here, coarse-grained calcareous sands and gravel are essentially composed of various bioclasts attributable to the reef structure, of lime-secreting algae, mollusks, echinoderms and large foraminifera.

Fine-grained calcareous mud is produced by green algae and benthic foraminifera as well as by the mechanical abrasion of shells of the macrobenthos. Considerable amounts of sediment is formed by bioerosion, through the action of boring, grazing and browsing, and predating organisms. Not all the details have been elucidated as to which measure the chemical and biological decomposition of the organic matter in biogenic hard materials might lead to the formation of primary skeletal chrystalites on the micrometer scale, and consequently contribute to the fine-grained calcareous mud formation.

It is obvious that the various calcareous-shelled groups, especially of those organisms who secrete aragonite and Mg-calcite, contribute significantly to the sediment formation in the shallow seas, whereas greater deposits of biogenic opal are rather absent in the shallow shelf seas. The isostatically over-deepened shelf region of Antarctica, where locally a significant accumulation of siliceous sponge oozes occurs, however, makes a remarkable exception. The relatively low opal concentration in recent shelf deposits does not result from an eventual dilution with terrigenous material. The reason is rather that recent tropical shallow waters have low silicate concentrations, from which it follows that diatoms and sponges are only capable of forming slightly silicified skeletons that quickly remineralize in markedly silicate-deficient waters.

With an increasing distance from the coastal areas, out toward the open ocean, the relevance of planktonic shells and tests in the formation of sediments increases as well (Fig. 1.6). Planktonic lime-secreting algae and silica-secreting algae, coccolithophorids and diatoms that dwell as primary producers in the photic zone which thickness measures approximately 100 m, as well as the calcareous foraminifers and siliceous radiolarians, and silicoflagellates (Table 1.2), are the producers of the by far most widespread and essential deep-sea sediments: the calcareous and siliceous biogenic oozes. Apart from the groups mentioned, planktonic mollusks, the aragonite-shelled pteropods, and some calcareous cysts forming dinoflagellates, also contribute to a considerable degree to sediment formation.

## 1.2.3    Hydrogenous Sediments

Hydrogenous sediments may be widely distributed, but as to their recent quantity they are relatively insignificant. They will be briefly mentioned in this context merely for reasons of being complete. According to Elderfield (1976), hydrogenous sediments can be subdivided into "precipitates", primary inorganic components which have precipitated directly from seawater, like sodium chloride, and "halmyrolysates", secondary components which are the reaction products of sediment particles with seawater, formed subsequent to *in-situ* weathering, but prior to diagenesis. Of these, manganese nodules give an example. In the scope of this book, these components are not conceived as being part of the "primary" solid phase sediment, but as "secondary" authigenic formations which only emerge in the course of diagenesis, as for instance some clay minerals like glauconite, zeolite, hydroxides of iron and manganese etc. In the subsequent Chapters 11 and 13 some aspects of these new formations will be more thoroughly discussed.

The distinction between detrital and newly formed, authigenic clay minerals is basically difficult to make on account of the small grain size and their amalgamation with quite similar detrital material. Yet it has been ascertained that the by far largest clay mineral proportion - probably more than 90% - located in recent to subrecent sediments is of detrital origin. (Chamley 1989, Hillier 1995). There are essentially three ways for smectites to be formed, which demand specific conditions as they are confined to local areas. Alterations produced in volcanic material is one way, especially by means of hydration of basaltic and volcanic glasses. This process is referred to as *palagonitization*. The probably best studied smectite formation consists in the vents of hydrothermal solutions and their admixture with seawater at the mid-oceanic mountain ranges. Should authigenic clay minerals form merely in recent surface sediments in very small amounts, their frequency during diagenesis (burial diagenesis), will demonstrate a distinct elevation. However, this aspect will not be considered any further beyond this point.

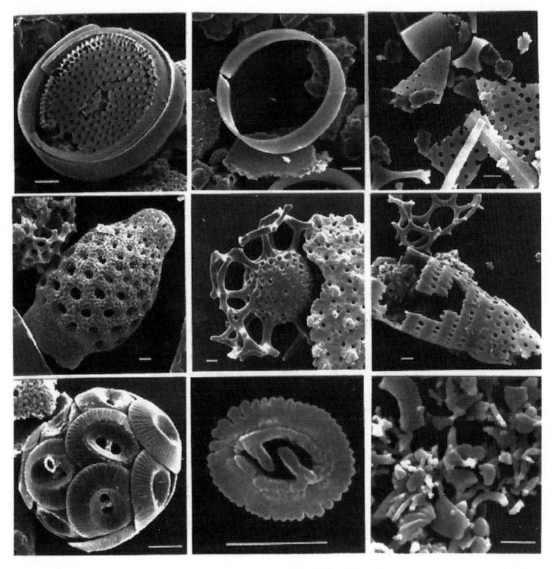

**Fig. 1.6**  SEM photographs of the important sediment-forming planktonic organisms in various stages of decay. *Above*: Siliceous centered diatoms. *Middle*: Siliceous radiolarian. *Below*: Calcareous coccolithophorid of single placoliths = coccoliths and lutitic abrasion of coccoliths. Identical scale of 5 μm.

## 1.3    Classification of Marine Sediments

As yet, there is no general classification scheme applicable to marine sediments that combines all the essential characteristics pertaining to a sediment. A large number of different schemes have been proposed in the literature which focus either on origin, grain-size distribution, chemical and mineralogical features of the sediment components, or the facial development of the sediments – all depending on the specific problem under study. The advantages and disadvantages of the various schemes will not be of any concern here. In the following, an attempt will be made to give a comprehensive concise summary, based on the combination of the various concepts.

Murray and Renard (1891) early introduced a basically simple concept which in its essentials differentiated according to the area of sediment deposition as well as to sediment sources. It distinguished (i) "shallow-water deposits from low-water mark to 100 fathoms", (ii) "terrigenous deposits in deep and shallow water close to land", i.e. the combined terrigenous deposits from the deep sea and the shelf seas, and (iii) "pelagic deposits in deep water removed from land". This scheme is very much appropriate to provide the basic framework for a simple classification scheme on the basis of terrigenous sediments, inclusive of the "shallow-water deposits" in the sense of Murray and Renard (1891), and the deep-sea sediments. The latter usually are subdivided into hemipelagic sediments and pelagic sediments.

### 1.3.1   Terrigenous Sediments

Terrigenous sediments, i.e. clastics consisting of material eroded from the land surface, are not only understood as nearshore shallow-water deposits on the shelf seas, but also comprise the deltaic foreset beds of continental margins, slump deposits at continental slopes produced by gravity transport, and the terrigenous-detrital shelf sediments redistributed into the deep sea by the activity of debris flows and turbidity currents.

The sand, silt, and clay containing shelf sediments primarily consist of terrigenous siliciclastic components transported downstream by rivers; they also contain various amounts of autochthonous biogenic shell material. Depending on the availability of terrigenous discharge, biogenic carbonate sedimentation might predominate in broad shelf regions. A great variety of grain sizes is typical for the sediments on the continental shelf, with very coarse sand or gravel accumulating in high-energy environments and very fine-grained material accumulating in low-energy environments. Coarse material in the terrigenous sediments of the deep sea is restricted to debris flow deposits on and near the continental margins and the proximal depocenters of episodic turbidites.

The development and the hydrodynamic history of terrigenous sediments is described in its essentials by the grain-size distribution and the derived sediment characteristics. Therefore, a classification on the basis of textural features appears suitable to describe the terrigenous sediments. The subdivision of sedimentary particles according the Udden-Wentworth scale encompasses four major categories: gravel (>2 mm), sand (2-0.0625 mm), silt (0.0625-0.0039 mm) and clay (<0.0039 mm), each further divided into a number of subcategories (Table 1.3). Plotting the percentages of these grain sizes in a ternary diagram results in a basically quite simple and clear classification of the terrigenous sediments (Fig.

1.7). Further subdivisions and classifications within the various fields of the ternary diagram are made quite differently and manifold, so that a commonly accepted standard nomenclature has not yet been established. The reason for this lies, to some extent, in the fact that variable amounts of biogenic components may also be present in the sediment, next to the prevalent terrigenous, siliclastic components. Accordingly, sediment classifications are likely to vary and certainly will become confusing as well; this is especially true of mixed sediments. However, a certain degree of standardization has developed owing to the frequent and identical usage of terms descriptive of sediment cores, as employed in the international Ocean Drilling Program (ODP) (Mazullo and Graham 1987).

Although very imprecise, the collective term "mud" is often used in literature to describe the texture of fine-grained, mainly non-biogenic sediments which essentially consist of a mixture of silt and clay. The reason for this is that the differentiation of the grain-size fractions, silt and clay, is not easy to manage and is also very time-consuming with regard to the applied methods. However, it is of high importance for the genetic interpretation of sediments to acquire this information. For example, any sediment mainly consisting of silt can be distinguished, such as a distal turbidite or contourite, from a hemipelagic sediment mainly consisting of clay. The classification of clastic sediments as proposed by Folk (1980) works with this distinction, providing a more precise definition of *mud* as a term (Fig. 1.8).

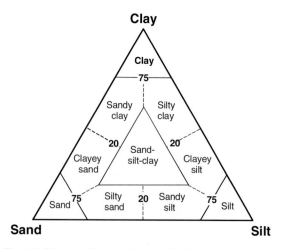

**Fig. 1.7** Ternary diagram of sand-silt-clay grain-size distribution showing principal names for siliciclastic, terrigeneous sediments (from Shepard 1954).

| mm | Udden-Wenthworth phi values ($\phi$) | Terminology |
|---|---|---|
| 1024 | -10 | Boulder |
| 512 | -9 | |
| 256 | -8 | Cobbles |
| 128 | -7 | |
| 64 | -6 | |
| 32 | -5 | Pebble |
| 16 | -4 | |
| 8 | -3 | |
| 4 | -2 | |
| 3,36 | -1,75 | Granule |
| 2,83 | -1,5 | |
| 2,38 | -1,25 | |
| 2,00 | -1,0 | |
| 1,68 | -0,75 | Very coarse sand |
| 1,41 | -0,5 | |
| 1,19 | -0,25 | |
| 1,00 | 0,0 | |
| 0,84 | 0,25 | Coarse sand |
| 0,71 | 0,50 | |
| 0,59 | 0,75 | |
| 0,50 | 1,00 | |
| 0,42 | 1,25 | Medium sand |
| 0,35 | 1,50 | |
| 0,3 | 1,75 | |
| 0,25 | 2,00 | |
| 0,21 | 2,25 | Fine sand |
| 0,177 | 2,50 | |
| 0,149 | 2,75 | |
| 0,125 | 3,00 | |
| 0,105 | 3,25 | Very fine sand |
| 0,088 | 3,50 | |
| 0,074 | 3,75 | |
| 0,0625 | 4,00 | |
| 0,053 | 4,25 | Coarse silt |
| 0,044 | 4,50 | |
| 0,037 | 4,75 | |
| 0,031 | 5,00 | |
| | 5,25 | Medium silt |
| | 5,50 | |
| | 5,75 | |
| 0,0156 | 6,00 | |
| | 6,25 | Fine silt |
| | 6,50 | |
| | 6,75 | |
| 0,0078 | 7,00 | |
| | 7,25 | Very fine silt |
| | 7,50 | |
| | 7,75 | |
| 0,0039 | 8,00 | |
| 0,00200 | 9,0 | **Clay** |
| 0,00098 | 10,0 | |
| 0,00049 | 11,0 | |

**Table 1.3** Grain-size scales and textural classification following the Udden-Wentworth US Standard. The phi values ($\phi$) according to Krumbein (1934, 1964); $\phi = - \log_2 d/d_0$ where $d_0$ is the standard grain diameter (i.e. 1 mm).

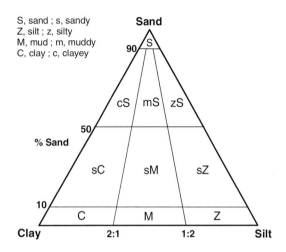

**Fig. 1.8** Textural classification of clastic sediments (modified after Folk 1980).

It needs to be stated in this context, that, in order to compare the quantitative reports of published grain sizes, the international literature does not draw the line between silt and clay at 0.0039 mm - as the U.S standard and the Udden-Wentworth scale does (Table 1.3), or the French AFNOR-norm, but sets the limit at 0.002 mm, a value very often found in German literature and complying with the DIN-standards. To add to further diversity, modern publications from Russia mark the silt-clay transition at a grain size of 0.01 mm (e.g. Lisitzin 1996).

### 1.3.2   Deep-sea Sediments

The sediments in the deep sea consist of only few basic types which in their manifold combinations are suited for the description of a varied facial pattern (Table 1.4). The characteristic pelagic deep-sea sediment far from coastal areas is deep-sea red clay, an extremely fine-grained (median < 1 μm) red-brown clay sediment which covers the oceanic deep-sea basins below the Calcite Compensation Depth (CCD). More than 90% is composed of clay minerals, other hydrogenous minerals, like zeolite, iron-manganese precipitates and volcanic debris. Such sediment composition demonstrates an authigenic origin. The small percentage of lithogenic minerals, such as quartz, feldspar and heavy minerals, confirms the existence of terrigenous components which in part should have originated from eolian transport processes. The biogenic oozes represent the most frequent type of deep-sea sediments; they mainly consist of shells and skeletal material from planktonic organisms living in the ocean where they drizzle from higher photic zones down to the ocean floor, like continuous rainfall, once they have died (Fig. 1.6). The fragments of the calcareous-shelled pteropods, foraminifera, and coccolithophorids constitute the calcareous oozes (pteropod ooze, foraminiferal ooze, or nannofossil ooze), whereas the siliceous radiolarians, silicoflagellates, and diatoms constitute the siliceous oozes (radiolarian ooze or diatomaceous ooze).

The hemipelagic sediments are basically made of the same components as the deep-sea red clay and the biogenic oozes, clay minerals and biogenic particles respectively, but they also contain an additional and sometimes dominating amount of terrigenous material, such as quartz, feldspars, detrital clay minerals, and some reworked biogenous components from the shelves (Table 1.4, Fig. 1.9).

Most deep-sea sediments can be described according to their composition or origin as a three-component system consisting of

(i)     biogenic carbonate,

(ii)    biogenic opal, and

(iii)   non-biogenic mineral constituents.

The latter group comprises the components of the deep-sea red clay and the terrigenous siliciclastics. On the basis of experiences made in the Deep Sea Drilling Project, Dean et al. (1985)

**Table 1.4**   Classification of deep-sea sediments according to Berger (1974).

**I.**  (Eu-)pelagic deposits (oozes and clays)
<25 % of fraction >5µm is of terrigenic, volcanogenic, and/or neritic origin
Median grain size <5µm (except in authigenic minerals and pelagic organisms)
A. Pelagic clays. $CaCO_3$ and siliceous fossils <30 %

    1. $CaCO_3$ 1-10 %. (Slightly) calcareous clay
    2. $CaCO_3$ 10-30 %. Very calcareous (or marl) clay
    3. Siliceous fossils 1-10 %. (Slightly) siliceous clay
    4. Siliceous fossils 10-30 %. Very siliceous clay

B. Oozes. $CaCO_3$ or siliceous fossils >30 %

    1. $CaCO_3$ >30 %. $<2/3$ $CaCO_3$: marl ooze. $>2/3$ $CaCO_3$: chalk ooze
    2. $CaCO_3$ <30 %. >30 % siliceous fossils: diatom or radiolarian ooze

**II.**  Hemipelagic deposits (muds)
>25 % of fraction >5 µm is of terrigenic, volcanogenic, and/or neritic origin
Median grain size >5µm (except in authigenic minerals and pelagic organisms)
A. Calcareous muds. $CaCO_3$ >30 %

    1. $<2/3$ $CaCO_3$: marl mud. $>2/3$ $CaCO_3$: chalk mud
    2. Skeletal $CaCO_3$ >30 %: foram ~, nanno ~, coquina ~

B. Terrigenous muds, $CaCO_3$ <30 %. Quarz, feldspar, mica dominant
   Prefixes: quartzose, arkosic, micaceous
C. Volcanogenic muds. $CaCO_3$ <30 %. Ash, palagonite, etc., dominant

**III.**  Special pelagic and/or hemipelagic deposits
    1. Carbonate-sapropelite cycles (Cretaceous).
    2. Black (carbonaceous) clay and mud: sapropelites (e.g., Black Sea)
    3. Silicified claystones and mudstones: chert (pre-Neogene)
    4. Limestone (pre-Neogene)

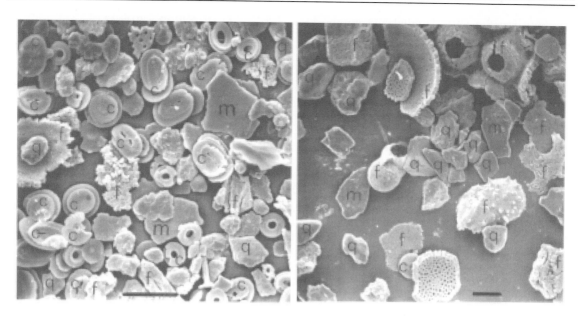

**Fig. 1.9** Hemipelagic sediment from Sierra Leone Rise, tropical North Atlantic. *Left*: Fine silt-size fraction composed of coccoliths (c), foraminiferal fragments (f) and of detrital quartz (q) and mica (m). *Right*: Coarse silt-sized fraction predominately composed of foraminiferal fragments (f) and some detrital quartz (q), scalebar 10 µm.

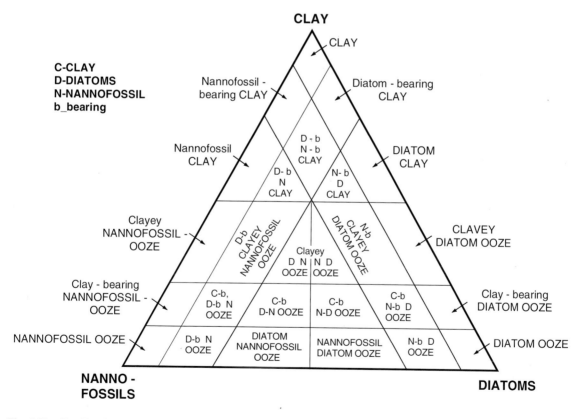

**Fig. 1.10** Classification of deep-sea sediments according to the main constituents, e.g. clay (non-biogenic), diatoms (siliceous biogenic), and nannofossils (calcareous biogenic). (modified from Dean et al. 1985).

have developed a very detailed and purely descriptive classification scheme (Fig. 1.10):

- The most frequently occurring component with a percentage higher than 50% determines the designation of the sediment. Non-biogenic materials are specified accordingly on the basis of the grain-size fractions: sand, silt or clay. Biogenic material is referred to as "ooze" and preceded by the most abundant biogenic component: nannofossil ooze, foraminifera ooze, diatom ooze, and radiolarian ooze respectively.

- Each component measuring between 25-50% is characterized by the following attributes: sandy, silty, clayey, or nannofossil, foraminiferal, diatomaceous, or radiolarian.

- Components with percentages between 10-25% are referred to by adding the suffix "-bearing", as in "clay-bearing", "diatom-bearing".

- Components with percentages below 10% are not expressed at all, but may be included by addition of the suffix "rich", as in "$C_{org}$-rich".

The thus established, four-divided nomenclature of deep-sea sediments with threshold limits of 10%, 25% and 50% easily permits a quite detailed categorization of the sediment which is adaptable to generally rare, but locally frequent occurrences of components, e.g. zeolite, eventually important for a more complete description.

## 1.4    Global Patterns of Sediment Distribution

The overall distribution pattern of sediment types in the world's oceans depends on few elementary factors. The most important factor is the relative amount with which one particle species contributes to sediment formation. Particle preservation and eventual dilution with other sediment components will modify the basic pattern. The formation and dispersal of terrigenous constituents derived from weathering processes on the continents, as well as autochthonous oceanic-biogenic constituents, both strongly depend on the prevalent climate conditions, so that, in the oceans, a latitude-dependent and climate-related global pattern of sediment distribution will be the ultimate result.

### 1.4.1    Distribution Patterns of Shelf Sediments

The particulate terrigenous weathering products mainly transported from the continents by rivers are not homogeneously distributed over the ocean floor, but concentrate preferentially along the continental margins, captured either on the shelf or the continental slope. Massive sediment layers are built where continental inputs are particularly high, and preferentially during glacial periods when sea-levels were low (Fig. 1.11).

Approximately 70% of the continental shelf surface is covered with relict sediment, i.e. sediment deposited during the last glacial period under conditions different from today's, especially at times when the sea-level was comparatively low (Emery 1968). It has to be assumed that there is a kind of textural equilibrium between these relict sediments and recent conditions. The fine-grained constituents of shelf sediments were eluted during the rise of the sea-level in the Holocene and thereafter deposited, over the edge of the shelf onto the upper part of the continental slope, so that extended modern shelf surface areas became covered with sandy relict sediment (Milliman et al. 1972, Milliman and Summerhays 1975).

According to Emery (1968), the sediment distribution on recent shelves displays a plain and distinctly zonal pattern (Fig. 1.12):

*Biogenic sediments* with coarse-grained calcareous sediments predominate at lower latitudes,

*Detrital sediments* with riverine terrigenous siliciclastic material at moderate latitudes, and

*Glacial sediments* of terrigenous origin transported by ice are limited to high latitudes.

In detail, this well pronounced pattern may become strongly modified by local superimpositions. Coarse-grained biogenic carbonate sediments will be found at moderate and high latitudes as well, at places where the riverine terrigenous inputs are very low (Nelson 1988).

Today, as a result of the post-glacial high sea-levels, most river-transported fine-grained material is deposited in the estuaries and on the flat inner shelves in the immediate proximity of river mouths. Only a small proportion is transported over the edge of the shelf onto the continental slope. These processes account for the development of mud belts on the shelf, of which 5 types

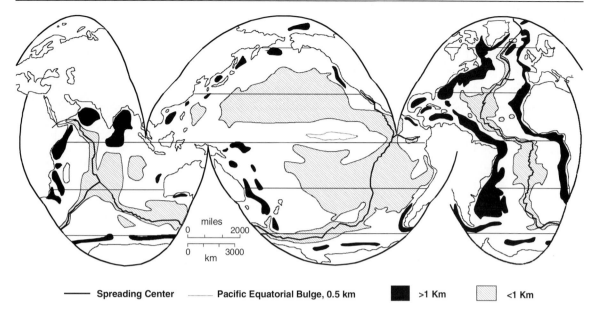

**Fig. 1.11**   Sediment thickness to acoustic basement in the world ocean (from Berger 1974).

of shelf mud accumulation can be distinguished (McCave 1972, 1985): "*muddy coasts, nearshore, mid-and outer shelf mud belts and mud blankets*" (Fig. 1.13). Mud belts depend on the amount of discharged mud load, the prevalent tidal and/or current system, or the distribution of the sus-

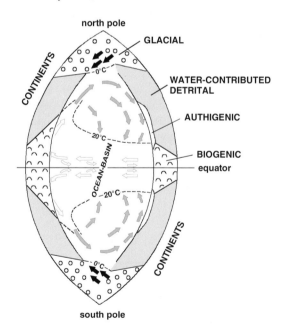

**Fig. 1.12**   Latitudinal distribution of sedimentary facies of the shallow marine environment of continental shelves in an idealized ocean. *Bold arrows* - cold water; *light arrows* - warm water; *grey arrows* - upwelling water (modified from Reineck and Singh 1973).

pended sediment. Muddy coasts will preferentially form near river mouths, whereas mid-shelf mud belts are characteristic of regions where wave and tidal activity are relatively lower than on the inner or outer shelf.

Especially in delta regions where the supply rates of terrigenous material are high – as in the tropics – even the entire shelf might become covered with a consistent blanket of mud, although the shelf represents a region of high energy conversion.

## 1.4.2    Distribution Patterns of Deep-sea Sediments

The two most essential boundary conditions in pelagic sedimentation are the nutrient content in the surface water which controls biogenic productivity and by this biogenic particle production, and the position of the calcite compensation depth (CCD) controlling the preservation of carbonate. The CCD, below which no calcite is found, describes a level at which the dissolution of biogenic carbonate is compensated for by its supply rate. The depth of the CCD is generally somewhere between 4 and 5 km below the surface, however, it varies rather strongly within the three great oceans due to differences in the water mass and the rates of carbonate production.

Calcareous ooze and pelagic clays are the predominant deep-sea sediments in offshore regions

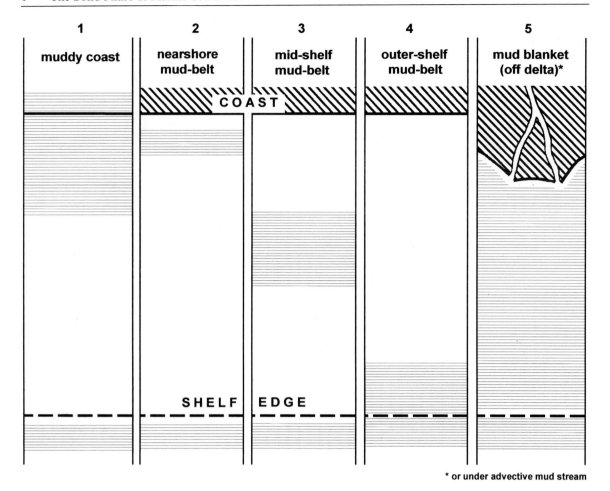

**Fig. 1.13**   Schematic representation of modern mud accumulation on continental shelves (modified from McCave (1972)).

(Table 1.5). The distribution of these sediments in the three great oceans shows a considerable degree of variation (Fig. 1.14). The distribution patterns strongly depend on the water depth, i.e. the position of the CCD. Calcareous ooze, primarily consisting of foraminiferal oozes and nannoplankton oozes, covers vast stretches of the sea-floor at water depths less than 3-4 km and roughly retraces the contours of the mid-oceanic ridges as well as other plateaus and islands, whereas pelagic clay covers the vast deep-sea plains in the form of deep-sea red clay. This particular pattern is especially obvious in the Atlantic Ocean.

**Table 1.5**   Relative areas of world oceans covered with pelagic sediments; area of deep-sea floor = $268.1 \cdot 10^6$ km$^2$ (from Berger (1976)).

| Sediments (%) | Atlantic | Pacific | Indian | World |
|---|---|---|---|---|
| Calcareous ooze | 65.1 | 36.2 | 54.3 | 47.1 |
| Pteropod ooze | 2.4 | 0.1 | --- | 0.6 |
| Diatom ooze | 6.7 | 10.1 | 19.9 | 11.6 |
| Radiolarian ooze | --- | 4.6 | 0.5 | 2.6 |
| Red clays | 25.8 | 49.1 | 25.3 | 38.1 |
| Relative size of ocean (%) | 23.0 | 53.4 | 23.6 | 100.0 |

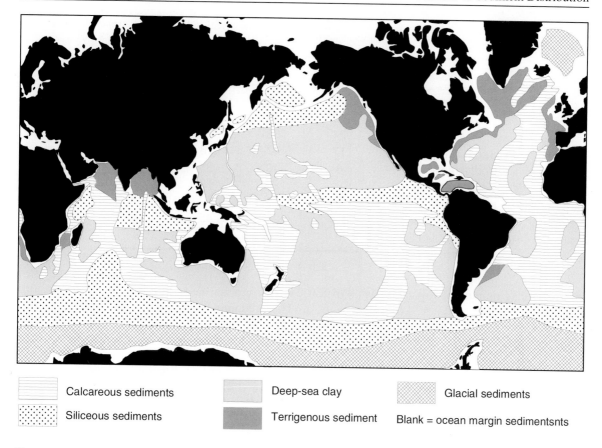

**Fig, 1.14** Distribution of dominant sediment types on the present-day deep-sea floor. The main sediment types are deep-sea clay and calcareous oozes which patterns are predominately depth-controlled. (from Davies and Gorsline (1976)).

Siliceous oozes, mostly consisting of diatom ooze, form a conspicuous ring around Antarctica which clearly marks the zone of the Polar Front in the Antarctic Circumpolar Current. A broad band of radiolarian ooze covers the Pacific ocean below the equatorial upwelling zone, whilst diatom ooze covers the oceanic margins of the Northern Pacific. Terrigenous sediments, especially in the form of mass flow deposits and turbidites cover vast stretches of the near-continental zones in the North Atlantic, the northeastern Pacific, and the broad deep-sea fans off the big river mouths in the northern Indian ocean. Glacio-marine sediments are restricted to the continental margins of Antarctica and to the high latitudes of the North Atlantic.

### 1.4.3    Distribution Patterns of Clay Minerals

Apart from the pelagic clays of the abyssal plains and the terrigenous sediments deposited along the

continental margins, primarily or exclusively consisting of clay minerals, various amounts of lithogenous clay minerals can be found in all types of ocean sediments (Table 1.6). The relative proportion of the various clay minerals in the sediments is a function of their original source, their mode of transport into the area of deposition - either by eolian or volcanic transport, or by means of water and ice - and finally the route of transportation (Petschick et al. 1996). In a global survey, it is easy to identify the particular interactions which climate, weathering on the continents, wind patterns, riverine transport, and oceanic currents have with regard to the relative distribution of the relevant groups of clay minerals (kaolinite, illite, smectite, and chlorite).

The distribution of kaolinite in marine sediments (Fig. 1.5) depends on the intensity of chemical weathering at the site of the rock's origin and the essential patterns of eolian and fluvial transport. Due to its concentration at equatorial and tropical latitudes, kaolinite is usually referred

**Table 1.6**  Average relative concentrations [%] of the principal clay mineral groups in the < 2 μm carbonate-free fraction in sediments from the major ocean basins (data from Windom 1976).

|  | Kaolinite | Illite | Smectite | Chlorite |
|---|---|---|---|---|
| North Atlantic | 20 | 56 | 16 | 10 |
| Gulf of Mexico | 12 | 25 | 45 | 18 |
| Caribbian Sea | 24 | 36 | 27 | 11 |
| South Atlanic | 17 | 47 | 26 | 11 |
| North Pacific | 8 | 40 | 35 | 18 |
| South Pacific | 8 | 26 | 53 | 13 |
| Indian Ocean | 16 | 30 | 47 | 10 |
| Bay of Bengal | 12 | 29 | 45 | 14 |
| Arabian Sea | 9 | 46 | 28 | 18 |

to as the "clay mineral of low latitudes" (Griffin et al. 1968).

Illite is the most frequent clay mineral to be found in ocean sediments (Fig. 1.16). It demonstrates a distinctly higher concentration in sediments at mid-latitudes of the northern oceans which are surrounded by great land masses. This follows particularly from its terrigenous origin and becomes evident when the Northern Pacific is compared with the Southern Pacific. The illite concentration impressively reflects the percentage and distribution of particles which were intro-

duced into marine sediments by fluvial transport. The predominance of illites in the sediments of the Pacific and Atlantic oceans at moderate latitudes, below the trajectories of the jet-stream, indicates the great importance of the wind system in the transport of fine-dispersed particulate matter.

The distribution pattern of smectite differs greatly in the three oceans (Fig. 1.17), and along with some other factors may be explained as an effect induced by dilution. Smectite is generally considered as an indicator of a "volcanic regime" (Griffin et al. 1968). Thus, high smectite concen-

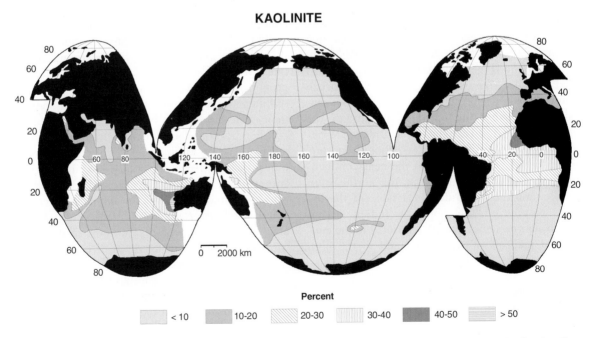

**Fig. 1.15**  Relative distribution of kaolinite in the world ocean, concentration in the carbonate-free < 2 μm size fraction (from Windom 1976).

## ILLITE

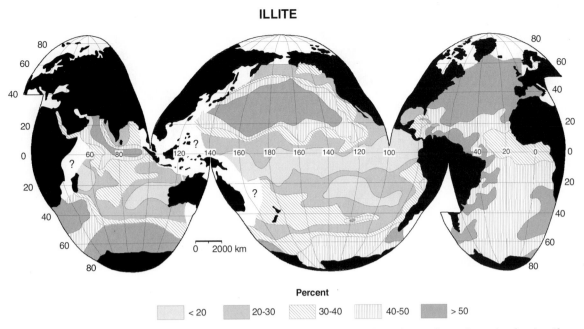

**Fig. 1.16** Relative distribution of illite in the world ocean, concentration in the carbonate-free < 2 μm size fraction (from Windom 1976).

trations are usually observed in sediments of the Southern Pacific, in regions of high volcanic activity, where the sedimentation rates are very low due to great distances from the shoreline, and where the dilution with other clay minerals is low as well. The low smectite concentration in the North Atlantic results from terrigenous detritus inputs which are rich in illites and chlorites.

The distribution of chlorite in deep-sea sediments (Fig. 1.18) is essentially inversely related

## SMECTITE

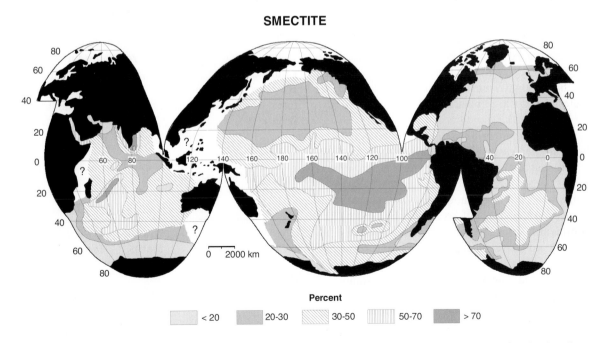

**Fig. 1.17** Relative distribution of smectite in the world ocean, concentration in the carbonate-free < 2 μm size fraction (from Windom 1976).

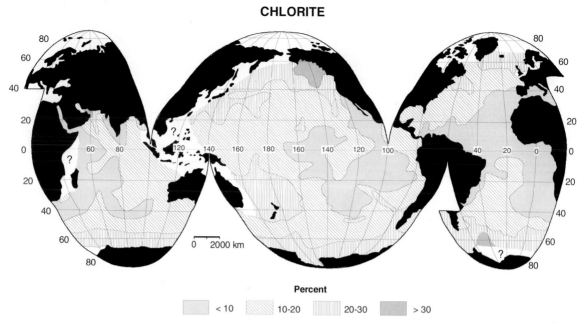

**Fig. 1.18**  Relative distribution of chlorite in the world ocean, concentration in the carbonate-free < 2 µm size fraction (from Windom 1976).

to the pattern of kaolinite. Although chlorite is distributed homogeneously over the oceans, its highest concentration is measured in polar regions and therefore is referred to as the "high latitude mineral" (Griffin et al. 1968).

### 1.4.4    Sedimentation Rates

As can be seen in Table 1.8, the sedimentation rates of typical types of deep-sea sediments show a strong geographical variability which is based on the regionally unsteady import of terrigenous material and a highly variable biogenic productivity in the ocean.

Basically, it can be stated that the sedimentation rate decreases with increasing distance from a sediment source, may this either be a continent or an area of high biogenic productivity. The highest rates of terrigenous mud formation are recorded on the shelf off river mouth's and on the continental slope, where sedimentation rates can amount up to several meters per one thousand years. Distinctly lower values are observed at detritus-starved continental margins, for example of Antarctica. The lowest sedimentation rates ever recorded lie between 1 and 3 mm/1000 yr. and are connected to deep-sea red clay in the offshore deep-sea basins (Table 1.7), especially in the central Pacific Ocean.

Calcareous biogenic oozes demonstrate intermediate rates which frequently lie between 10 and 40 mm/1000 yr. Their distribution pattern depends on the biogenic production and on the water depth, or the depth of the CCD. As yet, rate values between 2 and 10 mm/1000 yr. were considered as normal for the sedimentation of siliceous oozes. Recent investigations in the region of the Antarctic Circumpolar Current have revealed that a very high biogenic production, in connection with lateral advection and "sediment focusing", can even give rise to sedimentation rates of more than 750 mm per thousand years (Fig. 1.19).

**Table 1.7**   Sedimentation rates of red clay in various deep-sea basins of the world ocean (data from various sources, e.g. Berger 1974, Gross 1987).

|                | Rate (mm/1000 yr) | |
|                | Mean | Range |
|----------------|------|-------|
| North Atlantic | 1,8  | 0.5-6.2 |
| South Atlantic | 1,9  | 0.2-7.5 |
| North Pacific  | 1,5  | 0.4-6.0 |
| South Pacific  | 0,45 | 0.3-0.6 |

**Table 1.8**  Typical sedimentation rates of recent and subrecent marine sediments (data from various sources, e.g. Berger 1974, Gross 1987).

| Facies | Area | Average Sedimentation Rate (mm/1000 yr) |
|---|---|---|
| Terrigeneous mud | California Borderland | 50 - 2,000 |
| | Ceara Abyssal Plain | 200 |
| | Antarctic Continental Margin | 30-65 |
| Calcareous ooze | North Atlantic (40-50 °N) | 35-60 |
| | North Atlantic ( 5-20 °N) | 40-14 |
| | Equatorial Atlantic | 20-40 |
| | Caribbean | ~28 |
| | Equatorial Pacific | 5-18 |
| | Eastern Equatorial Pacific | ~30 |
| | East Pacific Rise (0-20 °N) | 20-40 |
| | East Pacific Rise (~30 °N) | 3-10 |
| | East Pacific Rise (40-50 °N) | 10-60 |
| Siliceous ooze | Equatorial Pacific | 2-5 |
| | Antarctic, Indian Sector | 2-10 |
| | Antarctic, Atlantic Sector | 25-750 |
| Red clay | Northern North Pacific (muddy) | 10-15 |
| | Central North Pacific | 1-2 |
| | Tropical North Pacific | 0-1 |
| | South Pacific | 0.3-0.6 |
| | Antarctic, Atlantic Sector | 1-2 |

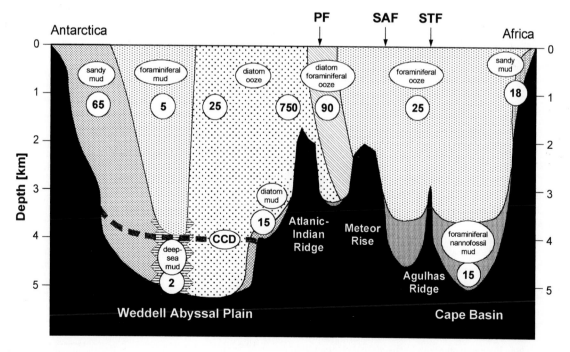

**Fig. 1.19**  Distribution of major sediment facies across the frontal system of the Antarctic Circumpolar Current (ACC) between Africa and Antarctica. Numbers are typical sedimentation rates in mm/1000 yr. PF = Polar Front, SAF = Sub Antarctic Front, STF = Subtropical Front.

This is contribution No 250 of the Special Research Program SFB 261 (*The South Atlantic in the Late Quaternary*) funded by the Deutsche Forschungsgemeinschaft (DFG).

# References

Berger, W.H., 1974. Deep - sea sedimentation. In: Burk, C.A. and Drake, C.L. (eds) The geology of continental margins. Springer Verlag, Berlin, Heidelberg, NY, pp 213-241.

Berger, W.H., 1976. Biogenic deep sea sediments: Production, preservation and interpretation. In: Riley, J.P. and Chester, R. (eds) Chemical Oceanography, Academic, 5. Press, London, NY, San Francisco, pp 266-388.

Berger, W.H. and Herguera, J.C., 1991. Reading the sedimentary record of the ocean's productivity. Pimary productivity and biogeochemical cycles in the sea. In: Falkowski, P.G. and Woodhead, E.D. (eds) Plenum Press, NY, London, pp 455-486.

Biscay, P.E., 1965. Mineraloy and sedimentation of recent deep-sea clay in the Atlantic Ocean and adjacent seas and oceans. Geol. Soc. Am. Bull., 76: 803-832.

Chamley, H., 1989. Clay sedimentology. Springer Verlag, Berlin, Heidelberg, NY, 623 pp.

Davies, T.A. and Gorsline, D.S., 1976. Oceanic sediments and sedimentary processes. In: Riley, J.P. and Chester, R. (eds) Chemical oceanography, 5. Academic Press, London, NY, San Francisco, pp 1-80.

Dean, W.E., Leinen, M. and Stow, D.A.V., 1985. Classification of deep-sea fine-grained sediments. Journal of Sediment Petrology, 55: 250-256.

Elderfield, H., 1976. Hydrogeneous material in marine sediments; excluding manganese nodules. In: Riley, J.P. and Chester, R. (eds) Chemical Oceanography, 5. Academic Press, London, NY, San Francisco, pp 137-215.

Emery, K.O., 1968. Relict sediments on continental shelves of the world. Bull. Am. Assoc. Petrol. Geologists, 52: 445-464.

Fabricius, F., 1977. Origin of marine ooids and grapestones. Contributions to sedimentology, 7, Schweizerbart, Stuttgart, 113 pp.

Flügel, E., 1978. Mikrofazielle Untersuchungs-methoden von Kalken. Springer Verlag, Berlin, Heidelberg, NY, 454 pp.

Folk, R.L., 1980. Petrology of sedimentary rocks. Hemphill. Publ. Co., Austin, 182 pp.

Garrels, R.M. and Mackenzie, F.T., 1971. Evolution of sedimentary rocks. Norton & Co, NY, 397 pp.

Gibbs, R.J., 1977. Transport phase of transition metals in the Amazon and Yukon rivers. Geological Society of America Bulletin, 88: 829-843.

Gingele, F., 1992. Zur Klimaabhängigen Bildung biogener und terrigener Sedimente und ihre Veränderungen durch die Frühdiagnese im zentralen und östlichen Südatlantic. Berichte, Fachbereich Geowissenschaften, Universität Bremen, 85, 202 pp.

Goldberg, E.D. and Griffin, J.J., 1970. The sediments of the northern Indian Ocean. Deep-Sea Research, 17: 513-537.

Griffin, J.J., Windom, H. and Goldberg, E.D., 1968. The distribution of clay minerals in the World Ocean. Deeep-Sea Research, 15: 433-459.

Grim, R.E., 1968. Clay Mineralogy, McGraw-Hill Publ., NY, 596 pp.

Gross, M.G., 1987. Oceanography: A view of the Earth. Prentice hall, Englewood Cliffs, NJ, 406 pp.

Hillier, S., 1995. Erosion, sedimentation and sedimentary origin of clays. In: Velde, B. (ed) Origin and mineralogy of clays. Springer Verlag, Berlin, Heidelberg, NY, pp 162-219.

Krumbein, W.C., 1934. Size frequency distribution of sediments. Journal of Sediment Petrology, 4: 65-77.

Krumbein, W.C. and Graybill, F.A., 1965. An introduction to statistical models in Geology. McGraw-Hill, NY, 328 pp.

Lisitzin, A.P., 1996. Oceanic sedimentation: Lithology and Geochemistry. Amer. Geophys. Union, Washington, D. C., 400 pp.

Mazullo, J. and Graham, A.G., 1987. Handbook for shipboard sedimentologists. Ocean Drilling Program, Technical Note, 8, 67 pp.

McCave, I.N., 1972. Transport and escape of fine-grained sediment from shelf areas. In: Swift, D.J.P., Duane, D.B. and Pilkey, O.H. (eds), Shelf sediment transport, process and pattern. Dowden, Hutchison & Ross, Stroudsburg, pp 225-248.

Milliman, J.D., Pilkey, O.H. and Ross, D.A., 1972. Sediments of the continental margins off the eastern United States. Geological Society of America Bulletin, 83: 1315-1334.

Milliman, J.D., 1974. Marine carbonates. Springer Verlag, Berlin, Heidelberg, NY, 375 pp.

Milliman, J.D. and Summerhays, C.P., 1975. Upper continental margin sedimentation of Brasil. Contribution to Sedimentology, 4, Schweizerbart, Stuttgart, 175 pp.

Milliman, J.D. and Meade, R.H., 1983. World-wide delivery of river sediment to the ocean. The Journal of Geology, 91: 1-21.

Milliman, J.D. and Syvitski, J.P.M., 1992. Geomorphic/tectonic control of sediment discharge to the ocean: The importance of small mountainous rivers. The Journal of Geology, 100: 525-544.

Murray, J. and Renard, A.F., 1891. Deep sea deposits - Report on deep-sea deposits based on specimens collected during the voyage of H.M.S. Challenger in the years 1873-1876. „Challenger" Reports, Eyre & Spottiswood, London; J. Menzies & Co, Edinburgh; Hodges, Figgis & Co, Dublin, 525 pp.

Nelson, C.S., 1988. Non-tropical shelf carbonates - Modern and Ancient. Sedimentary Geology, 60, 1-367 pp.

Petschick, R., Kuhn, G. and Gingele, F., 1996. Clay mineral distribution in surface sediments of the South Atlantic: sources, transport, and relation to oceanography. Marine Geology, 130: 203-229.

Prospero, J.M., 1996. The atmospheric transport of particles in the ocean. In: Ittekkot, V., Schäfer, P., Honjo, S. and Depetris, P.J. (eds), Particle Flux in the Ocean. Wiley & Sons, Chichester, NY, Brisbane, Toronto, Singapore, pp 19-52.

Pye, K., 1987. Aeolian dust and dust deposits. Academic Press, London, NY, Toronto, 334 pp.

Reineck, H.E. and Singh, I.B., 1973. Depositional sedimentary environments. Springer Verlag, Berlin, Heidelberg, NY, 439 pp.

Seibold, E. and Berger, W.H., 1993. The sea floor - An introduction to marine geology. Springer Verlag, Berlin, Heidelberg, NY, 356 pp.

Shepard, F., 1954. Nomenclature based on sand-silt-clay ratios. Journal of Sediment Petrology, 24: 151-158.

Straub, S.M. and Schmincke, H.U., 1998. Evaluating the tephra input into Pacific Ocean sediments: distribution in space and time. Geologische Rundschau, 87: 461-476.

Velde, B., 1995. Origin and mineralogy of clays. Springer Verlag, Berlin, Heidelberg, NY, 334 pp.

Windom, H.L., 1976. Lithogeneous material in marine sediments. In: Riley, J.P. and Chester, R. (eds), Chemical Oceanography. Academic Press, London, NY, pp 103-135.

25

# 2   Geophysical Properties in Marine Sediments

The traditional schism of Earth Science education and research into various specialties such as geophysics and geochemistry, although gradually fading, is still very much alive in most of the international and particularly the German academic community. Notwithstanding different experimental methods and often also different scientific objectives, isolated activities are at least ineffective if not a cul de sac in many fields. In recent years, the investigation of marine sediments has advanced to a highly successful example for the opposite strategy in multiple joined research efforts.

Modern marine geochemistry is a largely process-oriented subject, predominantly relying on aquatic chemistry rather than on customary analyses of element concentrations in the sediment matrix. It is now possible to study biogeochemical reactions in the pore water and to determine quantitatively the geochemical gradients and mass fluxes within unconsolidated sediments and across the sediment-water interface. With recently developed new core logging techniques marine geophysics achieves high-resolution data sets on various physical properties which in addition to understanding the sediment structure and stratigraphy are also indispensable to quantify biogeochemical processes of early diagenesis.

In many cases, marine geophysical and geochemical information obtained in this way are to the mutual advantage of both disciples. Some examples illustrating this approach are given below without any attempt at completeness.

- Detailed acoustic and resistivity records yield important data about the size and other relevant characteristics of the pore space in which aquatic-biogeochemical reactions and fluid flow take place. They also outline the nature and layering of the sediment solid phase which is a reaction partner for the fluid phase and, at the same time, a carrier of geochemical information.

- Diagenetic changes influence a variety of physical signals in the sediment as a result of dissolution, precipitation or re-crystallization of mineral components. Geophysical methods can help recognize these alterations and lead to a better understanding of the processes involved.

- Primary carriers of the magnetic signal in marine sediments are detrital and, depending on the respective depositional regime, biogenic minerals which change during the processes of early diagenesis in a controlled manner. While the fine grained magnetite fraction typically disintegrates, other specifically sulfur bearing minerals will be newly formed at fixed depths which are determined by biogeochemical zonations and reactions.

- Most all geophysical core logging tools are used on board ships today. Namely acoustic records give immediate insight to the cored materials delineating for example how unique or widely distributed are certain features of the sediment structure. This information may greatly help developing appropriate sampling schemes to study a specific geological or geochemical problem. Moreover, acoustic core logs, together with density logs can quantitatively be correlated to digital echosounder (Parasound) profiles providing a basis for the interpretation of single coring site results in a regional context.

- Applying orbital tuning techniques to physical property records, notably of sediment magnetic susceptibility or density, has repeatedly been proven to result in very reliable high-resolution chronostratigraphies. Such age data, often supplemented by paleomagnetic analyses, constitute an important framework for the interpretation of geochemical and other data.

- Most physical properties are particularly sensitive indicators of changes in (paleo-) environmental conditions. Various parameters have
been successfully established as suitable proxies for specific biogeochemical reactions and
geological or paleoceanographical processes.
Since they can be measured rapidly compared
to most geochemical parameters, such as carbonate contents or oxygen isotope ratios,
highly resolved physical property records may
efficiently be used for the interpolation of less
densely sampled geological or geochemical
sediment core logs. However, any calibration
for such purposes is typically valid only on a
regional scale and needs to be verified for
each individual setting.

# 2.1 Physical Properties of Marine Sediments

Monika Breitzke

Generally, physical properties of marine sediments are good indicators for the composition, microstructure and environmental conditions during and after the depositional process. Their study is of high interdisciplinary interest and follows various geoscientific objectives.

Physical properties provide a *lithological and geotechnical description* of the sediment. Questions concerning the composition of a depositional regime, slope stability or nature of seismic reflectors are of particular interest within this context. Parameters like P- and S-wave velocity and attenuation, elastic moduli, wet bulk density and porosity contribute to their solution.

As most physical properties can be measured very quickly in contrast to parameters like carbonate content, grain size distribution or oxygen isotope ratio, they often serve as *proxy parameters* for geological or paleoceanographical processes and changes in the environmental conditions. Basis for this approach are correlations between different parameters which allow to derive regression equations of regional validity. Highly resolved physical properties are used as input parameters for such equations to interpolate coarsely sampled geological, geochemical or sedimentological core logs. Examples are the P-wave velocity, attenuation and wet bulk density as indices for the sand content or mean grain size and the magnetic susceptibility as an indicator for the carbonate content or amount of terrigenous components. Additionally, highly resolved physical property core logs are often used for time series analysis to derive quaternary stratigraphies by orbital tuning.

In marine geochemistry, physical parameters which quantify the amount and distribution of pore space or indicate alterations due to diagenesis are important. Porosities are necessary to calculate flow rates, permeabilities describe how easy a fluid flows through a porous sediment and

increased magnetic susceptibilities can be interpreted by an increased iron content or, more generally, by a high amount of magnetic particles within a certain volume.

This high interdisciplinary value of physical property measurements results from the different effects of changed environmental conditions on sediment structure and composition. Apart from anthropogeneous influences variations in the environmental conditions result from climatic changes and tectonic events which affect the ocean circulation and related production and deposition of biogenic and terrigenous components. Enhanced or reduced current intensities during glacial or interglacial stages cause winnowing, erosion and redeposition of fine-grained particles and directly modify the sediment composition. These effects are subject of sedimentological and paleoceanographical studies. Internal processes like early diagenesis, newly built authigenic sediments or any other alterations of the solid and fluid constituents of marine sediments are of geochemical interest. Gravitational mass transports at continental slopes or turbidites in large submarine fan systems deposit as coarse-grained chaotic or graded beddings. They can be identified in high-resolution physical property logs and are often directly correlated with sea level changes, results valuable for sedimentological, seismic and sequence stratigraphic investigations. Generally, any changes in the sediment structure which modify the elastic properties of the sediment influence the reflection characteristics of the subsurface and can be imaged by remote sensing seismic or echosounder surveys.

## 2.1.1 Introduction

Physical properties of marine sediments depend on the properties and arrangement of the solid and fluid constituents. To fully understand the im-

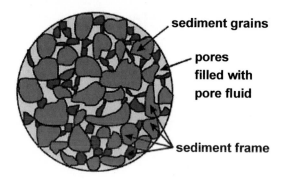

**Fig. 2.1** Components of marine sediments. The single particles are the sediment grains. The voids between these particles – the pores – are filled with pore fluid, usually sea water. Welded particles and sediment grains in close contact build the sediment frame.

age of geological and paleoceanographical processes in physical property core logs it is helpful to consider the single components in detail (Fig. 2.1).

By most general definition a sediment is a collection of particles - the sediment grains - which are loosely deposited on the sea floor and closely packed and consolidated under increasing lithostatic pressure. The voids between the sediment grains - the pores - form the pore space. In water-saturated sediments it is filled with pore water. Grains in close contact and welded particles build

the sediment frame. Shape, arrangement, grain size distribution and packing of particles determine the elasticity of the frame and the relative amount of pore space.

Measurements of physical properties usually encompass the whole, undisturbed sediment. Two types of parameters can be distinguished: (1) bulk parameters and (2) acoustic and elastic parameters. Bulk parameters only depend on the relative amount of solid and fluid components within a defined sample volume. They can be approximated by a simple volume-oriented model (Fig. 2.2a). Examples are the wet bulk density and porosity. In contrast, acoustic and elastic parameters depend on the relative amount of solid and fluid components and on the sediment frame including arrangement, shape and grain size distribution of the solid particles. Viscoelastic wave propagation models simulate these complicated structures, take the elasticity of the frame into account and consider interactions between solid and fluid constituents. (Fig. 2.2b). Examples are the velocity and attenuation of P- and S-waves. Closely related parameters which mainly depend on the distribution and capillarity of the pore space are the permeability and electrical resistivity.

Various methods exist to measure the different physical properties of marine sediments. Some

## Sediment Models

a)
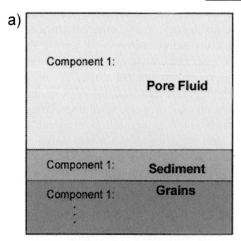

**Volume-Oriented Model**
for Bulk Parameters

b)
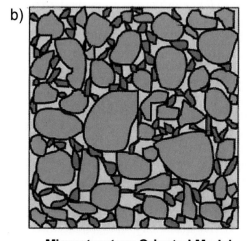

**Microstructure-Oriented Model**
for Acoustic, Elastic and Related
Parameters

**Fig. 2.2** Two types of sediment models. (a) The layered, volume-oriented model for bulk parameters only depends on the relative amount of solid and fluid components. (b) The microstructure-oriented model for acoustic and elastic parameters takes the complicated shape and geometry of the particle and pore size distribution into account and considers interactions between the solid and fluid constituents during wave propagation.

parameters can be measured directly, others indirectly by determining one or several related parameters and computing the desired property by empirical or model-based equations. For consolidated sedimentary, igneous and metamorphic rocks Schön (1996) described a large variety of such methods for all common physical properties. Some of these methods can also be applied to unconsolidated, water-saturated sediments, others have to be modified or are completely inappropriate. Principally, if empirical relations or sediment models are used, the implicit assumptions have to be checked carefully. An example is Archie's law (Archie 1942) which combines the electrical resistivities of the pore fluid and saturated sediment and with its porosity. An exponent and multiplier in this equation depends on the sediment type and composition and change from fine- to coarse-grained and from terrigenous to biogenic sediments. Another example is Wood's equation (Wood 1946) which relates P-wave velocities to porosities. This model approximates the sediment by a dilute suspension and neglects the sediment frame, any interactions between particles or particles and pore fluid and assumes a 'zero' frequency for acoustic measurements. Such assumptions are only valid for a very limited set of high porosity sediments so that for any comparisons these limitations should be kept in mind.

Traditional measuring techniques use small chunk samples taken from the split core. These techniques are rather time-consuming and could only be applied at coarse sampling intervals. The necessity to measure high-resolution physical property logs rapidly on a milli- to centimeter scale for a core-to-core or core-to-seismic data correlation, and the opportunity to use high-resolution physical property logs for stratigraphic purposes (e.g. orbital tuning) forced the development of non-destructive, automated logging systems. They record one or several physical properties almost continuously at arbitrary small increments under laboratory conditions. The most common tools are the multi sensor track (MST) system for P-wave velocity, wet bulk density and magnetic susceptibility core logging onboard of the *Ocean Drilling Program* research vessel JOIDES Resolution (e.g. Shipboard Scientific Party 1995), and the commercially available multi sensor core logger (MSCL) of GEOTEK™ (Schultheiss and McPhail 1989, Weaver and Schultheiss 1990, Gunn and Best 1998). Additionally, other core logging tools have simultaneously been developed for special research interests which for instance record electrical resistivities (Bergmann 1996) or full waveform transmission seismograms on sediment cores (Breitzke and Spieß 1993). They are particularly discussed in this paper.

While studies on sediment cores only provide information on the local core position, lateral variations in physical properties can be imaged by remote sensing methods like high-resolution seismic or sediment echosounder profiling. They facilitate core-to-core correlations over large distances and allow to evaluate physical property logs within the local sedimentation environment.

In what follows the theoretical background of the most common physical properties and their measuring tools are described. Examples for the wet bulk density and porosity can be found in Section 2.1.2. For the acoustic and elastic parameters first the main aspects of Biot-Stoll's viscoelastic model are summarized which computes P- and S-wave velocities and attenuations for given sediment parameters (Biot 1956a, b, Stoll 1974, 1977, 1989). Subsequently, analysis methods are described to derive these parameters from transmission seismograms recorded on sediment cores, to compute additional properties like elastic moduli and to derive the permeability as a related parameter by an inversion scheme (Sect. 2.1.4).

Examples from terrigenous and biogenic sedimentation provinces are presented (1) to illustrate the large variability of physical properties in different sediment types and (2) to establish a sediment classification which is only based on physical properties, in contrast to geological sediment classifications which mainly uses parameters like grain size distribution or mineralogical composition (Sect. 2.1.5).

Finally, some examples from high-resolution narrow-beam echosounder recordings present remote sensing images of terrigenous and biogenic sedimentation environments (Sect. 2.1.6).

### 2.1.2    Porosity and Wet Bulk Density

Porosity and wet bulk density are typical bulk parameters which are directly associated with the relative amount of solid and fluid components in marine sediments. After definition of both parameters this section first describes their traditional analysis method and then focuses on recently developed techniques which determine porosities and wet bulk densities by gamma ray attenuation and electrical resistivity measurements.

The porosity ($\phi$) characterizes the relative amount of pore space within a sample volume. It is defined by the ratio

$$\phi = \frac{volume\ of\ pore\ space}{total\ sample\ volume} = \frac{V_f}{V} \qquad (2.1)$$

Equation 2.1 describes the fractional porosity which ranges from 0 in case of none pore volume to 1 in case of a water sample. Multiplication with 100 gives the porosity in percent. Depending on the sediment type porosity occurs as *inter-* and *intraporosity*. Interporosity specifies the pore space between the sediment grains and is typical for terrigenous sediments. Intraporosity includes the voids within hollow sediment particles like foraminifera in calcareous ooze. In such sediments both inter- and intraporosity contribute to the total porosity.

The wet bulk density ($\rho$) is defined by the mass of a water-saturated sample per sample volume.

$$\rho = \frac{mass\ of\ wet\ sample}{total\ sample\ volume} = \frac{m}{V} \qquad (2.2)$$

Porosity and wet bulk density are closely related, and often porosity values are derived from wet bulk density measurements and vice versa. Basic assumption for this approach is a two-component model for the sediment with uniform grain and pore fluid densities ($\rho_g$ and $\rho_f$). The wet bulk density can then be calculated using the porosity as a weighing factor

$$\rho = \phi \cdot \rho_f + (1 - \phi) \cdot \rho_g \qquad (2.3)$$

If two or several mineral components with significantly different grain densities contribute to the sediment frame their densities are averaged in ($\rho_g$).

### 2.1.2.1  Analysis by Weight and Volume

The traditional way to determine porosity and wet bulk density is based on weight and volume measurements of small sediment samples. Usually they are taken from the center of a split core by a syringe which has the end cut off and a definite volume of e.g. 10 ml. While weighing can be done very accurately in shore-based laboratories measurements onboard of research vessels require special balance systems which compensate the shipboard motions (Childress and Mickel 1980).

Volumes are measured precisely by Helium gas pycnometers. They mainly consist of a sample cell and a reference cell and employ the ideal gas law to determine the sample volume. In detail, the sediment sample is placed in the sample cell, and both cells are filled with Helium gas. After a valve connecting both cells are closed, the sample cell is pressurized to ($P_1$). When the valve is opened the pressure drops to ($P_2$) due to the increased cell volume. The sample volume (V) is calculated from the pressure ratio ($P_1/P_2$) and the volumes of the sample and reference cell ($V_s$ and $V_{ref}$) (Blum 1997)

$$V = V_s + \frac{V_{ref}}{1 - P_1/P_2} \qquad (2.4)$$

While weights are measured on wet and dry samples having used an oven or freeze drying, volumes are preferentially determined on dry samples. A correction for the mass and volume of the salt precipitated from the pore water during drying must additionally be applied (Hamilton 1971, Gealy 1971) so that the total sample volume (V) consists of

$$V = V_{dry} - V_{salt} + V_f \qquad (2.5)$$

The volumes ($V_f$) and ($V_{salt}$) of the pore space and salt result from the masses of the wet and dry sample (m and $m_{dry}$) from the densities of the pore fluid and salt ($\rho_f = 1.024$ g cm$^{-3}$ and $\rho_{salt} = 2.1$ g cm$^{-3}$) and from the salinity (s)

$$V_f = \frac{m_f}{\rho_f} = \frac{m - m_{dry}}{(1 - s) \cdot \rho_f} \qquad (2.6)$$

$$\text{with}: \qquad m_f = \frac{m - m_{dry}}{1 - s}$$

$$V_{salt} = \frac{m_{salt}}{\rho_{salt}} = \frac{m_f - (m - m_{dry})}{\rho_{salt}} \qquad (2.7)$$

$$\text{with}: \qquad m_{salt} = m_f - (m - m_{dry})$$

Together with the mass (m) of the wet sample Equations 2.5 to 2.7 allow to compute the wet bulk density according to Equation 2.2.

Wet bulk density computations according to Equation 2.3 require the knowledge of grain densities ($\rho_g$). They can also be determined from weight and volume measurements on wet and dry samples (Blum 1997)

$$\rho_g = \frac{m_g}{V_g} = \frac{m_{dry} - m_{salt}}{V_{dry} - V_{salt}} \qquad (2.8)$$

Porosities are finally computed from the pore space and sample volume defined by the equations above.

### 2.1.2.2 Gamma Ray Attenuation

The attenuation of gamma rays passing radially through a sediment core is a widely used effect to analyze wet bulk densities and porosities by a non-destructive technique. Often [137]Cs is used as source emitting gamma rays of 662 keV energy. They are mainly attenuated by Compton scattering (Ellis 1987). The intensity (I) of the attenuated gamma ray beam depends on the source intensity ($I_0$), the wet bulk density of the sediment, the ray path length (d) and the specific Compton mass attenuation coefficient ($\mu$)

$$I = I_0 \cdot e^{-\mu \rho d} \qquad (2.9)$$

To determine the source intensity in practice and to correct for the attenuation in the liner walls first no core and then an empty core liner are placed between the gamma ray source and detector to measure the intensities ($I_{air}$) and ($I_{liner}$). The difference ($I_{air} - I_{liner}$) replaces the source intensity, and the ray path length is substituted by the measured outer core diameter ($d_{outside}$) minus the double liner wall thickness ($2d_{liner}$). With these corrections the wet bulk density can be computed according to

$$\rho = -\frac{1}{\mu \cdot (d_{outside} - 2d_{liner})} \cdot \ln\left(\frac{I}{I_{air} - I_{liner}}\right) \qquad (2.10)$$

Porosities are derived by rearranging Equation 2.3 and assuming a grain density ($\rho_g$).

The specific Compton mass attenuation coefficient ($\mu$) is a material constant. It depends on the energy of the gamma rays and on the ratio (Z/A) of the number of electrons (Z) to the atomic mass (A) of the material (Ellis 1987). For most sediment and rock forming minerals this ratio is about 0.5, and for a [137]Cs source the corresponding mass attenuation coefficient ($\mu_g$) is 0.0774 cm$^2$g$^{-1}$. However, for the hydrogen atom (Z/A) is close to 1.0 leading to a significantly different coefficient ($\mu_f$) of 0.0850 cm$^2$g$^{-1}$ in sea water (Gerland and Villinger 1995). So, in water-saturated sediments

the effective mass attenuation coefficient results from the sum of the mass weighted coefficients of the solid and fluid constituents (Bodwadkar and Reis 1994)

$$\mu = \phi \cdot \frac{\rho_f}{\rho} \cdot \mu_f + (1 - \phi) \cdot \frac{\rho_g}{\rho} \cdot \mu_g \qquad (2.11)$$

The wet bulk density in the denominator is defined by Equation 2.3. Unfortunately, Equations 2.3 and 2.11 depend on the porosity, a parameter which should actually be determined by gamma ray attenuation. Gerland (1993) used an average 'processing porosity' of 50% for terrigenous, 70% for biogenic and 60% for cores of mixed material to estimate the effective mass attenuation coefficient for a known grain density. Whitmarsh (1971) suggested an iterative scheme which improves the mass attenuation coefficient. It starts with an estimated 'processing porosity' and mass attenuation coefficient (equ. 2.11) to calculate the wet bulk density from the measured gamma ray intensity (equ. 2.10). Subsequently, an improved porosity (Eq. 2.3) and mass attenuation coefficient (Eq. 2.11) can be calculated for a given grain density. Using these optimized values in a second iteration wet bulk density, porosity and mass attenuation coefficient are re-evaluated. Gerland (1993) and Weber et al. (1997) showed that after few iterations (<5) the values of two successive steps differ by less than 0.1‰ even if a 'processing porosity' of 0 or 100% and a mass attenuation coefficient of 0.0774 cm$^2$g$^{-1}$ or 0.0850 cm$^2$g$^{-1}$ for a purely solid or fluid 'sediment' are used as starting values.

Gamma ray attenuation is usually measured by automated logging systems, e.g. onboard of the *Ocean Drilling Program* research vessel JOIDES Resolution (Boyce 1973, 1976), by the multi-sensor core logger of GEOTEK™ (Schultheiss and McPhail 1989, Weaver and Schultheiss 1990, Gunn and Best 1998) or by specially developed systems (Gerland 1993, Bodwadkar and Reis 1994, Gerland and Villinger 1995). The emission of gamma rays is a random process which is quantified in counts per second, and which are converted to wet bulk densities by appropriate calibration curves (Weber et al. 1997). To get a representative value for the gamma ray attenuation gamma counts must be integrated over a sufficiently long time interval. How different integration times and measuring increments influence the quality, resolution and reproducibility of gamma ray logs illustrates Figure 2.3. A 1 m long section of gravity core PS2557-1 from the South African continental margin was repeatedly measured with increments from 1 to 0.2 cm and integration times from 10 to 120 s. Generally, the dominant features, which can be related to changes in the lithology, are reproduced in all core logs. The prominent peaks are more pronounced and have higher amplitudes if the integration time increases (from A to D). Additionally, longer integration times reduce the scatter (from A to B). Fine-scale lithological variations are best resolved by the shortest measuring increment (D).

A comparison of wet bulk densities derived from gamma ray attenuation with those measured on discrete samples is shown in Figure 2.4a for two gravity cores from the Arctic (PS1725-2) and Antarctic Ocean (PS1821-6). Wet bulk densities, porosities and grain densities of the discrete samples were analyzed by their weight and volume. These grain densities and a constant 'processing porosity' of 50% were used to evaluate

the gamma counts. Displayed versus each other both data sets mostly differ by less than ±5% (dashed lines). The lower density range (1.20-1.65 g cm$^{-3}$) is mainly covered by the data of the diatomaceous and terrigenous sediments of core PS1821-6 while the higher densities (1.65-2.10 g cm$^{-3}$) are characteristic for the terrigenous core PS1725-2. According to Gerland and Villinger (1995) the scatter in the correlation of both data sets probably results from (1) a slight shift in the depth scales, (2) the fact that both measurements do not consider identical samples volumes, and (3) artifacts like drainage of sandy layers due to core handling, transportation and storage.

A detailed comparison of both data sets is shown in Figure 2.4b for two segments of core PS1725-2. Wet bulk densities measured on discrete samples agree very well with the density log derived from gamma ray attenuation. Additionally, the striking advantage of the almost continuous wet bulk density log clearly becomes obvious. Measured with an increment of 5 mm a lot of fine-scale variations are defined, a resolution

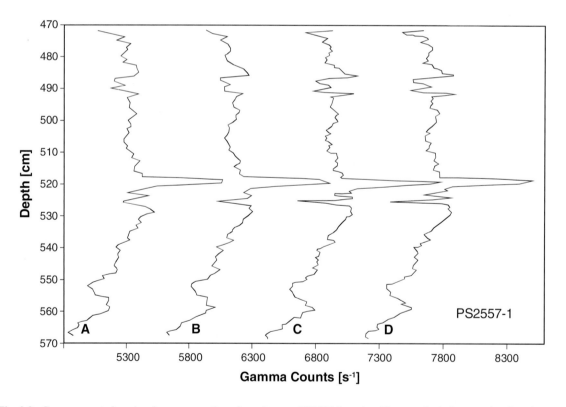

**Fig. 2.3** Gamma counts for a 1 m long core section of gravity core PS2557-1 used to illustrate the influence of various integration times and measuring increments. Curve A is recorded with an increment of 1 cm and an integration time of 10 s, curve B with an increment of 1 cm and an integration time of 20 s, curve C with an increment of 0.5 cm and an integration time of 60 s and curve D with an increment of 0.2 cm and an integration time of 120 s. To facilitate the comparison each curve is set off by 800 counts. Modified after Weber et al. (1997).

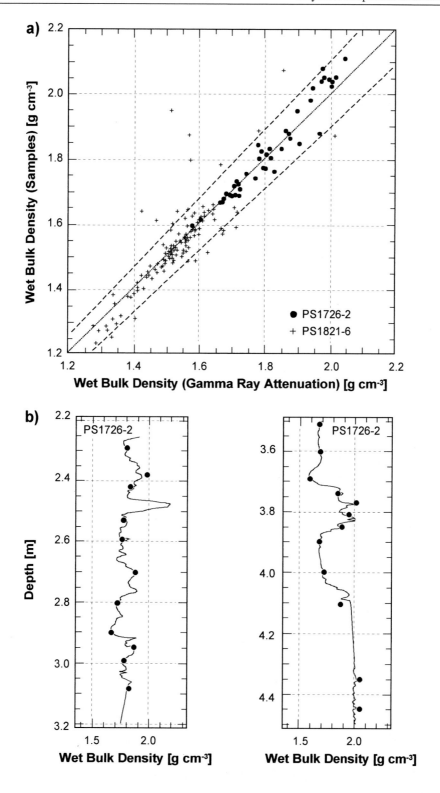

**Fig. 2.4** Comparison of wet bulk densities determined on discrete samples by weight and volume measurements and calculated from gamma ray attenuation. (a) Cross plot of wet bulk densities of gravity cores PS1821-6 from the Antarctic and PS1725-2 from the Arctic Ocean. The dashed lines indicate a difference of ±5% between both data sets. (b) Wet bulk density logs derived from gamma ray attenuation for two 1 m long core sections of gravity core PS1725-2. Superimposed are density values measured on discrete samples. Modified after Gerland and Villinger (1995).

which could never be reached by the time-consuming analysis of discrete samples (Gerland 1993, Gerland and Villinger 1995).

The precision of wet bulk densities can be slightly improved, if the iterative scheme for the mass attenuation coefficient is applied (Fig. 2.5a). For core PS1725-2 wet bulk densities computed with a constant mass attenuation coefficient are compared with those derived from the iterative procedure. Below 1.9 g cm$^{-3}$ the iteration produces slightly smaller densities than are determined with a constant mass attenuation coefficient and are thus below the dotted 1:1 line. Above 1.9 g cm$^{-3}$ densities based on the iterative procedure are slightly higher.

If both data sets are plotted versus the wet bulk densities of the discrete samples the optimization essentially becomes obvious for high densities (>2.0 g cm$^{-3}$, Fig. 2.5b). After iteration densities are slightly closer to the dotted 1:1 line.

While this improvement is usually small and here only on the order of 1.3% ($\approx$ 0.02 g cm$^{-3}$) differences between assumed and true grain density affect the iteration more distinctly (Fig. 2.5c). As an example wet bulk densities of core PS1725-2 were calculated with constant grain densities of 2.65, 2.75 and 2.10 g cm$^{-3}$, values which are typical for calcareous, terrigenous and diatomaceous sediments. Displayed as cross-plot, calculations with grain densities of 2.65 and 2.75 g/cm$^{3}$ almost coincide and follow the dotted 1:1 line. However, computations with a grain density of 2.10 g cm$^{-3}$ differ by 1.3-3.6% (0.02-0.08 g cm$^{-3}$) from those with a grain density of 2.65 g cm$^{-3}$. This result is particularly important for cores composed of terrigenous and biogenic material of

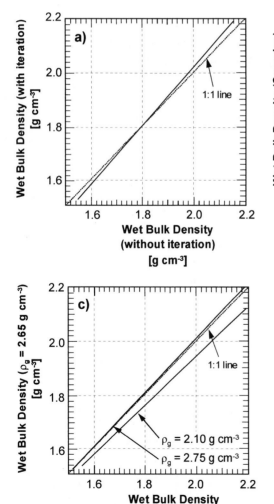

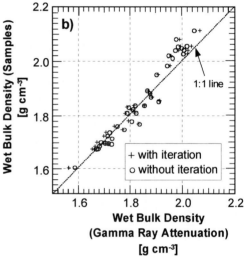

Fig. 2.5 Influence of an iterative mass attenuation coefficient determination on the precision of wet bulk densities. The gamma ray attenuation log of gravity core PS1725-2 was used as test data set. (a) Wet bulk densities calculated with a constant mass attenuation coefficient ('processing porosity' =50%) are displayed versus the data resulting from the iteration. A pore fluid density of 1.024 g cm$^{-3}$ and a constant grain density of 2.7 g cm$^{-3}$ were used, and the iteration was stopped if densities of two successive steps differed by less than 0.1‰ (b) Cross plot of wet bulk densities measured on discrete samples versus wet bulk densities calculated from gamma ray attenuation with a constant mass attenuation coefficient (O) and with the iterative scheme (+). (c) Influence of grain density on iteration. Three grain densities of 2.65, 2.75 and 2.1 g cm$^{-3}$ were used to calculate wet bulk densities. Modified after Gerland (1993).

significantly different grain densities like the terrigenous and diatomaceous components in core PS1821-6. Here, an iteration with a constant grain density of 2.65 g cm$^{-3}$ would give erroneous wet bulk densities in the diatomaceous parts. In such cores a depth-dependent grain density profile should be applied in combination with the iterative procedure, or wet bulk densities would be better calculated with a constant mass attenuation coefficient. However, if the grain density is almost constant downcore the iterative scheme could be applied without any problems.

### 2.1.2.3  Electrical Resistivity (Galvanic Method)

The electrical resistivity of water-saturated sediments depends on the resistivity of its solid and fluid constituents. However, as the sediment grains are poor conductors an electrical current mainly propagates in the pore fluid. The dominant transport mechanism is an electrolytic conduction by ions and molecules with an excess or deficiency of electrons. Hence, current propagation in water-saturated sediments actually transports material through the pore space, so that the resistivity depends on both the conductivity of the pore water and the microstructure of the sediment. The conductivity of pore water varies with its salinity and mobility and concentration of dissolved ions and molecules. The microstructure of the sediment is controlled by the amount and distribution of pore space and its capillarity and tortuosity. Thus, the electrical resistivity cannot be considered as a bulk parameter which strictly only depends on the relative amount of solid and fluid components, but as shown below, it can be used to derive porosity and wet bulk density as bulk parameters after calibration to a 'typical' sediment composition of a local sedimentation environment.

Several models were developed to describe current flow in rocks and water-saturated sediments theoretically. They encompass simple plane layered models (Waxman and Smits 1968) as well as complex approximations of pore space by self similar models (Sen et al. 1981), effective medium theories (Sheng 1991) and fractal geometries (Ruffet et al. 1991). In practice these models are of minor importance because often only few of the required model parameters are known. Here, an empirical equation is preferred which relates the resistivity of the wet sediment to its fractional porosity (Archie 1942)

$$F = \frac{R_s}{R_f} = a \cdot \phi^{-m} \qquad (2.12)$$

The ratio of the resistivity in sediment ($R_s$) to the resistivity in pore water ($R_f$) defines the formation (resistivity) factor (F). (a) and (m) are constants which characterize the sediment composition. As Archie (1942) assumed that (m) indicates the consolidation of the sediment it is also called cementation exponent (cf. Sect. 3.2.2). Several authors derived different values for (a) and (m). For an overview refer to Schön (1996). In marine sediments often Boyce's (1968) values (a = 1.3, m = 1.45), determined by studies on diatomaceous, silty to sandy arctic sediments, are applied. Nevertheless, these values can only be rough estimates. For absolutely correct porosities both constants must be calibrated by an additional porosity measurement, either on discrete samples or by gamma ray attenuation. Such calibrations are strictly only valid for that specific data set but, with little loss of accuracy, can be transferred to regional environments with similar sediment compositions. Wet bulk densities can then be calculated using Equation 2.3 and assuming a grain density.

Electrical resistivities are usually measured on split cores by a half-automated logging system (Bergmann 1996, Laser and Spieß 1998; Richter priv. comm.). It measures the resistivity ($R_s$) and temperature (T) by a small probe which is manually inserted into the upper few millimeters of the sediment. The resistivity of the interstitial pore water is simultaneously calculated from a calibration curve which defines the temperature-conductivity relation of standard sea water (35‰ salinity) by a fourth power law (Siedler and Peters 1986), (cf. App. B, Eq. 2.22).

The accuracy and resolution that can be achieved compared to measurements on discrete samples were studied on the terrigenous square barrel kastenlot core PS2178-5 from the Arctic Ocean. If both data sets are displayed as cross plots porosities mainly range within the dashed 10% error lines, while densities mainly differ by less than 5% (Fig. 2.6a). The core logs illustrate that the largest differences occur in the laminated sandy layers, particularly between 2.5 and 3.5 and between 4.3 and 5.1 m depth (Fig. 2.6b). Again, a drainage of the sandy layers, slight differences in the depth scale and different volumes considered by both methods probably cause this scatter. Nevertheless, in general both data sets agree very

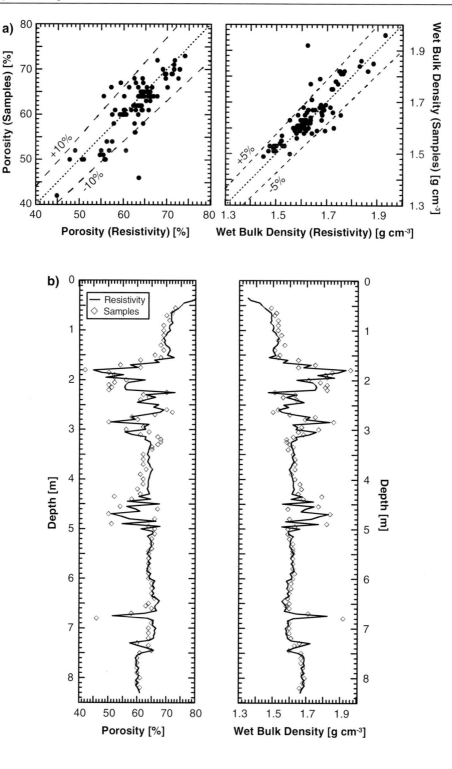

**Fig. 2.6** Comparison of porosities and wet bulk densities measured on discrete samples and by electrical resistivities. Boyce's (1968) values for the coefficients (a) and (m) and pore fluid and grain densities of 1.024 g cm$^{-3}$ and 2.67 g cm$^{-3}$ were used to convert formation factors into porosities and wet bulk densities. Wet and dry weights and volumes were analyzed on discrete samples. (a) Cross plots of both data sets for square barrel kastenlot core PS2178-5. The dashed lines indicate an error of 10% for the porosity and 5% for the density data. (b) Porosity and wet bulk density logs of core PS2178-5 derived from resistivity measurements. Superimposed are porosity and density values measured on discrete samples. Data from Bergmann (1996).

well having used Boyce's (1968) values for (a) and (m) and a typical terrigenous grain density of 2.67 g cm⁻³ for porosity and wet bulk density computations. This is valid because the sedimentation environment of core PS2178-5 from the Arctic Ocean is rather similar to the Bering Sea studied by Boyce (1968).

The influence of the sediment composition on the formation factor-porosity relation illustrates Figure 2.7 for six provinces in the South Atlantic. For each core porosities and formation factors were evaluated at the same core depths by wet and dry weights and volumes of discrete samples and by electrical resistivities. Displayed as a cross plot on a log-log scale the data sets of each core show different trends and thus different cementation exponents. Jackson et al. (1978) tried to relate such variations in the cementation exponent to the sphericity of sediment particles. Based on studies on artificial samples they found

higher cementation exponents if particles become less spherical. For natural sediments incorporating a large variety of terrigenous and biogenic particle sizes and shapes, it is difficult to verify similar relations. It is only obvious that coarse-grained calcareous foraminiferal oozes from the Hunter Gap show the lowest and fine-grained, diatom-bearing hemipelagic mud from the Congo Fan upwelling the highest cementation exponent. Simultaneously, the constant (a) decreases with increasing cementation exponent. Possibly, in natural sediments the amount and distribution of pore space are more important than the particle shape. Generally, cementation exponent (m), constant (a) and formation factor F together characterize a specific environment. Table 2.1 summarizes the values of a and m derived from a linear least square fit to the log-log display of the data sets and shortly describes the sediment compositions.

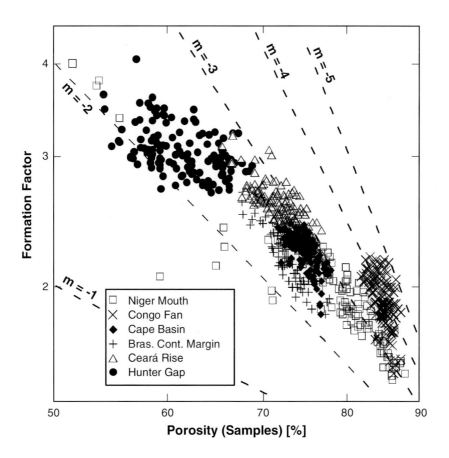

**Fig. 2.7** Formation factor versus porosity for six gravity cores retrieved from different sedimentation provinces in the South Atlantic. Porosities were determined on discrete samples by wet and dry weights and volumes, formation factors by resistivity measurements. The dashed lines indicate Archie's law for a = 1 and cementation exponents (m) between 1 and 5. For a description of the sedimentation provinces, core numbers, coring locations, sediment compositions, water depths and constants (a) and (m) derived from linear least square fits please refer to Table 2.1. Unpublished data from M. Richter, University Bremen.

With these values for (a) and (m) calibrated porosity logs are calculated which agree well with porosities determined on discrete samples (Fig. 2.8, black curves). The error in porosity that may result from using the 'standard' Boyce's (1968) values for (a) and (m) instead of those derived from the calibration appears in two different ways (gray curves). (1) The amplitude of the downcore porosity variations might be too large, as is illustrated by core GeoB2110-4 from the Brazilian Continental Margin. (2) The log might be shifted to higher or lower porosities, as is shown by core GeoB1517-1 from the Ceará Rise. Only if linear regression results in values for (a) and (m) close to Boyce's (1968) coefficients the error in the porosity log is negligible (core GeoB1701-4 from the Niger Mouth).

In order to compute absolutely correct wet bulk densities from calibrated porosity logs a grain density must be assumed. For most terrigenous and calcareous sediment cores this parameter is not very critical as it often only changes by few percent downcore (e.g. 2.6-2.7 g cm$^{-3}$). However, in cores from the Antarctic Polar Frontal Zone where an interlayering of diatomaceous and calcareous oozes indicates the advance and retreat of the oceanic front during glacial and interglacial stages grain densities may vary between about 2.0 and 2.8 g cm$^{-3}$.

Here, depth-dependent values must either be known or modeled in order to get correct wet bulk density variations from resistivity measurements. An example for this approach are resistivity measurements on *ODP* core 690C from the Maud Rise (Fig. 2.9). While the carbonate log (b) clearly indicates calcareous layers with high and diatomaceous layers with zero CaCO$_3$ percentages (O'Connell 1990), the resistivity-based porosity log (a) only scarcely reflects these lithological changes. The reason is that calcareous and diatomaceous oozes are characterized by high inter- and intraporosities incorporated in and between hollow foraminifera and diatom shells. In contrast, the wet bulk density log measured onboard of JOIDES Resolution by gamma ray attenuation ((c), gray curve) reveals pronounced variations. They obviously correlate with the CaCO$_3$-content and can thus only be attributed to downcore changes in the grain density. So, a grain density model (d) was developed. It averages the densities of carbonate ($\rho_{carb} = 2.8$ g cm$^{-3}$) and biogenic opal ($\rho_{opal} = 2.0$ g cm$^{-3}$; Barker, Kennnett et al., 1990) according to the fractional CaCO$_3$-content (C), $\rho_{model} = C \cdot \rho_{carb} + (1 - C) \cdot \rho_{opal}$. Based on this model wet bulk densities ((c), black curve) were derived from the porosity log which agree well with the gamma ray attenuation densities ((c), gray curve) and show less scatter.

**Table 2.1** Geographical coordinates, water depth, core length, region and composition of the sediment cores considered in Figure 2.7. The cementation exponent (m) and the constant (a) are derived from the slope and intercept of a linear least square fit to the log-log display of formation factors versus porosities.

| Core | Coordinates | Water Depth | Core Length | Region | Sediment Composition | Factor (a) | Cementation Exponent (m) |
|---|---|---|---|---|---|---|---|
| GeoB 1306-2 | 35°12.4'S 26°45.8'W | 4058 m | 6.97 m | Hunter Gap | foram.-nannofossil ooze, sandy | 1.6 | 1.3 |
| GeoB 1517-1 | 04°44.2'N 43°02.8'W | 4001 m | 6.89 m | Ceará Rise | nannofossil-foram. ooze | 1.3 | 2.1 |
| GeoB 1701-4 | 01°57.0'N 03°33.1'E | 4162 m | 7.92 m | Niger Mouth | clayey mud, foram. bearing | 1.4 | 1.4 |
| GeoB 1724-2 | 29°58.2'S 08°02.3'E | 5084 m | 7.45 m | Cape Basin | red clay | 1.0 | 2.8 |
| GeoB 2110-4 | 28°38.8'S 45°31.1'W | 3011 m | 8.41 m | Bras. Cont. Margin | pelagic clay. Foram. bearing | 0.8 | 3.3 |
| GeoB 2302-2 | 05°06.4'S 10°05.5'E | 1830 m | 14.18 m | Congo Fan | hemipelagic mud, diatom bearing, H$_2$S | 0.8 | 5.3 |

Though resistivities are only measured half-automatically including a manual insertion and removal of the probe, increments of 1 – 2 cm can be realized within an acceptable time so that more fine-scale structures can be resolved than by analysis of discrete samples. However, the real advantage compared to an automated gamma ray attenuation logging is that the resistivity measurement system can easily be transported, e.g. onboard of research vessels or to core repositories, while the transport of radioactive sources requires special safety precautions.

### 2.1.2.4   Electrical Resistivity (Inductive Method)

A second, non-destructive technique to determine porosities by resistivity measurements uses a coil as sensor. A current flowing through the coil induces an electric field in the unsplit sediment core while it is automatically transported through

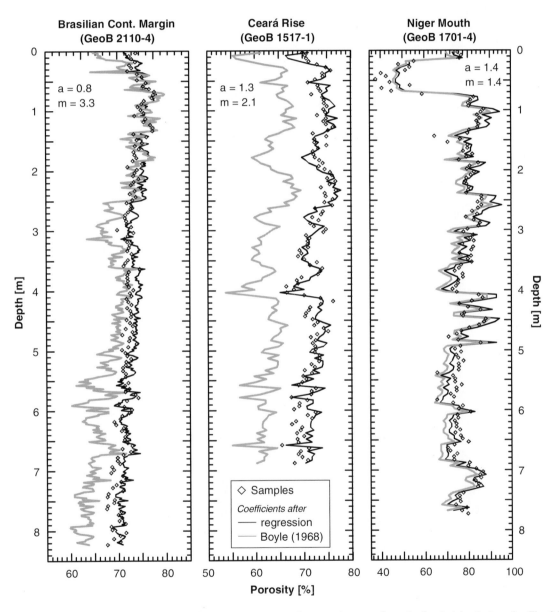

**Fig. 2.8** Porosity logs determined by resistivity measurements on three gravity cores from the South Atlantic (see also Fig. 2.7 and Table. 2.1). Gray curve: Boyce's (1968) values were used for the constants (a) and (m). Black curve: (a) and (m) were derived from the slope and intercept of a linear least square fit. These values are given at the top of each log. Superimposed are porosities determined on discrete samples by weight and volume measurements (unpublished data from P. Müller, University Bremen).

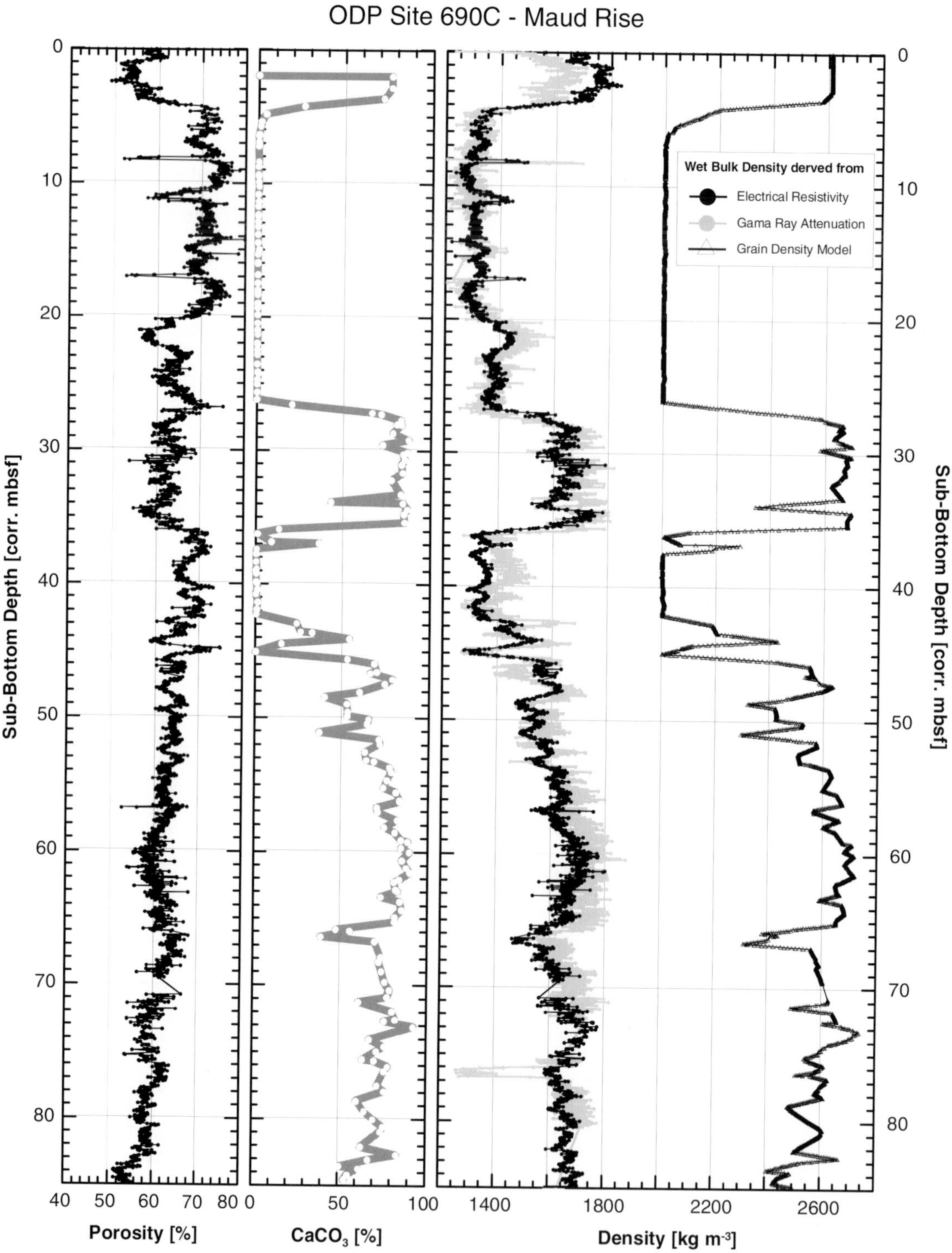

## ODP Site 690C - Maud Rise

**Wet Bulk Density derved from**
- ● Electrical Resistivity
- ● Gama Ray Attenuation
- △ Grain Density Model

Porosity [%]

CaCO₃ [%]

Density [kg m⁻³]

Sub-Bottom Depth [corr. mbsf]

**Fig. 2.9** Model based computation of a wet bulk density log from resistivity measurements on *ODP* core 690C. (a) Porosity log derived from formation factors having used Boyce's (1968) values for (a) and (m) in Archie's law. (b) Carbonate content (O'Conell 1990). (c) Wet bulk density log analyzed from gamma ray attenuation measurements onboard of JOIDES Resolution (gray curve). Superimposed is the wet bulk density log computed from electrical resistivity measurements on archive halves of the core (black curve) having used the grain density model shown in (d). Modified after Laser and Spieß (1998).

the center of the coil (Gerland et al. 1993). This induced electric field contains information on the magnetic and electric properties of the sediment.

Generally, the coil characteristic is defined by the quality value (Q)

$$Q = \omega \cdot \frac{L(\omega)}{R(\omega)} \qquad (2.13)$$

$(L(\omega))$ is the inductance, $(R(\omega))$ the resistance and $(\omega)$ the (angular) frequency of the alternating current flowing through the coil (Chelkowski 1980). The inductance depends on the number of windings, the length and diameter of the coil and the magnetic permeability of the coil material. The resistance is a superposition of the resistance of the coil material and losses of the electric field induced in the core. It increases with decreasing resistivity in the sediment. Whether the inductance or the resistance is of major importance depends on the frequency of the current flowing through the coil. Changes in the inductance can mainly be measured if currents of some kilohertz frequency or less are used. They simultaneously indicate variations in the magnetic susceptibility while the resistance is insensitive to changes in the resistivity of the sediment. In contrast, operating with currents of several megahertz allows to measure the resistivity of the sediment by

changes of the coil resistance while variations in the magnetic susceptibility do not affect the inductance. The examples presented here were measured with a commercial system (Scintrex CTU-2) which produces an output voltage that is proportional to the quality value (Q) of the coil at a frequency of 2.5 MHz, and after calibration is inversely proportional to the resistivity of the sediment.

The induced electric field is not confined to the coil position but extends over some sediment volume. Hence, measurements of the resistance integrate over the resistivity distribution on both sides of the coil and provide a smoothed, low-pass filtered resistivity record. The amount of sediment volume affected by the induction process increases with larger coil diameters. The shape of the smoothing function can be measured from the impulse response of a thin metal plate glued in an empty plastic core liner. For a coil of about 14 cm diameter this gaussian-shaped function has a half-width of 4 cm (Fig. 2.10), so that the effect of an infinitely small resistivity anomaly is smeared over a depth range of 10 – 15 cm. This smoothing effect is equivalent to convolution of a source wavelet with a reflectivity function in seismic applications and can accordingly be removed by deconvolution algorithms. How-

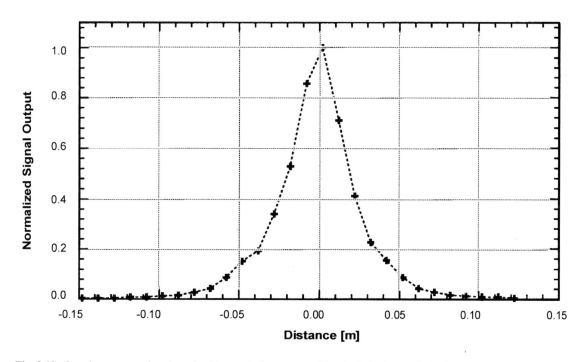

**Fig. 2.10**  Impulse response function of a thin metal plate measured by the inductive method with a coil of about 14 cm diameter. Modified after Gerland et al. (1993).

ever, only few applications from longcore paleo-magnetic studies are known up to now (Constable and Parker 1991, Weeks et al. 1993).

The low-pass filtering effect particularly becomes obvious if resistivity logs measured by the galvanic and inductive method are compared. Figure 2.11 displays such an example for a terrigenous core from the Weddell Sea (PS1635-1) and a biogenic foraminiferal and diatomaceous core from the Maud Rise (PS1836-3) in the Antarctic Ocean. Resistivities differ by maximum 15% (Fig. 2.11a), a rather high value which mainly results from core PS1635-1. Here, the downcore logs illustrate that the inductive methods produces lower resistivities than the galvanic method (Fig. 2.11b). In detail, the galvanic resistivity log reveals a lot of pronounced, fine-scale variations which cannot be resolved by induction measurements but are smeared along the core depth. For the biogenic core PS1836-3 this smoothing is not so important because lithology changes more gradually.

### 2.1.3    Permeability

Permeability describes how easy a fluid flows through a porous medium. Physically it is defined by Darcy's equation (cf. Sect. 3.5)

$$q = \frac{\kappa}{\eta} \cdot \frac{\partial p}{\partial x} \qquad (2.14)$$

which relates the flow rate (q) to the permeability ($\kappa$) of the pore space, the viscosity ($\eta$) of the pore fluid and the pressure gradient ($\partial p/\partial x$) causing the fluid flow. Simultaneously, permeability depends on the porosity and grain size distribution of the sediment, approximated by the mean grain size ($d_m$). Assuming that fluid flow can be simulated by an idealized flow through a bunch of capillaries with uniform radius ($d_m/2$) (Hagen Poiseuille's flow) permeabilities can for instance be estimated from Kozeny-Carman's equation (Carman 1956, Schopper 1982)

$$\kappa = \frac{d_m^2}{36k} \cdot \frac{\phi^3}{(1-\phi)^2} \qquad (2.15)$$

This relation is approximately valid for unconsolidated sediments of 30 - 80% porosity (Carman 1956). It is used for both geotechnical applications to estimate permeabilities of soil (Lambe and Whitman 1969) and seismic modeling of wave propagation in water-saturated sedi-

ments (Biot 1956a, b, Hovem and Ingram 1979, Hovem 1980, Ogushwitz 1985). ($\kappa$) is a constant which depends on pore shape and tortuosity. In case of parallel, cylindrical capillaries it is about 2, for spherical sediment particles about 5, and in case of high porosities $\geq 10$ (Carman 1956).

However, this is only one approach to estimate permeabilities from porosities and mean grain sizes. Other empirical relations exist, particularly for regions with hydrocarbon exploration (e.g. Gulf of Mexico, Bryant et al. 1975) or fluid venting (e.g. Middle Valley, Fisher et al. 1994) which compute depth-dependent permeabilities from porosity logs or take the grain size distribution and clay content into account.

Direct measurements of permeabilities in unconsolidated marine sediments are difficult, and only few examples are published. They confine to measurements on discrete samples with a specially developed tool (Lovell 1985), to indirect estimations by resistivity measurements (Lovell 1985), and to consolidation tests on ODP cores using a modified medical tool (Olsen et al. 1985). These measurements are necessary to correct for the elastic rebound (MacKillop et al. 1995) and to determine intrinsic permeabilities at the end of each consolidation step (Fisher et al. 1994). In Section 2.1.4.2 a numerical modeling and inversion scheme is described which estimates permeabilities from P-wave attenuation and dispersion curves.

### 2.1.4    Acoustic and Elastic Properties

Acoustic and elastic properties are directly concerned with seismic wave propagation in marine sediments. They encompass P- and S-wave velocity and attenuation and elastic moduli of the sediment frame and wet sediment. The most important parameter which controls size and resolution of sedimentary structures by seismic studies is the frequency content of the source signal. If the dominant frequency and bandwidth are high, fine-scale structures associated with pore space and grain size distribution affect the elastic wave propagation. This is subject of ultrasonic transmission measurements on sediment cores (Sects. 2.1.4 and 2.1.5). At lower frequencies larger scale features like interfaces with different physical properties above and below and bedforms like mud waves, erosion zones and channel levee systems are the dominant structures imaged by sediment echosounder and multi-channel seismic surveys (Sect. 2.1.6).

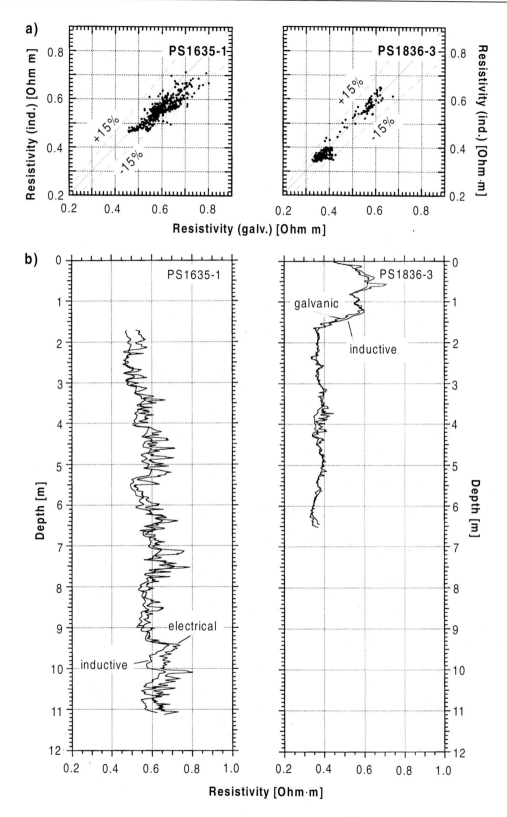

**Fig. 2.11** Comparison of electrical resistivities (galv.) measured with the small hand-held probe and determined by the inductive method (ind.) for the gravity cores PS1635-1 and PS1836-3. (a) Cross plots of both data sets. The dashed lines indicate a difference of 15%. (b) Downcore resistivity logs determined by both methods. Modified after Gerland et al. (1993).

In this chapter first Biot's viscoelastic model is summarized which simulates high- and low-frequency wave propagation in water-saturated sediments by computing phase velocity and attenuation curves. Subsequently, analysis techniques are introduced which derive P-wave velocities and attenuation coefficients from ultrasonic signals transmitted radially across sediment cores. Additional physical properties like S-wave velocity, elastic moduli and permeability are estimated by an inversion scheme.

### 2.1.4.1  Biot-Stoll Model

To describe wave propagation in marine sediments mathematically, various simple to complex models have been developed which approximate the sediment by a dilute suspension (Wood 1946) or an elastic, water-saturated frame (Gassmann 1951, Biot 1956a, b). The most common model which considers the microstructure of the sediment and simulates frequency-dependent wave propagation is based on Biot's theory (Biot 1956a, b). It includes Wood's suspension and Gassmann's elastic frame model as low-frequency approximations and combines acoustic and elastic parameters - P- and S-wave velocity and attenuation and elastic moduli - with physical and sedimentological parameters like mean grain size, porosity, density and permeability.

Based on Biot's fundamental work Stoll (e.g. 1974, 1977, 1989) reformulated the mathematical background of this theory with a simplified uniform nomenclature. Here, only the main physical principles and equations are summarized. For a detailed description refer to one of Stoll's publications or Biot's original papers.

The theory starts with a description of the microstructure by 11 parameters. The sediment grains are characterized by their grain density ($\rho_g$) and bulk modulus ($K_g$), the pore fluid by its density ($\rho_f$), bulk modulus ($K_f$) and viscosity ($\eta$). The porosity ($\phi$) quantifies the amount of pore space. Its shape and distribution are specified by the permeability ($\kappa$), a pore size parameter $a = d_m/3 \cdot \phi/(1-\phi)$, $d_m$ = mean grain size (Hovem and Ingram 1979, Courtney and Mayer 1993b), and structure factor $a' = 1 - r_0(1-\phi^{-1})$ $(0 \leq r_0 \leq 1)$ indicating a tortuosity of the pore space (Berryman 1980). The elasticity of the sediment frame is considered by its bulk and shear modulus ($K_m$ and $\mu_m$).

An elastic wave propagating in water-saturated sediments causes different displacements of the pore fluid and sediment frame due to their different elastic properties. As a result (global) fluid motion relative to the frame occurs and can approximately be described as Poiseuille's flow. The flow rate follows Darcy's equation and depends on the permeability and viscosity of the pore fluid. Viscous losses due to an interstitial pore water flow are the dominant damping mechanism. Intergranular friction or local fluid flow can additionally be included but are of minor importance in the frequency range considered here.

Based on generalized Hooke's law and Newton's 2. Axiom two equations of motions are necessary to quantify the different displacements of the sediment frame and pore fluid. For P-waves they are (Stoll 1989)

$$\nabla^2 (H \cdot e - C \cdot \zeta) = \frac{\partial^2}{\partial t^2} (\rho \cdot e - \rho_f \cdot \zeta)$$

(2.16a)

$$\nabla^2 (C \cdot e - M \cdot \zeta) = \frac{\partial^2}{\partial t^2} (\rho_f \cdot e - m \cdot \zeta) - \frac{\eta}{\kappa} \cdot \frac{\partial \zeta}{\partial t}$$

(2.16b)

Similar equations for S-waves are given by Stoll (1989). Equation 2.16a describes the motion of the sediment frame and Equation 2.16b the motion of the pore fluid relative to the frame.

$$e = div (\vec{u})$$  (2.16c)

and

$$\zeta = \phi \cdot div (\vec{u} - \vec{U})$$  (2.16d)

are the dilatations of the frame and between pore fluid and frame ($\vec{u}$ = displacement of the frame, $\vec{U}$ = displacement of the pore fluid). The term ($\eta/\kappa \cdot \partial \zeta / \partial t$) specifies the viscous losses due to global pore fluid flow, and the ratio ($\eta/\kappa$) the viscous flow resistance.

The coefficients (H), (C), and (M) define the elastic properties of the water-saturated model. They are associated with the bulk and shear moduli of the sediment grains, pore fluid and sediment frame ($K_g$), ($K_f$), ($K_m$), ($\mu_m$) and with the porosity ($\phi$) by

$$H = \frac{(K_g - K_m)^2}{D - K_m} + K_m + \frac{4}{3}\mu_m$$

(2.17a)

$$C = \frac{K_g (K_g - K_m)}{D - K_m}$$

(2.17b)

$$M = \frac{K_g^2}{D - K_m} \qquad (2.17c)$$

$$D = K_g \left(1 + \phi\left(K_g / K_f - 1\right)\right) \qquad (2.17d)$$

The apparent mass factor $m = a' \cdot \rho_f / \phi$ ($a' \geq 1$) in Equation 2.16b considers that not all of the pore fluid moves along the maximum pressure gradient in case of tortuous, curvilinear capillaries. As a result the pore fluid seems to be more dense, with higher inertia. ($a'$) is called structure factor and is equal to 1 in case of uniform parallel capillaries.

In the low frequency limit H - $4/3\mu_m$ (Eq. 2.17a) represents the bulk modulus computed by Gassmann (1951) for a 'closed system' with no pore fluid flow. If the shear modulus ($\mu_m$) of the frame is additionally zero, the sediment is approximated by a dilute suspension and Equation 2.17a reduces to the reciprocal bulk modulus of Wood's equation for 'zero' acoustic frequency ($K^{-1} = \phi/K_f + (1-\phi)/K_g$; Wood 1946).

Additionally, Biot (1956a, b) introduced a complex correction function (F) which accounts for a frequency-dependent viscous flow resistance ($\eta/\kappa$). In fact, while the assumption of an ideal Poiseuille flow is valid for lower frequencies, deviations of this law occur at higher frequencies. For short wavelengths the influence of pore fluid viscosity confines to a thin skin depth close to the sediment frame, so that the pore fluid seems be less viscous. To take these effects into account the complex function (F) modifies the viscous flow resistance ($\eta/\kappa$) as a function of pore size, pore fluid density, viscosity and frequency. A complete definition of (F) can be found in Stoll (1989).

The Equations of motions 2.16 are solved by a plane wave approach which leads to a 2 x 2 determinant for P-waves

$$\det \begin{pmatrix} H \cdot k^2 - \rho\omega^2 & \rho_f \omega^2 - C \cdot k^2 \\ C \cdot k^2 - \rho_f \omega^2 & m\omega^2 - M \cdot k^2 - i\omega F \eta / \kappa \end{pmatrix} = 0$$

$$(2.18)$$

and a similar determinant for S-waves (Stoll 1989). The variable $k(\omega) = k_r(\omega) + ik_i(\omega)$ is the complex wavenumber. Computations of the complex zeroes of the determinant result in the phase velocity $c(\omega) = \omega/k_r(\omega)$ and attenuation coefficient $\alpha(\omega) = k_i(\omega)$ as real and imaginary parts. Generally, the determinant for P-waves has two

and that for S-waves one zero representing two P- and one S-wave propagating in the porous medium. The first P- (P-wave of first kind) and the S-wave are well known from conventional seismic wave propagation in homogeneous, isotropic media. The second P-wave (P-wave of second kind) is similar to a diffusion wave which is exponentially attenuated and can only be detected by specially arranged experiments (Plona 1980).

An example of such frequency-dependent phase velocity and attenuation curves presents Figure 2.12 for P- and S-waves together with the slope (power n) of the attenuation curves ($\alpha = k \cdot f^n$). Three sets of physical properties representing typical sand, silt and clay (Table 2.2) were used as model parameters. The attenuation coefficients show a significant change in their frequency dependence. They follow an $\alpha \sim f^2$ power law for low and an $\alpha \sim \sqrt{f}$ law for high frequencies and indicate a continuously decreasing power (n) (from 2 to 0.5) near a characteristic frequency ($f_c$) = ($\eta\phi)/(2\pi\kappa\rho_f$). This characteristic frequency depends on the microstructure (porosity ($\phi$), permeability ($\kappa$)) and pore fluid (viscosity ($\eta$), density ($\rho_f$)) of the sediment and is shifted to higher frequencies if porosity increases and permeability decreases. Below the characteristic frequency the sediment frame and pore fluid move in phase and coupling between solid and fluid components is maximum. This behavior is typical for clayey sediments in which low permeability and viscous friction prevent any relative movement between pore fluid and frame up to several MHz, in spite of their high porosity. With increasing permeability and decreasing porosity the characteristic frequency diminishes so that in sandy sediments movements in phase only occur up to about 1 kHz. Above the characteristic frequency wavelengths are short enough to cause relative motions between pore fluid and frame.

Phase velocities are characterized by a low- and high-frequency plateau with constant values and a continuous velocity increase near the characteristic frequency. This dispersion is difficult to detect because it is confined to a small frequency band. Here, dispersion could only be detected from 1 – 10 kHz in sand, from 50 – 500 kHz in silt and above 100 MHz in clay. Generally, velocities in coarse-grained sands are higher than in fine-grained clays.

S-waves principally exhibit the same attenuation and velocity characteristics as P-waves. However, at the same frequency attenuation is

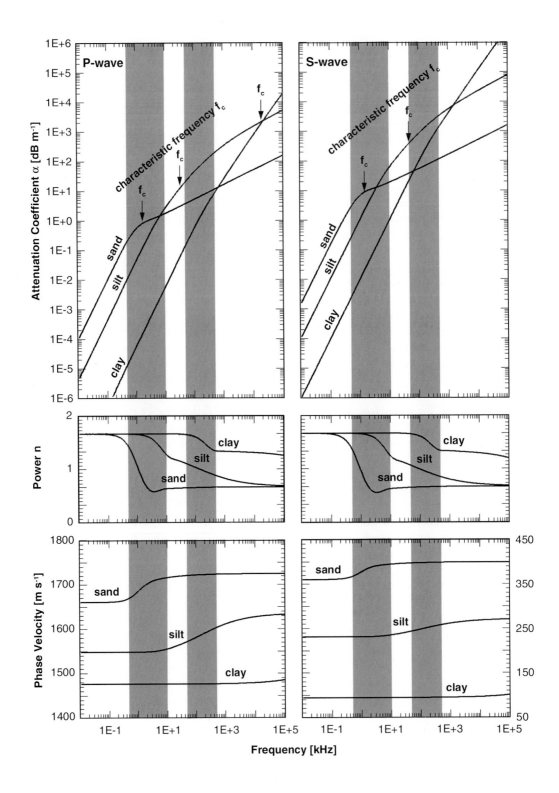

**Fig. 2.12** Frequency-dependent attenuation and phase velocity curves and power n of the attenuation law $\alpha = k \cdot f^{n}$ computed for P-and S-waves in typical sand, silt and clay. The gray-shaded areas indicate frequency bands typical for ultrasonic measurements on sediment cores (50-500 kHz) and sediment echosounder surveys (0.5-10 kHz).

about one order higher, and velocities are significantly lower than for P-waves. The consequence is that S-waves are very difficult to record due to their high attenuation, though they are of great value for identifying fine-scale variations in the elasticity and microstructure of marine sediments. This is even valid if the low S-wave velocities are taken into account and lower frequencies are used for S-wave measurements than for P-wave recordings.

The two gray-shaded areas in Figure 2.12 mark two frequency bands typical for ultrasonic studies on sediment cores (50-500 kHz) and sediment echosounder surveys (0.5-10 kHz). They are displayed in order to point to one characteristic of acoustic measurements. Attenuation coefficients analyzed from ultrasonic measurements on sediment cores cannot directly be transferred to sediment echosounder or seismic surveys. Primarily they only reflect the microstructure of the sediment. Rough estimates of the attenuation in seismic recordings from ultrasonic core measurements can be derived if ultrasonic attenuation is modeled, and attenuation coefficients are extrapolated to lower frequencies by such model curves.

### 2.1.4.2   Full Waveform Ultrasonic Core Logging

To measure the P-wave velocity and attenuation illustrated by Biot-Stoll's model an automated, PC-controlled logging system was developed which records and stores digital ultrasonic P-waveforms transmitted radially across marine sediment cores (Breitzke and Spieß 1993). These transmission measurements can be done at arbitrary small depth increments so that the resulting

**Table 2.2**   Physical properties of sediment grains, pore fluid and sediment frame used for the computation of attenuation and phase velocity curves according to Biot-Stoll's sediment model (Fig. 2.12).

| Parameter | Sand | Silt | Clay |
|---|---|---|---|
| *Sediment Grains* | | | |
| Bulk Modulus $K_g$ [$10^9$ Pa] | 38 | 38 | 38 |
| Density $\rho_g$ [g cm$^{-3}$] | 2.67 | 2.67 | 2.67 |
| *Pore Fluid* | | | |
| Bulk Modulus $K_f$ [$10^9$ Pa] | 2.37 | 2.37 | 2.37 |
| Density $\rho_f$ [g cm$^{-3}$] | 1.024 | 1.024 | 1.024 |
| Viscosity $\eta$ [$10^{-3}$ Pa·s] | 1.07 | 1.07 | 1.07 |
| *Sediment Frame* | | | |
| Bulk Modulus $K_m$ [$10^6$ Pa] | 400 | 150 | 20 |
| Shear Modulus $\mu_m$ [$10^6$ Pa] | 240 | 90 | 12 |
| Poisson Ratio $\sigma_m$ | 0.25 | 0.25 | 0.25 |
| *Pore Space* | | | |
| Porosity $\phi$ [%] | 50 | 60 | 80 |
| Mean Grain Size $d_m$ [$10^{-6}$ m] | 70 | 30 | 2 |
| Permeability $\kappa$ [m$^2$] | $5.4 \cdot 10^{-11}$ | $2.3 \cdot 10^{-12}$ | $7.1 \cdot 10^{-15}$ |
| Pore Size Parameter $a = d_m/3 \, \phi/(1-\phi)$ [$10^{-6}$ m] | 23 | 15 | 2.7 |
| Ratio $\kappa/a^2$ | 0.1 | 0.01 | 0.001 |
| Structure Factor $a' = 1 - r_0(1-\phi^{-1})$ | 1.5 | 1.3 | 1.1 |
| Constant $r_0$ | 0.5 | 0.5 | 0.5 |

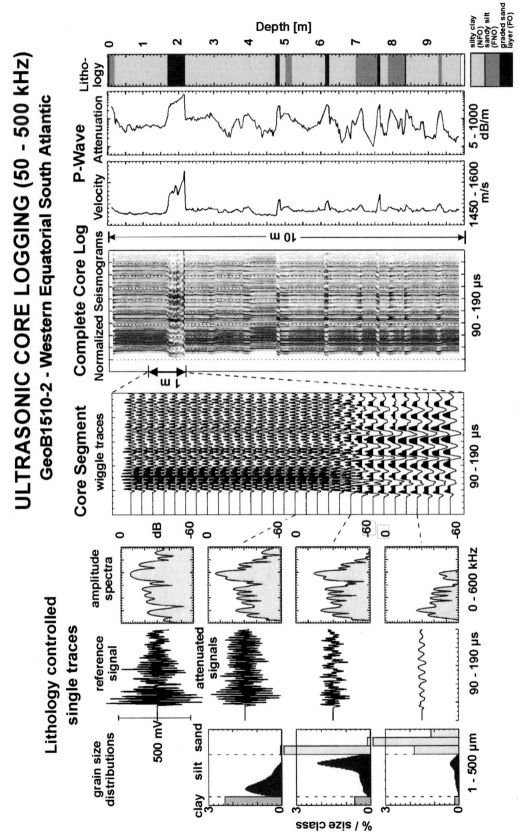

**Fig. 2.13** Full waveform ultrasonic core logging, from lithology controlled single traces to the gray shaded pixel graphic of transmission seismograms, and P-wave velocity and attenuation logs derived from the transmission data. The single traces on the left-hand side reflect true amplitudes while the wiggle traces of the core segment and the pixel graphic are normalized to maximum values. Data from Breitzke et al. (1996).

seismogram sections can be combined to an ultrasonic image of the core.

Figure 2.13 displays the most prominent effects involved in full waveform ultrasonic core logging. Gravity core GeoB1510-2 from the western equatorial South Atlantic serves as an example. The lithology controlled single traces and amplitude spectra demonstrate the influence of increasing grain sizes on attenuation and frequency content of transmission seismograms. Compared to a reference signal in distilled water, the signal shape remains almost unchanged in case of wave propagation in fine-grained clayey sediments (1st attenuated trace). With an increasing amount of silty and sandy particles the signal amplitudes are reduced due to an enhanced attenuation of high-frequency components (2nd and 3rd attenuated trace). This attenuation is accompanied by a change in signal shape, an effect which is particularly obvious in the normalized wiggle trace display of the 1 m long core segment. While the upper part of this segment is composed of fine-grained nannofossil ooze, a calcareous foraminiferal turbidite occurs in the lower part. The downward coarsening of the graded bedding causes successively lower-frequency signals which can easily be distinguished from the high-frequency transmission seismograms in the upper fine-grained part. Additionally, first arrival times are lower in the coarse-grained turbidite than in fine-grained nannofossil oozes indicating higher velocities in silty and sandy sediments than in the clayey part. A conversion of the normalized wiggle traces to a gray-shaded pixel graphic allows us to present the full transmission seismogram information on a handy scale. In this ultrasonic image of the sediment core lithological changes appear as smooth or sharp phase discontinuities resulting from the low-frequency waveforms in silty and sandy layers. Some of these layers indicate a graded bedding by downward prograding phases (1.60-2.10 m, 4.70-4.80 m, 7.50-7.60 m). The P-wave velocity and attenuation log analyzed from the transmission seismograms support the interpretation. Coarse-grained sandy layers are characterized by high P-wave velocities and attenuation coefficients while fine-grained parts reveal low values in both parameters. Especially, attenuation coefficients reflect lithological changes much more sensitively than P-wave velocities.

While P-wave velocities are determined online during core logging using a cross-correlation technique for the first arrival detection (Breitzke and Spieß 1993)

$$v_P = \frac{d_{outside} - 2d_{liner}}{t - 2t_{liner}} \qquad (2.19)$$

($d_{outside}$ = outer core diameter, $2\,d_{liner}$ = double liner wall thickness, $t$ = detected first arrival, $2t_{liner}$ = travel time across both liner walls), attenuation coefficients are analyzed by a post-processing routine. Several notches in the amplitude spectra of the transmission seismograms caused by the resonance characteristics of the ultrasonic transducers required a modification of standard attenuation analysis techniques (e.g. Jannsen et al. 1985, Tonn 1989, 1991). Here, a modification of the spectral ratio method is applied (Breitzke et al. 1996). It defines a window of bandwidth ($b_i = f_{ui} - f_{li}$) in which the spectral amplitudes are summed (Fig. 2.14a).

$$\overline{A}(f_{mi}, x) = \sum_{f_i = f_{li}}^{f_{ui}} A(f_i, x) \qquad (2.20)$$

The resulting value $\overline{A}(f_{mi}, x)$ is related to that part of the frequency band which predominantly contributes to the spectral sum, i.e. to the arithmetic mean frequency of the spectral amplitude distribution ($f_{mi}$) within the ith band. Subsequently, for a continuously moving window a series of attenuation coefficients ($\alpha(f_{mi})$) is computed from the natural logarithm of the spectral ratio of the attenuated and reference signal

$$\alpha(f_{mi}) = \ln\left[\frac{\overline{A}(f_{mi}, x)}{\overline{A}_{ref}(f_{mi}, x)}\right] \bigg/ x = k \cdot f^n \qquad (2.21)$$

Plotted in a log $\alpha$ - log f diagram the power (n) and logarithmic attenuation factor (log k) can be determined from the slope and intercept of a linear least square fit to the series of ($f_{mi}, \alpha(f_{mi})$) pairs (Fig. 2.14b). Finally, a smoothed attenuation coefficient $\alpha(f) = k \cdot f^n$ is calculated for the frequency (f) using these values for (k) and (n).

Figure 2.14b shows several attenuation curves ($\alpha(f_{mi})$) analyzed along the turbidite layer of core GeoB1510-2. With downward-coarsening grain sizes attenuation coefficients increase. Each linear regression to one of these curves provide a power (n) and attenuation factor (k), and thus one value $\alpha = k \cdot f^n$ on the attenuation log of the complete core (Fig. 2.14c).

## ATTENUATION ANALYSIS

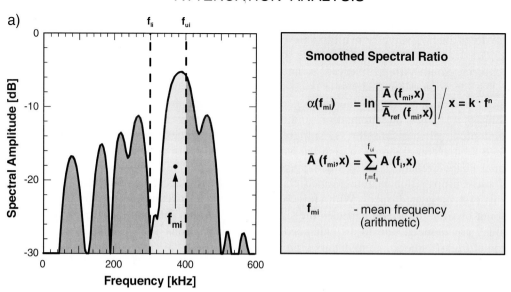

a)

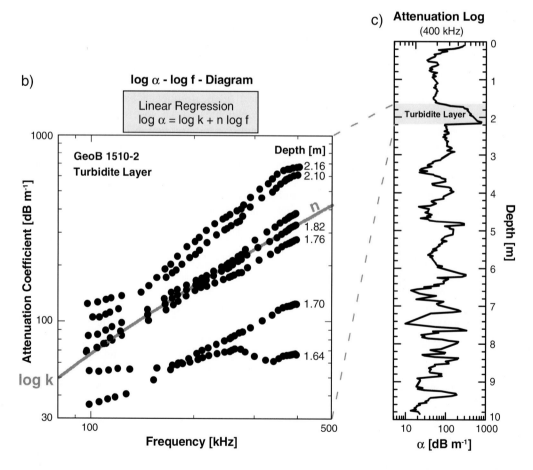

b)

c)

**Fig. 2.14** Attenuation analysis by the smoothed spectral ratio method. (a) Definition of a moving window of bandwidth $b_i = f_{ui} - f_{li}$, in which the spectral amplitudes are summed. (b) Seven attenuation curves analyzed from the turbidite layer of gravity core GeoB1510-2 and linear regression to the attenuation curve in 1.82 m depth. (c) Attenuation log of gravity core GeoB1510-2 for 400 kHz frequency. Data from Breitzke et al. (1996).

As the P-wave attenuation coefficient obviously depends on the grain size distribution of the sediment it can be used as a proxy parameter for the mean grain size, i.e. for a sedimentological parameter which is usually only measured at coarse increments due to the time-consuming grain size analysis methods. For instance, this can be of major importance in current controlled sedimentation environments where high-resolution grain size logs might indicate reduced or enhanced current intensities.

If the attenuation coefficients and mean grain sizes analyzed on discrete samples of core GeoB1510-2 are displayed as a cross plot a second order polynomial can be derived from a least square fit (Fig. 2.15a). This regression curve then allows to predict mean grain sizes using the attenuation coefficient as proxy parameter. The accuracy of the predicted mean grain sizes illus-

trates the core log in Figure 2.15b. The predicted gray shaded log agrees well with the superimposed dots of the measured data.

Similarly, P-wave velocities can also be used as proxy parameters for a mean grain size prediction. However, as they cover only a small range (1450-1650 m s$^{-1}$) compared to attenuation coefficients (20-800 dB m$^{-1}$) they reflect grain size variations less sensitively.

Generally, it should be kept in mind, that the regression curve in Figure 2.15 is only an example. Its applicability is restricted to that range of attenuation coefficients for which the regression curve was determined and to similar sedimentation environments (calcareous foraminiferal and nannofossil ooze). For other sediment compositions new regression curves must be determined, which are again only valid for that specific sedimentological setting.

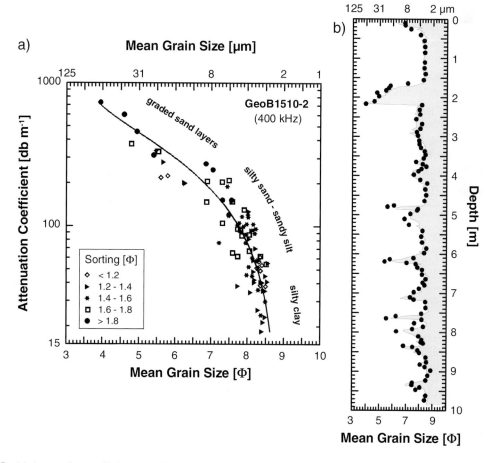

**Fig. 2.15** (a) Attenuation coefficients (at 400 kHz) of gravity core GeoB1510-2 versus mean grain sizes. The solid line indicates a second degree polynomial used to predict mean grain sizes from attenuation coefficients. (b) Comparison of the predicted mean grain size log (gray shaded) with the data measured on discrete samples (solid dots). Mean grain sizes are given in $\Phi = -log_2 d$, $d$ = grain diameter in mm. Modified after Breitzke et al. (1996).

Biot-Stoll's theory allows us to model P-wave velocities and attenuation coefficients analyzed from transmission seismograms. As an example Figure 2.16 displays six data sets for the turbidite layer of core GeoB1510-2. While attenuation coefficients were analyzed as described above frequency dependent P-wave velocities were determined from successive bandpass filtered transmission seismograms (Courtney and Mayer 1993a, Breitzke 1997). Porosities and mean grain sizes enter the modeling computations via the pore size parameter (a) and structure factor (a'). Physical properties of the pore fluid and sediment grains are the same as given in Table 2.2. A bulk and shear modulus of 10 and 6 MPa account for the elasticity of the frame. As the permeability is the parameter which is usually unknown but has the strongest influence on attenuation and velocity dispersion, model curves were computed for three constant ratios $\kappa/a^2 = 0.030$, 0.010 and 0.003 of permeability ($\kappa$) and pore size parameter (a). The resulting permeabilities are given in each diagram. These theoretical curves show that the attenuation and velocity data between 170 and 182 cm depth can consistently be modeled by an appropriate set of input parameters. Viscous losses due to a global pore water flow through the sediment are sufficient to explain the attenuation in these sediments. Only if the turbidite base is approached (188–210 cm depth) the attenuation and velocity dispersion data successively deviate from the model curves probably due to an increasing amount of coarse-grained foraminifera. An additional damping

mechanism which might either be scattering or resonance within the hollow foraminifera must be considered.

Based on this modeling computations an inversion scheme was developed which automatically iterates the permeability and minimizes the difference between measured and modeled attenuation and velocity data in a least square sense (Courtney and Mayer 1993b, Breitzke 1997). As a result S-wave velocities and attenuation coefficients, permeabilities and elastic moduli of water-saturated sediments can be estimated. They are strictly only valid if attenuation and velocity dispersion can be explained by viscous losses. In coarse-grained parts deviations must be taken into account for the estimated parameters, too. Applied to the data of core GeoB1510-2 in 170, 176 and 182 cm depth S-wave velocities of 67, 68 and 74 m s$^{-1}$ and permeabilities of $5 \cdot 10^{-13}$, $1 \cdot 10^{-12}$ and $3 \cdot 10^{-12}$ m$^2$ result from this inversion scheme.

### 2.1.5    Sediment Classification

Full waveform ultrasonic core logging was applied to terrigenous and biogenic sediment cores in order to analyze P-wave velocities and attenuation coefficients typical for the different settings. Together with the bulk parameters and the physical properties estimated by the inversion scheme they form the data base for a sediment classification which identifies different sediment types from their acoustic and elastic properties. Table 2.3 summarizes the cores used for this sediment classification.

**Table 2.3**  Geographical coordinates, water depth, core length, region and composition of the sediment cores considered for the sediment classification in Section 2.1.5.

| Core | Coordinates | Water Depth | Core Length | Region | Sediment Composition |
|------|-------------|-------------|-------------|--------|----------------------|
| 40KL | 07°33.1'N 85°29.7'E | 3814 m | 8.46 m | Bengal Fan | terrigenous clay, silt, sand |
| 47KL | 11°10.9'N 88°24.9'E | 3293 m | 10.00 m | Bengal Fan | terrigenous clay, silt, sand; foram. and nannofossil ooze |
| GeoB 2821-1 | 30°27.1'S 38°48.9'W | 3941 m | 8.19 m | Rio Grande Rise | foram. and nannofossil ooze |
| PS 2567-2 | 46°56.1'S 06°15.4'E | 4102 m | 17.65 m | Meteor Rise | diatomaceous mud/ooze; few foram. and nannofossil layers |

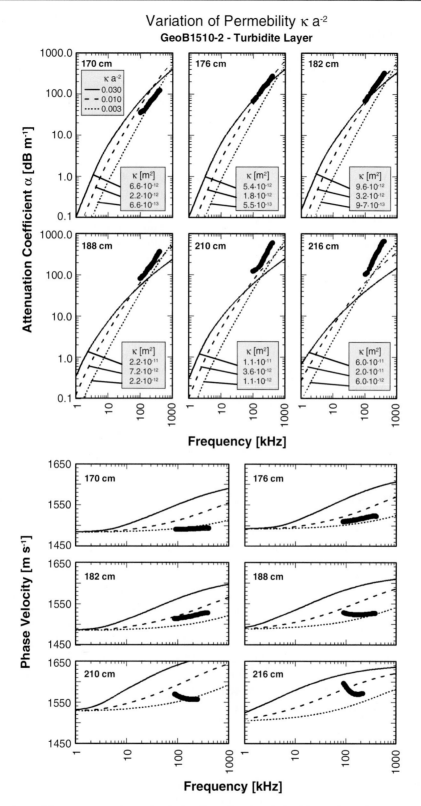

**Fig. 2.16** Comparison of P-wave attenuation and velocity dispersion data derived from ultrasonic transmission seismograms with theoretical curves based on Biot-Stoll's model for six traces of the turbidite layer of gravity core GeoB1510-2. Permeabilities vary in the model curves according to constant ratios $\kappa\,a^{-2}$ = 0.030, 0.010, 0.003 ($\kappa$ = permeability, a = pore size parameter). The resulting permeabilities are given in each diagram. Modified after Breitzke et al. (1996).

### 2.1.5.1  Full Waveform Core Logs as Acoustic Images

That terrigenous, calcareous and biogenic sili-ceous sediments differ distinctly in their acous-tic properties is shown by four transmission seismogram sections in Figure 2.17. Terrigenous sediments from the Bengal Fan (40KL, 47KL) are composed of upward-fining sequences of turbidites characterized by upward decreasing attenuations and P-wave velocities. Coarse-grained basal sandy layers can easily be located by low-frequency waveforms and high P-wave velocities.

Calcareous sediments from the Rio Grande Rise (GeoB2821-1) in the western South Atlantic also exhibit high- and low-frequency signals which scarcely differ in their P-wave velocities. In these sediments high-frequency signals indi-cate fine-grained nannofossil ooze while low-fre-quency signals image coarse-grained forami-niferal ooze.

The sediment core from the Meteor Rise (PS2567-2) in the Antarctic Ocean is composed of diatomaceous and foraminiferal-nannofossil ooze deposited during an advance and retreat of the Polar Frontal Zone in glacial and interglacial stages. Opal-rich, diatomaceous ooze can be iden-tified from high-frequency signals while fora-miniferal-nannofossil ooze causes higher attenua-tion and low-frequency signals. P-wave velocities again only show smooth variations.

Acoustic images of the complete core litholo-gies present the color-encoded graphics of the transmission seismograms, in comparison to the lithology derived from visual core inspection (Fig. 2.18). Instead of normalized transmission seismograms instantaneous frequencies are dis-played here (Taner et al. 1979). They reflect the dominant frequency of each transmission seismo-gram as time-dependent amplitude, and thus di-rectly indicate the attenuation. Highly attenuated low-frequency seismograms appear as warm red to white colors while parts with low attenuation and high-frequency seismograms are represented by cool green to black colors.

In these attenuation images the sandy turbidite bases in the terrigenous cores from the Bengal Fan (40KL, 47KL) can easily be distinguished. Graded beddings can also be identified from slightly prograding phases and continuously de-creasing travel times. In contrast, the transition to calcareous, pelagic sediments in the upper part (>5.6 m) of core 47KL is rather difficult to detect. Only above 3.2 m depth a slightly in-creased attenuation can be observed by slightly warmer colors at higher transmission times (>140 µs). In this part of the core (>3.2 m) sedi-ments are mainly composed of coarse-grained fora-miniferal ooze, while farther downcore (3.2-5.6 m) fine-grained nannofossils prevail in the pelagic sediments.

The acoustic image of the calcareous core from the Rio Grande Rise (GeoB2821-1) shows much more lithological changes than the visual core de-scription. Cool colors between 1.5-2.5 m depth indicate unusually fine grain sizes (Breitzke 1997). Alternately yellow/red and blue/black colors in the lower part of the core (> 6.0 m) reflect an inter-layering of fine-grained nannofossil and coarse-grained foraminiferal ooze. Dating by orbital tun-ing shows that this interlayering coincides with the 41 ky cycle of obliquity (von Dobeneck and Schmieder 1998) so that fine-grained oozes domi-nate during glacial and coarse-grained oozes dur-ing interglacial stages.

The opal-rich diatomaceous sediments in core PS2567-2 from the Meteor Rise are characterized by a very low attenuation. Only 2-3 calcareous layers with significantly higher attenuation (yel-low and red colors) can be identified as promi-nent lithological changes.

### 2.1.5.2  P- and S-Wave Velocity, Attenuation, Elastic Moduli and Permeability

As the acoustic properties of water-saturated sedi-ments are strongly controlled by the amount and distribution of pore space, cross plots of P-wave velocity and attenuation coefficient versus poros-ity clearly indicate the different bulk and elastic properties of terrigenous and biogenic sediments and can thus be used for an acoustic classification of the lithology. Additional S-wave velocities and elastic moduli estimated by least-square inversion specify the amount of bulk and shear moduli which contribute to the P-wave velocity.

The cross plots of the P-wave parameters of the four cores considered above illustrate that ter-rigenous, calcareous and diatomaceous sediments can uniquely be identified from their position in both diagrams (Fig. 2.19). In terrigenous sedi-ments (40KL, 47KL) P-wave velocities and at-tenuation coefficients increase with decreasing porosities. Computed S-wave velocities are very low ($\approx$ 60-65 m s$^{-1}$) and almost independent of

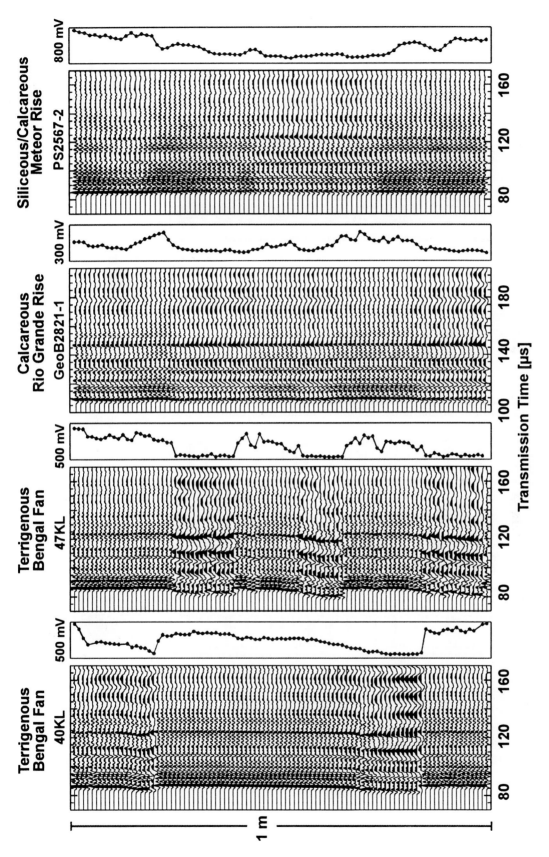

**Fig. 2.17** Normalized transmission seismogram sections of four 1 m long core segments from different terrigenous and biogenic sedimentation environments. Seismograms were recorded with 1 cm spacing. Core depths are 5.49-6.43 m (40KL), 6.03-6.97 m (47KL), 5.26-6.14 m (GeoB2821-1), 12.77-13.60 m (PS2567-2). Maximum amplitudes of the transmission seismograms are plotted next to each section from 0 to ... mV, as given at the top.

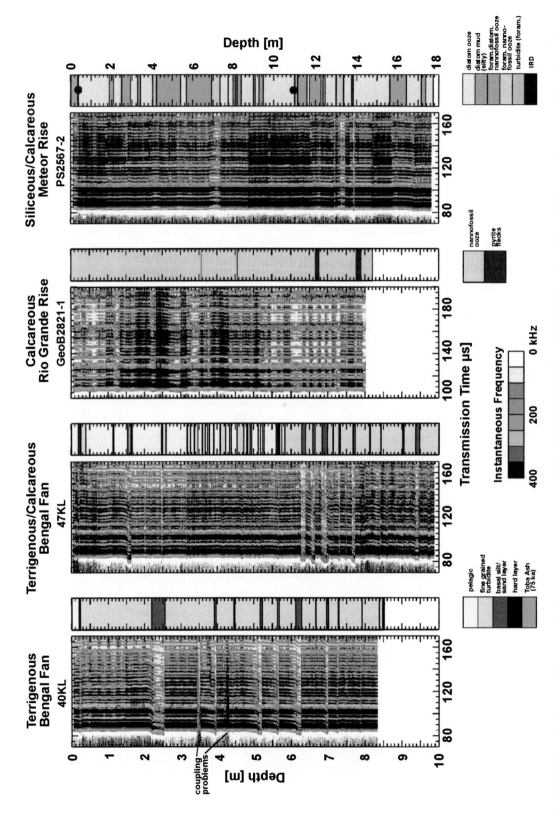

**Fig. 2.18** Ultrasonic images of the transmission measurements on cores 40KL, 47KL, GeoB2821-1 and PS2567-2 retrieved from different terrigenous and biogenic sedimentation environments. Displayed are the color-encoded instantaneous frequencies of the transmission seismograms and the lithology derived from visual core inspections.

porosity (Fig. 2.20). Accordingly, shear moduli are also low and do not vary very much so that higher P-wave velocities in terrigenous sediments mainly result from higher bulk moduli (Fig. 2.21).

If calcareous, particularly foraminiferal components (FNO) are added to terrigenous sediments porosities become higher ($\approx$ 70-80%, 47KL). P-wave velocities slightly increase from their minimum of 1475 m s$^{-1}$ at 70% porosity (Fig. 2.19a) due to an increase in the shear moduli (Fig. 2.21) . However, a much more pronounced increase can be observed in the P-wave attenuation coefficients (Fig. 2.19b). For porosities of about 80% they reach the same values as terrigenous sediments of about 55% porosity, but can easily be distinguished because of higher P-wave velocities in terrigenous sediments.

Calcareous foraminiferal and nannofossil oozes (NFO) show similar trends in both P-wave parameters as terrigenous sediments, but are shifted to higher porosities due to their additional intraporosities (GeoB2821-1).

In diatomaceous oozes P-wave velocities increase again (Fig. 2.19) though porosities are very high (>80%, PS2567-2). Here, the diatom shells build a very stiff frame which causes high shear moduli and S-wave velocities (Figs. 2.20, 2.21). It is this increase in the shear moduli which only accounts for the higher P-wave velocities while the bulk moduli remain almost constant and are close to the bulk modulus of sea water. P-wave attenuation coefficients are very low in these high-porosity sediments.

Permeabilities estimated from the least square inversion mainly reflect the attenuation characteristics of the different sediment types (Fig. 2.22). They reach lowest values of about $5 \cdot 10^{-14}$ m$^2$ in fine-grained clayey mud and nannofossil ooze. Highest values of about $5 \cdot 10^{-11}$ m$^2$ occur in diatomaceous ooze due to their high porosities. Nevertheless, it should be kept in mind that these permeabilities are only estimates based on the input parameters and assumptions incorporated in Biot-Stoll's model. For instance one of these assumptions is that only mean grain sizes are used, but the influence of grain size distributions is neglected. Additionally, the total porosity is usually used as input parameter for the inversion scheme without differentiation between inter- and intraporosities. Comparisons of these estimated permeabilities with direct measurements unfortunately do not exist up to know.

### 2.1.6    Sediment Echosounding

While ultrasonic measurements are used to study the structure and composition of sediment cores, sediment echosounders are hull-mounted acoustic systems which image the upper 10-200 m of sediment coverage by remote sensing surveys. They operate with frequencies around 3.5-4.0 kHz. The examples presented here were digitally recorded with the narrow-beam Parasound echosounder and ParaDIGMA recording system (Spieß 1993).

#### 2.1.6.1   Synthetic Seismograms

Figure 2.23 displays a Parasound seismogram section recorded across an inactive channel of the Bengal Fan. The sediments of the terrace were sampled by a 10 m long piston core (47KL). Its acoustic and bulk properties can either be directly compared to the echosounder recordings or by computations of synthetic seismograms. Such modeling requires P-wave velocity and wet bulk density logs as input parameters. From the product of both parameters acoustic impedances $I = v_P \cdot \rho$ are calculated. Changes in the acoustic impedance cause reflections of the normally incident acoustic waves. The amplitude of such reflections is determined by the normal incidence reflection coefficient $R = (I_2 - I_1) / (I_2 + I_1)$, with ($I_1$) and ($I_2$) being the impedances above and below the interface. From the series of reflection coefficients the reflectivity can be computed as impulse response function, including all internal multiples. Convolution with a source wavelet finally provides the synthetic seismogram. In practice, different time- and frequency domain methods exist for synthetic seismogram computations. For an overview, refer to books on theoretical seismology (e.g. Aki and Richards 1980). Here, we used a time domain method called 'state space approach' (Mendel et al. 1979).

Figure 2.24 compares the synthetic seismogram computed for core 47KL with the Parasound seismograms recorded at the coring site. The P-wave velocity and wet bulk density logs used as input parameters are displayed on the right-hand side, together with the attenuation coefficient log as grain size indicator, the carbonate content and an oxygen isotope ($\delta^{18}$O-) stratigraphy. An enlarged part of the gray shaded Parasound seismogram section is shown on the left-hand side. The comparison of synthetic and Parasound data indicates some core deformations. If the synthetic

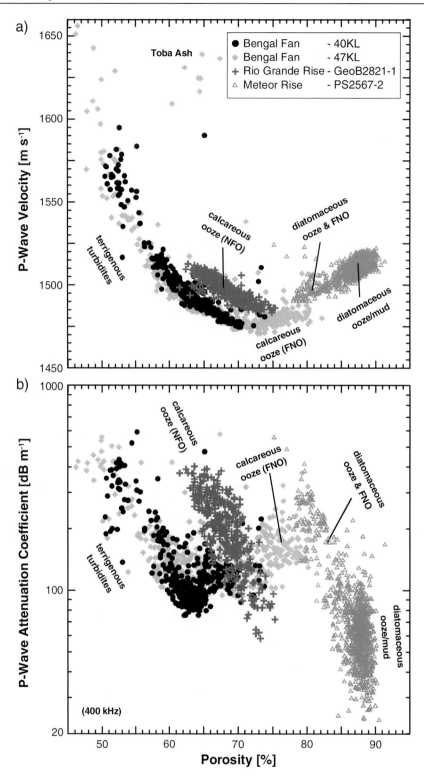

**Fig. 2.19**  (a) P-wave velocities and (b) attenuation coefficients (at 400 kHz) versus porosities displayed for cores 40KL, 47KL, GeoB2821-1 and PS2567-2.

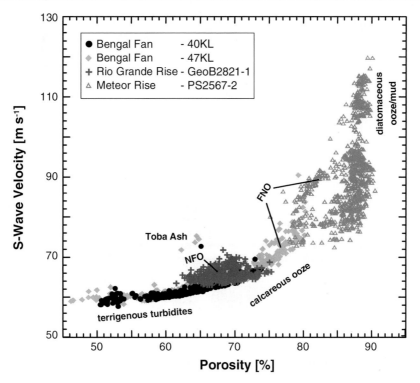

**Fig. 2.20** S-wave velocities estimated from least square inversion versus porosities for cores 40KL, 47KL, GeoB2821-1 and PS2567-2.

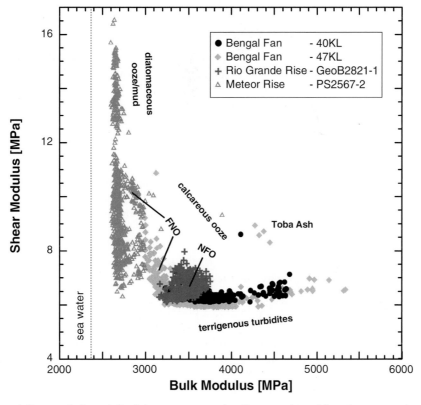

**Fig. 2.21** Shear moduli versus bulk moduli of the water-saturated sediments estimated from least square inversion for cores 40KL, 47KL, GeoB2821-1 and PS2567-2.

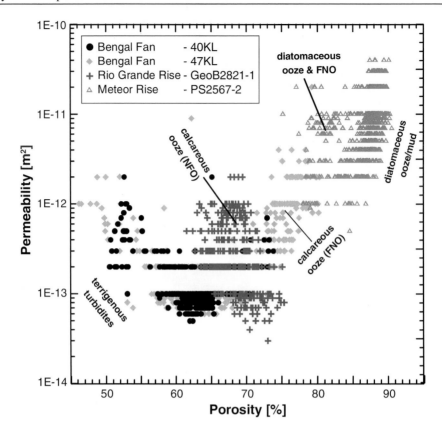

**Fig. 2.22** Permeabilities versus porosities estimated from least square inversion for cores 40KL, 47KL, GeoB2821-1 and PS2567-2.

seismograms are leveled to the prominent reflection caused by the Toba Ash in 1.6 m depth about 95 cm sediment are missing in the overlying younger part of the core. Deeper reflections, particularly caused by the series of turbidites below 6 m depth, can easily be correlated between synthetic and Parasound seismograms, though single turbidite layers cannot be resolved due to their short spacing. Slight core stretching or shortening are obvious in this lower part of the core, too.

From the comparison of the gray shaded Parasound seismogram section and the wiggle traces on the left-hand side with the synthetic seismogram, core logs and stratigraphy on the right-hand side the following interpretation can be derived. The first prominent reflection below sea floor is caused by the Toba Ash layer deposited after the explosion of volcano Toba (Sumatra) 75,000 years ago. The underlying series of reflection horizons (about 4 m thickness) result from an interlayering of very fine-grained thin terrigenous turbidites and pelagic sediments. They are younger than about 240,000 years

(oxygen isotope stage 7). At that time the channel was already inactive. The following transparent zone between 4.5-6.0 m depth indicates the transition to the time when the channel was active, more than 300,000 years ago. Only terrigenous turbidites causing strong reflections below 6 m depth were deposited.

Transferred to the Parasound seismogram section in Figure 2.23 this interpretation means that the first prominent reflection horizon in the levees can also be attributed to the Toba Ash layer (75 ka). The following transparent zone indicates completely pelagic sediments deposited on the levees during the inactive time of the channel (<240 ka). The suspension cloud was probably trapped between the channel walls so that only very thin, fine-grained terrigenous turbidites were deposited on the terrace but did not reach the levee crests. The high reflectivity part below the transparent zone in the levees is probably caused by terrigenous turbidites deposited here more than 300,000 years ago during the active time of the channel.

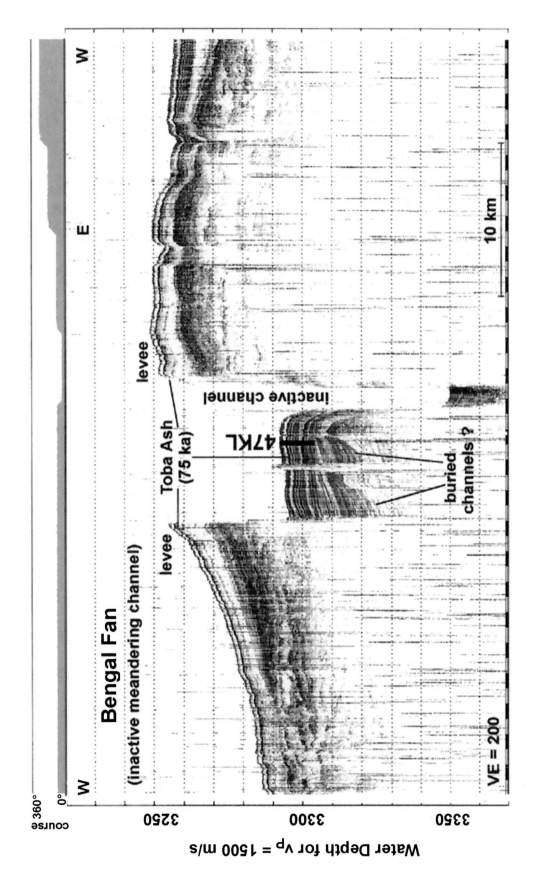

**Fig. 2.23** Parasound seismogram section recorded across an inactive meandering channel in the Bengal Fan. The sediments of the terrace were sampled by piston core 47KL marked by the black bar. Vertical exaggeration of sedimentary structures is 200. The ship's course is displayed above the seismogram section. The terrace exhibits features which might be interpreted as old buried channels.

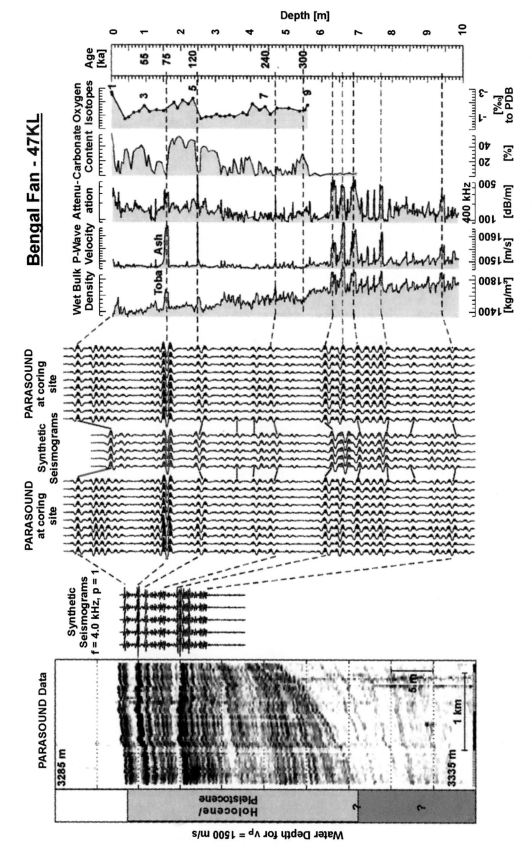

**Fig. 2.24** Coring site of 47KL. *From left to right:* Enlarged part of the Parasound seismogram section recorded on approaching site 47KL. Comparison of synthetic seismograms computed on basis of P-wave velocity and wet bulk density measurements on 47KL with Parasound seismograms recorded at the coring site. Physical properties and age model of piston core 47KL. Wet bulk densities, carbonate content and oxygen isotope stratigraphy (*G. ruber pink*) are taken from (Kudrass 1994; 1996).

### 2.1.6.2  Narrow-Beam Parasound Echosounder Recordings

Finally, some examples of Parasound echosounder recordings from different terrigenous and biogenic provinces are presented to illustrate that characteristic sediment compositions can already be recognized from such remote sensing surveys prior to the sediment core retrieval.

The first example was recorded on the Rio Grande Rise in the western South Atlantic (Bleil et al. 1994) and is typical for a calcareous environment (Fig. 2.25a). It is characterized by a strong sea bottom reflector caused by coarse-grained foraminiferal sands. Signal penetration is low and reaches only about 20 m.

In contrast, the second example from the Conrad Rise in the Antarctic Ocean (Kuhn 1998) displays a typical biogenic siliceous environment (Fig. 2.25b). The diatomaceous sediment coverage seems to be very transparent, and the Parasound signal penetrates to about 160 m depth. Reflection horizons are caused by calcareous foraminiferal and nannofossil layers deposited during a retreat of the Polar Frontal Zone during interglacial stages.

The third example recorded in the Weddell Sea/Antarctic Ocean (Kuhn 1998) indicate a deep sea environment with clay sediments (Fig. 2.25c). Signal penetration again is high (about 140 m), and reflection horizons are very sharp and distinct. Zones with upward curved reflection horizons might possibly be indicators for pore fluid migrations.

The fourth example from the distal Bengal Fan (Hübscher et al. 1997) displays terrigenous features and sediments (Fig. 2.25d). Active and older abandoned channel levee systems are characterized by a diffuse reflection pattern which indicates sediments of coarse-grained turbidites. Signal penetration in such environments is rather low and reaches about 30 - 40 m.

### Acknowledgment

We thank the captains, crews and scientists onboard of RV Meteor, RV Polarstern and RV Sonne for their efficient cooperation and help during the cruises in the South Atlantic, Antarctic and Indian Ocean. Additionally, special thanks are to M. Richter who provided a lot of unpublished electrical resistivity data and to F. Pototzki, B. Pioch and C. Hilgenfeld who gave much technical assistance and support during the development of the full waveform logging system. V. Spieß developed the ParaDIGMA data acquisition system for digital recording of the Parasound echosounder data and the program system for their processing and display. His help is greatly appreciated, too. H. Villinger critically read an early draft and improved the manuscript by many helpful discussions.

This is contribution No 251 of the Special Research Program SFB 261 (The South Atlantic in the Late Quaternary) funded by the Deutsche Forschungsgemeinschaft (DFG).

---

# Appendix

## A:  Physical Properties of Sediments Grains and Sea Water

The density ($\rho_g$) and bulk modulus ($K_g$) of the sediment grains are the most important physical properties which characterize the sediment type - terrigenous, calcareous and siliceous - and composition. Additionally, they are required as input parameters for wave propagation modeling, e.g. with Biot-Stoll's sediment model (Sect. 2.1.4.1). Table 2A-1 provides an overview on densities and bulk moduli of some typical sediment forming minerals.

Density ($\rho_f$), bulk modulus $1(K_f)$, viscosity ($\eta$), sound velocity ($v_P$), sound attenuation coefficient ($\alpha$) and electrical resistivity ($R_f$) of sea water depend on salinity, temperature and pressure. Table 2A-2 summarizes their values at laboratory conditions, i.e. 20°C temperature, 1 at pressure and 35‰ salinity.

## B:  Corrections to Laboratory and In Situ Conditions

Measurements of physical properties are usually carried out under laboratory conditions, i.e. room temperature (20°C) and atmospheric pressure (1 at). In order to correct for slight temperature variations in the laboratory and to transfer laboratory measurements to in situ conditions, usually temperature and in situ corrections are applied. Temperature variations mainly affect the pore fluid. Corrections to in situ conditions should -

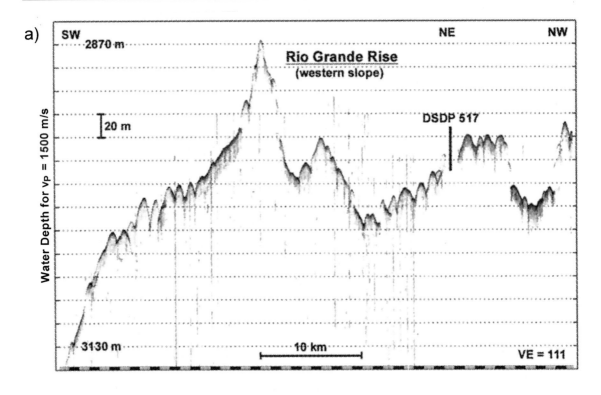

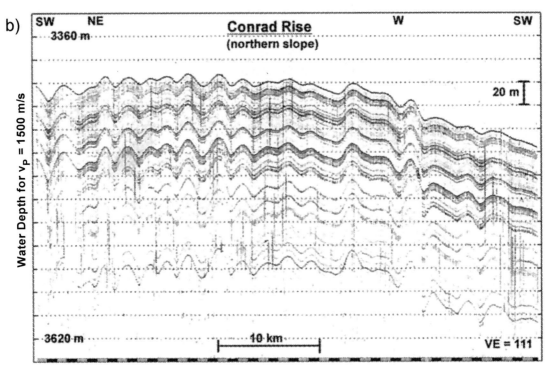

**Fig. 2.25a, b** Parasound seismogram sections representing typical calcareous (Rio Grande Rise) and opal-rich diatomaceous (Conrad Rise) environments. Lengths and heights of both profiles amount to 50 km and 300 m, respectively. Vertical exaggeration is 111.

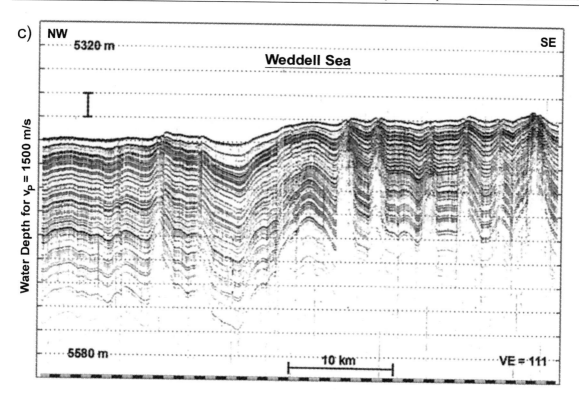

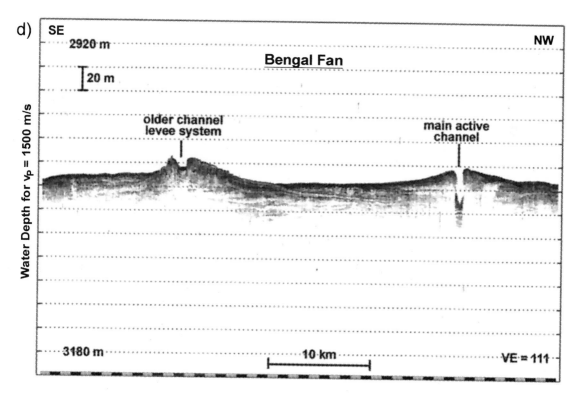

**Fig. 2.25c, d**  Parasound seismogram sections representing typical deep sea clay (Weddell Sea) and terrigenous sediments (Bengal Fan). Lengths and heights of both profiles amount to 50 km and 300 m, respectively. Vertical exaggeration is 111.

consider both the influence of reduced temperature and increased hydrostatic pressure at the sea floor.

*Porosity and Wet Bulk Density*

In standard sea water of 35‰ salinity density increases by maximum $0.3 \cdot 10^{-3}$ g cm$^{-3}$ per °C (Siedler and Peters 1986), i.e. by less than 0.1% per °C. Hence, temperature corrections can usually be neglected.

Differences between laboratory and in-situ porosities are less than 0.001% (Hamilton 1971) and can thus be disregarded, too. Sea floor wet bulk densities are slightly higher than the corresponding laboratory values due to the hydrostatic pressure and the resulting higher density of the pore water. For Central Pacific sediments of 75 - 85% porosity

**Table 2A-1**  Physical properties of sediment forming minerals, at laboratory conditions (20°C temperature, 1 at pressure). After [1]Wohlenberg (1982) and [2]Gebrande (1982).

| Mineral | Density[1] $\rho_g$ [g cm$^{-3}$] | Bulk Modulus[2] $K_g$ [$10^9$ Pa] |
|---|---|---|
| *terrigenous sediments* | | |
| quartz | 2.649 - 2.697 | 37.6 - 38.1 (trigonal) |
| biotite | 2.692 - 3.160 | 42.0 - 60.0 |
| muscovite | 2.770 - 2.880 | 43.0 - 62.0 |
| hornblende | 3.000 - 3.500 | 84.0 - 90.0 |
| *calcareous sediments* | | |
| calcite | 2.699 - 2.882 | 70.0 - 76.0 |
| *siliceous sediments* | | |
| opal | 2.060 - 2.300 | no data |

**Table 2A-2**  Physical properties of sea water, at laboratory conditions (20°C temperature, 1 at pressure, 35‰ salinity). After [1]Wille (1986) and [2]Siedler and Peters (1986).

| Parameter | Value |
|---|---|
| sound velocity[1] $v_P$ [m s$^{-1}$] | 1521 m s$^{-1}$ |
| sound attenuation coefficient[1] $\alpha$ [$10^{-3}$ dB m$^{-1}$] | |
| 4 kHz | ~0.25 |
| 10 kHz | ~0.80 |
| 100 kHz | ~ 40 |
| 400 kHz | ~ 120 |
| 1000 kHz | ~ 300 |
| density[2] $\rho_f$ [g/cm$^3$] | 1.024 |
| bulk modulus[2] $K_f$ [$10^9$ Pa] | 2.37 |
| viscosity[2] $\eta$ [$10^{-3}$ Pa·s] | 1.07 |
| electrical resistivity[2] $R_f$ [$\Omega$·m] | 0.21 |

Hamilton (1971) estimated a density increase of maximum 0.01 g cm$^{-3}$ for water depths between about 500-3000 m and of maximum 0.02 g cm$^{-3}$ for water depths between about 3000-6000 m. Thus, corrections to sea floor conditions are usually of minor importance, too. However, for cores of several hundred meter length the effect of an increasing lithostatic pressure has to be taken into account.

*Electrical Resistivity*

If porosities and wet bulk densities are determined by galvanic resistivity measurements (Sect. 2.1.2.3) varying sediment temperatures are considered by computation of the formation factor (F) (see Eq. 2.12). While the resistivity of the sediment (R$_s$) is determined by the small hand-held probe (cf. Sect. 2.2.3) the resistivity of the pore fluid (R$_f$) is derived from a calibration curve which describes the temperature (T) - conductivity (c) relation by a fourth power law (Siedler and Peters 1986)

$$R_f^{-1} = c_o + c_1 T + c_2 T^2 + c_3 T^3 + c_4 T^4 \qquad (2.22)$$

The coefficients $c_0$ to $c_4$ depend on the geometry of the probe and are determined by a least square fit to the calibration measurements in standard sea water.

*P-wave velocity and attenuation*

Bell and Shirley (1980) demonstrated that the P-wave velocity of marine sediments increases almost linearly by about 3 m s$^{-1}$ per °C while the attenuation is independent of sediment temperature, similar to the temperature dependence of sound velocity and attenuation in sea water. Hence, to correct laboratory P-wave velocity measurements to a reference temperature of 20°C Schultheiss and McPhail's (1989) equation

$$v_{20} = v_T + 3 \cdot (20 - T) \qquad (2.23)$$

can be applied, with v$_{20}$ = P-wave velocity at 20°C (in m s$^{-1}$), (T) = sediment temperature (in °C) and (v$_T$) = P-wave velocity measured at temperature T°C (in m s$^{-1}$).

To correct laboratory P-wave velocity measurements to in situ conditions a modified time-average equation (Wyllie et al. 1956) can be used (Shipboard Scientific Party 1995)

$$\frac{1}{v_{insitu}} = \frac{1}{v_{lab}} + \phi \cdot \left( \frac{1}{c_{insitu}} - \frac{1}{c_{lab}} \right) \qquad (2.24)$$

(v$_{lab}$) and (v$_{in-situ}$) are the measured laboratory (20°C, 1 at) and the corrected in situ P-wave velocities, (c$_{lab}$) and (c$_{in-situ}$) the sound velocity in sea water at laboratory and in situ conditions and the porosity ($\phi$). If laboratory and sea floor pressure, temperature and salinity are known from tables or CTD measurements (c$_{lab}$) and (c$_{in-situ}$) can be computed according to Wilson's (1960) equation

$$c = 1449.14 + c_T + c_P + c_s + c_{STP} \qquad (2.25)$$

(c$_T$), (c$_P$), (c$_S$) are higher order polynomials which describe the influence of temperature (T), pressure (p) and salinity (s) on sound velocity. (c$_{STP}$) depends on all three parameters. Complete expressions for (c$_T$), (c$_P$), (c$_S$), (c$_{STP}$) can be found in Wilson (1960). For T = 20°C, p = 1 at and s = 0.035 Wilson's equation results in a sound velocity of 1521 m s$^{-1}$.

# References

Aki, K. and Richards, P.G., 1980. Quantitative seismology. WH Freeman and Company, 932 pp.

Archie, G.E., 1942. The electrical resistivity log as an aid in determinig some reservoir characteristics. Transactions of the American Institute of Mineralogical, Metalurgical and Petrological Engineering, 146: 54-62.

Barker, P.F. and Kennett, J.P., 1990. Proceedings of the Ocean Drilling Program, Scientific Results. 113, Ocean Drilling Program, College Station (TX), 1033 pp.

Bell, D.W. and Shirley, D.J., 1980. Temperature variation of the acoustical properties of laboratory sediments. Journal of the Acoustical Society of America, 68: 277-231.

Bergmann, U., 1996. Interpretation digitaler Parasound Echolotaufzeichnungen im östlichen Arktischen Ozean auf der Grundlage physikalischer Sedimenteigenschaften (in German). Alfred-Wegener Institut für Polar- und Meeresforschung, 183, Bremerhaven, 164 pp.

Berryman, J.G., 1980. Confirmation of Biot's theory. Applied Physics Letters, 37: 382-384.

Biot, M.A., 1956a. Theory of propagation of elastic waves in a fluid-saturated porous solid. II. Higher frequency range. Journal of the Acoustical Society of America, 28: 179-191.

Biot, M.A., 1956b. Theory of wave propagation of elastic waves in a fluid-saturated porous solid. I. Low-frequency range. Journal of the Acoustical Society of America, 28: 168-178.

Bleil, U. and cruise participants, 1994. Report and preliminary results of Meteor cruise M29/2 Montevideo-Rio de Janeiro, 15.07-08.08.1994. Berichte, 59, Fachbereich Geowissenschaften, Universität Bremen, 153 pp.

Blum, P., 1997. Pysical properties handbook: a guide to the shipboard measurements of physical properties of the deep-sea cores. Technical Note, 26, Ocean Drilling Program, College Station, TX.

Bodwadkar, S.V. and Reis, J.C., 1994. Porosity measurements of core samples using gamma-ray attenuation. Nuclear Geophysics, 8: 61-78.

Boyce, R.E., 1968. Electrical resistivity of modern marine sediment from the Bering Sea. Journal of Geophysical Research, 73: 4759-4766.

Boyce, R.E., 1973. Appendix I. Physical properties-methods. In: Edgar, N.T., Sanders J.B. et al. (eds), Initial reports of the Deep Sea Drilling Project, U.S. Government Printing Office, Washington, 15: pp. 1115-1127.

Boyce, R.E., 1976. Definitions and laboratory techniques of compressional sound velocity parameters and wet-water content, wet-bulk density, and porosity parameters by gravity and gamma ray attenuation techniques. In: Schlanger, S.O., Jackson, E.D., et al. (eds), Initial reports of the Deep Sea Drilling Project, U.S. Government Printing Office, Washington, 33: pp 931-958.

Breitzke, M. and Spieß, V., 1993. An automated full waveform logging system for high-resolution P-wave profiles in marine sediments. Marine Geophysical Researches, 15: 297-321.

Breitzke, M., Grobe, H., Kuhn, G. and Müller, P., 1996. Full waveform ultrasonic transmission seismograms - a fast new methode for the determination of physical and sedimentological parameters in marine sediment cores. Journal of Geological Research, 101: 22123-22141.

Breitzke, M., 1997. Elastische Wellenausbreitung in marinen sedimenten - Neue Entwicklung der Ultraschall Sedimentphysik und Sedimenttechographie (in German). Berichte, 104, Fachbereich Geowissenschaften, Universität Bremen, 298 pp.

Bryant, W.R., Hottman, W. and Trabant, P., 1975. Permeability of unconsolidated and consolodated marine sediments, Gulf of Mexico. Marine Geotechnology, 1: 1-14.

Carman, P.C., 1956. Flow of gases through porous media. Butterworth Scientific Publications, London, 182 pp.

Chelkowski, A., 1980. Dilectric Physics. Elsevier, Amsterdam, 396 pp.

Childress, J.J. and Mickel, T.J., 1980. A motion compensated shipboard precision balance system. Deep Sea Research, 27: 965-970.

Constable, C. and Parker, R., 1991. Deconvolution of log-core paleomagnetic measurements-spline therapy for the linear problem. Geophysical Journal International, 104: 453-468.

Courtney, R.C. and Mayer, L.A., 1993a. Calculation of acoustic parameters by filter-correlation method. Journal of the Acoustical Society of America, 93: 1145-1154.

Courtney, R.C. and Mayer, L.A., 1993b. Acoustic properties of fine-grained sediments from Emerlad - Basin: Toward an inversion for physical properties using the Biot-Stoll model. Journal of the Acoustical Society of America, 93: 3193-3200.

Dobeneneck, v.T. and Schmieder, F., in press. Using rock magnetic proxy records for orbital tuning and extended time series analysis into the super- and sub- Milankovitch bands. In: Fischer, G. and Wefer, G. (eds), Use of proxies in paleoceanography: examples from the South Atlantic. Springer Verlag, Berlin.

Ellis, D.V., 1987. Well logging for earth scientists. Elsevier, Amsterdam, 532 pp.

Fisher, A.T., Fischer, K., Lavoie, D., Langseth, M. and Xu, J., 1994. Geotechnical and hydrogeological properties of sediments from Middle Valley, northern Juan de Fuca Ridge. In: Mottle, M.J., Davies, E., Fischer, A.T. and Slack, J.F. (eds), Proceedings of the Ocean Drilling Program, Scientific Results, 139, College Station (TX), pp. 627-647.

Gassmann, F., 1951. Über die elastizität poröser Medien. Vierteljahresschrift der Naturforschenden Gesellschaft in Zürich, 96: 1-23.

Gealy, E.L., 1971. Saturated bulk density, grain density and porosity of sediemt cores from western equatorial Pacific: Leg 7, Glomar Challenger. In: Winterer, E.L., et al (eds), Initial reports of the Deep Sea Drilling Project, 7, Washington, pp. 1081-1104.

Gebrande, H., 1982. Elastic wave velocities and constants of elasticity at normal conditions. In: Hellwege, K.H. (ed), Landolt-Börnstein. Numerical data and functional relationships in science and technology. Group V: Geophysics and space research 1, Physical Properties of Rocks, subvol. b. Springer Verlag, Berlin, pp. 8-35.

Gerland, S., Richter, M., Villinger, H. and Kuhn, G., 1993. Non-destructive porosity determination of Antarctic marine sediments derived from resistivity measurements with the inductive method. Marine Geophysical Researches, 15: 201-218.

Gerland, S., 1993. Non-destructive high resolution density measurements on marine sediments, Alfred-Wegener Institute for Polar and Marine Research, 123, Bremerhaven, 130 pp.

Gerland, S. and Villinger, H., 1995. Nondestructive density determinationon marine sediment cores from gamma-ray attenuation measurements. Geo-Marine-Letters, 15: 111-118.

Gunn, D.E. and Best, A.I., 1998. A new automated nondestructive system for high resolution multi-sensor logging of open sediment cores. Geo-Marine Letters, 18: 70-77.

Hamilton, E.L., 1971. Prediction of in situ acoustic and elastic properties of marine sediments. Geophysics, 36: 266-284.

Hovem, J.M. and Ingram, G.D., 1979. Viscous attenuation of sound in saturated sand. Journal of the Acoustical Society of America, 66: 1807-1812.

Hovem, J.M., 1980. Viscous attenuation of sound in suspensions and high-porosity marine sediments. Journal of the Acoustical Society of America, 67: 1559-1563.

Hübscher, C., Spieß, V., Breitzke, M. and Weber, M.E., 1997. The youngest channel-levee system of the Bengal Fan: results from digital echosounder data. Marine Geology, 141: 125-145.

Jackson, P.D., Taylor-Smith, D. and Stanford, P.N., 1978. Resistivity-porosity-particle shape relationships for marine sands. Geophysics, 43: 1250-1268.

Jannsen, D., Voss, J. and Theilen, F., 1985. Comparison of methods to determine Q in shallow marine sediments from vertical seismograms. Geophysical Propsecting, 33: 479-497.

Kudrass, H.R., 1994. SO93/1-3 Bengal Fan-Cruise report. Federal Institute for Geoscience and Natural Resources, Hannover.

Kudrass, H.R., 1996. Final Report Bengal Fan, Sonne Cruise SO93, Federal Institute for Geoscience and Natural Resource, Hannover.

Kuhn, G., in press. The expedition ANTARKTIS XI/4 of RV Polarstern in 1994. Rep. on polar research, Alfred-Wegener Institute for Polar and Marine Research, Bremerhaven.

Lambe, T.W. and Whitman, R.V., 1969. Soil mechanics. Wiley & Sons, NY, 553 pp.

Laser, B. and Spieß, V., subm. Comparision of high-resolution physical property core logs from ODP site 690 with digital Parasound data. Scientific Drilling.

Lovell, M.A., 1985. Thermal conductivity and permeability assessmentby electrical resitivity measurements in marine sediments. Marine Geotechnology, 6: 205-240.

MacKillop, A.K., Moran, K., Jarret, K., Farrell, J. and Murray, D., 1995. Consolidation properties of equatorial Pacific Ocean sediements and their relationship to stress history and offsets in the Leg 138 composite depth sections. In: Pisias, N.G., Mayer, L.A., Janecek, T.R., Palmer-Julson, A. and van Andel, T.H. (eds), Proceedings of the Ocean Drilling Program, Scientific Results, 138, College Station (TX), pp. 357-369.

Mendel, J.M., Nahi, N.E. and Chan, M., 1979. Synthetic seismograms using the state space approach. Geophysics,

44: 880-895.

O'Connell, S.B., 1990. Variation in upper cretaceous and cenozoic calium carbonate percentages, Maud Rise, Wedell Sea, Antarctica. In: Barker, P.F., Kennet, J.P., et al. (eds), Proceedings of the Ocean Drilling Program, Scientific Results, 113, College Station (TX), pp. 971-984.

Oguszwitz, P.R., 1985. Applicability of the Biot theory. II. Suspensions. Journal of the Acoustical Society of America, 77: 441-452.

Olsen, H.W., Nichols, R.W. and Rice, T.C., 1985. Low gradient permeability measurements in a triaxial system. Geotechnique, 35: 145-157.

Plona, T.J., 1980. Observation of a second bulk compressional wave in a porous medium at ultrasonic frequencies. Applied Physical Letters, 36: 159-261.

Ruffet, C., Guefuen, Y. and Darot, M., 1991. Complex conductivity measurements and fractal nature of porosity. Geophysics, 56: 758-768.

Schopper, J.R., 1982. Permability of rocks. In: Hellwege, K.H. (ed), Landolt-Börnstein. Numerical Data and Functional Relationships in Science and Technology, Group V: Geophysics and Space Research 1. Physical Properties of Rocks, subvol. a, Springer, Berlin, 278-303 pp.

Schön, J.H., 1996. Physical properties of rocks - fundamentals and principles of petrophysics. Handbock of Geophysical Exploration, 18, Section I, Seismic Exploration, Pergamon Press, Oxford, 583 pp.

Schultheiss, P.J. and McPhail, S.D., 1989. An automated P-wave logger for recording fine-scale compressional wave velocity structures in sediments. In: Ruddiman, W., Sarntheim, M., et al. (eds), Proceedings of the Ocean Drilling Program, Scientific Results, 108, College Station (TX), pp. 407-413.

Sen, P.N., Scala, C. and Cohen, M.H., 1981. A self-similar model from sedimentary rocks with application to dielectric constant of fused glass beads. Geophysics, 46: 781-795.

Sheng, P., 1991. Consistent modeling of electrical and elastic properties of sedimentary rocks. Geophysics, 56: 1236-1243.

Shipboard Scientific Party (1995) Explanatory Notes. In: Curry, W.B., Shackleton, N.J., Richter, C. et al (eds), Proceedings of the Ocean Drilling Program, Initial Reports, 154, College Station (TX), ODP, pp 11-38

Siedler, G. and Peters, H., 1986. Pysical properties (general) of seawater. In: Hellwege, K.H. and Madelung, O. (eds), Landolt-Börnstein. Zahlenwerte und Funktionen aus Naturwissenschaften und Technik. Group V: Geophysics and space research 3, Oceanography, subvol. a, Springer, Berlin, pp. 233-264.

Spieß, V., 1993. Digitale Sedimentechographie - Neue Wege zu einer hochauflösenden Akustostratigraphie (in German). Berichte, 35, Fachbereich Geowissenschaften, Universität Bremen, 199 pp.

Stoll, R.D., 1974. Acoustic waves in saturated sediments. In: Hampton, L. (ed) Physics of sound in marine sediments. Plenum Press, NY, pp. 19-39.

Stoll, R.D., 1977. Acoustic waves in ocean sediements. Geophysics, 42: 715-725.

Stoll, R.D., 1989. Sediment acoustics. Springer Verlag, Berlin, 149 pp.

Taner, M.T., Koehler, F. and Sheriff, R.E., 1979. Comlex seismic trace analysis. Geophysics, 44: 1041-1063.

Tonn, R., 1989. Comparision of seven methods for the computation of Q. Physics of the Earth and Planetary Interiors, 55: 259-268.

Tonn, R., 1991. The determiation of the seismic quality factor Q from VSP data: A comparision of different computational methods. Geophysical Propsecting, 39: 1-27.

Waxman, M.H. and Smits, L.J.M., 1968. Electrical conductivities in oil bearing shaly sandstones. Society of Petroleum Engineering, 8: 107-122.

Weaver, P.P.E. and Schultheiss, P.J., 1990. Current methods for obtaining, logging and splitting marine sediments cores. Marine Geophysical Researches, 12: 85-100.

Weber, M.E., Niessen, F., Kuhn, G. and Wiedicke, M., 1997. Calibration and application of marine sedimentary physical properties using a multi-sensor core logger. Marine Geology, 136: 151-172.

Weeks, R. et al., 1993. Improvements in long-core measurements techniques: applications in paleomagnetism and paleoceanography. Geophysical Journal International, 114: 651-662.

Whitmarsh, R.B., 1971. Precise sediment density determination by gamma-ray attenuation alone. Journal of Sedimentary Petrology, 41: 882-883.

Wille, P., 1986. Acoustical properties of the ocean. In: Hellwege, K.H. and Madelung, O. (eds), Landolt-Börnstein. Numerical Data and Functional Relationships in Science and Technology. Group V: Geophysics and Space Research 3, Oceanography, subvol. a, Springer, Berlin, pp. 265-382.

Wilson, W.D., 1960. Speed of sound in sea water as a function of temperature, pressure and salinity. Journal of the Acoustical Society of America, 32: 641-644.

Wohlenberg, J., 1982. Density of minerals. In: Hellwege, K.H. (ed), Landolt-Börnstein. Numerical Data and Functional Relationships in Science and Technology. Group V: Geophysics and Space Research 1, Physical Properties of Rocks, subvol. a, Springer, Berlin, pp. 66-113.

Wood, A.B., 1946. A textbook of sound. G. Bell and Sons, London, 578 pp.

Wyllie, M.R., Gregory, A.R. and Gardner, L.W., 1956. Elastic wave velocities in heterogeneous and porous media. Geophysics, 21: 41-70.

# 2.2 Sedimentary Magnetism

U L R I C H   B L E I L

## 2.2.1 Introduction

Magnetic studies of (marine) sediments traditionally focused on fossil natural remanent magnetization properties which have provided a wealth of information about the Earth's magnetic field history. Such magnetostratigraphic analyses are now routinely applied as a standard dating method. This 'magnetic view' has been substantially expanded during the recent two decades with the development of an entirely new research field, *Environmental Magnetism*'. Its basic premise is that composition, concentration and grain-size spectra of magnetic mineral assemblages in natural materials characterize depositional systems in relation to variable (paleo-)environmental regimes. Among others, Thompson and Oldfield (1986), Lund and Karlin (1990), King and Channell (1991), Verosub and Roberts (1995) and Frederichs et al. (1998) have compiled comprehensive reviews of the current research. Of the wide variety of environmental magnetic applications in geosciences, the authors particularly address paleoceanographic objectives such as

- regional lithostratigraphic correlations and orbitally tuned chronostratigraphies of sediment series using high-resolution records of rock magnetic parameters;
- identification of source areas and pathways of terrigenous magnetic mineral components in deep-sea sediments;
- reconstructions of biogeochemical settings and their temporal variability on the basis of rock magnetic proxy records delineating authigenic biomineralization or abiotic precipitation of ferrimagnetics and diagenetic alteration of magnetic mineral inventories.

This short contribution summarizes some major aspects of the last topic. For fundamentals in rock magnetism and modern magnetic analytical techniques refer to the new textbook by Dunlop and Özdemir (1997) and references therein.

## 2.2.2 Biogenic Magnetic Minerals in Marine Sediments

Very fine grained, so-called single-domain (SD) magnetite ($0.03 \leq \emptyset \leq 0.1$ µm) is the prime choice carrier of a natural remanent magnetization (NRM), because it preserves a stable magnetic information on geologic time scales. The discovery of fossil biogenic SD magnetite in deep-sea deposits (Petersen et al. 1986, Stolz et al. 1986) provoked a persistent debate over the relative importance of authigenic, biotic and abiotic formation of magnetic minerals for NRM and other sediment magnetic properties.

A series of subsequent studies (e.g., Chang et al. 1989, Vali et al. 1987, 1989, McNeill 1990, Stolz 1992) demonstrated that bacteria of diverse morphologies generate intracellurlarly single-domain magnetite crystals of different distinct regular shapes which are arranged in multi-particle chains bound by a membrane. They utilize these magnetosomes (Gorby et al. 1988) to control their locomotion in the chemically stratified pore space of unconsolidated sedimentary deposits. Observations in limnic and estuarin environments led to the assumption that magnetotactic bacteria are microaerophilic and could eventually indicate (paleo-)oxygen conditions (e.g., Blakemore et al. 1985, Chang et al. 1987, Lovley et al. 1987). In pelagic sediments of various ages fossil magnetosomes, preserved after decomposition of organic parts of the bacteria, were magnetically extracted and identified by transmission electron microscopy (TEM, Fig. 2.2.1).

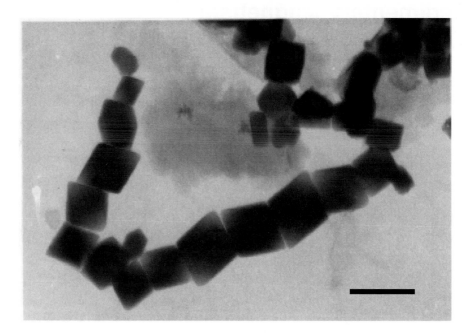

**Fig. 2.2.1** TEM image of magnetite magnetofossils extracted from Tertiary sediments recovered at Deep Sea Drilling Project Site 521 (Leg 73) in the Angola Basin, South Atlantic. Note very regular octahedral geometries and approximately uniform grain-sizes. Scale bar 0.1 μm. Courtesy of Petersen et al. (1986).

Living magnetotactic bacteria in hemipelagic sediments have been detected for the first time in the Santa Barbara Basin off California (Stolz et al. 1986). Petermann and Bleil (1993) and Petermann (1994) reported their widespread occurrence in open marine realms at the Southwest African and eastern South American continental margins as well as, there a rather rare exception, at three sites in the central South Atlantic. They found distinctive depth successions in the surface layers with maximum numbers a few centimeters below the water/sediment interface (Fig. 2.2.4). Population densities in hemipelagic deposits from the Benguela upwelling system off Namibia even exceeded those of high-productive shallow marine or brackish environments as for example intertidal mud flats of the North Sea. Down to about 2000 m water depth, magnetotactic bacteria were ubiquitous in the study areas, but their abundance varied by up to several orders of magnitude at any bathymetric level. Below 2000 m, only a very small number, frequently not a single individual was encountered. Altogether a pronounced decrease in maximum concentrations of live magnetotactic bacteria to deeper waters has been documented (Fig. 2.2.2). Approximately exponential regional trends reveal conspicuous differences between the African and

South American continental slopes. Directly corresponding to the actual findings, both their regressions suggest a lower limit of around 4000 m water depth for the occurrence of living magnetotactic bacteria. It's interesting to note that the deepest location, from where they could be retrieved so far, is in the central Subtropical Gyre (34.5°S / 21°W, 4016 m) of the eastern Argentine Basin. At other open ocean sites in waters deeper than 3000 m the search had regularly no success. Additional to locations shown in Figure 2.2.2, these observations were also made on the mid-Atlantic Ridge, the Ceará Rise and the lower Amazon Fan in the western equatorial South Atlantic. On the other hand, wherever surface sediments from these sites have been inspected by TEM, magnetosomes could positively be identified.

There is no rationale to argue for a *per se* dependence of living magnetotactic bacteria concentrations on water depth, whereas a link to nutrient supply and associated changes in geochemical milieu seems quite conceivable. Accumulation of organic matter at the ocean floor is a function of primary productivity in the surface waters (euphotic zone) and subsequent degradation processes in the water column. Using primary productivity data of Berger (1989) and applying the

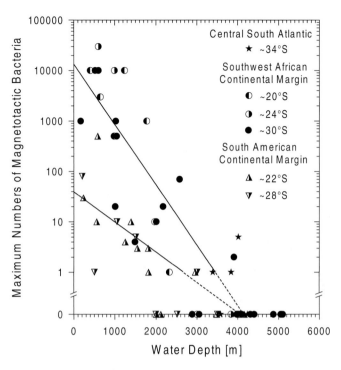

**Fig. 2.2.2** Maximum numbers of living magnetotactic bacteria in 50 µl of sediment versus water depth at the sampling sites. The solid lines are exponential least-square fits to the Southwest African and South American continental margin data sets using only locations, where living magnetotactic bacteria could be identified. Data from Petermann and Bleil (1993) and Petermann (1994).

empirical relationship of Betzer et al. (1984) for the marine flux of organic carbon ($C_{org}$) implies that the number of living magnetotactic bacteria is indeed largely controlled by $C_{org}$ accumulation (Fig. 2.2.3). Their living conditions apparently become continually favorable if more than about 5 g $C_{org}$ per square meter and year reach the ocean floor. Part of the high data scatter, specifically at low $C_{org}$ fluxes, may result from seasonal variations as the available primary productivity informations are several year mean estimates. Similar positive correlations also exist to $C_{org}$ and nitrate concentrations in the surface sediments as well as to the diffusive oxygen flux into the sediments (Petermann 1994).

Living magnetotactic bacteria have never been observed in the bottom water overlying the sediment. Where they are present, some individuals were always found immediately beneath the water/sediment interface. Maximum population densities typically occurred between 1 and 4 cm sub-bottom, at the South American continental margin sporadically as deep as 15 cm. Within a few centimeters further into the sediment column, numbers gradually declined to zero. Figure 2.2.4 shows two representative examples from the Benguela upwel-

ling system together with depth profiles of oxygen measured *in situ* using a microelectrode technique (Glud et al. 1994) and of nitrate concentrations in the pore water determined photometrically (Dahmke, unpublished data).

While oxygen is consumed, nitrate builds up in the sediment column by microbial degradation of organic material. With increasing depth nitrate will decompose. Oxidation of organic compounds provides energy for the microorganisms to maintain their metabolism. Froehlich et al. (1979) identified a systematic succession of terminal electron acceptors from oxygen in the uppermost layer to nitrate, Mn(IV) oxides, Fe(III) oxides and sulphate at depth. This sequence of oxidants reflects the accessible amount of energy per mole of organic carbon. Its depth extent critically depends on organic matter supply, sedimentation rate and the availability of reactants (Berner 1981). In the center of the Namibian upwelling cell, where the flux of organic material and thus the number of living magnetotactic bacteria is highest (station GeoB 1713, Fig. 2.2.4 left), the geochemical zonation is much narrower than at less productive peripheral locations (station GeoB 1719, Fig. 2.2.4 right).

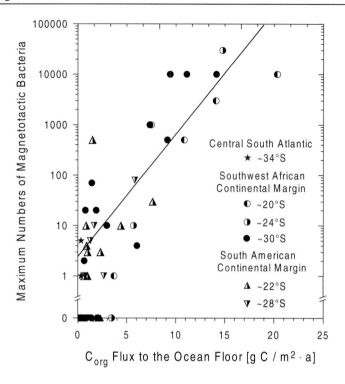

**Fig. 2.2.3** Maximum numbers of living magnetotactic bacteria in 50 μl of sediment versus flux of organic carbon to the ocean floor determined from primary productivity data of Berger (1989) and the empirical relationship of Betzer et al. (1984) for organic matter degradation in the water column. The solid line is an exponential least-square fit using only locations, where living magnetotactic bacteria could be identified. Data from Petermann (1994).

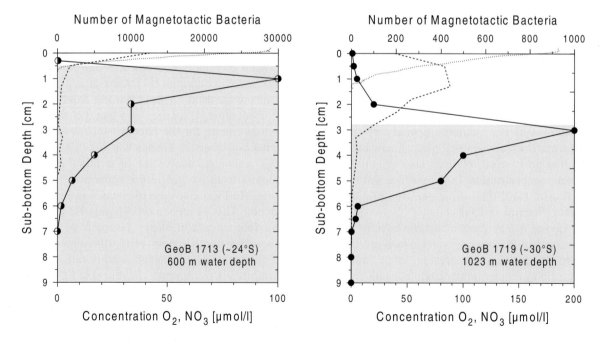

**Fig. 2.2.4** Depth distribution of living magnetotactic bacteria abundance in 50 μl of sediment at two sites in the Ben-guela upwelling region off Namibia together with oxygen profiles measured *in situ* (Glud et al. 1994, dotted line) and pore water nitrate concentrations (Dahmke unpublished data, dashed line). Grey shading indicates the sections, where iron was traced in the pore water. Data from Petermann and Bleil (1993).

Peak concentrations of living magnetotactic bacteria were always clearly located in the anaerobic nitrate respiration zone, compared to small numbers in the top aerobic horizon. At least in marine environments, the majority of magnetotactic bacteria thus appears to be reducers of nitrate or other nitrous oxides. This is consistent with studies on pure cultures (Bazylinski and Blakemore 1983, Bazylinski et al. 1988) using about ten times higher nitrate concentrations than observed under natural conditions. Repeatedly magnetotactic bacteria were encountered in very low nitrate concentrations ($\leq 3$ μmol/l). At present it remains an open question, whether they possibly use iron as terminal electron acceptor for their metabolic processes or microenvironmental effects are responsible for those findings. A most interesting discovery in this context was a type of non-mobile bacterium that combines the reduction of iron to the oxidation of organic matter to produce ultrafine grained (10 to 50 nm) magnetite extracellularly (Lovley et al. 1987). These magnetite crystals are comparatively irregular in shape and high proportions of superparamagnetic (SP) grains (Sparks et al. 1990), which are incapable to retain a stable remanent magnetization, imply that their purpose is not magnetotactic orientation. In sulphide rich water and sediment bacteria have been identified which synthesize intracellularly greigite ($Fe_3S_4$, Mann et al. 1990) and possibly also pyrrhotite ($Fe_7S_8$, Farina et al. 1990). In most sediments the occurrence of such ferrimagnetic iron sulphides is ascribed to reductive diagenesis of magnetite, however (e.g., Snowball and Thompson 1988, Reynolds et al., 1990). Taylor et al. (1987) have demonstrated that SP and SD magnetite also easily forms by abiotic oxidation of ferrous iron at ambient temperature and about neutral pH, conditions which are untypical for marine settings, but prevail in soils. Their erosion and fluvial or eolian transport to the oceans could thus be a significant source for fine grained magnetite in marine sediments (Maher and Taylor 1988).

## 2.2.3  Reduction Diagenesis of Magnetic Minerals in Marine Environments

Early diagenetic processes in marine sediments, which are primarily driven by microbially mediated degradation of organic matter, intensely affect the magnetic mineral assemblage. Dissolution of magnetite and formation of iron sulfides is a widespread phenomenon in suboxic to anoxic sediments (Canfield and Berner 1987, Karlin 1990a, b, Leslie et al. 1990). As a result, systematic changes in magnetic properties are observed (Karlin and Levi 1983, 1985). The following examples illustrate the potential of high-resolution rock magnetic analyses to contribute proxy parameters for geochemically induced alterations in marine sedimentary deposits.

Already a simple set of bulk sample measurements enables a rather detailed rock magnetic characterization of marine sediment series. All of them are non-destructive, can be performed in a short time on standard magnetic laboratory equipment and do not require laborious preparational techniques. Magnetic susceptibility $\kappa$ (per unit volume) or $\chi$ (per unit mass) gives a reliable estimate for the total concentration of magnetite minerals, isothermal ($M_{ir}$ or $\sigma_{ir}$) and anhysteretic ($M_{ar}$ or $\sigma_{ar}$) remanent magnetization for their coarser grained (pseudo-single-domain, PSD to multi-domain, MD) and finer grained (SD to PSD) portions, respectively. M refers to a magnetization per unit volume, $\sigma$ to a magnetization per unit mass (here, to unit dry mass to account for the variable water content in unconsolidated marine sediments). An ultrafine grained SP magnetite component can be identified from frequency dependent susceptibility ($\kappa_{fd}$ or $\chi_{fd}$). Back field remanent magnetization ($M_{-0.3T}$ or $\sigma_{-0.3T}$) allows to discriminate high-coercive minerals (antiferromagnetic hematite and/or goethite) versus low-coercive magnetite. All these concentration dependent parameters are more or less modulated by variations of the 'non-magnetic', biogenic or lithogenic major constituents of the sediment matrix. Their influence is eliminated by calculating suitable interparametric ratios which (predominantly) refer to individual fractions of the magnetite grain-size spectrum, $\kappa_{fd}/\kappa$ ($\chi_{fd}/\chi$) to SP, $M_{ar}/M_{ir}$ ($\sigma_{ar}/\sigma_{ir}$) and $M_{ar}/\kappa$ ($\sigma_{ar}/\chi$) to SD, $M_{ir}/\kappa$ ($\sigma_{ir}/\chi$) to MD (plus PSD) or quantify the hematite/goethite component ($S_{-0.3T}$). Note that unlike biogenetic magnetite, lithogenic magnetites ('titanomagnetites') contain in variable amounts titanium and also other cations which significantly modify their specific magnetic properties.

Magnetic hysteresis measurements require more sophisticated instrumentation and a quite elaborate sample preparation. Their complete data evaluation using appropriate mathematical algorithms (von Dobeneck 1996) provides a series of diagnostic key parameters, among others, ferrimagnetic (i.e., carried solely by magnetite) saturation magnetization ($M_s$ or $\sigma_s$) and saturation remanence ($M_{rs}$ or $\sigma_{rs}$), coercive field ($B_c$) and the coercivity of remanence ($B_{cr}$) delineating concentration and stability as well as, by the ratios $M_{rs}/M_s$ ($\sigma_{ar}/\sigma_{ir}$) and $B_{cr}/B_c$, magnetite domain states and grain-sizes. The total susceptibility signal ($\kappa_{tot}$ or $\chi_{tot}$) includes contributions of all mineral constituents, the non-ferromagnetic susceptibility ($\kappa_{nf}$ or $\chi_{nf}$) only those of the paramagnetic (Fe bearing silicates and clays) and diamagnetic (biogenic carbonates and opal) sediment matrix. The ratio $\kappa_{nf}/\kappa_{tot}$ ($\chi_{nf}/\chi_{tot}$) quantifies relative contributions of non-ferromagnetic constituents to total susceptibility.

Reducing conditions in the sediment column are common, where an intense primary biologic productivity in the surface waters supplies high fluxes of organic material to the sea floor. As discussed above, the Benguela upwelling system at the Southwest African continental margin is representative for such depositional regimes. In this region magnetosomes, the fossil remnants of magnetotactic bacteria, are the by far dominant carrier of the magnetic signal in recent sediments, also because of a very minor inflow of terrigenous ferrimagnetic mineral components compared to other hemipelagic environments in the South Atlantic (Frederichs et al. 1998, Däumler 1996). Within a few centimeters sub-bottom, sediment magnetic properties typically exhibit unusually prominent changes. As a typical example, Figure 2.2.5 shows data sets of an 18 cm surface sediment sequence recovered with a multicorer at site GeoB 1713 in the center of the high productivity zone (see also Fig. 2.2.4 left).

All concentration dependent parameters $\chi$, $\sigma_{ir}$, $\sigma_{ar}$ and $\sigma_s$ testify a drastically diminishing content

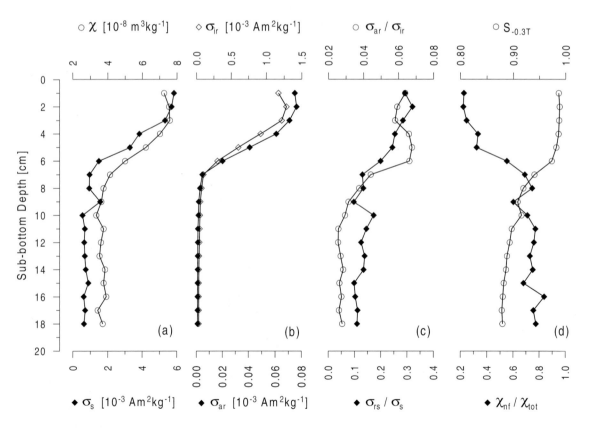

**Fig. 2.2.5** Rock magnetic depth profiles at site GeoB 1713 in the center of the Benguela upwelling system off Namibia. **(a)** Bulk magnetic susceptibility $\chi$ and ferrimagnetic saturation magnetization $\sigma_s$ delineating variable magnetite contents. **(b)** Isothermal remanent magnetization $\sigma_{ir}$ and anhysteretic remanent magnetization $\sigma_{ar}$ depicting variations in coarser grained (MD/PSD) and finer grained (SD/PSD) magnetite fractions, respectively. **(c)** Magnetite grain-size sensitive ratios $\sigma_{ar}/\sigma_{ir}$ and $\sigma_{rs}/\sigma_s$. **(d)** Hematite/goethite index $S_{-0.3T}$ and relative contributions of non-ferromagnetic sediment matrix constituents to total susceptibility $\chi_{nf}/\chi_{tot}$. Data from Däumler (1996) and Frederichs et al. (1998).

of ferrimagnetic components within the uppermost 10 centimeters (Fig. 2.2.5a, b). $\sigma_s$ indicates that magnetite concentration decreases by around 90% in a narrow depth interval between about 2 and 10 cm sub-bottom. Consequently, relative contributions of the paramagnetic sediment matrix to the susceptibility signal $\chi_{nf}/\chi_{tot}$ rise from approximately 20 to 80% (Fig. 2.2.5d). While the magnetic mineral assemblage contains about 25% highcoercive hematite/goethite in the top layer ($S_{-0.3T} \approx 0.99$), their relative proportion systematically increases downcore, reaching around 80% ($S_{-0.3T} \approx 0.88$) at the base. Note that the $S_{-0.3T}$ scale is non-linear and reflects extremely contrasting intrinsic magnetic properties of hematite/goethite and magnetite.

Grain-size sensitive parameters $\sigma_{ar}/\sigma_{ir}$ and $\sigma_{rs}/\sigma_s$ (Fig. 2.2.5c) clearly document a downward coarsening of the ferrimagnetic minerals from predominantly fine grained particles in the top centimeters to an on average much coarser fraction in deeper strata. High $\sigma_{ar}/\sigma_{ir}$ in the upper about 6 cm are indicative of substantial portions of single-domain magnetite vanishing to depth and leaving an essentially multi-domain ensemble. The ratios of saturation remanence to saturation magnetiza-

tion $\sigma_{rs}/\sigma_s$ and coercivity of remanence to coercive field $B_{cr}/B_c$ enable a more quantitative characterization of domain states and thus grain-size spectra of the magnetic mineral assemblage. Figure 2.2.6 illustrates these data sets in the standard diagram of Day et al. (1977). They reveal a distinct succession according to the depth position of the specimens. As typically observed for natural sample collections, the theoretical single-domain field ($\sigma_{rs}/\sigma_s \geq 0.5$) is not reached. In the present case a minor coarse grained terrigenous (titano-)magnetite MD fraction ($\varnothing \geq 1 - 10\ \mu m$) should be responsible. Ultrafine grained SP ($\varnothing \leq 0.03\ \mu m$) components (e.g., Gee and Kent 1995) or an interaction of SD particle clusters (Moskowitz et al. 1989) would cause the same effect, however. Interestingly, the smallest average grain-size is found just below the horizon of maximum numbers of living magnetotactic bacteria (see Fig. 2.2.4 left). From top to 6 cm sub-bottom the grain-size distribution is dominated by particles in the SD to fine PSD range, their relative quantity gradually decreases. Data from 7 to 18 cm sub-bottom fall into the coarse PSD to MD field.

A dissolution of the magnetic mineral inventory will successively reduce grain-sizes. Fine grained SD magnetite is most dramatically affected as it becomes superparamagnetic and does not contribute to the remanent magnetic properties any more. Yet, a SP component causes a drop in $\sigma_{rs}/\sigma_s$, while $B_{cr}/B_c$ rises (Jackson 1990, Gee and Kent 1995). The distinct increase of apparent grain-sizes in the upper about 10 cm (Fig. 2.2.6) thus should best be explained by the following two combined effects. Biogenic SD magnetite is converted to the superparamagnetic state. Initially, the fraction close to the SD/SP threshold simulates a coarsening of the grain-size spectrum, but with further dissolution these particles progressively lose influence. The coarse grained terrigenous MD fraction is comparatively resistant to dissolution and therefore steadily increases in relative concentration. The data scatter in the lower part of the core most likely reflects grain-size variations of the original terrigenous influx.

This interpretation is entirely consistent with transmission electron microscope analyses (Däumler 1996, Frederichs et al. 1998). In the sample from 2 - 3 cm depth magnetite grain-sizes and morphologies are typical for magnetosomes produced by magnetotactic bacteria. The crystals have perfect regular faces. In contrast, at 5 - 6 cm depth the magnetosomes exhibit notable corrosion and

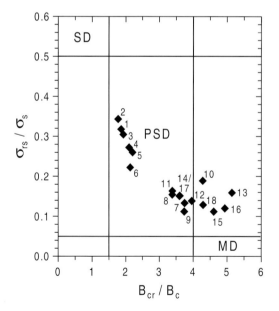

**Fig. 2.2.6** Depth variation of the avarage magnetite grain-size at site GeoB 1713 as inferred from ratios of saturation remanence to saturation magnetization $\sigma_{rs}/\sigma_s$ and coercivity of remanence to coercive field $B_{cr}/B_c$ (diagram following Day et al. 1977). SD/PSD boundary at ~ 0.1 μm, PSD/MD at ~ 1 - 10 μm. Data point labels refer to the depth position of the samples (in cm). Data from Däumler (1996) and Frederichs et al. (1998).

dissolution features. In deeper parts of the core no unequivocal identification of biogenic magnetites could be achieved. Direct measurements of particle dimensions revealed an about 20% reduction of average grain-sizes between 2 - 3 and 5 - 6 cm depth, shifting substantial fractions from the SD to the SP field (Butler and Banerjee 1975). Consequently, they become insignificant for remanent magnetic properties.

High fluxes of organic matter from an intense primary productivity in the surface waters of the Benguela upwelling center off Namibia sustain a very favorable living environment for magneto-tactic bacteria in the top few centimeters of the sediment column. Due to this extreme supply of organic matter, the early diagenetic zonation is evidently confined to a very shallow and narrow depth interval. The reducing conditions severely alter the magnetic mineral assemblage. While hematite/goethite and also coarse grained terrigenous (titano-)magnetite components apparently are not strongly affected, originally high concentrations of biogenic magnetite entirely disintegrate down to around 10 cm sub-bottom. It is certainly a paradox that in a marine setting, where ideal carriers of a sediment magnetization are presently produced in highest amounts, they ultimately do not contribute to the natural remanence at all. As biogenic magnetofossils have been found for example

in Tertiary deep-sea deposits of the Angola Basin (Petersen et al. 1986, see Fig. 2.2.1), their preservation hints at much less reductive paleoenvironments.

The Benguela upwelling example is likely to represent a steady state situation over the depth interval analyzed. On basis of magnetic hysteresis data Tarduno and Wilkison (1996) have postulated much deeper seated modern Fe redox boundaries in sediment series from the western equatorial Pacific Ocean. At three Ocean Drilling Program Leg 130 sites along a bathymetric transect from about 2500 to 3400 m water depth on the Ontong Java Plateau they found evidence for a systematic, 0.84 to 4.25 m sub-bottom shift of the boundary that is attributed to variable $C_{org}$ fluxes of between 0.04 and 0.005 g per square meter and year. Depth positions of the boundary were tentatively identified just beneath a sharp increase of coercitivity, interpreted to originate from an authigenic formation of biogenic SD magnetite. Under this narrow maximum coercitivities decay to lower averages than in the overlaying strata suggesting an overall coarsening of magnetite grain-sizes due to reductive dissolution effects. Combined with results of the South Atlantic living magnetotactic bacteria survey, these findings appear at least not implausible (Fig. 2.2.7). They need to be further confirmed by geochemical data, however, and

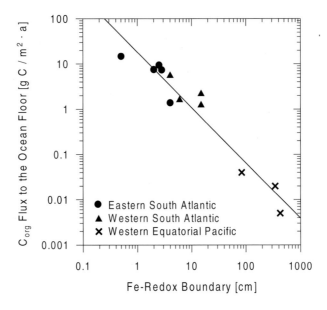

**Fig. 2.2.7** Sub-bottom depth positions of the Fe redox boundary versus flux of organic carbon to the ocean floor inferred from observations of living magnetotactic bacteria in South Atlantic (Petermann 1994) and rock magnetic analyses of equatorial Pacific sedimentary deposits (Tarduno and Wilkison 1996). The solid line is least-square fit to the data.

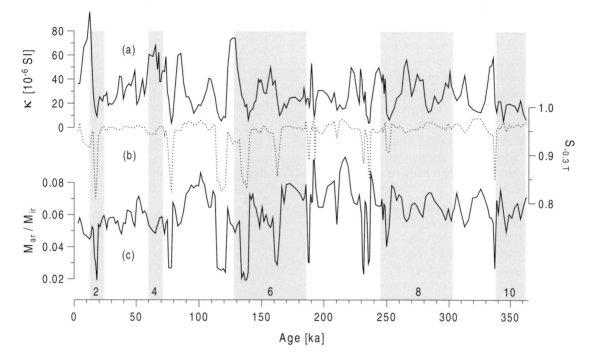

**Fig. 2.2.8** Rock magnetic records at site GeoB 2908 in the central equatorial Atlantic (0.1°S/23.3°W, 3809 m water depth). **(a)** Bulk magnetic volume susceptibility $\kappa$, **(b)** hematite/goethite index $S_{-0.3T}$ (dotted line), **(c)** magnetite grain-size sensitive ratio $M_{ar}/M_{ir}$. A series of diagenetic dissolution events is documented by the distinct coarsening of magnetite grain-sizes at $\kappa$ minima and maxima of relative hematite/goethite contents. Light shading indicates predominantly cold oxygen isotope stages 2 to 10. Data from Funk (1997).

specifically a convincing proof is required for a biogenic (?) formation of magnetite up to several meters deep in the sediment column. As pointed out by Tarduno and Wilkison (1996), such a process would have rather dramatic implications for the interpretation of the natural remanent magnetization pattern in marine sediment.

Alterations of sediment magnetic signals and particularly a delayed NRM acquisition due to biologically mediated and/or inorganic diagenesis resulting in dissolution and/or authigenesis of magnetic minerals have been discussed by various authors (e.g., van Hoof and Langereis 1991, Bloemendal et al. 1992, van Hoof et al. 1993, Dekkers et al. 1994, Tarduno 1995). They primarily relate these phenomena to non-steady state paleoenvironments, i.e., frequent changes from oxic to suboxic or anoxic conditions and vice versa. As an example Figure 2.2.8 shows rock magnetic records of a gravity core from the central equatorial Atlantic spanning the last about 350 ka.

As clearly demonstrated in other late Quaternary sediment sequences from this area (Frederichs et al 1998), the magnetic mineral contents are primarily controlled by climatically driven varia-

tions in eolian influx of terrigenous material from the African continent. However, and in contrast to numerous results from the western equatorial Atlantic, e.g., the Ceará Rise (von Dobeneck and Schmieder 1998), the magnetic volume susceptibility logs typically show very little similarities to oxygen isotope records over prolonged periods of time. A main reason are repeated extreme k minima. Their more detailed inspection by rock magnetic analyses reveals a simultaneous pronounced coarsening of the magnetite fraction (decreasing $M_{ar}/M_{ir}$ ratio) and a remarkable increase of the relative hematite (plus goethite) concentrations (decreasing magnetite/hematite ratio $S_{-0.3T}$), exactly the same characteristic features as seen in the presently active diagenetic system of the Benguela upwelling region. This striking analogy strongly suggests multiple suboxic to anoxic episodes which are likely to document enhanced $C_{org}$ fluxes to the ocean floor and hence periods of intensified primary productivity as proposed by Tarduno (1994).

The different topics briefly outlined above exemplify approaches of employing magnetic data sets as sensitive proxy parameters to characterize geochemical marine settings. Though *Environmental Magnetism* has not yet been exploited in its full range of capacities, the wide variety of rock magnetic measurements and parameters enable to trace the intense effects of diagenetic processes on magnetic mineral assemblages in appropriate details. Their application in future work should provide particularly valuable contributions to unravel paleoenvironmental condition in space and time. To this aim, further basic studies are needed for a more comprehensive understanding of modern analogues.

This is contribution No 252 of the Special Reseach Program SFB 261 (*The South Atlantic in the Late Quaternary*) funded by the Deutsche Forschungs-gemeinschaft (DFG).

# References

Bazylinski, D.A. and Blakemore, R.P., 1983. Denitrifica-tion and assimilatory nitrate reduction in *Aquaspiril-lum magnetotacticum*. Appl. Environ. Microbiol., 46: 1118-1124.

Bazylinski, D.A., Frankel, R.B. and Jannasch, H.W., 1988. Anaerobic magnetite production by a marine, magneto-tactic bacterium. Nature, 334: 518-519.

Berger, W.H., 1989. Global maps of ocean productivity. In: Berger, W.H., Smetacek, V.S. and Wefer, G. (eds) Produc-tivity of the Ocean: Present and Past. John Wiley & Sons, Chichester, pp. 429-455.

Berner, R.A., 1981. A new geochemical classification of sedi-mentary environments. J. Sed. Petrol., 51: 359-365.

Betzer, P.R., Showers, W.J., Laws, E.A., Winn, C.D., DiTullio, G.R. and Kroopnick, P.M., 1984. Primary pro-ductivity and particle fluxes on a transect of the Equator at 153°W in the Pacific Ocean. Deep-Sea Res., 31: 1-11.

Blakemore, R.P., Short, K.A., Bazylinski, D.A., Rosen-blatt, C. and Frankel, R.B., 1985. Microaerobic conditions are required for magnetite formation within *Aquaspirillum magnetotacticum*. Geomicrobiol. J., 4: 53-71.

Bloemendal, J., King, J.W., Hall, F.R. and Doh, S.-J., 1992. Rock magnetism of late Neogene and Pleisto-cene deep-sea sediments: relationship to sediment source, diagenetic processes, and sediment lithology. J. Geophys. Res., 97: 4361-4375.

Butler, R.F. and Banerjee, S.K., 1975. Theoretical single-do-main grain size range in magnetite and titanomagnetite. J. Geophys. Res., 80: 4049-4058.

Canfield, D.E. and Berner, R.A., 1987. Dissolution and pyritization of magnetite in anoxic marine sedi-ments. Geochim. Cosmochim. Acta, 51: 645-659.

Chang, S.R., Kirschvink, J.L. and Stolz, J.F., 1987. Biogenic magnetite as a primary remanence carrier in limestone de-posits. Phys. Earth. Planet. Inter., 46: 289-303.

Chang, S.R., Stolz, J.F., Awramik, S.M. and Kirschvink, J.L., 1989. Biogenic magnetite in stromatolites: occur-rence in ancient sedimentary environments. Precam-brian Res., 43: 305-315.

Däumler, K., 1996. Diagenetische Auflösung von bioge-nem Magnetit. Eine Untersuchung der magnetischen Eigenschaften von Oberflächensedimenten aus dem Benguela Auftriebsgebiet vor Namibia. Fachbereich Geowissenschaften, Universität Bremen (unpublished di-ploma thesis).

Day, R., Fuller, M. and Schmidt, V.A., 1977. Hysteresis prop-erties of titanomagnetites: grainsize and compo-sitional dependence. Phys. Earth Planet. Inter., 13: 260-267.

Dekkers, M.J., Langereis, C.G., Vriend, S.P., van Sant-voort, P.J.M. and de Lange, G.J., 1994. Fuzzy c-means cluster analysis of early diagenetic effects on natural remanent magnetisation acquisition in a 1.1 Myr piston core from the Central Mediterranean. Phys. Earth Planet. Inter., 85: 155-171.

von Dobeneck, T., 1996. A systematic analysis of natural magnetic mineral assemblages based on modelling hyster-esis loops with coercivity-related hyperbolic basis func-tions. Geophys. J. Int. 124: 675-694.

von Dobeneck, T. and Schmieder, F., 1998. Using rock mag-netic proxy records for orbital tuning and exten-ded time series analyses into the super- and sub-Milankovitch bands. In: Fischer, G. and Wefer, G. (eds) Proxies in Paleoceanography. Springer-Verlag, Berlin, Heidelberg, New York (in press).

Dunlop, D.J. and Özdemir, Ö., 1997. Rock Magnetism. Cam-bridge Studies in Magnetism, Cambridge University Press, 573 pp.

Farina, M., Esquivel, D.M.S. and Lins de Barros, H.G.P., 1990. Magnetic iron-sulphur crystals from a mag-netotactic microorganism. Nature, 343: 256-258.

Frederichs, T., Bleil, U., Däumler, K., von Dobeneck, T. and Schmidt, A., 1998. The magnetic view on the marine paleoenvironment: parameters, techniques, and potentials of rock magnetic studies as a key to paleoclimatic and paleoceanographic changes. In: Fischer, G. and Wefer, G. (eds) Proxies in Paleoocea-nography. Springer-Verlag, Berlin, Heidelberg, New York (in press).

Froehlich, P.N., Klinkhammer, G.B., Bender, M.L., Luedtke, N.A., Heath, G.R., Cullen, D., Hartman, B. and Maynard, V., 1979. Early oxidation of organic matter in pelagic sediments of the eastern equatorial Atlantic: suboxic diagenesis. Geochim. Cosmochim. Acta, 43: 1075-1090.

Funk, J., 1997 Sedimentologische, organisch-geochemi-sche und geophysikalische Untersuchungen am Kern 2908-7. Fachbereich Geowissenschaften, Universität Bremen (un-published diploma thesis).

Gee, J. and Kent, D.V., 1995. Magnetic hysteresis in young mid-ocean ridge basalts: dominant cubic anisotropy? Geophys. Res. Lett., 22: 551-554.

Glud, R.N., Gundersen, J.K., Jørgensen, B.B., Revsbech, N.P. and Schulz, H.D., 1994. Diffusive and total oxygen up-take of deep-sea sediments in the eastern South Atlantic Ocean: *in situ* and laboratory measurements. Deep-Sea Res., 41: 1767-1788.

Gorby, Y.A., Beveridge, T.J. and Blakemore, R.P., 1988. Characterizaton of the bacterial magnetosome membrane. J. Bacteriol., 170: 834-841.

van Hoof, A.A.M. and Langereis C.G., 1991. Reversal

records in marine marls and delayed acquisition of remanent magnetization. Nature, 351: 223-225.

van Hoof, A.A.M., Os, B.J.H., Rademakers, J.G., Lange-reis, C.G. and de Lange, G.J., 1993. A paleomagnetic and geochemical record of the upper Cochiti reversal and two subsequent precessional cycles from southern Sicily (Italy). Earth Planet. Sci. Lett., 117: 235-250.

Jackson, M., 1990. Diagenetic source of stable remanence in remagnetized Paleozoic cratonic carbonates. J. Geophys. Res., 95: 2753-2762.

Karlin, R., 1990a. Magnetite diagenesis in marine sedi-ments from the Oregon continental margin. J. Geo-phys. Res., 95: 4405-4420.

Karlin, R., 1990b. Magnetic mineral diagenesis in suboxic sediments at Bettis Site W-N, NE Pacific Ocean. J. Geophys. Res., 95: 4421-4436.

Karlin, R. and Levi, S., 1983. Diagenesis of magnetic miner-als in recent hemipelagic sediments. Nature, 303: 327-330.

Karlin, R. and Levi, S., 1985. Geochemical and sedimen-tological control of the magnetic properties of hemi-pelagic sediments. J. Geophys. Res., 90: 10373-10392.

King, J.W. and Channell, J.E.T., 1991. Sedimentary magnet-ism, environmental magnetism, and magnetostra-tigraphy. Rev. Geophys., 29: 358-370 (IUGG Report).

Leslie, B.W., Hammond, D.E., Berelson, W.M. and Lund, S.P., 1990. Diagenesis in anoxic sediments from the Cali-fornia continental borderland and its influence on iron, sulfur, and magnetite behavior. J. Geophys. Res., 95: 4453-4470.

Lovley, D.R., Stolz, J.F., Nord, J.G.L. and Philips, E.J.P., 1987. Anaerobic production of magnetite by a dis-similatory iron-reducing microorganism. Nature, 330: 252-254.

Lund, S.P. and Karlin, R., 1990. Introduction to the special section on physical and biogeochemical processes respon-sible for the magnetization of sediments. J. Geophys. Res., 95: 4353-4354.

Maher, B.A. and Taylor, R.M., 1988. Formation of ultra-fine-grained magnetite in soils. Nature, 336: 368-370.

Mann, S., Sparks, N.H.C., Frankel, R.B., Bazylinski, D.A. and Jannasch, H.W., 1990. Biomineralization of fer-rimagnetic greigite ($Fe_3S_4$) and iron pyrite ($FeS_2$) in a magnetotactic bacterium. Nature, 343: 258-261.

McNeill, D.F., 1990. Biogenic magnetite from surface Holocene carbonate sediments, Great Bahama Bank. J. Geophys. Res. 95: 4363-4372.

Moskowitz, B.M., Frankel, R.B., Bazylinski, D.A., Jan-nasch, H.W. and Lovley D.R., 1989. A comparison of magnetite particles produced anaerobically by mag-netotactic and dissimilatory iron-reducing bacteria. Geophys. Res. Lett., 16: 665-668.

Petermann, H., 1994. Magnetotaktische Bakterien und ihre Magnetosome in Oberflächensedimenten des Süd-atlantiks. Berichte, Fachbereich Geowissenschaften, Universität Bremen, 56: 1-134.

Petermann, H. and Bleil, U., 1993. Detection of live magnetotactic bacteria in South Atlantic deep-sea sediments. Earth Planet. Sci. Lett., 117: 223-228.

Petersen, N., von Dobeneck. T. and Vali, H., 1986. Fossil bac-terial magnetite in deep-sea sediments from the South At-lantic Ocean. Nature, 320: 611-615.

Reynolds, R.L., Fishman, N.S., Wanty, R.B. and Gold-haber, M.B., 1990. Iron sulphide minerals at Clement oil field, Oklahoma: implications for magnetic detection of oil fields. Geol. Soc. Amer. Bull., 102: 368-380.

Snowball, I. and Thompson, R., 1988. An occurrence of greigite in the sediments of Loch Lomond. J. Quat. Sci., 4: 121-125.

Sparks, N.H.C., Mann, S., Bazylinski, D.A., Lovley, D.R., Jannasch, H.W. and Frankel, R.B., 1990. Structure and morphology of anaerobically-produced magnetite by a marine magnetotactic bacterium and a dissimilatory iron-reducing bacterium. Earth Planet. Sci. Lett., 98: 14-22.

Stolz, J.F., 1992. Magnetotactic bacteria: biominera-lization, ecology, sediment magnetism, environmental indicator. In: Skinner H.G.W. and Fitzpatrick, R.W. (eds) Biomineralization, Processes of Iron and Manganese. Catena Verlag, Cremlingen, pp. 133-145.

Stolz, J.F., Chang, S.R. and Kirschwink, J.L., 1986. Mag-netotactic bacteria and single-domain magnetite in hemipelagic sediments. Nature, 321: 849-851.

Tarduno, J.A., 1994. Temporal trends of magnetic dissolution in the pelagic realm: gauging paleoproduc-tivity? Earth Planet. Sci. Lett., 123: 39-48.

Tarduno, J.A., 1995. Superparamagnetism and reduction diagenesis in pelagic sediments: enhancement or deple-tion? Geophys. Res. Lett., 22: 1337-1340.

Tarduno, J.A. and Wilkison, S.L., 1996. Non-steady state magnetic mineral reduction, chemical lock-in, and de-layed remanence acquisition in pelagic sediments. Earth Planet. Sci. Lett., 144: 315-326.

Taylor, R.M., Maher. B.A. and Self, P.G., 1987. Magne-tite in soils, I. The synthesis of single-domain and superparamagnetic magnetite. Clay Miner., 22: 411-422.

Thompson, R. and Oldfield, F., 1986. Environmental Magnet-ism. Allen & Unwin, London, 227 pp.

Vali, H., Förster, O., Amarantidis, G. and Petersen, N., 1987. Magnetotactic bacteria and their magnetofos-sils in sediments. Earth Planet. Sci. Lett., 86: 389-400.

Vali, H., von Dobeneck, T., Amarantidis, G., Förster, O., Morteani, G., Bachmann. L. and Petersen, N., 1989. Bio-genic and lithogenic magnetic minerals in Atlantic and Pacific deep-sea sediments and their paleomag-netic sig-nificance. Geol. Rundschau 78: 753-764.

Verosub, K.L. and Roberts, A.P., 1995. Environmental mag-netism: past, present, and future. J. Geophys. Res., 100: 2175-2192.

# 3 Quantification of Early Diagenesis: Dissolved Constituents in Marine Pore Water

HORST D. SCHULZ

A chapter on the pore water of sediments and the processes of early diagenesis, reflected by the concentration profiles therein, can certainly be structured in different ways. One possibility is, for example, to start with the sample-taking strategies, followed by the analytical treatment of the samples, then a model presentation to substantiate the processes, and finally a quantitative evaluation of the measured profiles with regard to material fluxes and reaction rates. The reader who would prefer this sequence is recommended to begin with the Sections 3.3, 3.4, 3.5 and to consult the Sections 3.1 and 3.2 later. However, in our opinion, marine geochemistry departed from its initial development stage of sample collection and anlysis some years ago. This book is therefore written with a different structure from earlier texts. The well understood theoretical knowledge on concentration profiles has to be introduced from the beginning. Only then do problems arise which necessitate the investigation of particular substances in pore water and the application of specific methods of sampling and analysis.

Particularly, two fundamental approaches are conceivable for investigating diagenesis processes, both having their advantages and disadvantages. The classical way leads to a sedimentological, mineralogical, and in the broadest sense geochemical examination of the solid phase. The advantages of this approach is that sample withdrawal and analytical procedures are generally easier to carry out and produce more reliable results. On the other hand, this approach in most cases will not provide us with any information on the timing at which specific diagenetical alterations in the sediment took place. Nothing will be learnt about reaction rates, reaction kinetics, and mostly we do not even know whether the processes, interpreted on account of certain measurements and/or observations, occur today or whether they were termi-

nated some time before. Furthermore, the distinction between the primary signals from the water column and the results of diagenetic reactions, often remains unclear.

If the processes of diagenesis and their time-dependency are to be recorded directly, then a second approach appears to be more reasonable; one that includes the measurement of concentration profiles in pore water. Figure 3.1 shows various concentration profiles of substances dissolved in the pore water of sediments taken at a depth of approximately 4000 m, off the estuary of the Congo River. The differently shaped curves shown here directly reflect the reactions that are happening today. The investigation of pore water thus constitutes the only way to determine the reaction rates and the fluxes of material which they produce. Yet, this procedure is rather laborious and unfortunately contains numerous opportunities for errors. Most errors are caused as soon as the sediment sample is taken, others are caused upon preparing the pore water from the sediment, or in the course of analyzing low concentrations obtained from small sample amounts. Actually, one would prefer to gain all results by *in situ* measurements at the bottom of the ocean, which is feasible, however, only in very few exceptional cases.

**GeoB 1401**

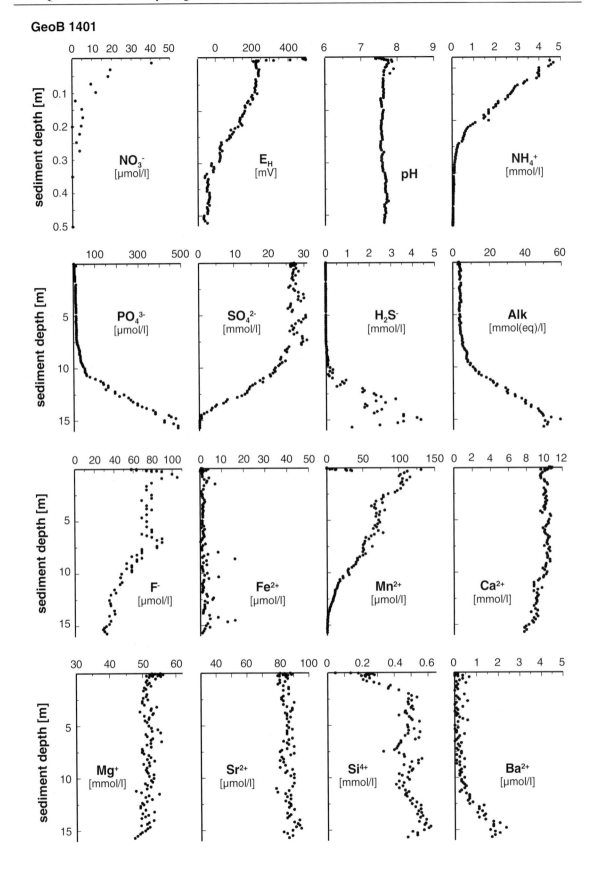

**Fig. 3.1** Concentration profiles in the pore water fractions of sediments obtained off the estuary of the River Congo, at a depth of approximately 4000 m. The sediments contain a relatively high amount of TOC. Values ranging from 1 to 3.5 wt. % indicate that this sediment is characterized by the high reaction rates of various early diagenesis processes. These processes are reflected by diffusion fluxes over gradients and by reaction rates determined by gradient changes (after Schulz et al. 1994).

# 3.1 Introduction: How to Read Pore Water Concentration Profiles

The processes of early diagenesis - irrespective of whether we are dealing with microbiological redox-reactions or predominantly abiotic reactions of dissolution and precipitation - are invariably reflected in the pore water of sediments. The aquatic phase of the sediments is the site where the reactions occur, and where they become visible as time-dependent or spatial distributions of concentrations. Thus, if the early diagenesis processes are to be examined and quantified in the young marine sediment, the foremost step will always consist of measuring the concentration profiles in the pore water fraction.

For reasons of simplification and enabling an easy understanding of the matter, only three fundamental processes will be considered in this section, along with their manifestation in the pore water concentration profiles:

- Consumption of a reactant in pore water (e.g. consumption of oxygen in the course of oxidizing organic material)

- Release of a substance from the solid phase into the pore water fraction (e.g. release of silica due to the dissolution of opal).

- Diffusion transport of dissolved substances in pore water and across the sediment/bottom water boundary.

Figure 3.2 shows schematic diagrams of some possible concentration profiles occurring in marine pore water. Here, the profile shown in Figure 3.2a is quite easy to understand. In the event of a substance that is not subject to any early diagenetic process, the pore water as the formation water contains exactly the same concentration as the

supernatant ocean water. If the concentration in the ocean floor water changes, this change will be reproduced in the pore water fraction on account of diffusion processes. After a definite time of non-stationary conditions, which depends on the depth of the profile under study and the required accuracy of the measurement, the concentration of the ocean floor water prevails again, stationary and constant within the entire profile. The details and the problems concerning stationary and non-

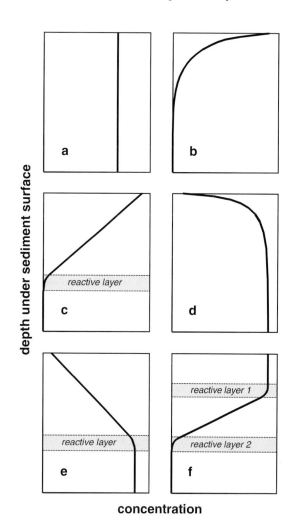

**concentration**

**Fig. 3.2** Principal concentration profile shapes as in measured pore water:

a: Profile of a non-reacting substance (e.g. chloride); b: profile of a substance that is depleted in the upper layers of the sediment (e.g. oxygen); c: profile of a substance that is consumed in a particular reactive layer; d: profile of a substance that is released into pore water in the upper layers of the sediment (e.g. silica); e: profile of a substance that is released into the pore water in specific reactive layers; f: profile of a substance that is released into pore water in one discrete depth (reactive layer 1), and, in another depth (reactive layer 2) is removed from the pore water by consumption (e.g. sulfate).

stationary conditions in pore water will be thoroughly discussed in Section 3.2.

Figure 3.2b shows the concentration profile of a substance which is consumed by early diagenetic reactions in the upper layers of the sediment. For example, the profiles of dissolved oxygen often look like this (cf. Chap. 6). Assuming steady-state conditions (cf. Sect. 3.2), the concentration gradient near the sediment surface will display its highest increases, because whatever is consumed in the deeper layers will arrive there by means of diffusion. With increasing depths the concentration gradient and the diffusive material transport will decline, until, at a particular depth, concentration, gradient, and diffusion simultaneously approach zero. This sediment depth, referred to as the penetration depth of a substance, often represents a steady-state condition composed of the diffusive reinforcement from above and the depletion of the substance in the sediment by the reactions of early diagenesis.

In principle, the situation depicted in Figure 3.2c is quite the same. Here, the process is just limited to one reactive layer. With an example of oxygen as the dissolved substance, this could be a layer containing easily degradable organic matter, or the sedimentary depth in which oxygen encounters and reacts with a reductive solute species coming from below (e.g. $Mn^{2+}$ or $Fe^{2+}$). At any rate, the space above the reactive layer is characterized by a constant gradient in the concentration profile and thus by a diffusive transport which is everywhere the same. Within the reactive layer the dissolved substance is brought to zero concentration by depletion.

The reverse case, which is in principle quite similar, is demonstrated in Figure 3.2d and e. Here, a substance is released anywhere into the layers near the sediment surface (Fig. 3.2d), whereas it might be released into the pore water only in one particular layer (Fig. 3.2e). The solute could be, for example, silica that often displays such concentration profiles on account of the dissolution of sedimentary opal.

The concentration profile as shown in Figure 3.2f and characterized by the presence of two different reactive layers is somewhat more complex. The concentration gradient and hence the diffusive material transport is practically zero above 'reactive layer 1' as well as below 'reactive layer 2'. The substance is released into the pore water within the upper 'reactive layer 1'; within the lower 'reactive layer 2' the substance is elimi-

nated from the pore water. Between both reactive layers the substance is transported by diffusion downward according to the gradient. Such concentration gradients are, for example, known for sulfate. In the case of sulfate profiles, methane coming from below reduces the sulfate in the lower 'reactive layer 2'. The liberated sulfide diffuses upwards until it is re-oxidized to sulfate 'in reactive layer 1'. Such a profile represents one half, the sulfate part, of the sulfur cycle in the sediment. Details as to these processes are dealt with in Chapter 8.

The described relations can be summarized by stating a few rules for reading and understanding of pore water concentration profiles:

- Diffusive material fluxes always occur in the form of concentration gradients; concentration gradients always represent diffusive material fluxes.

- Reactions occurring in pore water always constitute changes in the concentration gradient; changes in the concentration gradient always represent reactions occurring in pore water.

- A concave-shaped alteration in the concentration gradient profile (cf. Fig. 3.2 b, c) signifies the depletion of a substance from pore water; conversely, a convex-shaped concentration gradient profile (cf. Fig. 3.2 d, e) always depicts the release of a substance into the pore water.

In the following Section 3.2, the laws of diffusion and the particularities of their application to sediment pore waters will be treated in more detail. In this context, the problem of steady state and non-steady state situations will have to be covered, since the simple examples described above have anticipated steady state situations. Moreover, as a further simplification, the examples of this section have by necessity neglected advection, bioirrigation, and bioturbation. These processes will be discussed in detail in Section 3.6.

## 3.2 Calculation of Diffusive Fluxes and Diagenetic Reaction Rates

### 3.2.1 Steady State and Non-Steady State Situations

In the preceding section and especially in all the following chapters, one pair of terms will assume extraordinary importance for describing and understanding biogeochemical processes: the steady state and non-steady state situation.

Let us first consider the steady state situation as its description is more straight forward in a model concept. In Figure 3.2 c a concentration profile is shown in which a substance is continually consumed at a specific rate of reaction and within a reactive layer. At the same time, a con-

stant concentration is prevalent in the bottom water above the sediment surface, an infinite reservoir as compared to the consumption in the sediment. It follows therefore that a constant concentration gradient exists between the sediment surface and the reactive layer, and thus everywhere the same diffusive flux. Such a concentration gradient is referred to as being in steady state. It remains in this condition as long as its determining factors - turn-over rates in the reactive layer, concentration at the sea-floor, dimension and properties of the space between the reactive layer and sediment surface - are not changed.

Any change of the conditions that is liable to exert any kind of influence on the concentration profile, terminates the steady state situation. A non-steady state emerges, which is a time-dependent situation occurring in the pore water. If the system remains unperturbed in the changed situation for a sufficient length of time, a new steady state situation can become established, different from the first and reflecting the novel configuration of conditions.

Strictly speaking, there are no real steady-state situations in nature. Even the sun had begun to shine at a certain time, the earth and, upon her, the oceans have come into existence at a certain time, and all things must pass sooner or later. Hence, the term referring to the steady-state condition also depends on the particular stretch of time which is under study, as well as on the dimension of the system, and, not least, on the accuracy of the measurements with which we examine the parameters that describe the system.

A given concentration in the pore water of a pelagic sediment, several meters below the sediment surface, can be measured today, next month, next year, and after 10 years. Within the margins of reasonable analytical precision, we will always measure just about the same value and rightfully declare the situation to be steady-state. At the same time, pore water concentrations in sedimentary surface areas near the same pelagic sediment could be subject to considerable seasonal variation (such as residual deposits of algal bloom periods) and thus be classified as being in a non-steady state.

A calculated example for pore water concentrations in a non-steady state condition is shown in Figure 3.3. Details concerning the calculation procedure will not be discussed here. The conceptual model employed will be described in Section 3.2.4, a suitable computer model is described in

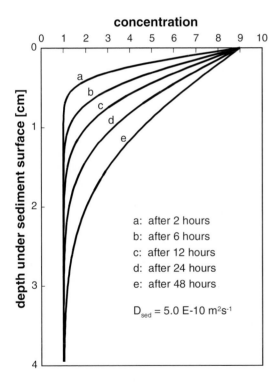

**concentration**

a: after 2 hours
b: after 6 hours
c: after 12 hours
d: after 24 hours
e: after 48 hours

$D_{sed} = 5.0\ E\text{-}10\ m^2 s^{-1}$

**Fig. 3.3** Calculated non-steady state concentration profiles in pore water of a young sediment. It was assumed that the concentration of '1' has been previously constant in the pore water of the sediment as well as in the supernatant bottom water for a long period of time, so that a steady state situation was prevalent. Then the concentration of bottom water changed shortly to '9'. The concentration profiles a to e are non-steady states after 2 and up to 48 hours. The calculation of such non-steady state concentration profiles can be performed , for example, with the aid of the model program CoTAM (Hamer and Sieger 1994) or CoTReM (cf. Chap. 14).

Chapter 14. A typical diffusion coefficient characteristic of young marine sediments and a characteristic porosity coefficient were used in the calculation.

In the calculated example shown in Figure 3.3 the assumption was made that a constant concentration of '1' has been prevalent for a long time in the pore water of the young sediment and in the bottom water above it. Then, the bottom water underwent a momentary change to yield a concentration of '9'. By means of diffusion, the new concentration gradually spread into the pore water. After 2 hours it reaches a depth of less than 1 cm, after 48 hours, respectively, a depth of about 3 to 4 cm below the sediment surface. If the concentration of '9' remains constant long enough in the bottom water above the sediment, this concentration will theoretically penetrate into an infinitely great depth. To what extent, and into what

depths of pore water, these concentrations are to be assigned to steady state or non-steady states can only be determined in each particular case, having its own concentration in the bottom water over a given period of time. At any rate, the allocation of one or the other state can only be done separately for each system, each parameter, and each time interval. Calculations as shown in Figure 3.3 are likely to produce valuable preliminary concepts, for evaluating real measured pore water profiles.

In the next example, the influence of seasonal variation in the bottom water lying above the sediment will be examined, as well as the resulting non-steady states in the pore water fraction. To this end, the following boundary conditions are selected: A substance concentration of '1' is supposed to be prevalent in the bottom water over one half year, afterwards the concentration changes to '10' for one half year, then it changes back to a value of '1', and so on, continually changing. Figure 3.4 shows the result of such an oscillatory situation after several years. The curve denoted 'a' demonstrates the situation in which the bottom concentration was '1' after half a year; the curve denoted 'b' reflects the situation in which the bottom concentration was '10' after half a year. It is evident that essential effects of such changes can only be observed down to a depth of less than 0.2 m below the sediment surface. In this model calculation, the effects of the preceding half year still remain visible in a depth between about 0.22 m and 0.35 m below the sediment surface. Below a depth of 0.4 m no further non-steady states can be seen, a constant and steady-state mean concentration of '5.5' prevails.

The conditions proposed in Figure 3.4 describe a very extreme situation. On the one hand, the half-yearly alternating concentrations differ by a whole order of magnitude, on the other hand, no transitory periods with intermediate concentrations were anticipated. Seasonal variations in nature usually display smaller concentration differences and are also likely to possess transitory intermediate concentrations. Both diminish the effects of non-steady state formation in sedimentary depths as were shown in the example of Figure 3.4.

Figure 3.4 also demonstrates clearly the effects of the analytical precision on differentiating steady state and non-steady states in pore water. In the theoretical calculation shown in Figure 3.4, the difference between both curves is still visible in form of a non-steady state at a depth between

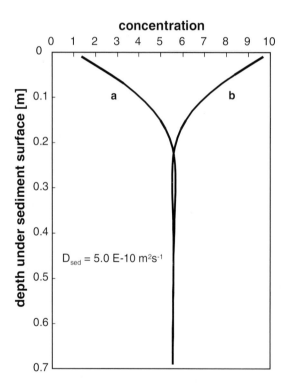

**Fig. 3.4** Calculated non-steady state concentration profiles in the pore water of a young sediment. The calculation assumes that concentrations of '1' and '10' were prevailing alternately in the bottom water over the sediment, each over a period of half a year. After several years, the curve a. reflects the situation of concentration '1' at the end of one half year, whereas curve b. reflects the situation of concentration '10' at the end of one half year. The calculation of such non-steady state concentration profiles can be performed, for example, with the aid of the model program CoTAM (Hamer and Sieger 1994) or CoTReM (cf. Chapter 14).

0.22 and 0.4 m. However, if the possible errors are taken into consideration that occur during sampling and in the analytic treatment of pore water samples (Sect. 3.3 - 3.5), then there is likely to be no current parameter with which these differences could be discovered practically. Below a depth of 0.2 m, one would always measure the same concentration and consequently judge the situation as stationary, or as a steady state.

### 3.2.2   The Steady State Situation and Fick's First Law of Diffusion

According to Fick's first law of diffusion, the diffusive flux (J) is directly proportional to concentration gradient ($\partial C/\partial x$) under steady state conditions. The factor of proportionality is the temperature-dependent and substance-dependent diffusion coefficient ($D^0$):

$$J = -D^0 \cdot \frac{\partial C}{\partial x} \qquad (3.1)$$

The negative sign indicates that the diffusive flux runs in opposition to the gradient's direction from high concentrations to lower concentrations. In these terms, increasing concentrations in greater depths will yield negative gradients in the sediment along with an upwards directed positive flux, and vice-versa.

The relations shown in Equation 3.1 are only valid for free solutions, thus without the 'disturbing' sedimentary solid phase. The diffusion coefficient $D^0$ is only valid for infinitely dilute solutions. Boudreau (1997) and Iversen and Jørgensen (1993) have summarized the current state of knowledge concerning the application of the diffusion laws to the pore water volume of sediments, accomplished on the basis of an extensive literature survey, and have thoroughly discussed the general problem. As nothing further needs to be added at this point, the calculation of the values shown in the following Tables 3.1 and 3.2 were performed by using the relations published by Boudreau (1997). Table 3.1 contains the diffusion coefficients for a number of ions, gases, and uncharged complex compounds dissolved in sea-water at various temperatures.

Looking at diffusion processes in the pore water volume of sediments, it must be taken into consideration that diffusion can only take place within the pore water volume (porosity $\phi$), hence a diffusive flux can only be proportionally effec-

tive with regard to this spatial compartment. Beyond this limitation, the diffusion coefficient is distinctly lower in the pore water volume of a sediment ($D_{sed}$) than in free solution. The diffusive flux in the sediment ($J_{sed}$) is calculated as:

$$J_{sed} = -\phi \cdot D_{sed} \cdot \frac{\partial C}{\partial x} \qquad (3.2)$$

The diffusion coefficient in the pore water volume of sediments differs from the diffusion coefficient of free solutions in such a manner that diffusion in the pore water volume cannot follow a straight course, but must take 'deviations' around each single grain. The degree of deviation around particles is called tortuosity ($\theta$). It describes the mean ratio between the real length of the pathway and the straight-line distance. Tortuosity can be quantified directly by measuring the electrical resistivity (R) (McDuff and Ellis 1979) and employing a related 'formation factor' (F).

$$F = R_s / R_f \qquad (3.3)$$

In this equation, $R_s$ is the specific electrical resistivity for the whole system composed of sediment and pore water, and $R_f$ denotes the electrical resistivity for pore water only. Since the electric current in the pore water volume of sediments is bound to the presence of charged ions in solution, the same deviations are valid (cf. Sect. 2.1.2). The tortuosity ($\theta$) is then calculated by applying the porosity ($\phi$) to the equation according to (Berner 1980):

$$\theta^2 = \phi \cdot F \qquad (3.4)$$

The diffusion coefficient in sediments ($D_{sed}$) can be calculated on the basis of a dimensionless tortuosity and the diffusion coefficient in free solutions of sea-water ($D^{sw}$ in Table 3.1):

$$D_{sed} = D^{sw} / \theta^2 \qquad (3.5)$$

If the tortuosity is not known on account of electric conductivity measurements and the 'formation factor' (F), its value may be estimated, a bit less accurately, by its empirical relationship to the porosity. About a dozen different empirical equations are known from literature. The most frequent is Archie's Law (Archie 1942):

$$\theta^2 = \phi^{(1-m)} \qquad (3.6)$$

**Table. 3.1** Diffusion coefficient in free solution for sea-water at various temperatures. The calculations were performed on the basis of relations derived from a comprehensive literature study carried out by Boudreau (1997).

| Ion | 0 °C | 5 °C | 10 °C | 15 °C | 20 °C | 25 °C |
|-----|------|------|-------|-------|-------|-------|
| | | | $D^{sw}$ [$m^2 s^{-1}$] | | | |
| $H^+$ | 5.17E-09 | 5.89E-09 | 6.60E-09 | 7.30E-09 | 8.01E-09 | 8.69E-09 |
| $D^+$ | 3.03E-09 | 3.68E-09 | 4.33E-09 | 4.97E-09 | 5.61E-09 | 6.24E-09 |
| $Li^+$ | 4.21E-10 | 5.34E-10 | 6.45E-10 | 7.56E-10 | 8.66E-10 | 9.74E-10 |
| $Na^+$ | 5.76E-10 | 7.15E-10 | 8.52E-10 | 9.88E-10 | 1.12E-09 | 1.26E-09 |
| $K^+$ | 9.08E-10 | 1.10E-09 | 1.29E-09 | 1.47E-09 | 1.66E-09 | 1.84E-09 |
| $Rb^+$ | 9.70E-10 | 1.17E-09 | 1.36E-09 | 1.56E-09 | 1.75E-09 | 1.94E-09 |
| $Cs^+$ | 9.79E-10 | 1.17E-09 | 1.36E-09 | 1.55E-09 | 1.74E-09 | 1.93E-09 |
| $Ag^+$ | 7.43E-10 | 9.11E-10 | 1.08E-09 | 1.24E-09 | 1.40E-09 | 1.57E-09 |
| $NH_4^+$ | 9.03E-10 | 1.10E-09 | 1.29E-09 | 1.47E-09 | 1.66E-09 | 1.85E-09 |
| $Ba^{2+}$ | 3.86E-10 | 4.68E-10 | 5.49E-10 | 6.30E-10 | 7.10E-10 | 7.88E-10 |
| $Be^{2+}$ | 2.44E-10 | 3.10E-10 | 3.75E-10 | 4.39E-10 | 5.03E-10 | 5.66E-10 |
| $Ca^{2+}$ | 3.42E-10 | 4.26E-10 | 5.09E-10 | 5.91E-10 | 6.72E-10 | 7.53E-10 |
| $Cd^{2+}$ | 3.15E-10 | 3.85E-10 | 4.56E-10 | 5.25E-10 | 5.95E-10 | 6.63E-10 |
| $Co^{2+}$ | 3.15E-10 | 3.85E-10 | 4.56E-10 | 5.25E-10 | 5.95E-10 | 6.63E-10 |
| $Cu^{2+}$ | 3.22E-10 | 3.96E-10 | 4.69E-10 | 5.41E-10 | 6.13E-10 | 6.84E-10 |
| $Fe^{2+}$ | 3.15E-10 | 3.84E-10 | 4.54E-10 | 5.22E-10 | 5.91E-10 | 6.58E-10 |
| $Hg^{2+}$ | 3.19E-10 | 4.17E-10 | 5.13E-10 | 6.09E-10 | 7.04E-10 | 7.98E-10 |
| $Mg^{2+}$ | 3.26E-10 | 3.93E-10 | 4.60E-10 | 5.25E-10 | 5.91E-10 | 6.55E-10 |
| $Mn^{2+}$ | 3.02E-10 | 3.75E-10 | 4.46E-10 | 5.17E-10 | 5.88E-10 | 6.57E-10 |
| $Ni^{2+}$ | 3.19E-10 | 3.80E-10 | 4.40E-10 | 4.99E-10 | 5.58E-10 | 6.16E-10 |
| $Sr^{2+}$ | 3.51E-10 | 4.29E-10 | 5.08E-10 | 5.85E-10 | 6.62E-10 | 7.38E-10 |
| $Pb^{2+}$ | 4.24E-10 | 5.16E-10 | 6.08E-10 | 6.98E-10 | 7.88E-10 | 8.77E-10 |
| $Ra^{2+}$ | 3.72E-10 | 4.65E-10 | 5.57E-10 | 6.48E-10 | 7.39E-10 | 8.28E-10 |
| $Zn^{2+}$ | 3.15E-10 | 3.85E-10 | 4.55E-10 | 5.24E-10 | 5.93E-10 | 6.60E-10 |
| $Al^{3+}$ | 2.65E-10 | 3.46E-10 | 4.26E-10 | 5.05E-10 | 5.83E-10 | 6.61E-10 |
| $Ce^{3+}$ | 2.80E-10 | 3.41E-10 | 4.02E-10 | 4.62E-10 | 5.22E-10 | 5.80E-10 |
| $La^{3+}$ | 2.64E-10 | 3.28E-10 | 3.91E-10 | 4.53E-10 | 5.15E-10 | 5.76E-10 |
| $Pu^{3+}$ | 2.58E-10 | 3.13E-10 | 3.69E-10 | 4.24E-10 | 4.79E-10 | 5.32E-10 |
| $OH^-$ | 2.46E-09 | 2.97E-09 | 3.48E-09 | 3.98E-09 | 4.47E-09 | 4.96E-09 |
| $OD^-$ | 1.45E-09 | 1.76E-09 | 2.07E-09 | 2.38E-09 | 2.68E-09 | 2.98E-09 |
| $Al(OH)_4^-$ | 4.24E-10 | 5.37E-10 | 6.50E-10 | 7.62E-10 | 8.73E-10 | 9.82E-10 |
| $Br^-$ | 9.51E-10 | 1.16E-09 | 1.36E-09 | 1.56E-09 | 1.76E-09 | 1.96E-09 |
| $Cl^-$ | 9.13E-10 | 1.12E-09 | 1.32E-09 | 1.52E-09 | 1.72E-09 | 1.92E-09 |
| $F^-$ | 5.98E-10 | 7.58E-10 | 9.17E-10 | 1.07E-09 | 1.23E-09 | 1.39E-09 |
| $HCO_3^-$ | 4.81E-10 | 6.09E-10 | 7.37E-10 | 8.63E-10 | 9.89E-10 | 1.11E-09 |
| $H_2PO_4^-$ | 3.82E-10 | 4.86E-10 | 5.90E-10 | 6.92E-10 | 7.94E-10 | 8.94E-10 |
| $HS^-$ | 9.89E-10 | 1.11E-09 | 1.24E-09 | 1.36E-09 | 1.49E-09 | 1.61E-09 |
| $HSO_3^-$ | 6.04E-10 | 7.34E-10 | 8.63E-10 | 9.91E-10 | 1.12E-09 | 1.24E-09 |
| $HSO_4^-$ | 5.69E-10 | 7.13E-10 | 8.55E-10 | 9.96E-10 | 1.14E-09 | 1.27E-09 |
| $I^-$ | 9.33E-10 | 1.13E-09 | 1.33E-09 | 1.53E-09 | 1.73E-09 | 1.92E-09 |

| Ion | 0 °C | 5 °C | 10 °C | 15 °C | 20 °C | 25 °C |
|---|---|---|---|---|---|---|
| | | | $D^{sw}$ [$m^2s^{-1}$] | | | |
| $IO_3^-$ | 4.43E-10 | 5.61E-10 | 6.78E-10 | 7.93E-10 | 9.08E-10 | 1.02E-09 |
| $NO_2^-$ | 9.79E-10 | 1.13E-09 | 1.28E-09 | 1.43E-09 | 1.58E-09 | 1.73E-09 |
| $NO_3^-$ | 9.03E-10 | 1.08E-09 | 1.26E-09 | 1.44E-09 | 1.62E-09 | 1.79E-09 |
| Acetate$^-$ | 4.56E-10 | 5.71E-10 | 6.84E-10 | 7.96E-10 | 9.08E-10 | 1.02E-09 |
| Lactate$^-$ | 4.19E-10 | 5.32E-10 | 6.44E-10 | 7.54E-10 | 8.64E-10 | 9.72E-10 |
| $CO_3^{2-}$ | 4.12E-10 | 5.04E-10 | 5.96E-10 | 6.87E-10 | 7.78E-10 | 8.67E-10 |
| $HPO_4^{2-}$ | 3.10E-10 | 3.93E-10 | 4.75E-10 | 5.56E-10 | 6.37E-10 | 7.16E-10 |
| $SO_3^{2-}$ | 4.31E-10 | 5.47E-10 | 6.62E-10 | 7.77E-10 | 8.91E-10 | 1.00E-09 |
| $SO_4^{2-}$ | 4.64E-10 | 5.72E-10 | 6.79E-10 | 7.86E-10 | 8.91E-10 | 9.95E-10 |
| $S_2O_3^{2-}$ | 4.58E-10 | 5.82E-10 | 7.06E-10 | 8.28E-10 | 9.50E-10 | 1.07E-09 |
| $S_2O_4^{2-}$ | 3.75E-10 | 4.63E-10 | 5.49E-10 | 6.35E-10 | 7.20E-10 | 8.04E-10 |
| $S_2O_6^{2-}$ | 5.04E-10 | 6.40E-10 | 7.75E-10 | 9.08E-10 | 1.04E-09 | 1.17E-09 |
| $S_2O_8^{2-}$ | 4.64E-10 | 5.89E-10 | 7.12E-10 | 8.35E-10 | 9.57E-10 | 1.08E-09 |
| Malate$^{2-}$ | 3.19E-10 | 4.05E-10 | 4.90E-10 | 5.74E-10 | 6.57E-10 | 7.39E-10 |
| $PO_4^{3-}$ | 2.49E-10 | 3.16E-10 | 3.82E-10 | 4.48E-10 | 5.13E-10 | 5.77E-10 |
| Citrate$^{3-}$ | 2.54E-10 | 3.22E-10 | 3.90E-10 | 4.57E-10 | 5.23E-10 | 5.89E-10 |
| $H_2$ | 1.92E-09 | 2.36E-09 | 2.81E-09 | 3.24E-09 | 3.67E-09 | 4.10E-09 |
| He | 2.74E-09 | 3.37E-09 | 4.00E-09 | 4.63E-09 | 5.24E-09 | 5.85E-09 |
| NO | 1.02E-09 | 1.26E-09 | 1.49E-09 | 1.72E-09 | 1.95E-09 | 2.18E-09 |
| $N_2O$ | 9.17E-10 | 1.13E-09 | 1.34E-09 | 1.55E-09 | 1.75E-09 | 1.96E-09 |
| $N_2$ | 8.69E-10 | 1.07E-09 | 1.27E-09 | 1.47E-09 | 1.66E-09 | 1.85E-09 |
| $NH_3$ | 9.96E-10 | 1.23E-09 | 1.45E-09 | 1.68E-09 | 1.90E-09 | 2.12E-09 |
| $O_2$ | 1.00E-09 | 1.23E-09 | 1.46E-09 | 1.69E-09 | 1.91E-09 | 2.13E-09 |
| CO | 1.00E-09 | 1.24E-09 | 1.47E-09 | 1.69E-09 | 1.92E-09 | 2.14E-09 |
| $CO_2$ | 8.38E-10 | 1.03E-09 | 1.22E-09 | 1.41E-09 | 1.60E-09 | 1.79E-09 |
| $SO_2$ | 6.94E-10 | 8.54E-10 | 1.01E-09 | 1.17E-09 | 1.33E-09 | 1.48E-09 |
| $H_2S$ | 9.17E-10 | 1.13E-09 | 1.34E-09 | 1.55E-09 | 1.75E-09 | 1.96E-09 |
| Ar | 8.65E-10 | 1.06E-09 | 1.26E-09 | 1.46E-09 | 1.65E-09 | 1.85E-09 |
| Kr | 8.38E-10 | 1.03E-09 | 1.22E-09 | 1.41E-09 | 1.60E-09 | 1.79E-09 |
| Ne | 1.31E-09 | 1.62E-09 | 1.92E-09 | 2.22E-09 | 2.51E-09 | 2.81E-09 |
| $CH_4$ | 7.29E-10 | 8.97E-10 | 1.06E-09 | 1.23E-09 | 1.39E-09 | 1.56E-09 |
| $CH_3Cl$ | 6.51E-10 | 8.01E-10 | 9.50E-10 | 1.10E-09 | 1.24E-09 | 1.39E-09 |
| $C_2H_6$ | 6.03E-10 | 7.42E-10 | 8.80E-10 | 1.02E-09 | 1.15E-09 | 1.29E-09 |
| $C_2H_4$ | 6.77E-10 | 8.33E-10 | 9.88E-10 | 1.14E-09 | 1.29E-09 | 1.44E-09 |
| $C_3H_8$ | 5.07E-10 | 6.23E-10 | 7.40E-10 | 8.54E-10 | 9.69E-10 | 1.08E-09 |
| $C_3H_6$ | 6.29E-10 | 7.74E-10 | 9.18E-10 | 1.06E-09 | 1.20E-09 | 1.34E-09 |
| $H_4SiO_4$ | 4.59E-10 | 5.67E-10 | 6.76E-10 | 7.84E-10 | 8.92E-10 | 1.00E-09 |
| $B(OH)_3$ | 5.14E-10 | 6.36E-10 | 7.57E-10 | 8.78E-10 | 9.99E-10 | 1.12E-09 |

The adaptation to specific data is done by means of (m) as long as parallel values are available for sediments obtained from direct measurements of their electrical conductivity. Boudreau (1997) shows, however, that this relation does not have any advantage as compared to:

$$\theta^2 = 1 - \ln(\phi^2) \tag{3.7}$$

This relation (Boudreau's law) has been used for the calculation of various porosity values prevalent in marine sediments as listed in Table 3.2.

By applying the contents of the Tables 3.1 and 3.2 as well as the relation expressed in Equation 3.5, the various examples of the following section are quantifiable. Notwithstanding, it should be emphasized that tortuosity values obtained by electrical conductivity measurements should always be, if available at all, favored to estimated values deduced on account of an empirical relation to porosity.

### 3.2.3  Quantitative Evaluation of Steady State Concentration Profiles

This section intends to demonstrate the application of Fick's first law of diffusion to some selected examples of concentration profiles which were derived from marine pore water samples. It needs to be stressed that all these calculations require that steady-state conditions are present. The calculation of non-steady state conditions will only be dealt with later in Section 3.2.4.

Figure 3.5 shows an oxygen profile measured *in-situ*. The part of concentration gradient exhibiting the highest inclination is clearly located directly below the sediment surface. This gradient of 22.1 mol m$^{-3}$m$^{-1}$ is identified in Figure 3.5 (cf. Sect. 12.2.1). A relatively high degree of porosity will have to be assumed for this sediment near the sediment surface. A porosity of $\phi = 0.80$ yields a tortuosity value ($\theta^2$) of 1.45, according to Table 3.2. The diffusion coefficient for oxygen dissolved in free seawater at 5 °C is shown in Table 3.1 to amount to D$^{sw}$ = 1.23$\cdot$10$^{-9}$ m$^2$s$^{-1}$. Applying Equation 3.5, it follows that the diffusion coefficient for oxygen in sediments is $D_{sed}$ = 8.5$\cdot$10$^{-10}$ m$^2$s$^{-1}$. This yields the diffusive oxygen flux from the bottom water into the sediment J$_{sed,\,oxygen}$ as:

$$J_{sed,oxygen} = -\,0.80 \cdot 8.5 \cdot 10^{-10} \cdot 22.1$$
$$= -\,1.5 \cdot 10^{-8}\ [\text{mol m}^{-2}\text{s}^{-1}] \tag{3.8}$$

To arrive at less complicated and more imaginable figures, and in order to compare this value with, for instance, sedimentological data, we multiply with the number of seconds in a year (365 · 24 · 60 · 60 = 31,536,000) and then we obtain:

$$J_{sed,oxygen} = -\,1.5 \cdot 10^{-8} \cdot 31,536,000$$
$$= -\,0.47\ [\text{mol m}^{-2}\text{a}^{-1}] \tag{3.9}$$

If we now assume that oxygen reacts in the sediment exclusively with C$_{org}$ in a ratio of 106:138 as Froelich et al. (1979) have indicated (cf. Sec-

**Table 3.2**  Tortuosity expressed in terms of a sediment's porosity. The calculation was performed by using the Equation 3.7 as published by Boudreau (1997).

| $\phi$ | $\theta^2$ | $\phi$ | $\theta^2$ | $\phi$ | $\theta^2$ |
|---|---|---|---|---|---|
| 0.20 | 4.22 | 0.44 | 2.64 | 0.68 | 1.77 |
| 0.22 | 4.03 | 0.46 | 2.55 | 0.70 | 1.71 |
| 0.24 | 3.85 | 0.48 | 2.47 | 0.72 | 1.66 |
| 0.26 | 3.69 | 0.50 | 2.39 | 0.74 | 1.60 |
| 0.28 | 3.55 | 0.52 | 2.31 | 0.76 | 1.55 |
| 0.30 | 3.41 | 0.54 | 2.23 | 0.78 | 1.50 |
| 0.32 | 3.28 | 0.56 | 2.16 | 0.80 | 1.45 |
| 0.34 | 3.16 | 0.58 | 2.09 | 0.82 | 1.40 |
| 0.36 | 3.04 | 0.60 | 2.02 | 0.84 | 1.35 |
| 0.38 | 2.94 | 0.62 | 1.96 | 0.86 | 1.30 |
| 0.40 | 2.83 | 0.64 | 1.89 | 0.88 | 1.26 |
| 0.42 | 2.74 | 0.66 | 1.83 | 0.90 | 1.21 |

tion 3.2.5), then we obtain the amount of $C_{org}$ which is annually oxidized per m²:

$$R_{ox,Corg} = 0.47 \cdot (106/138) \cdot 12$$
$$= 4.4 \; [gC \; m^{-2}a^{-1}] \qquad (3.10)$$

It must be pointed out that we have to differentiate very distinctly between two very different statements. On the one hand, the profile inescapably proves that 0.47 mol m⁻²a⁻¹ oxygen are consumed in the sediment. On the other hand, the calculation that 4.4 g m⁻²a⁻¹ $C_{org}$ is equivalent to this amount requires that all oxygen is, in fact, used in the oxidation of organic matter. However, it is imaginable that at least a fraction of oxygen is consumed by the oxidation of other reduced inorganic solute species (e.g. $Fe^{2+}$, $Mn^{2+}$, or $NH_4^+$; cf. also with example shown in Fig. 3.8). There is indeed evidence that, depending on the specific conditions of the various marine environments,

one or the other reaction contributes more or less to the consumption of oxygen. At any rate, this needs to be verified by other measurements, for instance, by recording the concentration profiles of the reducing solute species.

Figure 3.6 shows the concentration profile of dissolved sulfate obtained from the pore water of sediments sampled from the Amazon deep sea fan. The pore water was extracted by compression of sediment sampled with the gravity corer, and was immediately afterwards analyzed by ion-chromatography (compare Sects. 3.3 and 3.4). As compared to the previous example, a depth range comprising more than two orders of magnitude is dealt with here. The paths for diffusion are hence considerably longer in this example. Yet, the sulfate concentration in sea-water is also two orders of magnitude higher than the concentration of oxygen so that, in total, a similar gradient is formed nevertheless.

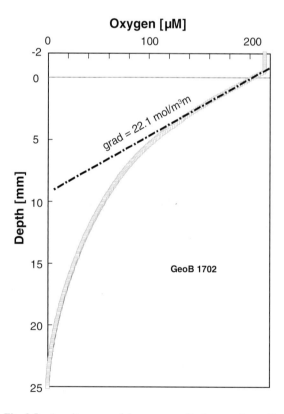

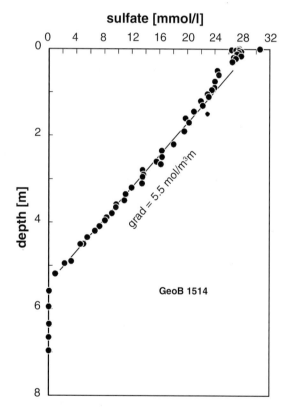

**Fig. 3.5** A quite successful oxygen profile in a marine sediment. This profile was measured by Glud et al. (1994) in highly reactive sediments off the western shoreline of Africa using the 'Profilur' lander in situ. The most pronounced concentration gradient (chain line) lies directly below the sediment surface. Down to a depth of only about 25 mm below the sediment surface, the oxygen dissolved in pore water is entirely depleted.

**Fig. 3.6** Sulfate profile in pore water from sediments of the Amazon deep sea fan at a water depth of about 3500 m. A linear concentration gradient can be distinctly derived from the sediment surface down to a depth of about 5.4 m. The gradient change, and thus a change in the diffusive flux, is strongly limited to a depth interval of at the most 10 to 20 cm (after Schulz et al. 1994).

In this case a concentration gradient of 5.5 mol m$^{-2}$a$^{-1}$ was derived as a mean value for the depths ranging from 0 to 5.4 m. Upon examining the curve in more detail, it is obvious that the gradient is less pronounced in the upper 2 meters which is probably explained by a somewhat higher porosity and a concurrently unchanged diffusive flux. As the discovered gradient within the sulfate profile is located distinctly deeper under the sediment surface than in the previous example, it is reasonable to assume a lower degree of porosity. A porosity of $\phi = 0.60$ yields, according to Table 3.2, a tortuosity value ($\theta^2$) of 2.02. On consulting Table 3.1 we find that the diffusion coefficient for sulfate in sea-water at 5 °C is $D^{sw} = 5.72 \cdot 10^{-10}$ m$^2$s$^{-1}$. Using Equation 3.5 it follows that the diffusion coefficient in the sediment amounts to $D_{sed} = 2.8 \cdot 10^{-10}$ m$^2$s$^{-1}$. Thus, the diffusive sulfate flux from the bottom water into the sediment corresponds to $J_{sed, sulfate}$:

$$J_{sed,sulfate} = -0.60 \cdot 2.8 \cdot 10^{-10} \cdot 5.5$$
$$= -9.2 \cdot 10^{-10} \text{ [mol m}^{-2}\text{s}^{-1}] \qquad (3.11)$$

To arrive at more relevant figures and for reasons of comparison with other sedimentologic data, we multiply this value with the number of seconds in one year (31,536,000) and obtain:

$$J_{sed,sulfate} = -9.2 \cdot 10^{-10} \cdot 31,536,000$$
$$= -0.029 \text{ [mol m}^{-2}\text{a}^{-1}] \qquad (3.12)$$

If we assume that sulfate reacts in the sediment exclusively with $C_{org}$, in a ratio of 106:53 as Froelich et al. (1979) have reported, then for the amount of $C_{org}$ that is annually oxidized per square meter amounts to:

$$R_{ox,Corg} = 0.029 \cdot (106/53) \cdot 12$$
$$= 0.70 \text{ [gC m}^{-2}\text{a}^{-1}] \qquad (3.13)$$

It should be indicated at this point as well that the calculated diffusive sulfate flux from the bottom water into the sediment, and from there into a depth of about 5.4 m, is the unequivocal consequence of the profile shown in Figure 3.6. It also follows that this sulfate is degraded in the depth of 5.4 m within a depth interval of at the most 10 to 20 cm thickness. The calculated $C_{org}$ amount that undergoes conversion again depends on the assumption made by Froelich et al. (1979) that indeed the whole of sulfate reacts with organic carbon. Several studies demonstrated that

this must not be generally the case. For sediments obtained from the Skagerak, Iversen and Jørgensen (1985) showed that an essential proportion of sulfate is consumed in the oxidation of methane. At different locations of the upwelling area off the shores of Namibia and Angola, Niewöhner et al. (1998) could even prove that the entire amount of sulfate is consumed due to the oxidation of methane which diffuses upwards in an according gradient.

Figure 3.7 shows a nitrate profile obtained from sediments of the upwelling area off Namibia which is rather characteristic of marine pore water. The processes behind such nitrate profiles are now well understood. The details of these reactions are described in the chapters 5 and 6; here, they will be discussed only briefly as much is necessary for the comprehension of calculated substance fluxes. The maximum at a specific

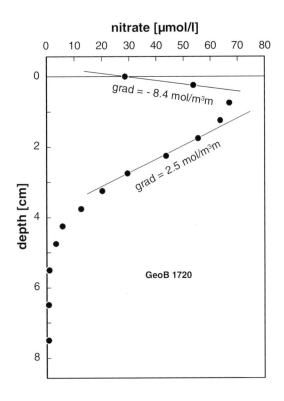

**Fig. 3.7**  Nitrate profile of pore water obtained from sediments of the upwelling area off the coast of Namibia. The profile displays the shape which is typical of nitrate profiles, with a maximum at a depth which is determined by the decomposition of organic material and the oxidized nitrogen released from it after having reacted with the dissolved oxygen. The gradients indicated document a flux upward into the bottom water and a flux downward into zones where nitrate functions as an electron acceptor in the oxidation of other substances.

depth below sea-level is the result of oxidation of organic material by the oxygen that diffuses into the sediment zone from above. Here, the nitrogen of the organic material is converted into nitrate. Mostly, a more pronounced gradient transports the major proportion of nitrate upwards into the bottom water. The smaller proportion travels downwards along a shallow gradient where it is finally consumed as an electron acceptor in the oxidation of other substances.

All these processes can be derived directly and quantitatively from Figure 3.7. Since they are bound to a sedimentary zone which lies very close to the surface, a high degree of porosity can be assumed. According to Table 3.2, the porosity $\phi = 0.80$ corresponds to a tortuosity $(\theta^2)$ of 1.45. Table 3.1 shows that the diffusion coefficient for nitrate in free sea-water at 5 °C amounts to $D^{sw} = 1.08 \cdot 10^{-9}$ m$^2$s$^{-1}$. Applying Equation 3.5 yields a sedimentary diffusion coefficient of $D_{sed} = 7.4 \cdot 10^{-10}$ m$^2$s$^{-1}$. Hence, the diffusive nitrate flux from the sediment to the bottom water compartment is calculated as:

$$
\begin{aligned}
J_{nitrate,up} &= -0.80 \cdot 7.4 \cdot 10^{-10} \cdot (-8.4) \\
&= 5.0 \cdot 10^{-9} \text{ [mol m}^{-2}\text{s}^{-1}] \quad (3.14)
\end{aligned}
$$

or as

$$
\begin{aligned}
J_{nitrate,up} &= 5.0 \cdot 10^{-9} \cdot 31,536,000 \\
&= 0.16 \text{ [mol m}^{-2}\text{a}^{-1}] \quad (3.15)
\end{aligned}
$$

The nitrate flux in a downward direction is calculated accordingly:

$$
\begin{aligned}
J_{nitrate,down} &= -0.80 \cdot 7.4 \cdot 10^{-10} \cdot 2.5 \\
&= -1.5 \cdot 10^{-9} \text{ [mol m}^{-2}\text{s}^{-1}] \quad (3.16)
\end{aligned}
$$

or as

$$
\begin{aligned}
J_{nitrate,down} &= -1.5 \cdot 10^{-9} \cdot 31,536,000 \\
&= -0.047 \text{ [mol m}^{-2}\text{a}^{-1}] \quad (3.17)
\end{aligned}
$$

The sum of both fluxes yields a minimal estimate value of the total nitrate concentration released from organic matter due to its reaction with oxygen. However, the real value for the total amount of released nitrate must be higher than the sum of both calculated fluxes. The gradient of the downward directed flux may be quite reliably calculated from numerous points, however, the more pronounced gradient of the flux leading upward into the bottom water consists only of two points, one of which merely represents the concentration in the bottom water whereas the other represents

the pore water of the uppermost 0.5 cm of sediment. It is probably correct to assume that the gradient is more pronounced in closer proximity to the sediment surface, and hence the flux should also prove to be more enhanced. A more accurate statement would only be possible under *in-situ* conditions with a depth resolution similar to the oxygen profile shown in Figure 3.5. As for measurements performed under *ex-situ* conditions, a better depth resolution than shown in Figure 3.7 is very hard to obtain.

If we assume that the released nitrate exclusively originates from the oxidation of organic matter, and that the organic matter is oxidized in a manner in which the C:N ratio corresponds to the Redfield-ratio of 106:16 (cf. Sect. 3.2.5), then we can also calculate the conversion rate of organic matter on the basis of the nitrate profile:

$$
R_{ox,Corg} = [\text{abs}(J_{nitrate,up}) + \text{abs}(J_{nitrate,down})] \cdot (106/16) \quad (3.18)
$$

or by employing the values of the above example:

$$
\begin{aligned}
R_{ox,Corg} &= (0.16 + 0.047) \cdot (106/16) \\
&= 1.37 \text{ [mol m}^{-2}\text{a}^{-1}] \quad (3.19)
\end{aligned}
$$

or

$$
R_{ox,Corg} = 16.5 \text{ [gC m}^{-2}\text{a}^{-1}] \quad (3.20)
$$

At any rate, such a value might serve only as a rough estimation, since apart from the aforementioned error, the calculation procedure implicitly contains some specific assumptions. It has been already mentioned that the C:N ratio of the oxidized material is supposed to be (106/16 = 6.625). More recent publications (Hensen et al. 1997), however, indicate that especially in sediments with a rich abundance of organic matter a distinctly lower ratio of almost 3 is imaginable (cf. Chap. 6). It must also be considered that the measured profiles are not just influenced by diffusion in the surface zones, but also by the processes of bioturbation and bioirrigation (cf. Sect. 3.6.2).

In principle, the shape of the manganese profile shown in Figure 3.8 is not dissimilar to the nitrate profile of Figure 3.7. Here, we again identify the zones of maximum concentrations – in this particular case about 0.5 m below the sea-floor level – as the site of dissolved manganese release into the pore water. Again, a pronounced negative gradient transports the dissolved manganese upwards away from the zone of its release. This time, however, the substance flux does not reach up to the bottom water above the sediment, instead, very low manganese concentrations are measured at a depth which is only few centimeters below the sediment surface. A flat positive gradient leads a smaller fraction of the released manganese to greater depth where it is withdrawn from the pore water at about 14 to 15 m below the sediment surface.

Here as well, similar to the previously discussed example, both fluxes can be calculated. The upwardly directed flux from the manganese release zone ($J_{manganese, up}$) is obtained from the gradient (-0.5 mol m$^{-3}$m$^{-1}$), where in the upper zone the sediment possesses an assumed porosity of = 0.80. Considering the values in Table 3.1 and

3.2, as well as the Equation 3.5, a sedimentary diffusion coefficient of $D_{sed} = 2.6 \cdot 10^{-10}$ m$^2$m$^{-1}$ yields the following manganese flux:

$$J_{manganese,up} = -0.80 \cdot 2.6 \cdot 10^{-10} \cdot (-0.5)$$
$$= 1.0 \cdot 10^{-10} \text{ [mol m}^{-2}\text{s}^{-1}] \quad (3.21)$$

or

$$J_{manganese,up} = 1.0 \cdot 10^{-10} \cdot 31,536,000$$
$$= 3.2 \cdot 10^{-3} \text{ [mol m}^{-2}\text{a}^{-1}] \quad (3.22)$$

Likewise, the downward-directed manganese flux is calculated, however, taking a porosity degree of $\phi = 0.60$ into account and a accordingly calculated $D_{sed} = 1.9$ E-10 m$^2$s$^{-1}$:

$$J_{manganese,down} = -0.60 \cdot 1.9 \cdot 10^{-10} \cdot 0.0084$$
$$= -9.6 \cdot 10^{-13} \text{ [mol}^1\text{m}^{-2}\text{s}^{-1}] \quad (3.23)$$

or

$$J_{manganese,down} = -9.6 \cdot 10^{-13} \cdot 31,536,000$$
$$= -3.0 \cdot 10^{-5} \text{ [mol m}^{-2}\text{a}^{-1}] \quad (3.24)$$

Again, both fluxes added together constitute the total release of dissolved manganese from the sediment into the pore water. Since the downward-directed flux is in this case almost two orders of magnitude lower than the upward-directed flux, it may be neglected considering the possible errors occurring in the determination of the upward stream. The release of manganese is equivalent to the conversion of oxidized tetravalent manganese into the soluble divalent manganese. Which substance is the electron donor in this reaction cannot be concluded from the manganese profile. It could be organic matter as proposed by Froelich et al. (1979). In this case we should not overlook the fact that the converted substance amounts are one order of magnitude lower than they are, for instance, in sulfate fluxes (Fig. 3.6), and more than two orders of magnitude lower than they are in the flux of oxygen. (Fig. 3.5). It should also be noted that upon reducing one mole of Mn(IV) to Mn(II) only two moles of electrons are exchanged, whereas it amounts to 4 moles of electrons per mole oxygen, and even 8 moles of electrons per mole sulfate. Even the 'impressive' gradient of the manganese profile shown in Figure 3.8 does not represent an essential fraction of the overall diagenetic processes in the sediment, involved in the oxidation of organic matter.

The upward-directed manganese flux does not reach into the bottom water, instead, the Mn$^{2+}$ is

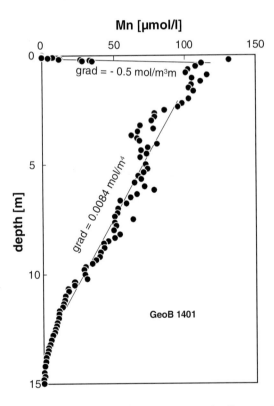

**Fig. 3.8** A manganese profile in pore water of sediments off the Congo River estuary, in a water depth of approx. 4000m. The profile is, in principle, quite similar to the profile of nitrate previously shown in Figure 3.7. Here, manganese is released into the pore water at a specific depth below the sediment surface. A gradient with a high negative slope leads most of the Mn$^{2+}$ upwards; another gradient, positive and more level, conveys manganese into a precipitation zone, which was just included in the lowest core meter (between 14 and 15 m).

re-oxidized in a depth of only few centimeters below the sediment surface. Generally, one would expect this re-oxidation to occur by the action of dissolved oxygen (cf. Chap. 11).

It is not within the scope of the present discussion to decide whether the variations in the concentration profile that were averaged upon calculating the gradient of 0.0084 mol m$^{-3}$m$^{-1}$ shown in Figure 3.8, merely reflect the inaccuracy pertinent to the sampling technique and/or the analytical procedure, or whether they represent discrete processes of their own. Since they do not vary independently, and since several points appear to constitute smaller minima and maxima, it may be suggested that these measurements do not simply represent analytical variations, but reasonable and true values. At any rate, the whole curve leads in its course to very low values that are reached at a depth of 14 m along with a distinct change of the gradient's slope. In many cases, a precipitation of Mn(II)-carbonate is to be expected. The identification of such diagenetic phases anticipated in geochemical modeling is discussed in more detail in Section 14.1.

### 3.2.4    The Non-Steady State Situation and Fick's Second Law of Diffusion

All examples of the preceding Sections 3.2.2 and 3.2.3 are strictly only valid under steady-state conditions when concentrations, and hence the gradients and diffusive fluxes, are constant over time. Fick's second law of diffusion is applicable in non-steady state situations:

$$\frac{\partial C}{\partial x} = D \cdot \frac{\partial^2 C}{\partial x^2} \qquad (3.25)$$

The diffusion coefficient in free solution (D) or the diffusion coefficient in sedimentary pore water ($D_{sed}$) are used due to the conditions of the system. In contrast to Fick's first law of diffusion, the time co-ordinate (t) appears here next to the local co-ordinate (x). The concentrations and thus the gradients and fluxes are variable for these co-ordinates. Such a partial differential equation cannot be solved without determining a specific configuration of boundary conditions. On the other hand, most of the known solutions are without great practical value for the geochemist due to their very specifically chosen sets of boundary conditions that are seldom related to real situations.

In the following only one solution will therefore be presented in detail. The following boundary conditions are to be considered as valid: In the whole sediment profile below the sediment surface the same diffusion coefficient ($D_{sed}$) is assumed to prevail, the same concentration ($C_0$) prevails in the sediment's pore volume. After a certain point in time (t=0) the bottom water attains another concentration ($C_{bw}$) via the sediment surface. The concentrations ($C_{x,t}$) in the depth profile (co-ordinate x = depth below the sediment surface) at a specific time-point (t) are for these conditions described as:

$$C_{x,t} = C_0 + (C_{bw} - C_0) \cdot erfc\left\{ x / \left( 2 \cdot \sqrt{D_{sed} \cdot t} \right) \right\}$$

$$(3.26)$$

The error function (erfc(a)), related to the Gauss-function, can be approximated according to Kinzelbach (1986) with the relation:

$$erfc(a) =$$

$$\exp(-a^2) \cdot (b_1 \cdot c + b_2 \cdot c^2 + b_3 \cdot c^3 + b_4 \cdot c^4 + b_5 \cdot c^5)$$

$$(3.27)$$

with:
$$b_1 = 0.254829592$$
$$b_2 = -0.284496736$$
$$b_3 = 1.421413741$$
$$b_4 = -1.453152027$$
$$b_5 = 1.061405429$$

and:
$$c = 1 / [1 + 0.327591117 \cdot abs(a)]$$

for negative values for (a) it follows:

$$erfc(a) = 2 - erfc(a)$$

Figure 3.9 shows the graphical representation of the error function according to the approximation published by Kinzelbach (1986). This is the function complementary to the error function of Boudreau (1997):

$$erfc(a) = 1 - erfc(a) \qquad (3.28)$$

With this analytical solution of Fick's second law of diffusion the components of Figure 3.3 can now be calculated. On doing this, one will find that the outcome is exactly the same as in the corresponding calculation with the numeric solution

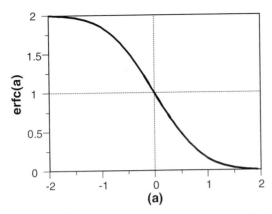

**Fig. 3.9** Graphical representation of the error function in the approximated form after Kinzelbach (1986). This is the function complementary to the error function of Boudreau (1997). Additionally, in the form published by Boudreau (1997), the range of negative values for (a) is omitted, as this has little relevance for sediments.

with the model CoTReM (cf. Chapter 14). Yet, the components of Figure 3.4, with the multiple change of the concentration in bottom water and the 'memory' of which is preserved over several cycles in the pore water fraction, is not accessible with this rather simple analytical solution.

Another example that can be assessed with this analytical solution results from the following considerations: At the beginning of the Holocene, about 10,000 years ago, the sea level rose more than 100 m as a result of thawing ice, which is equivalent to 3% of the entire water column. This means that sea water had been previously about 3% higher in concentration. If we assume a chloride concentration of 20,000 mg/l in the seawater today, and thus a mean concentration of 20,600 mg/l in sea-water of the ice age, then we are able to calculate the non-steady state chloride profile in pore water with the application of the analytical solution of equation 3.26.

From the result of this calculation (shown in Figure 3.10) it follows that we will find just about one half of the ice age seawater concentration (20,300 mg/l) at a depth of 12 m below the sediment surface. If we consider that the reliability of our analytical methods lies at best somewhere around 1.5%, the exemplary calculation reveals that the effect in pore water is almost at the limit of detection.

Other applications of such analytical solutions hardly make any sense, since, with the exception of chloride, practically all other parameters of pore water are strongly influenced by complex

biogeochemical processes. In order to retrace these processes appropriately, analytical solutions for non-steady states in pore water are usually not sufficiently flexible. Hence, numeric solutions are mostly employed. These will be discussed later in Chapter 14 with regard to connection to biogeochemical reactions.

### 3.2.5    The Primary Redox-Reactions: Degradation of Organic Matter

Nearly all biogeochemical processes in young marine sediments during early diagenesis are directly or indirectly connected with the degradation of organic matter. This organic matter is produced by algae in the euphotic zone of the water column by photosynthesis. Usually, only a small part of the primary production reaches the sediment surface and of which only a small part is incorporated into the sediment where it becomes the driving force for most of the primary diagenetic redox-reactions (cf. Fig. 12.1).

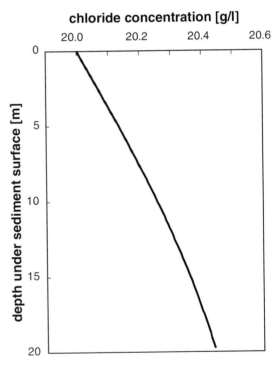

**Fig. 3.10** Calculated concentration profile in the pore water of a marine sediment according to an analytical solution of Fick's second law of diffusion. For reasons of simplification it was assumed that the seawater contained 3% less chloride concentration since the beginning of the Holocene as a result of thawing ice. This lower concentration (20,000 mg/l) had enough time over 10,000 years to replace the higher concentration (20,600 mg/l) from the sediment.

The conceptual model for the degradation of organic matter in marine sediments was first proposed by Froelich et al. (1979). Although many more details, variations and specific pathways of these redox-reactions have become known in the meantime, this 'Froelich-model' of the primary redox-reactions in marine sediments is still valid (Fig. 3.11). Usually, these reactions are based on a very simplified organic matter with a C:N:P-ratio of 106:16:1 ('Redfield-ratio') as described by Redfield (1958) (cf. Sects. 5.4.4 and 6.2). The standard free energies for these model reactions are also listed in Figure 3.11 (cf. Sect. 5.4.1). The reactions in this figure are listed in order of declining energy yield from top to bottom.

Based on these reactions, a succession of different redox zones is established:

- Close to the sediment surface, dissolved oxygen is usually transported from the bottom water into the sediment either by molecular diffusion or as a result of biological activity. In this upper zone (the oxic zone), dissolved oxygen is the electron acceptor for the degradation of organic matter. Products of this reaction are carbonate, nitrate and phosphate derived from the nitrogen and phosphorus in the organic matter. For details of these reactions, see Chapter 6.

- Below the oxic zone follows a zone where manganese(IV) oxides in the solid phase of the sediment serve as electron acceptors. Products of this reaction are usually carbonate, nitrogen, phosphate and dissolved $Mn^{2+}$-ions in the pore water. These dissolved $Mn^{2+}$-ions are usually transported either by diffusion or by bio-activities to the oxid zone where the manganese is re-oxidized and precipitated as manganese(IV) oxide. These re-

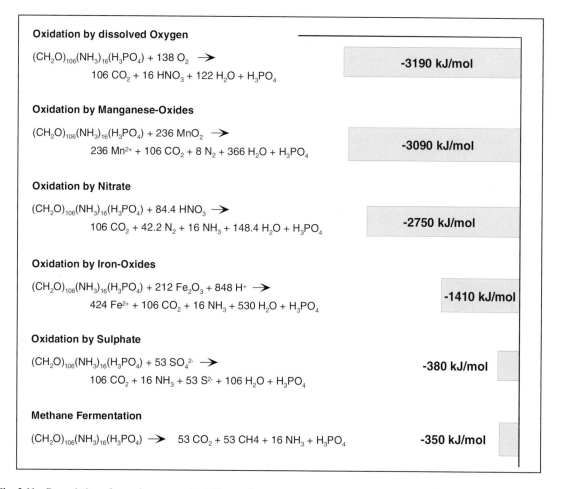

**Fig. 3.11** Degradation of organic matter with different electron acceptors (Froelich et al. 1979). The columns represent the different amounts of resulting energy.

actions belong to a manganese cycle, by which oxygen is transported into deeper parts of the sediment. For details of these reactions, see Chapter 11.

- Below this zone, nitrate serves as an electron acceptor, which is a product of the redox-reactions in the oxic zone. Carbonate, phosphate and nitrogen, as well as ammonia are produced. The oxygen for the oxidation of organic matter in this zone is derived from the nitrate, produced in the oxic zone. In most cases, this process oxidizes at least one order of magnitude less organic matter than the reactions in the oxic zone. For details of these reactions, see Chapter 6.

- Below this zone, iron(III) oxides or iron(III) hydroxides in the solid phase of the sediments act as electron acceptors. For details of these reactions, see Chapter 7.

- Below this zone, dissolved sulfate serves as electron acceptor for the oxidation of organic matter, according to the 'Froelich-model'. Recent publications however showed that, in most cases, not organic matter is oxidized in this zone, but predominantly methane, which diffuses up from the deeper parts of the sediment (Niewöhner et al. 1998). For details of these reactions, see Chapter 8.

- The reaction with the lowest yield of standard free energy is methane fermentation with the products carbonate, methane, ammonia and phosphate.

## 3.3    Sampling of Pore Water for *Ex-situ* Measurements

Ideally, one would prefer to analyze pore water exclusively under *in-situ* conditions, as will be explained in Section 3.5, for the parameters that permit such procedure. The pressure change, and frequently the change of temperature as well, are usually coupled to *ex-situ* measurement and exert a number of influences of varying potential. However, the *ex-situ* measurement will certainly remain a necessity for quite a long time, with regard to most of the substances dissolved in pore water and especially for great depths below the

sediment surface. This book is not the place to give a general review on sediment sampling techniques. Rather, the more common procedures for sediment sampling will be introduced with an emphasis put on pore water analysis. Then, the particularities, the possible errors as well as problems arising in the application of these sampling techniques, will be discussed.

The following section is concerned with the separation of the aquatic pore water phase from the solid sediment phase. As with all particularly problematic and error-inducing procedures, these various techniques have, depending on the case at hand, their specific advantages and disadvantages.

As a matter of course, one would want to analyze the obtained pore water as soon as it has been separated from the sediment to quantify the dissolved substances therein. In daily routine proceedings, compromises must be made since not all analyses can be carried out simultaneously, and since each and every analytical instrument is not present on board a ship. Thus, it will be necessary to describe the state of knowledge concerning pore water storage, transport and preservation.

### 3.3.1    Obtaining Samples of Sediment for the Analysis of Pore Water

A number of different techniques are available for the withdrawal of samples from the marine sediment. Depending on the scientific question under study, a tool may be chosen that is either capable of taking the sample without harming the sediment surface or disturbing the supernatant bottom water (e.g. multicorer, Rumohr-corer), or that punches out a large as possible sample from the sediment surface area (e.g. box corer), or one that yields cores from the upper less solid meters of the sediment which are long as possible and most unperturbed (gravity corer, piston corer, box corer).

*The Box Corer*

The generic term 'box corer' denotes a number of tools of different size and design used in sampling marine sediments, mostly lowered from ships by means of steel wire rope to the bottom. All have a square or rectangular metal box in common which is pressed into the sediment by their own weight, or perhaps by additionally mounted weights. Upon lifting the tool from the sea-floor the first pull on the steel ropes is used to close the box by means of a shovel while it is still situated in the sediment. At the same time, an opening at the top is shut, more or less tightly, so that the bottom water immediately above the sediment is entrapped. The lateral dimensions of small metal box corers are about 10 cm and they reach to that extent into the sediment. Larger box corers ('giant box corer') possess lateral dimensions and penetration depths of 50 cm, respectively.

Apart from the large sample volume that is collected, the giant box corer has the advantage that the *in-situ* temperature is kept stable at least in the central zones of the sample, even if the sample is raised at the equator, from a depth of several thousand meters where the sediment is about 2°C cold. This is about the only advantage the box corer has with regard to the geochemical pore water analysis, whereas various disadvantages are to be considered. The shutter on top of the box corer is often not very tightly sealed, and thus the entrapped bottom water might become uncontrollably adulterated upon being raised upwards through the water column. On hoisting the loaded, heavy box corer out of the water, and during its later transportation, onto the deck of a ship, the sediment surface is mostly destroyed to an extent that at least the upper 1-2 cm become worthless for the subsequent pore water analysis.

### The Multicorer

The multicorer is also employed from the ship using steel wire ropes and can also be used for all depths under water. On applying this tool, up to 12 plastic tubes (mostly acrylic polymers), each measuring a length of about 60 cm and about 5-10 cm in diameter, are simultaneously inserted approximately 30 cm into the sediment. As with the box corer, the first pull of the steel rope on lifting the appliance is used to seal the plastic tubes on both ends. These shutters are usually tight enough to ensure that the entrapped water will later represent the genuine bottom water.

Variation to the *in-situ* condition are caused in greater depths (about 1000 m and more) by the expansion of the pore water, which happens relative to the sediment, when the pressure diminishes. The uppermost millimeters of the profile are then distorted. Moreover, the temperature rise gives cause for disturbances upon raising the

samples upwards out of great depths, in the course of which microbial activity is activated within the sediment sample, which is distinctly higher than under *in-situ* conditions.

On the other hand, the multicorer provides, at present, the best solution for *ex-situ* sampling of sediments from the sediment/water interface. The nitrate profile shown in Figure 3.7 was measured in pore water extracted from a sediment sample which had been obtained by using the multicorer. It shows clearly that the concentration profile can be measured in pore water with an almost undisturbed depth resolution of 0.5 cm per each sample. The reliable sampling technique using the multicorer also becomes evident upon comparing the *in-situ* measured oxygen profiles (Holby and Riess 1996) with the *ex-situ* measured oxygen profiles of a multicorer sample (Enneking et al. 1996). Both measurements were conducted at the same location, at the same time, and lead to the same oxygen penetration depth and nearly identical oxygen concentration profiles. In both cases, the oxygen was measured with micro-electrodes.

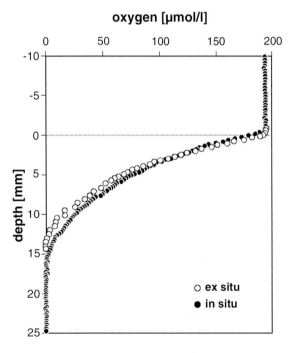

**Fig. 3.12** In-situ measured oxygen profile (dots, Holby and Riess 1996). Oxygen profile which was measured ex-situ by using a micro-electrode as well (circles, Enneking et al. 1996). The sample was obtained by using a multicorer tube. Both profiles were measured in the course of the Meteor M 34/2 expedition at the same time and at the same location in an upwelling area off Namibia, in a depth of approx. 1300 m below sea level.

*The Rumohr-Corer*

In case of the Rumohr corer, a plastic tube (mostly an acrylic polymerization product) with a diameter of about 6 cm and a variable weight, is pressed 0.5 m to 1 m deep into the sediment. Upon pulling the tube out, the top aperture is shut whereby the bottom water that just covers the sediment becomes entrapped. The tube is not sealed at the bottom because (mostly) the closure at the top suffices to prevent the sediment from falling out. But, as soon as the Rumohr corer is held out of the water, the bottom aperture of the tube must be sealed with an adequate rubber stopper.

The Rumohr corer is most frequently used in shallow waters, on the shelf and from the decks of smaller ships, since it does happen from time to time that the bulk sample slides out of the bottom prior to its closure with the rubber stopper, and owing to the fact that the corer has only one tube that is filled with sample of sediment. In greater depths, and whenever a larger ship is available, the multicorer will be used in preference. Otherwise, the quality of the samples is entirely comparable. Furthermore, the Rumohr corer is capable of extracting sample cores of up to 1 m in length.

*The Gravity Corer and the Piston Corer*

Upon using the gravity corer, a steel tube of about 13 cm in diameter is pressed between 6 m and 23 m deep into the sediment with the aid of a lead weight weighing about 3 - 4 t. Inside the steel tube, a plastic liner made of HD-PVC is situated that encompasses the core. Upon extraction from the sediment and raising the tube upwards through the water column, the tube's top aperture is kept shut by a valve. Below, another valve-like shutter - the core catcher - prevents the core material from falling out. Generally, cores of about 10 - 15 m in length are obtained in pelagic sediments, although at times they reach up to 20 m.

In a piston corer the same steel tubes and plastic liners are employed, but with a lighter lead weight. However, this corer is additionally equipped with a shear-action mechanism that moves a piston upwards through the tube when the tool is immersed into the sediment. The piston stroke produces a vacuum that facilitates the penetration of core into the tube. By using the piston corer, core lengths can be achieved which are frequently 20 - 30 m in length and thus several meters longer than the ones obtained with the gravity corer. Yet,

it is observed that the cores obtained with the gravity corer and with the piston corer display the same stratigraphical depth zones of the sediment. It is therefore assumed that the cores sampled with the gravity corer are 'somewhat compressed' especially in the lower parts, whereas the piston corer produces cores that are 'somewhat extended' in length. How this comes into effect and what conclusions are to be drawn with regard to the pore water samples is yet completely uncertain. In the examined pore water profiles of sediments extracted with the gravity corer, there were no indications as to any noticeable compression to date.

Both sampling tools can only be operated from the decks of larger ships that are equipped with steel ropes and winches designed for managing weights of 10 - 15 tons. The author would personally always prefer using the gravity corer because its handling is easier, safer and faster; and because, especially in low latitudes, every minute counts in which the core is unnecessarily exposed to high temperatures on board the deck of the ship.

At least the upper 10 to 30 cm of the core length obtained with either tool is usually adulterated in that it is not appropriate for pore water analysis. The multicorer, Rumohr corer, or at least the box corer should be employed in a parallel procedure to ensure that this layer will also be included as part of the sample. It should not be overlooked that, especially in the deep sea, sampling with two different tools 'at the same site' might imply a distance of several 100 m on the ocean floor. From this deviation considerable differences in pore water composition, and in some of the biogeochemical reactions close to sediment surface, are likely to result. Hence the specification as to 'same site' must be acknowledged with caution.

After the usage of either tool - the gravity corer and the piston corer - the sediment core obtained is immediately dissected into pieces of 1 m in length within the tube. Usually, the 1 m long tubular pieces, tightly sealed with caps, are stored at *in situ* temperature prior to the subsequent further processing which is to be carried out as quickly as possible.

*The Box-shaped Gravity Corer (Kastenlot)*

In principle, the box-shaped gravity corer is not dissimilar to the above mentioned gravity corer. Here, a metal box with a core length of about 10 m and a lateral dimension of about 0.1 m up to 0.3 m is used instead of a steel tube with a plastic liner. The core box is shut by valves upon being pulled out of the sediment and is then hoisted upwards through the water column. The advantage of using the box-shaped gravity corer consists in providing a large core diameter, less perturbations due to the metal box's thinner walls, and by removing one side of the box it allows the sediment stratigraphy to be examined. The essential disadvantage in examining pore water samples consists in the fact that the core is distinctly less accessible, and that processing of the core under an inert atmosphere (glove box) is hardly feasible.

*The Harpoon sampler*

Sayles et al. (1976) have described a device that allows collection of pore water samples under *in-situ* conditions. To achieve this, a tube is pressed about 2 meters deep into the sediment, like a harpoon. Then the pore water is withdrawn from various sections of the tube, through opened valves, and concomitantly passed through filters. The crucial step of separating sediment and pore water thus happens under *in-situ* conditions. The obtained water samples are then ready to be analyzed *ex-situ*. The concentrations which were measured by Sayles et al. (1976), however, do not differ greatly from those which were obtained 'from the same location' by expressing pore water under *ex-situ* conditions.

Jahnke et al. (1982) used the same (or, at least a quite similar) tool in 4450 m deep waters of the equatorial Pacific, at approximately 0° to 10° N, and approximately 140° W (MANOP-site). Here, the distances between the locations in which the harpoon sampler was used and in which, for reasons of comparison, pore water was obtained by compression after the sampling was carried out with the box corer, amounted to 300 m to up to 3000 m. In measurements, which were corroborated several times, Jahnke et al. (1982) found similar concentrations of nitrate, nitrite, silica, pH, and manganese. Yet, the concentrations of $\Sigma CO_2$, alkalinity and phosphate were distinctly higher and displayed statistical significance.

## 3.3.2 Pore Water Extraction from the Sediment

In the following section, the separation of pore water from the sediment will be described. The necessary procedures will be described in the sequence in which they are employed. This is to say, that we will begin with the filled tubes of the multicorer, the Rumohr corer, or the meter-sized pieces of the gravity corer or piston corer. The entire processing steps described in the following for the core material should be performed at temperatures which should be kept as close as possible to the temperatures prevailing under *in-situ* conditions.

In this context, the problem needs to be dealt with as to how long a tightly sealed core, which is exposed to the *in-situ* temperature all the while, can be left to itself before it is processed without the occurrence of any essential perturbation in the pore water fraction. In order to extract pore water from a sediment core, such as is shown in Figure 3.1, and subsequently analyze and preserve it, even a practiced team would require several days. In most cases, a compromise needs to be found therefore, between the highest number of samples and most rapid processing.

A special situation prevailed in the case of the core shown in Figure 3.1, as the ship was cruising for a relatively long time after the core had been taken. Then the experiment was conducted that led to the results shown in Figure 3.1. The core was analyzed with an almost unusual high sampling density. As these procedures afforded plenty of time, and since it was unsure whether the pore water of the sediment core, which was stored at *in-situ* temperature in the meantime, would change within the prospective processing time of ten days, the single pieces measuring one meter were intentionally *not* analyzed in a depth sequence, but in a *randomized* sequence. Thereby, the variations in the processing time had to become reflected as discontinuities in the corresponding data at the end of each meter interval. This had not been the case at either interval, from which it follows that a processing time of 10 days was obviously quite innocuous to the quality of the samples.

*Analysis of Dissolved Gases*

For some substances dissolved in pore water everything will be too late for a reliable analysis as soon as the sediment core lies freshly, but in a decompressed state, on the deck of the ship. This holds true predominantly for the dissolved gases.

In this respect, dissolved oxygen is relatively easy to manage, a circumstance which becomes evident upon comparing the *in-situ* oxygen profile with the *ex-situ* profile, both measured at the same sites (shown in Fig. 3.12). The reason for this similarity is that the relatively low concentrations, that are often below the saturation level, do not significantly assume a condition of over-saturation even after decompression. Notwithstanding, the results shown in Figure 3.12 reflect the situation too favourably, because differences of up to a factor of 2 were also observed between *in-situ* and *ex-situ* conditions with regard to the penetration depth of oxygen, and thus to the corresponding reaction rates as well. As to what extent these differences between *in-situ* and *ex-situ* conditions really exist, or whether such differences result from measurements carried out at not exactly identical sites, has not yet been sufficiently investigated.

The case is obviously similar for dissolved carbon dioxide whose concentration is essentially determined by the equilibrium of the aquatic carbonate phases. As for the alkalinity, no remarkable variations were found in measured values, even at high concentrations, when the measurements were performed successively on adjacent parts of the same sediment core (cf. alkalinity profile shown in Fig. 3.1).

The measurement of sulfide is much more troublesome, especially at high concentrations (It should be noted that $H_2S$ is strongly toxic and that one can become quickly accustomed to its smell after prolonged presence in the laboratory. When working with sulfide-containing core material the laboratory should be ventilated thoroughly at all times!). As a general rule, everything already perceived by its smell is already lost to analysis. Since sulfide is readily analyzed by various methods in aqueous solution, most errors arise from decompression of the core and the subsequent separation of pore water from the sediment. As soon as decompression begins, a great quantity starts to degas, initially forming a finely distributed effervescence. In this condition, measurements can mostly still be carried out, however with less satisfactory results, provided that a specific volume of water-containing degassed sediment is punched out with a syringe and immediately brought into an alkaline environment (SAOB = Sulfur Anti-Oxidizing Buffer with pH > 13, after Cornwell and Morse 1987).

Even more difficult is the sampling and the analysis of sediments with a marked content of methane gas. In this particular case, a considerable amount of degassing of the sample occurs immediately upon decompression, thus not per-mitting the measurement of the concentrations afterwards which are oversaturated under conditions of normal atmospheric pressure.

The results of measurements carried out on a core with high sulfide and methane concentrations are shown in Figure 3.13. The material was obtained in 1300 m deep water, in a high productivity zone of upwelling off the coast of Namibia. It can clearly be seen that in the depth zone of sulfate reduction, methane moving upwards meets with the downwards-diffusing sulfate, both substances displaying almost identical flux rates. These processes are dealt with more thoroughly in Chapter 8; at this point, just the sampling at the site of the core and the analytical sample treatment will be discussed.

As for both substances, sulfide and methane, concentrations found in the core are similar to that of sulfate. For the purpose of sampling, small 'windows' (2 x 3 cm) were cut into the plastic liners with a saw, immediately after the meter-long segments from the gravity corer were available. In each of these windows, 2-3 ml samples of fresh sediment were punched out with a syringe. For the determination of sulfide some of these sufficiently unperturbed partial samples were placed into a prepared alkaline milieu (see above, SAOB); for the determination of methane others were transferred directly to head space vials. The headspace vials comprising a volume of 50 ml contained 20 ml of a previously prepared solution of 1.2 M NaCl + 0.3 M $HgCl_2$. The analysis of sulfide was performed by measuring the sample with an ion-selective electrode, whereas methane analysis was done in the gaseous volume of the headspace vials by gas-chromatography.

It is noticeable that the values in the profiles hardly scatter at the somewhat lower concentrations (about 4 mmol/l for methane and 7 mmol/l for sulfide, respectively). The values of the higher concentrations obtained in greater depths scatter more strongly and are altogether far too low. This became evident since the highest methane concentration is to be found in a sample that was not obtained from an extra sawed-out 'window', but from one that was previously taken directly from the core catcher at the lower open end of the core. As to the question concerning the potential electron donor for sulfate reduction it was, however, of particular importance that methane could be reliably determined up to a concentration of 2-3 mmol/l, and that the upwards directed gradient reached into the zone of the reaction.

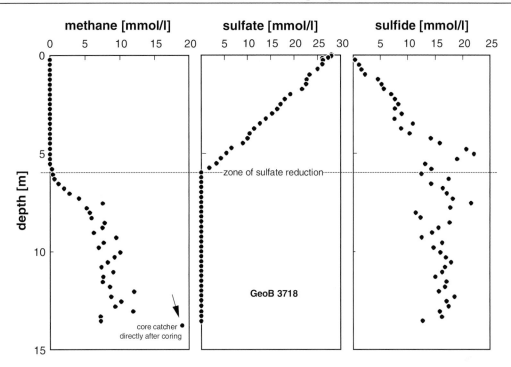

**Fig. 3.13**  Concentration profiles of pore water from anoxic sediments obtained from an upwelling area off Namibia at a water depth of approximately 1300 m. The analysis of sulfide and methane was carried out in samples that were punched out with syringes from small and quickly sawed-out 'windows' in the fresh sediment core. As for sulfide, these syringe-drawn samples were brought into an alkaline environment, whilst for methane analysis the samples were stored in head space vials for subsequent gas-chromatography analysis. The arrow points to a methane sample that originated from a sealed sediment core obtained by using a sample from the 'core catcher' (after Niewöhner et al. 1998).

*Opening and Sampling a Core under*
*Inert Gas inside the 'Glove Box'*

In order to extract the pore water from the sediment, the next step consists in opening, or rather cutting the meter-long segments previously excised from the sediment in a lengthwise fashion. In the case of multicorer or Rumohr corer, the cores are pushed upwards by using an appropriate piston. Caution should be taken that anoxic sediments (easy to identify: if uncertain, everything that is not distinctly and purely brown in color) must not, by any means, come into contact with an oxygen-containing atmosphere in the course of sampling. Therefore the opening of the core segments must happen in a inert atmosphere, or, as the case may be, the core must be pushed from below into the 'glove box' where such an atmosphere is maintained. A 'glove box' that suits the purpose and that can also be easily self-made and adapted to the requirements of the prevailing conditions is described by De Lange (1988).

The inert gases of higher molecular weight, as well as nitrogen, are suited for maintaining the desired inert gas atmosphere. As for nitrogen it must be considered that the least expensive version of the gas does not contain sufficiently low amounts of oxygen. Argon is often more useful as ordinary argon that is used for welding usually has sufficiently low oxygen. A slight excess pressure in the 'glove box' maintains that, in the event of small leakage, the dissipation of gas might increase, yet the invasion of oxygen is securely prevented.

*Measurement of $E_H$ and pH*
*with Punch-in Electrodes*

In the glove box, the redox potential ($E_H$-value) and the pH value should be measured by applying punch-in electrodes directly to the freshly cut core surface. Suitable electrodes are available today from most suppliers because they are, for example, quite often used in the examination of cheese. However, there is a risk that the electrodes become damaged when larger particles or solidified kernels are present in the sediment sample.

The measurement of the pH-value is relatively easy to do since the function of the electrode can be checked repeatedly with the aid of appropriate calibration solutions. However it must merely be taken into consideration that the comparison with *in situ* measurements close to the sediment surface (cf. Sect. 3.5) may not necessarily lead to corroborated values, because the decompression within the core material has an immediate effect on the pH-value as well.

The redox potential ($E_H$-value in mV) is considered as a quite troublesome and controversial parameter, with regard to the actual measurement and to its consequences as well. The electrodes available for this purpose cannot be checked with calibration solutions. The occasionally mentioned calibration solutions that function on the basis of divalent or trivalent Fe-ions are not satisfactory, since they actually just confirm the electrode's electric functioning and, at the same time, 'shock' the electrode to such a degree that it may remember the value of the calibration solution for several days to come. The method of choice can only consist of using well polished Pt-electrodes and of being continually aware of the plausibility and

variation of the measured values. Whenever necessary, the electrode must be substituted for another well-polished Pt-electrode. Useful recommendations as to the redox potential and its measurement are described by Kölling (1986) and by Seeburger and Käss (1989). A measured $E_H$-profile is shown in Figure 3.14 for the same core as was shown in the previous Figure 3.13. It can be seen that the values scatter by approximately +/- 30 mV, but that otherwise a reasonable profile has been measured that is likely to contribute to an understanding of the redox processes occurring in the sediment.

*Extraction of Pore water*

For extracting pore water from the sediment, various authors have developed appliances that all have a common design (e.g. Reeburgh 1967; De Lange 1988; Schlüter 1990). In each case, a sediment sample volume of 100 - 200 ml is transferred into a container made of PE or PTFE which has on the bottom side a piece of round filter foil (pore size: 0.1 - 0.2 μm) measuring 5-10 cm in diameter. On top, the sample is sealed with a rubber cover onto which argon or nitrogen is applied at a pressure of 5-15 atm. The sediment sample is thus compacted and a part of the pore water passes downwards through the filter layer. By this method, a pore water sample comprising 30 - 50 ml can easily be obtained from the water-rich and less solid layers lying close to the sediment surface.

It might make sense, in case of anoxic sediments, to load the extractor in an inert atmosphere, or even better, to carry out the whole procedure in the glove box under these conditions. This becomes inevitable if, for instance, divalent iron is to be analyzed. The question of how fast this pore water sample needs to be analyzed for single constituents, will be discussed further in Section 3.4.

Certain losses of $CO_2$ from the sample into the inert atmosphere of the glove box, and due to that, an increase of pH and a precipitation of calcite seem to be inevitable. Only the combined *in-situ* measurement of pH, $CO_2$ and $Ca^{2+}$ will lead to a reliable measurement of the calcite-carbonate-system.

It should be noted that the choice of the filter used in the extraction procedure will determine what will be evaluated as dissolved constituent and what will be evaluated as particle or solid

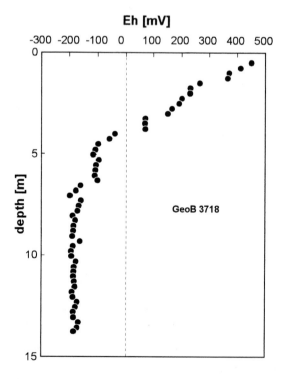

**Fig. 3.14**   Redox profile ($E_H$ in mV) for the same core as shown in Figure 3.13. Values varying approximately +/- 30 mV must be allowed, whereas an otherwise plausible profile is measured in coherence to the prevailing redox processes.

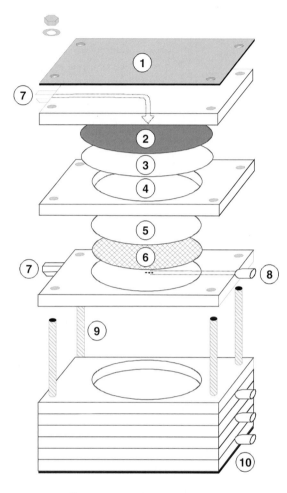

1. Top sheet (V2A)
2. Rubberdisk
3. Parafilm
4. Chamber for the sample
5. Filter membrane(0.2 μm)
6. Tissue
7. Pressure supply
8. Leak
9. Bottom sheet (V2A)
10. Thread (V2A)

**Fig. 3.15** Extractor used for the preparation of pore water from low density sediments (modified after Schlüter 1990).

sedimentary phase. The entire field of colloids which had not yet been investigated thoroughly will therefore be either evaluated as not existent, or as not dissolved, depending on the analytical method employed, for instance, whenever this fraction is eliminated by the addition of hydrochloric acid. With respect to the available and customary methods, a lot of work still remains to be done on this subject.

*Centrifugal Extraction of Pore water*

Centrifugal extraction of pore water from sediments is a procedure not very frequently performed although it appears rather easy at first examination. Anoxic sediments must be loaded into tightly sealed centrifugation tubes, to be filled and capped under the inert atmosphere conditions of a glove box. But as these plastic tubes are often not sufficiently air-tight, the invasion of ambient air might affect the sample. Moreover, a cooling centrifuge must be regularly employed, since heating the sediment above the *in situ* temperature is by no means desirable. According to the opinion of most workers, the operation of a centrifuge on board a ship is rather troublesome and, most frequently, leads in various ways to the reception of high repair bills.

In the course of centrifugation, the sediment will be compacted in such a manner that a supernatant of pore water can be decanted. Generally, less pore water is gained as compared to pressure exertion. At any rate, the water must be passed through a filter possessing a pore-size of 0.1 to 0.2 μm. As for anoxic sediments, all these steps must be carried out inside the glove box and under an inert atmosphere. The measurements comparing centrifugal extraction and pressing techniques, which have been done so far, indicate that both methods yield quite identical values (see Fig. 3.16).

*Whole Core Squeezing Method*

A simple method of pore water preparation from a sediment core that is still kept in a plastic liner has been described by Jahnke (1988). In this method, which is also suitable for employing on anoxic sediments, holes are drilled into the tubes used in the multicorer, Rumohr corer, or for a partial sample derived from the box corer. These holes are sealed prior to core sampling with plastic bolts and O-rings. As soon as the tube is filled

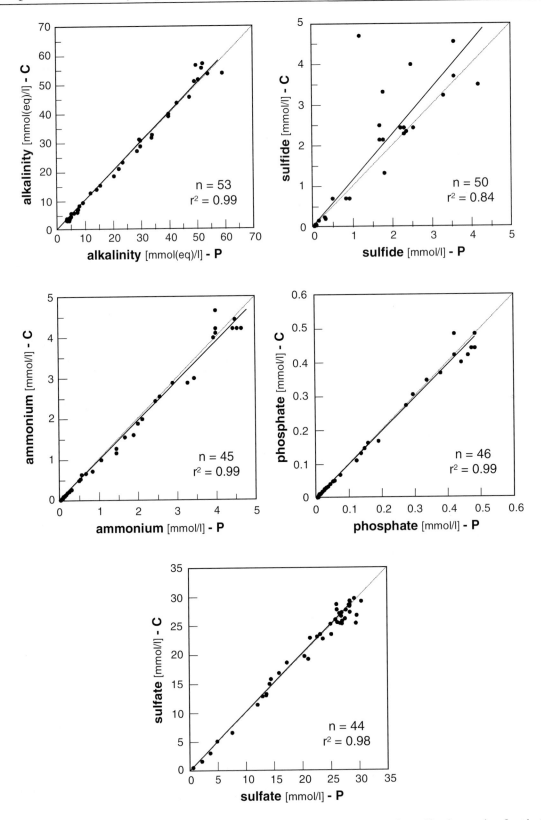

**Fig. 3.16** Comparison of analyses after the preparation of pore water by compression and centrifugal extraction. In principle, there are no observable differences. The larger variation of the sulfide values might be attributed to the decompression of the core and/or might have occurred upon mounting the sediment into the press or into the centrifuge (after Schulz et al. 1994).

with the sediment core, the device is mounted sol-
idly permitting a piston from above and below to
hold the core into position. Then, the prepared
drainage outlets are opened and sample ports
mounted to them, through which the pore water is
drained off and passed through a filter. If the core
is compressed either by the mechanical force of
the pistons, or by infusion of gas, the filtered pore
water runs off through the prefixed apertures and
can be collected without coming into contact with
the atmosphere, provided that such precautions
were taken. Jahnke (1988) compares concentration profile examples of pore water obtained by
the application of this method with pore water extracted centrifugally. Within the margins of at-
tainable precision, these profiles must be conceived as truly 'identical'.

*Sediments as yet Impossible to Sample*

Whatever has been said so far as to the separation
of pore water from sediments holds true only for
fine-grained material that is, in most cases, still
quite rich in water. Fortunately, this is what most
marine sediments and almost all deep-sea sediments are. However, sediments consisting of pure
sand might give rise to problems (e.g. sands from
the shelf or purely foraminiferal sands on the
mid-ocean ridge). The reason for this is, first of
all, that these sediments cannot be compacted by
pressure exertion or centrifugation. Consequently,
practically no water can be obtained from them at
all. It is of even more importance that these sediments possess a high degree of permeability so
that as soon as the corer is raised out of the water,
the pore water (and sometimes the core material
as well) is spilled out from the bottom. A simple
method for the extraction of pore water from
coarse, sandy material by centrifugation is described by Saager et al. (1990). However, this
technique does not solve the problem, how to
bring a sand together with its pore water undisturbed into a centrifuge.

   *In situ* measurements of pH-values and oxygen
with the aid of landers are certainly still possible
(cf. Section 3.5). But since the thin electrodes are
likely to break off upon examining such material,
these sediments are not very much favoured. Fortunately, this particular sediment type often contains only a small amount of decomposable organic matter, so that the biogeochemical processes of diagenesis do not display reaction rates
as high as other sediment types. Upon sparing

these sediments, the pore water geochemist usually suffers no great loss.

### 3.3.3  Storage, Transport and Preservation of Pore Water

In this section, the vessels in which the pressed or
centrifuged pore water is kept until the time for
analysis arrives will be briefly discussed. In Section 3.4 it will also be considered, in connection
with the analytical methods, how much time one
may allow to pass between pore water preparation
and analysis. Hence the decision depends on
whether immediate analysis is necessary, or
whether the analyses can be made at a later time,
provided that an appropriate preservation of the
material has been accomplished.

*Vessels, Bottles*

Bottles made of polyethylene have proven to be
quite useful for the storage of filtered pore water
samples. In most cases the disposable 20 ml PE-
bottles which are used in scintillation measurements are very much suited to the purpose. These
bottles need not be cleaned prior to use, but are
just briefly rinsed with the first milliliters that run
out of the press. A second use does not really
make sense considering the low price of these
bottles and the obligatory amount of work effort
and detergent to be invested in their cleaning. If
several bottles are to be used for the storage of
one and same sample, for instance when the
sample volume is rather large due to water-rich
sediments, precautions should be taken to be sure
that the entire sample amount is homogenized
prior to the analytical measurements, since differences in contrations between the first and the last
drop of pore water are possible.

*Preservation of Pore water Samples*

Only the alkali-metals, sulfate, and the halogenides do not require any preservation of the
samples prior to their analytical measurement at a
later date. Otherwise, pore water samples that are
not to be analyzed immediately can be preserved
with regard to some dissolved substances, allowing a delayed analysis.

   The concentrations of most cations - as long
as different valences of iron, for instance, are not
to be determined - are measurable after a longer
period of time when the samples' pH is lowered

by the addition of acid (mostly $HNO_3$) to a value of at least pH 2 or pH 3. Care should be taken that a sufficient amount of 'suprapur' quality acid is added. The required amounts can vary considerably depending on the alkalinity of the sample. Usually, concentrated acid will be used in order to prevent unnecessary dilution of the sample. It is also important that the sample has been previously passed through a filter of 0.1 - 0.2 μm pore-size, since the acid will dissolve all particles (and certainly the colloids that pass the filter as well).

As for the substances that are altered in their concentration by microbial activity, the toxification of sample is recommended. Mercury (II) salts are not very much favored since the handling of these extremely toxic substances - which are toxic also to man - requires special precaution measures. Chloroform is banned from most laboratories because it is considered as particularly carcinogenic. Thus, only the solvent TTE (trifluorotrichloro-ethane) remains which is frequently used in water analyses. The addition of one droplet usually suffices since only a small but sufficient proportion dissolves whereas the rest remains at the bottom phase of the bottle. Even though the compound is not as toxic in humans as those previously mentioned, no one would think of pipetting the sample thus (or otherwise) treated by mouth. It is anticipated that these toxified samples have a shelf-life of about 4 weeks.

Previously, water samples were often frozen. This procedure is not at all suited for marine pore water because mineral precipitations might occur. These precipitations do not become soluble upon thawing the samples, hence the pore water samples have become irreversibly damaged.

## 3.4    Analyzing Constituents in Pore Water, Typical Profiles

This section does not claim to re-write the analytical treatment of sea water anew, nor is it intended to present this aspect as a mere recapitulation of facts which will be far from being complete. This task would by far exceed the possibilities and, above all, the scope of this book. However, analytical methods are subject to continual improvement and relatively fast changes so that many of the methods to be introduced here would soon become outdated. Therefore, only a short account of the current methods will be given, especially of those that demand comment on the specificity to marine pore water.

The analytical details of methods that are currently used in the examination of marine pore water can be looked up in the internet under the following address:

http://www.geochemie.uni-bremen.de

We will take it upon ourselves to continually update the internet page, to evaluate the suggestions we receive for improvement, and eventually make this information available to other researchers.

If we take for granted that, in each single case, the determination of oxygen dissolved in marine pore water, as well as the measurement of the pH-value, $E_H$-value, and certainly the measurement of the temperature as well, has already been performed, the following sequence of single analytical steps is applied to the pressed or centrifuged pore water samples:

- The titration of alkalinity must be carried out as quickly as possible, i.e., within minutes or maximally within one hour, as a replacement to the direct determination of carbonate. Especially when high alkalinity values are present (maximal values up to 100 mmol(eq)/l are known to occur in pore water, whereas sea water contains only about 2.5 mmol(eq)/l), the adjustment to atmospheric $pCO_2$ might induce too low alkalinity values due to precipitation of calcite within the sample.

- If sampling for methane and sulfide has been performed as described above (see Sect. 3.3.2 'Analyses of Dissolved Gases'), the analytical measurement of the already preserved samples does not have to take place immediately within the first few minutes. Yet, the analysis should not be deferred for more than a couple of hours.

- Analyzing compounds that are sensitive to microbial action (phosphate, ammonium, nitrate, nitrite, silica) should always be completed approximately within one day, if the samples were not preserved by the addition of TTE (trifluorotrichloroethane). Even in its preserved state, it is not recommended to store the sample for longer than four weeks. Storage is best in the dark at about 4°C; freezing the samples is not recommended.

- The determination of sulfate and halogenides is not coupled to any particular time. It must be ascertained that no evaporation of the sample is likely to occur. Other instances of sample perturbation are not known. Storage is best at about 4°C; freezing the samples is not recommended.

*Photometrical Determinations, Auto-Analyzer*

Most procedures for photometric analysis of sea-water samples were described by Grasshoff et al. (1983). Owing to the quicker processing speed, and especially the need to apply lower amounts of sample solution, one would generally employ the auto-analyzer method to this end. The method is also described in the aforementioned textbook. The auto-analyzer can be either self-built or may be purchased from diverse commercial suppliers as a ready-to-use appliance. The analysis of marine pore water is generally characterized by small sample volumes and, as for some parameters, concentrations that are distinctly different from sea water. Allowing for these limitations, the recipes summarized under the above internet-address have proven quite useful for the determination of phosphate, ammonium, nitrate, nitrite, and iron(II) in pore water.

Figure 3.1 shows the characteristic profiles of ammonium and phosphate concentrations that were measured by photometrical analysis of pore water obtained from reactive sediments possessing a high amount of organic matter. Both parameters demonstrate quite similarly shaped curves compared to alkalinity. Here, the ratio of the concentrations derived from both profiles lies close to the C:N:P Redfield-ratio of 106:16:1 and clearly documents their release into the pore water due to the decomposition of organic matter. A typical photometric nitrate profile in reactive sediment zones near the sediment surface is shown in the quantitative evaluation of fluxes and reaction rates presented in Figure 3.7.

*Alkalinity*

Usually, alkalinity is actually the ultimate parameter that is subject to a procedure of 'genuinely chemical' titration, in this regard as a proxy parameter for carbonate. A new spectrophotometric method for the determination of alkalinity was proposed by Sarazin et al. (1999). Mostly, alkalinity will be, for reasons of simplification, set

equal to the total carbonate concentration, although a number of other substances in pore water will contribute to the titration of alkalinity as well. Most geochemical model programs (cf. Chap. 14) foresee the input of titrated alkalinity as an alternative to the input of carbonate. The model program will than calculate the proportion allocatable to the different carbonate species.

Another problem arising from titration of alkalinity in a pore water sample usually consists in the fact that samples with very small volumes cannot easily be used. Most ocean chemists will be accustomed to the titration of a volume of 100 ml, or at least 10 ml. Under certain circumstances, a pore water sample obtained from a definite depth might contain in total not more than 10 ml, hence, at best, merely 1 ml needs to be sacrificed for the titration of alkalinity. This requires that markedly pointed and thin (thus easily breakable) pH-electrodes are used. These are immersed, together with an electronically controlled micropipette, into a small vial so that a tiny magnetic stirrer bead still has enough room to fit inside as well.

Mostly, an adequate amount of 0.001 M hydrochloric acid will be added to a previously pipetted volume of 1 ml, thus the pH will be brought to a value of about 3.5. Since the dilution of the pre-pipetted volume must be considered, the alkalinity (Alk) is calculated according to the following equation:

$$Alk = [(V_{HCl} \cdot C_{HCl}) - 10^{-pH} \cdot (V_0 + V_{HCl}) \cdot f_{H+}^{-1}] \cdot V_0^{-1}$$

$$(3.29)$$

in which $V_{HCl}$ describes the volume of added hydrochloric acid, $C_{HCl}$ represents the molality of the added hydrochloric acid, pH denotes the pH value the solution attains after the addition of hydrochloric acid, $V_0$ is the pre-pipetted sample volume, and $f_{H+}$ the activity coefficient for $H^+$-Ions in solution. The activity coefficient $f_{H+}$ can be determined according to the method described by Grasshoff et al. (1983), or calculated by employing a geochemical model (e.g. PHREEQC, cf. Chap. 14). The Equation 3.29 is also formulated in the textbook published by Grasshoff et al. (1983) [equation 8-25 on page 108]. It must be pointed out that a very misleading error has unfortunately found its way into the equation in this reference. This error is hardly noticeable when low concentrations are prevalent in ocean water, or when the sample volumes are comparably large. In the case of high concentrations in combination with small sample volumes, however, alkalinity values will be calculated that are prone to an error of more than a factor of 2. For this reason, the equation is presented here in its correct form.

*Flow Injection Analysis*

A very interesting method to analyze total carbon dioxide and ammonium in pore water was introduced by Hall and Aller (1992). In this method, a sample carrier stream and a gas receiver stream flow past one another, separated only by a gas-permeable PTFE (Teflon®) membrane. For determining the total $CO_2$, the sample carrier stream consists of 10-30 mM HCl. To this stream, the sample in a volume of about 20 µl is added via a HPLC injection valve. The carbon dioxide traverses the PFTE membrane and enters the gas receiver stream which, in this case, consists of 10 mM NaOH. The $CO_2$ taken up by the gas receiver stream causes an electrical conductivity change that can be determined exactly in a micro-sized continous flow cell.

For the analysis of ammonium, the sample carrier stream consists of 10 mM NaOH + 0.2 M Na-citrate. This converts ammonium into $NH_3$-gas which penetrates the PTFE membrane and travels into the gas receiver stream, in this case consisting of 50 µm HCl. Here as well, an electric conductivity cell is used for the measurements. To provide a steady and impulse-free flow of the solutions, a multichannel peristaltic pump is used. The flow rates in both cases were set to approx. 1.4 ml min$^{-1}$.

These very reliable and easy to establish analytical procedures for the determination of these two important parameters of marine pore water are of special interest, because they require only small sample amounts and because a reliable analysis is obtained over a broad range of concentrations.

*Ion-selective Electrodes*

Ion-selective electrodes were accepted in water analysis only with great reluctance, despite of the advantages that had been expected on their introduction about 15 years ago. The reasons for this were manifold. The analytical procedure is inexpensive only at first sight. As has been confirmed for some electrodes (especially when its handling is not always appropriate), aging sets in soon after initial use which becomes noticeable with decreased sensitivity, low stability of measured values, and prolonged adjustment times.

With regard to marine pore water, ion-selective electrodes were successfully applied in the determination of fluoride (standard addition method) and in the analysis of sulfide within a mixture of sediment and pore water processed into SAOB-buffer (cf. Sect. 3.3.2). Figure 3.1 shows a typical profile of fluoride measured with an ion-selective electrode in pore water. The total sulfide profile shown in Figure 3.13 was measured in the way described above, by using an ion-selective electrode. The measured values occasionally display considerable variations, probably on account of the sediment's decompression and thereafter they might be produced upon withdrawing the sample from the core. They do not necessarily come about during the course of the analytical measurement.

*Ion-Chromatography*

The analysis of chloride and sulfate in marine pore water samples after an approximately 20-fold dilution is a standard method for normal ion-chromatography (HPLC), so that no further discussion is needed here. Considering the very dilute sample solution and the low sample amount required in ion-chromatography, the applied quantity of pressed or centrifuged pore water is negligibly small. However, the high background of chloride and sulfate which is due to the salt content in sea water prevents, with or without dilution, the determination of all other anion species by ion-chromatography. The sulfate profiles measured with this method are shown in the Figures 3.1, 3.6, and 3.13. Since the chloride profiles are mostly not very interesting, because they reflect practically no early diagenesis reactions at all, they can be consulted for control and eventually for correction of the sulfate profiles, whenever analytical errors have emerged in the course of their concomitant determination, e.g. due to faulty dilutions, or mishaps occurring in the injection valve of the machinery.

*ICP-AES, AAS, ICP-MS*

Generally the alkali metals (Li, Na, K, Rb, Cs), the alkali earth metals (Mg, Ca, Sr, Ba) and other metals (e.g. Fe, Mn, Si, Al, Cu, Zn, Cr, Co, Ni) are determined in the acid-preserved pore water samples. Analytical details depend on the special conditions provided by the applied analytical instruments ICP-AES (Inductively Coupled Plasma Atomic Emission Spectrometer), ICP-MS (Inductively Coupled Plasma Mass Spectrometer), or AAS (Atomic Absorption Spectrometer) of the respective laboratory. A discussion of details is not appropriate in this chapter. In most cases dilutions of 1:10 or 1:100 will be measured depending on the salt content in marine environments, so that the required sample amounts are rather low.

*Gas-Chromatography*

The quantification of methane with the aid of gas-chromatography (FID-detection) is an excellent standard method, which therefore does not need any further discussion. The main importance is the immediate withdrawal of a sediment/pore water sample as already mentioned in Section 3.3.2. This sample is placed into a 50 ml-headspace vial using a syringe in which the sample is combined with 20 ml of a prepared solution of 1.2 M NaCl + 0.3 M $HgCl_2$. After equilibrium is reached in the closed bottle between the methane concentration in the gaseous phase and its concentration present in aqueous solution (at least one hour), the gas phase is ready to be sampled by puncturing the rubber cap that seals the vial with a syringe.

## 3.5    *In-situ* Measurements

The desire to measure the properties of a natural system, or profiles of properties, *in-situ*, appears to be obvious whenever the system becomes subject to a change as a result of the sample collection procedure. In principle, this is applicable to many aquatic geosystems. Especially where solid and aquatic phases come together in a limited space, this boundary constitutes the site of essential reactions. Such a site is the surface boundary between bottom water and sediment. An essential biogeochemical process of early diagenesis is exemplified by the permeation of oxygen into the young sediment by diffusion, but also by the activity of organisms. The oxygen in the sediment reacts in many ways (cf. Chaps. 5 and 6). In part, the oxygen is depleted in the course of oxidizing

organic matter, in part, upon re-oxidizing various reduced inorganic species (e.g. $Fe^{2+}$, $Mn^{2+}$).

In organic-rich and highly reactive sediments, the depletion of oxygen previously imported by means of diffusion already occurs in the uppermost millimeters of the sediment (cf. Sect. 3.5). However, these organic-rich and young sediments are mostly extremely rich in water (porosity values up to 0.9) and very soft, so that every sample removal will most likely induce a perturbation of the boundary zone which is, in a crucial way, maintained by diffusion.

Moreover, the actively mediated import of oxygen into the young sediment, which runs parallel to diffusion, and which is transported by the organisms living therein, will function 'naturally' only as long as these organism are not 'unnaturally' treated. Any interference caused by sampling is certain to be regularly coupled to a change of temperature, pressure and the quality of the bottom water, and thus means a change in the 'normal' living conditions. Here as well, it is of special interest to measure the active import of oxygen into the sediment governed by biological activity *in-situ*. (cf. Sect. 3.6).

*Lander Systems*

In some very shallow waters the *in-situ* measurement at the sediment/bottom water boundary might be achieved from a ship, or by using light diving equipment. However, so-called 'Lander systems' have been developed for applications in greater depths and, above all, in the deep sea. Lander systems are deployed from the ship and sink freely, without being attached to the ship, down to the ocean floor where they softly 'land'. Standing on the bottom of the ocean they conduct measurements (profiles with microelectrodes), experiments (*in-situ* incubations), and/or remove samples from the sediment or the bottom water. When the designated tasks are completed, ballast is released and the whole system emerges back to the surface to be taken on board.

The lander technique described above is, especially in the deep sea, more easily outlined in brief theory than to put into operation, as the involved workers need to invest a considerable amount of construction effort and operational experience in the lander systems so far used.

The problems which cannot be discussed here in detail consist of, for example, deploying the device into the water without imparting any damage to it, conducting the subsequent performance of sinking and landing at the proper speed, keeping the lander in a proper position, recording and storing the measured values, the timing of ballast release and thus applying the correct measure of buoyancy all the way up to the ocean's surface, and finally, the lander's retrieval and ultimate recovery on board of the ship. The correct performance of the *in-situ* measurements is not even mentioned in this enumeration. In Figure 3.17 such an appliance is shown according to Jahnke and Christiansen (1989) which is very similar to the lander used by Gundersen and Jørgensen (1990) as well as Glud et al. (1994). A survey and comparative evaluation of 28 different lander systems is given by Tengberg et al. (1995) who also delivers an historical account on the development of landers. One of the first publications on the measurements of oxygen microprofiles in sediment/bottom water boundary layers of the deep sea is the report by Reimers (1978). Jørgensen and Revsbech (1985), subsequently, Archer et al. (1989) and Gundersen and Jørgensen (1990) were the first to measure a diffusive boundary layer using landers. The first *in-situ* incubations using a lander were carried out by Smith Jr. and Teal (1973).

*Profiles with Microelectrodes*

A profile measured *in-situ* with an oxygen microelectrode has already been presented and quantitatively evaluated in Figure 3.5. Such measurements became feasible only after oxygen electrodes had been developed that could measure oxygen concentrations at the sediment/bottom water boundary layers with a depth resolution of about 20 to 50 μm. It is also important to note that the measurement is carried out without being influenced by the oxygen depletion of the electrode. The development of the electrodes is closely linked to the name of N.P. Revsbech (Aarhus, Denmark) and is described in the publications of Revsbech et al. (1980), Jørgensen and Revsbech (1985), Revsbech and Jørgensen (1986) and Revsbech (1989).

The basic construction principle underlying the function of an oxygen microelectrode is shown in Figure 3.18, after a publication by Revsbech (1989). It is evident that such an electrode cannot be built without considerable practiced skill and experience - and they have to be self-made as they are not yet commercially avail-

able. It is furthermore evident that such microelectrodes are extremely sensitive and break easily as, when they either hit a rather solid zone in

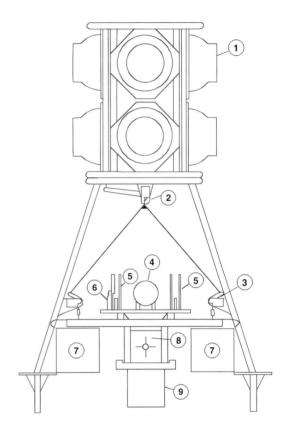

1. Glass floats (1 of 8)
2. Main ballast release
3. Secondary ballast release (1 of 3)
4. Electronic pressure case
5. Sampling racks
6. Syringes
7. Expendable ballast
8. Chamber lid and stirring mechanism
9. Chamber & chamber scoop

**Fig. 3.17** Schematic representation of a Lander System which is also used in the deep sea. The machinery sinks freely, without any attachment, to the ocean floor where it carries out measurements and returns back to the surface after the release of ballast. The depicted version stands about 2 -3 m in height. It is designed to conduct incubation experiments at the ocean floor, yet similar landers are used to record/monitor in situ microprofiles of the oxygen concentration through the sediment/bottom water boundary layer. The perspective shows only two of the three feet upon which the lander stands (after Jahnke and Christiansen 1989).

116

the sediment, contain slight imperfections from manufacturing, or are handled with the slightest degree of ineptitude.

A construction unit is shown in Figure 3.19 that permits the integration of microelectrodes in a lander along with the necessary mechanical system, the electronic control and data monitoring equipment. It is of utmost importance that this unit is built into the lander in such a manner that the measurements will begin above the sediment surface, then insert the electrodes deep enough into the sediment, allowing later derivations of the essential processes from the concentration profile. This requires, among other pre-requisites, the correct estimation as to how deep the feet of the lander will sink into the sediment. If the sediment is firmer than expected, only the bottom water might be measured; whereas, if the sediment is softer than expected, the profile might begin within the sediment. Thus, much experience and many aborted profiles stand behind those profiles which are now shown in the Figures 3.5 and 2.20. By using such electrodes it is now possible to de-

scribe a diffusive boundary layer at the transition zone between the sediment and the bottom water with a sufficient degree of depth resolution (Fig. 3.20). This layer characterizes a zone which lies directly over the sediment and is not influenced by any current. Instead, a diffusion controlled transport of material characterizes the exchange between bottom water and sediment. This layer which possesses a thickness of 0.2 - 1 mm still belongs to the bottom water with regard to its porosity (100% water), but due to its diffusion controlled material transport, it belongs instead to the sediment. In both oxygen profiles shown on the left in Figure 3.20, the location of the diffusive boundary layer is indicated. Above this layer, the bottom water is agitated, resulting in an almost constant oxygen concentration. A constant gradient is found within this layer because this is where only diffusive transport takes place, but no depletion of oxygen occurs. It may be of interest in this regard that the better known diffusion coefficient for water ($D^{sw}$ in Eq. 3.5 and in Table 3.1) must be applied for this diffusive transport,

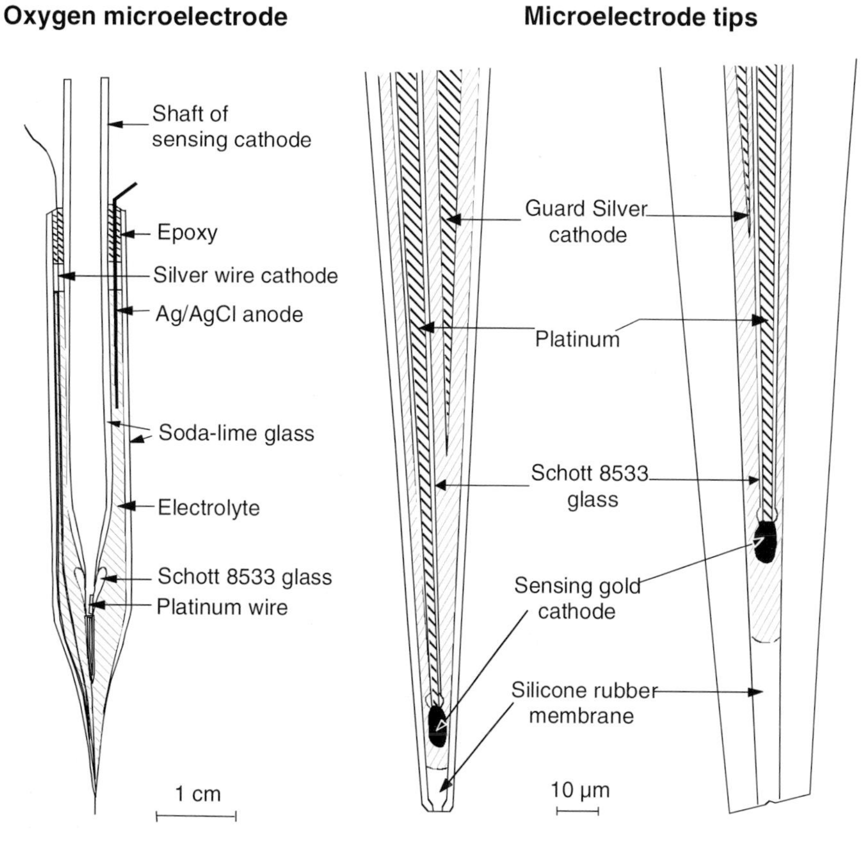

### Oxygen microelectrode

### Microelectrode tips

**Fig. 3.18**   Construction of an oxygen electrode equipped with a Guard-cathode (after Revsbech, 1989).

and not the diffusion coefficient for the sediment ($D_{sed}$). Since the entire diffusive transport from the bottom water to the sediment crosses this transitory layer, measurements within the layer allow for especially reliable estimations of the oxygen consumption in the sediment.

To date, extremely good results have been obtained with oxygen microelectrodes in sediments with high reaction rates, because the essential processes can be analyzed only here, with profiles most often reaching a few centimeters into the sediment. The lower ends of these profiles are frequently marked by breakage of the electrodes. In sediments displaying less vigorous turnovers, the measurement mostly includes just one single oxygen concentration gradient, not reaching into the penetration depth for oxygen.

Less frequently, pH-profiles are published in literature when compared to the oxygen profiles. The details as to their shapes and their underlying processes are not yet sufficiently understood. Additionally, a reliable calibration under the given pressure conditions has not been accomplished for *in-situ* pH-measurements. Consequently, only relative values are measured within one profile. Electrodes measuring $H_2S$ can also be built as microelectrodes. However, they are only suited for *in-situ* measurements when $H_2S$ reaches close enough to the sediment surface.

*Optodes*

Apart from the great advantages of electrodes - they have practically opened a domain of the aquatic environment which was previously not accessible to measurement - the disadvantages should be mentioned as well. On the one hand, only the three aforementioned types are applicable ($O_2$, pH, $H_2S$), on the other hand, these electrodes are not commercially available but must be self-made. The manufacturing of one electrode affords - if one has the necessary experience – half-a-day up to one day of labor. Additionally, the electrodes tend to break easily. Thus, in the experimental work that led to the publication by Glud et al. (1994) about 50 to 60 (!) microelectrodes were 'used up'.

It therefore appears to be reasonable to look for other methods and other measurement principles. Such an alternative principle possibly consists of the construction of the optode (by analogy: electrodes as functioning electrically, optodes functioning optically). The basic principle (Fig. 3.21) consists of the conductance of light possessing a specific wavelength (450 nm) via glass fibers to the site of measurement. The tip of the glass fiber is surrounded by a thin layer of epoxy resin that

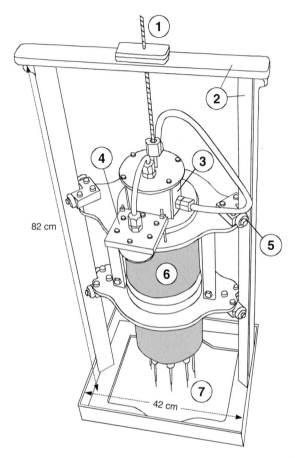

82 cm

42 cm

1. Thredded rod
2. Supporting Al - frame and tracks
3. DC - motor hausing
4. Oil - filled bladder
5. Radial ball bearings
6. Pressure cylinder
7. Microelectrodes

**Fig. 3.19** Schematic representation of the mechanic and electronic unit applied to microelectrodes in a lander system. The round pressure cylinder contains the electronic control equipment and the data monitoring device, to the bottom of which various vertically positioned microelectrodes are mounted. The entire cylinder is lowered with the aid of a stepper motor and a corresponding mechanical system in pre-adjustable intervals so that the electrodes begin measuring in the bottom water zone and then become immersed into the sediment (after Reimers 1978).

contains a dye (ruthenium(II)-tris-4,7-diphenyl-1,1-phenanthroline).

The dye emits a fluorescent light at a different wavelength (610 nm). The intensity of this fluorescence depends on the concentration of the substance to be measured that penetrates into the resin by diffusion, directly at the site of measurement. The construction of such microoptodes, and the first results of measuring oxygen at the sediment/bottom water boundary were described by Klimant et al. (1995). According to Klimant et al. (1995), the development of optodes is feasible for a number of substances (e.g. pH, $CO_2$, $NH_3$, $O_2$), yet, in the form of the microoptode, this new technology was only applied in the measurement of oxygen. The construction details and the alchemy of the fluorescent dyes will be presumably still the object of current development.

*in-situ Incubations*

The diffusive input of oxygen into the sediment contributes to only one part of the whole oxygen budget. An input of oxygen into the sediment conducted by macroscopic organisms must be considered to occur especially in the sediments of shallow waters, and when the proportion of organic matter in the sediment is high. These processes

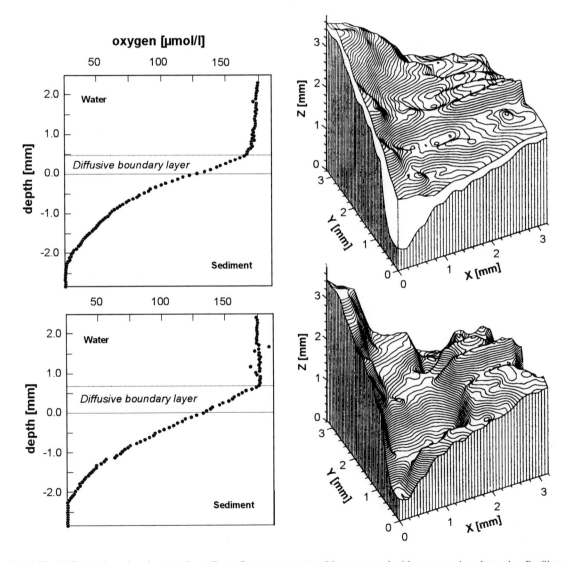

**Fig. 3.20** Diffusive boundary layer at the sediment/bottom water transition measured with oxygen microelectrodes. Profiles are shown in both representations on the left. The two representations on the right show the respective spatial patterns. In the above right corner, the surface is shown, whereas the subsurface of the diffusive boundary layer is shown below (after Gundersen and Jørgensen 1990).

will be discussed more thoroughly in the following Section 3.6 concerned with bioirrigation.

The total oxygen consumption at the transition between sediment and bottom water is measured in so-called incubation experiments which are best to be carried out under *in-situ* conditions. To this end, a distinct area of the sediment surface area (usually several dm³) is covered with a closed box in such a manner that a part of the bottom water is entrapped. Inside the closed box, the concentrations of oxygen and/or nutrients are repeatedly measured over a sufficient length of time. The total exchange of substances is deduced from the measured concentration changes in relation to the volume of entrapped bottom water and the area under study.

Figure 3.22 shows the result of such an *in-situ* experiment taken from a study conducted by Jahnke and Christiansen (1989). The sediment's uptake of the electron acceptors oxygen and nitrate is easy to quantify, as well as the concomitant release of phosphate, silica, and carbonate into the bottom water.

Glud et al. (1994) also carried out *in-situ* incubation experiments in the upwelling area off the coast of Namibia and Angola, in a parallel procedure to their recording diffusion controlled oxygen profiles by means of microelectrodes. Thus, they were able to obtain values for the sediment's total oxygen uptake and, at about the same site, simultaneously measured the exclusively diffusion-controlled oxygen uptake. The ratio of total uptake and diffusive uptake yielded values between 1.1 and 4. The higher ratios were coupled to the altogether higher reaction rates observable in organic-rich sediments at shallower depths. Similarly a good correlation between the difference of total oxygen uptake minus diffusive oxygen uptake on the one hand, and the abundance of macroscopic fauna on the other hand was observed.

*Peepers, Dialysis, and Thin Film Techniques*

At first glance, dialysis would appear to be a very useful technique for obtaining pore water samples under *in-situ* conditions which can be analyzed for nearly all constituents. Different types of 'peepers' are presently used which can be pressed into the sediment from the sediment surface. Such a peeper usually consists of a PE-column 0.5 - 2 m long with 'peepholes' on its outside. Cells filled with de-ionized water inside the peeper are allowed to equilibrate via dialysis through a filter membrane with the pore water outside the peeper. However, an equilibration time of many days or

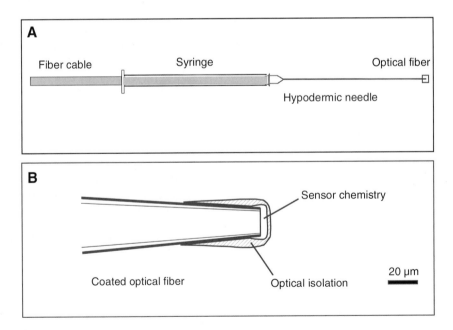

**Fig. 3.21**   Schematic representation of an oxygen microoptode. One end of the glass fiber can be mounted into a steel cannula for protection and stabilization (A). At the site of measurement, the tip of the glass fiber is surrounded by a thin layer of epoxy resin (B) which contains a fluorescent dye. The fluorescence properties of the dye depend on the concentration of the compound to be measured (after Klimant et al. 1995)

even a few weeks may be necessary if a volume of fluid greater than 5 - 10 cm³ is necessary for analysis and if a small peephole is required in order to achieve a certain depth resolution. This restricts the use of dialysis-peepers to shallow water environments where diving is possible and makes it especially useful in tidal areas. It should be pointed out that dialysis does not simply result in the sampling of pore water under *in-situ* conditions. Only those aquatic species that are small enough to pass through the membrane can find their way through the peephole into the cell. The species distribution is therefore dependent on the nature of the membrane used. A review of different techniques is given by Davison et al. (in press). More relevant to the study of deep sea sediments are *in-situ* techniques using the diffu-

sive equilibration in thin films (DET) and diffusive gradients in thin films (DGT) (Davison et al. 1991; Davison and Zhang 1994; Davison et al. 1997, Davison et al. in press).

Figure 3.23 shows concentration-depth profiles for Mn and Fe through the sediment/water interface with a depth resolution which would be impossible to obtain using any other technique. Measurements such as these have resulted in meaningful measurements of reaction rates within the uppermost layers of young sediments and of diffusive fluxes of various metals through the sediment/water interface for the first time. However, the preparation of the thin films, and their handling under field conditions is still quite troublesome and much work needs to be undertaken to further develop these techniques. None-

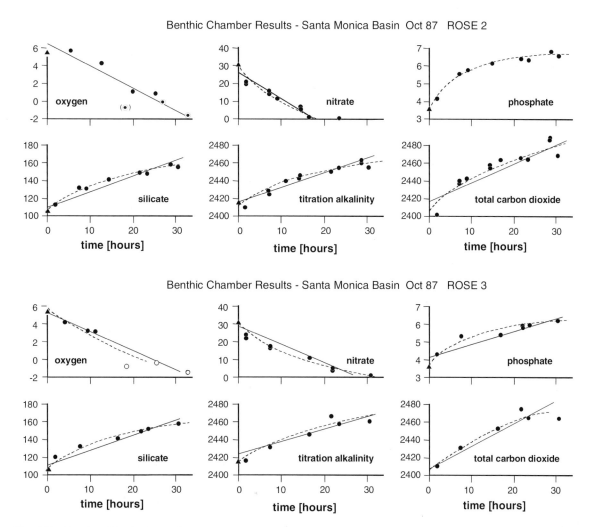

**Fig. 3.22**   Results of in situ incubation measurements. In both examples, a depletion of oxygen and nitrate and a concomitant release of phosphate, silica, and carbonate was observed in incubation time lasting 30 hours. The concentration unit is μmol/kg, alkalinity is expressed in μmol(eq)/kg. (after Jahnke and Christiansen 1989).

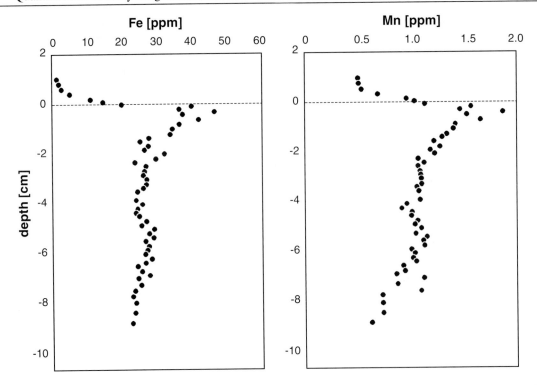

**Fig. 3.23** Concentration-depth profiles for Mn and Fe through the sediment/water interface. For these measurements, a DET gel was deployed for 24 hours in situ. (after Davison et al. 1994)

theless, this seems to be a most promising approach for obtaining reliable *in-situ* determinations of elements in sediment pore waters.

## 3.6    Influence of Bioturbation, Bioirrigation, and Advection

For the pure geochemist it would be most convenient if this chapter would close at this point. The concomitance of diffusive transport and reactions determinable on the basis of gradient alterations - although they are rather numerous - results in a sufficiently sophisticated system of pore water and sediment, quite well understood and calculable. For deeper zones below the sediment surface, beyond approximately 0.5 to 1 m, a description of the system in terms of diffusion and reactions probably seems to be quite acceptable, with a certain degree of accuracy obtained. The closer the zones under study are located to the sediment surface, the more inhomogeneities and non-steady state conditions become of quantitative importance. These are mainly determined by the activity of organisms living in the sediment.

Depending on whether the effect on the sediment or the pore water is considered, this phenomenon is referred to as bioturbation or bioirrigation, respectively. Advection (sometimes also called convection) is the term used for the motion of water coupled to a pressure gradient which partly overlaps with bioirrigation, when permeabilities are influenced by the habitats of the organisms. At the end of this section, model concepts will be outlined that allow for a quantitative description of this heterogeneous, yet very reactive domain of the sediment will be assessed.

*Bioturbation*

Bioturbation refers to the spatial rearrangement of the sediment's solid phase by diverse organisms, at least temporarily living in the sediment. This process implies that all sedimentary particles in the upper layers, which are inhabited by macro-organisms, are subject to a continual transfer, i.e. seen in the long run, are integrated into a permanent cycle. From the publications of many authors (e.g. Aller 1988, 1990, 1994; Dicke 1986) it can be concluded that this layer, essentially influenced by bioturbation, generally reaches 5 to 10 cm, rarely even up to 15 cm, below the sediment surface. Only when a sediment particle has reached a greater depth in the course of continued sedimentation, will the 'recycling' process become a rare event until it ultimately comes to a final standstill.

This continued recycling of the sediment's solid phase is of great importance for a number of geobiochemical processes. By the action of this cycle, the electron donor (organic matter) and the electron acceptors (e.g. iron and manganese oxides) of the biogeochemical redox processes in the sediment are continually procured from the sediment surface. The results obtained by Fossing and Jørgensen (1990) can only be understood when considerably more sulfate is chemically reduced in the upper layers of the sediment than is accounted for by the sedimentation rate of directly imported organic matter. The re-oxidation of once released sulfide back to sulfate, which is also of importance in this regard, can only proceed when the oxides of iron and manganese, the electron acceptors, are continually supplied from the surface in an oxidized state, and return there in a chemically reduced form. Such a cycle has been described for iron and manganese, for instance, by Aller (1990); Van Cappellen and Wang (1996) as well as Haese (1997).

*Bioirrigation*

The process in which living organisms in the sediment actively transport bottom water through their habitats is known as bioirrigation. In this process, oxygen-rich water is usually pumped into the sediment, and water with less oxygen is pumped out. Figure 3.24 which is derived from the publication by Glud et al. (1994) demonstrates an oxygen profile measured in situ with the aid of a microelectrode. The electrode recorded a normal profile reaching about 30 mm below the sediment surface. Beyond, from about 30 mm to 80 mm below the sediment surface, the oxygen concentration was practically zero, whereas an open cavity was detected between 80 and 90 mm flooded with oxygen-rich bottom water.

In the publications submitted by Archer and Devol (1992), a comparative study on the purely diffusive oxygen flux (based on measurements with microelectrodes) and the total oxygen flux (based on incubations) was conducted for the shelf and the continental slope off the State of Washington. In a similar study, Glud et al. (1994)

investigated the continental slope off Angola and Namibia. It was shown that the flux induced by bioirrigation was several times larger than the diffusive flux. However, such high fluxes only occur in densely populated sediments, mostly on the shelf. For deep sea conditions the total oxygen uptake was slightly higher then the diffusive oxygen uptake of the sediment.

*Advection of Pore water*

Except for bioirrigation, advection within the pore water fraction may only result from pressure gradients. Advection is sometimes referred to as convection, yet the terms are used rather synonymously. There are, in principle, three different potential causes leading to such pressure gradients, and thus to an advective flux:

- Sediment compaction and a resulting flow of water towards the sediment surface.

- Seafloor areas with warmer currents and a resulting upward-directed water flow, that corre-

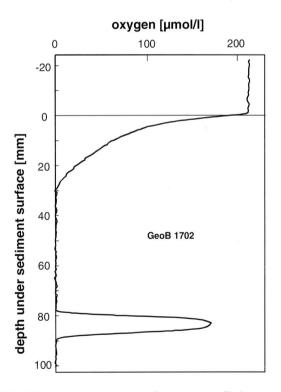

**Fig. 3.24** In situ measurement of an oxygen profile in a water depth of 3100 m, off the mouth of the Congo River. The microelectrode detected an open cavity flooded with oxygen-rich bottom water in a depth between 80 mm and 90 mm below the sediment surface (after Glud et al. 1994).

spondents with water flows directed down-wards at other locations.

- Currents in bottom water that induce pressure differences at luv and lee sides of uneven patches on the sedimentary surface.

The first case can be quite easily assessed quantitatively on the basis of porosity measurements: one might imagine a fresh and water-rich deposited sediment column and observe the compaction as a further descent of the solid phase relative to the water that remains at the same place. Hence, the upwards directed advective flux results as the movement of water relative to the further sinking sediment. If the sediment exhibits a water content of approximately 0.9 at its surface, and even a value of 0.5 in several meters depth, and provided that the boundary between sediment and bottom water moves upwards due to the accumulation of new sediment, then it will inevitably follow that the compaction induces an advective flux which will not exceed values similar to the rate of sedimentation. As for deep sea sediments, this process can produce values in the range of a few centimeters per millennium.

Some examples for the second case were provided by Schultheiss and McPhail (1986). By using an analytical instrument that freely sinks to the ocean floor, they were able to measure pressure differences between pore water, located 4 m below the sediment surface, and the bottom water directly. At some locations, deep sea sediments of the Madeira Abyssal Plain displayed pressure differences of about 0 Pa, at other locations, however, the pressure in pore water 4 m below the sediment surface was significantly lower (120 or 450 Pa) than in bottom water. With regard to the porosity ($\phi$) and the permeability coefficient of the sediment (k), the advective flux is calculated according to the following equation:

$$v_a = \left( k \cdot \frac{\Delta p}{\Delta x} \right) \Big/ \phi \qquad (3.30)$$

In this equation, known as the *Darcy equation*, and which is applied in hydrogeology for calculating advective fluxes in groundwater, $v_a$[m s$^{-1}$] denotes the velocity with which a particle/solute crosses a definite distance in aqueous sediments. $\Delta p$ refers to the pressure altitude measured in meters of water column ($10^5$ Pa = 1 bar $\approx$ 750 mm Hg $\approx$ 10.2 m water column); and $\Delta x$ [m] is the

distance across which the pressure difference is measured. In the example shown above ($\phi = 0.77$, k = 7·10ms$^{-1}$), this distance amounts to 4 m. Insertion into Equation 3.30 yields:

$$v_a = \left( 7 \cdot 10^{-9} \cdot \frac{0.012}{4} \right) \Big/ 0.77 = 2.7 \cdot 10^{-11} \text{ [m s}^{-1}]$$

$$(3.31)$$

or:

$$v_a = 0.86 \text{ [mm a}^{-1}]$$

Applying another value specified by Schultheiss and McPhail (1986) as 450 Pa, a velocity of $v_a$ = 3.2 [mm a$^{-1}$] ensues. (These authors end up with the same values, yet they do so by using a different, unnecessarily complicated procedure for calculation). Such values would thus be estimated as one, or rather two orders of magnitude greater than those resulting from compaction. Actually, such values should distinctly be reflected by the concentration profiles in pore water. However, despite of the huge number of published profiles, this consequence has not yet been demonstrated.

Upon applying the Darcy Equation 3.30 which is designed for permeable, sandy groundwaters, we must take into consideration that this equation is only applicable to purely laminar fluxes. Furthermore, it does not account for any forces effective near the grain's surface. A permeability value of k = 7·10$^{-9}$ m s$^{-1}$ implies that grain size of the sediment particles is almost identical to clay. Within such material, the forces working on the grain surfaces impose a limitation on the advective flux, so that the calculation of Equation 3.31, based on the pressure gradient, will surely lead to a considerable overestimation of the real advective flux.

The third case, a pressure gradient induced by the bottom current and a supposed advective movement of the pore water, is of importance mainly on the shelf where shallow waters, fast currents, an uneven underground, and high permeability values in coarse sediments are encountered. Ziebis et al. (1996) as well as Forster et al. (1996) were able to demonstrate by *in situ* measurements and in flume experiments that the influence of bottom currents may be indeed crucial for the superficial pore water of coarse sand sediments near to the coast.

At flow rates of about 10 cm s$^{-1}$ over an uneven sediment surface (mounds up to 1 cm high), the oxygen measured by means of microelectrodes had penetrated to a maximum depth of 40 mm, whereas a penetration depth of only 4 mm was measured under comparable conditions when the sediment surface was even. Huettel et al. (1996) were able to show in similar flume experiments that not only solutes, but, in the uppermost centimeters, even fine particulate matter was likewise transported into the pore water of coarsely grained sediments. Similar processes with marked advective fluxes are, however, not to be expected in the finely grained sediments predominant in the deep sea.

*Advection of Sediment*

An advection of the sediment's solid phase does not, at first sight, seem to make any sense, because it implies - in contrast to bioturbation - a movement of sediment particles which favor a specific direction. There is no known process that describes an advection of a solid phase, instead, the definition of such a process actually only results from the definition of the position of the system of coordinates, and thus already indicates the model boundary conditions which will be later dealt with in more detail. If the zero-point of the coordinate-system is set by definition to the boundary between sediment and bottom water, it follows that the coordinate-system will travel upwards with the velocity of the sedimentation rate. Concomitantly, this also implies that sediment particles move downwards relative to the coordinate-system with the velocity of the sedimentation rate. This definition may appear somewhat formalistic, at first, but it becomes more real and relevant from a quantitative point of view when diagenetic reactions are studied. Then most turnovers are limited by the solid phase components that are introduced into the system by this form of 'advection'.

*Conceptual Models*

Conceptual models and their applications will be discussed in more detail in Chapter 14 where computer models will be presented and examples taken from various fields of application will be calculated. At this point, conceptual models will just be mentioned in brief summary and will contain the most essential statements of this chapter.

- The description of the diffusive transport by Fick's first and second law of diffusion includes the transport of soluble substances in pore water, which is not mediated by macroorganisms. In this regard, diffusive fluxes are always produced by gradients; and fluxes are always reflected by gradients. Upon assessment of concentration profiles with respect to material fluxes, it needs to be considered whether the depth zones of the investigated profile are sufficiently (quasi) stationary with regard to the studied parameter. Non-steady state conditions are appropriately described only by Fick's second law of diffusion.

- If no biogeochemical processes can to be found in sediments at all, the pore water should display constant concentrations from the top down in a stationary way. Any reactions taking place in the pore water fraction, or between pore water and solid phase, will become visible as a change of the involved concentration gradients extending across the depth zone; any changes of the concentration gradients document the processes in which pore water has been involved.

- In principle, bioturbation as a macrobiological process should not be described in terms of easily manageable model concepts as can be done for molecular diffusion. Only if it is assured that the expansion of a given volume under study is large enough, and/or provided that the time-span necessary to make the observation is sufficiently long, can we conceive of the bioturbation processes, on average, as identically effective at all places. In such cases, a model concept of 'biodiffusion' is frequently applied in which Fick's laws of diffusion are applied upon introducing an analogous 'biodiffusion coefficient'. Aller and De Master (1984) reported on having determined $^{234}$Th and $^{210}$Pb and drew conclusions using a model concept of bioturbation. Figure 3.25 shows an empirically determined biodiffusion constant according to Tromp et al. (1995) in its dependence on the sedimentation rate. Deep sea sediments with sedimentation rates in the range of 0.001 and 0.01 cm a$^{-1}$ yield biodiffusion coefficients in the range of 0.1 and 1 cm$^2$a$^{-1}$, or, upon applying the dimensions contained in Table 3.1, values that lie between $3 \cdot 10^{-13}$ and $3 \cdot 10^{-12}$ m$^2$s$^{-1}$. Such biodiffusion coefficients are thus more than one order of magnitude smaller then the coefficients of molecular diffusion in sediment pore waters. The bioturbation process (= bio-diffusion) is therefore without immediate relevance for pore water because, if we judge the matter realistically, its effects are by far less than the accuracy of the molecular diffusion coefficient. Bioturbation (= biodiffusion) is, however, very important for the dynamics and the new organization of the sediment's solid phase. By this mechanism, and adjunct biogeochemical reactions, bioturbation consequently also exerts an indirect but nevertheless important influence on the distribution of concentrations in pore water.

- Upon introducing a related coefficient, Fick's laws of diffusion have also been applied to bioirrigation. However, this surely reflects a rather poor model concept. Figure 3.24 already showed this in two ways: firstly, this process is indeed of relevance and applicable to pore water. Secondly, both processes, molecular diffusion and biorrigation, take place simultaneously and are effective in parallel action at the same location. A workable model concept to deal with biorrigation is therefore still missing.

- The advection of pore water has the same order of magnitude as the sedimentation rate and is, therefore, compared to molecular diffusion, not of great importance. However, advection as a model concept in itself may be quite easily included into computer models by the employment of the Darcy Equation 3.30. The advection of the sediment's solid phase results from the formal logic that the zero-point of the system of coordinates is always defined at the sediment surface, and that the sediment particles stream downward relative to the co-ordination system. This process is of great importance for balancing the diagenetical effects in the sediment.

This is contribution No 253 of the Special Research Program SFB 261 (*The South Atlantic in the Late Quaternary*) funded by the Deutsche Forschungsgemeinschaft (DFG).

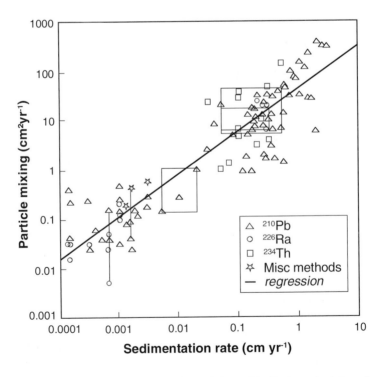

**Fig. 3.25** Biodiffusion coefficients (particle mixing) in relation to their empirically determined dependence on the rate of sedimentation (after Tromp et al. 1995).

# References

Aller, R.C. and DeMaster, D.J., 1984. Estimates of particle flux and reworking at the deep-sea floor using 234Th/238U disequilibrium. Earth and Planetary Science Letters, 67: 308-318.

Aller, R.C., 1988. Benthic fauna and biogeochemical processes in marine sediments: The role of burrow structures. In: Blackburn, T.H. and Sørensen, J. (eds), Nitrogen cycling coastal marine environments. SCOPE. Wiley & Sons Ltd., pp. 301-338.

Aller, R.C., 1990. Bioturbation and manganese cycling in hemipelagic sediments. Phil. Trans. R. Soc. Lond., 331: 51-68.

Aller, R.C., 1994. The sedimentary Mn cycle in Long Island Sound: Its role as intermediate oxidant and the influence of bioturbation, O2, and Corg flux on diagenetic reaction balances. Journal of Marine Research, 52: 259-293.

Archer, D., Emerson, S. and Smith, C.R., 1989. Direct measurements of the diffusive sublayer at the deep sea floor using oxygen microelectrodes. Nature, 340: 623-626.

Archer, D. and Devol, A., 1992. Benthic oxygen fluxes on the Washington shelf and slope: A comparison of in situ microelectrode and chamber flux measurements. Limnology and Oceanography, 37: 614 - 629.

Archie, G.E., 1942. The electrical resistivity log as an aid in determinig some reservoir characteristics. Trans. Am. Inst. Min. Metall., 146: 54 - 62.

Berner, R.A., 1980. Early diagenesis: A theoretical approach. Princton Univ. Press, Princton, NY, 241 pp.

Boudreau, B.P., 1997. Diagenetic models and their impletation: modelling transport and reactions in aquatic sediments. Springer, Berlin, Heidelberg, NY, 414 pp.

Cornwell, J.C. and Morse, J.W., 1987. The characterization of iron sulfide minerals in anoxic marine sediments. Marine Chemistry, 22: 193-206.

Davison, W., Grime, G.W., Morgan, J.A.W. and Clarke, K., 1991. Distribution of dissolved iron in sediment pore waters at submillimetre resolution. Nature, 352: 323-324.

Davison, W. and Zhang, H., 1994. In situ speciation measurements of trace components in natural waters using thin-film gels. Nature, 367: 546-548.

Davison, W., Zhang, H. and Grime, G.W., 1994. Performance characteristics of gel probes used for measuring the chemistry of pore waters. Environmental Science & Technology, 28: 1623-1632.

Davison, W., Fones, G.R. and Grime, G.W., 1997. Dissolved metals in surface sediment and a microbial mat at 100 µm resolution. Nature, 387: 885-888.

Davison, W., Fones, G., Harper, M., Teasdale, P. and Zhang, H., in press. Dialysis, DET and DGT: In situ diffusional techniques for studying water, sediments and soils. In: Buffle, J. and Horvai, G. (eds), In situ chemical measurements in aquatic systems. Wiley & Sons.

De Lange, G.J., 1988. Geochemical and early diagenetic aspects of interbedded pelagic/turbiditic sediments in two North Atlantic abyssal plains. Geologica Ultraiectina, Mededelingen van het Instituut vor Aardwetenschappen der Rijksuniversiteit te Utrecht, 57, 190 pp.

De Lange, G.J., Cranston, R.E., Hydes, D.H. and Boust, D., 1992. Extraction of pore water from marine sediments: A review of possible artifacts with pertinent examples from the North Atlantic. Marine Geology, 109: 53-76.

De Lange, G.J., 1992a. Shipboard routine and pressure-filtration system for pore-water extraction from suboxic sediments. Marine Geology, 109: 77-81.

De Lange, G.J., 1992b. Distribution of exchangeable, fixed, organic and total nitrogen in interbedded turbiditic/pelagic sediments of the Madeira Abyssal Plain, eastern North Atlantic. Marine Geology, 109: 95-114.

De Lange, G.J., 1992c. Distribution of various extracted phosphorus compounds in the interbedded turbiditic/pelagic sediments of the Madeira Abyssal Plain, eastern North Atlantic. Marine Geology, 109: 115-139.

Dicke, M., 1986. Vertikale Austauschkoeffizienten und Porenwasserfluß an der Sediment/Wasser Grenzfläche. Berichte aus dem Institut für Meereskunde an der Univ. Kiel, 155, pp 1-165.

Enneking, C., Hensen, C., Hinrichs, S., Niewöhner, C., Siemer, S. and Steinmetz, E., 1996. Poor water chemistry. In: Schulz, H.D. and cruise participants (eds), Report and preliminary results of Meteor cruise M34/2 Walvis Bay-Walvis Bay, 29.01.1996 - 18.02.1996. Berichte, Fachbereich Geowissenschaften, Univ. Bremen, 78, pp 87-102.

Forster, S., Huettel, M. and Ziebis, W., 1996. Impact of boundary layer flow velocity on oxygen utilisation in coastal sediments. Mar. Ecol. Prog. Ser., 143: 173-185.

Fossing, H. and Jørgensen, B.B., 1990. Oxidation and reduction of radiolabeled inorganic sulfur compounds in an estuarine sediment, Kysing Fjord, Denmark. Geochimica et Cosmochimica Acta, 54: 2731-2742.

Froelich, P.N., Klinkhammer, G.P., Bender, M.L., Luedtke, N.A., Heath, G.R., Cullen, D., Dauphin, P., Hammond, D. and Hartman, B., 1979. Early oxidation of organic matter in pelagic sediments of the eastern equatorial Atlantic: suboxic diagenesis. Geochimica et Cosmochimica Acta, 43: 1075-1090.

Glud, R.N., Gundersen, J.K., Jørgensen, B.B., Revsbech, N.P. and Schulz, H.D., 1994. Diffusive and total oxygen uptake of deep-sea sediments in the eastern South Atlantic Ocean: in situ and laboratory measurements. Deep-Sea Research, 41: 1767-1788.

Grasshoff, K., Kremling, K. and Ehrhardt, M., 1999. Methods of Seawater Analysis. Wiley-VCH, Weinheim, NY, 600 pp.

Gundersen, J.K. and Jørgensen, B.B., 1990. Microstructure of diffusive boundry layer and the oxygen uptake of the sea floor. Nature, 345: 604-607.

Haese, R.R., 1997. Beschreibung und Ouantifizierung frühdiagenetischer Reaktionen des Eisens in Sedimenten des Südatlantiks. Berichte, Fachbereich Geowissenschaften, Univ. Bremen, 99, 118 pp.

Hall, P.O.J. and Aller, R.C., 1992. Rapid, small-volume, flow injection analysis for SCO2 and NH4+ in marine and freshwaters. Limnology and Oceanography, 35: 1113-1119.

Hamer, K. and Seger, R., 1994. Anwendung des Modells CoTAM zur Simulation von Stofftransport und geochemischen Reaktionen. Ernst und Sohn, Berlin, 186 pp.

Hensen, C., Landenberger, H., Zabel, M., Gundersen, J.K., Glud, R.N. and Schulz, H.D., 1997. Simulation of early diagenetic processes in continental slope sediments in Southwest Africa: The computer model CoTAM tested. Marine Geology, 144: 191-210.

Holby, O. and Riess, W., 1996. In Situ Oxygen Dynamics and pH-Profiles. In: Schulz, H.D. and cruise participants (eds), Report and preliminary results of Meteor cruise M34/2 Walvis Bay-Walvis Bay, 29.01.1996 - 18.02.1996. Berichte, Fachbereich Geowissenschaften, Univ. Bremen, 78, pp. 85-87.

Huettel, M., Ziebis, W. and Forster, S., 1996. Flow-induced uptake of particulate matter in permeable sediments. Limnology and Oceanography, 41: 309-322.

Iversen, N. and Jørgensen, B.B., 1985. Anaerobic methane oxidation rates at the sulfate-methane transition in marine sediments from Kattegat and Skagerrak (Denmark). Limnology and Oceanography, 30: 944-955.

Iversen, N. and Jørgensen, B.B., 1993. Diffusion coefficients of sulfate and methane in marine sedimets: Influence of porosity. Geochimica et Cosmochimica Acta, 57: 571-578.

Jahnke, R.A., Heggie, D., Emerson, S. and Grundmanis, V., 1982. Pore waters of the central Pacific Ocean: nutrient results. Earth and Planetary Science Letters, 61: 233-256.

Jahnke, R.A., 1988. A simple, reliable, and inexpensive pore-water sampler. Limnology and Oceanography, 33: 483-487.

Jahnke, R.A. and Christiansen, M.B., 1989. A free-vehicle benthic chamber instrument for sea floor studies. Deep-Sea Research, 36: 625-637.

Jørgensen, B.B. and Revsbech, N.P., 1985. Diffusive boundary layers and the oxygen uptake of sediments and detritus. Limnology and Oceanography, 30: 111-122.

Jørgensen, B.B., Bang, M. and Blackburn, T.H., 1990. Anaerobic mineralization in marine sediments from the Baltic Sea-North Sea transition. Marine Ecology Progress Series, 59: 39-54.

Kinzelbach, W., 1986. Goundwater Modeling - An Introducion with Sample Programs in BASIC. Elsevier, Amsterdam, Oxford, NY, Tokyo, 333 pp.

Klimant, I., Meyer, V. and Kühl, M., 1995. Fiber-oxic oxygen microsensors, a new tool in aquatic biology. Limnology and Oceanography, 40: 1159-1165.

Kölling, M., 1986. Vergleich verschiedener Methoden zur Bestimmung des Redoxpotentials natürlicher Gewässer. Meyniana, 38: 1-19.

McDuff, R.E. and Ellis, R.A., 1979. Determining diffusion coefficients in marine sediments: A laboratory study of the validity of resistivity technique. American Journal of Science, 279: 66-675.

Niewöhner, C., Hensen, C., Kasten, S., Zabel, M. and Schulz, H.D., 1998. Deep sulfate reduction completely mediated by anaerobic methane oxidation in sediments of the upwelling area off Namibia. Geochimica et Cosmochimica Acta, 62: 455-464.

Redfield, A.C., 1958. The biological control of chemical factors in the environment. Am. Sci., 46: 206-226.

Reeburgh, W.S., 1967. An improved interstitial water sampler. Limnology and Oceanography, 12: 163-165.

Reimers, C.E., 1987. An in situ microprofiling instrument for measuring interfacial pore water gradients: methods and oxygen profiles from the North Pacific Ocean. Deep-Sea Research, 34: 2019-2035.

Revsbech, N.P., Jørgensen, B.B. and Blackburn, T.H., 1980. Oxygen in the sea bottom measured with a microelektrode. Science, 207: 1355-1356.

Revsbech, N.P. and Jørgensen, B.B., 1986. Microelectrodes: Their use in microbial ecology. Advances in Microbial Ecology, 9: 293-352.

Revsbech, N.P., 1989. An oxygen microsensor with a guard cathode. Limnology and Oceanography, 34: 474-478.

Saager, P.M., Sweerts, J.P. and Ellermeijer, H.J., 1990. A simple pore-water sampler for coarse, sandy sediments of low porosity. Limnology and Oceanography., 35: 747-751.

Sarazin, G., Michard, G. and Prevot, F., 1999. A rapid and accurate spectroscopic for alkalinity measurements in sea water samples. Wat. Res., 33: 290-294.

Sayles, F.L., Mangelsdorf, P.C., Wilson, T.R.S. and Hume, D.N., 1976. A sampler for the in situ collection of marine sedimentary pore waters. Deep-Sea Research, 23: 259-264.

Schlüter, M., 1990. Zur Frühdiagenese von organischem Kohlenstoff und Opal in Sedimenten des südlichen und östlichen Weddelmeeres. Berichte zur Polarforschung, Bremerhaven, 73, 156 pp.

Schultheiss, P.J. and McPhail, S.D., 1986. Direct indication of pore-water advection from pore pressure measurements in Madeira Abyssal Plain sediments. Nature, 320: 348-350.

Schulz, H.D., Dahmke, A., Schinzel, U., Wallmann, K. and Zabel, M., 1994. Early diagenetic processes, fluxes and reaction rates in sediments of the South Atlantic. Geochimica et Cosmochimica Acta, 58: 2041-2060.

Seeburger, I. and Käss, W., 1989. Grundwasser - Redoxpontentialmessung und Probennahmegeräte. DVWK-Schriften, Bonn, 84, 182 pp.

Smith, K.L.J. and Teal, J.M., 1973. Deep-sea benthic community respiration: An in-situ study at 1850 meters. Science, 179: 282-283.

Tengberg, A., De Bovee, F., Hall, P., Berelson, W., Chadwick, D., Ciceri, G., Crassous, P., Devol, A., Emerson, S., Gage, J., Glud, R., Graziottini, F., Gundersen, J., Hammond, D., Helder, W., Hinga, K., Holby, O., Jahnke, R., Khripounoff, A., Lieberman, S., Nuppenau, V., Pfannkuche, O., Reimers, C., Rowe, G., Sahami, A., Sayles, F., Schurter, M., Smallman, D., Wehrli, B. and De Wilde, P., 1995. Benthic chamber and profiling landers in oceanography - A review of design, technical solutions and function. Progress in Oceanography, 35: 253-292.

Tromp, T.K., van Cappellen, P. and Key, R.M., 1995. A global model for the early diagenetisis of organic carbon and organic phosphorus in marine sediments. Geochimica et Cosmochimica Acta, 59: 1259-1284.

Van Cappellen, P. and Wang, Y., 1996. Cycling of iron and manganese in surface sediments: a general theory for the coupled transport and reaction of carbon, oxygen, nitrogen, sulfur, iron, and manganese. American Journal of Science, 296: 197-243.

Ziebis, W. and Forster, S., 1996. Impact of biogenic sediment topography on oxygen fluxes in permeable seabeds. Mar. Ecol. Prog. Ser., 140: 227-237.

# 4 Organic Matter: The Driving Force for Early Diagenesis

Jürgen Rullkötter

## 4.1 The Organic Carbon Cycle

The organic carbon cycle on Earth is divided into two parts (Fig. 4.1; Tissot and Welte 1984). The *biological cycle* starts with photosynthesis of organic matter from atmospheric carbon dioxide or carbon dioxide/bicarbonate in the surface waters of oceans or lakes. It continues through the different trophic levels of the biosphere and ends with the metabolic or chemical oxidation of decayed biomass to carbon dioxide. The half-life is usually days to tens of years depending on the age of the organisms. The *geological organic carbon cycle* has a carbon reservoir several orders of magnitude larger than that of the biological organic carbon cycle ($6.4 \cdot 10^{15}$ t C compared with $3 \cdot 10^{12}$ t C in the biological cycle) and a turn-over time of millions of years. It begins with the incorporation

of biogenic organic matter into sediments or soils. It then leads to the formation of natural gas, petroleum and coal or metamorphic forms of carbon (e.g. graphite), which may be reoxidized to carbon dioxide after erosion of sedimentary rocks or by combustion of fossil fuels.

The tiny leak from the biological to the geological organic carbon cycle, particularly if seen from the point of view of a petroleum geochemist in the context of the formation of petroleum source rocks (see Littke et al. 1997a for a recent overview), is represented by the deposition and burial of organic matter into sediments. If looked at in detail, the transition from the biosphere to the geosphere is less well defined. The transformation of biogenic organic matter to fossil material starts immediately after the decay of living organisms. It may involve processes during transport, e.g. sinking through a water column, and al-

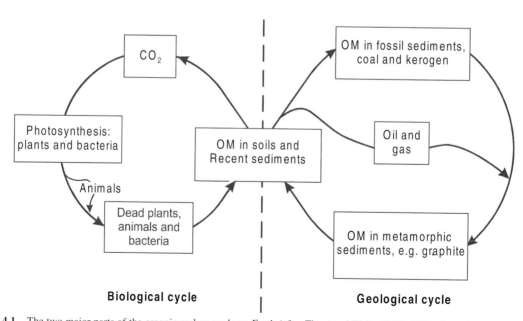

**Fig. 4.1** The two major parts of the organic carbon cycle on Earth (after Tissot and Welte 1984). OM = Organic matter.

teration at the sediment surface or in the upper sediment layers where epi- and endobenthic organisms thrive. Furthermore, it may extend deeply into the sedimentary column where bacteria were recently found to be still active at several hundreds of meters depth in layers deposited millions of years ago (Parkes et al. 1994).

## 4.2    Organic Matter Accumulation in Sediments

In the fossil record, dark-colored organic-matter-rich layers (*black shales*, *sapropels*, petroleum source rocks in general) witness periods of time when conditions for organic matter accumulation in sediments apparently were particularly favorable. As the other extreme, massive sequences of white- or red-colored sedimentary rocks are devoid of organic matter. Although these rocks contain abundant calcareous or siliceous plankton fossils, the organic matter of the organisms apparently was destroyed before it could be buried in the sediments.

Biogenic organic matter is considered labile (or metastable) under most sedimentary conditions due to its sensitivity to oxidative degradation, either chemically or biologically mediated. This is particularly true in well-oxygenated oceanic waters as they presently occur in the oceans almost worldwide. Thus, abundant accumulation of organic matter today is the exception rather than the rule. It is mainly restricted to the upwelling areas on the western continental margins and a few rather small oceanic deeps with anoxic bottom waters (such as the Cariaco Trench off Venezuela). In the geological past, more sluggish circulation in the deep ocean or in shallow epicontinental seas, accompanied by water column stratification, was probably one of the main causes leading to the deposition of massive organic-matter-rich rocks. A few examples are the Jurassic Posidonia Shales or Kimmeridge Clays in northwestern Europe, the Cretaceous black shales of the Atlantic Ocean and other oceanic areas of the world, and the Pliocene to Holocene sapropels of the Mediterranean Sea.

Stagnant oceanic bottom waters with a low concentration or absence of oxygen (*anoxia*) were considered for a long time as the main prerequisite for the accumulation of high amounts of organic matter in sediments (Demaison and Moore 1980). More recently, a controversy developed about two contrasting models to explain the deposition of organic-matter-rich sediments in the marine realm, either (1) by *preservation* under anoxic conditions in a static situation or (2) by high *primary productivity* in a dynamic system (Fig. 4.2; Pedersen and Calvert 1990, Demaison 1991, Pedersen and Calvert 1991). The relative importance of these two dominant controlling parameters is still being heavily debated, although Stein (1986) conceived that either one of these parameters could play a decisive role in different oceanographic situations. Another parameter brought into discussion more recently is the protective role of organic matter adsorption on mineral surfaces and its influence on organic matter accumulation in marine sediments (Keil et al. 1994a,b, Mayer 1994, Ransom et al. 1998).

### 4.2.1    Productivity *Versus* Preservation

Recognition of the sensitivity of organic matter towards oxidative destruction led to the idea that the concentration of free oxygen in the water column and particularly at the sediment/water interface is the most important factor determining the amount of organic matter that is incorporated into sediments (e.g. Demaison and Moore 1980). The stagnant basin or *Black Sea model* (Fig. 4.2A), developed from this idea, is based on the observation that lack of replenishment of oxygen by restricted circulation in the bottom part of larger water bodies can lead to longer-term oxygen-depleted (anoxic, suboxic; see Table 4.1 for definition) conditions in the water column and at the sediment/water interface. In the Black Sea (exceeding 2000m water depth in the center), this is caused by the development of a very stable halocline (preventing vertical mixing) at about 100m to 150m water depth. The surface layer is fed by relatively light riverine freshwater from the continent, and denser saline deep water is flowing in at a low rate from the Mediterranean Sea over the shallow Bosporus sill. Over time, oxidation of sinking remnants of decayed organisms consumed all of the free oxygen in the deeper water, which was not effectively replenished by Mediterranean water. Instead, the deep water in the Black Sea (also in Lake Tanganyika, an analogous contemporaneous example of a large stratified lake; Huc 1988) contains hydrogen sulfide restricting life to anaerobic microorganisms that are commonly thought to degrade organic matter less rapidly than aerobic bacteria, although there are also opposing views (see, e.g., discussion by Kristensen et al. 1995, Hulthe et al. 1998). Lack of intense organic matter degradation under anoxic conditions would then not necessarily require high surface water bioproductivity for high organic carbon concentrations to occur in the sediment.

The proponents of primary productivity as the decisive factor controlling organic matter accumulation (e.g. Calvert 1987, Pedersen and Calvert 1990, Bailey 1991) suggested that changes in primary productivity with time in different areas of the world, induced by climatic and related oceanographic changes, explain the distribution of Cretaceous black shales and more recent (Quaternary) organic-matter-rich sediments better than the occurrence of anoxic conditions in oceanic bottom waters. Reduced oxygen concentrations in

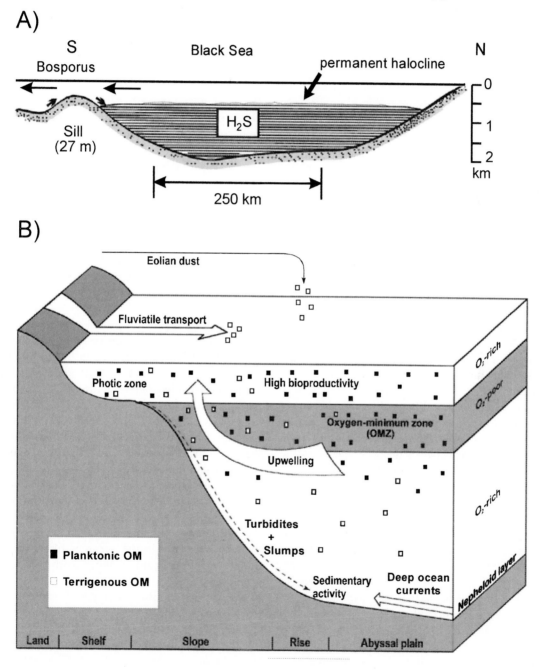

**Fig. 4.2**  Schematic models for organic matter accumulation in sediments. A) Stagnant basin or Black Sea model (after Demaison and Moore 1980 and Stein 1991); B) Productivity model (after Rullkötter et al. 1983). OM = Organic matter.

**Table 4.1** Terminology for regimes of low oxygen concentrations and the resulting biofacies according to Tyson and Pearson (1991)

| Oxygen (ml/l) | Environments | Biofacies | Physiological regime |
|---|---|---|---|
| 8.0-2-0 | Oxic | Aerobic | Normoxic |
| 2.0-0.2 | Dysoxic | Dysaerobic | Hypoxic |
| 2.0-1.0 | Moderate | | |
| 1.0-0.5 | Severe | | |
| 0.5-0.2 | Extreme | | |
| 0.2-0.0 | Suboxic | Quasi-anaerobic | |
| 0.0 (H$_2$S) | Anoxic | Anaerobic | Anoxic |

the water column, according to these authors, are a consequence of large amounts of decaying biomass settling toward the ocean bottom and consuming the dissolved oxygen.

The *productivity model* (Fig. 4.2B) is based on high primary bioproductivity in the photic zone of the ocean as it presently occurs in areas of *coastal upwelling* primarily on the western continental margins, along the equator and as a monsoon-driven phenomenon in the Arabian Sea. Upwelling brings high amounts of nutrients to the surface, which stimulates phytoplanktonic growth (e.g. Suess and Thiede 1983, Thiede and Suess 1983, Summerhayes et al. 1992). On continental margins, the formation of oxygen-depleted water masses (*oxygen minimum zones*) usually implies that they impinge on the ocean bottom where they create depositional conditions similar to those in a stagnant basin. To which extent the reduction in oxygen concentration at the sediment/water interface enhances, or is required for, the preservation of organic matter in the sediments, is the main subject of debate between the proponents of the productivity and the preservation models (e.g. Pedersen and Calvert 1990, 1991, Demaison 1991). For the geological past, e.g. the Cretaceous, it is also conceivable that the equable climate on our planet led to a more sluggish circulation of ocean water worldwide. The lack of turnover in the water column may have caused the development of anoxic bottom water conditions in some parts of the ocean that may explain the formation of Cretaceous black shales almost synchronously in different areas (Sinninghe Damsté and Köster 1998 and references therein). Also, transgression as a consequence of eustatic sea-level rise may have enhanced accumulation of organic matter in shelf areas (Wenger and Baker

1986) whereas times of regression may have promoted organic matter accumulation in prograding delta fans in deeper water. A detailed discussion of organic matter accumulation in different oceanic settings, including deep marine silled basins, progradational submarine fans, upwelling areas, anoxic continental shelves and fluviodeltaic systems was recently provided by Littke et al. (1997a).

### 4.2.2    Primary Production of Organic Matter and Export to the Ocean Bottom

The total annual *primary production* by photosynthetic planktonic organisms in the modern world oceans has been estimated to be in the order of $20\text{-}30 \cdot 10^9$ tons of carbon (Platt and Subba Rao 1975, Koblents-Mishke 1977, Berger et al. 1989). Oceanic carbon fixation is not evenly distributed, but displays zones of higher activity on continental margins (several hundred g $C_{org}$ m$^{-2}$yr$^{-1}$), whereas the central ocean gyres are mostly characterized by low primary production (about 25 g $C_{org}$ m$^{-2}$yr$^{-1}$); e.g. Romankevitch 1984, Berger 1989). Due to plate tectonics and the variable distribution of land masses together with climatic developments, the global amount and distribution of organic matter production in the ocean is likely to have undergone significant changes with geological time.

Of the total biomass newly formed in the photic zone of the ocean, only a very small portion reaches the underlying sea floor and is ultimately buried in the sediment (for reviews of water column processes see Emerson and Hedges 1988, Wakeham and Lee 1989). Most of the organic matter enters the biological food web in the surface waters and is respired or used for new heterotrophic biomass production. Because of this intense recycling, it is difficult to determine the organic matter flux at different levels of the photic zone. Oceanographers and biogeochemists usually consider only the flux of organic matter through the lower boundary of the photic zone and term it *'new' production* (Fig. 4.3). It equals 100% of export production (see below) and is not to be confused with net photosynthesis which is gross photosynthesis minus algal respiration. Below this boundary, the concentration of organic carbon in the water column decreases due to con-

sumption in the food web and to microbiological and chemical degradation as observed from the analysis of material in sediment traps deployed at different water depths.

The water depth-dependent flux is termed *export production* (Fig. 4.3). Export production decreases rapidly just below the photic zone. Then, there is mostly a quasi-linear, slower decline at greater water depths until the organic matter reaches the benthic boundary (nepheloid) layer close to the sediment/water interface where the activity of epibenthic organisms enhances organic matter consumption again. This enhanced con-

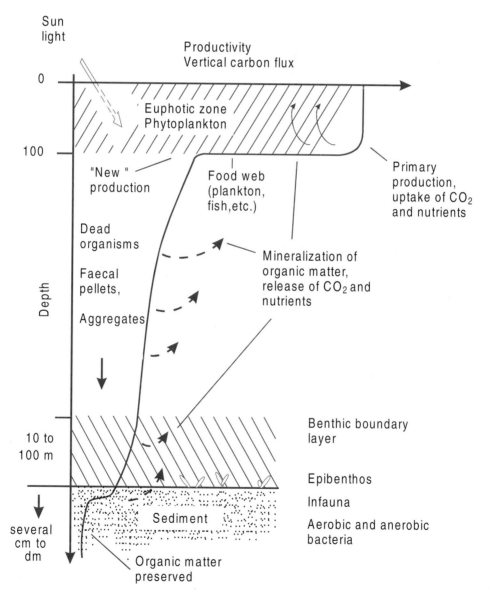

**Fig.4.3**    Schematic representation of organic matter flux to the ocean bottom.

sumption continues in the upper sediment layer where burrowing organisms depend on the supply from the water column. Organic matter degradation eventually extends deeply into the sediment pile as became evident from the detection of a so-called deep biosphere at several hundred meters below the sea floor (e.g. Parkes et al. 1994). However, the rate of organic matter degradation apparently decreases significantly with increasing depth of burial. Overall, it is estimated that only 1 to 0.01% of the primary production is buried deeply in marine sediments (cf. Fig. 12.1). The fraction strongly depends on a number of parameters including level of primary productivity, water depth, (probably) oxygen content in the water column and surface sediments (the latter affects benthic activity), particle size, adsorption to mineral surfaces and sediment porosity.

Although the fraction of organic matter buried in sediments is small relative to the amount produced by photosynthesis in oceanic surface waters, empirical relationships have been derived from the analysis of sedimentary organic matter to estimate oceanic *paleoproductivity* (PaP g C m$^{-2}$yr$^{-1}$) in the geological past:

$$PaP = C \cdot \rho (100 - \phi/100)/0.003 \cdot S^{0.3} \qquad (4.1)$$

In this equation of Müller and Suess (1979), the Pleistocene paleoproductivity of the (oxic) ocean is related to the organic carbon content of a sediment in percent dry weight (C), the density of the dry sediment ($\rho$; g cm$^{-3}$), its porosity ($\phi$; %) and the linear bulk sedimentation rate (S; cm of total sediment per 1000yr). The exponential factor was obtained from calibration with data from the present ocean. Because it was shown later that the organic carbon accumulation rate is a function of the carbon flux near the seafloor, which is related to both productivity and water depth, Stein (1986a, 1991) derived a more complex empirical relationship using flux data of Betzer et al. (1984):

$$PaP = 5.31[C(\rho_{WB} - 1.026\phi/100)]^{0.71} \cdot S^{0.07} \cdot D^{0.45}$$

$$(4.2)$$

with $\rho_{WB}$ being the wet bulk density of the sediment and D the water depth. Values from this equation are proportional, although numerically not identical, to the results obtained from a third empirical relationship developed by Sarnthein et al. (1987, 1988):

$$PaP = [15.9C \cdot S \cdot \rho(1-\phi)]^{0.66} \cdot S_{B-C}^{-0.71} \cdot D^{0.32} \quad (4.3)[1]$$

where $S_{B-C}$ is the organic-carbon-free linear bulk sedimentation rate. Given the complexity of the sedimentation and burial processes of organic matter, the wide range of chemical and physical properties of organic matter from different organisms and the effects of organic matter alteration during diagenesis after incorporation in the sediment, the equations can only be considered rough estimates. It can be expected that the investigation of more oceanic sediment profiles in the future will result in further modification of the equations. In particular, the extent of mixing of autochthonous marine organic matter with allochthonous organic matter from the continents (see Sect. 4.2.5) is difficult to estimate. Stein (1986a) used data from Rock-Eval pyrolysis, calibrated by organic petrographic data from microscopic analysis, to derive the marine organic matter fraction of a sediment from bulk pyrolysis measurements, but the correlation displays substantial scatter and can only be regarded as a crude approximation.

Müller and Suess (1979) applied their paleoproductivity relationship (Eq. 4.1) to sediments from the deep ocean off Northwest Africa. They found that the Pleistocene interglacial periods had about the same productivity as that measured in the present-day ocean. It was three times higher during glacial periods, probably due to a higher nutrient supply by more intense mixing of water masses or stronger coastal upwelling. The more sophisticated Equation 4.3 of Sarnthein et al. (1987, 1988) yielded essentially the same results for the last 500,000 years as those from Equation 4.1. Typical Pleistocene productivities in the upwelling area off Northwest Africa ranged between 150 and 300 g $C_{org}$ m$^{-2}$yr$^{-1}$, while the values were 20 to 50 g $C_{org}$ m$^{-2}$yr$^{-1}$ in the central Atlantic Ocean (Stein et al., 1989).

The paleoproductivity equations above describe the relationship between surface-water productivity and organic carbon accumulation, specifically under conditions of an oxic water column. A different relationship for anoxic depositional settings, suggested by Bralower and Thierstein (1987), implies that at least 2% of the organic

---

[1]    Numerical values in this equation were rounded because the author of this chapter believes that the number of decimals in the original publication suggests more accuracy than is both justified by, and required for, this empirical estimative approach.

carbon in the gross photosynthetic production is preserved in the sediments:

$$PaP = 5C \cdot S(\rho_{WB} - 1.026\phi/100) \qquad (4.4)$$

Stein (1986a) used this equation to calculate paleoproductivity in the Mesozoic Atlantic Ocean. Interpretation was considered preliminary due to the difficulty of obtaining precise age information, and thus sedimentation rate data, for the older and more compacted Mesozoic sediments lean in microfossils. The estimated productivity appeared to have been low off Northwest Africa in the Jurassic, to have increased during the Early Cretaceous and to have reached maximum values similar to those today during Aptian-Albian times (about 110 million years ago). Interestingly, low paleoproductivity was calculated for the time of deposition of black shales at the Cenomanian-Turonian boundary (90 million years ago) indicating that preservation may have played a more important role for organic matter accumulation than productivity.

The empirical relationships for paleo-productivity assessment illustrate how organic matter accumulation is related to primary productivity through factors such as organic carbon flux through the water column and bulk sedimentation rate. In addition, there is evidence that reduced oxygen concentrations in the water column enhance organic matter preservation. Thus, organic-carbon-rich sediments and sedimentary rocks are likely to be formed by the mutually enhancing effects of oxygen depletion (static or dynamic; anoxia), and productivity. In view of this, it appears to be too restrictive to assign a single controlling factor (Pederson and Calvert 1990). For example, an anoxic water column in the Holocene Black Sea is in itself apparently not sufficient to lead to black shale formation, whereas the enhanced primary productivity in equatorial upwelling areas is not reflected in a high organic carbon content of the underlying sediments due to oxidation in the deep oxic waters below the oxygen-minimum zone.

### 4.2.3 Transport of Organic Matter through the Water Column

The extent of degradation of particulate organic matter as it sinks through the water column is influenced by the residence time of organic matter particles in the water column. A measure of vertical transport is the sinking velocity (vs; m s$^{-1}$), which for a spherical particle follows Stokes' law:

$$vs = [(\rho_2 - \rho_1) \cdot g \cdot D^2]/18\eta \qquad (4.5)$$

$\rho_2$ and $\rho_1$ are the densities (g·cm$^{-3}$) of the particle and the water, respectively, g is the acceleration due to gravity (m s$^{-2}$), D is the particle diameter (cm) and $\eta$ is the dynamic viscosity (g cm$^{-1}$s$^{-1}$). Representative travel times of different idealized organic particles through a water column of 1000m cover a wide range (Table 4.2). They reflect the effects of different densities and diameters. Smaller, less dense particles clearly settle very slowly.

**Table 4.2** Sinking velocity and travel time for spherical particles in nonturbulent freshwater (after von Engelhardt 1973, Littke et al. 1997)

| Particle type | Sinking velocity (m·s$^{-1}$) | Travel time (1000 m water) |
|---|---|---|
| Large terrigenous organic matter particle $\varnothing = 100\ \mu m$, $\rho = 1.3\ g \cdot cm^{-3}$ | $1.6 \cdot 10^{-3}$ | 7.1 days |
| Small terrigenous organic matter particle $\varnothing = 10\ \mu m$, $\rho = 1.3\ g \cdot cm^{-3}$ | $1.6 \cdot 10^{-5}$ | 1.9 years |
| Fecal pellet $\varnothing = 1\ mm$, $\rho = 1.3\ g \cdot cm^{-3}$ | $1.6 \cdot 10^{-1}$ | 2 hours[a] |
| Quartz grain $\varnothing = 10\ \mu m$, $\rho = 2.6\ g \cdot cm^{-3}$ | $8.7 \cdot 10^{-5}$ | 133 days |

[a]Measured travel times for real faecal pellets: 4-20 days/1000 m (JK Volkman, pers. Com. 1998)

The vastly different travel times of the particles types range from hours to years. They would be only slightly higher in saline ocean water. The density values in the table imply some association of the organic matter with mineral matter, either biogenic calcareous or siliceous (plankton) frustules or detrital mineral matter like clay. Pure organic matter would have a lower density than water and not sink to the ocean bottom at all. Densities, e.g., of coal macerals usually vary between 1.1 and 1.7 $g\,cm^{-3}$ (van Krevelen 1961), and zooplankton fecal pellets often contain more mineral than organic matter. Degens and Ittekott (1987) strongly favored the transfer of organic matter by fecal pellets "which are jetted to the sea-floor at velocities of about 500 $cm\,day^{-1}$." Mineral-poor algal particles may have a very long residence time in the water column and a high chance of being metabolized before reaching the ocean floor. In deep oxic ocean water, the "fecal pellet express" may be an important mechanism of transporting marine organic matter to the sea-floor. Microscopic analysis often revealed that all of the labile marine organic matter in such sediments occurred as 'amorphous' degraded material in rounded bodies which were ascribed to fecal pellets (e.g. Rullkötter et al. 1987). On the other hand, Plough et al. (1997) measured rapid rates of mineralization of fecal pellets (relative to their sinking time towards the sea-floor). This is consistent, however, with the observation that only intact fecal pellets occur in deep-sea sediments (PK Mukhopadhyay, personal communication 1987). Obviously, lysis of fecal pellet walls leads to rapid mineralization of the entire organic content, and only those fecal pellets reach the sea-floor and are embedded in the sediment which escape this degradative process in the water column.

Other than noted in Table 4.2, real sinking velocities strongly depend on particle shape. For example, von Engelhardt (1973) showed that the sinking velocity of quartz grains of 10-100μm diameter is greater by a factor of about one hundred than that of muscovite plates of equal diameter. Most organic matter particles, apart from fecal pellets, are not spherical or well rounded and thus have a lower sinking velocity than indicated in Table 4.2. The typical shape of terrigenous organic particles in young open-marine sediments is irregularly cylindrical with the longest axis being about twice the length of the shortest axis (Littke et al. 1991a).

## 4.2.4    The Influence of Sedimentation Rate on Organic Matter Burial

Müller and Suess (1979) demonstrated the influence of sedimentation rate on organic carbon accumulation under oxic open-ocean conditions. They found that the organic carbon content of marine sediments increases by a factor of about two for every tenfold increase in sedimentation rate. The underlying mechanism was believed to be the more rapid removal and protection of organic matter from oxic respiration and benthic digestion at the sediment/water interface by increasingly rapid burial (cf. Sect. 12.3.3). Also conceivable, however, is that the relationship between sedimentation rate and organic carbon content is based on the protective effect of organic matter adsorption on mineral (particularly clay) surfaces, so that organic matter preservation increases with the increase of mineral surface available for adsorption (Keil et al. 1994a,b, Mayer 1994, Collins et al. 1995, Ransom et al. 1998).

There is general agreement on the positive relationship between sedimentation rate and organic carbon content (e.g. Heath et al. 1977, Ibach 1982, Stein 1986a,b, Bralower and Thierstein 1987, Littke et al. 1991b). However, in cases where biostratigraphy provides accurate time control, it has been noted by Tyson (1987) that deposition of marine sediments with very high organic matter contents (petroleum source rocks) often appears to be associated with low rather than high sedimentation rates. Very high sedimentation rates at some point may lead to low organic matter concentrations in sediments due to dilution even if much of the sinking organic matter is preserved (Note difference between linear sedimentation rate [$cm\,k^{-1}$] and sediment accumulation rate [$g\,cm^{-2}yr^{-1}$], i.e. in a highly diluted sediment with a moderate to low organic carbon content deposited at a high linear bulk sedimentation rate, organic matter accumulation (or preservation) with time may still be high).

According to Stein (1986b, 1990), the effects of oxic and anoxic conditions on marine organic matter preservation in oceanic sediments can be illustrated by a simple diagram of *sedimentation rate versus organic carbon content* (Fig. 4.4). Field A inside the diagonal lines represents the sedimentation rate-controlled accumulation of organic matter under open-marine oxic conditions. The hatched area B indicates anoxic or strongly oxygen-depleted conditions over a wide range of

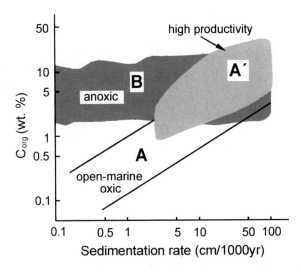

**Fig. 4.4** Correlation between marine organic carbon content and sedimentation rate (after Stein 1986b, 1990). The distinction between fields A, A' and B is based on data derived from Recent to Miocene sediments deposited in normal open-ocean environments (field A), upwelling high-productivity areas (field A') and anoxic environments (field B).

sedimentation rates with low rates being typical for stagnant basins like the Black Sea. The shaded area A', where areas A and B overlap at high sedimentation rates and high organic carbon contents, is typical of upwelling areas with high primary productivity on continental margins where the oxygen-minimum zone impinges on the shelf and upper slope. Strong dilution with mineral matter would place a sediment to the right of area A. Interestingly, the highly organic-matter-rich Atlantic Ocean black shales from a so-called "world-wide anoxic event" at the Cenomanian-Turonian boundary (about 90 million years ago; see Herbin et al. 1986) all fall in the left part of area B, i.e. they appear to have been deposited at low sedimentation rates under anoxic conditions (Stein 1986b).

### 4.2.5 Allochthonous Organic Matter in Marine Sediments

As schematically indicated in Figure 4.2B, marine sediments do not accumulate only organic matter from the (mainly planktonic) productivity in the overlying water column (*autochthonous* organic matter). *Allochthonous* organic matter originates from two other sources. One of them involves *redeposition* of marine sediments after erosion, often from a nearby location. Typical examples

are contour currents along continental margins or downslope transport events on (steep) continental margins. In these cases, sediment initially deposited at shallow(er) water depth is eroded by currents, mechanical instability (oversteepening), earthquakes or other tectonic movements. The eroded sediment is transported down the continental slope and redeposited at a deeper location. This may occur as a *turbidity current* by which sediment is suspended in the near-bottom water column and then settles again. This process often involves particle size fractionation, i.e. at the new site the larger particles are deposited first and become overlain by a sequence of progressively finer particles (Bouma series). Alternatively, a massive package of sediment material (*slump*) of variable size, from very small to cubic kilometers, may be redeposited as a whole, usually in a deep slope or continental rise setting. The effect on the organic matter is different in these two cases. Turbidity currents may expose the (fossil) organic matter to an oxygen-rich water mass causing further degradation during resettling. Compact slump masses may transport significant amounts of labile organic matter from an initial depositional setting, favorable for organic matter preservation (e.g. in an oxygen-minimum zone), to an oxygen-rich deep-water environment. There is no enhanced mineralization in this case due to the undisturbed embedding of organic matter in the sediment matrix (cf., e.g., Cornford et al. 1979, Rullkötter et al. 1982).

Redeposition may occur almost syn-sedimentarily, i.e. the redeposited allochthonous sediment will differ only very little in age from the underlying autochthonous sediment. The organic matter content of both may reflect more or less contemporaneous primary productivity with the only exception that the remains of shallow(er)-water species are relocated to a deep-water environment. This may be more evident in the mineral fossils than in the organic matter assemblage, however. Alternatively, redeposition may occur a long time after initial sedimentation took place. Deep-sea drilling on the Northwest African continental margin, for example, has recovered extended Miocene sediment series which evidently contained slumps a few centimeters to several meters thick and of various age, from Tertiary to Middle Cretaceous (von Rad et al. 1979, Hinz et al. 1984). Investigation of the organic matter on a molecular level clearly demonstrated the correspondence between slump clasts embedded in the

137

Lower Miocene sequence and underlying autochthonous Cenomanian series, while paleontological analysis showed a difference between outer shelf/upper slope species in the slumps and pelagic species in the autochthonous Cenomanian sediment (Rullkötter et al. 1984).

The second main source of allochthonous organic matter in marine sediments are the continents. Wind, rivers and glaciers transport large amounts of *land-derived organic matter* into the ocean (e.g. Romankevitch 1984). Again, two principal types of organic matter have to be distinguished: (1) fresh or (in geological terms) recently biosynthesized land plant material and (2) organic matter contained in older sediments that were weathered and eroded on the continent in areas ranging from mountains to coastal swamps. Organic matter in older sedimentary rocks that are being eroded may have had an extended history of geothermal heating at great burial depth into the stages of oil or bituminous coal formation and beyond. This organic matter carries a distinct signal of geochemical maturation that can easily be detected by bulk (e.g. atomic composition, pyrolysis yields, vitrinite reflectance; see Tissot and Welte 1984) and molecular geochemical parameters (e.g. compound ratios of specific geochemical fossils or biomarkers; see Peters and Moldowan 1983). Due to its advanced level of diagenetic alteration, even peat can be distinguished from fresh organic matter in marine sediments. Only when sediments have been buried to a depth corresponding to the temperature which the eroded organic matter earlier experienced do both fresh and prematured organic matter continue diagenesis or maturation at the same rate. Geochemical reactions are virtually stopped (i.e. reaction rates become very low) as soon as sediments (and their organic matter contents) in the course of tectonic uplift are cooled to a temperature of about 15°C lower than their previous maximum temperature. During transport to the ocean, oxidation of organic matter eroded on land - and of terrestrial organic matter in general - has an effect similar to maturation. The product is a highly refractory, inert material, which in organic petrography is termed inertinite and which is easily recognizable under the microscope by its high reflectance (see Stach et al. 1982 for details). Nevertheless, a substantial fraction of the terrigenous organic matter is reactive and metabolizable in the ocean (Hedges et al. 1997).

*Wind-driven dust and aerosols* carry terrigenous organic matter over long distances into the oceans and are estimated to contribute a total annual amount of $10^{14}$ t carbon each year (Romankevitch 1984). Entire organoclasts, like pollen and spores, are blown offshore as are lipids from the waxy coatings of plant cuticles adsorbed to mineral grains (e.g. Gagosian et al. 1987, Prahl 1992). This wind-blown terrestrial material may comprise the bulk of the organic matter in open-ocean sediments where very little of the primary marine organic matter reaches the deep ocean floor. The higher resistance of terrestrial organic matter toward oxidation has been invoked to explain this selective enrichment. Summerhayes (1981) estimated that most organic-matter-rich sediments in the Atlantic Ocean, including the Cretaceous and Jurassic black shales, contain a 'background level' of 1% terrestrial organic matter in total dry sediment.

Not all of this terrigenous material is brought into the ocean by winds. The most important contributors are the rivers draining into the ocean. Each year rivers transport approximately $0.4 \cdot 10^{15}$ t of organic carbon from continents to oceans (Schlesinger and Melack 1981), approximately four times as much as is transported by winds. About 60% of the *river run-off* derives from forested catchments, and the ratio of the contribution of tropical to temperate and boreal forests is about two to one (Schlesinger and Melack 1981). Much of the organic matter discharged by rivers appears to be soil-derived (Meybeck 1982, Hedges et al. 1986a,b) and highly degraded (Ittekott 1988, Hedges et al. 1994).

## 4.3   Early Diagenesis

### 4.3.1   The Organic Carbon Content of Marine Sediments

The concentration of organic carbon in marine surface sediments from different environments of the present-day oceans varies over several orders of magnitude depending on the extent of supply of organic matter, preservation conditions and dilution by mineral matter. The results of *organic carbon measurements* are usually expressed as TOC (total organic carbon) or $C_{org}$ values in percent of dry sediment. Romankevich (1984) compiled data from a great number of analyses of ocean bottom sediments by various authors and from all over the world. The TOC values range from 0.01% to more than 10% $C_{org}$ in a few cases. A statistical evaluation of data from the early phase of deep-sea drilling (Deep Sea Drilling Project Legs 1-31) showed deep-sea sediments to have a mean organic carbon content of 0.3%, with a median value of 0.1% (McIver 1975). However, the range of samples was certainly not representative, and the organic carbon contents are probably biased towards higher values. For example, in the vast abyssal plains and other deep-water regions far away from the continents, an organic carbon content as high as 0.05% is the exception rather than the rule.

In contrast to this, nearshore sediments on continental shelves and slopes usually have considerably higher organic carbon contents. Typical hemipelagic sediments on outer shelves and continental slopes range between 0.3 and 1% $C_{org}$. Sediments within the oxygen-minimum zone of upwelling areas contain several percent of organic carbon with exceptional values exceeding 10% $C_{org}$ where upwelling is very intense, e.g. off Peru and southwest Africa, and where the oxygen-minimum zone extends into the shallow waters of the shelf. Still, generalizations are difficult to make because sedimentation conditions are highly variable in space.

Generalizations are also difficult to make with respect to variation of organic carbon content with time. For long periods of the geological record the present-day conditions of organic carbon burial can be projected to the past. There were times, however, mostly relatively short and often termed an event, when high amounts of organic carbon were preserved, not only in shallow epicontinental seas, but also in deep-sea sediments. Examples are the Jurassic and particularly Cretaceous black shales of the Atlantic and Pacific Oceans with extreme organic carbon contents exceeding 20% (e.g. Herbin et al. 1986). Specific situations existed during the younger geological past in semi-enclosed basins like the Mediterranean Sea. Plio-Pleistocene sapropels in the eastern Mediterranean Sea were deposited at regular time intervals due to climatic changes induced by orbital forces, in this case the 23,000 year cycle of precession of the Earth's axis. Some of the Mediterranean sapropels, recovered during Leg 160 of the international Ocean Drilling Program, yielded more than 30% $C_{org}$ (Emeis et al. 1996).

The range of organic carbon contents in sediments and the associated variation in conditions for organic matter preservation imply that the amount of biogenic information incorporated in sediments as organic matter may vary drastically. In the same way, the extent to which the preserved organic matter is representative of the ecosystem in the water column above, may be vastly different. It is not surprising then that organic geochemists have preferentially investigated sediments with high organic carbon contents particularly when emphasis was on the formation of fossils fuels (petroleum or natural gas) or on molecular organic geochemical analysis which - at least in its early days - required relatively large amounts of material. It has to be kept in mind that this bias has certainly also influenced the choice of examples given throughout this chapter, although attempts are made to contrast case studies representing different environmental conditions in the oceanic realm.

Within a sediment, the organic carbon content decreases with increasing depth due to mostly microbiological remineralization, but possibly also due to abiological oxidation, during early (and later) diagenesis. The entire process takes place in a complex redox system where organic matter is the electron donor and a variety of substrates are electron acceptors. In other words, whenever organic matter is destroyed or altered by oxidation, a reaction partner has to be reduced. In an extended investigation of the biogeochemical cycling in an organic-matter-rich coastal marine basin, Martens and Klump (1984) schematically illustrated three independent approaches to quantify organic matter degradation in sediments with anoxic surface layers (Fig. 4.5). These involve (a) a mass balance of incoming, recycled and buried carbon fluxes, (b) kinetic modeling of the concentration/depth distribution of organic carbon and (c) measurement of degradation rates in the sediment column. The redox zones in the example given in Figure 4.5 are restricted to a depositional environment with anoxic surface sediment and comprise only sulfate reduction and methanogenesis. In the case of oxic conditions in the upper sediment layer, there would be additional oxygen, nitrate, Mn(IV) and Fe(III) reduction zones (Froelich et al. 1979; cf. also Fig. 4.6, where these zones are indicated, and Chap. 5).

Martens et al. (1992) applied the approaches in Figure 4.5 to study the composition and fate of organic matter in coastal sediments of Cape

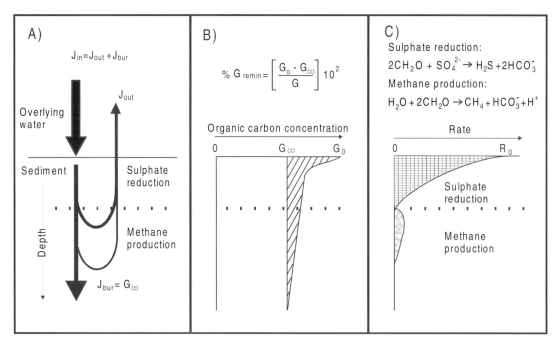

**Fig. 4.5**   Three independent approaches used to quantify organic matter degradation in anoxic coastal sediments: (A) carbon budget based on measurement of recycled ($J_{out}$; $CO_2$ and methane) and calculation of buried ($J_{bur}$) carbon fluxes; (B) kinetic modeling of the concentration/depth distribution of organic carbon (G; $G_0$ and $G_\infty$ are organic carbon concentrations at the top and the bottom of the studied depth interval, respectively); (C) calculated organic carbon remineralization based on modeled or measured rates of sulfate reduction and methanogenesis ($CH_2O$ stands for organic matter) (after Martens and Klump 1984).

Lookout Bight[2]. In separate studies it had previously been established that organic matter was mostly supplied from backbarrier island lagoons and marshes landward of the bight at a steady rate. Furthermore, the organic matter was extensively physically and biologically recycled in the lagoon before it ultimately accumulated in the sediments. Thus, systematic downcore decreases in amount of labile organic matter had to result from early diagenesis rather than variations of supply. The authors tried to answer the question of what fraction of the incoming particulate organic carbon (POC) is remineralized during early diagenesis under the conditions described by solving the simple mass balance equation

POC input = POC remineralized + POC buried

(4.6)

In their experience it has proven easiest to measure fluxes resulting from POC remineralization and burial and then to calculate POC input by adding these fluxes together. Numerical values of the fluxes are given in Figure 4.6. In this model, the incoming POC is either remineralized to $CO_2$, $CH_4$ and DOC (dissolved organic carbon) or buried. The $CO_2$, $CH_4$ and DOC produced during remineralization are either lost to the water column via sediment-water chemical exchange or buried as carbonate and dissolved components of sediment pore waters. Using $^{210}Pb$-based sedimentation rates, the POC burial rate was found to be $117\pm19$ mol C m$^{-2}$ yr$^{-1}$. Sediment-water chemical exchange accounts for losses of $40.6\pm6.6$ mol C m$^{-2}$yr$^{-1}$ as $CO_2$, $CH_4$ and DOC, whereas $7.0\pm1.1$ mol C m$^{-2}$yr$^{-1}$ of these species, including carbonate formed from $CO_2$, are buried (Fig. 4.6). The dissolved sediment-water exchange and burial fluxes sum to a total POC remineralization rate of $47.6\pm5.7$ mol C m$^{-2}$yr$^{-1}$. When this value is added to the POC burial rate, a total POC input of $165\pm20$ mol C m$^{-2}$yr$^{-1}$ can be calculated from Equation 4.6. From this result it follows that $29\pm5\%$ of the incoming POC is remineralized as an average over the first about

# Carbon Cycle in Cape Lookout Bight

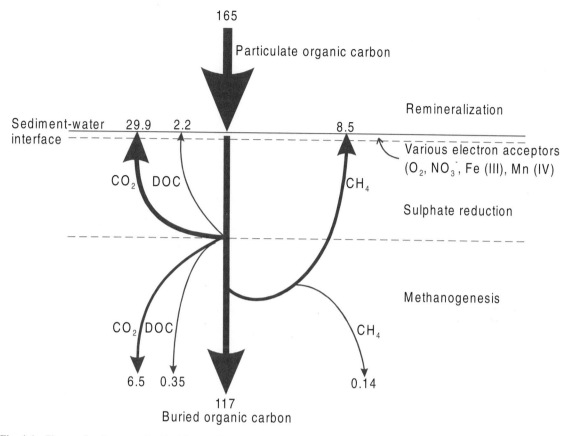

**Fig. 4.6**  Fluxes of carbon associated with organic matter supply, degradation and burial in Cape Lookout Bight sediments. The unit of all numerical flux values is moles C m$^{-2}$yr$^{-1}$ (after Martens et al. 1992).

ten years after sedimentation in Cape Lookout Bight.

Using a similar approach, Alperin et al. (1992) determined the POC remineralization rate for sediments of Skan Bay, Alaska, a pristine embayment with oxygen-depleted bottom water (<0.4 ml O$_2$/l water) and sulfidic surface sediments and with a shallow sill limiting advection of oxygen-rich water from the Bering Sea. Total sediment remineralization rate was calculated by three independent approaches: (1) the difference between POC deposition and preservation; (2) the quantity of carbon recycled to the water column and buried at depth; (3) depth-integrated rates of bacterial metabolism. The budget indicates that 84±3% of the organic carbon deposited is remineralized in the upper meter of the sediment column representing approximately 100 years. A steady state is nearly reached, however, at a depth of about 70 cm, i.e. remineralization is already very slow after ap-

proximately 70 years of burial in Skan Bay. Of the initial concentration of more than 9% organic carbon at the sediment surface, the content dropped to less than 2% of dry sediment at 0.7 to 1 m depth and most of that will survive deeper burial.

A third case study of organic carbon recycling and preservation in coastal environments, including a comprehensive budget of inorganic reactants and products, is from the Aarhus Bay (Denmark), a shallow embayment in the Kattegat that connects the North Sea with the Baltic Sea (Fig. 4.7; Jørgensen 1996). The bulk annual sedimentation rate in Aarhus Bay is about 2 mm yr$^{-1}$. Photosynthesis is in the upper mesotrophic range and annually produces organic matter corresponding to 21.8 mol C m$^{-2}$yr$^{-1}$. Planktonic oxygen respiration corresponds to mineralization of 68% of the primary productivity and 32% sedimentation, while direct sediment trap measurements accounted for 45% deposition. Of these 9.9 mol

C m$^{-2}$yr$^{-1}$, about 2.2 mol C m$^{-2}$yr$^{-1}$ are buried below the bioturbated zone. Metabolization in the sediment mainly occurs by oxygen and sulfate as electron acceptors, while nitrate, Mn(IV) and Fe(III) play a subordinate role. Methanogenesis was not included in the study of the carbon budget because only the water column and the shallow surface sediment were studied.

The three case studies show that organic carbon preservation and, thus, organic carbon contents strongly depend on the specific local environmental conditions. The extent of remineralization in these three cases ranges from about 30 to 85% in the upper sediment layers comprising, however, different time ranges. They correspond to the range quoted for continental shelf and estuarine sediments (20 to 90%) by Henrichs and Reeburgh (1987). Apparently, organic carbon flux, bulk sedimentation rate, water depth, oxygen concentration in the bottom water and related extent of bioturbation of surface sediments all have an influence on the intensity of organic mat-

ter remineralization during early diagenesis and on how much organic matter is buried to a sediment depth where further remineralization only proceeds very slowly. Only that organic matter fraction can be considered to become fossilized in a strict sense and to enter the geological organic carbon cycle.

### 4.3.2    Chemical Composition of Biomass

Apart from considering the fate of bulk organic matter (or organic carbon) during diagenesis, organic geochemistry has developed a more sophisticated understanding of diagenetic organic matter transformation down to the molecular level. Fundamental to this understanding is a comparison of the organic constituents of geological samples with the inventory of extant organisms. This was, and still partly is, hampered by the limited knowledge of the natural product chemistry particularly of unicellular marine algae, protozoans and bacteria.

The simplest way of describing the chemical nature of biomass is by its *elemental composition*.

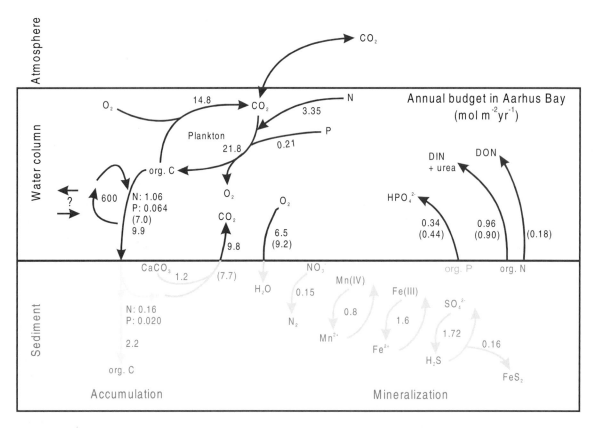

**Fig. 4.7**   Summary of fluxes and process rates measured in Aarhus Bay between May 1990 and May 1991. Numbers in parentheses were derived by difference while the others are based on independent rate measurements and calculations. Unit are given in mol m$^{-2}$yr$^{-1}$ for each component. DIN = dissolved inorganic nitrogen; DON = dissolved organic nitrogen (after Jørgensen 1996).

For marine phytoplankton as primary producers a relationship was found to the nutrients available in seawater which led to the definition of the Redfield ratio as C:N:P = 106:16:1 (Redfield et al. 1963). Derived from this is an average molecular formula of phytoplankton organic matter related to the general process of phytosynthesis (of which the reverse signifies remineralization):

$$106\ CO_2 + 106\ H_2O + 16\ NH_3 + H_3PO_4 \rightarrow$$
$$(CH_2O)_{106}(NH_3)_{16}H_3PO_4 + 106\ O_2$$

$$(4.7)$$

The formula of the organic matter product is often reduced to the summary version of $C_{106}H_{263}N_{16}O_{110}P$. It has no real chemical meaning in terms of molecular structure because it contains more hydrogen than the bonds of all the other atoms can account for. The reason is that the generalized average formula (i.e. the product in Eq. 4.7) is just the sum of separate neutral molecules which are involved in biosynthesis of organic matter. The formation of a molecular structure requires the formal loss of a number of molecules of water for condensation. While the formula does not represent the correct elemental organic matter composition of marine phytoplankton, at least not for hydrogen and oxygen, it has to be kept in mind that it is a crude generalisation in itself. It would certainly vary with nutrient conditions and planktonic species as has been observed, e.g., by Takahasi et al. (1985) in a study of plankton biomass from the Atlantic and Indian Oceans which resulted in a modified

Redfield ratio of 122(±18):16:1. There are quite a number of more recent studies that confirm this kind of deviation from the Redfield ratio (cf. Chap. 6).

Food chain and early diagenetic processes change the initial elemental composition drastically. Organic matter in sediments relative to the primary producers is enriched particularly in carbon and hydrogen, whereas it is depleted in oxygen (but the degree depends on the extent of oxidation of sedimentary organic matter), nitrogen and phosphorus. Depletion in phosphorus is due to the facile hydrolytic cleavage of bound phosphate groups. Loss of nitrogen occurs by preferential degradation of organic nitrogen compounds as discussed later (see Sect. 4.4 for a discussion of C/N ratios). Sulfur, not originally included in the general formula, would be equal to or less in content than phosphorus. The enrichment of sulfur in fossil organic matter is, however, not due to a relative enrichment in the course of preferential loss of other elements (as is the case for C and H). Sulfur enrichment rather is a consequence of diagenetic incorporation of reduced inorganic sulfur species (like $HS^-$ or corresponding polysulfides) which are formed from seawater sulfate by sulfate-reducing microorganisms in shallow sediments under anoxic conditions (see Chap. 8).

On the next higher level, the chemical composition of living organisms in the biosphere, despite their diversity, can be confined to a limited number of principal *compound classes*. Their proportions vary in the different groups of organisms as is evident from the estimates of Romankevitch

**Table 4.3** Biochemical composition of marine organisms (after Romankevitch 1984)

| Organism | Proteins (%) | Carbohydrates (%) | Lipids (%) | Ash (%) |
|---|---|---|---|---|
| Phytoplankton | 30 | 20 | 5 | 45 |
| Phytobenthos | 15 | 60 | 0.5 | 25 |
| Zooplankton | 60 | 15 | 15 | 10 |
| Zoobenthos | 27 | 8 | 3 | 62 |

**Table 4.4** The main chemical constituents of marine plankton in percent of dry weight (after Krey 1970)

| Organism | Proteins (%) | Carbohydrates (%) | Lipids (%) | Ash (%) |
|---|---|---|---|---|
| Diatoms | 24-48 | 0-31 | 2-10 | 30-59 |
| Dinoflagellates | 41-48 | 6-36 | 2-6 | 12-77 |
| Copepods | 71-77 | 0-4 | 5-19 | 4-6 |

(1984) for a few types of marine organisms (Table 4.3). Also, within the groups the compound class composition is highly variable (Table 4.4). It may even depend on the growth stage for a single species. The compound classes in turn comprise a very large number of single compounds with different individual chemical structures, although enzymatic systems limit the potential chemical diversity (that is why there are chemical biomarkers of taxonomic significance). Many of the compound classes are also represented in fossil organic matter, although not in the same proportions as they occur in the biosphere because of their different stabilities toward degradation and modification of original structures during sedimentation and diagenesis.

### Nucleic Acids and Proteins

Nucleic acids, as ribonucleic acids (RNA) or desoxyribonucleic acids (DNA), are biological macromolecules carrying genetic information. They consist of a regular sequence of phosphate, sugar (pentose) and a small variety of base units, i.e. nitrogen-bearing heterocyclic compounds of the purine or pyrimidine type. During biosynthesis, the genetic information is transcribed into sequences of amino acids, which occur as peptides, proteins or enzymes in the living cell. These macromolecules vary widely in the number of amino acids and thus in molecular weight. They account for most of the nitrogen-bearing compounds in the cell and serve in such different functions as catalysis of biochemical reactions and formation of skeletal structures (e.g. shells, fibers, muscles).

During sedimentation of decayed organisms, nucleic acids and proteins are readily hydrolyzed chemically or enzymatically into smaller, water-soluble units. Amino acids occur in rapidly decreasing concentrations in recent and subrecent sediments, but may also survive in small concentrations in older sediments, particularly if they are protected, e.g., by the calcareous frustules or shells of marine organisms. Nitrogen-bearing aromatic organic compounds in sediments and crude oils may relate to the purine and pyrimidine bases in nucleic acids, but this awaits unequivocal confirmation. A certain fraction of the nucleic acids and proteins reaching the sediment surface may be bound into the macromolecular organic matter network (humic substances, kerogen) of the sediments and there become protected against further rapid hydrolysis. Experiments in the laboratory have shown that kerogen-like material (melanoidins) can be obtained by heating amino acids with sugar.

### Saccharides, Lignin, Cutin, Suberin

Sugars are polyhydroxylated hydrocarbons that together with their polymeric forms (oligosaccharides, polysaccharides) constitute an abundant proportion of the biological material, particularly in the plant kingdom. Polysaccharides occur as supporting units in skeletal tissues (cellulose, pectin, chitin) or serve as an energy depot, for example, in seeds (starch). Although polysaccharides are largely insoluble in water, they are easily converted to soluble $C_5$ (pentoses) and $C_6$ sugars (hexoses) by hydrolysis and, thus, in the sedimentary environment will have a short-term fate similar to that of the proteins.

Lignin is a structural component of plant tissues where it occurs as a three-dimensional network together with cellulose. Lignin is a macromolecular condensation product of three different propenyl ($C_3$-substituted) phenols (one type of few biogenic aromatic compounds). It is preserved, even during transport from land to ocean and during sedimentation to the sea-floor where it occurs predominantly in humic organic matter of deltaic environments.

Cutin and suberin are lipid biopolymers of variable composition which are part of the protective outer coatings of all higher plants. Chemically, cutin and suberin are closely related polyesters composed of long-chain fatty and hydroxy fatty acid monomers. Both types of biopolymers represent labile, easily metabolizable terrigenous organic matter because they are sensitive to hydrolysis. After sedimentation, they have only a moderate preservation potential.

### Insoluble, Nonhydrolyzable Highly Aliphatic Biopolymers

Insoluble, nonhydrolyzable aliphatic biopolymers were discovered in algae and higher plant cell walls as well as in their fossil remnants in sediments (see de Leeuw and Largeau 1993, Largeau and de Leeuw 1995, for overviews). These substances are called algaenan, cutan or suberan according to their origin or co-occurrence with cutin and suberin in extant organisms. They consist of aliphatic polyester chains cross-linked with ether bridges (Blokker et al. 1998) which render them very stable toward degradation. Pyrolysis and other rigorous methods are needed to decompose these highly aliphatic biopolymers. This explains why they are preferentially preserved in sediments.

*Monomeric Lipids*

Biologically produced compounds that are insoluble in water but soluble in organic solvents such as chloroform, ether or acetone are called *lipids*. In a wider sense, these also include membrane components and certain pigments. They are common in naturally occurring fats, waxes, resins and essential oils. The low water solubility of the lipids derives from their hydrocarbon-like structures which are responsible for their higher survival rates during sedimentation compared to other biogenic compound classes like amino acids or sugars.

Various saturated and unsaturated *fatty acids* are the lipid components bound to glycerol in the triglyceride esters of fats (see Fig. 4.8 for examples of chemical structures of lipid molecules). Cell membranes consist to a large extent of fatty acid diglycerides with the third hydroxyl group of glycerol bound to phosphate or another hydrophilic group. In waxes, fatty acids are esterified with long-chain alcohols instead of glycerol. Plant waxes contain unbranched, long-chain saturated hydrocarbons (*n*-alkanes) with a predominance of odd carbon numbers (e.g. $C_{27}$, $C_{29}$, $C_{31}$) in contrast to the acids and alcohols which show an even-carbon-number predominance.

Isoprene (2-methylbuta-1,3-diene), a branched diunsaturated $C_5$ hydrocarbon, is the building block of a large family of open-chain and cyclic *isoprenoids* and *terpenoids* (Fig. 4.8). Essential oils of higher plants are enriched in monoterpenes ($C_{10}$) with two isoprene units. Farnesol, an unsaturated $C_{15}$ alcohol, is an example of a sesquiterpene with three isoprene units. The acyclic diterpene phytol is probably the most abundant isoprenoid on Earth. It occurs esterified to chlorophyll *a* and some bacteriochlorophylls and is, thus, widely distributed in the green pigments of aquatic and subaerial plants. Sesterterpenes ($C_{25}$) are of relatively minor importance except in some methanogenic bacteria (cf. Volkman and Maxwell 1986).

Cyclization of squalene (or its epoxide) is the biochemical pathway to the formation of a variety of *pentacyclic triterpenes* ($C_{30}$) consisting of six isoprene units. Triterpenoids of the oleanane, ursane, lupane and other less common types are restricted to higher plants, and in exceptional cases may dominate the extractable organic constituents of deep-sea sediments like in Baffin Bay (ten Haven et al. 1992). The geochemically most important and widespread triterpenes are from the hopane series, like diploptene which occurs in ferns, blue-green algae and bacteria. The predominant source of hopanoids are bacterial cell membranes, however, which contain bacteriohopanetetrol (and closely related molecular species) as rigidifiers. This $C_{35}$ compound has a sugar moiety attached to the triterpane skeleton *via* a carbon-carbon bond (Fig. 4.8). The widespread distribution of bacteria on Earth through time makes the hopanoids ubiquitous constituents of all organic-matter assemblages (Rohmer et al. 1992).

*Steroids* are tetracyclic compounds that are also biochemically derived from squalene epoxide cyclization, but have lost, in most cases, three methyl groups. Cholesterol ($C_{27}$) is the most important sterol of animals and of some plants as well. Higher plants frequently contain $C_{29}$ sterols (e.g. sitosterol) as the most abundant compound of this group. Steroids together with terpenoids are typical examples of biological markers (chemical fossils) because they contain a high degree of structural information that is retained in the carbon skeleton after sedimentation (e.g. Poynter and Eglinton 1991, Peters and Moldowan 1993) and often provides a chemotaxonomic link between the sedimentary organic matter and the precursor organisms in the biosphere.

*Carotenoids*, red and yellow pigments of algae and land plants, are the most important representatives of the tetraterpenes ($C_{40}$). Due to their extended chain of conjugated double bonds (e.g. β-carotene; Fig. 4.8) they are labile in most depositional environments and are found widespread but in low concentrations in marine surface sediments. Aromatization probably is one of the dominating diagenetic pathways in the alteration of the original structure of carotenoids in the sediment. Diagenetic intermolecular cross-linking by sulfur bridges may preserve the carotenoid carbon skeletons to a certain extent.

A second pigment type of geochemical significance are the *chlorophylls* and their derivatives that during diagenesis are converted into the fully aromatized porphyrins. Most porphyrins in sediments and crude oils are derived from the green plant pigment chlorophyll *a* and from bacteriochlorophylls.

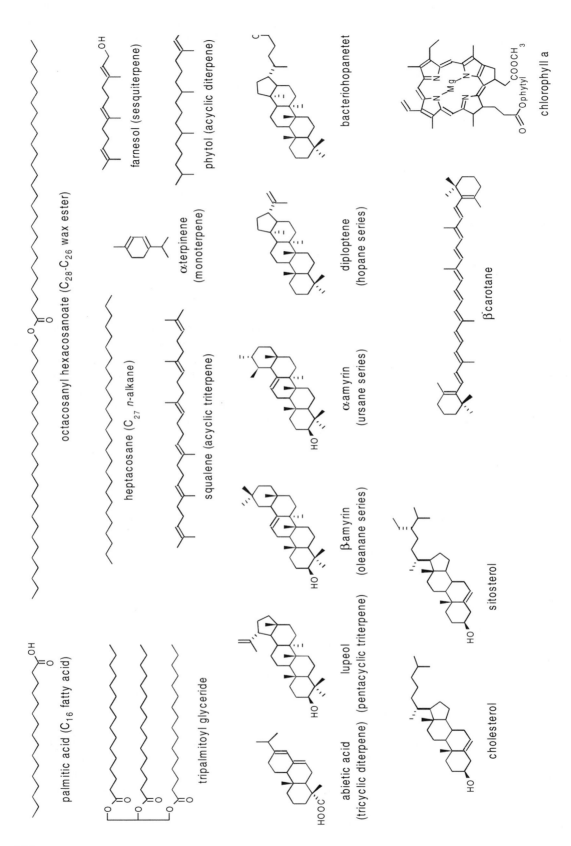

**Fig. 4.8**   Structural formulae of representative low-molecular-weight lipid compounds in living organisms and surface sediments (after Rullkötter 1992).

### 4.3.3 The Principle of Selective Preservation

Organic compounds and compound classes differ in their potential to be preserved in sediments and to survive early diagenesis. As a general rule, water-soluble organic compounds, or organic macromolecules, which are easily hydrolyzed to water-soluble monomers, have a low *preservation potential*. In contrast to this, compounds with a low solubility in water such as lipids and hydrolysis-resistant macromolecules are selectively enriched

in the sedimentary organic matter. Table 4.5 is a compilation of the source and preservation potential of some common organic compound types. It is based on anticipated chemical stabilities related to structures, reported biodegradability and reported presence in the geosphere, but not on mechanisms of preservation such as mechanical protection or bacteriostatic activities of certain chemicals in the (paleo)environment (de Leeuw and Largeau 1993).

The near-surface sediment layers represent the transition zone where biological organic matter is

**Table 4.5** Inventory of selected biomacromolecules, their occurrence in extant organisms, and their potential for survival during sedimentation and diagenesis (after Tegelaar et al. 1989 and de Leeuw and Largeau 1993; see there for chemical structures). The 'preservation potential' ranges from - (extensive degradation under all depositional conditions) to ++++ (little degradation under any depositional conditions).

| Biomacromolecules | Occurrence | 'Preservation potential' |
|---|---|---|
| Starch | Vascular plants; some algae; bacteria | - |
| Glycogen | Animals | - |
| Poly-β-hydroxyalkanoates | Eubacteria | - |
| Cellulose | Vascular plants; some fungi | -/+ |
| Xylans | Vascular plants; some algae | -/+ |
| Galactans | Vascular plants; algae | -/+ |
| Gums | Vascular plants | + |
| Alginic acids | Brown algae | -/+ |
| Dextrans | Eubacteria; fungi | + |
| Xanthans | Eubacteria | + |
| Chitin | Anthropods; copepods; crustacea; fungi; algae | + |
| Proteins | All organisms | -/+ |
| Mureins | Eubacteria | + |
| Teichoic acids | Eubacteria | + |
| Bacterial lipopolysaccharides | Gram-positive eubacteria | ++ |
| DNA, RNA | All organisms | - |
| Glycolipids | Plants; algae; eubacteria | +/++ |
| Polyisoprenoids (rubber, gutta) | Vascular plants | + |
| Polyprenols and dolichols | Vascular plants; bacteria; animals | + |
| Resinous polyterpenoids | Vascular plants | +/++ |
| Cutins, suberins | Vascular plants | +/++ |
| Lignins | Vascular plants | ++++ |
| Sporopollenins | Vascular plants | +++ |
| Algaenans | Algae | ++++ |
| Cutans | Vascular plants | ++++ |
| Suberans | Vascular plants | ++++ |

transformed into fossil organic matter. There are two slightly differing views about the nature of this process. The *classical view* (Fig. 4.9; Tissot and Welte 1984) implies that biopolymers are (mainly) enzymatically degraded into the corresponding biomonomers. The monomers then are either used by sediment bacteria and archaea to synthesize their own biomass or as a source of energy. Alternatively, they may randomly recombine by condensation or polymerization to geomacromolecules (see Sect. 4.3.4). The discovery of nonhydrolyzable, highly aliphatic biopolymers in extant organisms and geological samples has led to a reappraisal of the processes involved in the formation of geomacromolecules (Tegelaar 1989, de Leeuw and Largeau 1993). In a scheme modified from that of Tissot and Welte (1984), more emphasis is placed on the *selective preservation* of biopolymers (Fig. 4.10). This means that the role of consecutive and random polymerisation and polycondensation reactions of biomonomers formed by hydrolysis or other degradative pathways may be less important than previously thought. Support to this view is given by the detection of the close morphological relationship between some fossil 'ultralaminae' and the thin resistant outer cell walls of green microalgae (Largeau et al 1990).

To complete the modified view of geomacromolecule formation, the process of '*natural vul-*

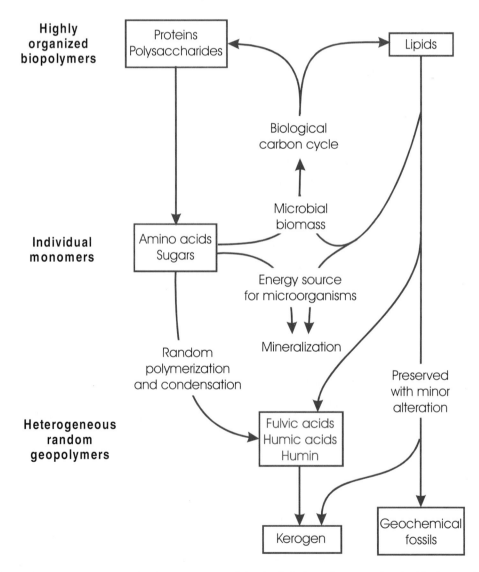

**Fig. 4.9**  From biomass to geomacromolecules - a summary of the classical view of processes involved in the transformation of biogenic organic matter into kerogen and geochemical fossils (after Tissot and Welte 1984).

*canisation*' (Fig. 4.10) has been proposed to play a major role under suitable conditions (e.g. Sinninghe Damsté et al. 1989a, 1990, de Leeuw and Sinninghe Damste 1990). Many marine sediments contain high-molecular-weight organic sulfur substances that are thought to be derived from intermolecular incorporation of inorganic sulfur species (HS⁻, polysulphides) into functionalized lipids during early diagenesis. This requires the reduction of seawater sulfate by sulfate-reducing microorganisms under anoxic conditions (see

Chap. 8). Sulfur incorporation into organic matter is further enhanced in depositional systems that are iron-limited, i.e. organosulfur compounds are particularly abundant in areas that receive little continental detritus with clays enriched in iron and where instead biogenic carbonate or opal is the dominant mineral component of the sediment.

As a consequence of the discussion of organo-mineral interaction for the preservation of organic matter in sediments (see Sect. 4.2), Collins et al. (1995) raised the question if sorption of organic

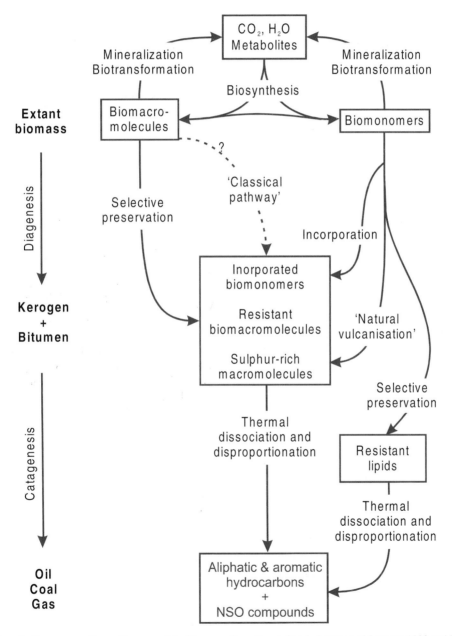

**Fig. 4.10** The selective preservation pathway model of kerogen formation (after de Leeuw and Largeau 1993 and Tegelaar et al. 1989).

matter on mineral surface did not lead to a rebirth of the classical condensation pathway for geomacromolecule formation. Neither adsorption nor condensation alone may be a satisfactory process for preservation of labile organic substances. Adsorption of monomers can merely retard their biodegradation, and condensation is not favored in (pore water) solution. However, if the processes operate in concert - adsorption promoting condensation and condensation enhancing adsorption of further reactants - a plausible mechanism for the preservation of organic matter arises. Condensation reactions between adsorbed compounds would lead to the formation of very strongly bound macromolecules resulting in a marked divergence in the diagenetic history of the adsorbed monolayer and nonmineral-bound organic matter. More direct evidence is, however, still required to establish the quantitative importance of this process relative to other processes, such as selective preservation.

### 4.3.4    The Formation of Fossil Organic Matter and its Bulk Composition

The geomacromolecular organic matter surviving microbial degradation in the early phase of diagenesis consists of three fractions, termed fulvic acids, humic acids and humin. They are ill-defined in their chemical structures, but are operationally distinguished by their solubilities in bases (humic and fulvic acids) and acids (fulvic acids only). All three are considered to be potential precursors of *kerogen*, which designates the type of geomacromolecules formed at a later stage of diagenesis. Kerogen is also only operationally defined as being insoluble in non-oxidizing acids, bases and organic solvents (Durand 1980, Tissot and Welte 1984). Besides this high-molecular-weight organic material, sediments contain low-molecular-weight organic substances that are collectively called *bitumen* and are extractable with organic solvents. Bitumen consists of nonpolar hydrocarbons and a variety of polar lipids with a great number of different functional groups, such as ketones, ethers, esters, alcohols, fatty acids and corresponding sulfur-bearing compounds.

Fulvic acids and subsequently humic acids decrease in their concentrations over time as a result of progressive combination reactions with increasing diagenesis. This process of kerogen formation concurrently involves the elimination of small molecules like water, carbon dioxide, am-

monia or hydrogen sulfide (Huc and Durand 1977). As a consequence, the degree of condensation of the macromolecular kerogen increases. In terms of elemental composition it becomes enriched in carbon and hydrogen and depleted in oxygen, nitrogen and sulfur. This carbon- and hydrogen-rich kerogen ultimately is the source material for the formation of petroleum and natural gas which starts when burial is deep and temperatures are so high that the carbon-carbon bonds in the kerogen are thermally cracked. The main phase of *oil formation* typically occurs between 90°C and 120°C, but the range largely depends on the heating rate (slow or rapid burial) and on the chemical structure of the kerogen. For example, kerogens rich in sulfur start oil formation earlier because carbon-sulfur bonds are broken more easily. The product, however, is a heavy oil (high density, high viscosity) rich in sulfur and economically less valuable than light and sulfur-lean crude oil. The difference between an oil shale and a rock actively generating petroleum is only the thermal history. An oil shale simply has not completed the phase of diagenetic release of small molecules described above. For more details on petroleum formation consult the textbooks of Tissot and Welte (1984), Hunt (1996) and Welte et al. (1997).

Molecular structural information about kerogen can be inferred from elemental analysis, spectroscopic methods and the results of pyrolysis and selected chemical degradation experiments (see Rullkötter and Michaelis 1990 for an overview). Yet, with the understanding that kerogen is a complex heterogeneous macro-molecular substance with contributions from a variety of organisms and a wide range of chemical alterations that occurred during diagenesis, it becomes clear that there will be no single molecular structure of kerogen, and only certain characteristic units can be described at the molecular level.

Kerogens have been classified into types derived from H/C and O/C atomic ratios in a van Krevelen diagram (Fig. 4.11). The types indicated are related to the hydrogen and oxygen richness, relative to carbon, of the biogenic precursor material. Roughly, kerogen Type I is related to hydrogen-rich organic matter as occurring, e.g., in waxes and algal mats, kerogen Type II represents typical oceanic plankton material and kerogen Type III is typical of land-derived organic matter as it occurs in coals. A kerogen type IV not indicated in the diagram has occasionally been defined to have very low H/C ratios and to represent highly oxidized, largely inert organic matter. The bold solid trend lines indicated in Figure 4.11 then represent the changes in elemental composition initially occuring during diagenesis (evolution grossly parallel to the x-axis due to loss of oxygen functionalities) and later during oil and gas formation (catagenesis; evolution grossly parallel to y-axis due to loss of hydrogen-rich petroleum hydrocarbons), until a carbon-rich residue is the ultimate product (near origin in xy diagram). For more details see Tissot and Welte (1984).

### 4.3.5 Early Diagenesis at the Molecular Level

A small portion of sedimentary organic matter is soluble in organic solvents and contains lipid compounds that are either directly inherited from the biological precursor organisms or cleaved by hydrolysis from larger cellular units like cell walls or membranes (cf. Figs. 4.9 and 4.10). The compounds include individual substances as well as homologous series of structurally related compounds. Most of them are functionalized polar

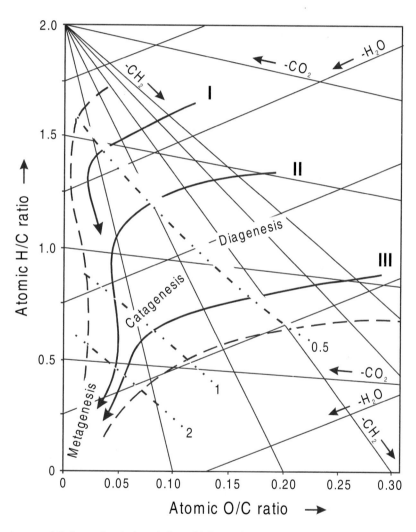

**Fig. 4.11** Kerogen types and their geochemical evolution with increasing burial (temperature increase) in a van Krevelen diagram of atomic H/C *versus* O/C ratios from elemental analysis (after Tissot and Welte 1984). Roman numbers indicate kerogen types, bold trend lines are idealized average values from a large number of data points. In organic geochemistry, diagenesis is the early (low-temperature) range of this evolution, catagenesis signifies the phase of petroleum and wet gas formation by thermal cracking, and metagenesis is the high-temperature range where still some dry gas (methane, ethane) is formed. Numbers associated with broken lines in the diagram (0.5, 1, 2) indicate approximate vitrinite reflectance ($R_o$) values commonly measured to indicate thermal maturity. Vitrinites are fossil woody organic particles of characteristic shape which increase their reflectivity as a function of geothermal history; the amount of reflected light can be quantitatively measured with a light microscope under defined conditions (see Sect. 4.5.3 and Stach et al. 1982 for more details).

lipids that undergo decarboxylation (organic acids) and dehydration reactions (alcohols) during early diagenesis to produce saturated and olefinic hydrocarbons, of which the latter are progressively hydrogenated into their saturated analogs during later diagenesis. Alternatively, aromatic hydrocarbons are formed by the loss of hydrogen. If these hydrocarbons essentially have the same carbon skeletons and steric configurations as their functionalized biogenic precursors, they are called *biological markers* (see Sects. 4.3.2 and 4.3.6). Parallel to retention of the biogenic carbon skeleton, structural rearrangements, catalyzed by clay minerals, partial cleavage of substituents or ring opening may occur as side reactions during diagenetic transformation of biogenic lipids. During the earlier phases of diagenesis, including processes occurring in the water column, such alterations appear to be mediated by microbial activity. With increasing burial they are more and

more driven by thermodynamic constraints as temperature increases.

The following discussion of biological marker reactions of course only applies to that fraction of lipid compounds that have escaped the highly efficient degradation in the uppermost sediment layer. It has recently been established through quantitative assessment of transformation reactions that degradation in this zone may occur over timescales of days and that reaction rates have been underestimated in the past by an order of magnitude (Canuel and Martens 1996). It could furthermore be demonstrated in this study that the degradation processes can be highly selective and depend on the origin of the compounds (marine, bacterial or terrestrial). Fatty acids apparently are particularly sensitive to degradation whereas sterols and hydrocarbons have a higher chance of entering the deeper sediment. As a consequence, quantitative assessment

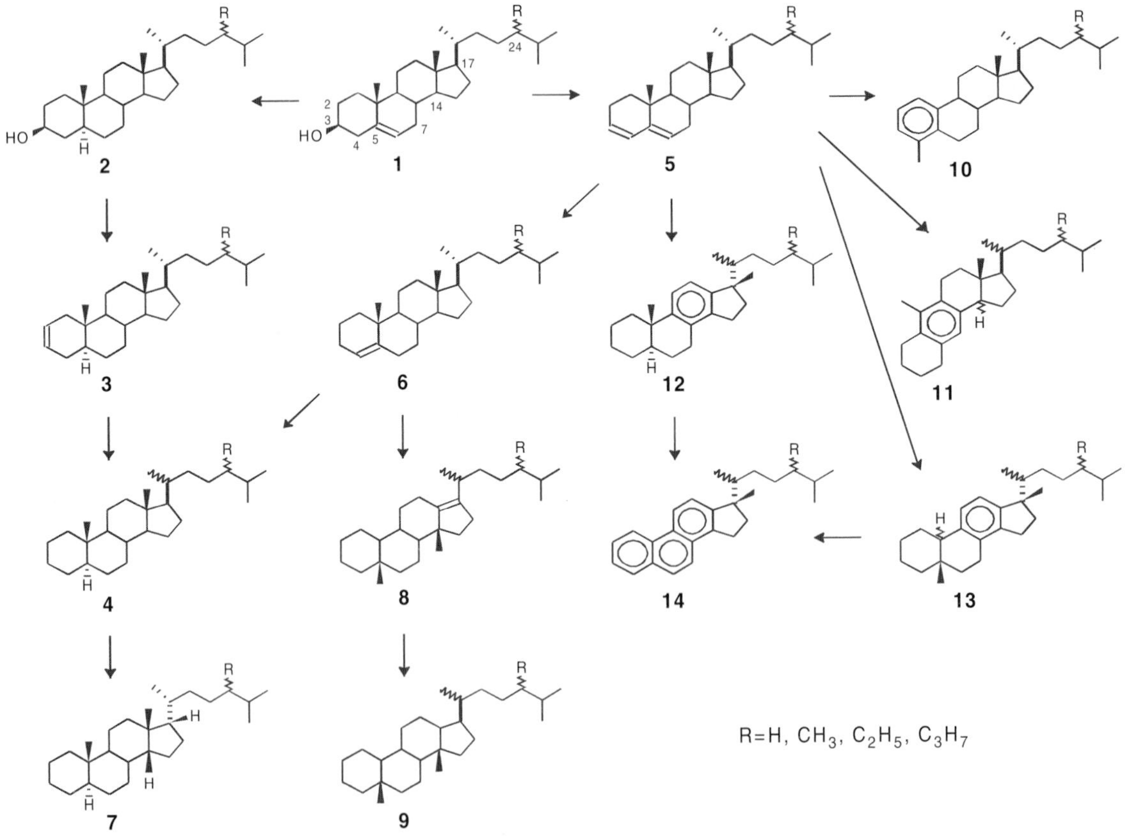

**Fig. 4.12** Diagenetic and catagenetic transformation of steroids. The precursor sterols are gradually transformed during diagenesis into saturated hydrocarbons by dehydration (elimination of water) and hydrogenation of the double bonds. At higher temperatures, during catagenesis, the thermodynamically most stable stereoisomers are formed. Alternatively, dehydration leads to aromatic steroid hydrocarbons of which are stable enough to occur in crude oils (after Rullkötter 1992). See text for detailed description of the reaction sequences.

of the source and fate of organic matter based on biological markers will be strongly limited as long as diagenetic effects cannot be separated from variations in organic matter supply.

## 4.3.6   Biological Markers (Molecular Fossils)

Molecules with a high degree of structural complexity are particularly informative and thus suitable for studying geochemical reactions because they provide the possibility of relating a certain product to a specific precursor. For example, specific biomarkers have been assigned to some common groups of microalgae. These compounds include long-chain ($C_{37}$-$C_{39}$) n-alkenones, highly-branched isoprenoid alkenes, long-chain n-alkandiols, 24-methylenecholesterol and dinosterol. They have been found to be unique for, or obviously be preferentially biosynthesized by, *haptophytes*, diatoms, *eustigmatophytes*, diatoms and dinoflagellates, respectively (see Volkman et al. 1998 for a recent comprehensive overview). Other long-chain n-alkyl lipids (e.g. Eglinton and Hamilton 1967), diterpenoids and 3-oxygenated triterpenoids (e.g. Simoneit 1986) are considered useful tracers for organic matter from vascular land plants. Although certain biological markers may be chemotaxonomically very specific, care has to be taken when using relative biomarker concentrations in geological samples to derive quantitative figures of the biological species that have contributed to the total organic matter. First of all, different types of biological markers may have different reactivities and, thus, may be selectively preserved during diagenesis (Hedges and Prahl 1993). In this respect, sequestering of reactive biomarkers by the formation of high-molecular-weight organic sulfur compounds may play an important role (e.g. Sinninghe Damsté et al 1989b; see Fig. 4.10). Furthermore, there may be a fractionation between high- and low-molecular-weight compounds. An extreme example is the (lacustrine) Messel oil shale. In its organic matter fraction, the residues of dinoflagellates are represented by abundant 4-methyl steroids in the bitumen while the labile cell walls were not preserved. On the other hand, certain green algae are clearly identifiable under the electron microscope due to the highly aliphatic biopolymers in their cell walls, but no biomarkers specific for green algae were found in the extractable organic matter (Goth et al. 1988).

The scheme in Figure 4.12 is an example of extensive and variable biomarker reactions after sedimentation. It illustrates the fate of sterols particularly during diagenesis. Although the scheme looks complex, it shows only a few selected structures out of more than 300 biogenic steroids and geochemical conversion products presently known to occur in sediments. The biogenic precursor chosen as an example in Figure 4.12 is cholesterol (structure 1, R=H), a widely distributed steroid in a variety of plants, but more typical of animals (e.g. zooplankton). Hydrogenation of the double bond leads to the formation of the saturated cholestanol (2). This reaction occurs in the uppermost sediment layers soon after deposition and is believed to involve microbial activity. Elimination of water gives rise to the unsaturated hydrocarbon 3. At the end of the diagenetic stage, the former unsaturated steroid alcohol 1 will have been transformed to the saturated steroid hydrocarbon 4 after a further hydrogenation step. An alternative route to the saturated sterane 4 is *via* dehydration of cholesterol (1, R=H), which yields the diunsaturated compound 5. Hydrogenation leads to a mixture of two isomeric sterenes (6; isomer with double bond in position 5 like in the starting material (1) not shown in Fig. 4.12). This compound cannot be formed from 3 as suggested for a long time, because such a double bond migration would require more energy than is available under the diagenetic conditions in sediments (de Leeuw et al. 1989). Further hydrogenation of 6 affords the saturated hydrocarbon 4. A change in steric configuration of this molecule, e.g. to form 7, occurs only during the catagenesis stage at elevated temperatures. A side reaction from sterene 6 is a skeletal rearrangement leading to diasterene 8 where the double bond has moved to the five-membered ring and two methyl groups (represented by the bold bonds) are now bound to the bottom part of the ring system. This reaction has been shown in the laboratory to be catalyzed by acidic clays. Thus, diasterenes (8) and the corresponding diasteranes (9), formed from diasterenes by hydrogenation during late diagenesis, are found in shales but not in those carbonates that lack acidic clays.

An additional alternative diagenetic transformation pathway of steroids leads to aromatic instead of saturated hydrocarbons. The diolefin 5 is a likely intermediate on the way to the aromatic steroid hydrocarbons 10-14. Compounds 10 and 11 are those detected first in the shallowest sediment layers. They obviously are labile and do not survive diagenesis. During late diagenesis, the aromatic steroid hydrocarbon 12 with the aromatic ring next to the five-membered ring co-occurs with compounds 10 and 11 in the sediments, but is also stable enough to survive elevated temperatures and thus to be found in crude oils. There is also a corresponding rearranged mono-aromatic steroid hydrocarbon (13). During catagenesis, monoaromatic steroid hydrocarbons are progressively transformed into triaromatic hydrocarbons (14) before the steroid record is completely lost by total destruction of the carbon skeleton at higher temperatures.

As a second example, Figure 4.13 shows five different diagenetic reaction pathways for pentacyclic triterpenoids of terrestrial origin that were found to be abundant, e.g., in Tertiary deep-sea sediments of the Baffin Bay (ten Haven et al. 1992). Diagenetic alteration with full retention of the carbon skeleton (e.g. in the case of β-amyrin; R=H) leads to an olefinic hydrocarbon after elimination of the oxygen functionality in the A-ring and later to the fully saturated hydrocarbon. If the substituent group is a hydroxyl or carboxylic acid group, oxidation would yield an unstable keto-carboxylic acid, which instantaneously eliminates $CO_2$ leading to a carbon skeleton with one carbon

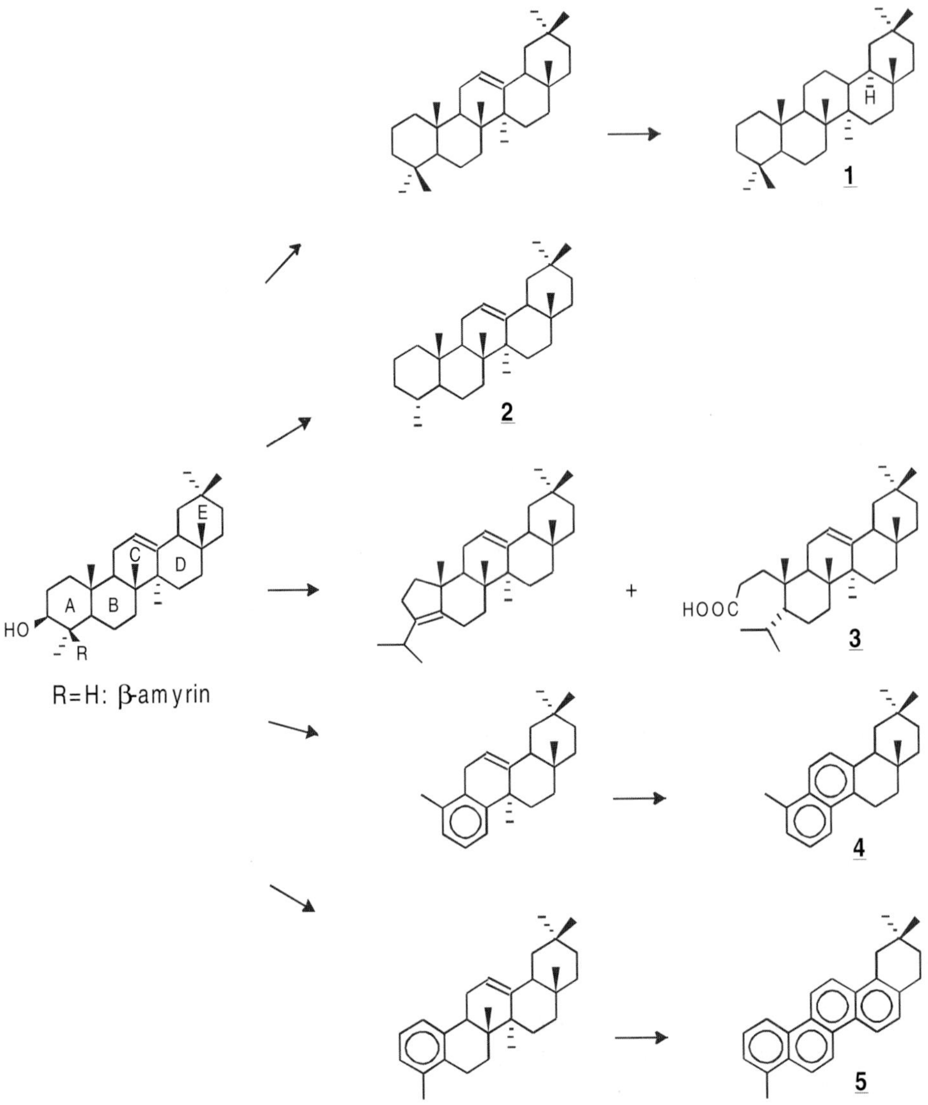

**Fig. 4.13** Schematic representation of five different (mostly oxidative) diagenetic reactions of triterpenoids from higher plants. R in the starting material should be an oxygen function, at least in the second pathway (after Rullkötter et al. 1994).

atom less than the starting molecule (second pathway; see Rullkötter et al. 1994). Direct chemical elimination of the hydroxyl group in ring A causes ring contraction, and eventually the ring is opened by oxidative cleavage of the double bond (third pathway). If the carbon atoms of the A-ring are completely lost during degradation, then subsequent aromatization may lead into the fourth pathway. Alternatively, aromatization may start with the intact carbon skeleton giving rise to a series of partly or fully aromatized pentacyclic hydrocarbons (fifth pathway). All these alterations are typical for terrigenous triterpenoids. They probably start soon after the decay of the organisms (or parts thereof, e.g. leaves) and continue during transport into the ocean. The compounds described and several others have been found in numerous marine sediments (see Corbet et al. 1980 and Rullkötter et al. 1994 for overviews).

## 4.4    Organic Geochemical Proxies

### 4.4.1    Total Organic Carbon and Sulfur

Organic carbon profiles in a sedimentary sequence, particularly if they are obtained with high stratigraphic resolution (e.g. Stein and Rack 1995), provide direct evidence for changes in depositional patterns. An in-depth interpretation, however, usually requires additional information on the quality of the organic matter, i.e. on its origin (marine *versus* terrigenous) and/or its degree of oxidation during deposition. The relationship between organic carbon and sedimentation rate may help to distinguish different depositional environments or to determine paleoproductivity as already discussed in Section 4.2.

Furthermore, the relationship between organic carbon and sulfur is also characteristic of the paleoenvironment. Leventhal (1983) and Berner and Raiswell (1983) observed an increase in pyrite sulfur content in marine sediments with increasing amount of total organic carbon (Fig. 4.14). The rationale behind this is that the amount of metabolizable organic matter available to support sulfate-reducing bacteria increases with the total amount of organic matter arriving at the sediment-water interface. As a consequence, the sedimentary pyrite sulfide content is positively correlated with the non-metabolized (resistant or unused) organic matter content (TOC). The trend-

line in Figure 4.14 is considered representative of normal marine (oxic) environments. Data from the Black Sea plot above the trendline (higher S/C ratios) because consumption of organic matter by sulfate-reducing bacteria leads to excess hydrogen sulfide, available for pyrite formation, in the water column. In contrast to this, freshwater sediments have very low S/C (or high C/S) ratios because of the low sulfate concentrations in most freshwater bodies. Although the trendline is based on pyrite sulfur, it is not important whether the reduced sulfur is present as metal sulfide (mostly pyrite) or bound to the organic matter. This is a question of iron limitation rather than sulfate reduction.

As outlined in Section 4.3.1 (and discussed more extensively in Chap. 8) there is a close connection between organic matter remineralization during early diagenesis and microbial sulfate reduction. If all of the sulfate reduction products were precipitated as pyrite or bound into (immobile) organic matter, measuring the amount of sulfur in these species would provide an easy way for determining the amount of organic matter that has been remineralized and was not preserved in the sediment. However, the main product of

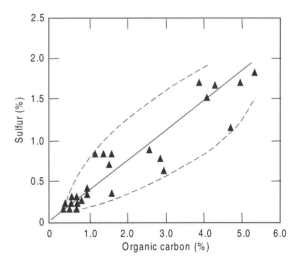

**Fig. 4.14** Plot of weight percent organic carbon *vs.* weight percent pyrite sulfur for normal-marine modern sediments. Each plotted point represents the average value of samples in a given core, taken at a sediment depth where concentrations of organic carbon and pyrite have attained quasi-steady-state values, i.e. where early diagenesis of carbon and sulfur is (essentially) complete. The dashed lines enclose data from a variety of other studies (after Berner and Raiswell 1983). Sediments deposited under anoxic (euxinic) conditions would plot above the trend line, freshwater sediments significantly below.

sulfate reduction, hydrogen sulfide, is volatile and can escape from the sediment, particularly when bioturbation of the surface sediment supports this transport.

Release of hydrogen sulfide from the sediment plays a less important role under strictly anoxic conditions where fine lamination indicates that the environment is hostile to burrowing organisms and bioturbation does not occur. It has been shown that in these cases the initial amount of organic matter deposited can be estimated by measuring concentrations of reduced sulfur in such sediments. Considering the amount of organic matter consumed by sulfate reduction, Lallier-Vergès (1993) defined a sulfate reduction index (SRI) as

$$SRI = \% \text{ initial organic carbon } /$$
$$\% \text{ preserved organic carbon}$$

$$(4.8)$$

The amount of initial organic carbon then is the sum of the preserved organic carbon (measured as TOC) and the metabolized organic carbon (determined from the sulfur content with stoichiometric correction for the sulfate reduction process). Furthermore, the diffusive loss has to be taken into account. With a correction factor of 0.75 and a term $1/(1-qH_2S)$ for the diffusive loss, Vetö et al. (1994) calculated the initial (or original) organic carbon content of a sediment before sulfate reduction as

$$TOC_{orig} = TOC + 0.75S \cdot 1/(1-qH_2S) \quad (4.9)$$

where TOC and S are the measured values of total organic carbon and total sulfur. The authors estimate that the diffusive $H_2S$ loss in non-bioturbated sediments usually is less than 45% and that this value can only be reached in cases of very high organic matter supply, high reactivity of this organic matter and iron limitation. Littke et al. (1991b) and Lückge et al. (1996) calculated that sulfate reduction consumed between 20 and 50% of the initially sedimented organic matter (or 1-3% of primary productivity) both from ancient rocks (Posidonia Shale) and Recent sediments (Oman Margin and Peru upwelling systems). In a study of the Pakistan continental margin in the northern Arabian Sea (Littke et al. 1997b), they clearly demonstrated that the carbon-sulfur relationship only holds in the laminated sections of the sediment profile whereas it fails (strongly underestimates sulfate reduction) in the intercalated homogeneous, i.e. bioturbated, sediments for the reason explained before.

### 4.4.2    Marine *Versus* Terrigenous Organic Matter

As pointed out in Section 4.2.5, even deep-sea sediments deposited in areas remote from continents may contain a mixture of marine and terrigenous organic matter. For any investigation of autochthonous marine organic matter preservation or marine paleoproductivity, these two sources of organic matter have to be distinguished. Furthermore, global or regional climate fluctuations have changed the pattern of continental run-off and ocean currents in the geological past (see Sect. 4.4.3). Being able to recognize variations in marine and terrigenous organic matter proportions may, thus, be of great significance in paleoclimatic and paleoceanographic studies.

A variety of parameters are used to assess organic matter sources. Bulk parameters have the advantage that they are representative of total organic matter, whereas molecular parameters address only part of the extractable organic matter, which in turn is only a small portion of total organic matter. Some successful applications of molecular parameters show that the small bitumen fraction may be representative of the total, but there are many other examples where this is not the case. On the other hand, oxidation of marine organic matter has the same effect on some bulk parameters as an admixture of terrigenous organic matter, because the latter is commonly enriched in oxygen through biosynthesis. It is, therefore, advisable to rely on more than one parameter, and to obtain complementary information.

*C/N Ratio*

Carbon/nitrogen (C/N) ratios of phytoplankton and zooplankton are around 6, freshly deposited marine organic matter ranges around 10, whereas terrigenous organic matter has C/N ratios of 20 and above (e.g. Meyers 1994, 1997 and references therein). This difference can be ascribed to the absence of cellulose in algae and its abundance in vascular plants and to the fact that algae are instead rich in proteins. Both weight and atomic ratios are used by various authors, but due to the small difference in atomic mass of carbon and nitrogen, absolute numbers of ratios do not deviate greatly.

Selective degradation of organic matter components during early diagenesis has the tendency to modify (usually increase) C/N values already in the water column. Still, these ratios are commonly sufficiently well preserved in shallow-marine sediments to allow a rough assessment of terrigenous organic matter contribution (e.g. Jasper and Gagosian 1990, Prahl et al. 1994). A different trend exists in deep oceanic sediments with low organic carbon contents. Inorganic nitrogen (ammonia) is released during organic matter decomposition and adsorbed to the mineral matrix (particularly clays) where it adds significantly to the total nitrogen. The C/N ratio is then changed to values below those of normal marine/terrigenous organic matter proportions (Müller 1977, Meyers 1994). This effect should be small in sediments containing more than 0.3% organic carbon. On the other hand, many sapropels from the eastern Mediterranean Sea and organic-matter-rich sediments underlying upwelling areas have conspicuously high C/N ratios (>15), i.e. well in the range of land plants despite a dominance of marine organic matter, for reasons yet to be determined (see Bouloubassi et al. 1999 for an overview).

*Hydrogen and Oxygen Indices*

Hydrogen Index (HI) values from Rock-Eval pyrolysis (see Sect. 4.5.2) below about 150 mg HC/g TOC are typical of terrigenous organic matter, whereas HI values of 300 to 800 mg HC/g TOC are typical of marine organic matter. Deep-sea sediments rich in organic matter usually show values of only 200-400 mg HC/g TOC, even if marine organic matter strongly dominates. Oxidation has lowered the hydrogen content of the organic matter in this case. It should also be mentioned that Rock-Eval pyrolysis was developed as a screening method for rapidly determining the hydrocarbon generation potential of petroleum source rocks (Espitalié et al. 1977) and that a range of complications may occur with sediments buried only to shallow depth. For example, unstable carbonates may decompose below the shut-off temperature of 390°C (cf. Sect. 4.5.2) which increases the Oxygen Index and falsely indicates a high oxygen content of the organic matter. Furthermore, Rock-Eval pyrolysis cannot be used for sediments with TOC<0.3% because of the so-

called mineral matrix effect. If sediments with low organic carbon contents are pyrolyzed, a significant amount of the products may be adsorbed to the sediment minerals and are not recorded by the flame ionization detector, thus lowering the Hydrogen Index (Espitalié et al. 1977).

*Maceral Composition*

If the morphological structure of organic matter is well preserved in sediments, organic petrographic investigation under the microscope is probably the most informative method to distinguish marine and terrestrial organic matter contributions to marine sediments by the relative amounts of macerals (organoclasts) derived from marine biomass and land plants (see Sect. 4.5.3). Many marine sediments, however, contain an abundance of non-structured organic matter (e.g. in upwelling areas; Lückge et al. 1996) which cannot easily be assigned to one source or the other. Furthermore, comprehensive microscopic studies are time-consuming. In his paleoproductivity assessments (see Sect. 4.2), Stein (1986a) calibrated Hydrogen Indices from Rock-Eval pyrolysis of marine sediments with microscopic data and suggested to use the more readily available pyrolysis data as a proxy for marine/terrigenous organic matter proportions.

*Stable Carbon Isotope Ratios*

Carbon isotope ratios are principally useful to distinguish between marine and terrestrial organic matter sources in sediments and to identify organic matter from different types of land plants. The stable carbon isotopic composition of organic matter reflects the isotopic composition of the carbon source as well as the discrimination (fractionation) between $^{12}C$ and $^{13}C$ during photosynthesis (e.g. Hayes 1993, cf. Sect. 10.4.2). Most plants, including phytoplankton, incorporate carbon into their biomass using the Calvin ($C_3$) pathway which discriminates against $^{13}C$ to produce a shift in $\delta^{13}C$ values of about -20‰ from the isotope ratio of the inorganic carbon source. Some plants use the Hatch-Slack ($C_4$) pathway which leads to an isotope shift of about -7‰. Other plants, mostly succulents, utilize the CAM (crassulacean acid metabolism) pathway, which more or less switches between the $C_3$ and $C_4$ pathways and causes the $\delta^{13}C$ values to depend on the growth dynamics.

Organic matter produced from atmospheric carbon dioxide ($\delta^{13}C \approx$ -7‰) by land plants using the $C_3$ pathway (including almost all trees and most shrubs) has an average $\delta^{13}C$ value of approximately -27‰ and by those using the $C_4$ pathway (many tropical savannah grasses and sedges) approximately -14‰. Marine algae use dissolved bicarbonate, which has a $\delta^{13}C$ value of approximately 0‰. As a consequence, marine organic matter typically has $\delta^{13}C$ values varying between -20‰ and -22‰. Isotopic fractionation also is temperature dependent which, e.g., in cold polar waters may lead to carbon isotope values for marine organic matter of -26‰ or lower (e.g. Rau et al. 1991).

The 'typical' difference of about 7‰ between organic matter of marine primary producers and land plants has been successfully used to trace the sources and distributions or organic matter in coastal oceanic sediments (e.g. Westerhausen et al. 1993, Prahl et al. 1994). Unlike C/N ratios, $\delta^{13}C$ values are not significantly influenced by sediment texture (Meyers 1997; cf. discussion of the effect of inorganic nitrogen on C/N ratios before), making them useful in reconstructing past sources of organic matter in changing depositional conditions. For varying continental run-off into the Pygmy Basin (northern Gulf of Mexico) during glacial-interglacial cycles this is illustrated in Figure 4.15 (Meyers 1997). Lowered sea-level increased delivery of land-derived material to the offshore area during the cold periods of Oxygen Isotope Stages 2 and 3. In the sediment sections representing this time, the C/N ratios are higher than in the interglacial stages 1 and 5. Simultaneously, the more negative $\delta^{13}C$ values of the organic matter also indicate a higher proportion of terrigenous organic matter. The global cooling process during the same period is illustrated by the more negative values of $\delta^{18}O$ determined on the carbonate frustules of foraminifera; the oxygen isotopes of the carbonate during growth of the foraminifera was in equilibrium with the sea water and reflect the fact that more of the heavy

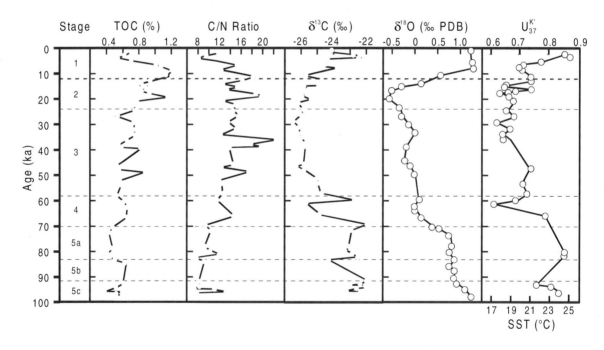

**Fig. 4.15** Glacial-interglacial changes in delivery of organic matter to sediments in the Pygmy Basin, northern Gulf of Mexico. Lowered sea-level increased delivery of land-derived material to this site during Oxygen Isotope Stages 2 and 3, as evidenced by high atomic C/N ratios and low $\delta^{13}C$ values. Sea-surface temperature simultaneously dropped by approximately 6°C according to $U^{K'}_{37}$ values and approximately 2°C according to $\delta^{18}O$ values of the planktonic foraminifera *Globigerina sacculifer* (from Meyers 1997, after Jasper and Gagosian 1989, 1990). The $U^{K'}_{37}$ index is based on the ratio of two constituents (long-chain alkenones) of phytoplanktonic algae. The ratio depends on the water temperature at which these organisms grow (see Sect. 4.4.3 for a more detailed explanation of this parameter). The $\delta^{18}O$ values of the carbonate frustules of planktonic foraminifera are sensitive to global climatic changes. Water bound in polar ice caps is enriched in $^{18}O$ and, correspondingly, ocean water during glacial periods is depleted in $^{18}O$. This leads to more negative values of the oxygen isotope ratio in the carbonate frustules due to oxygen exchange with seawater.

isotope was bound in the polar ice caps and glaciers during glacial periods. Local lower temperatures of the surface water in the Gulf of Mexico can be retraced in the sedimentary organic matter by measuring the $U^K{'}_{37}$ index, based on long-chain alkenones of haptophytes, and subsequent conversion to paleo-sea surface temperatures (see Sect. 4.4.3).

The stable carbon isotopic source information may be complicated in coastal areas that receive contributions of organic matter from marine algae and both $C_3$ and $C_4$ terrestrial plants (e.g. Fry et al. 1977). For this reason, a combination of carbon isotope data with other proxy parameters is advised. For example, a cross-plot of C/N and $\delta^{13}C$ values has been used to clearly distinguish four types of plant organic matter sources (Fig. 4.16; Meyers 1994).

The availability of dissolved $CO_2$ in ocean water has an influence on the carbon isotopic composition of algal organic matter because isotopic discrimination towards $^{12}C$ increases when the partial pressure of carbon dioxide ($pCO_2$) is high and decreases when it is low (see Fogel and Cifuentes 1993 for an overview). Organic-matter

$\delta^{13}C$ values, therefore, become indicators not only of origins of organic matter but also of changing paleoenvironmental conditions on both short- and long-term scales. For example, the $\delta^{13}C$ values of dissolved inorganic carbon (DIC; $CO_2$, bicarbonate, carbonate) available for photosynthesis in the cells varies over the year with the balance between photosynthetic uptake and respiratory production. During summer and spring, when rates of photosynthesis are high, the isotope ratio of remaining DIC is enriched in $^{13}C$. In fall, when respiration is the dominant process, the $\delta^{13}C$ of DIC becomes more negative because organic matter is remineralized.

Fluctuations that have been measured in the $\delta^{13}C$ values of sedimentary organic matter over the Earth's history (e.g. Schidlowski 1988) can thus be interpreted in terms of the productivity in the water column and the availability of DIC in a particular geological time period. In a study of sediments from the central equatorial Pacific Ocean spanning the last 255,000 years it has been demonstrated that the carbon isotopic composition of fossil organic matter depends on the exchange between atmospheric and oceanic $CO_2$.

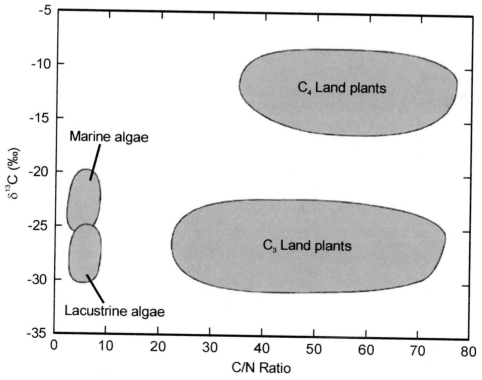

**Fig. 4.16** Elemental (atomic C/N ratios) and isotopic ($\delta^{13}C$ values) identifiers of bulk organic matter produced by marine algae, lacustrine algae, $C_3$ land plants and $C_4$ land plants (after Meyers 1994). Shaded area signify data scatter from a large number of analyses.

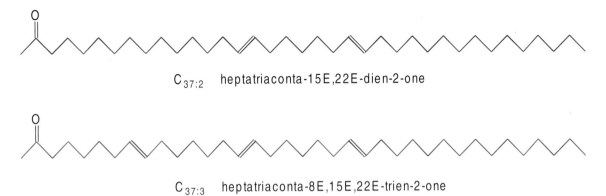

$C_{37:2}$   heptatriaconta-15E,22E-dien-2-one

$C_{37:3}$   heptatriaconta-8E,15E,22E-trien-2-one

**Fig. 4.17** Structural formulae of long-chain alkenones used for paleo-sea surface temperature assessment.

Changes with time can then be used to estimate past atmospheric carbon dioxide concentrations (Jasper et al. 1994).

### 4.4.3    Molecular Paleo-Seawater Temperature and Climate Indicators

*Past Sea-Surface Temperatures (SST) Based on Long-Chain Alkenones*

Paleoceanographic studies have taken advantage of the fact that biosynthesis of a major family of organic compounds by certain microalgae depends on the water temperature during growth. The microalgae belong to the class of *Haptophyceae* (often also named *Prymnesiophyceae*) and notably comprise the marine coccolithophorids *Emiliania huxleyi* and *Gephyrocapsa oceanica*. The whole family of compounds, which are found in marine sediments of Recent to mid-Cretaceous age throughout the world ocean, is a complex assemblage of aliphatic straight-chain ketones and esters with 37 to 39 carbon atoms and two to four double bonds (see Brassell 1993 for an overview and details), but principally only the $C_{37}$ methylketones with 2 and 3 double bonds are used for past sea-surface temperature assessment (Fig. 4.17).

It was found from the analysis of laboratory cultures and field samples that the extent of unsaturation (number of double bonds) in these long-chain ketones varies linearly with growth temperature of the algae over a wide temperature range (Marlowe et al. 1986, Prahl and Wakeham 1987). To describe this, an unsaturation index was suggested, which in its simplified form is defined by the concentration ratio of the two $C_{37}$ ketones:

$$U^{k'}_{37} = [C_{37:2}]/[C_{37:2} + C_{37:3}] \qquad (4.10)$$

Calibration was then made with the growth temperatures of laboratory cultures of different haptophyte species and with ocean water temperatures at which plankton samples had been collected. From these data sets, a number of different calibration curves evolved for different species and different parts of the world ocean (Fig. 4.18A) so that some doubts arose as to the universal applicability of the unsaturation index. In a major analytical effort, Müller et al. (1998) resolved the complications and arrived at a uniform calibration for the global ocean from 60°N to 60°S. The resulting relationship (Fig. 4.18B),

$$U^{K'}_{37} = 0.033T + 0.044 \qquad (4.11)$$

is identical within error limits with the widely used calibrations of Prahl and Wakeham (1987) and Prahl et al. (1988) based on *Emiliania huxleyi* cultures ($U^{K'}_{37} = 0.033T + 0.043$). Müller et al. (1998) also found that the best correlations were obtained using ocean water temperatures from 0 to 10 m water depth, suggesting that the sedimentary $U^{K'}_{37}$ ratio reflects mixed-layer temperatures and that the production of alkenones within or below the thermocline was not high enough to significantly bias the mixed-layer temperature signal. Regional variations in the seasonality of primary production also have only a negligible effect on the $U^{K}_{37}$ signal in the sediments. Furthermore, the strong linear relationships obtained for the South Atlantic and the global ocean indicate that $U^{K'}_{37}$ values of the sediments are neither affected to a measurable degree by changing species compositions nor by growth rate of algae and

A)

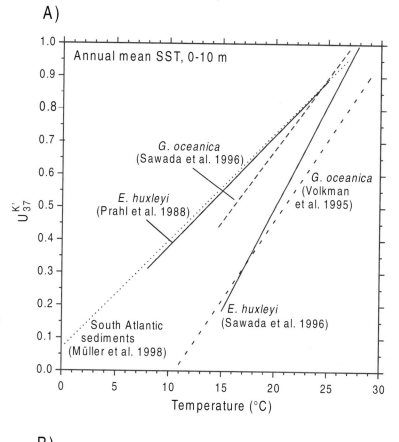

**Fig. 4.18** A) Comparison between $U^{K'}_{37}$-SST calibration for the eastern South Atlantic using annual mean SST at 0-10 m water depth (dotted line) and published culture calibrations for *E. huxleyi* (solid lines) and *G. oceanica* (dashed lines). The excellent agreement between the sediment-based calibration and the culture equation of Prahl et al. (1988) indicates that the North Pacific strain cultured by these authors is also representative for the alkenone-synthesizing algae in the South Atlantic (after Müller et al. 1998). B) Relationships between $U^{K'}_{37}$ and annual mean SST (0 m) for surface sediments from the global ocean between 60°N and 60°S (after Müller et al. 1998; see the original publication for regression parameters for the three oceans (dashed lines) and the global ocean (solid line)).

B)

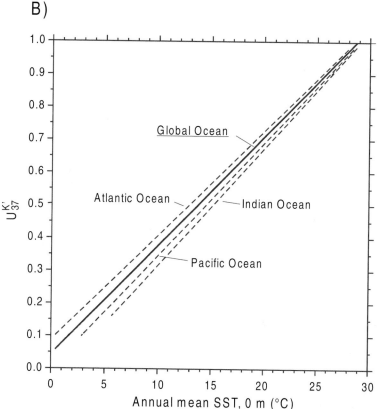

nutrient availability, other than expected from culture experiments.

### ACL Index Based on Land Plant Wax Alkanes

In marine sediments, higher-plant organic matter can be an indicator of climate variations both by the total amount indicating enhanced continental run-off during times of low sea-level or of humid climate on the continent and by specific marker compounds indicating a change in terrestrial vegetation as a consequence of regional or global climatic variations. Long-chain $n$-alkanes are commonly used as the most stable and significant biological markers of terrigenous organic matter supply (e.g. Eglinton and Hamilton 1967). The odd carbon-numbered $C_{27}$, $C_{29}$, $C_{31}$ and $C_{33}$ $n$-alkanes are major components of the epicuticular waxes of higher plants. These terrestrial biomarkers are often preferentially enriched in the marine environment, particularly under oligotrophic surface waters, because the compounds are protected by the resistant character of the plant particles and in part by the highly water-insoluble nature of the waxes themselves (Kolattukudy 1976).

The carbon number distribution patterns of $n$-alkanes in leaf waxes of higher land plants depend on the climate under which they grow. The distributions show a trend of increasing chain length nearer to the equator, i.e. at lower latitude (Gagosian et al. 1987), but they are also influenced by humidity (Hinrichs et al. 1998). In addition, waxes of tree leaves have molecular distributions different from those of grasses with either the $C_{27}$ or the $C_{29}$ $n$-alkane having the highest relative concentration in trees and the $C_{31}$ $n$-alkane in grasses (Cranwell 1973). Poynter (1989) defined an Average Chain Length (ACL) index to describe the chain length variations of $n$-alkanes,

$$ACL_{27-33} = (27[C_{27}] + 29[C_{29}] + 31[C_{31}] + 33[C_{33}])/([C_{27}] + [C_{29}] + [C_{31}] + [C_{33}])$$

$$(4.12)$$

in which $[C_x]$ signifies the concentration of the $n$-alkane with x carbon atoms. Poynter (1989) demonstrated the sensitivity of sedimentary $n$-alkane ACL values to past climatic changes. In Santa Barbara basin (offshore California) sediments from the last 160,000 years, Hinrichs et al. (1998) found the highest ACL values in the Eemian climate optimum (about 125,000 yr B.P.). Over the

entire sediment section, the ACL values were higher in homogeneous sediment layers deposited in periods of more humid climate than in laminated sediments deposited under a semi-arid continental climate like that of today (Fig. 4.19). The sedimentary ACL variations most probably recorded the climatic changes on the continent for the following two reasons:

Vegetation patterns on the continent responded rapidly to climatic oscillations, which were often characterized by drastic changes of temperature and precipitation. During relatively warm and dry periods when mainly laminated sediments were accumulated in the Santa Barbara basin, smaller proportions of grass-derived biomass may have contributed to the sedimentary organic matter.

Changes in continental precipitation significantly affected the degree of erosion and the transport of terrigenous detritus to the ocean, enhancing the proportion of biomass from other source areas (probably over longer distances from higher altitudes). This explains best the almost parallel and abrupt changes of oceanic conditions (e.g. bottom-water oxygen concentrations affecting sediment texture) and terrestrial signals recorded in the sediments (e.g. ACL indices).

## 4.5    Analytical Techniques

Organic geochemical analyses of marine sediments may range from the rapid determination of a few bulk parameters to a high level of sophistication if trace organic constituents are to be investigated at the molecular level. The analytical scheme in Figure 4.20 is one of many that are currently used in different laboratories. It is certainly not complete nor is it fully applied to each sample, particularly not when large series are studied. For example, the analysis of amino acids (e.g. Mitterer 1993), sugars (Cowie and Hedges 1984, Moers et al. 1993), lignin (Goñi and Hedges 1992), or humic substances (Brüchert 1998, Senesi and Miano 1994, Rashid 1985) requires a modification of the scheme in Figure 4.20. Other than in inorganic geochemical analysis, where modern instrumentation allows the simultaneous determination of a wide range of element concentrations, the millions of possible organic compounds require an *a priori* selectivity, but even the analysis of selected lipids can be quite time-consuming. In addition, the amount of sample material required for molecular

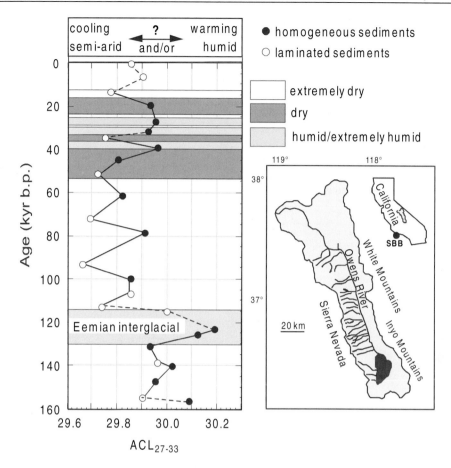

**Fig. 4.19** ACL time-series in relation to sediment texture (homogeneous *versus* laminated) for a core in the central Santa Barbara basin offshore California (after Hinrichs et al. 1998). Information on continental humidity from a sediment core in the Owens Lake basin was taken from Benson et al. (1996).

organic geochemical studies is higher than for many other types of analyses, and this limits stratigraphic (time) resolution.

### 4.5.1  Sample Requirements

Most organic geochemical methods require a well homogenized, pulverized sample aliquot. An exception is reflected-light microscopy in organic petrography where the association of organic matter with the sediment matrix can be quite informative and is, therefore, preserved. Before homogenization, the sample needs drying either at ambient temperature or by freeze-drying. Higher temperatures are to be avoided due to the thermal lability of the organic matter. For the same reason, sediments - particularly those of young age in which bacteria may still be active - should be stored deep-frozen (-18°C) between sampling and analysis. Grinding can be done by mortar and

pestle or in an electrical ball or disc mill, but excessive grinding should be avoided due to the associated rise in temperature in the sample.

### 4.5.2  Elemental and Bulk Isotope Analysis

The basic parameter determined in most organic geochemical studies is the total organic carbon (TOC, $C_{org}$) content. Most marine sediments and sedimentary rocks contain carbon both as carbonates ($C_{inorg}$, $C_{carb}$, $C_{min}$) and as organic matter. There are numerous methods for quantifying carbon, most of them are based on heating solid samples in an oxygen atmosphere and detection of the evolving $CO_2$ by coulometric or spectrometric techniques or by a thermal conductivity detector. Commonly used instruments are elemental analyzers, which determine carbon, nitrogen, hydrogen (only applicable to pure organic matter), and

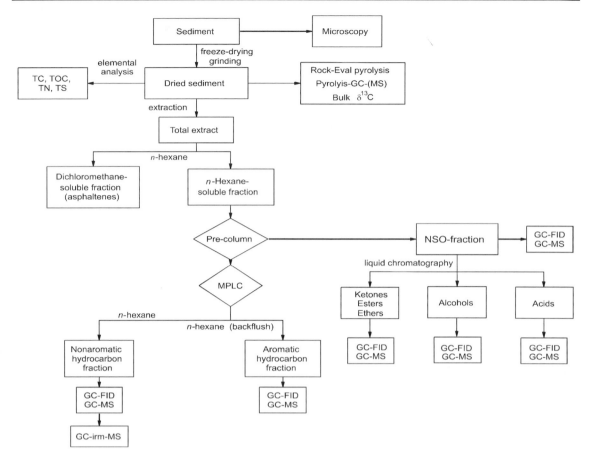

**Fig. 4.20** Example of an analytical scheme of organic geochemical analyses of marine sediments comprising bulk and molecular parameters.

sulfur (CHN, CNS, CS analyzers). Organic carbon is either determined directly, after destruction of carbonate carbon with mineral acids before combustion in the elemental analyzer, or as the difference between total carbon (combustion) and mineral carbon (measurement of $CO_2$ released upon acid treatment).

For the determination of bulk stable carbon isotope ratios ($^{13}C/^{12}C$) the organic matter is converted to carbon dioxide by oxidation following digestion of the sediment with mineral acid to remove carbonates. Traditionally, oxidation of organic matter was performed off-line, $CO_2$ was separated from other gaseous oxidation products, and the purified gas introduced into an isotope ratio mass spectrometer. Modern instruments provide on-line combustion isotope-ratio measurement facilities. In this case, an elemental analyzer is connected to an isotope ratio mass spectrometer *via* a special interface that allows removal of gases other than $CO_2$. This configuration can also

be used to separate sulfur and nitrogen oxides which, after on-line conversion to a suitable single species ($SO_2$ and $N_2$, respectively), can be used to determine stable sulfur ($^{34}S/^{32}S$) and nitrogen ($^{15}N/^{14}N$) isotope ratios. A special technical configuration of the mass spectrometer is required for hydrogen isotope ($^2H/^1H$) ratio measurement. Isotope ratios are not determined directly, but relative to an internationally accepted standard. The results are reported in the delta notation ($\delta^{13}C$, $\delta^{34}S$, $\delta^{15}N$, $\delta^2H$ [or commonly $\delta D$ for deuterium]) relative to this standard. For details see, e.g., Fogel and Cifuentes (1993) and references therein.

### 4.5.3    Rock-Eval Pyrolysis and Pyrolysis Gas Chromatography

Rock-Eval pyrolysis (Espitalié et al. 1985) is conducted using bulk sediment samples to determine, (1) the amount of hydrocarbon-type compounds already present in the sample (S1 peak [mg hydrocarbons per g sediment]; compounds released at low temperature and roughly equivalent to the amount of organic matter extractable with organic solvents), (2) the amount of hydrocarbon-type compounds generated by pyrolytic degradation of the macromolecular organic matter during heating up to 550°C (S2 peak [mg hydrocarbons per g sediment]), (3) the amount of carbon dioxide released from the organic matter up to 390°C, i.e. before carbonates decompose (S3 peak [mg $CO_2$ per gram sediment]), and (4) the temperature of maximum pyrolysis yield (Tmax [°C]). The Hydrogen Index (HI) and Oxygen Index (OI) derived from the S1 and S2 values correspond to the pyrolysis yield normalized to the content of organic carbon (mg hydrocarbons and mg $CO_2$ per g TOC, respectively). The results of Rock-Eval pyrolysis are usually displayed in a van-Krevelen-type diagram of HI *versus* OI values which roughly corresponds to an H/C *versus* O/C atomic ratio van Krevelen diagram (see Fig. 4.11; Tissot and Welte 1984).

One of the methods of studying the composition of macromolecular sedimentary organic matter in more detail is the molecular analysis of pyrolysis products. For this purpose, the pyrolysis products are transferred to a gas chromatographic column and analyzed as described for extractable organic matter in Section 4.5.4, with or without the combination with a mass spectrometer. Both flash pyrolysis (Curie-point pyrolysis; samples are heated on a magnetic wire by electrical induction almost instantaneously, e.g., to 610°C) or off-line pyrolysis at various heating rates have been applied to geological samples (see Larter and Horsfield 1993 for an overview of various pyrolysis techniques).

### 4.5.4    Organic Petrography

Organic petrography is the study of the macroscopically and, more importantly, microscopically recognizable organic matter components initially of coal, but meanwhile more generally in sediments and sedimentary rocks. Organic-matter-rich rocks and coal are usually studied as polished blocks in reflected light under a petrographic microscope. When sediments are lean in organic matter, the organic particles have to be concentrated by dissolution of the mineral matrix in consecutive treatments with hydrochloric and hydrofluoric acids. The concentrates are then analyzed

as smear slides in transmitted light or embedded in araldite® resin and subsequently studied as polished blocks similar to whole-rock samples.

Organic particles visible under the microscope (>1 µm) are called macerals. The most important groups in the order of increasing reflectance are liptinite, vitrinite and inertinite. Liptinites are lipid-rich parts of aquatic (e.g. alginite) or land plants (e.g. cutinite, suberinite, sporinite, resinite), the terms indicating the origin of these organoclasts. Many liptinites are probably related to nonhydrolyzable, highly aliphatic biopolymers found in algae (algaenan) and land plants (cutinan, suberan) (see Sect. 4.3.3). These biopolymers serve as cell wall components of the organisms and their stability allows the morphological shapes of plant material to be preserved after sedimentation and burial so that they can be identified under the microscope. Vitrinites derive from the woody parts of higher plants. Inertinites are highly reflecting particles of strongly oxidized or geothermally heated organic matter of various origin, most commonly from higher plants. Non-structured, often (incorrectly) called amorphous, organic matter is known as bituminite or sapropelinite. Lipid-rich organic matter, even if finely dispersed, can be recognized under the microscope after UV irradiation by its bright fluorescence.

In addition to maceral distribution, organic petrographers determine vitrinite reflectance as a measure of geothermal evolution of sedimentary organic matter. For more details of microscopic analysis see Stach et al. (1982).

### 4.5.5    Bitumen Analysis

The larger part of sedimentary organic matter is insoluble in organic solvents. The proportion of the soluble fraction (bitumen) can be relatively high in surficial sediments, then decreases in amount with increasing depth of burial due to formation of humic substances and kerogen. It only increases again when temperatures become high enough for thermal kerogen cracking to generate petroleum (Tissot and Welte 1984).

The most common solvent used for extraction of bitumen from dried sediments is dichloromethane ($CH_2Cl_2$) with a small admixture (e.g. 1%) of methanol, although more polar mixtures like chloroform/methanol or chloroform/toluene are also used. Occasionally, when very polar lipids from surficial sediments are to be extracted, wet sediment samples are preferred, and extraction starts with acetone or methanol or a mixture of these two because they mix with water. Extraction is then repeated with dichloromethane or a solvent of similar polarity. Extraction in a Soxhlet apparatus usually takes one to two days, while reasonably complete extracts can be obtained within minutes with the support of ultrasonication or blending. After filtration, the solvent is removed by rotary evaporation and the total extract yield determined gravimetrically, as is commonly done for any subfraction after further separation. Internal standards for quantitation of single compounds are added either before or after extraction, occasionally only after liquid chromatographic separation (see later).

The most polar and highest-molecular-weight extract components (called asphaltenes, a term derived from petroleum geochemistry, but often also applied to surficial sediments) may interfere with many subsequent separations and analyses. For this reason, these components are frequently removed from the total extract by dissolving it in a small amount of, e.g., dichloromethane and adding a large excess of a nonpolar solvent like *n*-hexane (or *n*-pentane, *n*-heptane). The *n*-hexane-soluble extract fraction can be further separated into compound classes (or fractions of similar polarities) either by column liquid chromatography, medium-pressure liquid chromatography (MPLC; Radke et al. 1980), high-performance liquid chromatography (HPLC), or thin-layer chromatography (TLC), depending on the sample quantity used and the sophistication of the separation required. In MPLC separation, as indicated in Figure 4.20, the polar lipids are withheld by a pre-column filled with deactivated silica gel, and only the hydrocarbons are separated on the main silica gel column into nonaromatic and aromatic hydrocarbon fractions. Because the main separation system is only operated with *n*-hexane, the difficult separation between the two hydrocarbon fractions is very reproducible, an important prerequisite for some applications, particularly in petroleum geochemistry. The polar lipids can be removed from the pre-column with a polar solvent. There are many variations of this separation scheme. The nature of the geological samples and the scientific objectives will determine the extent of separation required.

Total extracts and/or liquid chromatographic subfractions are then analysed by capillary column gas chromatography using a flame ionization

detector. Except for hydrocarbon fractions, derivatization is commonly applied to render polar lipids more volatile in order to reduce gas chromatographic retention times and to improve peak shape at the detector. Carboxylic acids are usually transformed into their methyl esters, and hydroxyl or amine groups into their trimethylsilyl ether derivatives. Alternatively, both acid and hydroxyl groups can be silylated. Acetate formation is another common derivatization method. A variety of derivatization reagents are commercially available for this purpose.

Only a few major compound series can be recognized by their molecular structure based on relative retention times and distribution patterns by gas chromatography alone. This applies to *n*-alkanes in the nonaromatic hydrocarbon fraction, *n*-fatty acids in the carboxylic acid fraction and in some cases *n*-alkanols in the neutral polar fraction. High abundances of a few single compounds (pristane, phytane, long-chain alkenones) sometimes also allow their direct identification from gas chromatograms.

The most powerful technique for assigning molecular structures to constituents of complex mixtures as they are found in the lipid extracts of geological samples is the combination of capillary column gas chromatography and mass spectrometry (GC-MS). Although the expression "identification" is frequently used, GC-MS alone is insufficient to fully characterize a new compound whose gas chromatographic and mass spectrometric behavior has not been described before. Normally, GC-MS analysis relies on a comparison with GC and MS data published in the (geochemical) literature or with data of standards, commercially available or synthesized in the laboratory, or on the interpretation of mass spectral fragmentation patterns following common empirical rules (e.g. McLafferty 1993). Unfortunately, there is no comprehensive compilation of mass spectra of geochemically relevant organic compounds. Peters and Moldowan (1993) have provided a detailed coverage of hydrocarbons of significance in petroleum geochemistry, and the mass spectral compilation of Philp (1985) may also be of some use, although the spectra are of variable quality.

The youngest, revolutionary development in analytical organic geochemistry is the on-line coupling of a gas chromatograph to an isotope ratio mass spectrometer *via* a combustion interface (GC-irm-MS; Hayes et al. 1990, Freeman et al.

1994). This instrument allows the determination of stable carbon isotope ratios of single organic compounds in complex mixtures provided they are gas chromatographically reasonably well separated. Chemotaxonomic relations to specific precursor organisms are then possible if these are distinct from other organisms in their carbon isotope fractionation behavior during photosynthesis and biosynthesis. Sample preparation and analysis are similar to those for GC-MS analysis with the provision that the isotopic composition of derivatizing agents is accounted for in data interpretation.

## 4.6 The Future of Marine Geochemistry of Organic Matter

The evolution of organic geochemistry has always been closely connected to the developments in instrumental techniques for the analysis of organic compounds in geological samples. Now that the instrumentation has reached a high level of sophistication and particularly sensitivity, one of the future targets will certainly be higher stratigraphic (time) resolution which is particularly important for climate research. Advancement in the fundamental understanding particularly of the early part of the geological organic carbon cycle will depend on the cooperation between organic geochemists and microbiologists. They will have to refine the knowledge of the biological effects on the early diagenesis of organic matter arriving at the sediment-water interface and becoming buried in the uppermost sediment. It will be necessary to broaden the natural product inventory of microorganisms, both of sedimentary bacteria and archaea and of unicellular algae, protozoans and other organisms at the lower end of the food chain in order to arrive at solid chemotaxonomic relationships between source organisms and molecular fossils.

Carefully designed laboratory simulation experiments together with high-resolution field studies will refine the mass balance approaches of organic matter exchange between sediment and water column related to early diagenetic processes. Finally, mathematical modeling of transport and reaction processes will become an increasingly important tool in marine geochemistry of organic matter.

## Acknowledgement

Philip A. Meyers (University of Michigan, Ann Arbor, USA), Peter Müller (University of Bremen, Germany) and Rüdiger Stein (Alfred Wegener Institute of Polar and Marine Research, Bremerhaven, Germany) kindly provided material for preparation of figures.

## References

Alperin, M.J., Reeburgh, W.S. and Devol, A.H., 1992. Organic carbon remineralization and preservation in sediments of Scan Bay, Alaska. In: Whelan, J.K. and Farrington, J.W. (eds), Organic matter: Productivity, accumulation and preservation in recent and ancient sediments. Columbia University Press, NY, pp. 99-122.

Bailey, G.W., 1991. Organic carbon flux and development of oxygen deficiency on the northern Benguela continental shelf south of 22°S: spatial and temporal variability. In: Tyson, R.V. and Pearson, T.H. (eds), Modern and ancient continental shelf anoxia, Geol. Soc. Spec. Publ., 58, Blackwell, Oxford, pp. 171-183.

Benson, L.V., Burdett, J.W., Kashgarian, M., Lund, S.P., Phillips, F.M. and Rye, R.O., 1996. Climatic and hydrologic oszillations in the Owens Lake basin and adjacent Sierra Nevada, California. Science, 274: 746-749.

Berger, W.H., Smetacek, V.S. and Wefer, G. (eds) 1989. Productivity of the ocean: Present and past. Dahlem Workshop Rep, Life Sci. Res. Rep., 44, Wiley, Chichester, 471 pp.

Berger, W.H., 1989. Global maps of ocean productivity. In: Berger, W.H., Smetacek, V.S., Wefer, G. (eds), Productivity of the ocean: Present and past; Dahlem Workshop Rep, Life Sci. Res. Rep., 44, Wiley, Chichester, pp. 429-456.

Berner, R. and Raiswell, R., 1983. Burial of organic carbon and pyrit sulfur in sediments over Phanerozoic time: a new theory. Geochim. Cosmochim. Acta, 47: 885-862.

Betzer, P.R., Showers, W.J., Laws, E.A., Winn, C.D., Ditullo, G.R. and Kroopnick, P.M. 1984. Primary productivity and particle fluxes on a transect of the equator at 153°W in the Pacific Ocean. Deep-Sea Research, 31: 1-11.

Blokker, P., Schouten, S., Sinninghe Damsté, J.S., van den Ende, H. and de Leeuw, J.W., 1998. Chemical structure of algaenans from fresh water algae *Tetraedon minimum*, *Scenedesmus communis* and *Pediastrum boryanum*. Org. Geochem., 29: 1453-1486.

Bouloubassi, I., Rullkötter, J. and Meyers, P.A., 1999. Origin and transformation of organic matter in Pliocene-Pleistocene Mediterranean sapropels: Organic geochemical evidence reviewed. Mar. Geol. 153: 177-199.

Bralower, T.J. and Thierstein, H.R., 1987. Organic carbon and metal accumulation in Holocene and mid-Cretaceous marine sediments: paleoceanogaphic significance.In: Brooks, J. and Fleet, A.J. (eds), Marine petroleum source rocks. Geol. Soc. Spec. Publ., 26, Blackwell, Oxford, pp. 345-369.

Brassel, S.C., Englinton, G., Marlowe, I.T., Pflaumann, U. and Sarnthein, M., 1986. Molecular stratigraphy: a new tool for climatic assessment. Nature, 320: 129-133.

Brassel, S.C., 1993. Application of biomarkers for delineating marine paleoclimatic fluctuations during the Pleistocene. In: Engel, M.H. and Macko, S.A. (eds), Organic geochemistry. Princibles and applications. Plenum Press, NY, pp. 699-738.

Brüchert, V., 1998. Early diagnosis of sulfur in estuarine sediments: The role of sedimentary humic and fulvic acids. Geochim. Cosmochim. Acta, 62: 1567-1586.

Calvert, S.E., 1987. Oceanic controls on the accumulation of organic matter in marine sediments. In: Brooks, J. and Fleet, A.J. (eds), Marine petroleum source rocks, Geol. Soc. Spec. Publ., 26, Blackwell, Oxford, pp.137-151.

Canuel, E.A. and Martens, C.S., 1996. Reactivity of recently deposited organic matter: Degradation of lipid compounds near the sediment - water interface. Geochim. Cosmochim. Acta, 60: 1793-1806.

Collins, M.J., Bishop, A.N. and Farrimond, P., 1995. Sorption by mineral surfaces: rebirth of the calssical condensation pathway for kerogen formation? Geochim. Cosmochim. Acta, 59: 2387-2391.

Corbet, B., Albrecht, A. and Ourisson, G., 1980. Photochemical or photomimetic fossil triterpenoids in sediments and petroleum. J. Amer. Chem. Soc., 78: 183-188.

Cornford, C., Rullkötter, J. and Welte, D., 1979. Organic geochemistry of DSDP Leg 47a, Site 397, eastern North Atlantic: Organic petrography and extractable hydrocarbons. In: von Rad, U., Ryan, W.B.F. et al. (eds), Init. Repts. DSDP, 47, US Government Printing Office, Washington, DC, pp. 511-522.

Cowie, G.L. and Hedges, J.L., 1984. Determination of neutral sugars in plankton, sediments, and wood by capillary gas chromatography of equilibrated isomeric mixtures. Anal Chem, 56: 497-504.

Cranwell, P.A., 1973. Chain-length distribution of n-alkanes from lake sediments in relation to postglacial environments. Freshw. Biol., 3: 259-265.

de Leeuw, J.W., Cox, H.C., van Graas, G., van de Meer, F.W., Peakman, T.M., Baas, J.M.A. and van de Graaf, B., 1989. Limited double bond isomerisation and selective hydrogenation of sterenes during early diagenesis. Geochim. Cosmochim. Acta, 53: 903-909.

de Leeuw, J.W. and Sinninghe Damsté, J.S., 1990. Organic sulphur compounds and other biomarkers as indicators of palaeosalinity: A critical evaluation. In: Orr, W.L. and White, C.M. (eds), Geochemistry of sulphur in fossil fuels, ACS symposium, 429, Washington, DC, pp. 417-443.

de Leeuw, J.W. and Largeau, C., 1993. A review of macromolecular organic compounds that comprise living organisms and their role in kerogen, coal and petroleum formation. In: Engel, M.H. and Macko, S.A. (eds), Organic geochemistry. Principles and applications. Plenum Press, NY, pp. 23-72.

Demaison, G.J. and Moore, G.T., 1980. Anoxic environments and oil source bed genesis. Bull. Am. Assoc. Petrol. Geol., 64: 1179-1209.

Demaison, G.J., 1991. Anoxiavs productivity: what controls the formation of organic-carbon-rich sediments and sedimentary rocks? Bull. Am. Assoc. Petrol. Geol., 75: 499.

Durand, B., 1980. Kerogen. Insoluble organic matter from the sedimentary rocks. Editions Technip., Paris, 519 pp.

Eglington, G. and Hamilton, R.J., 1967. Leaf epicuticular waxes. Science, 156: 1322-1335.

Emeis, K.-C., Robertson, A.H.F. and Richter, C., et al. (eds) 1996. Proceedings of the Ocean Drilling Program. Initial Reports., 160, ODP, College Station (TX), 972 pp.

Emerson, S. and Hedges, J.I., 1988. Processes controlling the organic carbon content of open ocean sediments. Paleoceanography, 3: 621-634.

Espitalié, J., Deroo, G. and Marquis, F., 1985. La pyrolyse Rock-Eval et ses applications. Rev. Inst. Fr. Pét., 40: 755 - 7´84.

Fogel, M.L. and Cifuentes, L.A., 1993. Isotope fractionation during primary production. In: Engel, M.H. and Macko, S.A. (eds), Organic geochemistry. Principles and applications. Plenum Press, NY, pp. 73-98.

Freeman, K.H., Boreham, C.J., Summons, R.E. and Hayes, J.M., 1994. The effect of aromatization on the isotopic compositions of hydrocarbones during early diagnesis. Org. Geochem., 21: 1037-1049.

Froelich, P.N., Klinkhammer, G.P., Bender, M.L., Luedtke, N.A., Heath, G.R., Cullen, D., Dauphin, P., Hammond, D., Hartmann, B. and Maynard, V., 1979. Early oxidation of organic matter in pelagic sediments of the eastern equatorial Atlantic: suboxic diagenesis. Geochim. Cosmochim. Acta, 43: 1075-1090.

Fry, B., Scalan, R.S. and Parker, P.L., 1977. Stable carbon isotope evidence for two sources of organic matter in coastel sediments: sea grasses and plankton. Geochim. Cosmochim. Acta, 41: 1875-1877.

Gagosian, R.B., Peltzer, E.T. and Merrill, J.T., 1987. Long-range transport of terrestrially derived lipids in aerosols from the South Pacific. Nature, 325: 800-803.

Goni, M.A. and Hedges, J.I., 1992. Lignin dimers: structures, distribution and potential geochemical applications. Geochim. Cosmochim. Acta, 56: 4025-4043.

Goth, K., de Leeuw, J.W., Püttmann, W. and Tegelaar, E.W., 1988. The origin of Messel shale kerogen. Nature, 336: 759-761.

Hayes, J.M., Freeman, K.H., Popp, B.N. and Hoham, C.H., 1990. Compound - specific isotope analyses, a novel tool for reconstruction of ancient biogeochemical processes. In: Duran, B. and Behar, F (eds), Advances in organic geochemistry 1989. Org. Geochem., 16, Pergamon Press, Oxford, pp. 1115-1128.

Hayes, J.M., 1993. Factors controlling $^{13}C$ contents of sedimentary organic compounds: Principles and evidence. Marine Geology, 113: 111-125.

Heath, G.R., Moore, T.C. and Dauphin, J.P., 1977. Organic carbon in deep-sea sediments. In: Anderson, N.R. and Malahoff, A. (eds), The fate of fossil fuel $CO_2$ in the oceans. Plenum Press, NY, 605-625 pp.

Hedges, J.I., Clark, W.A., Quay, P.D., Richey, J.E., Devol, A.H. and Santos, U.M., 1986a. Composition and fluxes of particulate organic material in the Amazonas River. Limnol. Oceanogr., 31: 717-738.

Hedges, J.I., Cowie, G.L., Quay, P.D., Grootes, P.M., Richey, J.E., Devol, A.H., Farwell, G.W., Schmidt, F.W. and Salati, E. 1986b. Organic carbon-14 in the Amazon river system. Nature, 321: 1129-1131.

Hedges, J.I. and Prahl, F.G., 1993. Early diagenesis: Consequence for applications of molecular biomarkers. Organic geochemistry. Principles and applications. Plenum Press, NY, 237-253 pp.

Hedges, J.I., Ertel, J.R., Richey, J.E., Quay, P.D., Benner, R., Strom, M. and Forsberg, B., 1994. Origin and processing of organic matter in the Amazo River as indicated by carbohydrates and amino acids. Limnol. Oceanogr., 39: 743-761.

Hedges, J.I., Keil, R.G. and Benner, R., 1997. What happens to terrestrial organic matter in the ocen? Org. Geochem., 27: 195-212.

Henrichs, S.M. and Reeburgh, W.S., 1987. Anaerobic mineralization of marine sediment organic matter: rates and the role of anaerobic processes in the oceanic carbon economy. Geomicrobiol. J., 5: 191-237.

Herbin, J.P., Montadert, L.O., Müller, C., Comez, R., Thurow, J. and Wiedmann, J., 1986. Organic-rich sedimentation at the Cenomanian/Turonian boundary in oceniac and coastal basins in the North Atlantic and Teths. In: Summerhayes C.P. and Shakleton N.J. (eds), North Atlantic paleoceanography, Geol. Soc. Spec. Publ., 21, Blackwell, Oxford, pp.389-422.

Hinrichs, K.-U., Rinna, J. and Rullkötter, J., 1998. Late Ouantenary paleoenvironmental conditions indicated by marine and terrestrial molecular biomarkers in sediments from the Santa Barbara basin. In: Wilson, R.C. and Tharp, V.L. (eds), Proceedings of the fourteenth annual Pacific climate (PACLIM) conference, april 6-9, 1997. Interagency Ecological Program, Technical Report, 57, California Department of Water Resources, Marysville (CA), pp. 125-133.

Hinz, K. and Winterer, E.L. et al. (eds), 1984. Initial Reports of the Deep Sea Drilling Project. 79, US Government Printing Office, Washington, DC, 934 pp.

Huc, A.Y. and Durand, B., 1977. Occurrence and significance of humic acids in ancient sediments. Fuel, 56: 73-80.

Huc, A.Y., 1988. Aspects of depositional processes of organic matter in sedimentary basins. In: Mattavelli, L. and Novelli, L. (eds), Advances in organic geochemistry 1987. Org. Geochem., 13, Pergamon Press, Oxford, pp. 263-272.

Hulthe, G., Hulth, S. and Hall, P.O.J., 1998. Effect of oxygen on degradation rate of refractory and labile organic matter in continental margin sediments. Geochim. Cosmochim. Acta, 62: 1319-1328.

Hunt, J.M., 1996. Petroleum geochemistry and geology. Freeman, NY, 743 pp.

Ibach, L.E., 1982. Relationships between sedimentation rate and total organic carbon content in ancient marine sediments. Bull. Am. Assoc. Petrol. Geol., 66: 170-188.

Ittekkot, V., 1988. Global trends in the nature of organic matter in river suspensions. Nature, 332: 436-438.

Jasper, J.P. and Gagosian, R.B., 1989. Alkenone molecular stratigraphy in an oceanic environment affected by glacial fresh-water events. Paleoceanography, 4: 603-614.

Jasper, J.P. and Gagosian, R.B., 1990. The source and deposition of organic matter in the Late Quantenary Pygmy Basin, Gulf of California. Geochim. Cosmochim. Acta, 54: 117-132.

Jasper, J.P., Hayes, J.M., Mix, A.C. and Prahl, F.G., 1994. Photosynthetic fractionation of $^{13}$C and concentrations of dissolved $CO_2$ in the central equatorial Pacific during the last 255,000 years. Paleoceanography, 9: 781-798.

Jørgensen, B.B., 1996. Case Study - Aarhus Bay. In: Jørgensen, B.B. and Richardson, K. (eds), Eutrophication in Coastal Marine Ecosystems. Coastal and Estuarine Studies, 52, American Geophysical Union, Washington, DC, pp. 137-154.

Keil, R.G., Tsamakis, E., Fuh, C.B., Giddings, J.C. and Hedges, J.I., 1994a. Mineralogical and textural controls on the organic matter composition of coastal marine sediments: hydrodynamic seperation using SPLITT-fractionation. Geochim. Cosmochim. Acta, 58: 879-893.

Keil, R.G., Montlucon, Prahl, F.G. and Hedges, J.I., 1994b. Sorptive preperation of labile organic matter in marine sediments. Nature, 370: 549-552.

Koblents - Mishke, O.I., 1977. Primary production. In: Vinogradow,

M.E. (ed) Oceanology and biology of the ocean 1 (in Russia). Nauka, Moscow, 62 pp.

Kolattukudy, P.E., 1976. Chemistry and biochemistry of natural waxes. Elsevier, Amsterdam, 459 pp.

Krey, J., 1970. Die Urproduktion des Meeres (in German) . In: Dietrich, G. (ed), Erforschung des Meeres. Umschau, Frankfurt, pp. 183-195.

Kristensen, E., Ahmed, S.A. and Devol, A.H., 1995. Aerobic and anaerobic decomposition of organic matter in marine sediments. Which is faster? Limnol. Oceanogr., 40: 1430-1437.

Lallier-Verges, E., Bertrand, P. and Desprairies, 1993. Organic matter composition and sulfate reduction intensity in Oman Margin sediments. Marine Geology, 112: 57-69.

Largeau, C., Derenne, S., Casadevall, E., Berkaloff, C., Corolleur, M., Lugardon, B., Raynaud, J.F. and Connan, J., 1990. Occurence and origin of "ultralaminar" structures in "amorphous" kerogens of various source rocks and oil shales. In: Durand, B. and Behar, F. (eds) Advances in organic geochemistry 1989. Org. Geochem., 16, Pergamon Press, Oxford, pp. 889-895.

Largeau, C. and de Leeuw, J.W., 1995. Insoluble, nonhydrolyzable, aliphatic macromolecular constituents of microbial cell walls. In: Jones, J.G. (ed), Advances in Microbial Ecology, 14, Plenum Press, NY, pp. 77-177.

Larter, S.R. and Horsfield, B., 1993. Determination of structural components of kerogens by the use of analytical pyrolysis methods. In: Engel, M.H. and Macko, S.A. (eds), Organic geochemistry. Prinsiples and applications. Plenum Press, NY, pp. 271-287.

Leventhal, J.S., 1983. An interpretation of carbon and sulphur relationships in Black Sea sediments as indicators of environments of depositions. Geochim. Cosmochim. Acta, 47: 133-138.

Littke, R., Rullkötter, J. and Schaefer, R.G., 1991a. Organic and carbonate carbon accumulation on Broken Ridge and Ninetyeast Ridge, central Indian Ocean. In: Weissel, J., Pierce, J., et al (eds), Proceedings of the Ocean Drilling Program. Sci. Res., 121, ODP, College Station (TX), pp. 467-487.

Littke, R., Baker, D.R., Leythaeuser, J. and Rullkötter, J., 1991b. Keys to the depositional history of the Posidonia Shale (Toarcian) in the Hils syncline, northern Germany. In: Tyson, R.V. and Pearson, T.H. (eds), Modern and ancient continental shelf anoxia. Geol. Soc. Spec. Publ., 58, The Geological Society, London, pp. 311-343.

Littke, R., Baker, D.R. and Rullkötter, J., 1997a. Deposition of petroleum source rocks. In: Welte, D.H., Horsfield, B. and Baker D.R. (eds), Petroleum and basin evolution. Springer Verlag, Heidelberg, pp. 271-333.

Littke, R., Lückge, A. and Welte, D.H., 1997b. Quantification of organic matter degradation by microbial sulphate reduction for Quaternary sediments from the northern Arabian Sea. Naturwissenschaften, 84: 312-315.

Lückge, A., Boussafir, M., Lallier - Vergès, E. and Littke, R., 1996. Comparative study of organic matter preservation in immature sediments along the continental margins of Peru and Oman. Part I: Results of petrographical and bulk geochemical data. Org. Geochem., 24: 437-451.

Martens, C.S. and Klump, J.V., 1984. Biogeochemical cycling in an organic-rich coastal marine basin-4. An organic carbon budget for sediments dominated by sulfate reduction and methanogenesis. Geochim. Cosmochim. Acta, 48: 1987-2004.

Martens, C.S., Haddad, R.I. and Chanton, J.P., 1992. Organic matter accumulation, remineralization, and burial in an anoxic coastal sediment. In: Whelan, J.K. and Farrington, J.W. (eds), Organic matter: Productivity, accumulation and preservation in recent and ancient sediments. Columbia University Press, NY, pp. 82-98.

Mayer, L.M., 1994. Surface area control of organic carbon accumulation in continental shelf sediments. Geochim. Cosmochim. Acta, 58: 1271-1284.

McIver, R., 1975. Hydrocarbon occurrences from JOIDES Deep Sea Drilling Project. Proc. 9th Petrol. Congr. (Tokyo), Applied Science Publishers, Barking, 2: 269-280.

McLafferty, F.W. and Turecek, F., 1993. Interpretation of mass spectra. University Science Books, Mill Valley (CA), 371 pp.

Meybeck, M., 1982. Carbon, nitrogen, and phosphorous transport by world rivers. Am. J. Sci., 282: 401-450.

Meyers, P.A., 1994. Preservation of elemental and isotopic identification of sedimentary organic matter. Chem. Geol., 144: 289-302.

Meyers, P.A., 1997. Organic geochemical proxies of paleoceanographic, paleolimnologie, and paleoclimatic processes. Org. Geochem., 27: 213-250.

Mitterer, R.M., 1993. The diagnosis of proteins and amino acids in fossil shells. In: Engel, M.H. and Macko, S.A. (eds), Organic geochemistry. Principles and applications. Plenum Press, NY, pp. 739-753.

Moers, M.E.C., Jones, D.M., Eakin, P.A., Fallick, A.E., Griffiths, H. and Larter, S.R. (eds), 1993. Carbohydrate diagenesis in hypersaline environments: application of GC-IRMS to the stable isotope analysis of derivatives of saccharides from surficial and buried sediments. Org. Geochem., 20: 927-933.

Müller, P.J., 1977. C/N ratios in Pacific deep-sea sediments: effect of inorganic ammonium and organic nitrogen compounds sorbed by clays. Geochim. Cosmochim. Acta, 41: 765-776.

Müller, P.J. and Suess, E., 1979. Productivity, sedimentation rate, and sedimentary organic matter in the oceans - Organic carbon preservation. Deep-Sea Res., 26: 1347-1362.

Müller, P.J., Kirst, G., Ruhland, G., von Storch, I. and Rosell - Melé, A., 1998. Calibration of the alkenone paleotemperature index $U^K_{37}$ based on coretops from the eastern South Atlantic and the global ocean (60°N-60°S). Geochim. Cosmochim. Acta, 62: 1757-1772.

Parkes, J.R., Cragg, B.A., Bale, S.J., Getliff, J.M., Goodman, K., Rochelle, P.C., Fry, J.C., Weightman, A.J. and Harvey, S.M., 1994. Deep bacterial biosphere in Pacific Ocean sediments. Nature, 371: 410-413.

Petersen, T.F. and Calvert, S.E., 1990. Anoxia versus productivity: What controls the formation of organic-carbon-rich sediments and sedimentary rocks? Bull. Am. Assoc. Petrol. Geol., 74: 454-466.

Pedersen, T.F. and Calvert, S.E., 1991. Anoxia vs. productivity: What controls the formation of organic-carbon-rich sediments and sedimentary rocks? Reply. Bull. Am. Assoc. Petrol. Geol., 75: 500-501.

Peters, K.E. and Moldowan, J.M., 1993. The biological marker guide. Interpreting molecular fossils in petroleum and ancient sediments. Prentice Hall, Englewood Cliffs (NJ), 363 pp.

Philp, R.P., 1985. Fossil fuel biomarkers - Applications and spectra. Methods in geochemistry and geophysics, 23, Elsevier, Amsterdam, 294 pp.

Platt, T. and Subba Rao, D.V., 1975. Primary production of marine microphytes. In: Cooper J.P. (ed) Photosynthesis and productivity in different environments. Cambridge University Press,

Cambridge, pp. 249-279 pp.

Plough, H., Kühl, M., Buchholz-Cleven, B. and Jørgensen, B.B., 1997. Anoxic aggregates - an ephemeral phenomenon in the pelagic environment? Aqu. Microb. Ecol., 13: 285-294.

Poynter, J., 1989. Molecular stratigraphy: The recongnition of paleoclimate signals in organic geochemical data. PhD Thesis, University of Bristol, 324 pp.

Poynter, J. and Eglinton, G., 1991. The biomarker concept - strengths and weakness. Fresenius J. Anal. Chem., 339: 725-731.

Prahl, F.G. and Wakeham, S.G., 1987. Calibration of unsaturation patterns in long-chain ketones compositions for paleotemperature assessment. Nature, 330: 367-369.

Prahl, F.G., Muelhausen, L.A. and Zahnle, D.L., 1988. Further evaluation of long-chain alkenones as indicators of paleoceanographic conditions. Geochim. Cosmochim. Acta, 52: 2303-2310.

Prahl, F.G., 1992. Prospective use of molecular paleontology to test for iron limitation on marine primary productivity. Mar. Chem., 39: 167-185.

Prahl, F.G., Ertel, J.R., Goñi, M.A., Sparrow, M.A. and Eversmeyer, B., 1994. Terrestrial organic carbonm contributions to sediments on the Washington margin. Geochim. Cosmochim. Acta, 58: 3035-3048.

Radke, M., Willsch, H. and Welte, D.H., 1980. Preperative hydrocarbon group type determination by automated medium pressure liquid chromatography. Anal Chem, 52: 406-411.

Ransom, B., Kim, D., Kastner, M. and Wainwright, S., 1998. Organic matter preservation on continental slopes: Importance of mineralogy and surface area. Geochim. Cosmochim. Acta, 62: 1329-1345.

Rashid, M.A., 1985. Geochemistry of marine humic compounds. Springer Verlag, Berlin, Heidelberg, NY, 300pp.

Rau, G.H., Takahashi, T., Des Marais, D.J. and Sullivan, C.W., 1991. Particulate organic matter d$^{13}$C variations across the Drake Passage. J. Geophys. Res., 96: 15131-15135.

Redfield, A.C., Ketchum, B.H. and Richards, F.A., 1963. The influence of organisms on the composition of sea-water. In: Hill, M.N. (ed), The sea, 2, Wiley Interscience, NY, pp. 26-77.

Rohmer, M., Bisseret, P. and Neunlist, S., 1992. The hopanoids, prokaryotic triterpenoids and precoursors of ubiquitous molecular fossil. In: Moldowan, J.M., Albrecht, P. and Philip, R.P. (eds), Biological markers in sediments and petroleum. Prentice Hall, Englewood Cliffs (NJ), pp. 1-17

Romankevich, E.A., 1984. Geochemistry of organic matter in the ocean. Springer, Heidelberg, 334 pp.

Rullkötter, J., Cornford, C. and Welte, D.H., 1982. Geochemistry and petrogaphy of organic matter in Northwest African continentalmagin sediments: quantity, provenance, depositional environment and temperature history. In: von Rad, U., Hinz, K., Sarntheim, M. and Seibold, E. (eds), Geology of the Northwest African continental margin. Springer Verlag, Heidelberg, pp. 686-703.

Rullkötter, J., Vuchev, V., Hinz, K., Winterer, E.L., Baumgartner, P.O., Bradshaw, M.L., Channel, J.E.T., Jaffrezo, M., Jansa, L.F., Leckie, R.M., Moore, J.M., Schaftenaar, C., Steiger, T.H. and Wiegand, G.E., 1983. Potential deep sea petroleum beds related to coastal upwelling. In: Thiede, J. and Suess, E. (eds), Coastal upwelling: Its sediment record, Part B: Sedimentary records of ancient coastal upwelling. Plenum Press, NY, pp. 467-483.

Rullkötter, J., Mukhopadhyay, P.K., Schaefer, R.G. and Welte, D.H., 1984. Geochemistry and petrography of organic matter in sediments from the Deep Sea Drilling Project Sites 545 and 547, Mazagan Escarpment. In: Hinz, K., Winterer, E.L., et al. (eds), Initial Reports DSDP, 79, US Government Printing Office, Washington, DC, pp. 775-806.

Rullkötter, J., Mukhopadhyay, P.K. and Welte, D.H., 1987. Geochemistry and petrography of organic matter from the Deep Sea Drilling Project Site 603, lower continental rise off Cape Hateras. In: van Hite, J.E., Wise, S.E., et al. (eds), Initial reports DSDP, 92, US Government Printing Office, Washington, DC, pp. 1163-1176.

Rullkötter, J. and Michaelis, W., 1990. The structure of kerogen and related materials. A review of recent progress and future trends. In: Durand, B. and Behar, F (eds), Advances in organic geochemistry 1989. Org. Geochem., 16, Pergamon Press, Oxford, pp. 829-852.

Rullkötter, J., 1992. Geochemistry, organic. In: Meyers, R.A. (ed), The encyclopedia of physical science and technology. 7, Academic Press, Orlando, pp. 165-192.

Sarnthein, M., Winn, K. and Zahn, R., 1987. Paleoproductivity of oceanic and the effect of atmospheric $CO_2$ and climatic change during deglaciation times. In: Berger, W.H. and Labeyrie, L.D. (eds), Abrupt climatic change. Reidel, Dordrecht, pp. 311-337.

Sarnthein, M., Winn, K., Duplessy, J.C. and Fontugne, M.R., 1988. Global variations of surface water productivity in low- and mid - latitudes: influence of $CO_2$ reservoirs of the deep ocean and atmosphere during the last 21,000 years. Paleoceanography, 3: 361-399.

Sawada, K., Handa, N., Shiraiwa, Y., Danbara, A. and Montani, S., 1996. Long-chain alkenones and alkyl alkenoates in the coastel and pelagic sediments of the northwest North Pacific, with special reference to the reconstruction of the *Emiliania huxleyi* and *Gephyrocapsa oceanica* ratios. Org. Geochem., 24: 751-764.

Schidlowski, M., 1988. A 3,800-million-year isotopic record of life from carbon in sedimentary rocks. Nature, 333: 313-318.

Schlesinger, W.H. and Melack, J.M., 1981. Transport of organic carbon in the world's rivers. Tellus, 33: 172-187 pp.

Senesi, N. and Miano, T.M., 1994. Humic substances in the global environment and implications on human health. Elsevier, Amsterdam.

Simoneit, B.R.T., 1986. Cyclic terpenoids of the geosphere. In: Johns, R.B. (ed), Biological markers in the sedimentary record. Elsevier, Amsterdam, pp. 43-99.

Sinninghe Damsté, J.S., Eglinton, T.I., de Leeuw, J.W. and Schenck, P.A., 1989a. Organic sulphur in macromolecular sedimentary organic matter I. Structure and origin sulphur-containing moieties in kerogen, asphaltenes and coal as revealed by flash pyrolysis. Geochim. Cosmochim. Acta, 53: 873-899.

Sinninghe Damsté, J.S., Rijpstra, W.I.C., Kock-van Dalen, A.C., de Leeuw, J.W. and Schenck, P.A., 1989b. Quenching of labile functionalized lipids by inorganic sulphur species: Evidence for the formation of sedimetary organic sulphur compounds at the early stages of diagenesis. Geochim. Cosmochim. Acta, 53: 1343-1355.

Sinninghe Damsté, J.S., Eglinton, T.I., Rijpstra, W.I.C. and de Leeuw, J.W., 1990. Characterization of organically bound sulphur in high - molecular - weight sedimentary organic matter using flash pyrolysis and Raney Ni desulphurisation. In: Orr, W.L. and White, C.M. (eds), Geochemistry of sulphur in fossil fuels, ACS symposium, 429, American Chemical Society, Washington, DC, pp.486-528.

Sinninghe Damsté, J.S. and Köster, J., 1998. A euxinic North Atlantic ocean during the Cenomanian/Turonian oceanic anoxic event. Earth Planet. Sci. Lett., 158: 165-173.

Stach, E., Mackowsky, M., Teichmüller, M., Taylor, G.H., Chandra, D. and Teichmüller, R., 1982. Stach's textbook of coal petrology. Gebrüder Bornträger, Stuttgart, 428 pp.

Stein, R., 1986a. Surface-water paleo-productivity as interrred from sediment deposits on oxic and anoxic deep-water environments of the Mesozoic Atlantic Ocean. Mitt. Geol.-Paläont. Inst. Univ. Hamburg, 60: 55-70

Stein, R., 1986b. Organic carbon and sedimentation rate - further evidence for anoxic deep-water conditions in the Cenomanian/ Turonian Atlantic Ocean. Marine Geology, 72: 199-209.

Stein, R., ten Haven, H.L., Littke, R., Rullkötter, J. and Welte, D.H., 1989. Accumulation of marine and terrigenous organic carbon at upwelling Site 658 and nonupwelling Sites 657 and 659: implications for the reconstruction of paleoenvironments in the eastern subtropical Atlantic through Late Cenozoic times. In: Rudiman, W., Sarnheim, M., et al. (eds), Proceedings of the Ocean Drilling Program, Sci. Res., 108, ODP, College Station (TX), pp. 361-385.

Stein, R., 1990. Organic carbon content/sedimentation rate relationship and its paleoenvironmental significance for marine sediments. Geo-Mar. Lett., 10: 37-44.

Stein, R., 1991. Accumulation of organic carbon in marine sediments. Lect. Notes Earth Science, 34: 1-217.

Stein, R. and Rack, F., 1995. A 160,000-year high-resolution record of quantity and composition of organic carbon in the Santa Barbara basin (Site 893). In: Kennett, J.P., Baldauf, J. and Lyle, M. (eds), Proceedings of the Ocean Drilling Program, Sci. Res., 146, ODP, College Station (TX), pp. 125-138.

Suess, E. and Thiede, J., 1983. Coastal upwelling: its sediment record. Part A: Responses of the sedimentary regime to present coastal upwelling. Plenum Press, NY, 604 pp.

Summerhayes, C.P., 1981. Organic facies of middle Cretaceous black shales in deep North Atlantic. Bull. Am. Assoc. Petrol. Geol., 65: 2364-2380.

Summerhayes, C.P., Prell, P.I. and Emeis, K.C. (eds), 1992. Upwelling systems: Evolution since the early Miocene. Geol. Soc. Spec. Publ., 64, Blackwell, Oxford, 519 pp.

Takahashi, T., Broeker, W.S. and Langer, S., 1985. Redfield ratio based on chemical data from isopycnal surface. J. Geophys. Res., 90: 6907-6924.

Tegelaar, E.W., de Leeuw, J.W., Derenne, S. and Largeau, C., 1989. A reappraisal of kerogen formation. Geochim. Cosmochim. Acta, 53: 3103-3107.

ten Haven, H.L., Peakman, T.M. and Rullkötter, J., 1992. Early diagnetic transformation of higher plant triterpenoids in deep sea sediments from Baffin Bay. Geochim. Cosmochim. Acta, 56: 2001-2024.

Thiede, J. and Suess, E. (eds), 1983. Coastal upwelling, its sediment record. Part B: Sedimentary records of ancient coastal upwelling. Plenum Press, NY, 610 pp.

Tissot, B.P. and Welte, D.H., 1984. Petroleum formation and occurrence. Springer Verlag, Heidelberg, 699 pp.

Tyson, R.V., 1987. The genesis and palynofacies characteristics of marine petroleum source rocks. In: Brooks, J. and Fleet, A.J. (eds), Marine petroleum source rocks. Geol. Soc. Spec. Publ., 58, Blackwell, Oxford, pp. 47-67.

Tyson, R.V. and Pearson, T.H. (eds), 1991. Modern and ancient continental shelf anoxia. Geol. Soc. Spec. Publ., 58, Blackwell,

Oxford, 470 pp.

van Krevelen, D.W., 1961. Coal typology-chemistry-physics-constitution. Elsevier, Amsterdam, 513 pp.

Vetö, I., Hetényi, M., Demény, A. and Hertelendi, E., 1994. Hydrogen index as reflecting intensity of sulphide diagenesis in non-bioturbated, shaly sediments. Org. Geochem., 22: 299-310.

Volkman, J.K. and Maxwell, J.R., 1986. Acyclic isoprenoides as biological markers. In: Johns, R.B. (ed), Biological markers in the sedimentary record. Elsevier, Amsterdam, pp 1-42.

Volkman, J.K., Barret, S.M., Blackburn, S.I. and Sikes, E.L., 1995. Alkenones in *Gephyrocapca oceanica*: Implications for studies of paleoclimate. Geochim. Cosmochim. Acta, 59: 513-520.

Volkman, J.K., Barret, S.M., Blackburn, S.I., Mansour, M.P., Sikes, E. and Gelin, F., 1998. Microalgal biomarkers: A review of recent research developments. Org. Geochem., 29: 1163-1179.

von Engelhardt, W., 1973. Sedimentpetrologie, Teil III: Die Bildung von Sedimenten und Sedimentgesteinen (in German). Schweizerbarth, Stuttgart, 378 pp.

von Rad, U. and Ryan, W.B.F., et al. (eds), 1979. Initial Reports of the Deep Sea Drilling Program, 47, US Government Printing Office, Washington, DC, 835pp.

Wakeham, S.G. and Lee, C., 1989. Organic geochemistry of particulate matter in the ocean: The role of particles in oceanic sedimentary cycles. Org. Geochem., 14: 83-96.

Welte, D.H., Horsfield, B. and Baker, D.R. (eds), 1997. Petroleum and basin evolution. Springer Verlag, Heidelberg, 535 pp.

Westerhausen, L., Poynter, J., Eglinton, G., Erlenkeuser, H. and Sarnthein, M., 1993. Marine and terrigenous origin of organic matter in modern sediments of the equatorial East Atlantic: the $d^{13}C$ and molecular record. Deep-Sea Res., 40: 1087-1121.

# 5 Bacteria and Marine Biogeochemistry

Bo Barker Jørgensen

Geochemical cycles on Earth follow the basic laws of thermodynamics and proceed towards a state of maximal entropy and the most stable mineral phases. Redox reactions between oxidants such as atmospheric oxygen or manganese oxide and reductants such as ammonium or sulfide may proceed by chemical reaction, but they are most often accelerated by many orders of magnitude through enzymatic catalysis in living organisms. Throughout Earth's history, prokaryotic physiology has evolved towards a versatile use of chemical energy available from this multitude of potential reactions. Biology, thereby, to a large extent regulates the rate at which the elements are cycled in the environment and affects where and in which chemical form the elements accumulate. By coupling very specifically certain reactions through their energy metabolism, the organisms also direct the pathways of transformation and the ways in which the element cycles are coupled. Microorganisms possess an enormous diversity of catalytic capabilities which is still only incompletely explored and which appears to continuously expand as new organisms are discovered. A basic understanding of the microbial world and of bacterial energy metabolism is therefore a prerequisite for a proper interpretation of marine geochemistry - a motivation for this chapter on biogeochemistry.

The role of microorganisms in modern biogeochemical cycles is the result of a long evolutionary history. The earliest fossil evidence of prokaryotic organisms dates back about 3.7 billion years (Schopf and Klein 1992). Only some 1.5 billion years later did the evolution of oxygenic photosynthesis apparently take place and it may have taken another 1.5 billion years before the oxygen level in the atmosphere and ocean rose to the present-day level, thus triggering the rapid evolutionary radiation of metazoans at the end of the Proterozoic era. Through the two billion years

that Earth was inhabited exclusively by prokaryotic microorganisms, the main element cycles and biogeochemical processes known today evolved. The microscopic prokaryotes developed the complex enzymatic machinery required for these processes and are even today much more versatile with respect to basic types of metabolism than plants and animals which developed over the last 600 million years. In spite of their uniformly small size and mostly inconspicuous morphology, the prokaryotes are thus physiologically much more diverse than the metazoans. In the great phylogenetic tree of all living organisms, humans are more closely related to slime molds than the sulfate reducing bacteria are to the methanogenic archaea. The latter two belong to separate domains of prokaryotic organisms, the *bacteria* and the *archaea*, respectively (the term 'prokaryote' is used rather than 'bacteria' when also the archaea are included). Animals and plants, including all the eukaryotic microorganisms, belong to the third domain, *eukarya*.

## 5.1 Role of Microorganisms

### 5.1.1 From Geochemistry to Microbiology – and back

Because of the close coupling between geochemistry and microbiology, progress in one of the fields has often led to progress in the other. Thus, analyses of chemical gradients in the pore water of marine sediments indicate where certain chemical species are formed and where they react with each other. The question is then, is the reaction biologically catalyzed and which microorganisms may be involved?

An example is the distribution of dissolved ferrous iron and nitrate, which in deep sea sedi-

173

ments often show a diffusional interface between the two species (Fig. 5.1A). Based on such gradients, Froelich et al. (1979), Klinkhammer (1980) and others suggested that $Fe^{2+}$ may be readily oxidized by nitrate, presumably catalyzed by bacteria. Marine microbiologists, thus, had the background information to start searching for nitrate reducing bacteria which use ferrous iron as a reductant and energy source. It took, however, nearly two decades before such bacteria were isolated for the first time and could be studied in pure culture (Straub et al. 1996, Benz et al. 1998; Fig. 5.1B). They appear to occur widespread in aquatic sediments but their quantitative importance is still not clear (Straub and Buchholz-Cleven 1998). The bacteria oxidize ferrous iron according to the following stoichiometry:

$$10FeCO_3 + 2NO_3^- + 24H_2O \rightarrow$$
$$10Fe(OH)_3 + N_2 + 10HCO_3^- + 8H^+ \quad (5.1)$$

Also the observation of a deep diffusional interface between sulfate and methane in marine sediments has led to a long-lasting search by microbiologists for methane oxidizing sulfate reducers. The geochemical data and experiments demonstrate clearly that methane is oxidized to $CO_2$ several meters below the sediment surface, at a depth where no other potential oxidant than sulfate seems to remain (Reeburgh 1969, Iversen and Jørgensen 1985, Alperin and Reeburgh 1985, Chaps. 3, 8 and 14). Attempts in several laboratories to isolate methane oxidizing sulfate reducers have, however, so far not led to confirmed results, and it is still not clear whether such organisms exist or do not exist. Alternatively, it has been suggested that methane oxidation is a reversal of the metabolic pathway of methane formation from $H_2$ and $CO_2$ in methanogenic organisms (Eq. 5.2). Thermodynamic calculations indicate that this would become the exergonic direction of reaction if the $H_2$ partial pressure were extremely low (Hoehler et al. 1994, 1998). The sulfate reducing bacteria are known to be highly efficient $H_2$ scavengers (Eq. 5.3). When methane diffuses into the sulfate zone, the $H_2$ partial pressure may thus be

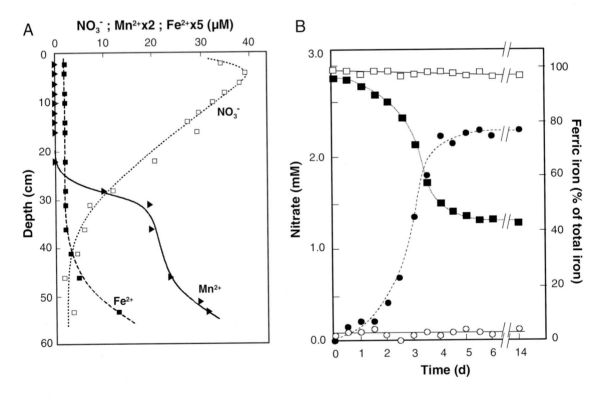

**Fig. 5.1**   A) Pore water gradients of nitrate, dissolved manganese, and iron in sediments from the eastern equatorial Atlantic at 5000 m depth. The gradients indicate that $Fe^{2+}$ is oxidized by $NO_3^-$, whereas $Mn^{2+}$ may be oxidized by $O_2$ (no data). (Data from Froelich et al. 1979; Station 10GC1). B) Anaerobic bacterial oxidation of ferrous to ferric iron with nitrate in an enrichment culture. Filled symbols show results from a growing culture, open symbols shows a control experiment with killed cells (no concentration changes). Symbols show ferric iron (● + ○) and nitrate (■ + □). (Data from Straub et al. 1996).

low enough that methane could be oxidized to $CO_2$:

$$CH_4 + 2H_2O \Leftrightarrow CO_2 + 4H_2 \qquad (5.2)$$

$$4H_2 + SO_4^{2-} + 2H^+ \rightarrow H_2S + 4H_2O \qquad (5.3)$$

There are also many examples of the reverse inspiration, that progress in microbiology has led to a new understanding of geochemistry. One such example was the discovery of a widespread ability among laboratory cultures of sulfate reducing and other anaerobic bacteria to disproportionate inorganic sulfur compounds of intermediate oxidation state (Bak and Cypionka 1987). By such a disproportionation, which can be considered an inorganic fermentation, elemental sulfur or thiosulfate may be simultaneously oxidized to sulfate and reduced to sulfide:

$$4S^0 + 4H_2O \rightarrow 3H_2S + SO_4^{2-} + 2H^+ \qquad (5.4)$$

$$S_2O_3^{2-} + H_2O \rightarrow H_2S + SO_4^{2-} \qquad (5.5)$$

A search for the activity of such bacteria, by the use of radiotracers and sediment incubation experiments, revealed their widespread occurrence in the sea bed and their great significance for the marine sulfur cycle (Jørgensen 1990, Jørgensen and Bak 1991, Thamdrup et al. 1993). Disproportionation reactions also cause a strong fractionation of sulfur isotopes, which has recently led to a novel interpretation of stable sulfur isotope signals in modern sediments and sedimentary rocks with interesting implications for the evolution of oxygen in the global atmosphere and ocean (Canfield and Teske 1996). The working hypothesis is that the large isotopic fractionations of sulfur, which in the geological record started some 600-800 million years ago, are the result of disproportionation reactions. From modern sediments we know that these conditions require two things: bioturbation and an efficient oxidative sulfur cycle. Both point towards a coupled evolution of metazoans and a rise in the global oxygen level towards the end of the Proterozoic and start of the Cambrian.

### 5.1.2 Approaches in Marine Biogeochemistry

The approaches applied in marine biogeochemistry are diverse, as indicated in Figure 5.2, and

range from pure geochemistry to experimental ecology, microbiology and molecular biology. Mineral phases and soluble constituents are analyzed and the data used for modeling of the diagenetic reactions and mass balances (Berner 1980, Boudreau 1997, Chap. 14). Dynamic processes are studied in retrieved sediment cores, which are used to analyze solute fluxes across the sediment-water interface or to measure process rates by experimental approaches using, e.g. radiotracers, stable isotopes or inhibitors. Studies are also carried out directly on the sea floor using advanced instrumentation such as autonomous benthic landers, remotely operated vehicles (ROV's), or manned submersibles. Benthic landers have been constructed which can be deployed on the open ocean from a ship, sink freely to the deep sea floor, carry out pre-programmed measurements while storing data or samples, release ballast and ascend again to the sea surface to be finally retrieved by the ship (Tengberg et al. 1995). Such in situ instruments may be equipped with A) samplers and analytical instruments for studying the benthic boundary layer (Thomsen et al. 1996), B) microsensors for high-resolution measurements of chemical gradients in the sediment (Reimers 1987; Gundersen and Jørgensen 1990), C) flux chambers for measurements of the exchange of dissolved species across the sediment-water interface (Smith et al. 1976, Berelson et al. 1987), D) coring devices for tracer injection and measurements of processes down to 0.5 meter sediment depth (Greeff et al. 1998).

Progress in microsensor technology has also stimulated research on the interaction between processes, bacteria and environment at a high spatial resolution (cf. Sect. 3.4). Microsensors currently used in marine research can analyze $O_2$, $CO_2$, pH, $NO_3^-$, $Ca^{2+}$, $S^{2-}$, $H_2S$, $CH_4$ and $N_2O$ as well as physical parameters such as temperature, light, flow or position of the solid-water interface (Kühl and Revsbech 1998). Sensors have mostly been based on electrochemical principles. However, microsensors based on optical fibers (optodes) or sensors based on enzymes or bacteria (biosensors) are gaining importance. Examples of microsensor data from marine sediments are given in Chapter 3.

Among the important contributions of microbiologists to biogeochemistry is to quantify the populations of bacteria, isolate new types of microorganisms from the marine environment, study their physiology and biochemistry in laboratory cultures, and thereby describe the microbial diversity and metabolic potential of natural microbial communities. There are currently about a thousand species of prokaryotes described from the sea and the number is steadily increasing. The rather time-consuming task of isolating and describing new bacterial species, however, sets a limit to the rate of progress. The classical microbiological approaches also have shortcomings, among others that only the viable or culturable bacteria are recognized, and that the counting procedures for these viable bacteria, furthermore, underestimate the true population size by some orders of magnitude.

In recent years, the rapid developments in molecular biology have opened new possibilities for the study of bacterial populations and their metabolic activity, even at the level of individual cells. A new taxonomy of the prokaryotic organisms has developed at a genetic level, based on sequence analyses of ribosomal RNA, which now allows a phylogenetic identification of microorganisms, even of those which have not yet been isolated and studied in laboratory cultures. This has revealed a much greater species diversity than had been anticipated just a decade ago and has for the first time enabled a true quantification of defined bacterial groups in nature. By the use of molecular probes, which bind specifically to RNA or DNA of selected target organisms, both cultured and uncultured bacteria can now be identified phylogenetically. When such probes are

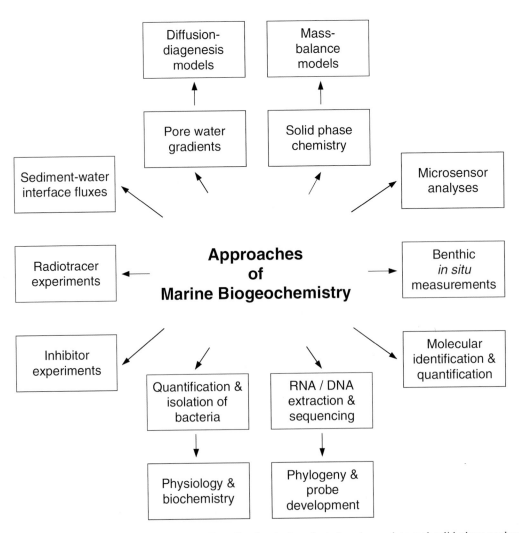

**Fig. 5.2**  Approaches of marine biogeochemistry. Top: Geochemical methods based on solute and solid phase analyses and modelling. Left: Experimental methods for the analyses of process rates. Bottom: Identification, quantification and characterization of the microbial populations. Right: High resolution and in situ methods for the analyses of microbial populations and their microenvironment. (Graphics by Niels Birger Ramsing and reproduced from Jørgensen 1996).

fluorescently labelled, a sediment sample may be stained by the probes and individual cells of the target bacteria can then be observed and counted directly under a fluorescense microscope. This technique is called Fluorescent In Situ Hybridization (FISH) and is rapidly gaining importance in biogeochemistry.

An example of such a FISH quantification is shown in Figure 5.3 (Llobet-Brossa et al. 1998). The total cell density of microorganisms in a sediment from the German Wadden Sea was up to $4 \cdot 10^9$ cells cm$^{-3}$, which is typical of coastal marine sediments. Of these cells, up to 73% could be identified as eubacteria and up to 45% were categorized to a known group of eubacteria. The sulfate reducing bacteria comprised some 10-15% of the identified microorganisms, while other members of the group proteobacteria comprised 25-30%. Surprisingly, members of the *Cytophaga-Flavobacterium* group were the most abundant in all sediment layers. These bacteria are specialized in the degradation of complex macromolecules and their presence in high numbers indicates their role in the initial hydrolytic degradation of organic matter.

Such studies on the spatial distribution of bacteria provides information on where they may be

actively transforming organic or inorganic compounds, and thereby indicate which processes are likely to take place. This is particularly important when chemical species are rapidly recycled so that the dynamics of their production or consumption is not easily revealed by geochemical analyses. The introduction of high resolution tools such as microsensors and molecular probes has helped to overcome one of the classical problems in biogeochemistry, namely to identify the relationship between the processes that biogeochemists analyze, and the microorganisms who carry them out. The magnitude of this problem may be appreciated when comparing the scale of bacteria with that of humans. The bacteria are generally 1-2 μm large, while we are 1-2 m. As careful biogeochemists we may use a sediment sample of only 1 cm$^3$ to study the metabolism of, e.g. methane producing bacteria. For the study of the metabolism of humans, this sample size would by isometric analogy correspond to a soil volume of 1000 km$^3$. Thus, it is not surprising that very sensitive methods, such as the use of radiotracers, are often necessary when bacterial processes are to be demonstrated over short periods of time (hours-days). It is also obvious, that a 1 cm$^3$ sample will include a great diversity of prokaryotic

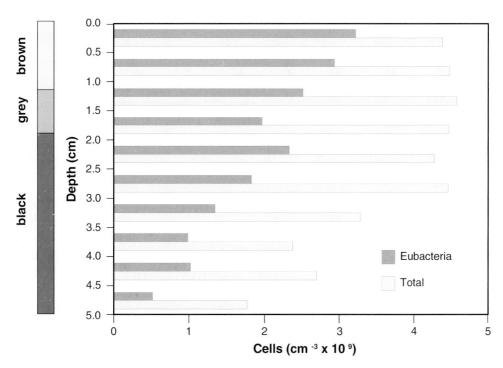

**Fig. 5.3** Distribution of microorganisms in sediment from the German Wadden Sea determined by total cell counts with DAPI staining and by fluorescent in situ hybridization of eubacteria. The color zonation of the sediment is indicated. (Data from Llobet-Brossa et al. 1998).

organisms and metabolic reactions and that a much higher resolution is required to sort out the activities of individual cells or clusters of organisms.

## 5.2    Life and Environments at Small Scale

The size spectrum of living organisms and of their environments is so vast that it is difficult to comprehend (Fig. 5.4). The smallest marine bacteria with a size of ca 0.4 μm are at the limit of resolution of the light microscope, whereas the largest whales may grow to 30 m in length, eight orders of magnitude larger. The span in biomass is nearly the third power of this difference, $<(10^8)^3$ or about $10^{22}$ (whales are not spherical), which is comparable to the mass ratio between humans and the entire Earth. It is therefore not surprising that the world as it appears in the microscale of bacteria is also vastly different from the world we humans can perceive and from which we have learnt to appreciate the physical laws of nature. These are the classical laws of Newton, relating mass, force and time with mass

movement and flow and with properties such as acceleration, inertia and gravitation. As we go down in scale and into the microenvironment of marine bacteria, these properties lose their significance. Instead, viscosity becomes the strongest force and molecular diffusion the fastest transport.

### 5.2.1    Hydrodynamics of Low Reynolds Numbers

Water movement on a large scale is characterized by inertial flows and turbulence. These are the dominant mechanisms of transport and mixing, yet they are inefficient in bringing substrates to the microorganisms living in the water. This is because the fundamental property required for turbulence, namely the inertial forces associated with mass, plays no role for very small masses, or at very small scales, relative to the viscosity or internal friction of the fluid. In the microscale, time is insignificant for the movement of water, particles or organisms and only instantaneous forces are important. Furthermore, fluid flow is rather simple and predictable, the water sticking to any solid surface and adjacent water volumes slipping past it in a smooth pattern of laminar

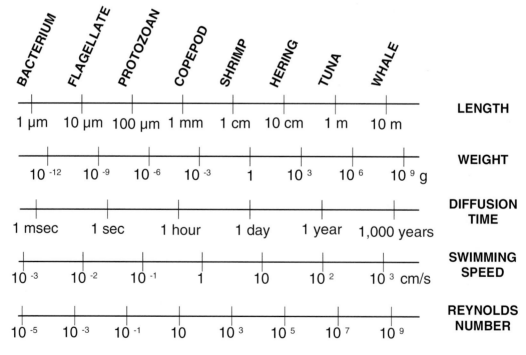

**Fig. 5.4**  Relationships between size, diffusion time and Reynolds number. Representative organisms of the different scales and their length and biomass are indicated. The diffusion times were calculated for small solutes with a molecular diffusion coefficient of $10^{-5}$ cm$^2$ s$^{-1}$. The Reynolds numbers were calculated for organisms swimming at a speed typical for their size or for water parcels of similar size and moving at similar speed. (From Jørgensen 1996).

flow. The transition from laminar to turbulent flow depends on the scale and the flow velocity and is described by the dimensionless Reynolds number, $Re$, which expresses the ratio between inertial and viscous forces affecting the fluid or the particle considered:

$$Re = uL/v \qquad (5.6)$$

where $u$ is the velocity, $L$ is the characteristic dimension of the water parcel or particle, and $v$ is the kinematic viscosity of the seawater (ca 0.01 cm$^2$ s$^{-1}$ at 20°C). The transition from low ($<1$) to high Reynolds numbers for swimming organisms, for sinking particles, or for hydrodynamics in general lies in the size range of 0.1-1 mm.

## 5.2.2   Diffusion at Small Scale

Diffusion is a random movement of molecules due to collision with water molecules which leads to a net displacement over time, a 'random walk'. When a large number of molecules is considered, the mean deviation, $L$ (more precisely: the root mean square of deviations), from the starting position is described by a simple but very important equation, which holds the secret of diffusion:

$$L = \sqrt{2Dt} \qquad (5.7)$$

where $D$ is the diffusion coefficient and $t$ is the time. This equation says that the distance molecules are likely to travel by diffusion increases only with the square root of time, not with time itself as in the locomotion of objects and fluids which we generally know from our macroworld. Expressed in a different way, the time needed for diffusion increases with the square of the distance:

$$t = \frac{L^2}{2D} \qquad (5.8)$$

These are very counter-intuitive relations which have some surprising consequences. From Equation 5.7 one can calculate the 'diffusion velocity', which is distance divided by time:

$$\text{'Diffusion velocity'} = L/t = \sqrt{2Dt} \qquad (5.9)$$

This leads to the curious conclusion that the shorter the period over which we measure diffusion, the larger is its velocity and vice versa. This is critical to keep in mind when working with

**Table 5.1**  Mean diffusion times for $O_2$ and glucose over distances ranging from 1 µm to 10 m.

| Diffusion distance | Time (10°C) | |
|---|---|---|
| | Oxygen | Glucose |
| 1 µm | 0.34 ms | 1.1 ms |
| 3 µm | 3.1 ms | 10 ms |
| 10 µm | 34 ms | 110 ms |
| 30 µm | 0.31 s | 1.0 s |
| 100 µm | 3.4 s | 10 s |
| 300 µm | 31 s | 100 s |
| 600 µm | 2.1 min | 6.9 min |
| 1 mm | 5.7 min | 19 min |
| 3 mm | 0.8 h | 2.8 h |
| 1 cm | 9.5 h | 1.3 d |
| 3 cm | 3.6 d | 12 d |
| 10 cm | 40 d | 130 d |
| 30 cm | 1.0 yr | 3.3 yr |
| 1 m | 10.8 yr | 35 yr |
| 3 m | 98 yr | 320 yr |
| 10 m | 1090 yr | 3600 yr |

diagenetic models, because the different chemical species have time and length scales of diffusion and reaction which vary over many orders of magnitude and which are therefore correspondingly difficult to compare.

Calculations from Equation 5.8 of the diffusion times of oxygen molecules at 10°C show that it takes one hour for a mean diffusion distance of 3-4 mm, whereas it takes a day to diffuse 2 cm and 1000 years for 10 meters (Table 5.1). For a small organic molecule such as glucose, these diffusion times are about three times longer. Over the scale of a bacterium, however, diffusion takes only 1/1000 second. Thus, for bacteria of 1 µm size, one could hardly envision a transport mechanism which would outrun diffusion within a millisecond. The transition between predominantly diffusive to predominantly advective or turbulent transport of solutes lies in the range of 0.1 mm for actively swimming organisms and somewhat higher for passively sinking marine aggregates (Fig. 5.4).

## 5.2.3   Diffusive Boundary Layers

The transition from a turbulent flow regime with advective and eddy transport to a small scale dominated by viscosity and diffusional transport is apparent when a solid-water interface such as the sediment surface is approached (Fig. 5.5). According to the classical eddy diffusion theory, the vertical component of the eddy diffusivity, $E$, decreases as a solid interface is approached according to: $E = A \, v \, Z^{3-4}$, where $A$ is a constant, $v$ is the kinematic viscosity, and $Z$ is the height above the bottom. An exponent of 3-4 shows that the eddy diffusivity drops very steeply as the sediment surface is approached. In the viscous sublayer, which is typically $=1$ cm thick in the deep sea, the eddy diffusivity falls below the kinematic viscosity of ca $10^{-2}$ cm$^2$s$^{-1}$. Even closer to the sediment surface, the vertical eddy diffusion coefficient for mass falls below the molecular diffusion coefficient, D, which is constant for a given solute and temperature, and which for small dissolved molecules is in the order of $10^{-5}$ cm$^2$s$^{-1}$. The level where E becomes smaller than D defines the diffusive boundary layer, $\delta_e$, which is typically about

0.5 mm thick. In this layer, molecular diffusion is the predominant transport mechanism, provided that the sediment is impermeable and stable.

The diffusive boundary layer plays an important role for the exchange of solutes across the sediment-water interface (Jørgensen 1998). For chemical species which have a very steep gradient in the diffusive boundary layer it may limit the flux and thereby the rate of chemical reaction. This may be the case for the precipitation of manganese on iron-manganese nodules (Boudreau 1988) or for the dissolution of carbonate shells and other minerals such as alabaster in the deep sea (Santschi et al. 1991). For chemical species with a weak gradient, such as sulfate, the diffusive boundary layer plays no role since the uptake of sulfate is totally governed by diffusion-reaction within the sediment. The function of the diffusive boundary layer as a barrier for solute exchange is also reflected in the mean diffusion time of molecules through the layer, which is about 1 min over a 0.5 mm distance (Table 5.1).

The existence of a diffusive boundary layer is apparent from microsensor measurements of oxygen and other solutes at the sediment-water interface. In a Danish coastal sediment, the concentration of oxygen dropped steeply over the 0.5 mm thick boundary layer (Fig. 5.5) which consequently had a significant influence on the regulation of oxygen uptake in this sediment of high organic-matter turnover. Figure 5.6 shows, as another example, oxygen penetration down to 13 mm below the surface in a fine-grained sediment . The profile was measured in-situ in the sea bed by a free-falling benthic lander operating with a 100 μm depth resolution, which was just sufficient to resolve the diffusive boundary layer. Based on the boundary layer gradient, the vertical diffusion flux of oxygen across the water-sediment interface can be calculated (cf. Chap. 3).

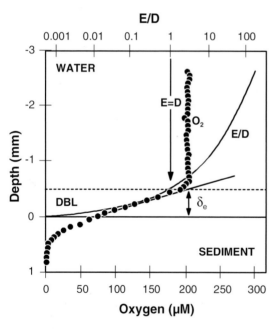

**Fig. 5.5** Oxygen microgradient (n) at the sediment-water interface compared to the ratio, E/D (logarithmic scale), between the vertical eddy diffusion coefficient, E, and the molecular diffusion coefficient, D. Oxygen concentration was constant in the overflowing seawater. It decreased linearly within the diffusive boundary layer (DBL), and penetrated only 0.7 mm into the sediment. The DBL had a thickness of 0.45 mm. Its effective thickness, $\delta_e$ is defined by the intersection between the linear DBL gradient and the constant bulk water concentration. The diffusive boundary layer occurs where E becomes smaller than D, i.e. where E/D = 1 (arrow). Data from Aarhus Bay, Denmark, at 15 m water depth during fall 1990 (Gundersen et al. 1995).

## 5.3   Regulation and Limits of Microbial Processes

Bacteria and other microorganisms are the great biological catalysts of element cycling at the sea floor. The degradation and remineralization of organic matter and many redox processes among inorganic species are dependent on bacterial catalysis, which may accelerate such processes up to $10^{20}$-fold relative to the non-biological reaction

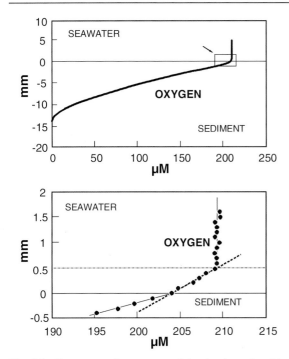

**Fig. 5.6** Oxygen gradient measured in-situ by a benthic lander in Skagerrak at the transition between the Baltic Sea and the North Sea at 700 m water depth. Due to the high depth resolution of the microelectrode measurements it was possible to analyze the O₂ micrograient within the 0.5-mm thick diffusive boundary layer. The framed part in the upper graph is blown up in the lower graph. (Data from Gundersen et al. 1995)

rate. It is, however, important to keep in mind that this biological catalysis is based on living organisms, each of which has its special requirements and limits for its physical and chemical environment, its supply of nutrients, growth rate and mortality, interaction with other organisms, etc..

### 5.3.1    Substrate Uptake by Microorganisms

Among the many reasons for the importance of prokaryotic organisms in biogeochemical cycling are:

A) Their metabolic versatility, especially the many types of anaerobic metabolism and the ability to degrade complex polymeric substances or to catalyze reactions among inorganic compounds.

B) Their small size which allows them to inhabit nearly any environment on the surface of the Earth and which strongly enhances the efficiency of their catalytic activity.

C) The wide range of environmental conditions under which they thrive, including temperature, salinity, pH, hydrostatic pressure, etc.

The metabolic versatility of prokaryotic organisms is discussed in Section 5.4. The small size of the individual cells is related to their nutrition exclusively on solutes such as small organic molecules, inorganic ions or gases. The uptake of food by the prokaryotic cells thus takes place principally by molecular diffusion of small molecules to the cell surface and their transport through the cytoplasmic membrane into the cell. This constrains the relationships between cell size, metabolic rate, substrate concentration and molecular diffusion coefficients (D) (e.g. Koch 1990, 1996, Karp-Boss et al. 1996). The concentration gradient around the spherical cell is:

$$C_r = (R/r) \cdot (C_0 - C_\infty) + C_\infty \; , r > R \quad (5.10)$$

where $C_r$ is the concentration at the radial distance, $r$, $C_0$ is the substrate concentration at the cell surface, $C_\infty$ is the ambient substrate concentration, and $R$ is the radius of the cell (Fig. 5.7). The maximal substrate uptake rate of a cell is reached when the substrate concentration at the cell surface is zero. The total diffusion flux, $J$, to the cell is then:

$$J = 4 \pi DR C_\infty \quad (5.11)$$

This flux provides the maximal substrate supply to the diffusion-limited cell, which has a volume of $^4/_3 \pi R^3$. The flux thus determines the maximal specific rate of bacterial metabolism of the substrate molecules, i.e. the metabolic rate per volume of biomass:

Specific metabolic rate =
$(4 \pi DR C_\infty)/(^4/_3 \pi R^3) = (3D/R^2)C_\infty \quad (5.12)$

Equation 5.12 shows that the biomass-specific metabolic rate of the diffusion-limited cell varies inversely with the square of its size. This means that the cell could potentially increase its specific rate of metabolism 4-fold if the cell diameter were only half as large. The smaller the cell, the less likely it is that its substrate uptake will reach diffusion limitation. Thus, at the low substrate concentrations normally found in marine environments, microorganisms avoid substrate limitation by forming small cells of <1 µm size. Thereby, the bacteria become limited by their transport efficiency of molecules across the cell membrane rather than by diffusion from their surroundings (Fig. 5.7). In the nutrient-poor seawater, where substrates are available only in sub-micromolar and even nanomolar concentrations, free-living bacteria may have an impressively high substrate uptake efficiency which corresponds to clearing a seawater volume for substrate each second which is several hundred times their own volume. In sediments, on the other hand, bacteria often form microcolonies or their diffusion supply is impeded by sediment structures so that the microorganisms may at times be diffusion limited in their substrate uptake. It is not clear, however, how this affects the kinetics of substrate turnover in sediments.

The theoretical limit for substrate availability required to sustain bacterial growth and survival is of great biogeochemical significance, e.g. in relation to the exceedingly slow degradation of organic material in million-year old sediments and oil reservoirs deep under the sea floor (Stetter et al. 1993, Parkes et al. 1994). The recent discovery of this 'deep biosphere' has added a new perspective to the limits of life and the regulation of organic carbon burial in marine sediments. It appears, that the gradual geothermal heating of sediments as they become buried to several hundred

meters depth enhances the availability of even highly refractory organic material to microbial attack and leads to further mineralization with the ultimate formation of methane (Wellsbury et al. 1997). The methane slowly diffuses upwards from the deep deposits and becomes oxidized to $CO_2$ as it reaches the bottom of the sulfate zone (Chap. 8), or it accumulates as gas hydrate within the upper sediment strata (Borowski et al. 1996).

## 5.3.2   Temperature as a Regulating Factor

The sea floor is mostly a cold environment with 85% of the global ocean having temperatures below 5°C. At the other extreme, hydrothermal vents along the mid-oceanic ridges have temperatures reaching above 350°C. In each temperature range from the freezing point to an upper limit of around 110°C there appear to be prokaryotic organisms which are well adapted and even thrive optimally at that temperature. Thus, extremely warm-adapted (hyperthermophilic) methane producing or elemental-sulfur reducing bacteria, which live at the boiling point of water, are unable to grow at temperatures below 60-70°C because it is too cold for them (Stetter 1996).

It is well known that chemical processes as well as bacterial metabolism are slowed down by low temperature. Yet, the biogeochemical recycling of deposited organic material in marine sediments does not appear to be less efficient or less complete in polar regions than in temperate or tropical environments. Sulfate reduction rates in marine sediments of below 0°C around Svalbard at 78 degrees north in the Arctic Ocean are comparable to those at similar water depths along the European coast (Sagemann et al. 1998). Figure 5.8A shows the short-term temperature dependence of sulfate reduction rates in such Svalbard sediments. As is typical for microbial processes, the optimum temperature of 25-30°C is high above the in-situ temperature, which was -1.7°C during summer. Above the optimum, the process rate dropped steeply, which is due to enzymatic denaturation and other physiological malfunctioning of the cells and which shows that this is a biologically and not a chemically catalyzed process. Although the sulfate reduction was slow at <0°C this does not mean that the bacteria do not function well at low temperature. On the contrary, the organisms had their highest growth efficiency (i.e. highest biomass production per amount of substrate consumed) at around 0°C, in contrast to

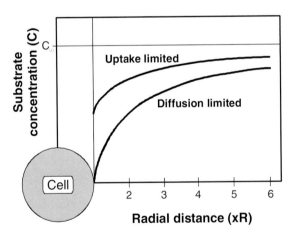

**Fig. 5.7**   Theoretical concentration gradient of substrate molecules around a spherical cell at different radial distances from its center (R is the radius of the cell). The concentration of substrate in the bulk water is $C_\infty$ and concentration curves were calculated from Equation 5.10 for a cell limited in its substrate uptake by external diffusion ('Diffusion limited') or by the uptake capacity across its own cell membrane ('Uptake limited'). Note how the substrate concentration only gradually approaches the bulk concentration with increasing distance from the cell.

sulfate reducers from temperate environments which have their highest growth efficiency at the warm temperatures experienced during summer (Isaksen and Jørgensen 1996, C. Knoblauch, unpublished).

The radiotracer method of measuring sulfate reduction in sediments (see Sect. 5.6) is a sensitive tool to demonstrate the temperature strains among the environmental microorganisms. In ca. 100°C warm sediment from the hydrothermal sediments of the Guaymas Basin, Gulf of California, three main groups of sulfate reducers could be discriminated from their temperature optima: a) mesophiles with an optimum at 35°C, b) thermophiles with optimum at 85°C and, c) hyperthermophiles with optimum at 105°C (Fig. 5.8B). The mesophiles comprise the sulfate reducers known best from pure cultures, whereas thermophiles with optimum at 85°C are known among the genus, *Archaeoglobus*. Such prokaryotic organisms have been isolated from hydrothermal environments and have also been found in oil reservoirs 3000 m below the sea bed where the temperature is up to 110°C and the hydrostatic pressure up to 420 bar. Hyperthermophilic sulfate reducing bacteria which can metabolize at >100°C have never been isolated and are still not known.

Surprisingly, sulfate reducers adapted to the normal low temperatures of the main sea floor were also unknown until recently. Such cold-adapted (psychrophilic) bacteria are generally scarce among the culture collections in spite of their major biogeochemical significance. This is partly because of their slow growth which makes them difficult to isolate and cumbersome to study. A number of psychrophilic sulfate reducers were recently isolated which have temperature optima down to 7°C (C. Knoblauch, unpublished).

These examples demonstrate a general problem in marine microbiology, namely that only a very small fraction (one percent?) of the bacterial species in the ocean are known to science today. Many of the unknown microorganisms may be among the biogeochemically very important species. The estimate is based on recent methodological advances in molecular biology which have made it possible to analyze the diversity of natural prokaryotic populations, even of the many which have still not been isolated or studied.

### 5.3.3    Other Regulating Factors

The major area of the sea floor lies in the deep sea below several thousand meters of water where an enormous hydrostatic pressure prevails. Since the pressure increases by ca 1 bar (1 atm) for every 10 meters, bacteria living at 5000 m depth must be able to withstand a pressure of 500 bar. Microorganisms isolated from sediments down to 3000-4000 m have been found to be preferentially barotolerant, i.e. they grow equally well at sea surface pressure as at their in-situ pressure. At depths exceeding 4000 m the isolated bacteria become increasingly barophilic, i.e. they grow opti-

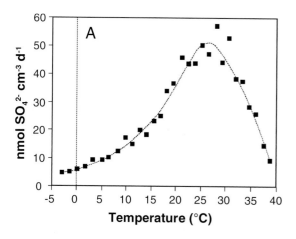
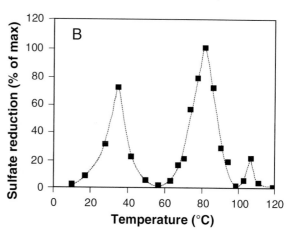

**Fig. 5.8** Temperature regulation of bacterial sulfate reduction in different marine sediments. The data show the rates of sulfate reduction in a homogenized sample from the upper 5-10 cm of sediment measured by short-term incubations in a temperature gradient block with radiolabelled sulfate as a tracer. A) Arctic sediment from 175 m depth in Storfjorden on Svalbard at 78°N where the in situ temperature was -1.7°C. B) Sediment from the hydrothermal area of the Guaymas Basin at 2000 m depth. The hydrothermal fluid here seeps up through fine-grained, organic-matter rich sediment which was collected at 15-20 cm depth where the in situ temperature was 90-125°C. (Data from Sagemann et al. 1998, and Jørgensen et al. 1992).

mally at high pressures, and bacteria isolated from deep-sea trenches at 10,000 m depth were found to grow optimally at 700-1000 bar (Yayanos 1986). Barophilic bacteria are also psychrophilic which is in accordance with the low temperature of 2-3°C in the deep sea. They appear to grow relatively slowly which may be an adaptation to low temperature and low nutrient availability rather than a direct effect of high pressure. The degradation of organic material appears to be just as efficient in the deep sea as in shallower water, since only a few percent of sedimenting detritus resist mineralization and are buried deep down into the sediment.

## 5.4    Energy Metabolism of Prokaryotes

Microorganisms can be considered 1 μm large bags of enzymes in which the important biogeochemical processes are catalyzed. This analogy, however, is too crude to understand how and why these microscopic organisms drive and regulate the major cycles of elements in the ocean. This requires a basic knowledge of their energy metabolism and physiology.

The cells use the chemical energy of organic or inorganic compounds for cell functions which require work: growth and division, synthesis of macromolecules, transport of solutes across the cell membrane, secretion of exoenzymes or exopolymers, movement etc. The organisms catalyze redox processes from which they conserve a part of the energy and couple it to the formation of a proton gradient across the cytoplasmic membrane. The socalled *proton motive force* established by this gradient is comparable to the electron motive force of an ordinary battery. It is created by the difference in electrical charge and $H^+$ concentration between the inside (negative and low $H^+$, i.e. alkaline) and the outside (positive and high $H^+$) of the membrane. By reentry of protons into the cell through membrane-bound ATPase protein complexes, it drives the formation of high energy phosphate bonds in compounds such as *ATP* (adenosine triphosphate), which functions as a transient storage of the energy and is continuously recycled as the phosphate bond is cleaved in energy-requiring processes. ATP is utilized very widely in organisms as the fuel to drive energy requiring processes, and a fundamental question in all cellular processes is, how much ATP do they produce or consume?

Reduction-oxidation (redox) processes, whether biological or chemical, involve a transfer of one or more electrons between the chemical reactants. An example is the oxidation of ferrous iron to ferric iron by oxygen at low pH:

$$2Fe^{2+} + \tfrac{1}{2}O_2 + 2H^+ \rightarrow 2Fe^{3+} + H_2O \qquad (5.13)$$

By this reaction, a single electron is transferred from each $Fe^{2+}$ ion to the $O_2$ molecule. The ferrous iron is thereby oxidized to ferric iron, whereas the oxygen is reduced to water. In the geochemical literature, $O_2$ in such a reaction is termed the *oxidant* and $Fe^{2+}$ the *reductant*. In the biological literature, the terms *electron acceptor* for $O_2$ and *electron donor* for $Fe^{2+}$ are used.

### 5.4.1    Free Energy

Chemical reactions catalyzed by microorganisms yield highly variable amounts of energy and some are directly energy consuming. The term *free energy*, G, of a reaction is used to express the energy released per mol of reactant, which is available to do useful work. The change in free energy is conventionally expressed as $\Delta G^0$, where the symbol $\Delta$ should be read as 'change in' and the superscript $^0$ indicate the following standard conditions: pH 7, 25°C, a 1 M concentration of all reactants and products and a 1 bar partial pressure of gasses. The value of $\Delta G^0$ for a given reaction is expressed in units of kilojoule (kJ) per mol of reactant. If there is a net decrease in free energy ($\Delta G^0$ is negative) then the process is *exergonic* and may proceed spontaneously or biologically catalyzed. If $\Delta G^0$ is positive, the process is *endergonic* and energy from ATP or from an accompanying process is required to drive the reaction. The change in free energy of several metabolic processes in prokaryotic organisms is listed in Table 5.2.

A thorough discussion of the theory and calculation of $\Delta G^0$ for a variety of anaerobic microbial processes is given by Thauer et al. (1977). As a general rule, processes for which the release of energy is very small, < ca 7 kJ mol$^{-1}$, are insufficient for the formation of ATP and are thus unable to serve the energy metabolism of microorganisms. It is important to note, however, that standard conditions are seldom met in the marine environment and that the actual conditions of pH,

**Table 5.2**  Pathways of organic matter oxidation, hydrogen transformation and fermentation in the sea floor and their standard free energy yields, $\Delta G^0$, per mol of organic carbon. $\Delta G^0$ values according to Thauer et al. (1977), Fenchel et al. (1998), and Conrad et al. (1986) (cf. Fig.3.11).

| Pathway and stoichiometry of reaction | $\Delta G^0$ (kJ mol-1) |
|---|---|
| **Oxic respiration:** | |
| $\quad CH_2O + O_2 \rightarrow CO_2 + H_2O$ | -479 |
| **Denitrification:** | |
| $\quad 5CH_2O + 4NO_3^- \rightarrow 2N_2 + 4HCO_3^- + CO_2 + 3H_2O$ | -453 |
| **Mn(IV) reduction:** | |
| $\quad CH_2O + 3CO_2 + H_2O + 2MnO_2 \rightarrow 2Mn^{2+} + 4HCO_3^-$ | -349 |
| **Fe(III) reduction:** | |
| $\quad CH_2O + 7CO_2 + 4Fe(OH)_3 \rightarrow 4Fe^{2+} + 8HCO_3^- + 3H_2O$ | -114 |
| **Sulfate reduction:** | |
| $\quad 2CH_2O + SO_4^{2-} \rightarrow H_2S + 2HCO_3^-$ | -77 |
| $\quad 4H_2 + SO_4^{2-} + H^+ \rightarrow HS^- + 4H_2O$ | -152 |
| $\quad CH_3COO^- + SO_4^{2-} + 2H^+ \rightarrow 2CO_2 + HS^- + 2H_2O$ | -41 |
| **Methane production:** | |
| $\quad 4H_2 + HCO_3^- + H^+ \rightarrow CH_4 + 3H_2O$ | -136 |
| $\quad CH_3COO^- + H^+ \rightarrow CH_4 + CO_2$ | -28 |
| **Acetogenesis:** | |
| $\quad 4H_2 + 2CO_3^- + H^+ \rightarrow CH_3COO^- + 4H_2O$ | -105 |
| **Fermentation:** | |
| $\quad CH_3CH_2OH + H_2O \rightarrow CH_3COO^- + 2H_2 + H^+$ | 10 |
| $\quad CH_3CH_2COO^- + 3H_2O \rightarrow CH_3COO^- + HCO_3^- + 3H_2 + H^+$ | 77 |

temperature and substrate/product concentrations must be known before the energetics of a certain reaction can be realistically calculated. Several processes, which under standard conditions would be endergonic, may be exergonic in the normal marine sediment.

An important example of this is the formation of $H_2$ in several bacterial fermentation processes which is exergonic only under low $H_2$ partial pressure. The hydrogen cycling in sediments is therefore dependent on the immediate consumption of $H_2$ by other organisms, such as the sulfate reducing bacteria, which keep the partial pressure of $H_2$ extremely low. The $H_2$-producing and the $H_2$-consuming bacteria thus tend to grow in close proximity to each other, thereby facilitating the diffusional transfer of $H_2$ at low concentration from one organism to the other, a socalled 'interspecies hydrogen transfer' (Conrad et al. 1986).

### 5.4.2    Reduction-Oxidation Processes

The electron transfer in redox processes is often accompanied by a transfer of protons, $H^+$. The simplest example is the oxidation of $H_2$ with $O_2$ by the socalled 'Knallgas-bacteria', which occur widespread in aquatic sediments:

$$H_2 + \tfrac{1}{2}O_2 \rightarrow H_2O \qquad (5.14)$$

In the energy metabolism of cells, an intermediate carrier of the electrons (and protons) is commonly required. Such an *electron carrier* is, for instance, $NAD^+$ (nicotinamid adenin dinucleotide), which formally accepts two electrons and one proton and is thereby reduced to NADH. The NADH may give off the electrons again to specialized electron acceptors and the protons are released in the cell sap. Thereby, the NADH, which must be used repeatedly, is recycled.

A redox process such as Equation 5.14 formally consists of two reversible half-reactions. The first is the oxidation of $H_2$ to release electrons and protons:

$$H_2 \Leftrightarrow 2e^- + 2H^+ \qquad (5.15)$$

The second is the reduction of oxygen by the transfer of electrons (and protons):

$$\frac{1}{2}O_2 + 2e^- + 2H^+ \Leftrightarrow H_2O \qquad (5.16)$$

Compounds such as $H_2$ and $O_2$ vary strongly in their tendency to either give off electrons and thereby become oxidized ($H_2$) or to accept electrons and thereby become reduced ($O_2$). This tendency is expressed as the redox potential, $E_0'$ of the compounds. This potential is expressed in volts and is measured electrically in reference to a standard compound, namely $H_2$. By convention, redox potentials are expressed for half reactions which are reductions, i.e. 'oxidized form + $e^- \rightarrow$ reduced form'. When protons are involved in the reaction, the redox potential in the biological literature is expressed at pH 7, because the interior of cells is approximately neutral in pH. A reducing compound such as $H_2$, with a strong tendency to give off electrons (Eq. 5.15), has a strongly negative $E_0'$ of -0.414 V. An oxidizing compound such as $O_2$, with a strong tendency to accept electrons (Eq. 5.16), has a strongly positive $E_0'$ of +0.816 V. The free energy yield, $\Delta G^0$, of a process is proportional to the difference in $E_0'$ of its two half-reactions, and the redox potentials of electron donor and electron acceptor in a microbial energy metabolism therefore provide important information on whether they can serve as useful substrates for energy metabolism:

$$\Delta G^0 = -n\, F\, \Delta E_0' \; [kJ\; mol^{-1}] \qquad (5.17)$$

where n is the number of electrons transferred by the reaction and $F$ is Faraday's constant (96,485 Coulomb $mol^{-1}$).

The redox potentials of different half-reactions, which are of importance in biogeochemistry, are shown in Figure 5.9. The two 'electron towers' show the strongest reductants (most negative $E_0'$) at the top and the strongest oxidants at the bottom. Reactions between electron donors of more negative $E_0'$ with electron acceptors of more positive $E_0'$ are exergonic and may provide the basis for biological energy metabolism. The larger the drop in $E_0'$ between electron donor and acceptor, the more energy is released. As an example, the oxidation of ferrous iron in the form of $FeCO_3$ with $NO_3^-$ is shown, a process which was discussed in Section 5.1.1 in relation to geochemical pore water gradients in deep-sea sediments (Fig. 5.1, Straub et al. 1996). The oxidation of organic compounds such as lactate or acetate by $O_2$ releases maximum amounts of energy, and aerobic respiration is, accordingly, the basis for metazoan life and for many microorganisms.

### 5.4.3    Relations to Oxygen

Before discussing the biological reactions further, a few basic concepts should be clarified. One is the relation of living organisms to $O_2$. Those organisms, who live in the presence of oxygen, are termed *aerobic* (meaning 'living with oxygen'), whereas those who live in the absence of oxygen are *anaerobic*. This discrimination is both physiologically and biogeochemically important. In all aerobic organisms, toxic forms of oxygen, such as peroxide and superoxide, are formed as by-products of the aerobic metabolism. Aerobic organisms rapidly degrade these aggressive species with enzymes such as catalase, peroxidase or superoxide dismutase. Many organisms, which are obligately anaerobic, lack these enzymes and are not able to grow in the presence of oxygen for this and other reasons. Some organisms need oxygen for their respiration, yet they are killed by higher $O_2$ concentrations. They are *microaerophilic* and thrive at the lower boundary of the $O_2$ zone, between the *oxic* (containing oxygen) and the *anoxic* (without oxygen) environments. Other, non-obligately (facultatively) anaerobic bacteria may live both in *oxidized* and in *reduced* zones of the sediment. The oxidized sediment is characterized by redox potentials, as measured by a naked platinum electrode, of $E_H > 0$-100 mV and up to about +400 mV. Oxidized sediments are generally brown to olive because iron minerals are in oxidized forms. In most marine sediments the $O_2$ penetration is small relative to the depth of oxidized iron minerals. Most of the oxidized zone is therefore anoxic and has been termed *suboxic* (Froelich et

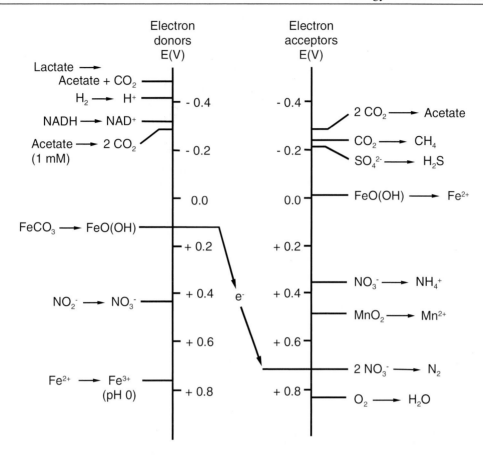

**Fig. 5.9** The 'electron towers' of redox processes in biogeogeochemistry. By the half-reaction on the left side, electrons are released from an electron donor and are transferred to an electron acceptor in the half-reaction on the right side. The drop in redox potential between donor and acceptor is a measure of the chemical energy released by the process. The redox potentials are here calculated for standard conditions at pH 7 and 1 mM concentrations of substrates and products. As an example, the electron transfer (arrow between electron towers) is shown for the oxidation of ferrous carbonate with nitrate (see text).

al. 1979). The reducing sediment below the sub-oxic zone has $E_H$ below 0-100 mV and down to -200 mV and may be black or gray from different forms of iron sulfide minerals.

### 5.4.4    Definitions of Energy Metabolism

We are now ready to explore the main types of energy metabolism. Microbiologists use a termi-nology for the different types based on three cri-teria: a) the energy source, b) the electron donor and c) the carbon source (Table 5.3). To the term for each of these criteria or their combination is added '-*troph*', meaning 'nutrition'. For each of the three criteria there are two alternatives, thus lead-ing to $2^3 = 8$ possibilities. The combinations are, however, partly coupled and only 4-5 of them are of biogeochemical significance.

**Table 5.3**    Microbiological terminology for different types of energy metabolism.

| Energy source | | Electron donor | | Carbon source | |
|---|---|---|---|---|---|
| Light: | *photo-* | Inorganic: | *litho-* | $CO_2$: | *auto-troph* |
| Chemical: | *chemo-* | Organic: | *organo-* | Organic C: | *hetero-troph* |

187

The energy source of the *phototrophic* organisms such as plants, algae and photosynthetic bacteria is light. For the *chemotrophic* organisms such as animals, fungi and many bacteria it is chemical energy, e.g. of glucose, methane or ammonium. Many phototrophic microorganisms have the capacity to switch between a phototrophic mode of life in the light and a heterotrophic mode, where they take up dissolved organic substrates or catch prey, or they may do both.

The electron donor of green plants is an inorganic compound, $H_2O$, and the final electron acceptor is $CO_2$:

$$CO_2 + H_2O \rightarrow [CH_2O] + O_2 \qquad (5.18)$$

where $CH_2O$ symbolizes the organic matter in plant biomass. A more accurate stoichiometry, originally suggested by Redfield (1958) for marine phytoplankton assimilating also nitrate and phosphate as nutrients, is:

$$106CO_2 + 16NO_3^- + HPO_4^{2-} + 122H_2O + 18H^+ \rightarrow$$
$$C_{106}H_{263}O_{110}N_{16}P + 138O_2 \qquad (5.18a)$$

which shows that the biomass, in particular due to the lipid fraction, is more reduced than $CH_2$.

The electron donor of many biogeochemically important, phototrophic or chemotrophic bacteria is also inorganic, such as $H_2$, $Fe^{2+}$ or $H_2S$. All these organisms are called *lithotrophic* from 'lithos' meaning rock. In contrast, *organotrophic* organisms such as animals use an organic electron donor:

$$[CH_2O] + O_2 \rightarrow CO_2 + H_2O \qquad (5.19)$$

Finally, organisms such as the green plants, but also many lithotrophic bacteria, are *autotrophs*, i.e. they are able to build up their biomass from $CO_2$ and other inorganic nutrients. Animals and many bacteria living on organic substrates instead incorporate organic carbon into their biomass and are termed *heterotrophs*. Among the prokaryotes, there is not a strict discrimination between autotrophy and heterotrophy. Thus, heterotrophs generally incorporate 4-6% $CO_2$ (mostly to convert $C_3$ compounds to $C_4$ compounds in anaplerotic reactions, in order to compensate for $C_4$ compounds which were consumed for biosynthetic purposes in the cell). This heterotrophic $CO_2$ assimilation has in fact been used to estimate the total heterotrophic metabolism of

bacterioplankton based on their dark $^{14}CO_2$ incorporation. As another example, sulfate reducing bacteria living on acetate are in principle heterotrophs but derive about 30% of their cell carbon from $CO_2$. Aerobic methane oxidizing bacteria, which strictly speaking are also living on an organic compound, incorporate 30-90% $CO_2$.

The three criteria in Table 5.3 can now be used in combination to specify the main types of energy metabolism of organisms. Green plants are photolithoautotrophs while animals are chemoorganoheterotrophs. This detail of taxonomy may seem an exaggeration for these organisms which are in daily terms called photoautotrophs and heterotrophs. However, the usefulness becomes apparent when we need to understand the function of, e.g. chemolithoautotrophs or chemolithoheterotrophs for the cycling of nitrogen, manganese, iron or sulfur in marine sediments (see Sect. 5.5).

### 5.4.5    Energy Metabolism of Microorganisms

The basic types of energy metabolism and representative organisms of each group are compiled in Table 5.4. Further information can be found in Fenchel et al. (1998), Ehrlich (1996), Madigan et al. (1997), and other textbooks. Most photoautotrophic organisms use water to reduce $CO_2$ according to the highly endergonic reaction of Equation 5.18. Since water has a high redox potential it requires more energy than is available in single photons of visible light to transfer electrons from water to an electron carrier, $NADP^+$, which can subsequently reduce $CO_2$ through the complex pathway of the Calvin-Benson cycle. Modern plants and algae, which use $H_2O$ as an electron donor, consequently transfer the electrons in two steps through two photocenters. In photosystem II, electrons are transferred from $H_2O$ which is oxidized to $O_2$. In photosystem I these electrons are transferred to a highly reducing primary acceptor and from there to $NADP^+$. More primitive phototrophic bacteria, which predominated on Earth before the evolution of oxygenic photosynthesis, have only one photocenter and are therefore dependent on an electron donor of a lower redox potential than water. In the purple and green sulfur bacteria or cyanobacteria, $H_2S$ serves as such a low-$E_H$ electron donor. The $H_2S$ is oxidized to $S^0$ and mostly further to sulfate. Some purple bacteria are able to use $Fe^{2+}$ as an electron donor to reduce $CO_2$. Their existence had been suggested many years ago, but they were discovered and isolated only recently (Widdel et al. 1993, Ehrenreich and Widdel 1994). This group is geologically interesting, since direct phototrophic Fe(II) oxidation with $CO_2$ opens the theoretical possibility for iron oxidation in the absence of $O_2$ some 2.0-2.5 billion years ago, at the time when the great deposits of banded iron formations were formed on Earth.

Some of the purple and green bacteria are able to grow photoheterotrophically, which may under some environmental conditions be advantageous as they cover their energy requirements from light but can assimilate organic substrates instead of spending most of the light energy on the assimilation of $CO_2$. The organisms may either take up or excrete $CO_2$ in order to balance the redox state of their substrate with that of their biomass.

### 5.4.6    Chemolithotrophs

The chemolithotrophs comprise a large and diverse group of exclusively prokaryotic organisms, which play important roles for mineral cycling in marine sediments (Table 5.4). They conserve energy from the oxidation of a range of inorganic compounds. Many are autotrophic and assimilate $CO_2$ via the Calvin cycle similar to the plants. Some are heterotrophic and assimilate organic carbon, whereas some can alter between these carbon sources and are termed *mixotrophs*. Many respire with $O_2$, while others are anaerobic and use nitrate, sulfate, $CO_2$ or metal oxides as electron acceptors.

It is important to note that the autotrophic chemolithotrophs have a rather low growth yield (amount of biomass produced per mol of substrate consumed). For example, the sulfur bacteria use

**Table 5.4**   Main types of energy metabolism with examples of representative organisms.

| Metabolism | | Energy source | Carbon source | Electron donor | Organisms |
|---|---|---|---|---|---|
| Photoautotroph | | | $CO_2$ | $H_2O$ | Green plants, algae, cyanobacteria |
| | | Light | | $H_2S$, $S^0$, $Fe^{2+}$ | Purple and green sulfur bact. (*Chromatium*, *Chlorobium*), Cyanobacteria |
| Photoheterotroph | | | Org. C $\pm$ $CO_2$ | | Purple and green non-sulfur bact. (*Rhodospirillum*, *Chloroflexus*) |
| Chemo-lithotroph | Chemolitho-autotroph | Oxidation of inorganic compounds | $CO_2$ | $H_2S$, $S^0$, $S_2O_3^{2-}$, $FeS_2$ $NH_4^+$, $NO_2^-$ $H_2$ $Fe^{2+}$, $Mn^{2+}$ | Aerobic: Colorless sulfur bact. (*Thiobacillus, Beggiatoa*) Nitrifying bact. (*Thiobacillus, Nitrobacter*) Hydrogen bact. (*Hydrogenomonas*) Iron bact. (*Ferrobacillus, Shewanella*) |
| | | | | $H_2 + SO_4^{2-}$ $H_2S/S^0/S_2O_3^{2-} + NO_3^-$ $H_2 + CO_2 \rightarrow CH_4$ $H_2 + CO_2 \rightarrow$ acetate | Anaerobic: Some sulfate reducing bact. (*Desulfovibrio spp.*) Denitrifying sulfur bact. (*Thiobac. denitrificans*) Methanogenic bact. Acetogenic bact. |
| | Mixotroph | | $CO_2$ or Org. C | $H_2S$, $S^0$, $S_2O_3^{2-}$ | Colorless sulfur bact. (some *Thiobacillus*) |
| | Chemolitho-heterotroph | | Org. C | $H_2S$, $S^0$, $S_2O_3^{2-}$ $H_2$ | Colorless sulfur bact. (some *Thiobacillus, Beggiatoa*) Some sulfate reducing bact. |
| Heterotroph (=chemoorganotroph) | | Oxidation of organic compounds | Org. C (max. 30% $CO_2$) | Org. C | Aerobic: Animals, fungi, many bacteria Anaerobic: Denitrifying bacteria Mn- or Fe-reducing bacteria Sulfate reducing bacteria Fermenting bacteria |
| | | | Org. $C_1$ (30-90% CO2) | $CH_4$ | Methane oxidizing bact. |

most of the electrons and energy from sulfide oxidation to generate ATP. They need this transient energy storage for $CO_2$ assimilation via the Calvin cycle, which energetically is a rather inefficient pathway in spite of its widespread occurrence among autotrophic organisms. Only some 10-20% of the electrons derived from $H_2S$ oxidation flow into $CO_2$ and are used for autotrophic growth (Kelly 1982). As a consequence, it has a more limited significance for the overall organic carbon budget of marine sediments whether sulfide is oxidized autocatalytically without bacterial involvement or biologically with a resulting formation of new bacterial biomass (Jørgensen 1987). In spite of this low growth yield, filamentous sulfur bacteria such as *Beggiatoa* or *Thioploca* may often form mats of large biomass at the sediment surface and thereby influence the whole community of benthic organisms as well as the chemistry of the sediment (Fossing et al. 1995).

Among the aerobic chemolithotrophs, colorless sulfur bacteria may oxidize $H_2S$, $S^0$, $S_2O_3^{2-}$ or $FeS_2$ to sulfate. Many $H_2S$ oxidizers, such as *Beggiatoa*, accumulate $S^0$ in the cells and are conspicuous due to the highly light-refractive $S^0$-globules which give *Beggiatoa* mats a bright white appearence. Nitrifying bacteria consist of two groups, those which oxidize $NH_4^+$ to $NO_2^-$ and those which oxidize $NO_2^-$ to $NO_3^-$. The hydrogen oxidizing 'Knallgas-bacteria' mentioned before (Eq. 5.14) oxidize $H_2$ to water. They have a higher efficiency of ATP formation than other chemolithotrophs because their electron donor, $H_2$, is highly reducing which provides a large energy yield when feeding electrons into the respiratory chain (see below). A variety of iron and manganese oxidizing bacteria play an important role for metal cycling in the suboxic zone, but this group of organisms is still very incompletely known. Much research has been done on the important acidophilic forms, e.g. *Thiobacillus ferrooxidans*, associated with acid mine drainage, due to their significance and because it is possible to grow these in a chemically stable medium without exessive precipitation of iron oxides.

The anaerobic chemolithotrophs use alternative electron acceptors such as nitrate, sulfate or $CO_2$ for the oxidation of their electron donor. Thus, several sulfur bacteria can respire with nitrate (denitrifiers) and thereby oxidize reduced sulfur species such as $H_2S$, $S^0$ or $S_2O_3^{2-}$ to sulfate. Also the above mentioned iron oxidizing nitrate reducing bacteria belong in this group. Many

sulfate reducing bacteria can use $H_2$ as electron donor and thus live as anaerobic chemolithotrophs. Two very important groups of organisms are the methanogens and the acetogens. The former, which are archaea and not bacteria, form methane from $CO_2$ and $H_2$:

$$CO_2 + 4H_2 \rightarrow CH_4 + H_2O \qquad (5.20)$$

while the latter form acetate:

$$2CO_2 + 4H_2 \rightarrow CH_3COO^- + H^+ \qquad (5.21)$$

In marine sediments below the sulfate zone, methanogenesis is the predominant terminal pathway of organic carbon degradation. Methane may also be formed from acetate or from organic $C_1$-compounds such as methanol or methylamines:

$$CH_3COO^- + H^+ \rightarrow CH_4 + CO_2 \qquad (5.22)$$

## 5.4.7  Respiration and Fermentation

The best known type of energy metabolism in the sea bed is the aerobic respiration by heterotrophic organisms such as animals and many bacteria. Heterotrophic bacteria take up small organic molecules such as glucose, break them down into smaller units, and ultimately oxidize them to $CO_2$ with oxygen. The first pathway inside the cell, called *glycolysis*, converts glucose to pyruvate and conserves only a small amount of potential energy in ATP (Fig. 5.10). A few electrons are transferred to $NAD^+$ to form reduced NADH. Through the complex cyclic pathway called the *tricarboxylic acid cycle (TCA cycle)*, the rest of the available electrons of the organic substrate are transferred, again mostly to form NADH, and $CO_2$ is released. The electron carrier, NADH is recycled by transferring its electrons via a membrane-bound *electron transport chain* to the *terminal electron acceptor*, $O_2$, which is thereby reduced to $H_2O$. Through the electron transport chain, the proton-motive force across the cell membrane is maintained and most of the energy carrier, ATP, is generated. The energy yield of aerobic respiration is large, and up to 38 mol of ATP may be formed per mol of glucose oxidized. This corresponds to a 43% efficiency of energy utilization of the organic substrate, the rest being lost as entropy (heat). By the transfer of electrons from the electron carrier, NADH, to the terminal electron acceptor, $O_2$, the reduced form of the carrier, NAD$^+$, is regenerated and the electron flow can proceed in a cyclic manner.

Many heterotrophic prokaryotes are anaerobic and have the ability to use terminal electron acceptors other than O$_2$ for the oxidation of their organic substrates (Fig. 5.10). The denitrifying bacteria respire with NO$_3^-$ which they reduce to N$_2$ and release into the large atmospheric nitrogen pool. The sulfate reducing bacteria can use a range of organic molecules as substrates and oxidize these with the concomitant reduction of SO$_4^{2-}$ to H$_2$S. Some bacteria can use manganese(III, IV) or iron(III) oxides as electron acceptors in their heterotrophic metabolism and reduce these to Mn$^{2+}$ or Fe$^{2+}$.

Many bacteria do not possess a complete TCA-cycle and electron transport chain and are thus not able to use an external electron acceptor as terminal oxidant. This prevents the formation of large amounts of ATP via the electron transport chain and thus leads to a low energy yield of the metabolism. The organisms may still form a small amount of ATP by so-called substrate level phosphorylation through the glycolysis, and this is sufficient to enable growth of these *fermentative* organisms. The cells must, however, still be able to recycle the reduced electron carrier, NADH, in order to continue the glycolytic pathway. In principle, they do this by transferring the electrons from NADH to an intermediate such as pyruvate (CH$_3$COCOO$^-$). An example of a fermentation reaction is the formation of lactate by lactic acid bacteria (brackets around the pyruvate indicate that this is an intermediate and not an external substrate being assimilated and transformed):

$$[CH_3COCOO^-] + NADH + H^+ \rightarrow$$
$$CH_3CHOHCOO^- + NAD^+ \qquad (5.23)$$

It is seen that fermentation does not require an external electron acceptor and there is no net oxidation of the organic substrate, glucose. There is rather a reallocation of electrons and hydrogen atoms within the cleaved molecule, whereby a small amount of energy is released. Fermentations often involve a cleavage of the C$_3$ compound to a C$_2$ compound plus CO$_2$. The pyruvate is then coupled to coenzyme-A and cleaved in the form of acetyl-CoA and energy is subsequently conserved by the release of acetate.

$$[CH_3COCOO^-] \rightarrow CH_3COO^- + CO_2 + H_2 \quad (5.24)$$

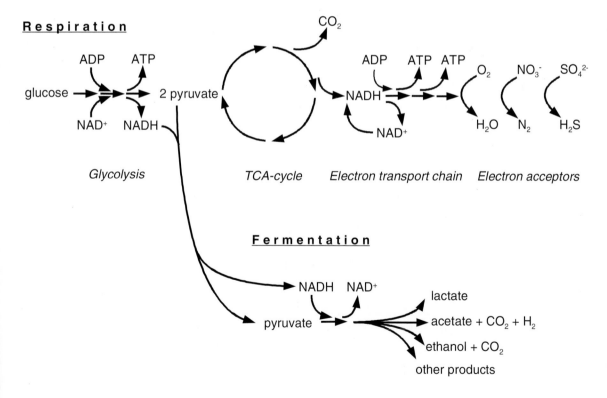

**Fig. 5.10**   Main pathways of catabolic metabolism in respiring and fermenting heterotrophic organisms (see text).

The $H_2$ is formed by a transfer of electrons from pyruvate via the enzyme ferredoxin to $NAD^+$:

$$NADH + H^+ \rightarrow NAD^+ + H_2 \qquad (5.25)$$

The degradation of organic matter via fermentative pathways to small organic molecules such as lactate, butyrate, propionate, acetate, formate, $H_2$ and $CO_2$ is very important in marine sediments, since these compounds are the main substrates for sulfate reduction and partly for methane formation. A form of inorganic fermentation of sulfur compounds was discovered in recent years in several sulfate reducers and other anaerobic bacteria (Bak and Cypionka 1987). These organisms may carry out a disproportionation of $S^0$, $S_2O_3^{2-}$ or $SO_3^{2-}$ by which $H_2S$ and $SO_4^{2-}$ are formed simultaneously (Eqs. 5.4 and 5.5). Disproportionation reactions have turned out to play an important role in the sulfur cycle of marine sediments and for the isotope geochemistry of sulfur (Jørgensen

1990, Thamdrup et al. 1993, Canfield and Teske 1996).

In comparison to all other heterotrophs, the microorganisms oxidizing methane and other $C_1$ compounds such as methanol, have a unique metabolic pathway which involves oxygenase enzymes and thus requires $O_2$. Only aerobic forms have thus been isolated and studied in laboratory culture, yet methane oxidation in marine sediments is known to take place mostly anaerobically at the deep transition to the sulfate zone. Based on the high activation energy and the apparent requirement for molecular oxygen to attack methane in all known methane oxidizers, it is still questionable whether sulfate reducers may directly oxidize methane (see Sect. 5.1).

Although the different groups of prokaryotes may be catagorized in the scheme of Table 5.4, it should be noted that they are often very diverse and flexible and may thus fit into different categories according to their immediate environmen-

**Table 5.5**  Diversity of sulfur metabolism among the sulfate reducing bacteria. The changes in free energy, $\Delta G^0$, have been calculated for standard conditions. The data show that the disproportionation of elemental sulfur is an endergonic process under standard conditions and therefore requires an efficient removal of the formed $HS^-$ to pull the reaction and make it exergonic. After Cypionka (1994).

| *Pathway* and stoichiometry | $\Delta G^0$ (kJ mol$^{-1}$) |
|---|---|
| *Reduction of sulfur compounds:* | |
| $SO_4^{2-} + 4H_2 + H^+ \rightarrow HS^- + 4H_2O$ | -155 |
| $SO_3^{2-} + 3H_2 + H^+ \rightarrow HS^- + 3H_2O$ | -175 |
| $S_2O_3^{2-} + 4H_2 \rightarrow 2HS^- + 3H_2O$ | -179 |
| $S^0 + H_2 \rightarrow HS^- + H^+$ | -30 |
| *Incomplete sulfate reduction:* | |
| $SO_4^{2-} + 2H_2 + H^+ \rightarrow S_2O_3^{2-} + 5H_2O$ | -65 |
| *Disproportionation:* | |
| $S_2O_3^{2-} + H_2O \rightarrow SO_4^{2-} + HS^- + H^+$ | -25 |
| $4SO_3^{2-} + H^+ \rightarrow 3SO_4^{2-} + HS^-$ | -236 |
| $4S^0 + 4H_2O \rightarrow SO_4^{2-} + 3HS^- + 5H^+$ | +33* |
| *Oxidation of sulfur compounds:* | |
| $HS^- + 2O_2 \rightarrow SO_4^{2-} + H^+$ | -794 |
| $HS^- + NO_3^- + H^+ + H_2O \rightarrow SO_4^{2-} + NH_4^+$ | -445 |
| $S_2O_3^{2-} + 2O_2 + H_2O \rightarrow 2SO_4^{2-} + 2H^+$ | -976 |
| $SO_3^{2-} + {}^1/_2 O_2 \rightarrow SO_4^{2-}$ | -257 |

tal conditions and mode of life. A good example are the sulfate reducing bacteria which were previously assumed to be obligate anaerobes specialized on the oxidation of a limited range of small organic molecules with sulfate (Postgate 1979, Widdel 1988). Recent studies have revealed a great diversity in the types of sulfur metabolism in these organisms as listed in Table 5.5 (Bak and Cypionka 1987, Dannenberg et al. 1992, Krekeler and Cypionka 1995). Different sulfate reducing bacteria may, alternatively to $SO_4^{2-}$, use $SO_3^{2-}$, $S_2O_3^{2-}$ or $S^0$ as electron acceptors or may disproportionate these in the absence of an appropriate electron donor such as $H_2$. They may even respire with oxygen or nitrate and may oxidize reduced sulfur compounds with oxygen. There is also evidence from marine sediments and pure cultures that sulfate reducing bacteria may reduce oxidized iron minerals (Coleman et al. 1993) and are even able to grow with ferric hydroxide as the sole electron acceptor (C. Knoblauch, unpublished observations; cf. Sect. 7.4.2.2):

$$8Fe(OH)_3 + CH_3COO^- \rightarrow$$
$$8Fe^{2+} + 2HCO_3^- + 5H_2O + 15OH^- \quad (5.26)$$

# 5.5    Pathways of Organic Matter Degradation

Organic material is deposited on the sea floor principally as aggregates which sink down through the water column as a continuous particle rain (Chap. 4). This particulate organic flux is related to the primary productivity of the overlying plankton community and to the water depth through which the detritus sinks while being gradually decomposed. As a mean value, some 25-50% of the primary productivity reaches the sea floor in coastal seas whereas the fraction is only about 1% in the deep sea (Suess 1980; cf. Fig. 12.1). Within the sediment, most organic material remains associated with the particles or is sorbed to inorganic sediment grains. Freshly arrived material, in particular after the sedimentation of a phytoplankton bloom, often forms a thin detritus layer at the sediment-water interface. This detritus is a site of high microbial activity and rapid organic matter degradation, even in the deep sea where the bacterial metabolism is otherwise strongly limited by the low organic influx (Lochte and Turley 1988). Due to the activities of

detritus-feeding benthic invertebrates and to bioturbation, the particulate organic material is gradually buried into the sediment and becomes an integral part of the sedimentary organic matter.

## 5.5.1    Depolymerization of Macromolecules

The organic detritus is a composite of macromolecular compounds such as structural carbohydrates, proteins, nucleic acids and lipid complexes. Prokaryotic organisms are unable to take up particles or even larger organic molecules and are restricted to a molecular size less than ca 600 daltons (Weiss et al. 1991). A simple organic molecule such as sucrose has a molecular size of 342 daltons. Thus, although some bacteria are able to degrade and metabolize fibers of cellulose, lignin, chitin or other structural polymers, they must first degrade the polymers before they can assimilate the monomeric products (Fig. 5.11). This depolymerization is caused by exoenzymes produced by the bacteria and either released freely into their environment or associated with the outer membrane or cell wall. The latter requires a direct contact between the bacterial cells and the particulate substrate and many sediment bacteria are indeed associated with solid surfaces in the sediment. The excretion of enzymes is a loss of carbon and nitrogen and thus an energy investment of the individual bacterial cells, but these have strategies of optimizing the return of monomeric products (Vetter et al. 1998) and may regulate the enzyme production according to the presence of the polymeric target compound in their environment ('substrate induction'). Thus, there is a positive correlation between the availability of specific polymeric substances in the sediment and the concentration of free enzymes which may degrade them (Boetius and Damm 1998, Boetius and Lochte 1996).

The depolymerization of sedimentary organic matter is generally the rate-limiting step in the sequence of mineralization processes. This may be concluded from the observation, that the monomeric compounds and the products of further bacterial degradation do not accumulate in the sediment but are rapidly assimilated and metabolized. Thus, the concentrations of individual free sugars, amino acids or lipids in the pore water are low. Dissolved organic matter (DOM) is a product of partial degradation and may be released from the sediment to the overlying water. It preferentially consists of complex dissolved polymers and oligomers rather than of monomers. Dissolved organic molecules may not only be taken up by bacteria but may also be removed from the pore water by adsorption onto sediment particles or by condensation reactions, such as monosaccharides and amino acids forming melanoidins (Sansone et al. 1987, Hedges 1978). Organic matter, which is not mineralized, may be adsorbed to mineral surfaces and transformed to 'geomacromolecules' which are highly resistant to enzymatic hydrolysis and bacterial degradation and which become more permanently buried with the sediment (Henrichs 1992, Keil et al. 1994). This buried fraction also contains some of the original organic matter which was deposited as refractory biomolecules such as lignins, tannins, algaenan, cutan and suberan (Tegelaar et al. 1989). The burial of organic matter deep down into the sediment shows a positive correlation with the rate of deposition and constitutes in the order of 5-20% of the initially deposited organic matter in shelf sediments and 0.5-3% in deep-sea sediments (Henrichs and Reeburgh 1987, cf. Fig. 12.1)

### 5.5.2  Aerobic and Anaerobic Mineralization

The particulate detritus consumed by metazoans, and the small organic molecules taken up by microorganisms in the oxic zone, may be mineralized completely to $CO_2$ through aerobic respiration within the individual organisms. The aerobic food chain consists of organisms of very diverse feeding biology and size, but of a uniform type of energy metabolism, namely the aerobic respiration. The oxic zone is, however, generally only mm-to-cm thick in shelf sediments as shown by measurements with $O_2$ microsensors (Chap. 3). In slope and deep-sea sediments the oxic zone expands to many cm or dm depth (Reimers 1987, Glud et al. 1994).

Much of the organic mineralization thus takes place within the anoxic sediment. This anoxic world is inhabited primarily by prokaryotic organisms having a high diversity of metabolic types (Fig. 5.11). There are denitrifiers and metal oxide reducers in the suboxic zone which can utilize a wide range of monomeric organic substances and can respire these to $CO_2$. With depth into the sediment, however, the energy yield of bacterial metabolism becomes gradually smaller and the organisms become narrower in the spec-

trum of substrates which they can use. While denitrifiers are still very versatile with respect to usable substrates, the sulfate reducers are mostly unable to respire, for example sugars or amino acids. Instead, these monomeric compounds are taken up by fermenting bacteria and converted into a narrower spectrum of fermentation products which include primarily volatile fatty acids such as formate, acetate, propionate and butyrate, as well as $H_2$, lactate, some alcohols and $CO_2$. Through a second fermentation step the products may be focused even further towards the key products, acetate, $H_2$ and $CO_2$. The sulfate reducers depend on these products of fermentation which they can respire to $CO_2$. Several well-known sulfate reducers such as *Desulfovibrio* can only carry out an incomplete oxidation of substrates such as lactate, and they excrete acetate as a product. Other sulfate reducers have specialized on acetate and catalyze the complete oxidation to $CO_2$. The degradative capacity of anaerobic bacteria seems, however, to be broader than formerly expected and new physiological types continue to be discovered. As an example, sulfate reducing bacteria able to oxidize aromatic and aliphatic hydrocarbons have now been isolated, which shows that anaerobic prokaryotes are able to degrade important components of crude oil in the absence of oxygen (Rueter et al. 1994, Rabus et al. 1996).

The methane forming (methanogenic) archaea can use only a narrow spectrum of substrates, primarily $H_2$ plus $CO_2$ and acetate (Eqs. 5.20 and 5.22). Within the sulfate zone, which generally penetrates several meters down into the sea bed, the sulfate reducing bacteria compete successfully with the methanogens for these few substrates and methanogenesis is, therefore, of little significance in the sulfate zone. Only a few 'non-competitive' substrates such as methylamines are used only by the methanogens and not by the sulfate reducers (Oremland and Polcin 1982). Below the sulfate zone, however, there are no available electron acceptors left other than $CO_2$, and methane accumulates here as the main terminal product of organic matter degradation.

### 5.5.3  Depth Zonation of Oxidants

The general depth sequence of oxidants used in the mineralization of organic matter is $O_2 \rightarrow NO_3^- \rightarrow Mn(IV) \rightarrow Fe(III) \rightarrow SO_4^{2-} \rightarrow CO_2$. This sequence corresponds to a gradual decrease

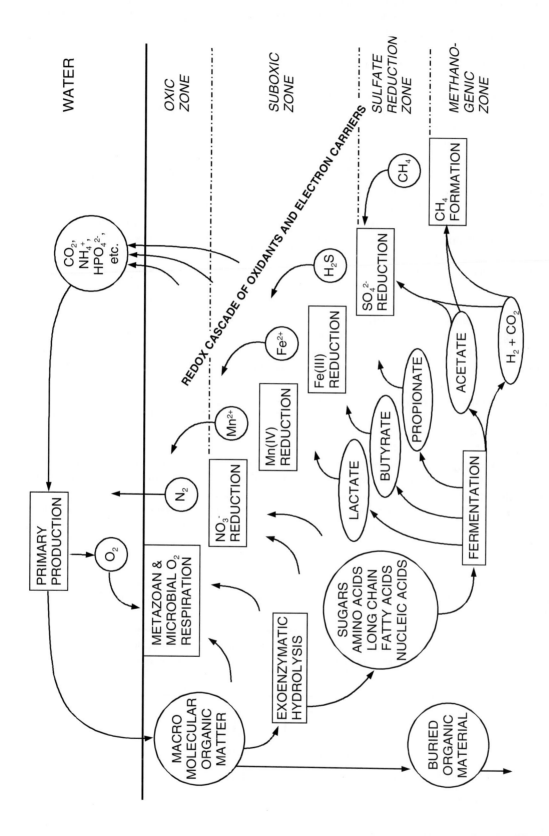

**Fig. 5.11** Pathways of organic carbon degradation in marine sediments and their relation to the geochemical zonations and the consumption of oxidants. (After Fenchel and Jørgensen 1977).

in redox potential of the oxidant (Fig. 5.9) and thus to a decrease in the free energy available by respiration with the different electron acceptors (Table 5.2; cf. Sect. 3.2.5). The $\Delta G^0$ of oxic respiration is the highest, -479 kJ mol$^{-1}$, and that of denitrification is nearly as high. Table 5.2 shows that respiration of one mol of organic carbon with sulfate yields only a small fraction of the energy of respiration with oxygen. The remaining potential chemical energy is not lost but is mostly bound in the product, $H_2S$. This energy may become available to other microorganisms, such as chemolithotrophic sulfur bacteria, when the sulfide is transported back up towards the sediment surface and comes into contact with potential oxidants.

The quantitative importance of the different oxidants for mineralization of organic carbon has been studied intensely over the last few decades, both by diagenetic modeling and by incubation experiments. It is generally found that oxygen and sulfate play the major role in shelf sediments, where 25-50% of the organic carbon may be mineralized anaerobically by sulfate reducing bacteria (Jørgensen 1982). With increasing water depth and decreasing organic influx down the continental slope and into the deep sea, the depth of oxygen penetration increases and sulfate reduction gradually looses significance. Nitrate seems to play a minor role as an oxidant of organic matter. Manganese oxides occur in shelf sediments mostly in lower concentrations than iron oxides and, expectedly, the Mn(IV) reduction rates should be lower than those of Fe(III) reduction. Manganese, however, is recycled nearer to the sediment sur-

face and relatively faster than iron. These solid-phase oxidants are both dependent on bioturbation as the mechanism to bring them from the sediment surface down to their zone of reaction. The shorter this distance, the faster can the metal oxide be recycled.

Table 5.6 shows an example of process rates in a coastal marine sediment. In this comparison, it is important to keep in mind that the different oxidants are not equivalent in their oxidation capacity. When for example the iron in Fe(III) is reduced, the product is Fe(II) and the iron atoms have been reduced only one oxidation step from +3 to +2. Sulfur atoms in sulfate, in contrast, are reduced eight oxidation steps from +6 in $SO_4^{2-}$ to -2 in $H_2S$. One mol of sulfate, therefore, has 8-fold higher oxidation capacity than one mol of iron oxide. In order to compare the different oxidants and their role for organic carbon oxidation, their reduction rates were recalculated to carbon equivalents on the basis of their change in oxidation step. It is then clear that oxygen and sulfate were the predominant oxidants in this sediment, sulfate oxidizing about 44% of the organic carbon in the sediment. Of the $H_2S$ formed, 15% was buried as pyrite, while the rest was reoxidized and could potentially consume one third of the total oxygen uptake. This redox balance is typical for coastal sediments where up to half of the oxygen taken up by the sediment is used for the direct or indirect reoxidation of sulfide and reduced manganese and iron (Jørgensen 1982).

Manganese and iron behave differently with respect to their reactivity towards sulfide. Iron

Table 5.6 Annual budget for the mineralization of organic carbon and the consumption of oxidants in a Danish coastal sediment, Aarhus Bay, at 15 m water depth. The basic reaction and the change in oxidation step is shown for the elements involved. The rates of processes were determined for one $m^2$ of sediment and were all recalculated to carbon equivalents. From data compiled in Jørgensen (1996).

| Reaction | Δ Oxidation steps | Measured rate | Estimated carbon equivalents |
|---|---|---|---|
| | | mol m$^{-2}$ yr$^{-1}$ | |
| $CH_2O \rightarrow CO_2$ | C:  $0 \rightarrow +4 = 4$ | 9,9 | 9,9 |
| $O_2 \rightarrow H_2O$ | O:  $2 \cdot 0 \rightarrow -2 = 4$ | 9,2 | 9,2 |
| $NO_3^- \rightarrow N_2$ | N:  $+5 \rightarrow 0 = 5$ | 0,15 | 0,19 |
| $Mn(IV) \rightarrow Mn^{2+}$ | Mn:  $+4 \rightarrow +2 = 2$ | 0,8 | 0,4 |
| $Fe(III) \rightarrow Fe^{2+}$ | Fe:  $+3 \rightarrow +2 = 1$ | 1,6 | 0,4 |
| $SO_4^{2-} \rightarrow HS^-$ | S:  $+6 \rightarrow -2 = 8$ | 1,7 | 3,4 |

binds strongly to sulfide as FeS or $FeS_2$, whereas manganese does not. Furthermore, the reduced iron is more reactive when it reaches up near the oxic zone, and $Fe^{2+}$ generally does not diffuse out of the sediment, although it may excape by advective pore water flow (Huettel et al. 1998). The $Mn^{2+}$, in contrast, easily recycles via the overlying water column and can thereby be transported from the shelf out into deeper water. This mechanism leads to the accumulation of manganese oxides in continental slope sediments (see Chap. 11). The important role of manganese and iron in slope sediments is evident from Figure 5.12. Manganese constituted about 5% of the total dry weight of this sediment. Below the manganese zone, iron oxides and then sulfate took over the main role as oxidants. In the upper few cm of the sediment, manganese oxide was the dominant oxidant below the $O_2$ zone (Canfield 1993).

The consecutive reduction of oxidants with depth in the marine sediment and the complex reoxidation of their products constitute the 'redox cascade' (Fig. 5.11). An important function of this sequence of reactions is the transport of electrons from organic carbon via inorganic species back to oxygen. The potential energy transferred from the organic carbon to the inorganic products of oxidation is thereby released and may support a variety of lithotrophic microorganisms. These may make a living from the oxidation of sulfide with Fe(III), Mn(IV), $NO_3^-$ or $O_2$. Others may be involved in the oxidation of reduced iron with Mn(IV), $NO_3^-$

or $O_2$ etc. The organisms responsible for these reactions are only partly known and new types continue to be isolated.

The depth sequence of electron acceptors in marine sediments, from oxygen to sulfate, is accompanied by a decrease in the degradability of the organic material remaining at that depth. This decrease can be formally considered to be the result of a number of different organic carbon pools, each with its own degradation characteristics and each being degraded exponentially with time and depth. This is the 'multi-G model' of Westrich and Berner (1984). The consecutive depletion of these pools leads to a steep decrease in organic carbon reactivity, which follows the sum of several exponential decays, and which can be demonstrated from the depth distribution of oxidant consumption rates. Thus, the $O_2$ consumption rate per volume in the oxic zone of a coastal sediment was found to vary between 3,000 and 30,000 nmol $O_2$ $cm^{-3}d^{-1}$ as the oxygen penetration depth varied between 5 mm in winter and 1 mm in summer (Gundersen et al. 1995). Some five cm deeper into the sediment, where sulfate reduction predominated and reached maximal activity, the rates of carbon mineralization decreased 100-fold to 25-150 nmol $cm^{-3}d^{-1}$ between winter and summer (Thamdrup et al. 1994). A few meters deep into the sediment, where methanogenesis predominated below the sulfate zone, carbon mineralization rates were again about a 100-fold lower. Although the rates down there may seem insig-

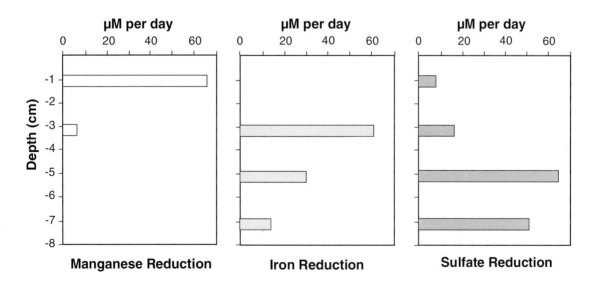

**Fig. 5.12** Depth zonation of reduction rates for the oxidants, Mn(IV), Fe(III) and $SO_4^{2-}$ in a marine sediment from Skagerrak at 700 m water depth. Data from Canfield (1993).

nificant, the slow organic matter decomposition to methane proceeds to great depth in the sediment and is therefore important on an areal basis. This methane diffuses back up towards the sulfate zone where it is oxidized to $CO_2$ at the expense of sulfate (Chap. 8). It may consume 10% or more of the total sulfate reduction in coastal sediments, depending on the depth of sulfate penetration (Iversen and Jørgensen 1985, Niewöhner et al. 1998, Chap. 8).

## 5.6    Methods in Biogeochemistry

A diversity of approaches, of which only a few examples can be discussed here, is used in the study of marine biogeochemistry (Fig. 5.2). Among the important goals is to quantify the rate at which biological and chemical processes take place in different depth zones of the sediment. These are often fast processes for which the reactants may have turnover times in the order of days or hours or even minutes. Data on dissolved species in the pore water and on solid-phase geochemistry can be used in diagenetic models to calculate such rates, as discussed in Chapter 3 (e.g. Schulz et al. 1994). The dynamic process is then derived from the chemical gradients in the pore water and from knowledge about the diffusion coefficient of the chemical species. Such diffusion-diagenesis models work best for either very steep and shallow gradients, e.g. of oxygen for which diffusion is rapid, or very deep gradients, e.g. of sulfate which penetrates deep below the zone affected by biological transport of pore water (bioirrigation) or of solid-phase sediment (bioturbation). For intermediate depth scales, e.g. in the suboxic zone, the burrowing fauna influences the transport processes so strongly that advection and bioirrigation tend to dominate over molecular diffusion. If the transport factor is enhanced to an unknown degree, a rate calculation based on molecular diffusion would be correspondingly wrong. In such cases, it may be more realistic to model the solid phase combined with estimates of the rate of burial and of mixing by the infauna (bioturbation). Burial and mixing rates are most often estimated from the vertical distribution and decay of natural radionuclides such as [210]Pb in the sediment.

Another problem is that many compounds are not only consumed but also recycled in a chemi-cal zone. Thus, sulfate in the upper sediment layers is both consumed by sulfate reduction and produced by sulfide oxidation so the net sulfate removal may be insignificant relative to the total reduction rate. The net removal of sulfate in the whole sediment column is determined by the amount of sulfide trapped in pyrite etc. and is often only 10% of the gross sulfate reduction. A modeling of the sulfate gradient would in that case lead to a 10-fold underestimate of the total reduction rate. In the upper part of the reducing zone, where rates of sulfate reduction are highest, there may be no net depletion of sulfate and the underestimation would be 100%.

### 5.6.1    Incubation Experiments

To overcome this problem, process rates can be measured experimentally in sediment samples by following the concentration changes of the chemical species with time, either in the pore water or in the solid phase. By such incubation experiments it is critical that the physico-chemical conditions and the biology of the sediment remain close to the natural situation. Since gradual changes with time are difficult to avoid, the duration of the experiment should preferably be restricted to several hours up to a day. This corresponds roughly to one doubling time of normal sediment bacteria. A careful way of containing the sediment samples is to distribute the sediment from different depths into gas-tight plastic bags under an atmosphere of nitrogen and keep these at the natural temperature. The sediment can then be mixed without dilution or contact with a gas phase, and subsamples can be taken over time for the analyses of chemical concentrations. Such a technique has been used to measure the rates of organic matter mineralization from the evolution of $CO_2$ or $NH_4^+$ or to measure the reduction rates of manganese and iron oxides (Fig. 5.12, Canfield et al. 1993).

### 5.6.2    Radioactive Tracers

Radioactive tracers are used in a different manner to identify and measure processes in marine sediments, mostly those carried out by microorganisms, but also purely chemical or physical processes. Radiotracers are mostly applied if chemical analysis alone is too insensitive or if the pathway of processes is more complex or cyclic. Thus, if a process is very slow relative to the pool

**Table 5.7**   Radio-isotopes most often used as tracers in biogeochemistry. The measurement of radioactivity is based on the emission of β-radiation (high-energy electrons) or γ-radiation (electromagnetic) with the specified maximum energy.

| Isotope | Emission mode, $E_{max}$ | Half-life | Examples of application |
|---------|--------------------------|-----------|------------------------|
| $^3$H | β, 29 keV | 12 years | Turnover of org. compounds, autoradiography |
| $^{14}$C | β, 156 keV | 5730 years | Turnover of org. compounds, $CO_2$ fixation |
| $^{32}$P | β, 1709 keV | 14 days | Phosphate turnover & assimilation |
| $^{35}$S | β, 167 keV | 87 days | Sulfate reduction, thiosulfate transformations |
| $^{55}$Fe | γ, 6 keV | 2.7 years | Iron oxidation and reduction |
| $^{59}$Fe | β, 466 keV | 45 days | do. |

sizes of the reactants or products, a long-term experiment of many days or months would be required to detect a chemical change. Since this would lead to changing sediment conditions and thus to non-natural process rates, higher sensitivity is required. By the use of a radiotracer, hundred- or thousand-fold shorter experiment durations may often be achieved. An example is the measurement of sulfate reduction rates (see Sect. 5.6.3). Sulfate can be analyzed in pore water samples with a precision of about ±2%, and a reduction of 5-10% or more is, therefore, required to determine its rate with a reasonable confidence. If radioactive, $^{35}$S-labelled sulfate is used to trace the process, it is possible to detect the reduction of only 1/100,000 of the $SO_4^{2-}$ by analyzing the radioactivity of the sulfide formed. The sensitivity by using radiotracer is thus improved >1000-fold over the chemical analysis.

The radioisotopes most often used in biogeochemistry are shown in Table 5.7, together with examples of their application. $^3$H is a weak ß-emitter, whereas $^{32}$P is a hard ß-emitter. $^{14}$C and $^{35}$S have similar intermediate energies. The $^{14}$C is most widely used, either to study the synthesis of new organic biomass through photo- or chemosynthesis ($^{14}CO_2$-assimilation), or to study the transformations and mineralization of organic material such as plankton detritus or specific compounds such as glucose, lactate, acetate and other organic molecules. The $^3$H-labelled substrates have in particular been used in connection with autoradiography, where the incorporation of label is quantified and mapped at high spatial resolution by radioactive exposure of a photographic film emulsion. By this technique it may be possible to demonstrate which microorganisms are ac-

tively assimilating the labeled compound. Since phosphorus does not undergo redox processes in sediments, $^{32}$P (or $^{33}$P) is mostly used to study the dynamics and uptake of phosphate by microorganisms. $^{35}$S has been used widely to study processes of the sulfur cycle in sediments, in particular to measure sulfate reduction.

Although radiotracer techniques may offer many advantages, they also have inherent problems. For example, applications of $^{35}$S to trace the pathways of $H_2S$ oxidation have been flawed by isotope exchange reactions between the inorganic reduced sulfur species such as elemental sulfur, polysulfides and iron sulfide (Fossing and Jørgensen 1990, Fossing et al. 1992, Fossing 1995). This isotope exchange means that the sulfur atoms swap places between two compounds without a concomitant net chemical reaction between them. If a $^{35}$S atom in $H_2S$ changes place with a $^{32}$S atom in $S^0$, the resulting change in radioactivity will appear as if $^{35}$S-labelled $H_2S$ were oxidized to $S^0$, although there may be no change in the concentrations of $H_2S$ or $S^0$. When the distribution of $^{35}$S radiolabel is followed with time the results can hardly be distinguished from a true net process and may be incorrectly interpreted as such. There is no similar isotopic exchange with sulfate at normal environmental temperatures which would otherwise prevent its use for the measurement of sulfate reduction rates.

The radioisotopes of iron or manganese, $^{55}$Fe, $^{59}$Fe and $^{54}$Mn, have had only limited application as tracers in biogeochemical studies. Experiments with $^{55}$Fe as a tracer for Fe(III) reduction in marine sediments showed problems of unspecific binding and possibly isotope exchange (King 1983, Roden and Lovley 1993), which has discouraged other researchers from the use of this isotope. Similar problems complicate experiments with manganese as a radiotracer. The iron isotope, [57]Fe, has been used in a completely different manner for the analysis of iron speciation in marine sediments. The isotope, [57]Fe, is added to a sediment and is allowed to equilibrate with the iron species. It can then be used to analyze the oxidation state and the mineralogy of iron by Mössbauer spectroscopy and thus to study the oxidation and reduction of iron minerals.

It has been a serious draw-back in studies of nitrogen transformations that a useful radioisotope of nitrogen does not exist. The isotope, [13]N, is available only at accelerator facilities and has a half-life of 5 min, which strongly limits its applicability. Instead, the stable isotope, [15]N, has been used successfully as a tracer for studies of nitrogen transformations in the marine environment. For example, the use of [15]NO$_3^-$ in a recently developed 'isotope pairing' technique has offered possibilities to study the process of denitrification of [15]NO$_3^-$ to N$_2$ (Nielsen 1992). The [15]NO$_3^-$ is added to the water phase over the sediment and is allowed to diffuse into the sediment and gradually equilibrate with [14]NO$_3^-$ in the pore water. By analyzing the isotopic composition of the formed N$_2$, i.e. [14]N[14]N, [14]N[15]N and [15]N[15]N, it has been possible not only to calculate the rate of denitrification, but even to discriminate between denitrification from nitrate diffusing down from the overlying seawater and denitrification from nitrate

formed by nitrification within the sediment. The results have shown that at the normal, low concentrations of nitrate in seawater, <10-20 μM, the main source of nitrate for denitrification is the internally formed nitrate derived from nitrification.

### 5.6.3   Example: Sulfate Reduction

As an example of a radio-tracer method, the measurement of sulfate reduction rates using [35]SO$_4^{2-}$ according to Jørgensen (1978) and Fossing and Jørgensen (1989) is described in brief (Fig. 5.13). The [35]SO$_4^{2-}$ is injected with a microsyringe into whole, intact sediment cores in quantities of a few microliters which contain ca 100 kBq (1 becquerel = 1 radioactive disintegration per second; 37 kBq = 1 microcurie). After 4-8 hours the core is sectioned, and the sediment samples are fixed with zinc acetate. The zinc a) binds the sulfide and prevents its oxidation and, b) kills the bacteria and prevents further sulfate reduction. The reduced [35]S is then separated from the sediment by acidification in a distillation apparatus, and the released H$_2$[35]S is transferred in a stream of N$_2$ to traps containing zinc acetate where Zn[35]S precipitates. The radioactivities of the Zn[35]S and of the remaining [35]SO$_4^{2-}$ are then analyzed by liquid scintillation counting. In a separate sediment core the concentration gradient of sulfate in the pore water is analyzed, and the rates of sulfate reduction can then be calculated according to the equation:

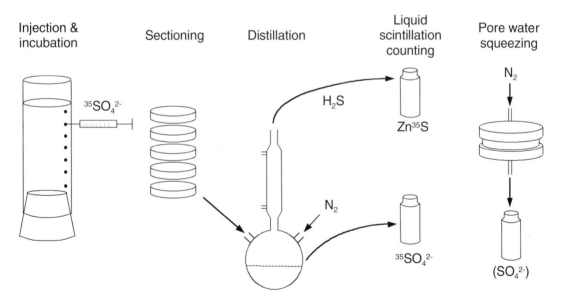

**Fig. 5.13**   Principle of sulfate reduction measurements in sediments using [35]SO$_4^{2-}$ as a tracer (see text).

Sulfate reduction rate =

$$\phi(SO_4^{2-}) \cdot \frac{H_2^{35}S}{^{35}SO_4^{2-}} \cdot \frac{24}{t} \cdot 1.06 \text{ nmol } SO_4^{2-}cm^{-3}day^{-1}$$

(5.27)

where $\phi$ is porosity, $(SO_4^{2-})$ is sulfate concentration in the pore water ($\phi$ $(SO_4^{2-})$ is then sulfate concentration per volume sediment), $H_2^{35}S$ is radioactivity of total reduced sulfur, $^{35}SO_4^{2-}$ is radioactivity of added sulfate tracer, $t$ is experiment time in hours, and 1.06 is a correction factor for the small dynamic isotope discrimination by the bacteria against the heavier isotope. This formula is a good approximation as long as only a small fraction of the labeled sulfate is reduced during incubation, a condition normally fulfilled in marine sediments.

Sulfate reduction rates in marine sediments commonly lie in the range of 1-100 nmol $SO_4^{2-}$ $cm^{-3}day^{-1}$ (Jørgensen 1982). Since the sulfate concentration in the pore water is around 28 mM or 20-25 µmol $cm^{-3}$, the turn-over time of the sulfate pool is in the order of 1-100 years. A purely chemical experiment would thus require a month to several years of incubation. This clearly illustrates why a radiotracer technique is required to measure the rate within several hours.

Similar principles as described here are used in measurements of, e.g. the oxidation of $^{14}C$-labelled methane or the formation of methane from $^{14}C$-labelled $CO_2$, acetate or methylamine. Often there are no gaseous substrates or products, and it is not possible to separate the radioisotopes as efficiently as by the measurement of sulfate reduction or methane formation and oxidation. This is the case for radiotracer studies of intermediates in the mineralization of organic matter in sediments, e.g. of $^{14}C$-labelled sugars, amino acids, or volatile fatty acids, studies which have been important for the understanding of the pathways of organic degradation (Fig. 5.11). These organic compounds, however, occur at low concentrations and have a fast turn-over of minutes to hours, so the sensitivity is less important. More important is the fact that these compounds are at a steady-state between production and consumption, which means that their concentration may not change during incubation in spite of their fast turnover. A chemical experiment would thus not immediately be able to detect their natural dynamics, but a radiotracer experiment may.

### 5.6.4  Specific Inhibitors

This limitation of chemical experiments is often overcome by the use of a specific inhibitor (Oremland and Capone 1988). The principle of an inhibitor technique may be, a) to block a sequence of processes at a certain step in order to observe the accumulation of an intermediate compound, b) to inhibit a certain group of organisms in order to observe whether the process is taken over by another group or may proceed auto-catalytically, c) to let the inhibitor substitute the substrate in the process and measure the transformation of the inhibitor with higher sensitivity. Ideally, the inhibitor should be specific for only the relevant target organisms or metabolic reaction. In reality, most inhibitors have side effects

**Table 5.8**  Some inhibitors commonly used in biogeochemistry and microbial ecology.

| Inhibitor | Process | Principle of function |
|---|---|---|
| BES | Methanogenesis | Blocks $CH_4$ formation (methyl-CoM reductase) |
| $MoO_4^{2-}$ | Sulfate reduction | Blocks $SO_4^{2-}$ reduction (depletes ATP pool) |
| Nitrapyrin | Nitrification | Blocks autotrophic $NH_4^+$ oxidation |
| Acetylene | Denitrification | Blocks N2O → N2 (also blocks nitrification) |
| Acetylene | N2-fixation | C2H2 is reduced to C2H4 instead of N2 → NH4+ |
| DCMU | Photosynthesis | Blocks electron flow between Photosystem II → I |
| Cyanide | Respiration | Blocks respiratory enzymes |
| β-fluorolactate | Lactate metabol. | Blocks heterotrophic metabolism of lactate |
| Chloramphenicol | Growth | Blocks prokaryotic protein synthesis |
| Cycloheximide | Growth | Blocks eukaryotic protein synthesis |

and it is necessary through appropriate control experiments to determine these and to find the minimum inhibitor dose required to obtain the required effect. Some examples are given in Table 5.8.

BES (2-bromoethanesulfonic acid) is a structural analogue of mercaptoethanesulfonic acid, also known as coenzyme-M in methanogenic bacteria, a coenzyme associated with the terminal methylation reactions in methanogenesis. BES inhibits this methylation and thus the formation of methane. It belongs to the near-ideal inhibitors because its effect is specific to the target group of organisms. BES has been used to determine the substrates for methane formation in aquatic sediments and to show that some substrates such acetate and $H_2$ are shared in competition between the methanogens and the sulfate reducers, whereas others such as methylamines are 'non-competitive' substrates which are used by the methano-

gens alone (Oremland and Polcin 1982, Oremland et al. 1982).

Similarly, molybdate ($MoO_4^{2-}$) together with other group VI oxyanions are analogues of sulfate and inhibit sulfate reduction competitively. Molybdate specifically interferes with the initial 'activation' of sulfate through reaction with ATP and tends to deplete the ATP pool, thus leading to cell death of sulfate reducing bacteria. Also molybdate has been important for clearing up the substrate interactions between methanogens and sulfate reducers in sediments. Molybdate has been used to demonstrate quantitatively which substrates play a role for sulfate reduction in marine sediments (Fig. 5.14). When molybdate is added to a sediment at a concentration similar to that of seawater sulfate, 20 mM, sulfate reduction stops. The organic substrates, which were utilized by the sulfate reducers in the uninhibited sediment and which were kept at a minimum concen-

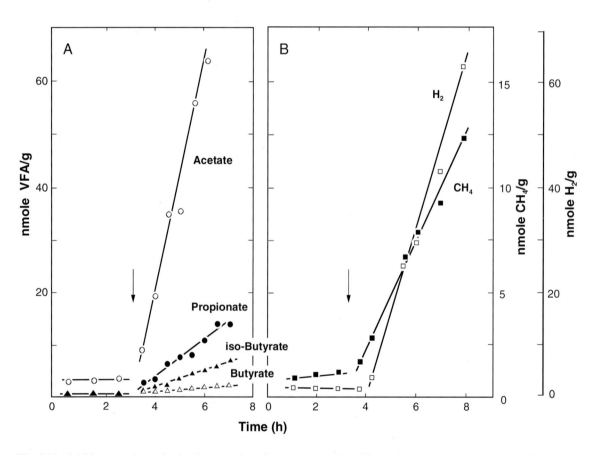

**Fig. 5.14** Inhibitor experiment for the demonstration of substrates used by sulfate reducing bacteria in a marine sediment. The concentrations of volatile fatty acids, hydrogen and methane are followed during a time-course experiment over 8 hours. At 3.5 hours (arrow) molybdate was added and the substrates accumulate at a rate corresponding to their rate of consumption before inhibition. The formation of methane shows the release of competition for the common substrates for methanogenesis and sulfate reduction ($H_2$ and acetate). Data from Sørensen et al. (1981).

tration as long as they were active, then suddenly start to accumulate because they are no longer consumed. Since the bacterial processes leading to the formation of these substrates are not inhibited, the substrates will accumulate at a similar rate at which they were used by the sulfate reducers before inhibition. Such experiments have demonstrated that acetate, propionate, butyrate, isobutyrate and $H_2$ are among the most important substrates for sulfate reducing bacteria in marine sediments (Sørensen et al. 1981; Christensen 1984).

Nitrapyrin (N-serve) was first introduced in agriculture as a means to inhibit the conversion of ammonium fertilizer to nitrate with subsequent wash-out of the nitrate. The nitrapyrin blocks the copper-containing cytochrome oxidase involved in the initial enzymatic oxidation of ammonium to hydroxylamine. The rate of ammonium accumulation after nitrapyrin inhibition is thus a measure of the nitrification rate before inhibition (Henriksen 1980).

Acetylene ($C_2H_2$) has been used to study the process of denitrification in which it blocks the last step from $N_2O$ to $N_2$ (Sørensen 1978). It causes an accumulation of $N_2O$ which can be analyzed by a gas chromatograph equipped with an electron capture detector or by a $N_2O$ microelectrode (Revsbech et al. 1988). The $N_2O$ accumulation rate is equal to the denitrification rate before inhibition. Acetylene is also used for studies of $N_2$ fixation. Acetylene has a triple bond analogous to $N_2$ (HC≡CH vs. N≡N) and can substitute $N_2$ competitively. Instead of reducing $N_2$ to $NH_4^+$, the key enzyme of nitrogen fixation, nitrogenase, reduces acetylene to ethylene ($H_2C=CH_2$). The formation rate of ethylene, which is easily analyzed in the headspace by a gas chromatograph with flame ionization detector, is thus a measure of the nitrogen fixation rate.

Other inhibitors are applied in studies of photosynthesis (DCMU) and respiration (cyanide) or are used to distinguish activities of prokaryotic (chloramphenicol) versus eukaryotic (cycloheximide) microorganisms (Table 5.8).

### 5.6.5    Other Methods

Information on the hydrolytic activity in marine sediments has been obtained from the use of model substrates labeled with fluorescent dyes such as methylumbelliferone (MUF) or fluorescein. These substrates may be small dimeric molecules, the hydrolytic cleavage of which re-

leases the fluorescence signal, which is then indicative of the activity of specific enzymes such as glucosidase, chitobiase, lipase, aminopeptidase or esterase (Chrost 1991). Also large fluorescently labeled polymers such as the polysaccharides laminarin or pullulan have been used in experiments to demonstrate the mechanism and kinetics of bacterial degradation (Arnosti 1996).

This is contribution No 254 of the Special Research Program SFB 261 (*The South Atlantic in the Late Quaternary*) funded by the Deutsche Forschungsgemeinschaft (DFG).

## References

Alperin, M.J. and Reeburgh, W.S., 1985. Inhibition Experiments on Anaerobic Methane Oxidation. Applied and Environmental Microbiology, 50: 940-945.

Arnosti, C., 1996. A new method for measuring polysaccharide hydrolysis rates in marine environments. Organic Geochemistry, 25: 105-115.

Bak, F. and Cypionka, H., 1987. A novel type of energy metabolism involving fermentation of inorganic sulphur compounds. Nature, 326: 891-892.

Benz, M., Brune, A. and Schink, B., 1998. Anaerobic and aerobic oxidation of ferrous iron and neutral pH by chemoheterotrophic nitrate-reduction bacteria. Arch. Microbiology, 169: 159-165.

Berelson, W.M., Hammond, D.E., Smith, K.L. Jr; Jahnke, R.A., Devol, A.H., Hinge, K.R., Rowe, G.T. and Sayles, F. (eds), 1987. In situ benthic flux measurement devices: bottom lander technology. MTS Journal, 21: 26-32.

Berner, R.A., 1980. Early diagenesis: A theoretical approach. Princton Univ. Press, Princton, NY, 241 pp.

Boetius, A. and Lochte, K., 1996. Effect of organic enrichments on hydrolytic potentials and growth of bacteria in deep-sea sediments. Marine Ecology Progress Series, 140: 239-250.

Boetius, A. and Damm, E., 1998. Benthic oxygen uptake, hydrolytic potentials and microbial biomass at the Arctic continental slope. Deep-Sea Research I, 45: 239-275.

Borowski, W.S., Paull, C.K. and Ussler, W., 1996. Marine pore-water sulfate profiles indicate in situ methane flux from underlying gas hydrate. Geology, 24: 655-658.

Boudreau, B.P., 1988. Mass-transport constraints on the growth of discoidal ferromanganese nodules. American Journal of Science, 288: 777-797.

Boudreau, B.P., 1997. Diagenetic models and their impletation: modelling transport and reactions in aquatic sediments. Springer, Berlin, Heidelberg, NY, 414 pp.

Canfield, D.E., 1993. Organic matter oxidation in marine sediments. In: Wollast, R., Mackenzie, F.T. and Chou, L. (eds), Interactions of C, N, P and S biogeochemical cycles. NATO ASI Series, 14. Springer, Berlin, Heidelberg, NY, pp. 333-363.

Canfield, D.E., Jørgensen; B.B., Fossing, H., Glud, R.N., Gundersen, J., Ramsing, N.B., Thamdrup, B., Hansen, J.W., Nielsen, L.P. and Hall, P.O.J., 1993b. Pathways of organic carbon oxidation in three continental margin sediments. Marine Geology, 113: 27-40.

Canfield, D.E. and Teske, A., 1996. Late Proterozoic rise in atmospheric oxygen concentration inferred from phylogenetic and sulphur-isotope studies. Nature, 382: 127-132.

Christensen, D., 1984. Determination of substrates oxidized by sulfate reduction in intact cores of marine sediments. Limnology and Oceanography, 29: 189-192.

Chrost, R.J., 1991. Microbial enzymes in aquatic environments. Springer, Berlin, Heidelberg, NY, 317 pp.

Coleman, M.L., Hedrick, D.B., Lovley, D.R., White, D.C. and Pye, K., 1993. Reduction of Fe(III) in sediments by sulphate-reducing bacteria. Nature, 361: 436-438.

Conrad, R., Schink, B. and Phelps, T.J., 1986. Thermodynamics of H2-consuming and H2-producing metabolic reactions in diverse methanogenic environments under in situ conditions. FEMS Microbiology Ecology, 38: 353-360.

Cypionka, H., 1994. Novel matabolic capacities of sulfate-reducing bacteria, and their activities in microbial mats. In: Stal, L.J. and Caumette, P.(eds), Microbial mats, NATO ASI Series, 35, Springer, Berlin, Heidelberg, NY, pp. 367-376.

Dannenberg, S., Kroder, M., Dilling, W. and Cypionka, H., 1992. Oxidation of H2, organic compounds and inorganic sulfur compounds coupled to reduction of O2 or nitrate by sulfate-reducing bacteria. Archives of Microbiology, 158: 93-99.

Ehrenreich, A. and Widdel, F., 1994. Anaerobic oxidation of ferrous iron by purple bacteria, a new type of phototrophic metabolism. Applied and Environmental Microbiology, 60: 4517-4526.

Ehrlich, H.L., 1996. Geomicrobiology. Marcel Dekker, NY, 719 pp.

Fenchel, T.M. and Jørgensen, B.B., 1977. Detritus food chains of aquatic ecosystems: The role of bacteria. In: Alexander, M. (ed), Advances in Microbial Ecology, 1, Plenum Press, NY, pp. 1-58.

Fenchel, T., King, G.M. and Blackburn, T.H., 1998. Bacterial biogeochemistry: The ecophysiology of mineral cycling. Academic Press, London, 307 pp.

Fossing, H. and Jørgensen, B.B., 1989. Measurement of bacterial sulfate reduction in sediments: evaluation of a single-step chromium reduction method. Biogeochemistry, 8: 205-222.

Fossing, H. and Jørgensen, B.B., 1990. Isotope exchange reactions with radiolabeled sulfur compounds in anoxic seawater. Biogeochemistry, 9: 223-245.

Fossing, H., Thode-Andersen, S. and Jørgensen, B.B., 1992.

Sulfur isotope exchange between 35S-labeled inorganic sulfur compounds in anoxic marine sediments. Marine Chemistry, 38: 117-132.

Fossing, H., Gallardo, V.A., Jørgensen, B.B., Hüttel, M., Nielsen, L.P., Schulz, H., Canfield, D.E., Forster, S., Glud, R.N., Gundersen, J.K., Küfer, J., Ramsing, N.B., Teske, A., Thamdrup, B. and Ulloa, O., 1995. Concentration and transport of nitrate by the mat-forming sulphur bacterium Thioploca. Nature, 374: 713-715.

Fossing, H., 1995. 35S-radiolabeling to probe biogeochemical cycling of sulfur. In:. Vairavamurthy, M.A and. Schoonen, M.A.A (eds), Geochemical transformations of sedimentary sulfur. ACS Symposium Series, 612, American Chemical Society, Washington, DC, pp. 348-364.

Froelich, P.N., Klinkhammer, G.P., Bender, M.L., Luedtke, N.A., Heath, G.R., Cullen, D., Dauphin, P., Hammond, D., Hartman, B. and Maynard, V., 1979. Early oxidation of organic matter in pelagic sediments of the eastern equatorial Atlantic: suboxic diagenesis. Geochimica et Cosmochimica Acta, 43: 1075-1088.

Glud, R.N., Gundersen, J.K., Jørgensen, B.B., Revsbech, N.P. and Schulz, H.D., 1994. Diffusive and total oxygen uptake of deep-sea sediments in the eastern South Atlantic Ocean: in situ and laboratory measurements. Deep-Sea Research, 41: 1767-1788.

Greeff, O., Glud, R.N., Gundersen, J., Holby, O. and Jørgensen, B.B., in press. A benthic lander for tracer studies in the sea bed: in situ measurements of sulfate reduction. Continental Shelf Research.

Gundersen, J.K. and Jørgensen, B.B., 1990. Microstructure of diffusive boundary layer and the oxygen uptake of the sea floor. Nature, 345: 604-607.

Gundersen, J.K., Glud, R.N. and Jørgensen, B.B., 1995. Oxygen turnover of the sea floor (in Danish). Marine research from the danisch environmental agency, 57, Danish Ministry of Environment and Energy, Copenhagen, 155 pp.

Hedges, J.I., 1978. The formation and clay mineral reactions of melanoidins. Geochimica Cosmochimica Acta, 42: 69-76.

Henrichs, S.M. and Reeburgh, W.S., 1987. Anaerobic mineralization of marine sediment organic matter: rates and the role of anaerobic processes in the oceanic carbon economy. Geomicrobiological Journal, 5: 191-237.

Henrichs, S.M., 1992. Early diagenesis of organic matter in marine sediments: progress and perplexity. Marine Chemistry, 39: 119-149.

Henriksen, K., 1980. Measurement of in situ rates of nitrification in sediment. Microbial Ecology, 6: 329-337.

Hoehler, T.M., Alperin, M.J., Albert, D.B. and Martens, C.S., 1994. Field and laboratory studies of methane oxidation in an anoxic marine sediment: Evidence for a methanogen-sulfate reducer consortium. Global Biogeochemical Cycles, 8: 451-463.

Hoehler, T.M., Alperin, M.J., Albert, D.B. and Martens, C.S., 1998. Thermodynamic control on hydrogen concentrations in anoxic sediments. Geochimica et

Cosmochimica Acta, 62: 1745-1756.

Huettel, M., Ziebis, W., Forster, S. and Luther, G.W., 1998. Advective transport affecting metal and nutrient distributions and interfacial fluxes in permeable sediments. Geochimica et Cosmochimica Acta, 62: 613-631.

Isaksen, M.F. and Jørgensen, B.B., 1996. Adaptation of psychrophilic and psychrotrophic sulfate-reducing bacteria to permanently cold marine environments. Applied and Environmental Microbiology, 62: 408-414.

Iversen, N. and Jørgensen, B.B., 1985. Anaerobic methane oxidation rates at the sulfate-methane transition in marine sediments from Kattegat and Skagerrak (Denmark). Limnology and Oceanography, 30: 944-955.

Jørgensen, B.B., 1978. A comparison of methods for the quantification of bacterial sulfate reduction in coastal marine sediments. I. Measurement with radiotracer techniques. Geomicrobiology Journal, 1: 11-27.

Jørgensen, B.B., 1982. Mineralization of organic matter in the sea bed-the role of sulphate reduction. Nature, 296: 643-645.

Jørgensen, B.B., 1987. Ecology of the sulphur cycle: Oxidative pathways in sediments. In: Cole, J.A. and Ferguson, S. (eds), The nitrogen and sulphur cycles. Society for General Microbiology Symposium, 42, Cambridge University Press, pp. 31-63.

Jørgensen, B.B., 1990. A thiosulfate shunt in the sulfur cycle of marine sediments. Science, 249: 152-154.

Jørgensen, B.B. and Bak, F., 1991. Pathways and microbiology of thiosulfate transformation and sulfate reduction in a marine sediment (Kattegat, Denmark). Applied and Environmental Microbiology, 57: 847-856.

Jørgensen, B.B., Isaksen, M.F. and Jannasch, H.W., 1992. Bacterial sulfate reduction above 100°C in deep-sea hydrothermal vent sediments. Science, 258: 1756-1757.

Jørgensen, B.B., 1996. The micro-world of marine bacteria (in German). Naturwissenschaften, 82: 269-278.

Jørgensen, B.B., in press. Microbial life in the diffusive boundery layer. In: Boudreau, B.P. and Jørgensen, B.B. (eds), The benethic boundary layer: transport processes and biogeochemistry. Oxford University Press, Oxford.

Karp-Boss, L., Boss, E. and Jumars, P.A., 1996. Nutrient fluxes to planktonic osmotrophs in the presence of fluid motion. Oceanographic Marine Biology Annual Reviews, 34: 71-107.

Keil, R.G., Montlucon, Prahl, F.G. and Hedges, J.I., 1994b. Sorptive preperation of labile organic matter in marine sediments. Nature, 370: 549-552.

Kelly, D.P., 1982. Biochemistry of the chemolithotrophic oxidation of inorganic sulphur. Philosophic Transactions of the Royal Society of London, 298: 499-528.

King, G.M., 1983. Sulfate reduction in Georgia salt marsh soils: An evaluation of pyrite formation by use of 35S and 55Fe tracers. Limnology and Oceanography, 28: 987-995.

Klinkhammer, G.P., 1980. Early diagenesis in sediments from the eastern equatorial Pacific, II. Pore water metal results. Earth and Planetary Science Letters, 49: 81-101.

Koch, A.L., 1990. Diffusion, the crucial process in many aspects of the biology of bacteria. In: Marshall, K.C. (ed), Advances in microbial ecology, 11, Plenum, NY, pp. 37-70.

Koch, A.L., 1996. What size should a bacterium be? A question of scale. Annual Reviews of Microbiology, 50: 317-348.

Krekeler, D. and Cypionka, H., 1995. The preferred electron acceptor of Desulfovibrio desulfuricans CSN. FEMS Microbiology Ecology, 17: 271-278.

Kühl, M. and Revsbech, N.P., in press. Microsensors for studies of interfacial biogeochemical processes. In: Boudreau, B.P. and Jørgensen, B.B. (eds), The benthic boundary layer: transport processes and biogeochemistry. Oxford University Press, Oxford.

Llobet-Brossa, E., Roselló-Mora, R. and Ammann, R., 1998. Microbial community composition of Wadden Sea sediments as revealed by fluorescent in situ hybridization. Applied and Environmental Microbiology, 64: 2691-2696.

Lochte, K. and Turley, C.M., 1988. Bacteria and cyanobacteria associated with phytodetritus in the deep sea. Nature, 333: 67-69.

Madigan, M.T., Martinko, J.M. and Parker, J., 1997. Biology of microorganisms. Prentice Hall, London, 986 pp.

Nielsen, L.P., 1992. Denitrification in sediment determined from nitrogen isotope pairing. FEMS Microbiology Ecology, 86: 357-362.

Niewöhner, C., Hensen, C., Kasten, S., Zabel, M. and Schulz, H.D., 1998. Deep sulfate reduction completely mediated by anaerobic methane oxidation in sediments of the upwelling area off Namibia. Geochimica et Cosmochimica Acta, 62: 455-464.

Oremland, R.S. and Polcin, S., 1982. Methanogenesis and sulfate reduction: Competitive and noncompetetive substrates in estuarine sediments. Applied and Environmental Microbiology, 44: 1270-1276.

Oremland, R.S., Marsh, L.M. and Polcin, S., 1982. Methane production and simultaneous sulphate reduction in anoxic saltmarsh sediments. Nature, 296: 143-145.

Oremland, R.S. and Capone, D.G., 1988. Use of „specific" inhibitors in biogeochemistry and microbial ecology. In: Marshall, K.C. (ed), Advances in microbial ecology, 10, Plenum Press, NY, pp. 285-383.

Parkes, J.R., Cragg, B.A., Bale, S.J., Getliff, J.M., Goodman, K., Rochell, P.A., Fry, J.C., Weightman, A.J. and Harvey, S.M., 1994. Deep bacterial biosphere in Pacific Ocean sediments. Nature, 371: 410-413.

Postgate, J.R., 1979. The sulfate-reduction bacteria. Cambridge University Press, Cambridge, 208 pp.

Rabus, F., Fukui, M., Wilkes, H. and Widdel, F., 1996. Degradative capacities and 16S rRNA-targeted whole-cell hybridization of sulfate-reducing bacteria in an anaerobic enrichment culture utilizing alkylbenzenes from crude oil. Applied and Environmental Microbiology, 62: 3605-3613.

Redfield, A.C., 1958. The biological control of chemical factors in the environment. Am. Scientist, 46: 206-222.

Reeburgh, W.S., 1969. Observations of gases in Chesapeake Bay sediments. Limnology and Oceanography, 14: 368-375.

Reimers, C.E., 1987. An in situ microprofiling instrument for measuring interfacial pore water gradients: methods and oxygen profiles from the North Pacific Ocean. Deep-Sea Research, 34: 2019-2035.

Revsbech, N.P., Nielsen, L.P., Christensen, P.B. and Sørensen, J., 1988. Combined oxygen and nitrous oxide microsensor for denitrification studies. Applied and Environmental Microbiology, 54: 2245-2249.

Roden, E.E. and Lovley, D.R., 1993. Evaluation of 55Fe as a tracer of Fe(III) reduction in aquatic sediments. Geomicrobiological Journal, 11: 49-56.

Rueter, P., Rabus, R., Wilkes, H., Aeckersberg, F., Rainey, F.A., Jannasch, H.W. and Widdel, F., 1994. Anaerobic oxidation of hydrocarbons in crude oil by new types of sulphate-reducing bacteria. Nature, 372: 455-458.

Sagemann, J., Jørgensen, B.B. and Greeff, O., 1998. Temperature dependence and rates of sulfate reduction in cold sediments of Svalbard, Arctic Ocean. Geomicrobiological Journal, 15: 83-98.

Sansone, F.J., Andrews, C.C. and Okamoto, M.Y., 1987. Adsorption of short-chain organic acids onto nearshore marine sediments. Geochimica et Cosmochimica Acta, 51: 1889-1896.

Santschi, P.H., Anderson R.F., Fleisher, M.Q. and Bowles, W., 1991. Measurements of diffusive sublayer thicknesses in the ocean by alabaster dissolution, and their implications for the measurements of benthic fluxes. Journal of Geophysical Research, 96: 10.641-10.657.

Schopf, J.W. and Klein, C. (eds), 1992. The proterozoic biosphere. Cambridge University Press, Cambridge, 1348 pp.

Schulz, H.D., Dahmke, A., Schinzel, U., Wallmann, K. and Zabel, M., 1994. Early diagenetic processes, fluxes and reaction rates in sediments of the South Atlantic. Geochimica et Cosmochimica Acta, 58: 2041-2060.

Smith, K.L.Jr., Clifford, C.H. Eliason, A.h., Walden, B., Rowe, G.T. and Teal, J.M., 1976. A free vehicle for measuring benthic community metabolism. Limnology and Oceanography, 21: 164-170.

Sørensen, J., 1978. Denitrification rates in a marine sediment as measured by the acetylene inhibition technique. Applied and Environmental Microbiology, 35: 301-305.

Sørensen, J., Christensen, D. and Jørgensen, B.B., 1981. Volatile fatty acids and hydrogen as substrates for sulfate-reducing bacteria in anaerobic marine sediment. Applied and Environmental Microbiology, 42: 5-11.

Stetter, K.O., Huber, R., Blöchl, E., Knurr, M., Eden, R.D., Fielder, M., Cash, H. and Vance, I., 1993. Hyperthermophilic archaea are thriving in deep North Sea and Alaskan oil reservoirs. Nature, 365: 743-745.

Stetter, K.O., 1996. Hyperthermophilic procaryotes. FEMS Microbiology Revue, 18: 149-158.

Straub, K.L., Benz, M., Schink, B. and Widdel, F., 1996. Anaerobic, nitrate-dependent microbial oxidation of ferrous iron. Applied and Environmental Microbiology, 62: 1458-1460.

Straub, K.L. and Buchholz-Cleven, B.E.E., 1998. Enumeration and detection of anaerobic ferrous iron-oxidizing, nitrate-reducing bacteria from diverse European sediments. Applied and Environmental Microbiology, 64: 4846-4856.

Suess, E., 1980. Particulate organic carbon flux in the oceans-surface productivity and oxygen utilization. Nature, 288: 260-263.

Tegelaar, E.W., de Leeuw, J.W., Derenne, S. and Largeau, C., 1989. A reappraisal of kerogen formation. Geochimica et Cosmochimica Acta, 53: 3103-3106.

Tengberg, A., de Bovee, F., Hall, P, Berelson, W., Chadwick, D., Ciceri, G., Crassous, P., Devol, A., Emerson, s., Gage, J., Glud, R., Graziottin, F., Gundersen, J., Hammond, D., Helder, W., Hinga, K., Holby, O., Jahnke, R., Khripounoff, A., Lieberman, S., Nuppenau, V., Pfannkuche, O., Reimers, C., Rowe, G., Sahami, A., Sayles, F., Schurter, M., Smallman, D., Wehrli, B. and de Wilde, P., 1995. Benthic chamber and profiling landers in oceanography - A review of design, technical solutions and function. Progress in Oceanography, 35: 253-292.

Thamdrup, B., Finster, K., Hansen, J.W. and Bak, F., 1993. Bacterial disproportionation of elemental sulfur coupled to chemical reduction of iron or manganese. Applied and Environmental Microbiology, 59: 101-108.

Thamdrup, B., Fossing, H. and Jørgensen, B.B., 1994. Manganese, iron, and sulfur cycling in a coastal marine sediment, Aarhus Bay, Denmark. Geochimica et Cosmochimica Acta, 58: 5115-5129.

Thauer, R.K., Jungermann, K. and Decker, K., 1977. Energy conservation in chemotrophic anaerobic bacteria. Bacterial Reviews, 41: 100-180.

Thomsen, L., Jähmlich, S., Graf, G., Friedrichs, M., Wanner, S. and Springer, B., 1996. An instrument for aggregate studies in the benthic boundary layer. Marine Geology, 135: 153-157.

Vetter, Y.A., Deming, J.W., Jumars, P.A. and Kriegerbrockett, B.B., 1998. A predictive model of bacterial foraging by means of freely released extracellular enzymes. Microbiology Ecology, 36: 75-92.

Weiss, M.S., Abele, U., Weckesser, J., Welte, W. und Schulz, G.E., 1991. Molecular architecture and electrostatic properties of a bacterial porin. Science, 254: 1627-1630.

Wellsbury, P., Goodman, K., Barth, T., Cragg, B.A., Barnes, S.P. and Parkes R.J., 1997. Deep marine biosphere fuelled by increasing organic matter availability during burial and heating. Nature, 388: 573-576.

Westrich, J.T. and Berner, R.A., 1984. The role of sedimentary organic matter in bacterial sulfate reduction: The G model tested. Limnology and Oceanography, 29: 236-249.

Widdel, F., 1988. Microbiology and ecology of sulfate-and sulfur-reduction bacteria. In: Zehnder, A.J.B. (ed). Biology of anaerobic microorganisms. Wiley & Sons, NY, 469-585 pp.

Widdel, F., Schnell, S., Heising, S., Ehrenreich, A., Assmus, B. and Schink, B., 1993. Ferrous iron oxidation by anoxygenic phototrophic bacteria. Nature, 362: 834-836.

Yayanos, A.A., 1986. Evolutional and ecological implications of the properties of deep-sea barophilic bacteria. Proc. Natl. Acad. Sci., 83: 9542-9546.

# 6 Early Diagenesis at the Benthic Boundary Layer: Oxygen and Nitrate in Marine Sediments

CHRISTIAN HENSEN AND MATTHIAS ZABEL

## 6.1 Introduction

All particles settling to the sea floor continuously undergo diagenetic alteration due to physical and chemical processes in the sediment (e.g. particle mixing, compaction, redox reactions). One of the most intensely studied topic in marine geology and geochemistry is the early diagenesis of organic material deposited in marine sediments. Marine sediments are the primary long-term repository of organic matter and the analysis of the controls on the input of organic particles and the processes those undergo until they are finally buried is a major prerequisite for the reconstruction of biogeochemical cycles in the ocean. This chapter mainly focuses on processes occurring when fresh, bio-available organic material reaches the sea floor and is subject to intense bacterially mediated oxidation. Oxygen and nitrate are treated together because they are thermodynamically the most favorable electron acceptors in the diagenetic sequence of organic matter decomposition and their pathways are coupled through oxidation of reduced nitrogen species (nitrification) in oxic systems (cf. Sect. 3.2.3). Generally, their involvement in the biogeochemical cycles of the ocean is much more complex than only seen from this point of view and therefore, the combined examination of both parameters is for reasons of convenience and follows the general concept of this book. Both oxygen and nitrate pathways are very important and inherent for the understanding of the oceanic carbon cycle. Oxygen is introduced into surface waters by photosynthesis and, even more important, by exchange with the atmosphere. Conversely, it is consumed in the course of the degradation of organic matter. The latter occurs throughout the water column and in the sediments resulting in the release of carbon dioxide and nitrate. On the one hand, nitrate itself is used as the "next" suitable electron acceptor for organic matter degradation in environments where oxygen availability is limited, such as oxygen minimum zones or below the oxygen penetration depth in the sediments. On the other hand, the availability of nitrate as an oxidant in geochemical processes is restricted since it is - together with phosphate - an important limiting nutrient for primary productivity. In the following sections, we will first give a short overview concerning the distribution and the geochemistry of oxygen and nitrate in the modern oceans followed by a more detailed description of the processes in marine sediments.

## 6.2 Oxygen and Nitrate Distribution in Seawater

The distribution of dissolved oxygen in seawater results from the interaction of different factors. Those are (a) the input of oxygen across the atmosphere-ocean interface and the oxygen production by phytoplankton, (b) the microbially catalyzed degradation of organic matter and oxidation of other reduced substances, and (c) physical transport and mixing processes in the ocean. Theoretically, the oxygen concentration in seawater is limited by its solubility, but in fact the saturation concentration is only reached in surface waters. Sometimes, surface waters are even supersaturated with respect to oxygen which is thought to be caused by an entrapment of air bubbles (cf. Chester 1990). Below the productive mixed layer oxygen is depleted by bacterial respiration processes. Numerous studies show that this process is most intense within the upper 1000 m of the water column which marks approximately the lower end of the permanent thermocline. As a consequence significant oxygen minimum zones can be observed in areas with high surface water productivity. Such areas exist where upwelling water masses significantly enhance the supply of nutrients, mainly at the west coasts of the continents (trade wind belts of America and Africa), but also along the equatorial divergence zones and related to the Polar Fronts. These patterns are generated in compilations of the global ocean primary productivity (e.g. Antoine et al. 1996, Behrenfeld and Falkowski 1997; see Fig. 12.5). Generally, the distribution of oxygen in the water column is dependent on how much the oxygen depletion exceeds the supply by vertical and lateral diffusion and advection at a certain depth. Since organic carbon oxidation is the main reason for oxygen depletion, a syngenetically increase of dissolved nutrients (nitrate and phosphate) and a decrease of particulate organic carbon with increasing water depth is the typical feature. Figure 6.1 shows idealized depth profiles of oxygen, nitrate and alkalinity. Alkalinity, in this case, has to be understood as a sum parameter for dissolved carbon species which increase with depth due to the continued decay of organic material. These patterns, however, may deviate significantly between the ocean basins depending on the oceano-graphic setting (currents, mixing of water masses), the particle-transport through the water column, and the composition of mineralized organic matter. Figure 6.2 shows some examples of the nitrate distribution in different ocean basins. The deep waters of the Pacific and the Indian Ocean are enriched in nitrate relative to the North Atlantic due to deep-water transport through the ocean basins (see below).

The investigation of the effective processes has largely benefited from the invention of sediment traps measuring the particle flux through the water column over long time periods. A number of researchers have attempted to quantify export fluxes of organic carbon from surface waters and their transition to the bottom by empirical formulations (cf. Sect. 4.2; e.g. Betzer et al. 1984, Berger et al. 1987, Martin et al. 1987). Generally, these formulations are exponential and power equations predicting about 90% of the remineralization within the upper hundreds of meters of the water column (cf. Fig. 12.1). To illustrate this process Figure 6.3 shows the application of three different equations to predict the vertical (transit) carbon flux. In all cases a surface primary productivity of 250 gC m$^{-2}$yr$^{-1}$ is assumed. At a depth horizon of 1000 m less than 5% of primarily produced organic carbon remain.

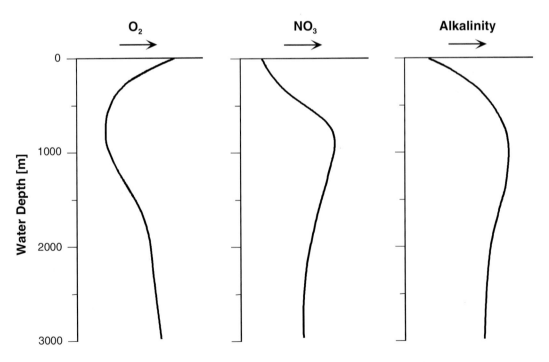

**Fig. 6.1** Idealized vertical sections of dissolved oxygen, nitrate and alkalinity through the water column. Nitrate and CO$_2$ are produced while oxygen is consumed during organic matter decomposition.

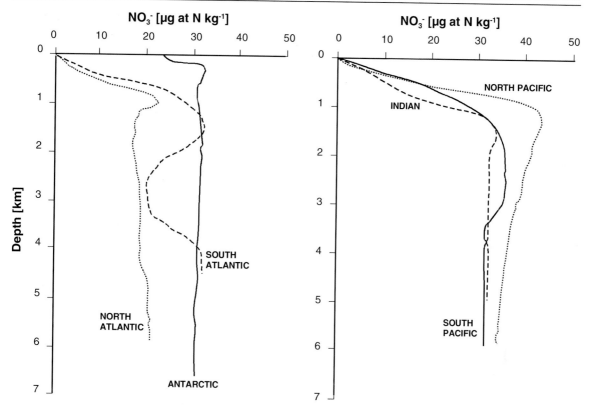

**Fig. 6.2**   Vertical sections of nitrate in different ocean basins. Surface waters are depleted in nutrients compared to deeper water layers. Older water masses in the Pacific and Indian Ocean are generally more enriched in nitrate - as well as in phosphate and alkalinity or $\Sigma CO_2$, respectively - (redrawn from Sverdrup et al 1942).

Comparative studies of the relation between primary productivity and benthic mineralization processes (Jahnke et al. 1990, Rowe et al. 1994, Hensen et al. 1998) show, however, that the use of these empirical formulations is restricted to regional use. A further important factor in this regard is the reaction stoichiometry of organic matter degradation, since it determines the proportional release of $CO_2$, $NO_3$ and $PO_4$. Based on planktonic decomposition studies, Redfield et al. (1963) suggested an overall $C/N/P/O_2$ ratio of 106/16/1/138 (see Eq. 6.1). However, although widely used, subsequent investigations have put this formulation into question. Deviating ratios were formulated as 140/16/1/172 (Takahashi et al. 1985) or 117/16/1/170 (Anderson and Sarmiento 1994) implying that there is still debate on the general validity of the use of one Redfield ratio for all ocean basins and all water depths. Instead, C/N/P ratios seem to be subject to regional variation.

Oxygen depth profiles show an opposite trend to alkalinity and nitrate resulting from low mineralization rates in the deep ocean waters and input of oxygen-rich water masses by advective transport. Figure 6.4 shows a meridional transect of oxygen concentration through the Atlantic Ocean compiled by Reid (1994). The most prominent pattern is the southward flow of oxygen-rich North Atlantic Deep Water raising the oxygen concentration along its flow path into the equatorial South Atlantic at water depths of 3000-4000 m. Less prominent, but still significant, is also the northward flow of oxygen-rich Antarctic Intermediate Water which is marked by elevated oxygen concentrations on a south-north path (50°S - 20°S) between 0-1000 m water depth. The distribution of oxygen in ocean water is therefore strongly dependent on large scale circulation patterns. The same is valid for the distribution of nutrients: Vertical concentration profiles are always a mixture of *in-situ* decomposition and advective transport processes. The global deep water circulation pattern indicates a general flow path from the North Atlantic to the North Pacific and Indian Oceans. As a result, "older" deep waters in the Pacific and Indian Oceans are depleted of oxygen and enriched in nutrients (Fig. 6.2) and $CO_2$ (Broecker and Peng 1982, Kennett 1982, Chester 1990; see Chap. 9).

When studying the early diagenesis of deep-sea sediments, it is very important to consider all features of the oceanic environment and consequently of the composition of the overlying deep-ocean water, since it determines the availability of any solute, in this case oxygen and nitrate, as possible oxidants for organic carbon. However, there is at present, no oxidant limitation in the most part of the sea floor below the big central gyres, resulting from a very limited supply of degradable material to the sediments, which was, however, not always the case in Earth history (cf. Sect. 4.1).

## 6.3    The Role of Oxygen and Nitrate in Marine Sediments

To estimate the role of oxygen and nitrate we have to describe the general processes occurring close to the sediment water interface, the methods how to measure concentrations, fluxes, and consumption rates, and how to relate them to organic matter degradation or other processes. Subsequently, we will show examples from case studies to characterize the magnitude of fluxes and different environments in deep-sea sediments. The early diagenetic processes at the sediment-water interface are of special interest in global biogeochemical cycles because it is decided at this separation line between ocean water and sediment whether any substance is recycled or buried for a long period of time in a geological sense.

### 6.3.1    Respiration and Redox Processes

#### 6.3.1.1  Nitrification and Denitrification

In principle, the sequence of oxidants is determined by the energy yield for the microorganisms. When oxygen and nitrate are depleted reduction of Mn- and Fe-(oxo-)hydroxides and sulfate as well as methane fermentation follow in the sequence with decreasing yield of energy (Froelich et al. 1979; cf. Sect. 3.2.5). This sequence is generally valid, even though numerous studies have identified an overlap of carbon oxidation pathways within the sediment resulting from competition between microbial populations

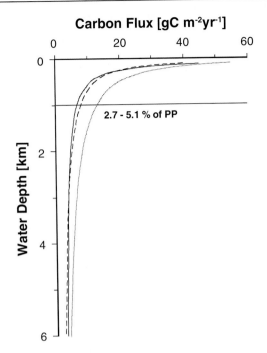

**Carbon Flux [gC m⁻²yr⁻¹]**

2.7 - 5.1 % of PP

**Fig. 6.3**  Example calculations for the transit flux of organic carbon after empirical equations assuming a surface primary productivity of 250 gC m⁻²yr⁻¹. More than 95% of the organic carbon is oxidized above the 1000 m horizon.

Solid line:    $J_{Corg}$ [gC m⁻²yr⁻¹] = 17 PP/z + PP/100 (Berger et al., 1987)

Broken line:  $J_{Corg}$ [gC m⁻²yr⁻¹] = 9 PP/z + 0.7 PP/z$^{0.5}$ (Berger et al., 1987)

Dotted line:  $J_{Corg}$ [gC m⁻²yr⁻¹] = 0.409 PP$^{1.41}$/z$^{0.628}$ (Betzer et al., 1984)

(Canfield 1993) and the presence of microenvironments (e.g. Jørgensen 1977; cf. Chaps. 7, 8, 12).

The general equations of coupled oxic respiration and nitrification (Eq. 6.1) and denitrification (Eq. 6.2) describing the "top" of the diagenetic sequence are given as:

*Nitrification*

$$(CH_2O)_{106}(NH_3)_{16}(H_3PO_4) + 138\ O_2 \rightarrow$$

$$106\ CO_2 + 16\ HNO_3 + H_3PO_4 + 122\ H_2O$$

$$\Delta G^0 = -\ 3190\ kJ\ mol^{-1} \qquad (6.1)$$

*Denitrification*

$$(CH_2O)_{106}(NH_3)_{16}(H_3PO_4) + 94.4\ HNO_3 \rightarrow$$

$$106\ CO_2 + 55.2\ N_2 + H_3PO_4 + 177.2\ H_2O$$

$$\Delta G^0 = -\ 3030\ kJ\ mol^{-1} \qquad (6.2)$$

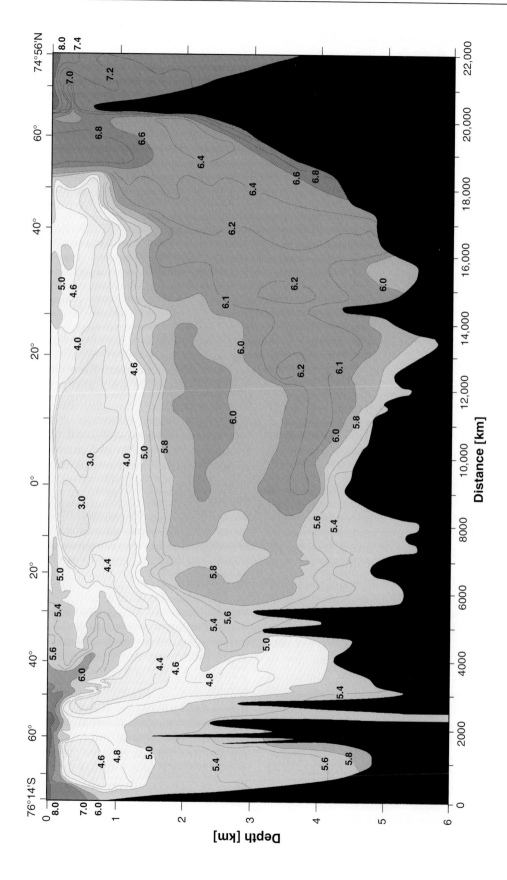

**Fig. 6.4** North-South section of dissolved oxygen (ml/l) across the Atlantic Ocean (after Reid 1994). See text for explanation.

The $\Delta G^0$-values indicate the higher energy yield for using oxygen rather than nitrate as the terminal electron acceptor. In these equations, it is assumed that organic matter with a C/N/P ratio of 106/16/1 is oxidized. Equation 6.1 requires that ammonia released during oxic respiration is quantitatively oxidized to nitrate. The process of nitrification is, however, more complex than described above and is known as a stepwise oxidation of nitrogen species by different microbes of the family Nitrobacteraceae. These bacteria can be grouped into ammonia oxidizers and nitrite oxidizers. Since high ammonia concentrations are toxic for nitrite oxidizers they strictly depend on the presence of the ammonia oxidizers (Schlegel 1985). Important ammonia oxidizers are *Nitrosomonas* and *Nitrosococcus*. The oxidation of ammonia to nitrite is also a two step process in which hydroxylamine is formed as an intermediate product (Eq. 6.3). The second step yields the biogeochemically useful energy (Eq. 6.4):

$$2 \, NH_3 + O_2 \rightarrow 2 \, NH_2OH \qquad (6.3)$$

$$NH_2OH + O_2 \rightarrow NO_2^- + H_2O + H^+ \qquad (6.4)$$

In the following step nitrite is oxidized by autotrophs like *Nitrobacter* or *Nitrococcus* to nitrate (Eq. 6.5):

$$NO_2^- + \tfrac{1}{2} \, O_2 \rightarrow NO_3^- \qquad (6.5)$$

In summary, nitrifying bacteria are considered to be strictly aerobic and therefore depend on adequate oxygen supply for their energy gain (Painter 1970). Experimental results of Henriksen et al. (1981) on control factors of nitrification rates in shallow water sediments from Denmark revealed that nitrification is strongly dependent on temperature, oxygen availability (oxygen penetration depth), ammonia supply and the number of nitrifying bacteria. These complex interactions are more thoroughly discussed in comprehensive reviews on nitrification in coastal marine environments by Kaplan (1983) and Henriksen and Kemp (1988).

Denitrification starts when oxygen is almost depleted (below the oxygen penetration depth) by inducing an enzymatic system of nitrate reductase and nitrite reductase by facultative aerobic bacteria which can only use nitrogenous oxides if oxygen is - nearly - absent. Measurements carried out with a combined microsensor for $O_2$ and $N_2O$ in-

dicated that denitrification is restricted to a thin anoxic layer below the oxic zone (Christensen et al. 1989). Denitrification is the only biological process that produces free nitrogen (Eq. 3.1). It removes fixed nitrogen compounds and, therefore, exerts a negative feedback on eutrophication making it a crucial process for the preservation of life on Earth. For example, denitrification in rivers and coastal environments decreases by about 40% of the continentally derived nitrogen transported to the oceans (Seitzinger 1988). The reduction of nitrate to dinitrogen occurs first as a reduction of nitrate to nitrite (Eq. 6.6) and then a stepwise reduction to nitrogen oxide, dinitrogen oxide (Eq. 6.7) and dinitrogen (Eq. 6.8):

$$NO_3^- + 2 \, H^+ + 2 \, e^- \rightarrow NO_2^- + H_2O \qquad (6.6)$$

$$2 \, NO_2^- + 6 \, H^+ + 4 \, e^- \rightarrow N_2O + 3 \, H_2O \quad (6.7)$$

$$N_2O + 2 \, H^+ + 2 \, e^- \rightarrow N_2 + H_2O \qquad (6.8)$$

The last reduction step from nitrous oxide to dinitrogen (Eq. 6.8) is not always completed so that the final product of denitrification is not necessarily dinitrogen. Nitrous oxygen might therefore be produced or consumed during denitrification. A compilation of Seitzinger (1998) for coastal marine environments shows, however, that in most cases the net ratios between $N_2O:N_2$ production rates are usually very small (between 0.0002-0.06). The total amount of dinitrogen produced obviously depends on the partial pressure of oxygen (higher oxygen contents seem to be suitable for the production of $N_2O$; Jørgensen et al. 1984), the pH, and the presence of $H_2S$.

The major prerequisite for denitrification is the availability of nitrate (including nitrite). For the marine environment the dominant source is nitrate produced during nitrification (Middelburg et al., 1996a), but also nitrate from the overlying water might be mixed into the sediment by bioturbation, bioirrigation or diffusion (see Sect. 6.3.2). Furthermore, denitrification is strongly dependent on temperature, but also the oxygen concentration and the availability of organic matter are limiting for the process (Middelburg et al. 1996a). There is also evidence that denitrification might be reduced at high sulfate reduction rates, because low sulfide concentrations completely inhibit nitrification which in turn is necessary for denitrification (Seitzinger 1988). Generally, most suitable conditions for denitrification are obtained at intermediate carbon availability levels when carbon is not limiting for oxic respiration and nitrification (nitrate supply), but sulfate reduction is still low or absent. Studies of Jørgensen and Sørensen (1985) along a salinity gradient in a Danish estuary revealed evidence that denitrification and nitrate reduction (Eq. 6.9) became a more important pathway as sulfate became limited due to increased freshwater input close to the river outlet. The ability of some bacteria to reduce nitrite further to ammonia is a process called nitrate reduction or ammonification (Jørgensen and Sørensen 1985, Sørensen 1987):

$$NO_2^- + 8\,H^+ + 6\,e^- \rightarrow NH_4^+ + 2\,H_2O \quad (6.9)$$

Figure 6.5 shows typical pore water profiles of oxygen and nitrate measured in organic rich sediments off Namibia summing up the net reactions described above. Due to nitrification, the highest nitrate concentrations are reached approximately at the oxygen penetration depth. At about 3 cm depth, nitrate is consumed in the process of denitrification. The nitrate profile indicates an upward flux into the bottom water and a downward flux to the zone of denitrification. Both profiles are verified by the application of Equations 6.1 and 6.2 within the computer model CoTAM/CoTReM (cf. Chap. 14) as indicated by the solid and dashed lines.

It could be shown that the above described processes have an essential influence on the distribution of oxygen and nitrate in the sediment column and can directly be related to the degradation of organic matter. The reduction by any other reduced species (e.g. $H_2S$, $Fe^{2+}$, $Mn^{2+}$) can, however, also be an important pathway (cf. Chaps. 7, 8). Generally, most of these processes are microbially mediated (cf. Chap. 5; e.g. Chapelle 1993, Stumm and Morgan 1996). This also includes the (re-)oxidation of upward diffusing reduced nitrogen species (mainly ammonia) during oxic respiration. In the example of Figure 6.5 the C/N ratio of decomposed organic matter had to be decreased to a factor of 3.7 (instead of 6.625; cf. Eq. 6.1) to reproduce the measured nitrate concentration profile. This is a reasonable explanation because fresh organic matter could be supplied in this specific marine environment (cf. Sect. 4.4), but the oxidation of upward diffusing reduced nitrogen species and/or artificially increased nitrate concentrations (see Sect. 6.3.2) have to be considered for interpretation.

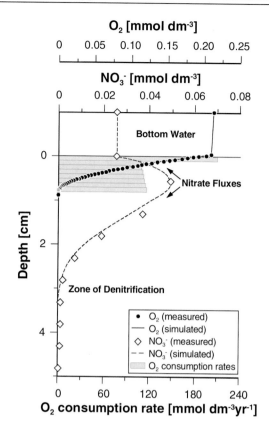

**Fig. 6.5** Measured and simulated profiles of oxygen and nitrate of station GeoB 1711 from the continental slope off Namibia at a water depth of approximately 2000 m. Degradation of organic matter with a C/N ratio of 3.7 was assumed for simulation. Bars indicate oxygen consumption rates required for model fit (after Hensen et al. 1997).

The general reaction principle and the coupling of all processes described in this section is illustrated in Figure 6.6. All dissolved species created during nitrification and denitrification might either escape into the bottom water, mainly by diffusion, or are involved in subsequent redox processes. Ammonium is generally re-oxidized in oxygenated sediments, but it might also diffuse into the bottom water, if oxygen is either limited or absent.

### 6.3.1.2 Coupling of Oxygen and Nitrate to other Redox Pathways

Below the oxygenated zone anoxic diagenesis is stimulated, if biodegradable organic matter is sufficiently available. Anoxic mineralization processes result in a release of reduced species, like ammonia (Fig. 6.6), into pore water which might be re-oxidized again when they diffuse back to the surface layers. Stimulation of anoxic diagenesis requires a high input of organic matter to the sediment surface which is usually combined with high sediment advection restricted to coastal and high production areas adjacent to continental margins. The general coupling between sediment advection, deposition (burial) and mineralization of organic matter is depicted in Figure 6.7 and has been stated in numerous studies (cf. Sections 4.2, 4.3; e.g. Henrichs and Reeburgh 1987, Henrichs 1992, van Cappellen et al. 1993, Tromp et al. 1995).

The amount of fresh organic matter arriving at the sediment surface also constitutes a control parameter for the population density of benthic macrofauna responsible for the biological mixing of the sediment. Biological mixing is known to be much more important for the transport of labile organic particles to deeper sediment layers than sedimentation (cf. Sect. 7.4.3). The strong correlation between sedimentation rate and particle mixing was recently compiled by Tromp et al. (1995) and is shown in Figure 3.24. Based on this compilation Tromp et al. (1995) derived the following regression equation:

$$\log D_{bio} = 1.63 + 0.85 \log \omega \qquad (6.10)$$

where $D_{bio}$ is the bioturbation coefficient ($cm^2 \, yr^{-1}$) and $\omega$ is the sedimentation rate ($cm \, yr^{-1}$). A closer look at the data shown in Figure 3.24 and Equation 6.10 makes clear that there is a difference between both parameters of up to three orders of magnitude. The correlation is, however, only applicable when other environmental factors, like bottom water anoxia, extreme sedimentation rates, or current action, are not effective. Bottom water deficient of oxygen (below 20% sat.) has been shown to seriously decrease the bioturbation intensity (Rhoads and Morse 1971) and below an oxygen saturation of about 5% nearly no macrofauna will survive (Baden et al. 1990).

For some practical reasons, oxygen is often used as a measure for total respiration of a sediment, mainly in the marine environment (cf. Jahnke et al. 1996; Sect. 12.5.1; Figs. 12.12, 12.13). Because of all subsequent mineralization processes occurring below, this is of course not strictly appropriate. Rather a complete net-reoxidation it required –ultimately by means of oxygen - of all reduced species produced during anoxic diagenesis (Pamatmat 1971). Any fixation and burial of reduced species (e.g. the formation of sulfides or carbonates; pyrite, siderite,...) or the escape of reduced solutes across the sediment-wa-

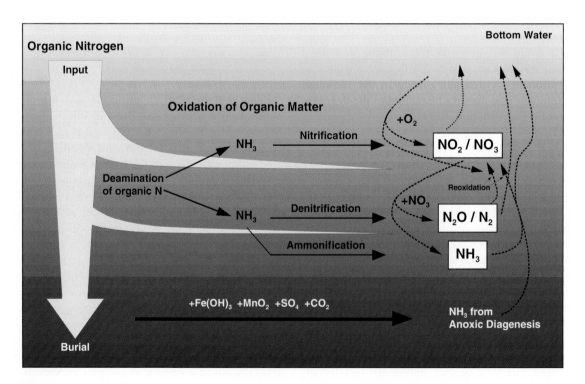

**Fig. 6.6** Pathways of nitrogen in marine surface sediments. Arrows: black, organic matter degradation; gray, particulate organic nitrogen; dotted, diffusion of solutes.

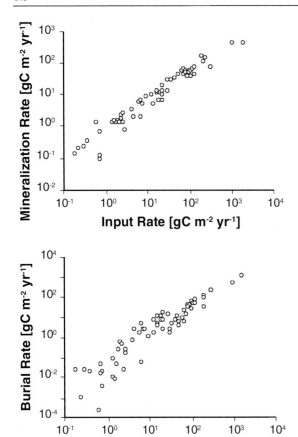

**Fig. 6.7** Input rates of organic carbon plotted against carbon mineralization and carbon burial from different data sources (after Henrichs 1992).

gator. Reimers et al. (1992) calculated for rapidly accumulating sediments on the continental margin off California that aerobic respiration is the major pathway of organic matter oxidation and more than 90% of the oxygen flux into the sediments is used for organic carbon oxidation. Since respiration by anoxic processes is estimated to exceed 30% of the total mineralization in the sediment this indicates an incomplete reoxidation cycle. Results of Canfield et al. (1993 a,b) and Greeff et al. (sub.) from continental shelf sediments of the Baltic Sea indicate that oxygen consumption by reoxidation processes is quantitatively more important than aerobic respiration. In these environments sulfate reduction seems to be the dominant respiration process (Greeff et al. sub., Thamdrup et al. 1994), although metal oxides can also play a significant role in particular cases. Canfield et al. (1993 a,b) have shown that the importance of metal oxides in the diagenetic sequence is strongly dependent on bioturbation activity. Mineralization rates simply derived from pore water gradients might underestimate the true rates by one order of magnitude (Haese 1997). Such complex interactions between different pathways of organic matter decomposition and redox reactions are restricted to coastal marine environments and highly accumulating upwelling regions. In more oligotrophic regions of the deep sea, 100 to 1000-times more organic carbon is oxidized by oxygen than by sulfate reduction and other pathways (Canfield 1989). For the major part of the world oceans the oxygen flux into the sediment provides, therefore, a good approximation to the total rate of organic carbon oxidation.

### 6.3.2 Determination of Consumption Rates and Benthic Fluxes

#### 6.3.2.1 Fluxes and Concentration Profiles Determined by *In-Situ* Devices

One reason of determining changes of oxidant concentrations or consumption / production rates in the pore water fraction of a sediment is to quantify the underlying respiration processes and to define the reactive horizons. Until today, however, there is no method to determine oxic respiration directly. Total oxic respiration has to be calculated from the difference in the oxygen demand of a sediment and the amount of oxygen consumed by oxidation of reduced species (see above). There are two main methods to determine

ter interface (e.g. $CH_4$, $NH_4$; $N_2O$, $N_2$, $Mn^{2+}$, $Fe^{2+}$; Bartlett et al. 1987, Seitzinger 1988, Devol 1991, Tebo et al. 1991, Johnson et al. 1992, Thamdrup and Canfield 1996, Greeff et al. sub.) results in an underestimation of the total respiration. The evaluation whether a reoxidation is complete is generally very difficult and is limited by the availability of measurements of all key species. The main questions are: (1) How big is the contribution of each pathway compared to the total mineralization? (2) To which amount and by which processes are these pathways interrelated? Since in most studies a lack of information on certain parameters remains, or different methods are applied to determine one pathway (e.g. differences resulting from *in-situ/ex-situ* determination of a species, or different methods to determine for example denitrification and sulfate reduction rates; see Sects. 6.3.2 and 12.2.3), the conclusion is at least to some extent arbitrary to the investi-

diffusive or total oxygen uptake rates in a deep-sea sediment. These are (1) the application of microelectrodes and optodes to obtain one- or even two-dimensional (planar optodes) depth profiles and (2) benthic chambers to reveal total areal uptake rates (cf. Sect. 12.2). Clark-type microelectrodes as they are commonly in use since more than a decade (e.g. Revsbech et al. 1980, Revsbech and Jørgensen 1986, Revsbech 1989, Gundersen and Jørgensen 1990) and the more recently invented optodes (Klimant et al. 1995) have become a driving force in performing measurements of oxygen consumption and penetration depths in any kind of soft sediment. A new instrument, a planar optode for the measurement of two-dimensional oxygen distribution was recently introduced by Glud et al. (1996) showing an excellent correlation with measurements performed with microelectrodes. For the general principles of microelectrodes and optodes and their application in the deep sea we refer to the description in Section 3.5 and the references above.

Within the last decade benthic lander systems have been increasingly applied in the deep sea (e.g. Berelson et al. 1987, Reimers et al. 1992, Jahnke et al. 1997) to avoid artefacts resulting from sediment recovery (see discussion below). Some results obtained by these devices have already been shown in Chapter 3. The microelectrodes or optodes provide information about the oxygen penetration depth and its depth-dependent distribution, whereas the benthic chambers measure total fluxes across the sediment water interface. There is generally good agreement of fluxes obtained by both methods (Jahnke et al. 1990; Fig. 12.7), but total fluxes might exceed calculated diffusive fluxes from oxygen profiles. While diffusive transport of pore water is the dominant process in the large area of the oligotrophic oceans where the input of degradable organic matter to the sea floor is low (Sayles and Martin 1995), this is not the case adjacent to continental margins. Glud et al. (1994) found that the total oxygen uptake was always larger than the diffusive uptake in continental slope sediments off Southwest Africa. The results shown in Figure 6.8a indicate a good correlation between the dry weight of macrofauna and the total oxygen uptake. Subtracting the diffusive oxygen uptake - measured with microelectrodes - from the total uptake rates reveals values close to zero for stations with a low dry weight of macrofauna obviously increasing with increasing population density (Fig. 6.8b). On one hand, the difference between both fluxes provides a measure of bio-irrigation meaning that there is additional transport across the sediment-water interface due to active pumping of organisms and a higher oxygen demand due to an increased surface area at additional sites where oxygen is consumed (worm burrows etc.). On the other hand, the respiration by the macro- or meiofauna itself increases the

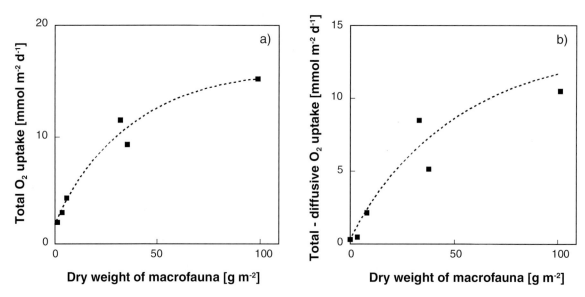

**Fig. 6.8** Correlation of (a) total oxygen fluxes and (b) total-diffusive oxygen fluxes with the dry weight of organic macrofauna indicating the effect of macrobenthic activity for benthic respiration processes (from Glud et al. 1994). Broken lines are fitted by eye.

oxygen consumption (Glud et al. 1994, Heip et al. 1995, Soetaert et al. 1997). Since the chamber system provides information on total mineralization rates and fluxes, and the profiling lander on the depth distribution of solutes and redox processes, the application of both lander systems is required to properly investigate the oxygen demand and the pathways of oxygen in a sediment.

The determination of *in-situ* fluxes of nitrate is comparatively limited. Although there are electrodes for the determination of micro-concentration profiles in form of biosensors (Larsen et al. 1996), they are not yet suitable to be used on lander systems in the deep sea. The measurement of total nitrate fluxes by benthic chambers is possible, but only a limited number of flux measurements have been published until today. Figure 3.22 shows flux measurements in continental slope sediments off California measured by Jahnke and Christiansen (1989). There, nitrate fluxes are directed into the sediment indicating strong denitrification supported by very low oxygen concentrations in the overlying bottom water. Under normal deep-sea oxygen conditions, the nitrate flux is generally directed out of the sediment due to nitrification and lower denitrification as shown by results reported by Hammond et al. (1996) from measurements in the central equatorial Pacific (Fig. 6.9). It is further indicated that fluxes of oxygen, silicate, or $\Sigma CO_2$ are easier to determine than nitrate or, particularly, phosphate, because of the magnitude of concentration change over time. Phosphate in Figure 6.9 shows large scattering and does not allow a reliable flux calculation. The differences in the order of magnitude result from the reaction stoichiometry of organic matter decomposition (Eq. 6.1). Additional processes, like the dissolution of biogenic opal and calcium carbonate in the sediments, is of further significance for the parameters silicate and alkalinity. Whereas silicate fluxes depend on the amount and the surface area of soluble opal, the degree of silica undersaturation in pore waters, and the content of terrigenous components (cf. Sect. 12.3.3), variations of alkalinity and $\Sigma CO_2$ fluxes are controlled by respiratory production of $CO_2$ and dissolution of calcium carbonate (cf. Sect. 9.3.2).

The result of the phosphate measurement shown in Figure 6.9 indicates that flux chamber measurements are restricted to a certain number of measurable parameters. Even more, this is true for profiling lander systems. Apart from oxygen,

only pH- and $pCO_2$-electrodes have been successfully employed *in-situ* until today (Cai et al. 1995, Hales and Emerson 1997). There is, however, another promising microelectrode technique described by Brendel and Luther (1995) which has, however, not yet been tested *in-situ*. This voltammetric microelectrode technique allows the simultaneous determination of the most characteristic redox species: oxygen, manganese, iron and sulfide. Recently published results from continental slope sediments of northeast Canada displayed redox sequences which were not disturbed by different vertical depth resolutions, which usually occurs when different methods are applied (Luther et al. 1997, 1998).

### 6.3.2.2 *Ex-Situ* Pore Water Data from Deep-Sea Sediments

Mostly, solutes are still determined *ex-situ* by extraction of pore water from multicorer or box corer samples (see Chap. 3). Additionally, numerous *ex-situ* measurements of oxygen and nitrate exist which were carried out by onboard core incubations or microelectrode measurements. Generally, the determination of a dissolved species in water samples does not pose a problem, if sampling is handled carefully and, in specific cases, contact with atmospheric oxygen is avoided. Manifold problems arise, however, when a sediment sample is retrieved from some thousand meters below the sea surface and subsequently during the extraction of pore water aboard a ship which leads to changes in pore water concentrations compared to *in-situ* conditions. Such effects can be caused either by decompression and/or transient heating of a sample during its transport through the water column. These problems arise because of the large temperature gradient between deep water and surface water in most areas of the global ocean. With regard to oxygen a possible intensification of organic matter decay, triggered by the temperature increase or the lysis of cells, might result in a reduction of the oxygen penetration depth due to a higher oxygen consumption. Glud et al. (1994) showed that there might be significant discrepancies between *in-situ* and *ex-situ* measured fluxes and oxygen penetration depths. This effect is illustrated in Figure 6.10 and shows that the discrepancy obviously increases with increasing water depth. Above 1000 m water depth the sampling effect seems to be more or less negligible.

In analogy to oxygen, such artefacts are believed to exist also for nitrate profiles determined *ex situ*. The occurrence of increased (compared to Redfield stoichiometry in Eq. 6.1) subsurface nitrate concentrations has been described in a number of studies (e.g. Hammond et al. 1996, Martin and Sayles 1996), but is mostly attributed to the centrifugation method in pore water extraction (see Sect. 3.3.2). Artificially increased subsurface nitrate concentration would consequently lead to an overestimation of benthic fluxes of nitrate. Berelson et al. (1990) and Hammond et al. (1996) found evidence for increased nitrate fluxes after pore water centrifugation compared to

lander measurements which were up to a factor of 3, but also agreement between both methods for quite a number of stations was found. However, the possibility of low C/N organic matter or oxidation of reduced nitrogen species (diffusing upwards from deeper layers) can be important natural factors increasing nitrate concentrations deviating from the general expected stoichiometry. Following results of Luther et al. (1997) the amount of subsurface nitrate production due to nitrification can also be regulated by the manganese oxide content of the sediment. As discussed in Section 6.3.2.3 high $MnO_2$ concentrations favor a catalytic reduction of nitrate to $N_2$ already

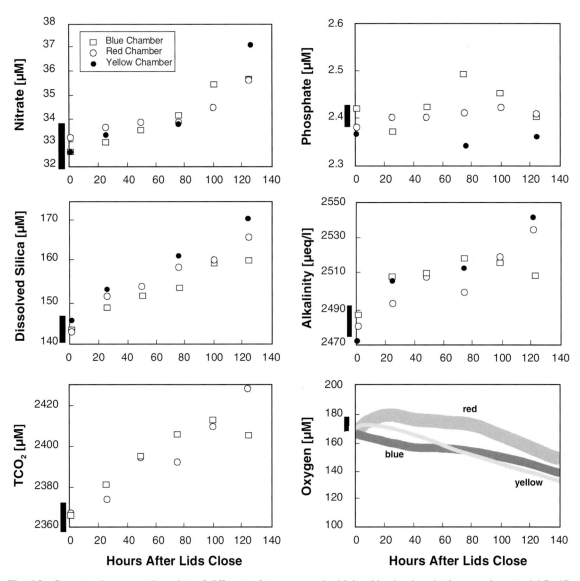

**Fig. 6.9** Concentration versus time plots of different solutes measured with benthic chambers in the central equatorial Pacific (from Hammond et al. 1996). Black vertical bars indicate bottom water values. Blue, red and yellow indicate different chambers.

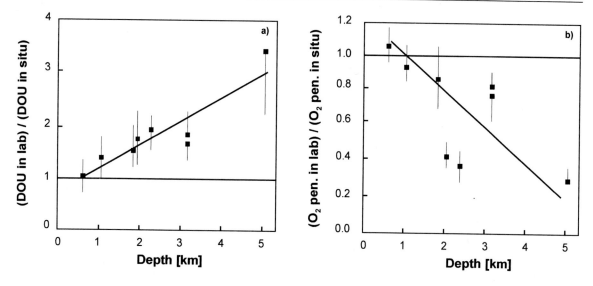

**Fig. 6.10** Plots of (a) the ratio between diffusive oxygen uptake rates (DOU) *ex situ* and *in situ* and (b) oxygen penetration depth *ex situ* and *in situ* versus water depth from stations off the continental slope off Southwest Africa (from Glud et al. 1994). Solid lines indicate linear regressions. With increasing water depth fluxes appear to be overestimated and oxygen penetration underestimated when measured *ex situ*.

in the oxic zone of the sediment so that nitrate peaks only occur when the solid-phase manganese content is low.

In general, fluxes which are calculated on the basis of *ex-situ* data should be interpreted with caution, the overall result, however, in most cases reveals a reasonable approximation to real conditions (cf. Sect.12.2).

### 6.3.2.3  Determination of Denitrification Rates

The downward flux of nitrate (e.g Fig. 6.5) indicates the depth of active denitrification, but is no measure for the total rate of denitrification. As mentioned above, denitrifying bacteria are facultative anaerobic and denitrification is generally located directly below the oxic zone (Christensen et al. 1989). The shape of a nitrate profile depends on the total rate of nitrification and the denitrification rate. Considering the example shown in Figure 6.5 , the fit of the measured nitrate profile was achieved by determining an indirect nitrification rate (coupled oxidation and nitrification as described by Equation 6.1 and the C/N ratio) and denitrification occurring with a distinct rate at a depth of about 3 cm. A reduction of the denitrification rate would result in a greater nitrate penetration depth, i.e. not all the nitrate can be consumed in this depth zone. On the other hand, an increase of the denitrification rate would

lead to a depression of the nitrate maximum and therefore reduce upward and downward fluxes, and finally induce a total nitrate flux from the bottom water into the sediment. To clarify these interactions we plotted three nitrate profiles from different regions of the South Atlantic in Figure 6.11. Two profiles indicate high respiration rates with a nitrate penetration depth of approximately 3 cm, but a distinct peak is visible only at one station. The station with the nearly linear gradient into the sediment is indicative for strong denitrification. The third profile shows a low gradient into the bottom water which is due to nitrification and remains nearly constant with depth indicating that denitrification does not occur close to the sediment surface.

Depletion of nitrate and the formation of dinitrogen is, however, not exclusively coupled to denitrification, since the microbially mediated reduction utilizing reduced species like $Mn^{2+}$ or $Fe^{2+}$ might occur. Based on field observations, a number of studies invoke the reduction of nitrate by $Mn^{2+}$ to form $N_2$ instead of organic matter respiration (Aller 1990, Schulz et al. 1994, Luther et al. 1997,1998).

The direct determination of denitrification rates has been carried out by a number of different methods which all have certain advantages and restrictions. It is beyond the scope of this text to describe all of these methods in detail and so we will only give a short summary of the general principles. A more detailed overview of this subject can be found, for example, in Koike and Sørensen (1988), Seitzinger (1988), and Seitzinger et al. (1993).

The most important methods of measuring denitrification are (1) the detection of totally produced $N_2$ gas by incubation, (2) isotope-labeling methods with $^{15}N$ and $^{13}N$, and (3) the acetylene ($C_2H_2$) inhibition technique.

(1) The first method aims at measuring the total production of $N_2$ as equal to the rate of denitrification (Eq. 6.6-6.8) by sediment incubation (Seitzinger et al. 1984, Devol 1991). At present, this is thought to be the best method to accurately determine rates of overall denitrification, although the contamination with atmospheric $N_2$ is possible during long incubation periods and thought to be the main restriction (Koike and Sørensen 1988, Seitzinger 1988). More recently, however, Kana et al. (1994, 1998) have developed a technique of membrane inlet mass spectrometry which is able to measure small changes of dissolved nitrogen with high temporal resolution and without perturbation of the sediment.

(2) The intention of the isotope methods is to add $^{15}NO_3$ or $^{13}NO_3$ to the nitrate pool of the supernatant water during incubation and to measure the $^{15}N$ and $^{13}N$ content of the total amount of produced $N_2$. As the half-life of $^{13}N$ is 10 minutes, this method is only of limited use. The $^{15}N$ method already applied by Goering and Pamatmat (1970) to marine sediments off Peru is, in contrast, widely accepted and was used in a number of recent incubation studies of shallow marine and freshwater environments (e.g. Nielsen 1992, Rysgaard et al. 1994). This so-called ion-pairing method (Nielsen 1992) allows the determination of the total rate of denitrification and its dependence on the nitrate concentration in bottom water. Additionally, a number of authors (e.g. Nielsen 1992, Rysgaard et al. 1994, Sloth et al. 1995) believe that it also enables to distinguish between the source of nitrate, either as coming directly from the bottom water or from nitrification. This, however, has caused an intense discussion regarding the potential of the method and the benefits of its performance (Middelburg et al. 1996bc, Nielsen et al. 1996).

(3) At last, the $C_2H_2$ inhibition technique takes advantage of the property of acetylene to block the reduction of $N_2O$ to $N_2$ after it is injected into the sediment. The total amount of $N_2O$ produced is then the measure for the denitrification rate as it is easy to determine by gas chromatography (Andersen et al. 1984) or by microsensors (Christensen et al. 1989). The advantage of this method is that analyses can be carried out rapidly and sensitively. Problems are: (a) $N_2O$ reduction is sometimes incomplete, (b) a homogenous distribution of $C_2H_2$ in the pore water is difficult to maintain, (c) $C_2H_2$ inhibits nitrification in the sediment meaning that the coupled system (nitrification / denitrification) might be seriously affected due to the applied method, and (d) might lose its inhibitory properties in the presence of low sulfide concentrations (Sørensen et al. 1987, Koike and Sørensen 1988, Hynes and Knowles 1978, Seitzinger 1988).

The application of acetylene and nitrogen labeling methods usually results in the determination of lower denitrification rates when compared to total nitrogen fluxes (Seitzinger et al. 1993; Section 6.3.3). Apart from the above restrictions inherent to the methods themselves, the recently published concept of Luther et al. (1997) provides

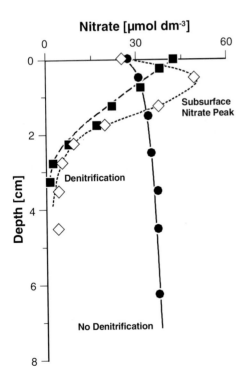

**Fig. 6.11** Typical nitrate concentration profiles of surface sediments from different productivity regions in the South Atlantic Ocean. The profiles from the continental slope of the Argentine and the Cape Basin indicate denitrification at about 3 cm. The profile from an oligotrophic equatorial site shows no denitrification.

an additional explanation for the observed discrepancy. They tested the thermodynamic properties of several redox reactions and found evidence for a catalytic short circuit for the coupled process of nitrification-denitrification within the oxic zone (Eq. 6.11.). Organic nitrogen and ammonia released during oxic respiration is therefore oxidized to $N_2$ by $MnO_2$, instead of being further oxidized to $NO_3$.

$$2 \, NH_3 + 3 \, MnO_2 + 6 \, H^+ \rightarrow 3 \, Mn^{2+} + N_2 + 6 \, H_2O$$

$$(6.11)$$

This process may outweigh nitrification in manganese-rich surface sediments and circumvent denitrification. Elevated $N_2$ fluxes without increasing denitrification rates and even $N_2$ production in oxidized sediments as observed by Seitzinger (1988) may be explained by this process.

### 6.3.3 Oxic Respiration, Nitrification and Denitrification in Different Marine Environments

After the description of the general biogeochemical processes controlling the distribution of oxygen and nitrate in marine sediments, including the possibilities and limitations of determining these inorganic compounds in the deep sea, the following section will give an overview of the dimensions of their fluxes and their distribution in different marine environments.

### 6.3.3.1 Quantification of Rates and Fluxes

As stated above, the basic mechanism inducing microbial activity is the supply of organic matter to the sea floor and this is generally coupled to surface water productivity. Most of the highly productive areas in the global ocean are adjacent to the continents, so that we can expect a decrease of respiration intensity from the coastal marine environments over the continental shelves and slopes into the deep sea. This becomes evident when we look at the data compiled by Middelburg et al. (1993) which indicate that 83% mineralization and 87% burial in marine sediments occurs in the coastal zone occupying only ~9% of the total ocean area. This means that the sediments with the highest respiration rates also have the highest burial efficiency in marine environments. For a

more detailed discussion of this subject see also reviews by (Henrichs and Reeburgh 1987, Henrichs 1992, Canfield 1993). As shown in Figure 6.3 the organic matter supply is not only a function of productivity, but also of water depth generally amplifying this gradient between shallow and deep water environments.

Fluxes of oxygen and nitrate, therefore, vary over several orders of magnitude between oligotrophic open ocean areas and continental shelf and slope areas. This is about 50 to 6000 mmol m$^{-2}$yr$^{-1}$ for oxygen and -600 to 380 mmol m$^{-2}$yr$^{-1}$ for nitrate (e.g. Devol and Christensen 1993, Glud et al. 1994, Berelson et al. 1994, Hammond et al. 1996, Luther et al. 1997, Hensen et al. 1998) where negative nitrate fluxes indicate fluxes into the sediment. The above minimum and maximum values do not permit differentiation between total and diffusive or in-situ and ex-situ fluxes. Figure 6.12 reveals the distribution of nitrate fluxes released from sediments below 1000 m water depth in the South Atlantic (Hensen et al. 1998) based on about 180 ex-situ nitrate concentration profiles. Averaged fluxes vary between 10-180 mmol m$^{-2}$yr$^{-1}$. Based on this compilation the total annual release for the whole area (about one tenth of the global deep ocean) is about $1.6 \cdot 10^{12}$ mol $NO_3^-$ yr$^{-1}$.

Global denitrification in marine sediments was recently calculated to be about $1.64$-$2.03 \cdot 10^{13}$ mol N yr$^{-1}$ with a contribution of $0.71 \cdot 10^{13}$ mol N yr$^{-1}$ of shelf sediments (Middelburg et al. 1996a). This model-based re-estimation produces values which are up to 3 to 20 times higher than those previously estimated by a number of authors in the mid-1980s. The predicted denitrification rates, however, are in the range of those derived from literature, and the total contribution of organic matter mineralization of 7-11% corresponds to results of estimates as well (see compilation in Middelburg et al. 1996a). Total denitrification rates vary between 0.4 mmol m$^{-2}$yr$^{-1}$ in the deep sea (Bender and Heggie 1984) and 1200 mmol m$^{-2}$yr$^{-1}$ (measured by the $N_2$ method) in continental margin sediments (Devol 1991) which is up to a factor of two higher than otherwise indicated by the highest fluxes of nitrate into the sediments. For estuarine and coastal areas Seitzinger et al. (1988) have summarized common rates between 440-2200 mmol m$^{-2}$yr$^{-1}$ with highest rates up to 9000 mmol m$^{-2}$yr$^{-1}$. However, considering the suggestions of Luther et al. (1997) a high amount of $N_2$ fluxes may be due to ammonia oxidation by

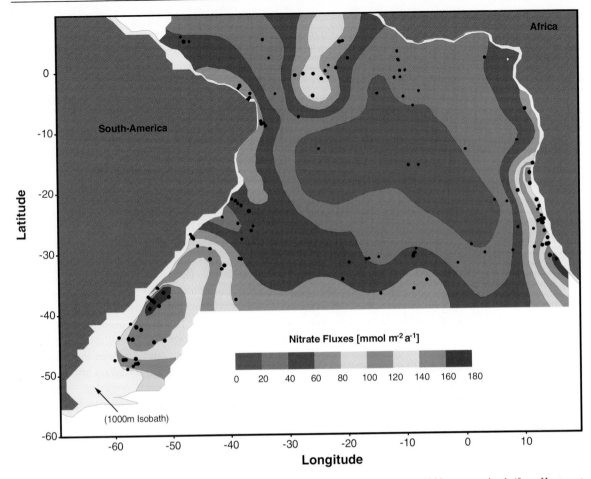

**Fig. 6.12**   Regional distribution of benthic nitrate release rates in the South Atlantic below 1000 m water depth (from Hensen et al. 1998).

$MnO_2$ in the oxic zone of the sediment bypassing denitrification and thus organic matter decay. If their estimate is correct that this process could contribute to up to 90% of $N_2$ production in continental margin sediments a careful evaluation and possibly a re-estimation of published denitrification rates for this environment is required.

### 6.3.3.2   Variation in Different Marine Environments: Case Studies

We already emphasized the importance of oxic respiration over other pathways in the deep sea. Denitrification (or related processes) only account for a few percent of the carbon oxidation rate by oxic respiration, provided that oxygen is sufficiently available. In oxygen depleted waters, the proportions can be dramatically shifted and denitrification might become an important pathway. Figure 6.13 represents data from different

oceanic regions as compiled by Canfield (1993) where the ratio of carbon oxidation by oxygen and nitrate is plotted as a function of the oxygen concentration in bottom water. It clearly shows that denitrification becomes more important than oxic respiration below oxygen concentrations of about 20 $\mu$mol l$^{-1}$.

To illustrate the general trend of decreasing respiration processes from the continental margin to the deep sea, Figure 6.14 shows the results of oxygen microelectrode measurements across the continental slope in the upwelling area off Namibia. Both the diffusive oxygen flux and the oxygen penetration depth at these stations can be clearly described as a function of water depth, even though scattering data points indicate some regional variability. The oxygen penetration depth at the 5000 m station had to be extrapolated from the measured gradient, since microelectrode measurements are limited to a few centimeters of

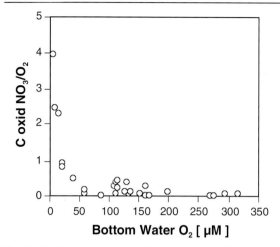

**Fig. 6.13** Ratio of carbon oxidation by denitrification and oxic respiration as a function of bottom water oxygen content (from Canfield 1993).

gree of small scale variability related to sediment surface topography and the inhomogeneous distribution of easily degradable organic matter. Such variability can be observed when several oxygen microprofiles are recorded on a given surface area of about 10 cm² (the area is determined by the construction of the device; cf. Fig. 3.20) during lander deployments. This often reveals conspicuous differences in the shape of the profiles as well as the oxygen penetration depth. A more sophisticated method used to the determine vertical and lateral oxygen distribution in sediments on a millimeter scale is provided by planar optodes (Glud et al. 1996). Their data show an excellent resolution of oxygen distribution within the sediment and the diffusive boundary layer. Furthermore, it shows that variations are due to differences in the surface topography. Another reason might be that other pathways, like denitrification or sulfate reduction, become more important in areas characterized by high sediment accumulation rates which are associated with a large input of degradable organic matter (see above).

sediment depth. To investigate deep-oxygen penetration in oligotrophic areas, the new optode techniques allow measurements up to several decimeters into the sediment (Fig. 6.15). The example is from a station located in the oligotrophic western equatorial Atlantic.

There is a number of studies that confirm the trend indicated in Figure 6.14, but on a regional scale there is much more variability, so that a simple relation between oxygen or nitrate fluxes and water depth cannot be found. The problem of regional flux variability will be dealt with further in Chapter 12, but generally, there is a high de-

Recently, Cai and Reimers (1995) compiled oxygen flux data across the Northeast Pacific continental margin into the deep sea. They related the oxygen fluxes to the availability of oxygen in the bottom water (as the limiting oxidant) and to the organic carbon content in the surface sediments (as limiting phase for respiration processes). The highest oxygen fluxes were measured on the lower continental slope where the

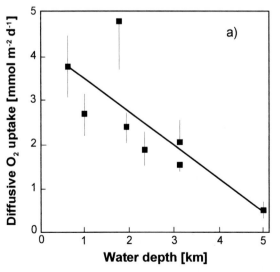

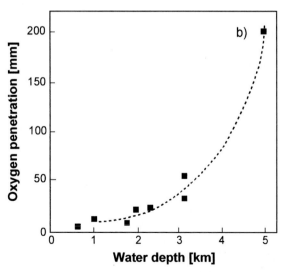

**Fig. 6.14** Diffusive oxygen fluxes (a) and oxygen penetration depths (b) at stations across the continental margin off Namibia (from Glud et al. 1994). Solid line indicates linear regression, broken line indicates fit by eye. Oxic respiration decreases with increasing water depth resulting in higher oxygen penetration into the sediment.

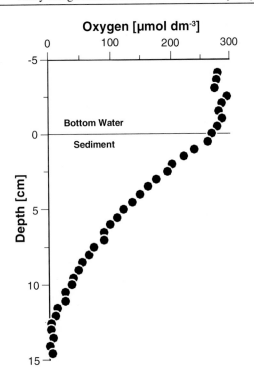

**Fig. 6.15** Oxygen concentration profile measured with an *in-situ* optode technique (after Wenzhöfer et. al. sub.).

tal rate of carbon oxidation is to be avoided which is probably higher in the upper slope sediments. As stated previously, the mineralization rate in a sediment as well as the burial rate is correlated to the rate of organic carbon input, so that areas with the highest input of organic matter consequently display the highest mineralization rates, but also the highest burial potential as well (Fig. 6.6). A further constraint for the standardization of empirical relations as given by Equation 6.12 is that temporal constancy, namely

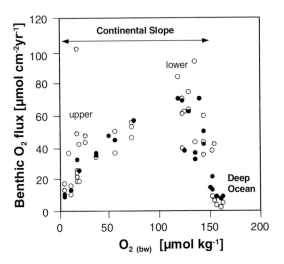

conditions for oxic respiration were optimal, because of the quantitative ratio between oxidant and organic matter availability (Fig. 6.16). On the upper slope, organic matter is available in excess, but oxygen is the limiting phase and reduces the total oxygen uptake in this area. The opposite situation can be observed for the deep ocean. Cai and Reimers (1995) also developed an empirical equation representing this obvious relationship between oxygen flux on the one hand and oxygen bottom water concentration and surface organic carbon content on the other hand with:

$$FO_2 = \frac{\pi C_{org}\,[O_2]_{BW}}{(126 + [O_2]_{BW})} \qquad (6.12)$$

This quantitative approach revealed a good agreement with the measured data. This formulation was also applied to a different data set recorded in the Atlantic Ocean, which also revealed a good agreement with measured data. However, the general applicability needs to be tested on a global scale with larger data sets.

Looking at the above situation, we need to emphasize the fact that any confusion with the to-

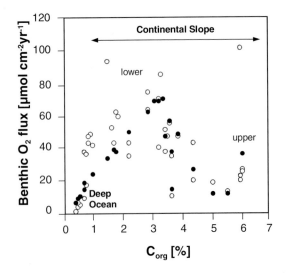

**Fig. 6.16** Distribution of benthic oxygen fluxes across the continental slope in the Northeast Pacific related to (a) oxygen bottom water concentration and (b) organic carbon content in the surface sediments. Highest oxygen respiration occurs at the lower continental slope. Open circles are measured fluxes, solid circles were calculated by applying Equation 6.12 (after Cai and Reimers, 1995).

steady-state conditions, are required. The time dependent variability of early diagenetic processes, however, has been identified even in deep-sea sediments (e.g. Smith and Baldwin 1984). One important question for the above case is: How fast does oxic respiration react to the input of labile organic matter? If the reaction constant is high, organic matter will be quickly recycled at the sediment surface. If furthermore, the input of organic matter is episodic or seasonal, a highly variable oxygen flux at a given time interval might occur. Since organic particles are subject to burial, mixing, and other respiratory processes the surface content will always result in a more or less time-integrated value that does not necessarily reflect the oxygen flux at the time of a single measurement.

More recently, Soetaert et al. (1996) produced striking evidence for the dependence of degradation rates of different mineralization pathways, as well as oxygen, nitrate, and other material fluxes on seasonal variations in organic matter deposition and its reaction rate. Some of the model results compared to the measured carbon flux to the sediment and the oxygen uptake rates are plotted in Figure 6.17. The study was based on data derived from box corer, and benthic chamber deployments in the abyssal Pacific and covered a time span of more than two years. The carbon flux function was derived from sediment trap data and sedimentation rates, whereas oxygen fluxes were obtained by the adaptation of total mineralization rates of organic material arriving at the sediment surface (and a number of other input parameters). A higher rate would account for varia-

tion as reflected by the carbon flux curve, whereas a low rate would continually reduce the seasonal variability.

A situation different from that in the Northeast Pacific which does not comply to a general relationship can be found in the Argentine Basin. As shown in Figures 6.12 and 12.14, distribution maps of nutrient release from deep-sea sediments in the South Atlantic indicate high mineralization rates in this area. Figure 6.18a shows diffusive nitrate fluxes on several transects across the continental slope in front of the Rio de la Plata mouth. The highest release rates of nitrate were detected at intermediate and low depths of the slope. In contrast to the situation in the Northwest Pacific, there is no oxygen limitation in the bottom water of the Argentine Basin, suggesting that other processes must be responsible for the observed flux distribution. In this case, it is assumed that intense downslope transport processes deliver large amounts of sediments and organic matter to the

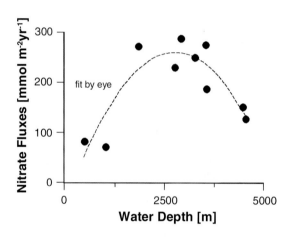

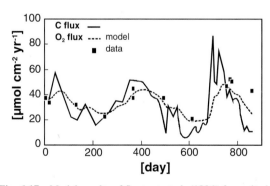

**Fig. 6.17** Model results of Soetaert et al. (1996) for a site in the abyssal Pacific. The curve of mineralization rates (oxygen fluxes) is smoother and shows a slight shift compared to the sedimentary carbon flux caused by the effective reaction kinetics. Squares indicate oxygen fluxes determined on the basis of benthic chamber and box corer data (redrawn from Soetaert et al., 1996)

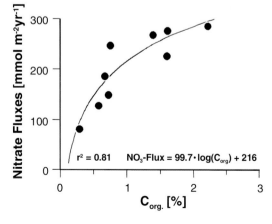

**Fig. 6.18** Plot of diffusive benthic nitrate fluxes against (a) water depth and (b) organic carbon content in surface sediments off the Rio de la Plata mouth (Argentine Basin).

lower slope where most of the degradable material is deposited, whereas the upper slope surface sediments are partly depleted of organic carbon. This is indicated by the good correlation of nitrate fluxes with organic carbon content in the surface sediments (Fig. 18b).

The above examples have shown that there are large discrepancies between benthic biogeochemical processes in areas of high and low productivity, but there are no simple relationships or master variables to correlate benthic fluxes and mineralization rates with primary productivity or sediment parameters. However, the quantitative coupling between these processes is a main objective of future research.

## 6.4    Summary

In this chapter we have briefly summarized the general availability of oxygen and nitrate in the oceans and aspects of their significance in the biogeochemical cycles of marine sediments. Oxic respiration is by far the most important pathway of organic carbon in deep-sea sediments and significantly determines the recycling of organic matter introduced onto the sediment surface. Even if some uncertainty remains, benthic oxygen fluxes reveal probably a reasonable approximation to the total carbon oxidation rate under deep-marine conditions. However, benthic fluxes vary considerably in the various ocean basins and the interactions between all parameters controlling benthic biogeochemical activities are not yet completely understood. Above, we showed some examples in which benthic activity deviates from the generally expected situation as suggested by export fluxes from the surface waters. We will further elaborate these concepts in Chapter 12 where we will summarize existent approaches for the regionalization of benthic fluxes and present methods how to access mineralization processes in surface sediments of the deep sea globally and arrive at a definition of benthic biogeochemical provinces.

This is contribution No 255 of the Special Research Program SFB 261 (*The South Atlantic in the Late Quaternary*) funded by the Deutsche Forschungsgemeinschaft (DFG).

## References

Aller, R.C., 1990. Bioturbation and manganese cycling in hemipelagic sediments. Phil. Trans. R. Soc. Lond., 331: 51-68.

Andersen, T.K., Jensen, M.H. and Sørensen, J., 1984. Diurnal variation of nitrogen cycling in coastal marine sediments: 1. Dentrification. Marine Biology, 83: 171-176.

Anderson, L.A. and Sarmiento, J.L., 1994. Redfield ratios of remineralization determined by nutrient data analysis. Global Biogeochemical Cycles, 8: 65-80.

Antoine, D., André, J.-M. and Morel, A., 1996. Oceanic primary production; 2. Estimation at global scale from satellite (coastal zone color scanner) chlorophyll. Global Biogechemical Cycles, 10: 57-69.

Baden, S.P., Loo, L.O., Pihl, L. and Rosenberg, R., 1990. Effects of eutrophication on benethic communities including fish: Swedish west coast. Ambio., 19: 113-122.

Bartlett, K.B., Bartlett, D.S., Harriss, R.C. and Sebacher, D.I., 1987. Methane emission along a salt marsh salinity gradient. Biogeochemistry, 4: 183-202.

Behrenfeld, M.J. and Falkowski, P.G., 1997. Photosynthetic rates derived from satellite-based chlorophyll concentration. Limnology and Oceanography, 42: 1-20.

Bender, M.L. and Heggie, D.T., 1984. Fate of organic Carbon reaching the deep-sea floor: a status report. Geochimica et Cosmochimica Acta, 48: 977-986.

Berelson, W.M., Hammond, D.E. and Johnson, K.S., 1987. Benthic fluxes and the cycling of biogenic silica and carbon in two southern California borderland basins. Geochimica et Cosmochimica Acta, 51: 1345-1363.

Berelson, W.M. et al., 1990. Benthic fluxes and pore water studies from sediments of the central equatorial north Pacific: Nutrient diagenesis. Geochimica et Cosmochimica Acta, 54: 3001-3012.

Berelson, W.M., Hammond, D.E., McManus, J. and Kilgore, T.E., 1994. Dissolution kinetics of calcium carbonate in equatorial Pacific sediments. Global Biogeochemical Cycles, 8: 219-235.

Berger, W.H., Fischer, K., Lai, C. and Wu, G., 1987. Ocean producitivity and organic carbon flux. I. Overview and maps of primary production and export production. University California, San Diego, SIO Reference, 87-30, 67 pp.

Betzer, P.R., Showers, W.J., Laws, E.A., Winn, C.D., DiTullio, G.R. and Kroopnick, P.M., 1984. Primary productivity and particle fluxes on a transect of the equator at 153°W in the Pacific Ocean. Deep-Sea Research, 31: 1-11.

Brendel, P.J. and Luther III, G.W., 1995. Development of a gold amalgam votametric microelectrode for the determination of dissolved iron, manganese O2, and S(-II) in pore waters of marine and freshwater sediments. Enviromental Science & Technology, 29: 751-761.

Broecker, W.S. and Peng, T.H., 1982. Tracer in the Sea. Palisades, NY, 690 pp.

Cai, W.-J., Reimers, C.E. and Shaw, T., 1995. Microelectrode studies of organic carbon degradation and calcite dissolution at a California Continental rise site. Geochimica et Cosmochimica Acta, 59: 497-511.

Canfield, D.E., 1989. Sulfate reduction and oxic respiration in marine sediments: implications for organic carbon preservation in euxinic environments. Deep-Sea Research, 36: 121-138.

Canfield, D.E., 1993. Organic matter oxidation in marine sediments. In: Wollast, R., Chou, L. and Mackenzie, F. (eds) Interactions of C,N,P and S biogeochemical cycles. NATO ASI Series, Springer, Berlin, Heidelberg, NY, pp 333-363.

Canfield, D.E., Jørgensen, B.B., Fossing, H., Glud, R.N., Gundersen, J.K., Ramsing, N.B., Thamdrup, B., Hansen, J.W. and Hall, P.O.J., 1993a. Pathways of organic carbon oxidation in three continental margin sediments. Marine Geology, 113: 27-40.

Canfield, D.E., Thamdrup, B. and Hansen, J.W., 1993b. The anaerobic degradation of organic matter in Danish coastal sediments: Iron reduction, manganese reduction, and sulfate reduction. Geochimica et Cosmochimica Acta, 57: 3867-3883.

Chapelle, F.H., 1993. Ground - Water Microbiology and Geochemistry. Wiley, NY, 424 pp.

Chester, R., 1990. Marine Geochemistry. Chapman & Hall, London, 698 pp.

Christensen, J.P., Nielsen, L.P., Revsbech, N.P. and Sørensen, J., 1989. Microzonation of denitrification activity in stream sediments as studied with a combined oxygen and nitrous oxide microsensor. Appl. Environ. Microbiology, 55: 1234-1241.

Devol, A.H., 1991. Direct measurement of nitrogen gas fluxes from continental shelf sediments. Nature, 349: 319-321.

Devol, A.H. and Christensen, J.P., 1993. Benthic fluxes and nitrogen cycling in sediments of the continental margin of the eastern North Pacific. Journal of Marine Research, 51: 345-372.

Froelich, P.N., Klinkhammer, G.P., Bender, M.L., Luedtke, N.A., Heath, G.R., Cullen, D. and Dauphin, P., 1979. Early oxidation of organic matter in pelagic sediments of the eastern equatorial Atlantic: suboxic diagenesis. Geochimica et Cosmochimica Acta, 43: 1075-1090.

Glud, R.N., Gundersen, J.K., Jørgensen, B.B., Revsbech, N.P. and Schulz, H.D., 1994. Diffusive and total oxygen uptake of deep-sea sediments in the eastern South Atlantic Ocean: in situ and laboratory measurements. Deep-Sea Research, 41: 1767-1788.

Glud, R.N., Ramsing, N.B., Gunderson, J.K. and Klimant, I., 1996. Planar Optrodes: A new tool for fine scale measurements of two-dimensional $O_2$ distribution in benthic communities. Mar. Ecol. Prog. Ser., 140: 217-226.

Goering, G.G. and Pamatmat, M.M., 1970. Denitrification in sediments of the sea of Peru. Invest. Pesq., 35: 233-242.

Greeff, O., Rieß, W., Wenzhöfer, F., Weber, A., Holby, O. and Glud, R.N., submitted. Pathways of carbon oxidation in Gotland Basin, Baltic Sea, measured in situ by use of benthic landers. Continental Shelf Research.

Gundersen, J.K. and Jørgensen, B.B., 1990. Microstructure of diffusive boundry layer and the oxygen uptake of the sea floor. Nature, 345: 604-607.

Haese, R.R., 1997. Beschreibung und Quantifizierung frühdiagenetischer Reaktionen des Eisens in Sedimenten des Südatlantiks. Berichet, Fachbereich Geowissenschaften, Universität Bremen, 99, 118 pp.

Hales, B. and Emerson, S., 1997a. Calcite dissolution in sediments of the Ceara Rise: In situ measurements of porewater $O2$, pH, and $CO2(aq)$. Geochimica et Cosmochimica Acta, 61: 501-514.

Hammond, D.E., McManus, J., Berelson, W.M., Kilgore, T.E. and Pope, R.H., 1996. Early diagenesis of organic material in equatorial Pacific sediments: stoichiometry and kinetics. Deep-Sea Research, 43: 1365-1412.

Heip, C.H.R., Goosen, N.K., Herman, P.M.J., Kromkamp, J., Middelburg, J.J. and Soetart, K., 1995. Production and Consumption of Biological Particles in Temperate Tidal Estuaries. Oceanography and Marine Biology, 33: 1-149.

Henrichs, S.M. and Reeburgh, W.S., 1987. Anaerobic mineralization of marine sediment organic matter: rates and the role of anaerobic processes in the oceanic carbon economy. Geomicrobiol. Journal, 5: 191-237.

Henrichs, S.M., 1992. Early diagenesis of organic matter in marine sediments: progress and perplexity. Marine Chemistry, 39: 119-149.

Henriksen, K., Hansen, J.I. and Blackburn, T.H., 1981. Rates of nitrification, distribution of nitriffying bacteria and nitrate fluxes in different types of sediment from Danish waters. Marine Biology, 61: 299-304.

Henriksen, K. and Kemp, W.M., 1988. Nitrification in estuarine and coastal marine sediments. In: T.H. Blackburn and J. Sørensen (eds), Nitrogen cycling in coastal marine enviroments. SCOPE Reports, Wiley & Sons, Chichester, pp. 207-249.

Hensen, C., Landenberger, H. Zabel, M. Gundersen, J.K., Glud, R.N. and Schulz, H.D., 1997. Simulation of early diagenetic processes in continental slope sediments in Southwest Africa: The computer model CoTAM tested. Marine Geology, 144: 191-210.

Hensen, C., Landenberger, H., Zabel, M. and Schulz, H.D., 1998. Quantification of diffusive benthic fluxes of nitrate, phosphate and silicate in the Southern Atlantic Ocean. Global Biogeochemical Cycles, 12: 193-210.

Hynes, R.K. and Knowles, R., 1978. Inhibition by acetylene of ammonia oxidation in Nitrosomonas europaea. FEMS Microbiol. Letters, 4: 319-321.

Jahnke, R.A. and Christiansen, M.B., 1989. A free-vehicle benthic chamber instrument for sea floor studies. Deep-Sea Research, 36(4): 625-637.

Jahnke, R.A., Reimers, C.E. and Craven, D.B., 1990. Intensification of recycling of organic matter at the sea floor near ocean margins. Nature, 348: 50-54.

Jahnke, R.A., Craven, D.B., McCorkle, D.C. and Reimers, C.E., 1997. CaCO3 dissolution in California continental margin sediments: The influence of organic matter remineralization. Geochimica et Cosmochimica Acta, 61: 3587-3604.

Johnson, K.S., Berelson, W.M., Coale, K.H., Coley, T.L., Elrod, V.A., Fairey, W.R., Iams, H.D., Kilgore, T.E. and Nowicki, J.L., 1992. Manganese flux from continental margin sediments in a transect through the oxygen minimum. Science, 257: 1242-1245.

Jørgensen, B.B., 1977. Bacterial sulfate reduction within reduced microniches of oxidized marine sediments. Marine Biology, 41: 7-17. Jørgensen, K.S., Jensen, H.B. and Sørensen, J., 1984. Nitrouse oxide production from nitrification and denitrification in marine sediment at low oxygen concentrations. Canadian Journal of Microbiology, 30: 1073-1078.

Jørgensen, B.B. and Sørensen, J., 1985. Seasonal cycles of O2, NO3- and SO42- reduction in estuarine sediments: the significance of an NO3- reduction maximum in spring. Mar. Ecol. Prog. Ser., 24: 65-74.

Kaplan, W.A., 1983. Nitrification. In: Carpenter, J.E. and Capone, D.G. (eds) Nitrogen in the marine environment. Academic Press, NY, pp. 139-190.

Kennett, J.P., 1982. Marine Geology. Prentice Hall, New Jersey, 813 pp.

Klimant, I., Meyer, V. and Kühl, M., 1995. Fiber-oxic oxygen microsensors, a new tool in aquatic biology. Limnology and Oceanography, 40: 1159-1165.

Koike, I. and Sørensen, J., 1988. Nitrate reduction and denitrification in marine sediments. In: Blackburn, T.H. and Sørensen, J. (eds) Nitrogen cycling in coastal marine environments. Wiley & Sons, pp 251-273.

Larsen, L.H., Revsbech, N.P. and Binnerup, S.J., 1996. A microsensor for nitrate based on immobilized denitrifying bacteria. Appl Environm. Microbiol., 62: 1248-1251.

Luther III, G.W., Sundby, B., Lewis, B.L., Brendel, P.J. and Silverberg, N., 1997. Interactions of manganese with the nitrogen cycle: Alternative pathways for dinitrogen formation. Geochimica et Cosmochimica Acta, 61: 4043-4052.

Luther III, G.W., Brendel, P.J., Lewis, B.L., Sundby, B., Lefrançois, L. Silverberg, N. and Nuzzio, D.B., 1998. Simultaneous measurement of O2, Mn, Fe, I-, and S(-II) in marine pore waters with a solid-state voltammetric microelectrode. Limnology and Oceanography, 43: 325-333.

Martin, J.H., Knauer, G.A., Karl, M. and Broenkow, W.W., 1987. VERTEX: carbon cycling in the northeast Pacific. Deep-Sea Research, 34(2): 267 - 285.

Martin, W.R. and Sayles, F.L., 1996. CaCO3 dissolution in sediments of the Ceara Rise, western equatorial Atlantic. Geochimica et Cosmochimica Acta, 60: 243-263.

Middelburg, J.J., Vlug, T., Jaco, F. and Van der Nat, W.A., 1993. Organic matter mineralization in marine systems. Glob. Planet. Change, 8: 47-58.

Middelburg, J.J., Soetaert, K., Herman, P.J.M. and Heip, C.H.R., 1996a. Denitrification in marine sediments: A model sudy. Global Biogeochemical Cycles, 10: 661-673.

Middelburg, J.J., Soetaert, K. and Herman, P.M.J., 1996b. Evaluation of the nitogen isotpe-pairing for measuring benthic denitrification: A simulation analysis. Limnology and Oceanography, 41: 1839-1844.

Middelburg, J.J., Soetaert, K. and Herman, P.M.J., 1996c. Reply to the comment by Nielson et al. Limnology Oceanography, 41: 1846-1847.

Nielsen, L.P., 1992. Denitrification in sediment determined from nitrogen isotope pairing. FEMS Microbiol. Ecol., 86: 357-362.

Nielsen, L.P., Risgaard-Petersen, N., Rysgaard, S. and Blackburn, T., H., 1996. Reply to the note by Midelburg et al. Limnology andceanography, 41: 1845-1846.

Painter, H.A., 1970. A review of literature on inorganic nitrogen metabolism in microorganisms. Water Res., 4: 393-450.

Pamatmat, M.M., 1971. Oxygen consumption by the seabed

IV.shipboard and laboratory experiments. Limnology and Oceanography, 16: 536-550.

Redfield, A.C., Ketchum, B.H. and Richards, F.A., 1963. The influence of organisms on the composition of sea-water. In: Hill, N.N. (ed) The sea. Interscience, NY, pp. 26-77.

Reid, J.L., 1994. On the total geostrophic circulation of the North Atlantic Ocean: Flow patterns, tracers, and transports. Prog. Oceanogr., 33: 1-92.

Reimers, C.E., Jahnke, R.H. and McCorkle, D.C., 1992. Carbon fluxes and burial rates over the continental slope and rise off central California with implications for the global carbon cycle. Global Biogeochemical Cycles, 6: 199-224.

Revsbech, N.P., Jørgensen, B.B. and Blackburn, T.H., 1980. Oxygen in the sea bottom measured with a microelektrode. Science, 207: 1355-1356.

Revsbech, N.P. and Jørgensen, B.B., 1986. Microelectrodes: Their use in microbial ecology. Adv. Microb. Ecol., 9: 293-352.

Revsbech, N.P., 1989. An oxygen electrode with a guard cathode. Limnology and Oceanography, 34: 474-478.

Rhoads, D.L. and Morse, J.W., 1971. Evolutionary and ecologic significance of oxygen deficient marine basins. Lethaia, 4: 413-428.

Rowe, G.T., Boland, G.S., Phoel, W.C., Anderson, R.F. and Biscaye, P.E., 1994. Deep-sea floor respiration as an indication of lateral input of biogenic detritus from continental margins. Deep-Sea Research, 41: 657-668.

Rysgaard, S., Risgaard-Petersen, N., Sloth, N.P., Jensen, K. and Nielsen, L.P., 1994. Oxygen regualtion of nitrification and denitrification in sediments. Limnology and Oceanography, 39: 1643-1652.

Sayles, F.L. and Martin, W.R., 1995. In Situ tracer studies of solute transport across the sediment-water interface at the Bermuda Time Series site. Deep-Sea Research, 42: 31-52.

Schlegel, H.G., 1985. Allgemeine Mikrobiologie. Thime, Stuttgard, 571 pp.

Schulz, H.D., Dahmke, A., Schinzel, U., Wallmann, K. and Zabel, M., 1994. Early diagenetic processes, fluxes and reaction rates in sediments of the South Atlantic. Geochimica et Cosmochimica Acta, 58: 2041-2060.

Seitzinger, S.P., Nixon, S.W. and Pilson, M.E.Q., 1984. Denitification and nitrous oxide production in a coastal marine ecosystem. Limnology and Oceanography, 29: 73-83.

Seitzinger, S.P., Nielson, L.P., Caffrey, J. and Christensen, P.B., 1993. Denitrification measurements in aquatic sediments: A comparision of three methods. Biogeochem., 23: 147-167.

Seitzinger, S.P., 1998. Denitrification in freshwater and coastal marine environments: Ecological and geochemical significance. Limnology and Oceanography, 33: 702-724.

Sloth, N.P., Blackburn, T., H., Hansen, L.S., Risgaard - Petersen, N. and Lomstein, B.A., 1995. Nitrogen cycling in sediments with different organic loading. Mar. Ecol Prog. Serv., 116: 163-170.

Smith, K.L.J. and Baldwin, R.J., 1984. Seasonal fluctuations in deep-sea sediment community oxygen consumption: central and eastern North Pacific. Nature, 307: 624-626.

Soetaert, K., Herman, P.M.J. and Middelburg, J.J., 1996. A model of early diagenetic processes from the shelf to abyssal depth. Geochimica et Cosmochimica Acta, 60: 1019-1040.

Soetaert, K. et al., 1997. Nematode distribution in ocean margin sediments of the Goban Spur (northeast Altantic) in relation to sediment geochemistry. Deep-Sea Research, 44: 1671-1683.

Sørensen, J., 1987. Nitrate reduction in marine sediment: pathways and interactions with iron and sulfur cycling. Geomicrobiol., 5: 401-421.

Sørensen, J., Rasmussen, L.K. and Koike, I., 1987. Micromolar sulfide concentrations alleviate blockage of nitous oxide reduction by denitrifying Pseudomonas fluorescens. Canadian Journal of Microbiology, 33: 1001-1005.

Stumm, W. and Morgan, J.J., 1996. Aquatic Chemistry. Wiley & Sons, NY, 1022 pp.

Sverdrup, H.U., Johnson, M.W. and Fleming, R.H., 1942. The oceans. Prentice Hall, NY.

Takahashi, T., Broeker, W.S. and Langer, S., 1985. Redfield ratio based on chemical data from isopycnal surface. J. Geophys. Res., 90: 6907-6924.

Tebo, B.M., Rosson, R.A. and Nealson, K.H., 1991. Potential for manganese (II) oxydation and manganese (IV) reduction to co-occur in the suboxic zone of the Black Sea. In: Izdar, E. and Murray, W. (eds) Black Sea Oceanography. Kluwer Academic Publishers, Dordrecht, pp. 173-185.

Thamdrup, B., Fossing, H. and Jörgensen, B.B., 1994. Manganese, iron, and sulfur cycling in a coastal marine sediment, Aarhus Bay, Denmark. Geochimica et Cosmochimica Acta, 58(23): 5115-5129.

Thamdrup, B. and Canfield, D.E., 1996. Pathways of carbon oxidation in continental margin sediments off central Chile. Limnology Oceanography, 41(8): 1629-1650.

Tromp, T.K., van Cappellen, P. and Key, R.M., 1995. A global model for the early diagenetisis of organic carbon and organic phosphorus in marine sediments. Geochimica et Cosmochimica Acta, 59(7): 1259-1284.

van Cappellen, P., Gaillard, J.-F. and Rabouille, C., 1993. Biogeochemical transformations in sediments: kinetic models of early diagenesis. Interactions of C, N, P and S biogeochemical cycles and global change - NATO ASI Series, 4. Springer-Verlag, Berlin, Heidelberg, 401-445 pp.

Wenzhöfer, F., Kohls, O. and Holby, O., Deep oxigen penetration in deep sea sediments measured in situ by optodes. .

# 7  The Reactivity of Iron

Ralf R. Haese

## 7.1  Introduction

For our understanding of interactions between living organisms and the solid earth it is fascinating to investigate the reactivity of iron at the interface of the bio- and geosphere. Similar to manganese (Chap. 11), iron occurs in two valence states as oxidized ferric iron, Fe(III), and reduced ferrous iron, Fe(II). Two principal biological processes are of importance: Microorganisms such as magnetotactic bacteria or phytoplankton (Chap. 2.2 and Sect. 7.3) depend on the uptake of iron as a prerequisite for their cell growth (assimilation). Others conserve energy from the reduction of Fe(III) to maintain their physiology (dissimilation). In this case, ferric iron serves as an electron acceptor which is also termed oxidant. Apart from biotic reactions, manifold abiotic reactions occur depending on thermodynamic and kinetic conditions. Due to redox-reactions, dissolution and precipitation of iron-bearing minerals may result which has great influence on the sorption/desorption and co-precipitation/release behavior of various components such as phosphate and trace metals. From a geologic point of view, it is striking to find discrete iron-enriched layers such as black shales or strata of the banded iron formation, which challenge geochemists to reconstruct the environmental conditions of their formation.

## 7.2  Pathways of Iron Input to Marine Sediments

Within the continental crust, iron is the fourth most abundant element with a concentration of 4.32 wt% (Wedepohl 1995). It is transported to marine sediments by three major regimes: fluvial, aeolian and submarine hydrothermal input. For the investigation of iron reactivity, it is important to differentiate regions of predominant input regimes since characteristic reactions occur at the interface of the transport regime and the marine environment. For the chemistry of hydrothermal fluids and reactions during mixing with seawater refer to Chapter 13.

### 7.2.1  Fluvial Input

In Figure 7.1, averaged concentrations of dissolved and particulate iron, major fluxes and the respective predominant reactions are shown. Average dissolved iron concentrations clearly show lower solubilities in marine relative to river water. In contrast, the concentration of particulate iron does not change significantly and is similar to the average continental crust concentration. This is also reflected in the conservative behavior of iron under chemical weathering conditions. Along with Al, Ti and Mn, Fe belongs to the refractive elements (Canfield 1997). Note that in Figure 7.1, the given value for particulate iron concentration in marine sediments is derived from pelagic clay sediments. Biogenic constituents such as carbonate and opal may significantly dilute the terrigineous fraction (cf. Chap. 1) and thus decrease the iron concentration.

The decrease of dissolved iron concentrations can be tracked within estuarine mixing zones. Within river water, dissolved iron is mainly present as Fe(III)oxyhydroxide, which is stabilized in colloidal dispersion by high-molecular-weight humic acids (Hunter 1983). Due to increasing salinity and therefore ionic strength, the colloidal dispersion becomes electrostatically and chemically destabilized which results in the coagulation of the fluvial colloids. This process is reflected in Figure 7.2 showing the distribution of dissolved iron on a transect from the mouth of the Congo (formerly: Zaire) river towards the open ocean.

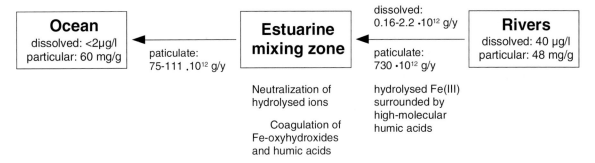

**Fig. 7.1**  Averaged concentrations and global fluxes of iron from rivers to the open ocean. The estuarine mixing zone serves as a sink for particulate iron and causes precipitation. Fluxes and concentrations are given according to Bewers and Yeats (1977), Yeats and Bewers (1982), Martin and Whitfield (1983).

With increasing salinity, the concentration of the dissolved iron decreases exponentially. These results also point out the operatively defined differentiation of dissolved and particulate phase. The concentration of dissolved iron depends on the pore size of the used filters. Commonly, particles smaller than 0.4 μm are considered to be 'dissolved'.

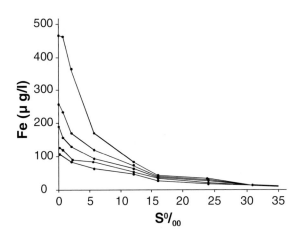

**Fig. 7.2**  Dissolved iron concentrations of surface water from a transect off the Congo River towards the open ocean. Pore sizes of 1.2, 0.45, 0.22, 0.05 and 0.025 μm (according to graphs from above downward) were used to separate particulate from dissolved phase (redrawn after Figuères et al. 1978).

Apart from being a zone of coagulation, the estuarine and coastal zones represent the most important sinks for particulate iron of fluvial origin. Since approximately 90% of the gross fluvial input is retained in estuarine and coastal zones (Chester 1990), only 10% of particulate iron coming from fluvial input reaches the open ocean (Fig. 7.1).

## 7.2.2    Aeolian Input

In contrast to the fluvial transport regime, the aeolian transport is highly efficient for the deposition of terrigenous matter in the deep sea. Donaghay et al. (1991) have published a global map of atmospheric iron flux to the ocean (Fig. 7.3). Among others, the North African desert and Asia can be identified as major dust sources. Asian loess is transported across the northern Pacific making up to 100% of the terrigenous fraction of pelagic sediments (Blank et al. 1985) and North African dust is spread out over the north equatorial Atlantic eventually reaching the Caribbean Sea (Carlson and Prospero 1972) and the northern coast of Brazil (Prospero et al. 1981). Iron settles out of wind-driven air layers by wet deposition which can be tracked in equatorial areas (Murray and Leinen 1993) where a distinctively high humidity occurs within the Inner Tropical Convergence Zone (ITCZ). A characteristically high deposition of iron therefore occurs at the equatorial Atlantic (Fig. 7.3). Fluxes of dissolved and particulate iron from atmospheric transport are compared to fluvial transport in Table 7.1.

As iron is discussed as a limiting nutrient for phytoplankton productivity of distinct oceanic regions (Sect. 7.3) it is important to investigate the flux, degree of solubility and the bioavailability of the introduced iron into open ocean surface water.

It has long been known that sorption onto suspended particles is highly efficient in the removal of trace metals from solution (Krauskopf 1956) resulting in a decrease of dissolved concentration relative to the thermodynamic saturated concentration. Chester (1990) listed the

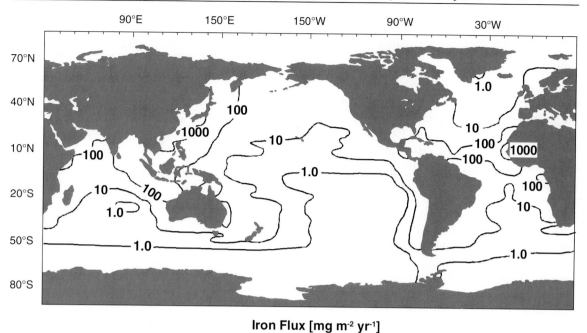

**Iron Flux [mg m⁻² yr⁻¹]**

**Fig. 7.3** Global map atmospheric iron fluxes to the deep sea (redrawn after Donaghay et al. 1991).

solubility of several trace metals in coastal and open ocean surface water and specified the solubility of aeolian transported iron with $\leq 7\%$. This is in agreement with a short review of published results provided by Zhuang et al. (1990) revealing a range of iron solubility between 1 and 10%. By contrast, Zhuang et al. (1990) determined a solubility of ~50% of atmospheric iron suggesting that all the dissolved iron in North Pacific surface water is provided by atmospheric input. Zhuang and Duce (1992) showed that the concentration of suspended particles is the prime variable controlling the adsorbed fraction, yet, an increase in aeolian deposition would also result in a net increase of dissolved iron.

The reason for the dissolution of solid phase iron in surface water is the photochemical reduc-

tion of Fe(III). The produced $Fe^{2+}$ is subsequently reoxidized, but the photoreduction causes a net increase of dissolved iron. Model and experimental results by Johnson et al. (1994) reflect the dependence of dissolved iron on irradiated light (Fig. 7.4). During incubations diurnal cycles and gradually decreasing dissolved iron concentrations result due to induced daily irradiation and a gradual uptake by phytoplankton.

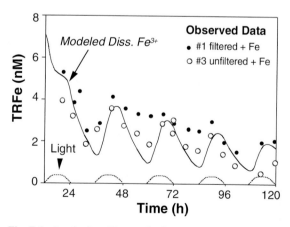

**Fig. 7.4** Incubation (dots and triangles) and model (solid line) results of dissolved iron within surface ocean water. The artificial light intensity (dashed line) drives the photochemical reduction. The gradual decrease of dissolved iron is caused by the uptake by phytoplankton (redrawn after Johnson et al. 1994).

**Table 7.1** Comparison of input fluxes of dissolved and particulate Fe from the atmosphere and rivers ([1] Duce et al. 1991, [2] GESAMP 1987).

|  | Dissolved | Particulate |
|---|---|---|
|  | **Fe [10¹² g yr⁻¹]** | |
| Atmosphere [1] | ~ 3 | ~ 29 |
| Rivers [2] | ~ 1 | ~ 110 |

## 7.3    Iron as a Limiting Nutrient for Primary Productivity

The growth of phytoplankton in the world ocean is undoubtedly one of the driving influences for the global carbon cycle and thus for the present and past climate (Berger et al. 1989, de Baar and Suess 1993, cf. Chap. 12.). This primary productivity (PP) is limited by the availability of nutrients. Apart from the major nutrients, nitrogen, phosphorous and silica, so-called micronutrients - especially iron - have long been speculated to have a limiting control on PP (Hart 1934). However, detailed investigations concerning their importance for the carbon cycle have only been possible for the past decade as a result of limitations in analytical methods. Virtually all microorganisms require iron for their respiratory pigments, proteins and many enzymes. Dissolved iron therefore shows a similar vertical profile in the water column to nitrate and is reduced to near zero within the surface layer where PP takes place and is increased within the oxygen minimum zone as a result of the mineralization of iron bearing organic matter (Fig. 7.5).

Three major oceanic regions (20% of the world's open ocean) have been found to be distinguished by high-nitrate and low-chlorophyll (HNLC) concentrations. The PP of the Southern Ocean (Broecker et al. 1982), the equatorial Pacific (Chavez and Barber 1987) and the Gulf of Alaska (McAllister et al. 1960) are obviously not limited by nitrate. Alternatively, as atmospheric dust loads in the Antarctic and equatorial Pacific are the lowest in the world (Prospero 1981, Uematsu et al. 1983), the importance of iron as limiting micronutrient for PP has become increasingly discussed.

The effect of added atmospheric dust to clean sea water from HNLC-regions were studied in bottle experiments by Martin et al. (1991) demonstrating increased PP by a factor of 2-4 (Fig. 7.6). Since the role of large grazers living on phytoplankton was not considered by such experiments, doubts about the use of these laboratory results for statements on large-scale environmental processes remained. Martin and colleagues therefore conducted a large-scale iron enrichment experiment south of the Galapagos Islands in the equatorial Pacific (Martin et al. 1994). A total volume of 15,600 l of iron solution (450 kg Fe) were distributed by ship over an area of 64 km$^2$ which in-

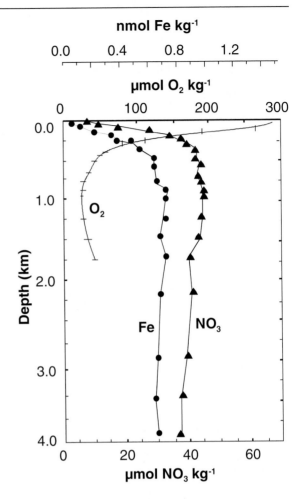

**Fig. 7.5** Vertical distribution of $NO_3^-$, dissolved iron and oxygen at a station of the Gulf of Alaska (redrawn after Martin et al. 1989).

creased the original dissolved iron concentration of 0.06 nM to ~4 nM. Iron concentrations, various parameters monitoring primary productivity and an inert tracer were continuously analyzed for 10 days. As a result, an increase of PP by a factor of 2-4 within the iron fertilized open ocean patch was observed (Martin et al. 1994) which gave evidence for the importance of iron as limiting micronutrient within HNLC-areas.

If iron is a limiting micronutrient for present-day PP, it may be an important link in explaining the glacial-interglacial climatic cycles of the past. Martin (1990) postulated the 'iron hypothesis' which explains decreased atmospheric $CO_2$ concentrations during glacial times with increased iron deposition by aeolian input resulting in increased PP and thus increased $CO_2$-uptake by the oceans. A comparison of iron and $CO_2$ concentra-

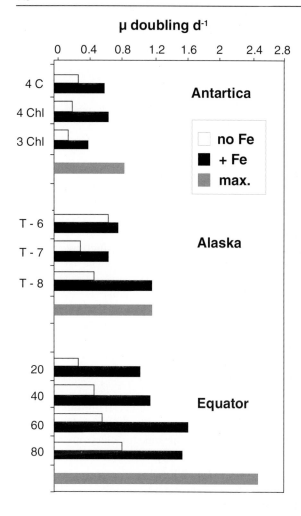

**Fig. 7.6** The effect of iron addition to surface water of high-nutrient, low-chlorophyll (HNLC) regions. The doubling rate, μ, is an expression for the increase of primary productivity and maximum values, Max., depend on light intensity and temperature. An increase of primary productivity by a factor of 2-4 is resulting due to the addition of atmospheric iron (redrawn after Martin et al. 1991).

depth the iron concentrations always remain constant at ~0.6 nM. Other elements with such short residence time (100 to 200 years) typically decrease continuously with depth and age. This suggests a substantial decrease in the iron removal rate below this concentration. As organic ligands with a binding capacity of 0.6 nM Fe have been found (Rue and Bruland 1995, Wu and Luther 1995), iron-organic complexes are regarded as being of great importance for the distribution of dissolved iron. Additional evidence for the interaction with dissolved organic molecules comes from the study of the mechanisms of iron uptake by organisms. To make dissolved iron more accessible, microorganisms have acquired the ability to synthesize chelators which complex ferric iron of solid phase. These chelators are commonly called siderophores and consist of a low-molecular-mass compound with a high affinity for ferric iron. Siderophores are secreted out of the microorganism where they form a complex with ferric iron. After transport into the cell, the chelated ferric iron is enzymatically reduced and released from the siderophore, which is secreted again for further complexation. For the open ocean environment, ferric iron availability by the secretion of siderophores was shown for phytoplankton (Trick et al. 1983), as well as for bacteria (Trick 1989). However, this extracellular substance has only been found in few open-ocean microorganisms and thus their importance for PP seems to be unlikely. In contrast, Kuma et al. (1994) showed that total natural organic $Fe^{3+}$-

tions of the past 160,000 years recorded within an Antarctic ice core revealed a strikingly negative correlation (Fig. 7.7) which supports the 'iron hypothesis'.

So far, dissolved iron has been discussed with respect to the assimilation by phytoplankton. The chemical state of the bioavailable dissolved species is presently a matter of intensive studies and discussions. Due to thermodynamic reasons, the concentrations of free ions of dissolved iron are extremely low under oxic and pH-neutral conditions. A discussion paper by Johnson et al. (1997) reviews regional distributions and depth profiles of dissolved iron and points out that at greater

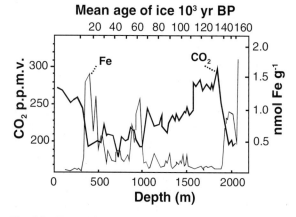

**Fig. 7.7** Fe and $CO_2$ concentrations of the Antarctic Vostok ice core for the past 160,000 years (redrawn after De Angelis et al. 1987). Measured Al concentrations were converted to Fe concentrations according to the average continental crust composition. The negative correlation of $CO_2$ and Fe supports the 'iron hypothesis' (see text).

chelators are abundant in open ocean regions of the eastern Indian Ocean and the western North Pacific Ocean where they control the dissolved iron concentration. Future studies should clarify the importance of natural organic chelators for the bioavailability of iron for phytoplankton and thus for the PP.

## 7.4   The Early Diagenesis of Iron in Sediments

The fundamental work by Froelich et al. (1979) established a conceptual model for the organic matter respiration in marine sediments which has been modified, verified and extended in numerous aspects since then (cf. Chaps. 3 and 5). Froelich and colleagues found a succession of electron acceptors used by dissimilatory bacteria according to their energy gain. Consequently, a biogeochemical zonation of the sediment results where $O_2$, $NO_3^-$, bioavailable Mn(IV) and Fe(III) and $SO_4^{2-}$ diminish successively with depth (cf. Chap. 3). Apart from the consumption of electron acceptors a production of reduced species such as $NH_4^+$, $Mn^{2+}$, $Fe^{2+}$, $HS^-$ and $CH_4$ occurs. These components may be reoxidized abiotically under given thermodynamic conditions. As will be shown, these reoxidation reactions can also be microbiologically catalyzed. It is important to note that, for the investigation of iron reactivity, a differentiation of biotic and abiotic reactions is inherently important but often very difficult to achieve. Another considerable question with respect to the iron reactivity in sediments concerns the bioavailable fraction of iron-bearing minerals. So far, it could be shown that ferric iron of iron oxides as well as certain sheet silicates can be used by dissimilatory iron reducing bacteria but their quantities vary significantly. A final discussion of this section will compare different depositional environments with respect to the importance of dissimilatory iron reduction, chemical reduction and the availability of ferric iron.

### 7.4.1   Dissimilatory Iron Reduction

Iron (and manganese) can be reduced enzymatically by various pathways which have been summarized in great detail by Lovley (1991). The following equations will be considered as representative for a series of reactions within each group.

1. Fermentative $Fe^{3+}$-reduction:

$$C_6H_{12}O_6 + 24Fe^{3+} + 12H_2O$$
$$\Rightarrow 6HCO_3^- + 24Fe^{2+} + 30H^+ \qquad (7.1)$$

2. Sulfur-oxidizing $Fe^{3+}$-reduction:

$$S^0 + 6Fe^{3+} + 4H_2O$$
$$\Rightarrow HSO_4^- + 6Fe^{2+} + 7H^+ \qquad (7.2)$$

3. Hydrogen-oxidizing $Fe^{3+}$-reduction:

$$H_2 + 2Fe^{3+} \Rightarrow 2H^+ + 2Fe^{2+} \qquad (7.3)$$

4. Organic-acid-oxidizing $Fe^{3+}$-reduction:

$$acetate^- + 8Fe^{3+} + 4H_2O$$
$$\Rightarrow 2HCO_3^- + 8Fe^{2+} + 9H^+ \qquad (7.4)$$

5. Aromatic-compound-oxidizing $Fe^{3+}$-reduction:

$$phenol + 28Fe^{3+} + H_2O$$
$$\Rightarrow 6HCO_3^- + 28Fe^{2+} + 34H^+ \qquad (7.5)$$

For the case of fermentative Fe(III)-reduction it is important to note that during fermentation Fe(III) only serves as a minor sink for electrons and that organic substrates are primarily transformed to organic acids or alcohols.

A review of organisms reducing Fe(III), the respective electron donors and the applied forms of Fe(III) is given by Lovley (1987) and Lovley et al. (1997). In Figure 7.8, a typical result of ferric iron reduction mediated by a distinct ferric iron-reducing bacteria (i.e. *Geobacter metallireducens*) and an appropriate electron donor (i.e.

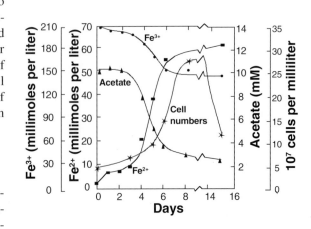

**Fig. 7.8**  Iron reduction by GS-15 with acetate as electron donor (redrawn after Lovley and Philips 1988).

acetate) is shown. Note that the reproduction (cell numbers) ceases even though acetate and ferric iron are still present. This is a general observation which is explained by the limited bioavailability of the ferric iron fraction (Sect. 7.4.2.2). The bacteria need direct contact with solid phase ferric iron (Munch and Ottow 1982). Figure 7.9 represents a model for the degradation of natural organic matter with ferric iron as the sole electron acceptor (Lovley 1997). The primary organic matter is hydrolyzed to smaller compounds such as sugars, amino and fatty acids, and aromatic compounds through the activity of hydrolytic enzymes which are produced by a variety of microorganisms. Subsequently, these intermediate products can be used for the dissimilatory ferric iron reduction.

### 7.4.2 Solid Phase Ferric Iron and its Bioavailability

Solid phase ferric iron is present in various iron bearing minerals and amorphous phases of marine sediments. Two major groups can be distinguished: Iron oxides (including iron oxyhydroxide) and (sheet) silicates. As the iron bearing mineral assemblage varies considerably among different depositional environments the microbiological availability of specific Fe(III)-bearing compounds has been assessed for a long time.

#### 7.4.2.1 Properties of Iron Oxides

During terrestrial weathering a minor amount of the released iron from silicates (biotite, pyroxene, amphibole, olivine) is incorporated into clay minerals and a major fraction serves for the formation of iron oxides. Among iron oxides goethite ($\alpha$-FeOOH) and hematite ($\alpha$-$Fe_2O_3$) are the most abundant and are mostly associated with each other. Lepidocrocite ($\gamma$-FeOOH), maghemite ($\gamma$-$Fe_2O_3$) and magnetite ($Fe^{2+}Fe_2^{3+}O_4$) are generally less abundant. However, magnetite and maghemite are of great importance with regard to the magnetization of the sediment due to their ferrimagnetic character (cf. Chap. 2). Iron phases precipitating from solution are collectively called iron oxyhydroxide. By means of mineralogical identification methods (X-ray diffraction, infrared and Mössbauer spectroscopy), the precipitate may be completely amorphous representing a ferric gel or it may be a poorly crystallized, water contain-

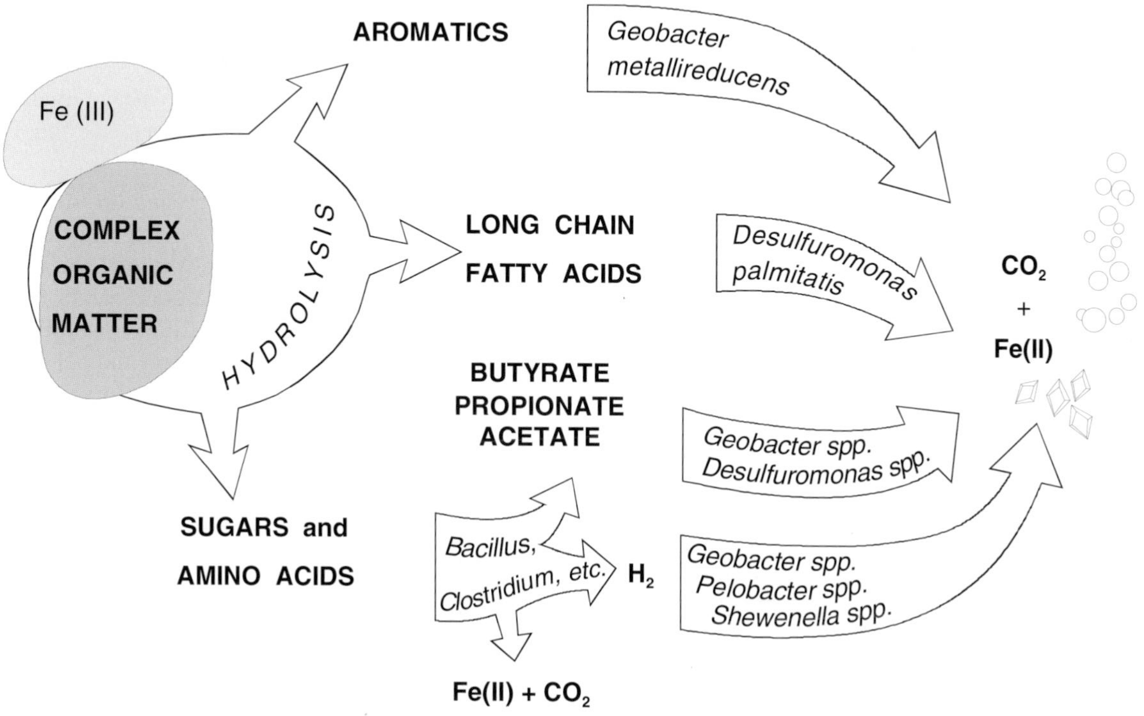


**Fig. 7.9** Schematic model for the oxidation of complex organic matter with ferric iron as sole electron acceptor (redrawn after Lovley 1997).

ing phase such as ferrihydrite ($5Fe_2O_3 \cdot 9H_2O$). Formerly, this fraction was called 'amorphous Fe(III)hydroxide' (Böhm 1925). Aging causes these earliest precipitates to increase their crystallinity which means an increase in the ordering of the crystal lattice. Depending on the pH-value of ambient water, the relative proportion of goethite to hematite varies. Hematite is favored under seawater conditions (Schwertmann and Murad 1983, Fig. 7.10). The effect of aging also explains the observation of a decrease in the easy reducible Fe-oxide fraction with water depth and distance to the coast, although the total fraction of Fe-oxides and total iron concentration increases (Haese et al. submitted). With respect to the adsorption of anions and cations as well as to organic ligand formation (see below), it is important to know an approximate dimension of the specific surface area of iron oxides/oxyhydroxide. Crosby et al. (1983) determined specific surface areas for iron oxyhydroxide synthesized under natural conditions to be 159-234 $m^2g^{-1}$ for precipitates from $Fe^{3+}$ solutions and 97-120 $m^2g^{-1}$ for precipitates from $Fe^{2+}$ solutions. Natural samples showed a range from 6.4-164 $m^2g^{-1}$.

Physical, chemical and mineralogical properties of the iron oxides can be variable. Aluminium (Al(III)) may substitute isomorphically for Fe(III) due to very similar ionic radii (Fe(III): 0.73 Å; Al(III): 0.61 Å) and the same valence. Norish and Taylor (1961) found a maximum of one-third mole percent Al substitution for Fe within goethite and a maximum of one-sixth substitution was determined for hematite in soils (Schwertmann et al. 1979). As a product of terrestrial weathering magnetites usually contain minor amounts of Ti(IV) (ionic radius: 0.69 Å) as a substitution for Fe which is then call titano-magnetite. In contrast, bio-mineralized magnetite (cf. Chap. 2.2) is a pure iron oxide. Different colors of iron oxides/oxyhydroxide are immediately apparent in nature. This holds true for different pure iron oxides, for different grain sizes of one oxide (e.g. goethite) as well as for distinct substitutions for Fe e.g. by Mn or Cr (Schwertmann and Cornell 1991). The color of synthetic minerals and natural sediment can be quantitatively determined by reflectance spectroscopy (e.g. Morris et al. 1985).

In the presence of water, the surface of an iron oxide is completely hydroxylated. This can be understood as a two-step reaction which is shown schematically in Figure 7.11.

As iron oxides/oxyhydroxide have a very high affinity for the adsorption of anions as well as cations under natural conditions, the early diagenetic reactivity of iron is often of great significance for the behavior of compounds such as trace metals, phosphate and organic acids. The adsorption on iron oxides is caused by the hydroxylation of the mineral surface (S-OH). Depending on the pH of ambient water, protonation or deprotonation occurs according to the equations

$$S\text{-}OH + H^+ \quad \Leftrightarrow \quad S\text{-}OH_2^+ \tag{7.6}$$

$$S\text{-}OH + OH^- \Leftrightarrow \quad S\text{-}O^- + H_2O \tag{7.7}$$

As a result, the surface charge and the surface potential vary depending on the concentration of $H^+$ ions in solution. Apart from the pH, the surface charge is influenced by the concentration of the electrolyte and the valence of ions in solution. Where the net surface charge is zero, the pH cor-

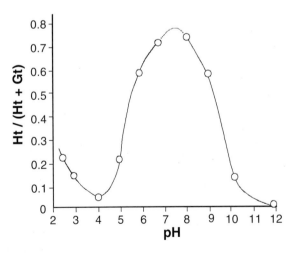

**Fig. 7.10** Recrystallization products of a ferrihydrite suspension after 441 days. Aging causes a formation of hematite (Ht) and goethite (Gt) depending on the pH of ambient water (redrawn after Schwertmann and Murad 1983).

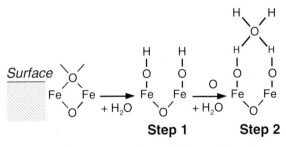

**Fig. 7.11** In the presence of water the iron oxide surface is hydroxylated (step 1) and subsequently $H_2O$ is adsorbed (step 2) (redrawn from Schwertmann and Taylor 1989).

responds to the zero pointy charge ($pH_{zpc}$). For pure synthetic iron oxides, these vary between 7 and 9. An excess of positive or negative charge is balanced by the equivalent amount of anions ($A^-$) or cations ($C^+$). As representative for the adsorption of cations (e.g. $Cu^{2+}$), anions (e.g. $H_2PO_4^-$) and organic compounds (e.g. oxalate) the following reactions are given:

$$\begin{array}{l} \text{-S-OH} \\ | \qquad + Cu^{2+} \\ \text{-S-OH} \end{array} \Leftrightarrow \begin{array}{l} \text{-S-O} \\ | \quad Cu \quad + 2H^+ \\ \text{-S-O} \end{array} \qquad (7.8)$$

$$\begin{array}{l} \text{-S-OH} \\ | \qquad + H_2PO_4^- \\ \text{-S-OH} \end{array} \Leftrightarrow \begin{array}{l} \text{-S-O} \quad O^- \\ | \quad P \quad + 2H_2O \\ \text{-S-O} \quad O \end{array} \qquad (7.9)$$

$$\text{-S-OH} + H_2PO_4^- \Leftrightarrow \text{-S} \begin{array}{l} O\text{-}C=O^- \\ | \quad + H_2O \\ O\text{-}C=O \end{array} \qquad (7.10)$$

As the complexation of cations causes a release of $H^+$ and anions compete with surface bound hydroxyl groups ('ligand exchange'), the adsorption is strongly pH-dependent (see above). As a result, anions are preferably adsorbed at lower pH-values whereas cations are primarily adsorbed at higher pH values (Fig. 7.12).

A maximum of 2.5 - 2.8 µmol $m^{-2}$ of adsorbed phosphate on iron oxides were found (Goldberg and Sposito 1984, Pena and Torrent 1984). In or-

der to approximate a maximum adsorbed phosphate concentration in the sediment, one can assume 1 $cm^3$ of sediment with a porosity of 75%, a dry weight density of 2,65 g $cm^{-3}$, 50 µmol $g_{Sed}^{-1}$ Fe bound to iron oxides and an iron oxide specific surface area of 120 $m^2g^{-1}$ Fe (see this section above). For the wet sediment, we can calculate a Fe concentration of 33 µmol $cm^{-3}$ which is bound to iron oxides. This fraction has a specific surface area of ~0,22 $m^2$ within 1 $cm^3$ of wet sediment which may then adsorb up to ~0,57 µmol P (assuming an adsorption capacity of 2.6 µmol P per square meter of Fe oxide) . For a comparison of adsorbed and dissolved phosphate concentration we can furthermore assume a concentration of 5 µM phosphate within the interstitial water which is equivalent to 0.0037 µmol $cm^{-3}$ of wet sediment. Consequently, the adsorbed fraction of phosphate can be more than two orders of magnitude greater than the dissolved fraction due to the presence of iron oxides. In reality, the ratio of Fe bound to poorly crystalline Fe oxides and P bound by these phases has been determined to be ~10Fe : 1P for coastal and shelf sediments (Slomp et al. 1996a). This differs significantly from the given theoretical sample in which we find a ratio of 58 (33 µmol Fe: 0,57 µmol(P) $cm^{-3}$). Either P is additionally bound in the crystalline lattice of the iron oxides (e.g. Torrent et al. 1992) or the adsorption capacity for P in shelf sediments is much greater than derived from the calculated example. In the latter case, one must conclude that the specific surface area of Fe oxides is higher than assumed as another quantitatively important adsorbent of P in sediments is not likely. Instead of 120 $m^2g^{-1}$ Fe oxide, a specific surface area of 650-700 $m^2g^{-1}$ is required to justify such high P adsorption.

### 7.4.2.2  Bioavailability of Iron Oxides

As dissimilation is a biological process used to gain electrons for cell energetic functions, the energy gain from the induced reactions is noteworthy. The following reactions and their respective $\Delta G^o$ values are given to provide an overview of the energy gain due to the reaction with various iron oxyhyxdroxide/oxides.

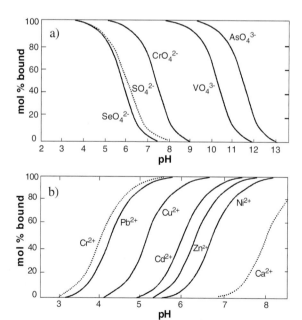

**Fig. 7.12** The pH dependence of anion (a) and cation (b) sorption on hydrous ferric oxide (redrawn after Stumm and Morgan 1996)

$$4Fe(OH)_3 + CH_2O + 7H^+$$
$$\Rightarrow 4Fe^{2+} + HCO_3^- + 10H_2O \qquad (7.11)$$
$$\Delta G^0: (-376) - (-228) \text{ kJ mol}^{-1}$$

$$4FeOOH + CH_2O + 7H^+$$
$$\Rightarrow 4Fe^{2+} + HCO_3^- + 6H_2O \qquad (7.12)$$

$\Delta G^0$: $(-387) - (-239)$ kJ mol$^{-1}$

$$2\alpha\text{-}Fe_2O_3 + CH_2O + 7H^+$$
$$\Rightarrow 4Fe^{2+} + HCO_3^- + 4H_2O \qquad (7.13)$$

$\Delta G^0$: $-236$ kJ mol$^{-1}$

$$2Fe_3O_4 + CH_2O + 11H^+$$
$$\Rightarrow 6Fe^{2+} + HCO_3^- + 6H_2O \qquad (7.14)$$

$\Delta G^0$: $-328$ kJ mol$^{-1}$

$\Delta G^0$ values were adopted from Stumm and Morgan (1996). For Fe(OH)$_3$ and FeOOH a range of free energy is given as the solubility products $K_s = [Fe^{3+}] [OH^-]^3$ range from $10^{-37.3}$ to $10^{-43.7}$ depending on the mode of preparation, age and molar surface. For example, aged goethite ($\alpha$-FeOOH) reveals a $\Delta G^0$ of $-489$ kJ mol$^{-1}$ whereas for freshly precipitated amorphous FeOOH a value of $-462$ kJ mol$^{-1}$ was determined (Stumm and Morgan 1996). The variability of FeOOH thermodynamic properties is a result of the metastability of freshly precipitated ferric iron phases (Sect. 7.4.2.1, Fig. 7.10).

By means of microbial cultures growing on the various iron oxides as terminal electron acceptors, it has been shown that they all can be used for the purpose of dissimilation (e.g. Lovley 1991, Kostka and Nealson 1996). However, it was generally found that the ferric iron phases were reduced at different rates and to varying degrees. Consequently, Ottow (1969) determined the following sequence of biological availability: FePO$_4$·4H$_2$O > Fe(OH)$_3$ > lepidocrocite ($\gamma$-FeOOH) > goethite ($\alpha$-FeOOH) > hematite (Fe$_2$O$_3$). Within natural soil samples a preferential reduction of amorphous to crystalline iron oxides was found (Munch and Ottow 1980). Most recently, Roden and Zachara (1996) discovered the importance of the solid phase specific surface area for the degree of reducibility. The greater the specific surface area, the more reduced ferric iron was determined (Fig. 7.13). Additionally, they could show that, after rinsing the solid phase, which could not be reduced any further, dissimilatory reduction of the original ferric iron continued. They concluded that the reduction of iron oxides is limited by the adsorption of some components, possibly Fe$^{2+}$, which inhibits further microbial access. This find-

ing also explains the results of Munch and Ottow (1980; see above) because amorphous and poorly crystalline phases are usually characterized by a higher specific surface area than well crystallized phases, and the greater the specific surface area the more can be adsorbed.

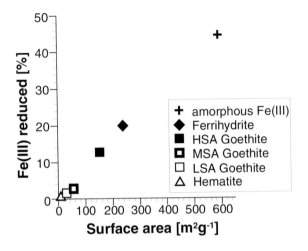

**Fig. 7.13** The dependence of solid phase ferric iron reducibility from the specific surface area of iron oxyhydroxide / oxides (redrawn from Roden and Zachara 1996). HSA, MSA, and LSA represent high, medium, and low specific surface area.

### 7.4.2.3  The Bioavailability of Sheet Silicate Bound Ferric Iron

As early as in 1972 Ruth and Tullock published results on the chemical reduction of smectites. Subsequently, Rozenson and Heller-Kallai (1976a,b) studied the potential of reduction and reoxidation in various dioctahedral smectites with the aid of different reducing and oxidizing agents. Concurrent with the change of the intercrystalline redox state, a change of smectite color was observed. Oxidized smectites were white or yellowish whereas the reduced smectites were greenish-grey or black. To balance the intercrystalline charge, (de-) protonation was postulated. Additionally, as a consequence of smectite reduction, a decrease of the specific surface area and the swellability in water, as well as an increase of nonexchangeable Na$^+$ was found (Lear and Stucki 1989). The potential importance of smectite redox reactivity in natural sediments was first pointed out by Lyle (1983) for sediments of the eastern equatorial Pacific where a distinctive color transition from tan (above) and green-gray (below) was found in near-surface sediments. König et al. (1997) conducted a high resolution Mössbauer-spectroscopy study for one core from the Peru Basin which revealed a present-day reduction of 42% of total iron which can be differentiated into an immobile fraction (36%) consisting of smectite bound iron and a mobile fraction (16%) which diffuses back into the oxidized upper sediment layer. Evidence for the bioavailability of ferric iron bound to smectites was given by Kostka et al. (1996). During culturing of an iron reducing bacteria (*Shewanella putrefaciens* strain MR-1) on smectite as a sole electron acceptor a reduction of $Fe^{3+}$ by 15% within the first 4 hours and a total of 33% after two weeks occurred.

### 7.4.3    Iron and Manganese Redox Cycles

Processes of early diagenesis can only be understood by integrating microbiological induced/ chemical reactions and modes of transport in the sediment. With respect to the quantification of iron and manganese reactions, molecular diffusion and bioirrigation need to be considered for the dissolved phase whereas bioturbation and advection are relevant for the particulate transport (cf. Chap. 3). Bioirrigation is the term for solute exchange between the bottom water and tubes in which macro-benthic organisms actively pump water. For iron and manganese a recent study (Hüttel et al. 1998) points out the importance of solute transport in the sediment and across the sediment/bottom water interface due to pressure gradients induced by water flow over a rough sediment topography. Advection in the context of particulate transport describes the downward

transport of particles relative to the sediment surface due to sedimentation. Strictly speaking, bioturbation (sometimes more generally termed mixing) also induces a vertical transport of dissolved phase. However, since molecular diffusive transport is usually much greater than dissolved transport by bioturbation, the latter is usually neglected.

The cycling of reduced and oxidized iron and manganese species are discussed together in this chapter since the driving processes are principally the same. In Figure 7.14 the operating modes of transport are shown schematically along with a redox boundary. Above this boundary the reactive fraction of total solid phase iron or manganese is present as oxidized species whereas below the reduced species occur. Note that at this boundary a build-up in the pore water occurs if no immediate precipitation (e.g. FeS) or adsorption inhibits a release into ambient water. The change in the redox state implies oxidation above and reduction below by some electron donor/acceptor. In case of dissimilatory iron/manganese reduction the organic carbon serves as electron donor, the other most important oxidants and reductants will be discussed in the Sections 7.4.4.1 and 7.4.4.2.

The intensity of the redox cycling and thus the importance for oxidation and reduction reactions in the sediment is terminated by either one of the following conditions: 1. Where no efficient oxidant (e.g. $O_2$) is present in the upper-most layer or bottom water, no oxidation will occur and the redox cycling cannot be maintained. 2. Where no reactive fraction (bioavailable or 'rapidly' reducible by HS⁻; Sect. 7.4.4.1) is present in the lower layer, no reduction will occur and the redox cycle

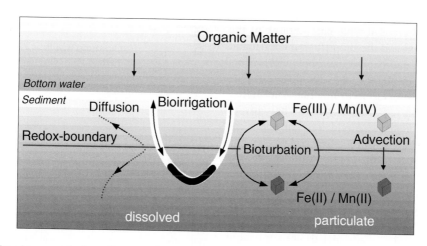

**Fig. 7.14**   Modes of transport in the sediment: molecular diffusion, bioirrigation, bioturbation, and advection.

will cease. 3. A vertical transport mode must be maintained between the zone of oxidation and the zone of reduction. Since advection is usually very much slower than the downward transport by bioturbation the intensity of bioturbation, terminates the transport between the redox-zones.

For the most simple assumption of an homogeneously mixed layer, the intensity of bioturbation is expressed by the biodiffusion (or mixing) coefficient, $D_b$, which can be deduced appropriately along with the sedimentation rate with the aid of natural radioactive isotopes. According to Nittrouer et al. (1983/1984), the general advection-diffusion equation can be rearranged to calculate the sedimentation rate (SR):

$$SR = \frac{\lambda x}{\ln(C_0/C_x)} - \frac{D_b}{x} \cdot \ln(C_0/C_x) \qquad (7.15)$$

with $\lambda$: decay constant $[y^{-1}]$, x: depth interval between two levels [cm], $C_0$, $C_x$: activity at an upper sediment level and at a lower level with the distance x below $C_0$ [decays per minute, dpm] $D_b$: biodiffusion coefficient $[cm^2 y^{-1}]$. If mixing is negligible ($D_b = 0$), then the above equation can be simplified:

$$SR = \frac{\lambda x}{\ln(C_0/C_x)} \qquad (7.16)$$

In case of a very low sedimentation rate relative to mixing ($SR^2 \ll \lambda \cdot D_b$) Equation 7.15 can be rearranged to calculate the biodiffusion coefficient, $D_b$:

$$D_b = \lambda \left( \frac{x}{\ln C_0/C_x} \right)^2 \qquad (7.17)$$

The above restrictions for the calculations of the sedimentation rate and the biodiffusion coefficient imply the use of radioactive isotopes with different half-lifes ($t_{1/2} = 0.693 \cdot \lambda^{-1}$) for different purposes and depositional environments. The higher the sedimentation rate, the shorter the half-life of the radioactive isotope should be. The more intense the bioturbation in the surface layer is, the shorter the half-life of the applied radioactive isotope should be. For coastal and shelf sediments sedimentation rates of several decimeters to few meters per 1000 years are typical and can be determined by $^{210}Pb$ ($t_{1/2} = 22.3$ yr). Shorter lived isotopes (e.g. $t_{1/2}$ of $^{234}Th = 24.1$ d) are applicable for the determination of the mixing intensity.

$^{230}Th$ ($t_{1/2} = 75,200$ yr) is a commonly used radioactive isotope in oceanographic sciences to trace processes over longer periods of times. The above isotopes are rapidly scavenged by particles once they are formed from the decay of some parent isotopes and settle to the sea floor. Limited by the analytical methods, post-depositional processes can be traced for a time period of 4 to 5 times of the radioactive half-life which is approximately 100 years in case of $^{210}Pb$.

In Figure 7.15, a typical $^{210}Pb$-activity depth profile from the Washington shelf is shown. The uppermost 9 cm are characterized by constant $^{210}Pb$ activity implying intensive mixing in the surface layer. Below, $^{210}Pb$ activity decreases linearly (on a log-scale) indicating no mixing and continuous decay. The lowest part of the profile is characterized by constantly very low values resulting from the long-term decay of $^{226}Ra$ to $^{210}Pb$ in the sediment. This background value is subtracted from the above activities which are then termed excess-$^{210}Pb$ or unsupported-$^{210}Pb$. The depth interval showing a linear decrease on a log-activity scale is often used to calculate a sedimen-

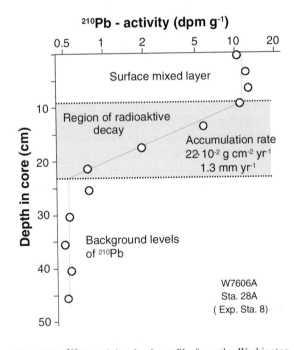

**Fig. 7.15** $^{210}Pb$-activity depth profile from the Washington shelf (redrawn after Nittrouer 1983/1984). The sediment surface layer is mixed as can be deduced from homogenous $^{210}Pb$ values. Below, constantly decreasing values imply no or hardly any mixing. This gradient can be used to calculate a sedimentation rate. The deepest part is characterized by a homogenous background activity resulting from the decay of $^{226}Ra$ in the sediment.

tation rate. However, a strong overestimation of the true sedimentation rate is possible as slight deep bioturbation in this part may not be seen exclusively by $^{210}Pb$ as was shown by Aller and DeMaster (1984). The investigation of an additional, shorter-lived isotope within this depth interval will reveal a potential influence of bioturbation.

As mentioned in the beginning of this section, bioturbation and advection by sedimentation cause particle transport in the sediment. In Figure 7.16, three scenarios are schematically shown representing constant molecular diffusive and advective transport while bioturbation varies. As a result, the shape of the solid phase profile varies distinctively. In case of no bioturbation, the molecular diffusive flux ($J_{diff.}$) from the zone of dissolution into the zone of precipitation causes a thin, sharp peak (enrichment) (Fig. 7.16a) whereas slight bioturbation and thus vertical up- and down-transport of particles ($J_p$) broadens the enrichment (Fig. 7.16b). In case of very intense particle transport relative to the molecular diffusive transport ($J_{p.} \gg J_{diff.}$, Fig. 7.16c) hardly any or no enrichment will be formed although a distinctive depth of precipitation is still present. In summary, we can conclude that the solid phase profile is a result of the dissolution within the lower part of

the enrichment, as well as of the sedimentation rate and of the bioturbation (mixing) intensity. This can be expressed mathematically by a one-dimensional transport-reaction model according to Aller (1980). If constant bioturbation and sedimentation with depth (no compaction), as well as steady-state conditions (see Chap. 3) can be assumed and solid phase decreases linearly over the interval of dissolution, then

$$P = \frac{(C_1 - C_2)}{(x_2 - x_1)} \cdot D_b + SR \cdot (C_1 - C_2) \qquad (7.18)$$

can be calculated according to Sundby and Silverberg (1985). P resembles the production (or dissolution) rate and the other variables are according to Eqation 7.15 except that C is given as concentration per volume [$\mu mol\ cm^{-3}$] as the depth distribution of solid phase strongly depends on the porosity.

In addition to the dissolution rate, one can calculate the burial rate of unreactive phase and the input rate to the sediment surface once the sedimentation rate is known. Based on these independently calculated fluxes Sundby and Silverberg (1985) developed a depth-zonated flux model for manganese in the St. Lawrence estuary. Their depth-dependent reactive zones were surface wa-

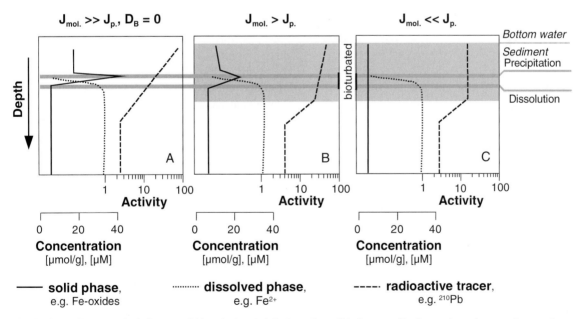

**Fig. 7.16** To illustrate the influence of bioturbation (mixing) on the solid phase profile three schematic scenarios are drawn. For all scenarios the same molecular diffusive transport ($J_{mol.}$) and sedimentation is assumed, yet the particulate transport ($J_{p.}$) by bioturbation is varied. A: As no bioturbation occurs a distinctive, thin solid phase enrichment is formed in the zone of precipitation. B: The enrichment broadens up- and downwards as slight bioturbation is present. C: In case of a much higher particulate transport relative to the diffusive transport ($J_p \gg J_{mol.}$) hardly any enrichment will be formed.

ter, bottom water, sediment depth of precipitation (oxidation), sediment depth of dissolution (reduction) and depth of eventually buried sediment. One example of their Mn-cycling results is given in Figure 7.17.

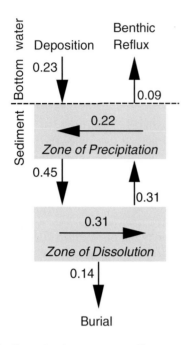

**Fig. 7.17** Example of manganese cycling across the sediment/bottom water interface and within the sediment (redrawn after Sundby and Silverberg 1985). The applied depth-dependent flux model is described in the text. Depositional, burial, and molecular diffusive fluxes as well as the reduction rate within the zone of dissolution were calculated independently.

The cycling of elements in bioturbated surface sediments can also be expressed in terms of turnover times defined as period of time required for a complete oxidation - reduction cycle of the reactive fraction. Additional consideration of the bio-

turbation depth and the sedimentation rate then reveals the number of redox-cycles before ultimate burial. In Table 7.2, representative results for estuarine (coastal) and slope sediments are given.

### 7.4.4    Iron Reactivity towards S, $O_2$, Mn, $NO_3^-$, P, $HCO_3^-$, and Si-Al

In Section 7.4.3, the importance of iron and manganese transport in surface sediments is shown. A continuous transport is only maintained if oxidation and reduction at different depth levels is permanently occurring. In this section, the reactions with major oxidants and reductants will be introduced. The (microbial) dissimilatory iron reduction was shown already in Section 7.4.1. Additionally, the interactions of iron and phosphorus, as well as the formation of siderite and iron-bearing sheet silicates will be pointed out briefly to show the variety of reactions in marine sediments coupled to the reactivity of iron.

### 7.4.4.1   Iron Reduction by HS- and Ligands

Apart from the dissimilatory iron reduction (Sect. 7.4.1), iron oxides can be dissolved by protons, ligands and reductants. Dissolution reactions by protons and ligands are generally considered to be the rate-determining step for weathering processes. For iron-bearing minerals in marine sediments, proton-promoted dissolution is of no importance due to prevailing neutral or slightly alkaline conditions. Ligands (e.g. oxalates and citric acid) are by-products of biological decomposition and dissolve iron oxides by primary surface complexation onto the iron oxide surface resulting in a weakening of the Fe-O bond which is followed by a detachment of $Fe^{3+}$-ligand. Reductive dissolution (e.g. by HS-, ascorbate, and dithionite)

**Table 7.2**   Calculated turn-over times and times of redox cycling before burial in coastal and slope sediments. The dynamic of redox cycling becomes evident by envisaging a complete oxidation - reduction cycle on a 2 - 6 months time scale. ([1] Sundby and Silverberg 1985, [2] Aller 1980, [3] Canfield et al. 1993a, [4] Thamdrup and Canfield 1996)

| Location | Fe/Mn | Turn-over time [d] | Times cycled before burial |
|---|---|---|---|
| St. Lawrence estuary [1] | Mn | 43 - 207 | |
| Long Island Sound estuary [2] | Mn | 60 - 100 | |
| Skagerrak [3] | Fe/Mn | 70 - 250 | 130 - 300 |
| Slope off Chile [4] | Fe | 70 | 31 - 77 |

is characterized by primary surface complexation followed by an electron transfer from the reductant to ferric iron and detachment of $Fe^{2+}$. The three pathways of Fe(III) (hydr)oxide dissolution are shown in Figure 7.18.

Experimental determinations of reduction rates (e.g. Pyzik and Sommer 1981, Dos Santos Afonso and Stumm 1992, Peiffer et al. 1992) reveal rates under well defined conditions and information on the reaction kinetics. Under natural conditions, the composition of dissolved compounds in solution is complex and may have significant influence on the dissolution rate. Biber et al. (1994) demonstrated the inhibition of reductive dissolution by $H_2S$ and the ligand-promoted dissolution by EDTA due to the presence of oxoanions (e.g. phosphate, borate, arsenate) (Fig. 7.19 a,b, cf. Chap. 5).

In a combined laboratory and field study Canfield et al. (1992) established a sequence of mineral reactivity towards sulfide. In Table 7.3

the reactivity towards sulfide is expressed as rate constant ($\lambda$) and the respective half-life ($t_{1/2}$).

The presence of iron minerals and their respective reactivity towards sulfide is of greatest importance for the pore water chemistry and the limitation for pyrite formation (see Chap. 8). In case of reactive iron rich sediments dissolved iron may build-up in pore water and dissolved sulfide is hardly present although sulfate reduction occurs (cf. Chap. 8). In contrast, dissolved sulfide can build-up instead of dissolved iron in sediments characterized by a low content of reactive iron (Canfield 1989). The degree of pyritisation (DOP) was originally defined by Berner (1970) as

$$DOP = \frac{Fe_{Pyrite}}{Fe_{Pyrite} + Fe_{acid-soluble}} \quad (7.19)$$

and has been modified since then with respect to the determination of the acid-soluble fraction

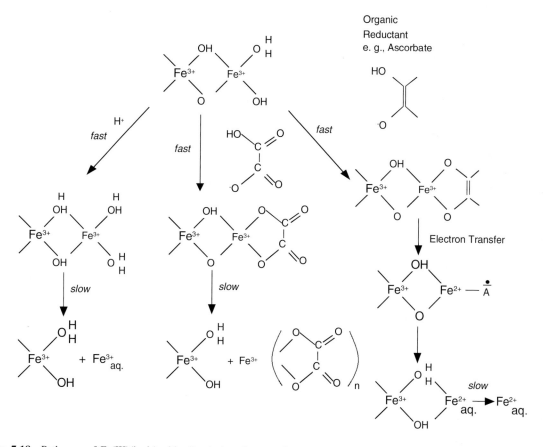

**Fig. 7.18** Pathways of Fe(III)(hydr)oxide dissolution. Proton-, ligand- (here: oxalate), and reductant- (here: ascorbate) promoted dissolution is initiated by surface complexation. The subsequent step of detachment ($Fe^{3+}_{aq.}$, $Fe^{3+}$-ligand, $Fe^{2+}_{aq.}$) is rate determining. Note that the shown pathways of dissolution are fundamental for the described extractions (Sect. 7.5) (redrawn after Stumm and Morgan 1996).

a)                                          b)

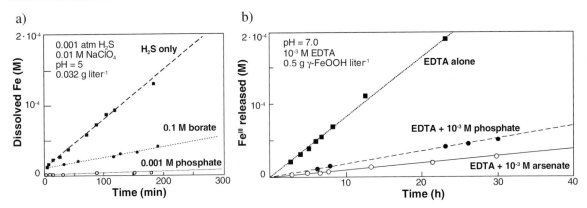

**Fig. 7.19** Inhibition of reductive- and ligand-promoted dissolution of iron oxides by oxoanions. a: The dissolution of goethite by $H_2S$ is inhibited by borate and phosphate. b: The dissolution of lepidocrocite by EDTA is inhibited by phosphate and arsenate (redrawn after Stumm and Morgan 1996; original data from Biber et al. 1994).

(Leventhal and Taylor 1990, Raiswell et al. 1994). A DOP-value of 1 means a complete pyritisation of reactive iron, which has been found in sediments overlain by anoxic-sulfidic bottom water (Raiswell et al. 1988). In contrast, it was shown for sediments, which have been exposed to sulfide for more than one million years (Raiswell and Canfield 1996) that a fraction of poorly reactive iron is not yet completely turned into pyrite. This clearly states the range of silicate iron reactivity towards sulfide, which is influenced by the mineral assemblage, degree of crystallinity and grain size.

### 7.4.4.2   Iron Oxidation by $O_2$, $NO_3^-$, and $Mn^{4+}$

The reaction with oxygen is known to be fast and its rate was determined in seawater by Millero et al. (1987):

$$\frac{d[Fe(II)]}{dt} = \frac{k_H[O_{2aq.}]}{[H^+]^2}\cdot[Fe(II)] \qquad (7.20)$$

at 20°C, $k_H = 3\cdot10^{-12}$ mol min$^{-1}$liter$^{-1}$. For a temperature of 5°C, the rate decreases by about a factor of 10. As the oxidation rate (-d[Fe]/dt) is inversely correlated with the power of the proton concentration ($[H^+]^2$) the importance of the pH becomes obvious. The lower the pH, the lower is the rate of ferrous iron oxidation. Within a ferrous iron solution with a very low pH-value, e.g. acidified with HCl, the reaction is therefore so slow that oxidation under atmosphere is negligible over weeks. Under natural conditions, this reaction is so fast that dissolved iron may only escape from the sediment into the bottom water if

the oxygen penetration depth is very little or even anoxic bottom water conditions are given (cf. Chap. 6). The effects on iron, manganese, phosphate and cobalt fluxes during a controlled decrease of oxygen bottom water concentration and the importance of the diffusive boundary layer within a benthic flux-chamber (see Chap. 3) were studied by Sundby et al. (1986). They demonstrated that manganese release increased prior to iron according to thermodynamic predictions as a result of a decrease of diffusive oxygen flux into the sediment (Balzer 1982). Stirring within a flux-chamber controls the thickness of the benthic boundary layer and thus the diffusive flux of oxy-

**Table 7.3**   Reactivity of iron minerals towards sulfide according to [1] Canfield et al. (1992) and [2] Raiswell and Canfield (1996). The 'poorly-reactive silicate fraction' was determined operationally as: $(Fe_{HCl, boiling} - Fe_{Dithionite}) / Fe_{total}$

| Iron Mineral / fraction | Rate constant $\lambda[yr^{-1}]$ | Half life $t_{1/2}$ |
|---|---|---|
| Ferrihydrite [1] | 2,200 | 2.8 hours |
| Lepidocrocite [1] | > 85 | < 3 days |
| Goethite [1] | 22 | 11.5 days |
| Hematite [1] | 12 | 31 days |
| Magnetite [1] | $6.6\cdot10^{-3}$ | 105 years |
| Sheet silicates [1] | $1.0\cdot10^{-5}$ | 10,000 years |
| poorly-reactive Silicate fraction [2] | $0.29\cdot10^{-3}$ | $2.4\cdot10^{6}$ years |

gen into the sediment. A decrease or even an interruption of stirring results in a significant increase of benthic efflux of redox-sensitive constituents such as iron and manganese.

Buresh and Moraghan (1976) showed the thermodynamic potential of ferrous iron oxidation by nitrate, yet the reaction is not spontaneous. In the presence of solid phase Cu(II), Ag(I), Cd(II), Ni(II), and Hg(II) serving as catalysts, ferrous iron can reduce nitrate rapidly (Ottley et al. 1997). Similarly, the formation of a Fe(II)-lepidocrocite (γ-FeOOH) surface complex was found to catalyze chemodenitrification (Sørensen and Thorling 1991). The potential significance of microorganisms inducing ferrous iron oxidation was pointed out within the last years (Widdel et al. 1993, Straub et al. 1996). Ferrous iron was found to serve as electron donor in cultures of nitrate-reducing bacteria (cf. Chap. 6). Even in the presence of acetate as typical electron donor ferrous iron was additionally oxidized. This implies that iron oxide formation typically occurring at the interface of nitrate and iron bearing pore water is at least in parts microbially mediated. Ehrenreich and Widdel (1994) have described a microbial mechanism of pure anaerobic oxidation by iron oxidizing photoautotrophs. This exciting observation challenges the conviction that the earliest iron oxidation on Earth occurred during the build-up of free oxygen. One may now speculate that the accumulation of the Banded Iron Formation (Archaic age, ~3 billions years B.P.) was microbiologically induced under suboxic/anoxic conditions. Similarly, manganese oxidation rates in natural environments were determined to be considerable higher than determined in laboratory studies under abiotic conditions implying a microbially-mediated manganese oxidation (Thamdrup et al. 1994, Wehrli et al. 1995).

Ferrous iron oxidation by manganese oxide was found to be especially fast as long as no iron oxyhydroxide precipitates, which presumably blocks reactive sites on the manganese oxide surface (Postma 1985). The interaction of ferrous iron and manganese oxide was suggested to be important for the interpretation of pore water profiles (Canfield et al. 1993a, Haese et al. submitted). In Figure 7.20 pore water profiles of iron and manganese reveal concurrent liberation of the two elements which was attributed to dissimilatory iron reduction and subsequent iron reoxidation by manganese oxides which in turn results in the production of $Mn^{2+}_{aq}$. Obviously, iron reduction exceeds the rate of reoxidation and consequently $Fe^{2+}_{aq}$ is released, too.

Similar to the possibility of concurrent reduction of sulfate and ferric iron by a culture of a single bacteria (Coleman et al. 1993, Sect. 7.4.4.4), other iron reducing bacteria were found to additionally maintain dissimilation with more than one electron acceptors under suboxic conditions (Lovley and Phillips 1988) or even under oxic

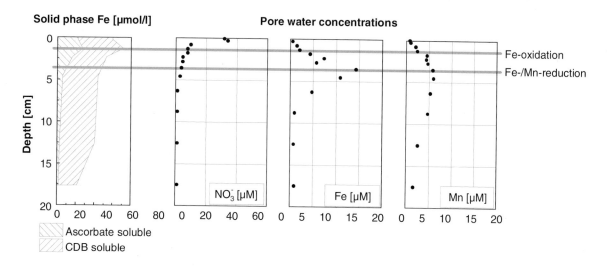

**Fig. 7.20** Pore water and extraction results from hemipelagic sediments off Uruguay (redrawn from Haese et al., sub.). Dissolution and precipitation of Fe is reflected by the easy reducible iron oxyhydroxide fraction whereas less reducible iron oxides soluble by subsequent citrate/dithionite/bicarbonate (CDB) extraction remain constant. A concurrent liberation of Mn and Fe indicates dissimilatory iron reduction and subsequent iron reoxidation by manganese oxides, which results in the build-up of $Mn^{2+}$. Under these conditions the actual dissimilatory iron reduction rate is higher than deduced from iron pore water gradients.

conditions (Myers and Nealson 1988a). In the presence of Fe(III) and Mn(IV), strain MR-1 was found to reduce both but additional manganese reduction occurred due to the immediate abiotic reaction with released $Fe^{2+}$ (Myers and Nealson 1988b). The interactions of biotic and abiotic reactions are shown in Figure 7.21.

### 7.4.4.3  Iron-Bound Phosphorus

In Section 7.4.2.1 the theoretical significance of phosphate adsorption onto iron oxides was illustrated. Numerous field studies suggest that iron oxides control phosphate pore water and solid phase concentrations and drive the sedimentary phosphate cycle (Krom and Berner 1980, Froelich et al. 1982, Sundby et al. 1992, Jensen et al. 1995, Slomp et al. 1996a,b). A generalized representation of major phosphorous cycles in the water column and in the sediment is shown in Figure 7.22. Apart from the Fe-bound P, organic P and carbonate fluorapatite (CFA) representing authigenic P are the principal carriers of solid phase P. $HPO_4^{2-}$ is the predominant dissolved P species under sea water conditions (Kester and Pytkowicz 1967).

In Figure 7.23, typical depth distributions of dissolved and particulate phosphorus and iron concentrations are shown from sediments off southwest Africa (Zabel and Steinmetz sub.). Phosphorus is bound by adsorption (soluble by 0.35 M NaCl) as well as within crystalline iron oxides (soluble by citrate/dithionite/bicarbonate). An active cycling of both constituents is apparent

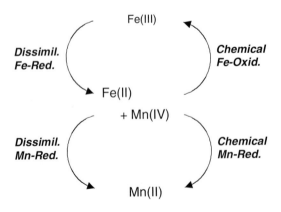

**Fig. 7.21**  Interaction of dissimilatory Fe / Mn reduction and abiotic reaction of $Fe^{2+}$ with Mn(IV). Note that additional interactions with species and microbial processes typically occuring in surface sediments (e.g. sulfate reduction and subsequent reactions of HS-) are not considered and that Mn(IV) is not replenished.

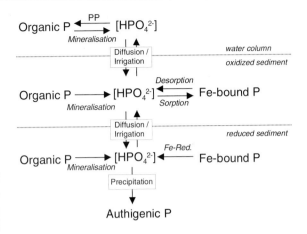

**Fig. 7.22**  Generalized phosphorus cycles in the water column and in the sediment (modified after Slomp et al. 1996b).

due to a release into pore water at the lower part of the solid phase enrichment and transfer into solid phase at the depth of the peak-maximum.

### 7.4.4.4  The Formation of Siderite

In the marine environment, siderite ($FeCO_3$) is hardly found relative to iron sulfides because it is thermodynamically not stable in the presence of even low dissolved sulfide activities. Due to the complex molecular and redox chemistry of sulfur, the interaction pathways of iron and sulfur are still a matter of intensive investigation. To study sulfur geochemistry, the reader is refered to Chapter 8 and for iron sulfur interactions an up-to-date summary is given by Rickard et al. (1995).

Postma (1982) proved the calculation of the solubility equilibrium between siderite and ambient pore water chemistry to be a reliable approach for the investigation of present-day siderite formation. In salt marsh sediments where the influence of salt and fresh water varies temporarily and spatially, the formation of siderite and pyrite are closely interlinked. Recently, Mortimer and Coleman (1997) demonstrated that siderite precipitation is microbiologically induced. They showed that $\delta^{18}O$ values of siderite precipitated during the culturing of one specific iron-reducing microorganism, *Geobacter metallireducens*, were distinctively lower than expected according to equilibrium fractionation between siderite and water (Carothers et al. 1988) which is a contradiction to a pure thermodynamically induced reaction.

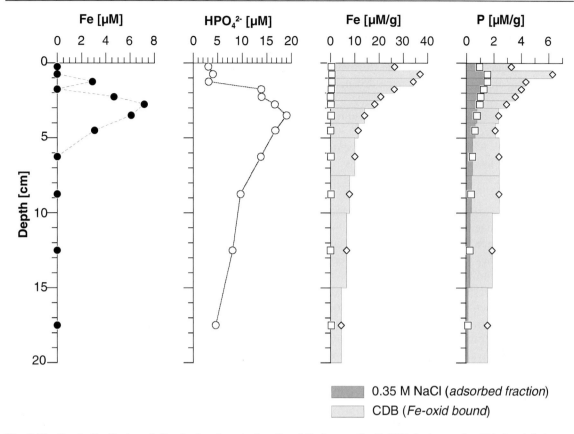

**Fig. 7.23** Depth distribution of dissolved and particulate P and Fe in core GeoB 3702 (redrawn after Zabel and Steinmetz sub.). Dissolved and particulate P depends on the dissolution and precipitation of Fe due to the high adsorption capacity of iron oxide.

In fully marine systems, siderite formation probably occurs below the sulfate reduction zone where dissolved sulfide is absent, if reactive iron is still present and the Fe/Ca-ratio of pore water is high enough to stabilize siderite over calcite (Berner 1971). The coexistence of siderite and pyrite in anoxic marine sediments was shown by Ellwood et al. (1988) and Haese et al. (1997). Both studies attribute this observation to the presence of microenvironments resulting in different characteristic early diagenetic reactions next to each other within the same sediment depth (cf. Sect. 12.2.4). It appears that in one microenvironment sulfate reduction and the formation of pyrite is predominant whereas in another dissimilatory iron reduction and local supersaturation with respect to siderite occurs. The importance of microenvironments has similarly been pointed out for various other processes (Jørgensen 1977, Bell et al. 1987, Canfield 1989, Gingele 1992).

Apart from microenvironments, an explanation for the concurrent dissimilatory sulfate and iron reduction was provided by Postma and

Jakobsen (1996). They demonstrated that the stabilities of iron oxides are decisive with respect to iron and/or sulfate reduction assuming that the fermentative step and not the overall energy yield is overall rate limiting. Additionally, it shall be noted that the typical sulfate reducing bacteria *Desulfovibrio desulfuricans* was found to reduce iron oxide enzymatically contemporarily or optionally (Coleman et al. 1993). When only very small concentrations of $H_2$ as sole electron donor were available iron oxide instead of sulfate was used as electron acceptor by *D. desulfuricans*.

### 7.4.4.5  The Formation of Iron Bearing Aluminosilicates

In 1966 Mackenzie and Garrels (1966) pointed out the possibility for aluminosilicates to be formed in marine environments. This process has potential significance for the oceanic chemistry and global elemental cycles. As elements are transferred into solid phase and thus become insoluble, this process is referred to 'reverse weathering'. Within the scope of this chapter, only a brief overview of the major processes and conditions of formation is intended to be outlined.

Four major pathways for the formation of iron-bearing aluminosilicates can be distinguished:

1.    from weathered basalt and volcanic ashes
2.    as glauconite
3.    in the vicinity of hydrothermal vents
4.    under low temperature conditions

The first two pathways will not be discussed here as they were found to be only of local/regional importance and are not considered to be of major importance for early diagenetic reactions. Iron-bearing clay mineral formation under high-temperature conditions near a hydrothermal system of the Red Sea was studied by Bischoff (1972). A direct precipitation of an iron-rich smectite (nontronite) within the metalliferous sediments was found. This pathway of clay mineral formation was shown to occur at temperatures typically ranging between 70°C and 150°C under oxic and anoxic conditions (Cole and Shaw 1983). Yet, lowest temperatures of formation were deduced to be ~20°C (McMurtry et al. 1983, Singer et al. 1984). Experimental studies by Decarreau et al. (1987) demonstrated the synthesis of dioctahedral smectite, containing Fe(III) within the octahedral sheet, only under strictly oxic conditions.

Experimental results by Harder (1976, 1978) gave evidence for the potential of iron-bearing clay mineral formation under low temperature deep-sea floor conditions. Subsequent findings of sediments rich in montmorillonites in the north equatorial Pacific (Hein et al. 1979) and nontronite in the Bauer Deep of the eastern equatorial Pacific (Cole and Shaw 1983, Cole 1985) were attributed to authigenic aluminosilicate formation. For the formation of low-temperature iron-bearing aluminosilicates, the deposition of skeletal opal (e.g. radiolarian) and iron oxyhydroxide (e.g. precipitation products of hydrothermal activity) as well as a low carbonate content are considered (Cole and Shaw 1983). Enhanced opal dissolution due to the presence of high iron oxide concentrations were reported (Mayer et al. 1991). However, kinetic reasoning of this observation remains unclear. As the skeletal opal closely associated with the iron oxyhydroxide becomes buried, it dissolves and forms an amorphous Fe(III)-silica complex at the skeleton surface which subse-

quently recrystallizes to form nontronite on the surface of the partially dissolved skeletons (Cole 1985). By this analogy, Harder (1976,1978) also found an amorphous Fe(III)-silicate precipitate as a precursor which developed during aging under suboxic conditions into a crystalline iron-rich clay mineral. The presence of $Fe^{2+}$ was a prerequisite for the synthesis of clay minerals under experimental conditions and therefore partial reduction of iron oxyhydroxide within the microenvironment of an opal skeleton must be assumed. The oxygen isotopic composition ($\delta^{18}O$) of the authigenic mineral can be used to reconstruct the prevailing temperature during formation by applying the geothermometric equation of Yeh and Savin (1977). For the aluminosilicates from the north equatorial Pacific and the Bauer Deep, formation temperatures of ~3-4°C were deduced representing authigenic formation under low-temperature conditions in deep-sea sediments. Similar to the above described deep-sea conditions, the Amazon delta represents an iron- and silicate-rich depositional environment. Incubation experiments with sediments from the Amazon delta revealed substantial K-Fe-Mg-clay mineral formation within 1-3 years under low-temperature conditions (Michalopoulos and Aller 1995) implying significant elemental transfer into solid phase within the estuarine mixing zone.

### 7.4.5    Discussion: The Importance of Fe- and Mn-Reactivity in Various Environments

The above sections of this chapter have shown the high variability of iron-input modes, fluxes and reactivity towards oxidized and reduced species in marine sediments. Within this section the importance of iron and manganese reactivity with respect to the mineralization of organic matter as well as to the chemical oxidation (reduction) of reduced (oxidized) species within different depositional environments will be discussed and hopefully inspire further considerations.

In order to investigate the importance of iron and manganese reduction and oxidation processes one needs to determine their rates and compare these with other early diagenetic reaction pathways. For example, rates of organic carbon mineralization by each electron acceptor ($O_2$, $NO_3^-$, Mn(IV), Fe(III), $SO_4^{2-}$) are – strictly speaking – necessary to make a statement on the relative contribution of dissimilatory iron and manganese reduction. For the case of iron and manganese, the determination of such rates is problematic as both electron acceptors may be reduced by reduced species as well as by microbial respiration. Additionally, pore water fluxes usually strongly underestimate the true reduction rate as adsorption and precipitation of $Fe^{2+}/Mn^{2+}$-bearing minerals buffer the build-up within pore-water. 'Rates of dissimilatory Fe-oxide and Mn-oxide reduction are therefore the least well quantified of the carbon oxidation pathways' (Canfield 1993).

An overview of methods to determine the various organic carbon oxidation pathways is provided by Canfield (1993). As these methods are technically highly demanding and time-consuming, only very few sediments have been investigated with respect to the contribution of the different organic carbon oxidation pathways. Among such studies, different methods have been applied which result in additional uncertainties. A present-day discussion on the importance of iron and manganese must therefore to some extent be speculative. In Table 7.4, the results of different studies concerning the relative contribution of dissimilatory iron and manganese reduction are summarized.

Note that the results by Wang and Van Cappellen (1996) are model results for which some results of Canfield et al. (1993 a,b) were used for the basic data set. A comparison of different locations reveals significant variabilities in

the biodiffusion coefficient. For open ocean sediments, one can expect even much lower deposition rates and biodiffusion coefficients. Similarly, the proportion of dissimilatory reduction (relative to dissimilatory plus chemical reduction) as well as the proportion of organic carbon mineralization by iron and manganese reduction (relative to total organic carbon oxidation) varies significantly.

The above sites of investigation are distinct by different depositional environments. Skagerrak sediments were retrieved from water depths of 200 (S4), 400 (S6), and 700 (S9) meters and site S9 is located in the central Norwegian Trough where solid phase manganese content made up to 3.5-4 wt%. The Panama Basin site is ~4000 m deep and is located in the equatorial upwelling region as well as in the vicinity of hydrothermal activity causing a delivery of large amounts of reactive organic matter and manganese to the sea floor. Sites from the continental slope off Chile were investigated during a period of intense upwelling. The intensity of redox-cycling is controlled by the intensity of bioturbation, the input and reduction of reactive Fe(III)/Mn(IV), as well as by the oxidation rate (Sect. 7.4.3). However, the great significance of Fe/Mn cycling for the decomposition of organic matter may only result if reduction is predominantly coupled to dissimilation and chemical reduction is unimportant. This is the case for the manganese-dominated sites, Skagerrak S9 and Panama Basin, where nei-

**Table 7.4** Summary of results quantifying the relative contribution of dissimilatory iron and manganese reduction for the decomposition of organic matter.
([1] Wang and Van Cappellen 1996, [2] Canfield et al. 1993b, [3] Canfield et al. 1993a, [4] Aller 1990, [5] Thamdrup and Canfield 1996)

| Location | | Net-Fe/Mn deposition $[\mu mol\ cm^{-2}y^{-1}]$ | Biodiff. coeff. $[cm^{-2}y^{-1}]$ | Dissimilatory Fe/Mn reduction [%] | Total org. C decomposition [%] |
|---|---|---|---|---|---|
| Skagerrak, S4/S6 | Fe | 6 - 14 [3] | 80 - 87 [2] | | 32 - 51 [2] |
| | Fe | 13 - 24 [1] | | 71 - 84 [1] | |
| | Mn | 1 - 15 [1] | | 3 - 24 [1] | 0 [2] |
| Skagerrak, S9 | Mn | 5 - 10 [3] | 19 [2] | | 90 [2] |
| | Mn | 13 [1] | | 100 [1] | |
| | Fe | 14 [1] | | 0 [1] | 0 [2] |
| Panama Basin [4] | Mn | 'high' | 100 | 100 | 100 |
| Cont. Slope Chile [5] | Fe | 5.1 | 9; 29 | 46; 84 | 12; 29 |

ther the production of HS⁻ nor of $Fe^{2+}$ was found. $O_2$ consumption was (almost) completely attributed to the reoxidation of $Mn^{2+}$ in these cases. In contrast, Skagerrak sites S4 and S6 as well as sediments off Chile resemble situations where iron is reduced chemically (HS⁻) by up to ~50% and only smaller amounts of ferric iron are available for the dissimilation. A third situation can be inferred for typical open ocean sediments where low Fe/Mn and organic matter deposition along with low bioturbation intensity occurs. Here, organic matter decomposition is restricted to aerobic dissimilation and denitrification. Iron and manganese reduction rates are presumably negligible, yet over a long period of time a significant proportion of manganese is redistributed and the composition of iron oxide phases changes (Haese et al. 1998).

An extensive study in a shallow water, estuarine mixing zone elucidated ideal conditions for efficient manganese cycling (Aller 1994). Manganese turn-over was found to be most intense during warm periods with intensive bioturbation, well-oxygenated bottom water, and moderate organic matter input. Under such conditions, manganese reoxidation consumed 30-50% of the benthic oxygen flux and manganese reduction was mainly induced by HS⁻, FeS and $FeS_2$. As soon as bottom water became $O_2$-depleted the sedimentary Mn-cycle was reduced as dissolved Mn escaped out of the sediment. The influence of bioirrigation has not yet been explicitly investigated but modeling results by Wang and Van Cappellen (1996) imply enhanced metal cycling efficiency with increasing irrigation due to more rapid $Fe^{2+}$- / $Mn^{2+}$-oxidation.

A conceptual model describing the importance of iron and manganese reactivity in different environments is only sketchy as results are scarce and manifold aspects deserve further investigations. A general complication concerns the differentiation of dissimilatory and chemical reduction of Mn-oxide when concurrent iron reduction is apparent and further investigations need to discriminate each reaction pathway. The reactivity of smectite-bound iron has only been shown qualitatively (Kostka et al. 1996, König et al. 1997). Quantification with respect to dissimilatory and/or chemical reactions is still not available. An important role of adsorbed $Fe^{2+}$ / $Mn^{2+}$ is indicated (Sørensen and Thorling 1991, Roden and Zachara 1996) but not proved by sediment studies. Investigations focusing on the importance of metal cy-

cling and its influence on pathways of organic matter decomposition are scarce (Canfield et al. 1993a, Wang and Van Cappellen 1996), but are necessary to understand an important link of the carbon cycle.

## 7.5    The Assay for Ferric and Ferrous Iron

In order to study iron reactivity qualitatively and quantitatively, it is essential to quantify the ferrous and ferric iron fractions of the present minerals or mineral groups. With respect to the determination of iron speciation, the principal problem is the rapid oxidation of ferrous iron. Atmospheric oxygen diffuses into pore water where it oxidizes dissolved ferrous iron 'immediately' and starts oxidizing FeS and $FeS_2$. Reduced smectites may also become oxidized under atmosphere within hours. If dissolved iron and iron speciation of solid phase are to be determined, samples therefore need to be conserved under inert gas atmosphere. No extra care is needed for the determination of the total iron content of the solid phase.

Above a pH of 3 and in the absence of chelators, dissolved iron is only present as ferrous iron under natural conditions. The colored complex that results from the reaction between ferrous iron and a reagent can therefore be analysed colorimetrically and correlates with the concentration of total dissolved iron. Most conveniently, one can mix a drop of Ferrozine® solution (Stookey 1970), one drop of $H_2SO_4$ (diluted 1:4) and 1 ml of pore water in the glove box, wait until complex formation is completed (20-30 minutes) and quantify the iron concentration by the intensity of the color at a wavelength of 562 nm. To avoid matrix effects, standards should be prepared with artificial seawater.

The assay of ferrous and ferric iron in the solid phase has a long tradition due to the early interest in soil chemistry. Publications on extraction/leaching conditions and results from varying soils and sediments are extensive and thus, within the scope of a textbook, only important principals and a description of the (subjectively) most important extractions can be given. Since a great variety of the quantitatively most important iron bearing minerals / mineral groups is present in marine sediments, a series of different extractions is necessary in order to achieve a complete distribution of ferrous and ferric iron. In general, the extraction conditions applied to natural sediments result from experiments conducted in advance proving the dissolution of individual minerals or mineral groups. Because the grain size, degree of crystallinity, ionic substitution within minerals and varying matrix constituents influence the dissolution kinetics during the extraction a clear-cut mineral specific determination is usually not pos-

sible with this approach. Yet, extractions have been successfully applied to show patterns of mineral (group) dissolution and precipitation and to deduce reaction rates. For a comparison with results of other studies exactly the same extraction conditions (reagent composition, sediment : solution ratio, contact time) must be applied.

The following table (Table 7.5) gives an overview of experimentally derived dissolution be-

**Table 7.5** Solubility of iron bearing minerals derived under experimental conditions. + / - imply a solubility of $\geq 97\ \%$ / $\leq 3\ \%$, values indicate a percentage of release.

[1] Schwertmann (1964): 0.2 M $NH_4^+$-oxalate / 0.2 M oxalic acid; pH: 2.5, 2 h in darkness

[2] Ferdelman (1988): 10 g Na-citrate + 10 g Na-bicarbonate mixed in 200 ml distilled and deionized water, deaerated, before 4 g ascorbic acid are added; pH: 7.5, 24 h

[3] Lord (1980): 0.35 M acetate / 0.2 M Na-citrate + 1.0 g Na-dithionite for each sample (~ 1 g wet sediment in 20 ml solution); pH: 4.8, 4 hours.

[4] Chao and Zhou (1983): 1 M HCl, 30 min; Canfield (1988): 1 M HCl 20-23 h; Cornwell and Morse (1987): 1 M HCl, 45 min; Kostka and Luther (1994): 0.5 M, 1 h.

[5] Haese et al. (1997): 1 ml distilled and deionized water + 1 ml conc. $H_2SO_4$ + 2 ml HF were added to ~ 250 mg of wet sediment under inert gas atmosphere and constant stirring over few minutes.

[6] Canfield et al. (1986): 15 ml of $O_2$-free 1 M $CrCl_2$ in 0.5 M HCl + 10 ml of 12 M HCl under inert gas atmosphere.

[a] Chou and Zhou 1983, [b] Canfield 1988, [c] Kostka and Luther 1994, [d] Ruttenberg 1992, [e] Mehra and Jackson 1960, [f] Cornwell and Morse 1987, [g] Haese et al. 1997, [h] Canfield et al. 1986)

| Mineral | Oxalate [1] | Ascorbate [2] | Dithionite [3] | HCl [4] | HF/H₂SO₄ [5] | Cr(II)/ HCl [6] |
|---|---|---|---|---|---|---|
| am. Fe(OH)₃ | + [a] | | | 34 - 72 [a]  + [b] | | |
| Ferrihydrite | + [b] | + [c] | + [b,c,d] | + [b,c] | | |
| Lepidocrocite | + [b] | | + [b] | 7 [b] | | |
| Goethite | - [a,b] | - [c] | 91 [c]  + [b] | - [a,b,c] | + [g] | |
| Hematite | - [a,b] | - [c] | 63 [c]  + [b,d,e] | - [a,b,c] | + [g] | |
| Magnetite | 60 [c] | - [c] | 90 [c]  - [b] | - [a,b,c] | + [g] | |
| (am.) FeS | + [c] | | | + [f] | | + [h] |
| Pyrite (FeS₂) | | | | - [b,f] | | + [h] |
| Chlorite | - [b,c] | - [c] | 5 - 7 [c,b] | 27 [c]  32 [b] | 10 - 100 [g] | |
| Nontronite | - [b] | | 27 [b] | 7 [b] | | |
| Glauconite | - [b] | | 10 [b] | 10 [b] | | |
| Garnet | - [b] | | - [b] | - [b] | | |

havior of iron bearing minerals under some selected extraction conditions:

The investigation of total-Fe from ascorbate
and dithionite solution can be determined by ICP-
AES or flame-AAS. Ferrous and ferric iron from
non-reducing or -oxidizing extractions can be determined by polarographic methods (Wallmann et
al. 1993) or colorimetrically with and without the
addition of a reducing agent (e.g. hydroxylamine
hydrochloride; Kostka and Luther 1994). During
the acidic extractions evolving sulfide can be
trapped in a separate alkaline solution (e.g. Sulfur
Antioxidant Buffer, SAOB; Cornwell and Morse
1987) where it can be determined polarographi-
cally, by precipitation titration with Pb or by a
standard ion sensitive electrode. Sulfide evolving
from HCl extraction is called Acid Volatile Sulfur
(AVS).

The leaching with $HF/H_2SO_4$ as described
above and the subsequent polarographic determination of ferrous and ferric iron is based on work
by Beyer et al. (1975) and Stucki (1981) in order
to quantify the silicate bound ferrous and ferric
iron. This extraction has hardly been applied with
respect to questions of early diagenesis so far,
yet, the silicate bound iron fraction is quantitatively very important in marine sediments and
even a small reactive fraction of this pool may be
of overall significance for the iron reactivity. As a
complementary method to the commonly applied
extractions (Table 7.5) it renders the calculation
of the total iron speciation in the sediment which
may then be compared to Mössbauer-spectro-
scopic results (Haese et al. 1997; Haese et al. submitted).

One of the pitfalls in the interpretation of extraction results from natural sediments is caused
by the fact that the presence of $Fe^{2+}$ complexed
by carboxylic acid catalyzes the reduction of crystalline iron oxides such as hematite (Sulzberger et
al. 1989), magnetite (Blesa et al. 1989) and
goethite (Kostka and Luther 1994). In order to
avoid this catalytic dissolution of well-crystal-
lized iron oxides by $Fe^{2+}$ during the oxalate extraction Thamdrup and Canfield (1996) air-dried
the sediment in advance, thereby oxidizing FeS
and $FeCO_3$ to ferrihydrite. In addition, they applied the anoxic oxalate extraction and subtracted
the released amount of $Fe^{2+}$ from the amount of
$Fe^{3+}$ determined from the oxic extraction to calculate the poorly crystallized iron oxide fraction as
intended according to Table 7.5.

## Acknowledgements

The author wishes to thank Caroline Slomp for
her critical review of the manuscript. This is contribution No 256 of the Special Research Program
SFB 261 (*The South Atlantic in the Late Quaternary*) funded by the Deutsche Forschungsgemeinschaft (DFG).

# References

Aller, R.C., 1980. Diagnetic processes near the sediment-water interface of Long Island Sound. 2. Fe and Mn. Advances in Geophysics, 22: 351-415.

Aller, R.C. and DeMaster, D.J., 1984. Estimates of particle
flux and reworking at the deep-sea floor using 234TH/
238U disequilibrium. Earth and Planetary Science Letters,
67: 308-318.

Aller, R.C., 1990. Bioturbation and manganese cycling in
hemipelagic sediments. Philosophical Transactions of the
Royal Society of London, 331: 51-68.

Aller, R.C., 1994. The sedimentary Mn cycle in Long Island
Sound: Its role as intermediate oxidant and the influence
of bioturbation, O2, and Corg. flux on diagenetic reaction
balances. Journal of Marine Research, 52: 259-295.

Balzer, W., 1982. On the distribution of iron and manganese
at the sediment/water interface: thermodynamic versus kinetic control. Geochimica et Cosmochimica Acta, 46:
1153-1161.

Bell, P.E., Mills, A.L. and Herman, J.S., 1987.
Biogeochemical conditions favoring magnetite formation
during anaerobic iron reduction. Applied and Environmental Microbiology, 53: 2610-2616.

Berger, W.H., Smetacek, V.S. and Wefer, G., 1989. Ocean productivity and paleoproductivity-an overview. In: Berger,
W.H., Smetacek, V.S. and Wefer, G. (eds), Productivity of
the ocean: present and past. Wiley & Sons, Chichester,
pp. 1-34.

Berner, R.A., 1970. Sedimentary pyrite formation. American
Journal of Science, 268: 1-23.

Berner, R.A., 1971. Principals of chemical sedimentology.
McGraw-Hill, New York, 240 pp.

Bewers, J.M. and Yeats, P.A., 1977. Oceanic residence times
of trace metals. Nature, 268: 595-598.

Beyer, M.E., Bond, A.M. and McLaughlin, R.J.W., 1975. Simultaneous polarographic determination of ferrous, ferric
and total iron in standard rocks. Analytical Chemistry, 47:
479-482.

Biber, M.V., Dos Santos Afonso, M. and Stumm, W., 1994.
The coordination chemistry of weathering: IV. Inhibition
of the dissolution of oxide minerals. Geochimica et
Cosmochimica Acta, 58: 1999-2010.

Bischoff, J.L., 1972. A ferroan nontronite from the red Sea
geothermal system. Clays and Clay Minerals., 20: 217-223.

Blank, M., Leinen, M. and Prospero, J.M., 1985. Major Asian
aelian inputs indicated by the mineralogy of aerosols and

sediments in the western North Pacific. Nature, 314: 84-86.

Blesa, M.A., Marinovich, H.A., Baumgartner, E.C. and Marota, A.J.G., 1987. Mechanism of dissolution of magnetite by oxalic acid-ferrous ion solution. Inorganic Chemistry, 26: 3713-3717.

Böhm, J., 1925. Über Aluminium und Eisenoxide I (in German). Zeitschrift der Anorganischen Chemie, 149: 203-218.

Broecker, W.S., Spencer, D.W. and Craig, H., 1982. GEOSECS Pacific Expedition: Hydrographic Data. U.S. Govermaent Printing Office, Washington, DC, 3: 137 pp.

Buresh, R.J. and Moraghan, J.T., 1976. Chemical reduction of nitrate by ferrous iron. Journal of Environmental Quality, 5: 320-325.

Canfield, D.E., Raiswell, R., Westrich, J.T., Reaves, C.M. and Berner, R.A., 1986. The use of chromium reduction in the analysis of reduced inorganic sulfur in sediments and shales. Chemical Geology, 54: 149-155.

Canfield, D.E., 1988. Sulfate reduction and the diagenesis of iron in anoxic marine sediments. Ph.D. thesis, Yale Univ., 248pp.

Canfield, D.E., 1989. Reactive iron in marine sediments. Geochimica et Cosmochimica Acta, 51: 619-632.

Canfield, D.E., Raiswell, R. and Bottrell, S., 1992. The reactivity of sedimentary iron minerals toward sulfide. American Journal of Science, 292: 659-683.

Canfield, D.E., 1993. Organic matter oxidation in marine sediments. In: Wollast, R., Mackenzie, F.T. and Chou, L. (eds). Interactions of C, N, P and S biogeochemical cycles and global change. NATO ASI Series, 4, Springer, Berlin, Heidelberg, NY, pp. 333-363.

Canfield, D.E., Thamdrup, B. and Hansen, J.W., 1993a. The anaerobic degradation of organic matter in Danish coastal sediments: Iron reduction, manganese reduction, and sulfate reduction. Geochimica et Cosmochimica Acta, 57: 3867-3883.

Canfield, D.E., Jørgensen, B.B., Fossing, H., Glud, R., Gundersen, J., Ramsing, N.B., Thamdrup, B., Hansen, J.W., Nielsen L.P. and Hall, P.O.J., 1993b. Pathways of organic carbon oxidation in three continental margin sediments. Marine Geology, 113: 27-40.

Canfield, D.E., 1997. The geochemistry of river particles from the continental USA: Major elements. Geochimica Cosmochimica Acta, 61: 3349-3365.

Carlson, T.N. and Prospero, J.M., 1972. The large-scale movement of Saharan air outbreaks over the northern equatorial Atlantic. Journal of Applied Meteorology, 11: 283-297.

Carothers, W.W., Adami, L.H. and Rosenbauer, R.J., 1988. Experimental oxygen isotope fractionation between siderite-water and phosphoric acid liberated CO2-siderite. Geochimica et Cosmochimica Acta, 52: 2445-2450.

Chavez, F.P. and Barber, R.T., 1987. An estimate of new production in the equatorial PAcific. Deep-Sea Research, 34: 1229-1243.

Chester, R., 1990. Marine Geochemistry. Chapman & Hall, London, 698 pp.

Chou, T.T. and Zhou, L., 1983. Extraction techniques for selective dissolution of amorphous iron oxides from soils and sediments. Soil Science Society American Journal, 47: 225-232.

Cole, T.G. and Shaw, W.F., 1983. The nature and origin of authigenic smectites in some recent marine sediments. Clay Minerals, 18: 239-252.

Cole, T.G., 1985. Composition, oxygen isotope geochemistry, and orgin of smectite in the metalliferous sediments of the Bauer Deep, southeast Pacific. Geochimica et Cosmochimica Acta, 49: 221-235.

Coleman, M.L., Hedrick, D.B., Lovley, D.R., White, D.C. and Pye, K., 1993. Reduction of Fe(III) in sediments by sulphate-reducing bacteria. Nature, 361: 436-438.

Cornwell, J.C. and Morse, J.W., 1987. The characterization of iron sulfide minerals in anoxic marine sediments. Marine Chemistry, 22: 193-206.

Crosby, S.A., Glasson, D.R., Cuttler, A.H., Butler, I., Turner, D.R., Whitfield, M. and Millward, G.W., 1983. Surface areas and porosities of Fe(III)- and Fe(II)-derived oxyhydroxides. Environmental Science and Technology, 17: 709-713.

De Angelis, M., Barkov, N.I. and Petrov, V.N., 1987. Aerosol concentration over the last climatic cycle (160 kyr) from an Antarctic ice core. Nature, 325: 318-321.

De Baar, H.J.W. and Suess, E., 1993. Ocean carbon cycle and climate change - An introduction to the interdisciplinary union symposium. Global and Planetery Change, 8: VII-XI.

Decarreau, A., Bonnin, D., Badauth-Trauth, D., Couty, R. and Kaiser, P., 1987. Synthesis and crystallogenesis of ferric smectite by evolution of Si-Fe coprecipitates in oxidizing conditions. Clay Minerals, 22: 207-223.

Donaghay, P.L., 1991. The role of episodic atmospheric nutrient inputs in chemical and biological dynamics of oceanic ecosystems. Oceanography, 4: 62-70.

Dos Santos Afonso, M. and Stumm, W., 1992. The reductive dissolution of iron (III) (hydr) oxides by hydrogen sulfide. Langmuir, 8: 1671-1676.

Duce, R.A., Liss, P.S., Merrill, J.T., Atlas, E.L., Buat-Menard, P., Hicks, B.B., Miller, J.M., Prospero, J.M., Arimoto, R., Church, T.M., Ellis, W., Galloway, J.N., Hansen, L., Jickells, T.D., Knap, A.H., Reinhardt, K.H., Schneider, B., Soudine, A., Tokos, J.J., Tsunogai, S., Wollast, R. and Zhou, M., 1991. The atmospheric input of trace species to the world ocean. Global Biogeochemical Cycles, 5: 193-259.

Ehrenreich, A. and Widdel, F., 1994. Anaerobic oxidation of ferrous iron by purple bacteria, a new type of phototrophic metabolism. Applied and Environmental Microbiology, 60: 4517-4526.

Ellwood, B.B., Chrzanowski, T.H., Hrouda, F., Long, G.J. and Buhl, M.L., 1988. Siderite formation in anoxic deep-sea sediments: A synergetic bacterially controlled process with important implications in palaeomagnetism. Geology, 16: 980-982.

Ferdelman, T.G., 1980. The distribution of sulfur, iron, manganese, copper, and uranium in a salt marsh sediment core as determined by a sequential extraction method. Masters thesis, University Delaware.

Figuères, G., Martin, J.M. and Meybeck, M., 1978. Iron behaviour in the Zaire estuary. Netherlands Journal of Sea Research, 12: 329-337.

Froelich, P.N., Klinkhammer, G.P., Bender, M.L., Luedtke, N.A., Heath, G.R., Cullen, D., Dauphin, P., Hammond, D.

and Hartman B, 1979. Early oxidation of organic matter in pelagic sediments of the eastern equatorial Atlantic: suboxic diagenesis. Geochimica et Cosmochimica Acta, 43: 1075-1090.

Froelich, P.N., Bender, M.L., Luedtke, N.A., Heath, G.R. and DeVries, T., 1982. The marine phosphorus cycle. American Journal of Science, 282: 474-511.

GESAMP (Group of Experts on the Scientific Aspects of Marine Pollution), 1987. Land/sea boundary flux of contaminants: Contributions from rivers. GESAMP Rep. Stud., 32: 172 pp.

Gingele, F., 1992. Zur Klimaabhängigen Bildung biogener und terrigener Sedimente und ihre Veränderungen durch die Frühdiagenese im zentralen und östlichen Südatlantik (in German). Berichte, 26, Fachbereich Geowissenschaften, Universität Bremen, 202 pp.

Goldberg, E.D. and Sposito, G., 1984. A chemical model od phosphate adsorption by soils I. Reference oxide minerals. Soil Science Society American Journal, 48: 772-778.

Haese, R.R. Wallmann, K., Kretzmann, U., Müller, P.J. and Schulz, H.D., 1997. Iron species determination to investigate early diagenetic reactivity in marine sediments. Geochimica et Cosmochimica Acta, 61: 63-72.

Haese, R.R., Petermann, P., Dittert, L. and Schulz, H.D., 1998. The early Diagenesis of iron in pelagic sediments-a multidisciplinary approach. Earth and Planetary Science Letters, 157: 233-248.

Haese, R.R., Schramm, J., Rutgers van der Loeff, M.M. and Schulz, H.D., in press. A comparative study of iron and manganese diagenesis in continental slope and deep sea basin sediments of Uruguay (SW Atlantic). Geologische Rundschau.

Harder, H., 1976. Nontronite synthesis at low temperatures. Chemical Geology, 18: 169-180.

Harder, H., 1978. Synthesis of iron layer silicate minerals under natural conditions. Clays and Clay Minerals, 26: 65-72.

Hart, T.J., 1934. On the phytoplankton of the south-west Atlantic and the Bellinghausen Sea, 1929-31. Discovery Reports, VIII.

Hein, J.R., Yeh, H-W. and Alexander, E., 1979. Origin of iron-rich montmorillonite from the manganese nodule belt of the north eqatorial Pacific. Clays and Clay Minerals, 27: 185-194.

Hunter, K.A., 1983. On the estuarine mixing of dissolved substances in relation to colloid stability and surface properties. Geochimica et Cosmochimica Acta, 47: 467-473.

Hüttel, M., Ziebis, W., Forster, S. and Luther, G.W. III., 1998. Advective transport affecting metal and nutrient distributions and interfacial fluxes in permeable sediments. Geochimica Cosmochimica Acta, 62: 613-631.

Jensen, H.S., Mortensen, P.B., Andersen, F.Ø., Rasmussen, E. and Jensen, A., 1995. Phosphorus cycling in a coastal marine sediment, Aarhus Bay, Denmark. Limnology and Oceanography, 40: 908-917.

Johnson, K.S., Coale , K.H., Elrod, V.A. and N.W., T., 1994. Iron photochemistry in seawater from equatorial Pacific. Marine Chemistry, 46: 319-334.

Johnson, K.S., Gordon, R.M. and Coalae, K.H., 1997. What controls dissolved iron concentrarions in the world ocean? Marine Chemistry, 57: 137-161.

Jørgensen, B.B., 1977. Bacterial sulfate reduction within reduced microniches of oxidized marine sediments. Marine Biology, 41: 7-17.

Kester, D.R. and Pytkowicz, R.M., 1967. Determination of apparent dissociation constants of phosphoric acid in sea water. Limnology and Oceanography, 12: 243-252.

Kostka, J.E. and Luther, G.W. III., 1994. Partitioning and speciation of solid phase iron in saltmarsh sediments. Geochimica et Cosmochimica Acta, 58: 1701-1710.

Kostka, J.E. and Nealson, K.H., 1995. Dissolution and reduction of magnetite by bacteria. Environmental Science and Technology, 29: 2535-2540.

Kostka, J.E., Nealson, K.H., Wu, J. and Stucki, J.W., 1996. Reduction of the structural Fe(III) in smectite by a pure culture of the Fe-reducing bacterium, Shewanella putrefaciens strain MR-1. Clays and Clay Minerals, 44: 522-529.

Köning, I., Drodt, M., Suess, E. and Trautwein, A.X., 1997. Iron reduction through the tan-green color transition in deep-sea sediments. Geochimica et Cosmochimmica Acta, 61: 1679-1683.

Krauskopf, K.B., 1956. Factors controlling the concentration of thirteen trace metals in seawater. Geochimica et Cosmochimica Acta, 12: 331-334.

Krom, M.D. and Berner, R.A., 1980. Adsorption of phosphate in anoxic marine sediments. Limnology and Oceanography, 25: 797-806.

Kuma, K., Nishioka, J. and Matsunaga, K., 1994. Controls on iron(III) hydroxide solubility in seawater: The influence of pH and natural organic chelators. Limnology and Oceanography, 41: 396-407.

Lear, P.R. and Stucki, J.W., 1989. Effects of iron oxidation state on the specific surface area of Nontronite. Clays and Clay Minerals, 37: 547-552.

Leventhal, J. and Taylor, C., 1990. Comparison of methods to determine degree of pyritization. Geochimica et Cosmochimica Acta, 54: 2621-2625.

Lord, C.L. III., 1980. The chemistry and cycling of iron, manganese, and sulfur in salt marsh sediments. Ph.D. thesis, University Delaware, 177 pp.

Lovley, D.R., 1987. Organic matter mineralization with the reduction of ferric iron: A review. Geomicrobiology Journal, 5: 375-399.

Lovley, D.R. and Phillips, E.J.P., 1988. Novel mode of microbial energy metabolism: Organic carbon oxidation coupled to dissimilatory reduction of iron and manganese. Applied and Environmental Microbiology, 54: 1472-1480.

Lovley, D.R., 1991. Dissimilatory Fe(III) and Mn(IV) Reduction. Microbiology Reviews, 55: 259-287.

Lovley, D.R., 1997. Microbial Fe(III) reduction in subsurface environments. FEMS Microbiological Reviews, 20: 305-313.

Lovley, D.R., Coates, J.D., Saffarini, D. and Loneran, D.J., 1997. Diversity of dissimilatory Fe(III)-reducing bacteria. In: Winkelman, G. and Carrano, C.J. (eds), Iron and related transition metals in microbial metabolism. Harwood Academic Publishers, Switzerland, pp. 187-215.

Lyle, M., 1983. The brown-green color transition in marine sediments: A marker of the Fe(III)-Fe(II) redox boundary. Limnology and Oceanography, 28: 1026-1033.

Mackenzie, F.T. and Garrels, R.M., 1966. Chemical mass balance between rivers and oceans. America Journal of Science, 264: 507-525.

Martin, J.M. and Whitfield, M., 1983. The significance of the river input of chemical elements to the ocean. In: Wong, C.S., Boyle, E., Bruland, K.W., Burton, J.D. and Goldberg, E.D. (eds), Trace metals in sea water. Plenum Press, NY, pp. 265-296.

Martin, J.H., Gordon, R.M., Fitzwater, S.E. and Broenkow, W.W., 1989. VERTEX: phytoplankton/iron studies in the gulf of Alaska. Deep-Sea Research, 36: 649-680.

Martin, J.H., 1990. Glacial-interglacial CO2 change: The iron hypothesis. Paleoceanography, 5: 1-13.

Martin, J.H., Gordon, R.M. and Fitzwater, S.E., 1991. The case for iron. In: Chisholm, S.W. and Morel, F.M.M. (eds). What controls phytoplankton production in nutrient-rich areas of the open sea?. ASLO Symposium, Lake San Marcos, California, Feb. 22-24, Allen Press, Lawrence.

Martin, J.H., Coale, K.H., Johnson, K.S., Fitzwater, S.E., 1994. Testing the iron hypothesis in ecosystems of the equatorial Pacific Ocean. Nature, 371: 123-129.

Mayer, L.M., Jorgensen, J. and Schnitker, D., 1991. Enhancement of diatom frustule dissolution by iron oxides. Marine Geology, 99: 263-266.

McAllister, C.D., Parsons, T.R. and Strickland, J.D.H., 1960. Primary producivity and fertility at station „P" in the north-east Pacific Ocean. Journal du Conseil, 25: 240-259.

McMurtry, G.M., Chung-Ho, W. and Hsueh-Wen, Y., 1983. Chemical and isotopic investigation into the origin of clay minerals from the Galapagos hydrothermal mound field. Geochimica et Cosmochimica Acta, 47: 291-300.

Mehra, O.P. and Jackson, M.L., 1960. Iron oxide removal from soils and clays by a dithionite-citrate system bufferd with sodium carbonate. Proceedings of the national conference on clays and clay mineralogy, 7: 317-327.

Michalopoulos, P. and Aller, R.C., 1995. Rapid clay mineral formation in Amazon delta sediments: Reverse weathering and oceanic elemental cycles. Science, 270: 614-617.

Millero, F.J., Sotolongo, S. and Izaguirre, M., 1987. The oxidation kinetics of Fe(II) in seawater. Geochimica et Cosmochimica Acta, 51: 793-801.

Morris, R.V., Lauer, H.V. Jr, Lawson, C.A., Gibson, E.K. Jr., Nace, G.A. and Stewart, C., 1985. Spectral and other physicochemical properties of submicron powders of hematite (a-Fe2O3), maghemite (g-Fe2O3), magnetite (Fe3O4), goethite (a -FeOOH), and lepidocrocite (g-FeOOH). Journal of Geophysical Research, 90: 3126-3144.

Mortimer, R.J.G. and Coleman, M.L., 1997. Microbial influence on the oxygen isotopic composition of diagenetic siderite. Geochimica et Cosmochimica Acta, 61: 1705-1711.

Munch, J.C. and Ottow, J.C.G., 1980. Preferential reductions of amorphous to crystalline iron oxides by bacterial activity. Journal of Soil Science, 129: 15-21.

Munch, J.C. and Ottow, J.C.G., 1982. Einfluß von Zellkontakt und Eisen (III) oxidform auf die bakterielle Eisenreduktion (in German). Zeitschrift der Pflanzenernährung und Bodenkunde, 145: 66-77.

Murray, R.W. and Leinen, M., 1993. Chemical transport to the seafloor of the equatorial Pacific Ocean across a latitudinal transect at 135°W: Tracking sedimentary major, trace, and rare earth element fluxes at the Equator and the Intertropical Convergence Zone. Geochimica et Cosmochimica Acta, 57: 4141-4163.

Myers, C.R. and Nealson, K.H., 1988a. Bacterial manganese reduction and growth with manganese oxide as the sole electron acceptor. Science, 240: 1319-1321.

Myers, C.R. and Nealson, K.H., 1988b. Microbial reduction of manganese oxides: Interactions with iron and sulfur. Geochimica et Cosmochimica Acta, 52: 2727-2732.

Nittrouer, C.A., DeMaster, D.J., McKee, B.A., Cutshall, N.H. and Larsen, I.L., 1983/1984. The effect of sediment mixing on Pb-210 accumulation rates for the Washington continental shelf. Marine Geology, 54: 201-221.

Norrish, K. and Taylor, R.M., 1961. The isomorphous replacement of iron by aluminium in soil goethites. Journal of Soil Sciences, 12: 294-306.

Ottley, C.J., Davison, W. and Edmunds, W.M., 1997. Chemical catalysis of nitrate reduction by iron(II). Geochimica et Cosmochimica Acta, 61: 1819-1828.

Ottow, J.C.G., 1969. Der Einfluß von Nitrat, Chlorat, Sulfat, Eisenoxidform und Wachstumsbedingungen auf das Ausmaß der bakteriellen Eisenreduktion (in German). Zeitschrift der Pflanzenernährung und Bodenkunde, 124: 238-253.

Peiffer, G., Dos Santos Afonso, M., Werhli, B. and Gächter, R., 1992. Kinetics and mechanism of the reaction of H2S with lepidocrocite. Environmental Science and Technology, 26: 2408-2412.

Pena, F. and Torrent, J., 1984. Relationships between phosphate sorption and iron oxides in alfisols from a river terrace sequence of mediterranean Spain. Geoderma, 33: 283-296.

Postma, D., 1982. Pyrite and siderite formaton brackish and freshwater swamp sediments. American Journal of Science, 282: 1151-1183.

Postma, D., 1985. Concentration of Mn and separation from Fe in sediments. Kinetics and stoichiometry of the reaction between birnessite and dissolved Fe(II) at 10°C. Geochimica et Cosmochimica Acta, 49: 1023-1033.

Postma, D. and Jakobsen, R., 1996. Redox zonation: Equilibrium constraints on the Fe(III)/SO4-reduction interface. Geochimica et Cosmochimica Acta, 60: 3169-3175.

Prospero, J.M., 1981. Eolian transport to the world ocean. In: Emiliani, C. (ed), The sea. 7, Wiley, NY, pp. 801-874.

Prospero, J.M., Glaccum, R.A. and Nees, R.T., 1981. Atmospheric transport of soil dust from Africa to South America. Nature, 289: 570-572.

Pyzik, A.J. and Sommer, S.E., 1981. Sedimentary iron monosulfides: kinetics and mechanism of formation. Geochimica et Cosmochimica Acta, 45: 687-698.

Raiswell, R., Buckley, F., Berner, R.A. and Anderson, T.F., 1988. Degree of pyritisation as a paleoenvironmental indicator of bottom water oxygenation. Journal of Sedimentary Petrology, 58: 812-819.

Raiswell, R., Canfield, D.E. and Berner, R.A., 1994. A comparison of iron extraction methods for the determination of degree of pyritisation and the recognition of iron-limited pyrite formation. Chemical Geology, 111: 101-110.

Raiswell, R. and Canfield, D.E., 1996. Rates of reaction between silicate iron and dissolved sulfide in Peru Margin sediments. Geochimica et Cosmochimica Acta, 60: 2777-2787.

Rickard, D., Schoonen, M.A.A. and Luther III., G.W., 1995. Chemistry of iron sulfides in sedimentary environments. In: Vairavamurthy, M.A. and Schoonen, M.A.A. (eds), Geochemical transformations of sedimentary sulfur. ACS Symposium Series, 612, Washington, DC,pp. 168-194.

Roden, E.E. and Zachara, J.M., 1996. Microbial reduction of crystalline iron (III) oxides: Influence of oxides surface area and potential for cell growth. Environmental Science and Technology, 30: 1618-1628.

Roth, C.B. and Tullock, R.J., 1972. Deprotonation of nontronite resulting from chemical reduction of strucktural ferric iron. Proceedings of the International Clay Conference, Madrid, 89-98.

Rozenson, I. and Heller-Kallai, L., 1976a. Reduction and oxidation of Fe3+ in dioctahedral smectites - 1: Reduction with Hydrazine and Dithionite. Clays and Clay Minerals, 24: 271-282.

Rozenson, I. and Heller-Kallai, L., 1976b. Reduction and oxidation of Fe3+ in dioctahedral smectites - 2: Reduction with sodium sulphide solutions. Clays and clay minerals, 24: 283-288.

Rue, E.L. and Bruland, K.W., 1995. Comlexation of Fe(III) by natural organic ligands in the central North Pacific as determined by a new competitive ligand equilibration/ adsorptive cathodic stripping voltametric method. Marine Chemistry, 50: 117-138.

Ruttenberg, K.C., 1992. Development of a sequential extraction method for different forms of phosphorus in marine sediments. Limnology and Oceanography, 37: 1460-1482.

Schwertmann, U., 1964. Differenzierung der Eisenoxide des Bodens durch photochemische Extraktion mit saurer Ammoniumoxalat-Lösung (in German). Zeitschrift zur Pflanzenernährung und Bodenkunde, 195: 194-202.

Schwertmann, U., Fitzpatrick, R.W., Taylor, R.M. and Lewis, D.G., 1979. The influence of aluminium on iron oxides. Part II. Preperation and properties of Al substituted hematites. Clays and Clay Minerals, 11: 189-200.

Schwertmann, U. and Murad, E., 1983. Effect of pH on the formation of goethite and haematite from ferrihydrite. Clays and Clay Minerals, 31: 277-284.

Schwertmann, U. and Taylor, R.M., 1989. Iron oxides. In: Dinauer, R.C. (ed) Minerals in soil environment. Soil Science Society of America, Book Series, 1, Madison, WI, pp. 379-438.

Schwertmann, U. and Cornell, R.M., 1991. Iron oxydes in the laboratory. VCH Verlagsgesellschaft mbH, Weinheim, 137 pp.

Singer, A., Stoffers, P., Heller-Kallai, L. and Szafranek, D., 1984. Nontronite in a deep-sea ore from the south Pacific. Clays and Clay Minerals, 32: 375-383.

Slomp, C.P., Van der Gaast, S.J. and Van Raaphorst, W., 1996a. Phosphorus binding by poorly cristalline iron oxides in North Sea sediments. Marine Geochemistry, 52: 55-73.

Slomp, C.P., Epping, E.H.G., Helder, W. and Van Raaphorst, W., 1996b. A key role for iron-bound phosphorus in authigenic apatite formation in North Atlantic continental platform sediments. Journal of Marine Research, 54:

1179-1205.

Sørensen, J. and Thorling, L., 1991. Stimulation by lepidocrocite (g-FeOOH) of Fe(II)-dependent nitrite reduction. Geochimica et Cosmochimica Acta, 55: 1289-1294.

Stookey, L.L., 1970. Ferrozine-A new spectrophotometric reagent for iron. Analytical Chemistry, 42: 779-781.

Straub, K.L., Benz, M., Schink, B. and Widdel, F., 1996. Anaerobic, nitrate-dependent microbial oxidation of ferrous iron. Applied and Environmental Microbiology, 62: 1458-1460.

Stucki, J.W., 1981. The quantitative assay of minerals for Fe2+ and Fe3+ using 1,10-Phenanthroline: II. A photochemical Method. Soil Science Society of America Journal, 45: 638-641.

Stumm, W. and Morgan, J.J., 1996. Aquatic Chemistry. Wiley & Sons, London, 1022 pp.

Sulzberger, B., Suter, S., Siffert, C., Banwart, S. and Stumm, W., 1989. Dissolution of Fe(III) hydroxides in natural waters; Laboratory assessment on the kinetics controlled by surface coordination. Marine Chemistry, 28: 127-144.

Sundby, B. and Silverberg, N., 1985. Manganese fluxes in the benthic boundary layer. Limnology and Oceanography, 30: 372-381.

Sundby, B., Anderson, L.G., Hall, P.O.J., Iverfeldt, Å., Rutgers van der Loeff, M. and Westerlund S.F.G., 1986. The effect of oxygen on release and uptake of cobalt, manganese, iron and phosphate at the sediment-water interface. Geochimica et Cosmochimica Acta, 50: 1281-1288.

Sundby, B., Gobeil, C., Silcerberg, N. and Mucci, A., 1992. The phosphorus cycle in coastal marine sediments. Limnology and Oceanography, 37: 1129-1145.

Thamdrup, B., Glud, R.N. and Hansen, J.W., 1994b. Manganese oxidation and in situ manganese fluxes from a coastal sediment. Geochimica et Cosmochimica Acta, 58: 2563-2570.

Thamdrup, B. and Canfield, D.E., 1996. Pathways of carbon oxidation in continental margin sediments off central Chile. Limnology and Oceanography, 41: 1629-1650.

Torrent, J., Barrón, V. and Schwertman, U., 1992. Fast and slow phosphate sorption by goethite-rich natural materials. Clays and Clay Mineralogy, 40: 14-21.

Trick, C.G., Andersen, R.J., Gillam, A. and Harrison, P.J., 1983. Prorocentrin: An extracellular siderophore produced by the marine dinoflagellate Prorocentrum minimum. Science, 219: 306-308.

Trick, C.G., 1989. Hydroxomate-siderophore production and utilization by marine eubacteria. Current Microbiology, 18: 375-378.

Uematsu, M., Duce, R.A., Prospero, J.M., Chen, L., Merrill, J.T. and McDonald, R.L., 1983. Transport of mineral aerosol from Asia over the North Pacific Ocean. Journal of Geophysical Research, 88: 5343-5352.

Wallmann, K., Hennies, K., König, I., Petersen, W. and Knauth, H.-D., 1993. A new procedure for the determination of 'reactive' ferric iron and ferrous iron minerals in sediments. Limnologogy and Oceanography, 38: 1803-1812.

Wang, Y. and Cappellen, P.v., 1996. A multicomponent reactive transport model of early diagenesis: Application to redox cycling in coastal marine sediments. Geochimica et

Cosmochimica Acta, 60: 2993-3014.

Wedepohl, K.H., 1995. The composition of the continental crust. Geochimica et Cosmochimica Acta, 59: 1217-1232.

Wehrli, B., Friedl, G. and Manceau, A., 1995. Reaction rates and products of manganese oxidation at the sediment-water interface. In: Huang, C.P., O'Melia, C.R. and Morgan, J.J. (eds) Aquatic chemistry: Interfacial and interspecies processes. ACS Advances in Chemistry, 244, pp. 111-134.

Widdel, F., Schnell, S., Heising, S., Ehrenrech, A., Assmus, B. and Schink, B., 1993. Ferrous iron oxidation by anoxygenic phototrophic bacteria. Nature, 362: 834-836.

Wu, J. and Luther, G.W., 1995. Complexation of Fe(III) by natural organic ligands in the Northwest Atlantic Ocean by competitve ligand equilibration method and kinetic approach. Marine Chemistry, 50: 159-177.

Yeats, P.A. and Bewers, J.M., 1982. Discharge of metals from the St. Lawrence River. Canadian Journal of Earth Science, 19: 982-992.

Yeh, H.W. and Savin, S.M., 1977. Mechanism of burial metamorphism of argillaceaous sediments: 3. O-isotope evidence. Bulletin of the Geological Society of America, 88: 1321-1330.

Zabel, M. and Steinmetz, E., subm. Phosphorus forms in surficial sediments off Namibia-indicator for the benthic particle cycling. Marine Geology.

Zhuang, G., Duce, R.A. and Kester, D.A., 1990. The dissolution of atmospheric iron in surface seawater of the open ocean. Journal of Geophysical Research, 59: 16207-16216.

Zhuang, G. and Duce, R.A., 1993. The adsorption of dissolved iron on marine aerosol particles in surface waters of the open ocean. Deep-Sea Research, 40: 1413-1429.

# 8 Sulfate Reduction in Marine Sediments

Sabine Kasten and Bo Barker Jørgensen

The present chapter deals with the biogeochemical transformations of sulfur within marine sediments during early diagenesis. The term 'early diagenesis' refers to the whole range of post-depositional processes that take place in aquatic sediments coupled either directly or indirectly to the degradation of organic matter. We focus on the processes that drive sulfate reduction together with the manifold associated biotic and abiotic reactions that make up the sedimentary sulfur cycle. Furthermore, we will give an overview of the quantitative significance of microbial sulfate reduction in the remineralization of organic matter and oxidation of methane in different depositional environments and discuss the often observed discrepancy between sulfate reduction rates deduced from radiotracer methods and those calculated from pore-water concentration profiles and/or solid-phase sulfur data. As sedimentary pyrite represents one of the two most important sinks for seawater sulfate, the different mechanisms of pyrite formation are presented. The formation of organic sulfur is the second major sink for oceanic sulfur, but will not be discussed here. For a detailed overview of the processes and pathways involved in the incorporation of sulfur into organic matter, we refer to Orr and White (1990), Krein and Aizenshtat (1995) or Schouten et al. (1995).

## 8.1 Introduction

With $1.3 \cdot 10^9$ teragrams (Tg = $10^{12}$ g) of sulfur present as sulfate the oceans represent one of the largest sulfur pools (Vairavamurthy et al. 1995). The main input of sulfur into the oceans occurs via river water carrying the products of mechanical and chemical weathering of continental rocks. In contrast to this fluvial input, the atmospheric transport of sulfur is of minor importance. It mainly consists of recycled oceanic sulfate from seaspray, volcanic sulfur gases, $H_2S$ released by sulfate-reducing bacteria, organic S-bearing compounds released into seawater and subsequently into the atmosphere by phytoplankton and anthropogenic emissions of sulfur dioxide. Due to the oxic conditions that prevail in most parts of the world's oceans, the dominant sulfur species in seawater is by far the sulfate ion ($SO_4^{2-}$).

Marine sediments are the main sink for seawater sulfate which demonstrates that the sedimentary sulfur cycle is a major component of the global sulfur cycle. The most important mechanisms for removing sulfate from the oceans to the sediments are (1) the bacterial reduction of sulfate to hydrogen sulfide, which reacts with iron to form sulfide minerals, particularly pyrite ($FeS_2$), (2) the formation of organic sulfur, i.e. the incorporation of sulfur into sedimentary organic matter during early diagenesis, and (3) the precipitation of calcium sulfate minerals in evaporites (Vairavamurthy et al. 1995). With respect to the relative importance of each pathway, Vairavamurthy et al. (1995) point out that although marine evaporites were important sinks for sulfate from the Late Precambrian to the Recent, their rate of formation in today's oceans, in the last few million years, is quantitatively insignificant. Thus, they conclude that - based on recent research - the burial of sulfide minerals and organic sulfur (both formed during early diagenesis) represents the major sinks for oceanic sulfur in the modern oceans.

## 8.2    Sulfate Reduction and the Degradation of Organic Matter

The sedimentary sulfur cycle here starts with the bacterial reduction of seawater sulfate to hydrogen sulfide ($H_2S$). Two types of sulfate reduction can be distinguished: (1) assimilatory sulfate reduction which serves for the biosynthesis of organic-sulfur compounds, and (2) dissimilatory sulfate reduction from which microorganisms conserve energy. We will focus here on the latter process.

As described in detail in Chapter 5, many microorganisms gain the energy necessary for their life processes from the oxidation of organic matter with an external electron acceptor. Thermodynamically, oxygen is the most favorable electron acceptor. The supply of oxygen from seawater into the sediment is, however, transport-limited. In general, a high organic matter flux or low oxygen content of bottomwater reduce the thickness of the oxic surface layer of the sediment. Thus, in coastal marine sediments or in ocean areas of high productivity, e.g. upwelling regions, a high flux of organic material to the seafloor leads to rapid depletion of $O_2$ in the sediment. In these areas, oxygen penetrates only a few mm or cm into the sediment.

Where oxygen has been consumed by aerobic respiration, the sediment is anoxic, i.e. oxygen-free, and microorganisms utilize other terminal electron acceptors for the mineralization of organic matter. Listed in an order of decreasing energy gain these are: nitrate ($NO_3^-$), manganese oxides (represented by $MnO_2$), iron oxides (represented by $Fe_2O_3$) and sulfate ($SO_4^{2-}$). Although $NO_3^-$, $MnO_2$ and $Fe_2O_3$ are energetically more favorable than sulfate, they are usually less important biogeochemically because of their limited supply to the sediments. The high concentration in seawater makes sulfate a dominant electron acceptor in the anaerobic degradation of organic matter (Henrichs and Reeburgh 1987). With an average concentration of about 29 mmol/l (Vairavamurthy et al. 1995), sulfate concentrations in seawater are two orders of magnitude higher than in freshwater (Bowen 1979). The mineralization of organic material by sulfate in marine sediments is, therefore, much more important than in freshwater.

Most of the known dissimilatory sulfate reducers are bacteria, but also some thermophilic archaea belong to this group (Stetter et al. 1993). For an overview of the most common sulfate-reducing bacteria in marine sediments we refer to Widdel (1988). Besides sulfate there are also other oxidized sulfur compounds, e.g. thiosulfate ($S_2O_3^{2-}$) and elemental sulfur ($S^0$), that can serve as the terminal electron acceptor (Ehrlich 1996). They are, however, not of similar quantitative importance.

Dissimilatory sulfate reduction can be described by the following net equation (e.g. Coleman and Raiswell 1995):

$$2CH_2O + SO_4^{2-} \rightarrow 2HCO_3^- + H_2S \qquad (8.1)$$

A number of estimates have been made of the overall significance of dissimilatory sulfate reduction for the degradation of organic matter in different marine depositional environments. They suggest that the combined processes of oxic respiration and sulfate reduction account for most of the organic carbon oxidation (Jørgensen 1983, Henrichs and Reeburgh 1987, Canfield 1989). Whereas oxic respiration dominates in deep-sea sediments with relatively low depositional rates, sulfate reduction and oxic respiration may be of similar magnitude in near-shore sediments (Jørgensen 1982, Canfield 1989). Moreover, in sediments of the upwelling region off Chile, sulfate reduction is the dominant pathway of organic carbon oxidation (Thamdrup and Canfield 1996). Sulfide produced by sulfate reduction reacts with iron to form iron sulfides within the sediments. Up to 90% of these reduced sulfur compounds are transported up to the oxidized sediment layers by bioturbation, where they are reoxidized to sulfate (Jørgensen 1982). In organic-rich coastal marine environments typically 25-50% of the sediment oxygen consumption is used either directly or indirectly for the reoxidation of sulfide (Jørgensen 1982).

Direct measurements of sulfate reduction rates using radiotracer methods have revealed that dissimilatory sulfate reduction is the most important anaerobic pathway of organic matter decomposition in most continental margin sediments (Jørgensen 1977, Canfield et al. 1993). Determinations of sulfate reduction rates by this technique, however, have mainly been carried out in shallower waters, so that the data base for sediments below 500 m water depth is very limited. One of the most extensive, radiotracer-based data sets of sulfate reduction rates for continental slope sediments is presented by Ferdelman et al. (in press) for the continental margin of southwest Africa. This area forms part of the Benguela upwelling system (stations between 855 and 4766 m water depth). They demonstrated that the depth-integrated sulfate reduction rates over the upper 20 cm of the sediment strongly correlated with the concentrations of organic carbon within the surface sediments. They further estimated that for the stations at 1300 m water depth sulfate reduction accounts for between 20 to 90% of total oxygen consumption (which is taken as the total oxidation of organic carbon to $CO_2$). This indicates that a significant fraction of organic matter in these sediments is degraded anaerobically through sulfate reduction.

In Table 8.1 we give an overview of the quantitative significance of dissimilatory sulfate reduction in different marine environments. The different methods for the determination of sulfate reduction rates in marine sediments will be described below and in more detail in Chapter 5.

Sulfate is also reduced in connection with anaerobic methane oxidation (e.g. Reeburgh 1976, Devol and Ahmed 1981, Reeburgh and Alperin 1988). Methane is generated below the sulfate zone of the sediments as the end-product of anaerobic degradation of organic matter. Anaerobic methane oxidation integrates the whole degradation of buried organic carbon from the lower boundary of the sulfate zone to very deep sediment layers which were deposited many thousands or millions of years ago. The coupled sulfate-methane reaction has been proposed to proceed according to the following net equation assuming a one to one stoichiometry (e.g. Murray et al. 1978, Devol and Ahmed 1981):

$$CH_4 + SO_4^{2-} \rightarrow HCO_3^- + HS^- + H_2O \quad (8.2)$$

Most of the dissimilatory sulfate reduction takes place close to the sediment surface (Iversen and Jørgensen 1985, Ferdelman et al. in press), depending on the quantity and quality of the organic matter transported into the sediment. Sulfate reduction based on anaerobic methane oxidation takes place in a distinct zone typically located one to several meters below the sediment surface (Fig. 8.1). Within this reaction zone, referred to as the 'base of sulfate reduction zone', the 'sulfate-methane interface' or the 'sulfate-methane transition zone', pore-water methane and sulfate are both consumed to depletion. Figure 8.1 shows examples of the sulfate-methane transition zone in sediments of the Kattegat and Skagerrak (Denmark), the two seas connecting the Baltic Sea with the North Sea. Both sampling locations are characterized by high sedimentation rates (Station B: 0.16 cm yr$^{-1}$; Station C: 0.29 cm yr$^{-1}$); the sediments consist of fine-grained silt and clay, with average organic matter contents of 12.1% at station B and 10.3% at station C (Iversen and Jørgensen 1985).

**Table 8.1** Quantitative significance of sulfate reduction for the total degradation of organic matter in different depositional environments. Highest significance is reached in areas of oxygen-depleted bottomwater, e.g. upwelling systems and stagnant basins.

| Location | Rate of $C_{org}$ degradation [mmol m$^{-2}$ d$^{-1}$] | % $C_{org}$ remineralized by sulfate reduction | Reference |
|---|---|---|---|
| Limfjord (Denmark) | 36 | 53 | Jørgensen (1977) |
| Inner shelf | 10 | 50 | Jørgensen (1983) |
| Outer shelf | 2.8 | 20 - 30 | Jørgensen (1983) |
| Gulf of St. Lawrence | 7.5 - 10 | 10 - 50 | Edenborn et al. (1987) |
| Washington shelf | 20 | 72 | Christensen (1989) |
| Skagerrak | 3 - 8 | 28 - 51 | Canfield et al. (1993) |
| Chilean shelf | 10 | 56 - 79 | Thamdrup and Canfield (1996) |
| Svalbard shelf | 6 - 20 | 28 - 42 | Sagemann et al. (1998) |
| Benguela upwelling area | 2.3 - 15.5 | 3 - 90 | Ferdelman et al. (in press) |
| Gotland Deep (Baltic Sea) | 6 - 23 | 30 - 96 | Greeff et al. (sub.) |

The process of anaerobic methane oxidation in marine deposits is well documented geochemically by measured pore-water concentration profiles of sulfate and methane (e.g. Martens and Berner 1974, Reeburgh 1980, Iversen and Blackburn 1981, Devol 1983, Iversen and Jørgensen 1985, Hoehler et al. 1994, Blair and Aller 1995). Despite this strong geochemical evidence, the

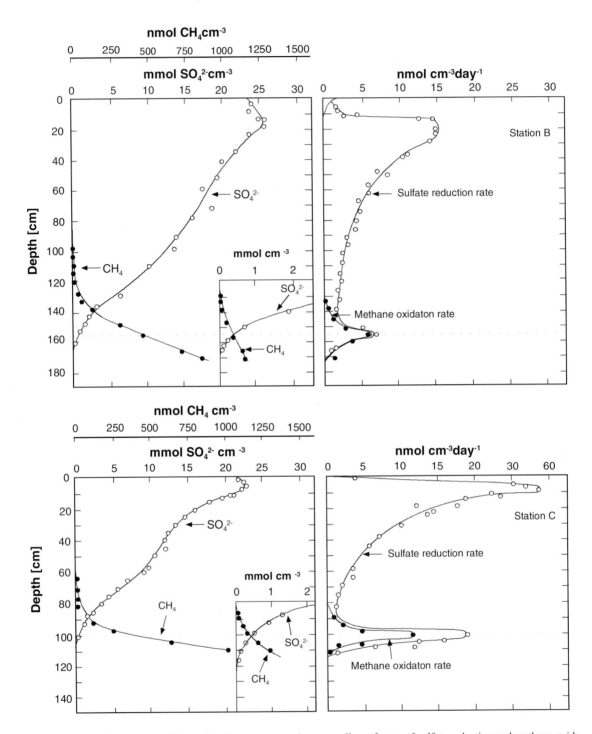

**Fig. 8.1** Profiles of pore-water sulfate and methane concentrations as well as of rates of sulfate reduction and methane oxidation for sediment cores recovered from the Kattegat (Station B; 65 m water depth) and the Skagerrak (Station C; 200 m water depth). The broken horizontal line denotes the depth where sulfate and methane were at equimolar concentrations - indicating the sulfate-methane interface. Refer to text for details. Modified from Iversen and Jørgensen (1985).

mechanism for the reaction and especially the microorganisms involved have, however, not been identified although repeated attempts were made to isolate the responsible bacteria.

As a possible mechanism for the coupled reactions of sulfate reduction and methane oxidation, Hoehler et al. (1994) suggested that anaerobic methane oxidation is mediated by a consortium of methanogenic and sulfate-reducing bacteria at the base of the sulfate zone. They proposed that in the sulfate-methane transition zone sulfate reducers consume hydrogen and thus limit methanogenesis. In the deeper part of the transition zone, the methanogens obtain sufficient hydrogen, presumably through fermentation reactions, to produce methane. This deeper part belongs to the methanogenic zone where sulfate is present at very low concentrations and methane builds up in the pore water. The rapid increase in hydrogen concentrations as sulfate concentrations decrease from 1.2 mmol/l to 10 µmol/l has also been shown in a recent study by Hoehler et al. (1998). Within the transition zone, methanogens may oxidize methane and produce carbon dioxide and hydrogen through 'reverse methanogenesis':

$$CH_4 + 2\,H_2O \rightarrow CO_2 + 4\,H_2 \qquad (8.3)$$

Hoehler et al. (1994) assumed that net methane oxidation occurs because hydrogen is maintained at sufficiently low concentrations through oxidation by sulfate reducers:

$$HSO_4^- + 4\,H_2 \rightarrow HS^- + 4\,H_2O \qquad (8.4)$$

Previous studies have shown that anaerobic methane oxidation is restricted to a narrow depth interval at the sulfate-methane transition zone - just above the depth of sulfate depletion. This has been confirmed by (1) a shift in $\Sigma CO_2$-gradients (Reeburgh 1976, Reeburgh and Heggie 1977), (2) a $\delta^{13}CO_2$-minimum due to oxidation of isotopically light methane with $\delta^{13}C$ values ranging from -40 to -90‰ to $\Sigma CO_2$ more $^{12}C$-enriched than $CO_2$ resulting from the mineralization of other organic substances (Claypool and Kaplan 1974, Reeburgh 1982, Reeburgh and Alperin 1988, Blair and Aller 1995) and (3) peaks of *in-situ* methane oxidation and sulfate reduction rates at the same depth (Iversen and Blackburn 1981, Devol 1983, Iversen and Jørgensen 1985; c.f. Fig. 8.1).

In order to determine anaerobic methane oxidation rates in the sulfate-methane transition zone, Iversen and Jørgensen (1985) carried out radiotracer measurements in sediment cores from the Kattegat and Skagerrak (Denmark). They found that the integrated rates of methane oxidation in the transition zone accounted for 61 to 89% of sulfate reduction at this depth (see Fig. 8.1). Based on pore-water concentration profiles, Niewöhner et al. (1998) calculated the diffusive flux of sulfate and methane into the transition zone in sediments of the Benguela upwelling area. Their calculations revealed that anaerobic methane oxidation accounted for 100% of the deep sulfate reduction within the sulfate-methane transition zone, i.e. it could consume the total diffusive sulfate flux. These findings demonstrate that methane is the primary electron donor for sulfate reduction in the sulfate-methane transition zone.

The fraction of total sulfate reduction (per unit area throughout the sulfate zone), which is fueled by anaerobic methane oxidation, varies among locations (Devol 1983, Alperin and Reeburgh 1985, Iversen and Jørgensen 1985). The particular method used for the determination of sulfate reduction rates also has a strong influence on the estimated contribution of anaerobic methane oxidation to the total sulfate reduction. Percentages of anaerobic methane oxidation in total sulfate reduction deduced from pore-water concentration profiles generally overestimate the significance of anaerobic methane oxidation as will be demonstrated in more detail in the last part of this chapter.

From interstitial flux calculations, Reeburgh (1976, 1982) estimated that approximately 50% of the net downward sulfate flux at a Cariaco Trench station – an anoxic basin – could be accounted for by methane oxidation. For Saanich Inlet sediments, 75% of the downward sulfate flux were attributed to anaerobic methane oxidation according to simple box model calculations (Murray et al. 1978). Devol et al. (1984) obtained lower percentages of 23 to 40% of the downward sulfate flux consumed by methane oxidation for these same sediments using a coupled reaction diffusion model. Iversen and Jørgensen (1985) reported that in Kattegat and Skagerrak sediments methane oxidation accounted for 10% of the electron donor requirement for sulfate reduction measured by radiotracer techniques in the entire sulfate zone.

Under molecular diffusion conditions, the proportion of sulfate consumed by organic matter degradation in comparison to the proportion consumed by anaerobic methane oxidation may be calculated from the shape of sulfate profiles. Borowski et al. (1996) inferred that the linear sulfate pore-water concentration profiles found in Carolina Rise and Blake Ridge sediments imply that anaerobic methane oxidation is the dominant sulfate-consuming process. They calculated the upward methane flux from measured sulfate pore-water profiles assuming that downward sulfate flux is stoichiometrically balanced by upward methane flux. In this way they used the sulfate concentra-

tion profiles as 'proxy measurements' of methane flux from the underlying methane gas hydrates which presence was documented by seismics. (Methane gas hydrates are ice-like solids - generally composed of water and methane - which occur naturally in sediments under conditions of high pressure, low temperature and high methane concentrations). Due to the relatively low solubility of methane and the large pressure difference between sea surface and *in-situ* depth, methane often bubbles out of the pore water during recovery of sediment cores. For this reason, true methane concentration gradients above atmospheric pressure are difficult to determine. Although oxidation of organic matter demonstrably occurs throughout the sulfate zone (see below), the dominant pore water signal is that of anaerobic methane oxidation occurring below. Niewöhner et al. (1998) found similar linear sulfate concentration profiles in sediments of the Benguela upwelling area (see Fig. 8.4) and showed, based on pore-water data, that anaerobic methane oxidation accounted for 100% of deep sulfate reduction.

These studies demonstrate that the sulfate-consuming process occurring in the deeper sediments, i.e. anaerobic methane oxidation, is the dominant factor determining the shape of the sulfate pore-water profiles. Linear sulfate profiles can, therefore, be used to calculate the upward methane flux (e.g. Devol and Ahmed 1981, Borowski et al. 1996, Niewöhner et al. 1998) but they do not give accurate sulfate reduction rates occurring in surface-near sediments. Concave-down pore-water profiles develop when methane-dependent sulfate reduction in deeper sediment layers is less important as is schematically illustrated in Figure 8.2.

With respect to the global significance of anaerobic methane oxidation for the total sulfate reduction at different depositional settings, available data show that in near-shore to continental margin sediments anaerobic methane oxidation can account for 5-20% of total sulfate reduction.

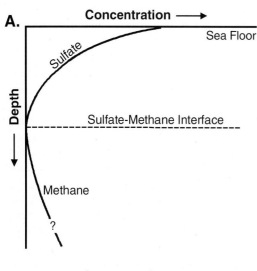

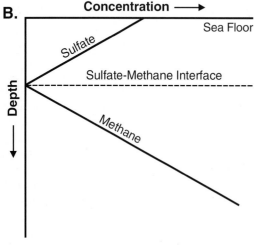

**Fig. 8.2** Differences in the shape of sulfate profiles under diffusive conditions: (a) concave-down curvature of the sulfate pore-water profile when anaerobic methane oxidation is less important and sulfate is primarily consumed by the degradation of organic matter; (b) linear sulfate pore-water profile when anaerobic methane oxidation controls sulfate depletion at the sulfate-methane interface. Modified from Borowski and Paull (1997).

## 8.3    Biotic and Abiotic Processes coupled to Sulfate Reduction

In recent years numerous studies have demonstrated the importance of inorganic sulfur transformations in the early diagenesis of marine sediments. Rather than being a simple cycle, composed of anaerobic bacterial reduction of sulfate to hydrogen sulfide and aerobic reoxidation of $H_2S$ to $SO_4^{2-}$, the transformations of sulfur in aquatic sediments include a combination of inter-mediate cycles or shunts (Jørgensen 1990, Luther and Church 1991, Thamdrup et al. 1993) sche-matically illustrated in Figure 8.3. Within this complex cycle, sulfur compounds occur in oxida-tion states ranging from -2 ($H_2S$) to +6 ($SO_4^{2-}$) (Table 8.2).

By bacterial sulfate reduction $H_2S$ is produced as the extracellular end-product (Widdel and Hansen 1991). During the oxidation of $H_2S$, oxic or anoxic, chemical or biological, compounds such as zero-valent sulfur (in elemental sulfur, poly-sulfides, or polythionates), thiosulfate ($S_2O_3^{2-}$), and sulfite ($SO_3^{2-}$) are produced (Cline and Richards 1969, Pyzik and Sommer 1981, Kelly 1988, Dos Santos Afonso and Stumm 1992). These intermediates may then be further trans-formed by one or several of the following proc-esses (Thamdrup et al. 1994a):

- Respiratory bacterial reduction to $H_2S$,
- bacterial or chemical oxidation,
- chemical precipitation (e.g. FeS formation), or
- bacterial disproportionation to $H_2S$ and $SO_4^{2-}$.

Bacterial disproportionation can be regarded as an inorganic fermentation process whereby $H_2S$ and $SO_4^{2-}$ are produced concurrently without participation of an external electron acceptor or donor (Bak and Pfennig 1987, Thamdrup et al. 1993). The following three equations give the known disproportionation reactions of inorganic sulfur species as well as the stoichiometry of $H_2S$ and $SO_4^{2-}$ production:

Sulfite:

$$4HSO_3^- \rightarrow H_2S + 3SO_4^{2-} + 2H^+ \qquad (8.5)$$
$$H_2S{:}SO_4^{2-} = 1{:}3$$

Thiosulfate:

$$S_2O_3^{2-} + H_2O \rightarrow H_2S + SO_4^{2-} \qquad (8.6)$$
$$H_2S{:}SO_4^{2-} = 1{:}1$$

Elemental sulfur:

$$4\ S^0 + 2\ H_2O + 2OH^- \rightarrow 3H_2S + SO_4^{2-} \quad (8.7)$$
$$H_2S{:}SO_4^{2-} = 3{:}1$$

**Table 8.2** Biogeochemically important forms of sulfur and their oxidation states. Modified from Ehrlich (1996).

| Compound | Formula | Oxidation state(s) of sulfur |
|---|---|---|
| Sulfide | $S^{2-}$ | - 2 |
| Polysulfide | $S_n^{2-}$ | - 2, 0 |
| Sulfur[a] | $S_8$ | 0 |
| Hyposulfite (dithionite) | $S_2O_4^{2-}$ | + 3 |
| Sulfite | $SO_3^{2-}$ | + 4 |
| Thiosulfate[b] | $S_2O_3^{2-}$ | - 1, + 5 |
| Dithionate | $S_2O_6^{2-}$ | + 5 |
| Trithionate | $S_3O_6^{2-}$ | - 2, + 6 |
| Tetrathionate | $S_4O_6^{2-}$ | - 2, + 6 |
| Pentathionate | $S_5O_6^{2-}$ | - 2, + 6 |
| Sulfate | $SO_4^{2-}$ | + 6 |

[a] Occurs in an octagonal ring in crystalline form.

[b] Outer sulfur has an oxidation state of - 1; the inner sulfur has an oxidation state of + 5.

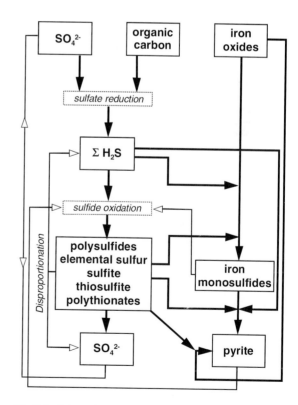

**Fig. 8.3** Schematic representation of the sedimentary sulfur cy-cle as it applies to the formation of iron sulfide minerals. Modi-fied from Cornwell and Sampou (1995) and Vairavamurthy et al. (1995).

and thus represents an important component of seafloor hydrothermal activity, although the actual chemical fluxes through the ridge flanks have not yet been documented. According to Wheat and Mottl (1994) and Ginster et al. (1994), up to 90% of the heat in the neovolcanic axial rift areas may also be removed as a result of diffuse discharge which originates from subseafloor mixing of high-temperature hydrothermal fluids with cold seawater. The significance of axial and off-axial diffuse discharge indicates that the chemical fluxes to the ocean cannot be simply calculated from the composition and venting rates of high-temperature hydrothermal fluids at the ridge crests (cf. Alt 1995).

Following the initial discovery of ore-forming hydrothermal systems in the Red Sea (Miller et al. 1966), black smokers, polymetallic massive sulfide deposits, and vent biota were located at the Galapagos Spreading Center (Corliss et al. 1979) and the East Pacific Rise at 21°N (Francheteau et al. 1979, Spiess et al. 1980). This initiated an intensive investigation of the mid-ocean ridge systems in the Pacific, Atlantic and Indian Ocean, which resulted in the delineation of numerous new sites of hydrothermal activity. In 1986, the first inactive hydrothermal sites were found at the

active back-arc spreading center of the Manus Basin in the Southwest Pacific (Both et al. 1986). Subsequently, active hydrothermal systems and associated sulfide deposits were reported from the Marianas back-arc (Craig et al. 1987, Kastner et al. 1987), the North Fiji back-arc (Auzende et al. 1989), the Okinawa Trough (Halbach et al. 1989), and the Lau back-arc (Fouquet et al. 1991). Today, more than 100 sites of hydrothermal mineralization are known on the modern seafloor (Rona 1988, Rona and Scott 1993, Hannington et al. 1994) including at least 25 sites with high-temperature (350-400°C) black smoker venting.

## 13.1  Hydrothermal Convection and Generation of Hydrothermal Fluids at Mid-Ocean Ridges

At mid-ocean ridges, seawater penetrates deeply into layers 2 and 3 of the newly formed oceanic crust along cracks and fissures, which are a response to thermal contraction and seismic events typical for zones of active seafloor spreading (Fig. 13.1). The seawater circulating through the oceanic crust at seafloor spreading centers is con-

**Table 13. 1**  Conversion of seawater to a hydropthermal fluid through water/rock interaction above a high-level magma chamber.

| temperature: | 2°C | $\rightarrow$ | > 400°C | magma chamber (1200°C) |
|---|---|---|---|---|
| pH (acidity): | 7.8 | $\rightarrow$ | < 4 | $H_2O \rightarrow OH^- + H^+$ |
| | | | | $2OH^- + Mg^{2+} \rightarrow Mg(OH)_2$ (> 350°C: Ca instead of Mg) |
| | | | | $Mg(OH)_2$ fixed in      smectite (<200°C)      chlorite (>200°C) |
| | | | | excess $H^+$ = pH decrease |
| $E_H$ (redox state): | + | $\rightarrow$ | - | $Fe^{2+} \rightarrow Fe^{3+}$ (in basalt) |
| | | | | seawater $SO_4^{2-} \rightarrow S^{2-}$ ($H_2S$) at > 250°C |
| | | | | note: a significant amount of the reduced $S^{2-}$ in the hydrothermal fluid results from leaching of sulfide inclusions in the basalt |

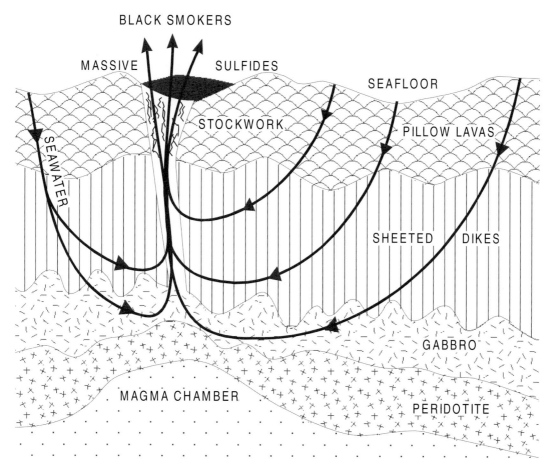

**Fig. 13.1**  Model showing a seawater hydrothermal convection system above a subaxial magma chamber at an oceanic spreading center. Radius of a typical convection cell is about 3-5 km. Depth of the magma chamber usually varies between 1.5 and 3.5 km (see text for details).

verted into an ore-forming hydrothermal fluid in a reaction zone which is situated close to the top of a subaxial magma chamber. Major physical and chemical changes in the circulating seawater include (i) increasing temperature, (ii) decreasing pH, and (iii) decreasing $E_H$ (Table. 13.1).

The increase in temperature from about 2°C to values >400°C (Richardson et al. 1987, Schöps and Herzig 1990) is a result of conductive heating of a small percentage of seawater close to the frozen top of a high-level magma chamber (Cann and Strens 1982). This drives the hydrothermal convection system and gives rise to black smokers at the seafloor. High-resolution seismic reflection studies have indicated that some of these magma reservoirs may occur only 1.5-3.5 km be-

low the seafloor (Detrick et al. 1987, Collier and Sinha 1990; Fig. 13.2). The crustal residence time of seawater in the convection system has been constrained to be 3 years or less (Kadko and Moore 1988). Data from water/rock interaction experiments indicate that, with increasing temperatures, the $Mg^{2+}$ dissolved in seawater (about 1,280 ppm) combines with OH-groups (which originate from the dissociation of seawater at higher temperatures) to form $Mg(OH)_2$. The $Mg^{2+}$ is incorporated in secondary minerals such as smectite (<200°C) and chlorite (>200°C) (Hajash 1975, Seyfried and Mottl 1982, Seyfried et al. 1988, Alt 1995). The removal of OH-groups creates an excess of $H^+$ ions, which is the principal acid-generating reaction responsible for the drop

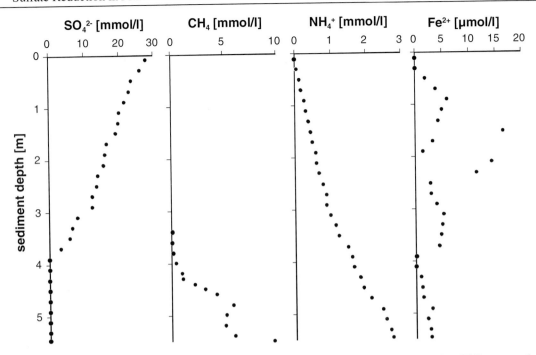

**Fig. 8.5**   Pore-water concentration profiles for gravity core GeoB 4417-7 from the Amazon deep-sea fan (3510 m water depth). Modified from Kasten et al. (1998).

mation is also possible with hydrogen sulfide itself acting as an oxidant for metastable iron sulfides according to the following equation:

$$FeS + H_2S \rightarrow FeS_2 + H_2 \qquad (8.12)$$

In this reaction pyrite is formed in the absence of partially oxidized sulfur such as $S^0$. Rickard (1997) found that this process is by far the most rapid of the pyrite-forming reactions hitherto identified and suggested that it represents the dominant pyrite forming pathway in strictly anoxic systems.

Direct precipitation of pyrite without intermediate iron sulfide precursors was reported for salt marsh sediments, where pore waters were undersaturated with respect to amorphous FeS (Howarth 1979, Giblin and Howarth 1984). In these sediments the oxidizing activity favored the formation of elemental S and polysulfides which were thought to react directly with $Fe^{2+}$. The direct reaction pathway may proceed within hours, resulting in the formation of small, single euhedral pyrite crystals (Rickard 1975, Luther et al. 1982). Framboidal pyrite – apart from that formed by the mechanism presented by Drobner et al. (1990) and Rickard (1997) - is formed slowly (over years) via intermediate iron sulfides (Sweeney and Kaplan 1973, Raiswell 1982).

### 8.3.2    Effects of Sulfate Reduction on Sedimentary Solid Phases

Sulfate reduction – either occurring due to the oxidation of methane or the mineralization of other organic material – can lead to a pronounced overprinting or modification of the primary sediment composition by dissolution of minerals and precipitation of authigenic minerals. The formation of authigenic iron sulfides has already been described above. Besides the precipitation of mineral phases, dissolution of a wide range of minerals initially supplied to the seafloor can occur in sediments that undergo sulfate reduction, and this may complicate the interpretation of the primary sedimentary record. Two examples for sediment constituents relevant to paleoceanographic research, which are subject to dissolution under sulfate-reducing conditions, are iron (hydr)oxides and barite ($BaSO_4$).

Magnetic iron (hydr)oxides, especially magnetite ($Fe_3O_4$), are the main carriers of remanent magnetization in sediments. Magnetostratigraphy of deep-sea sediment cores can be used as a valuable method of dating and comparing sedimentary records. Dissolution of these magnetic minerals under sulfate- and iron-reducing conditions and/or subsequent precipitation of authigenic minerals at different sediment levels may, however, alter the initial remanent magnetization and seriously compromise the interpretation of the sedimentary geomagnetic record (e.g. Karlin and Levi 1983, 1985) (compare also Chap. 2).

A second sedimentary component important for paleoceanographic reconstructions, which is prone to dissolution under conditions of sulfate reduction, is the barium sulfate mineral, barite ($BaSO_4$). Since a correlation has been detected between barite and the flux of organic matter through the water column, the concentration of barite in sediments is considered a promising parameter to reconstruct past changes in ocean productivity. Since dissolution of barite occurs in the

sediment zone depleted of sulfate, the use of barite as a geochemical tracer for paleoproductivity is limited in such sediments (e.g. Gingele and Dahmke 1994). This effect of barite dissolution in the zone of sulfate depletion, and the subsequent reprecipitation of barite at higher sediment levels is illustrated in Figure 8.6 for a sediment core recovered from the continental margin off Angola.

A particularly pronounced overprinting of the sedimentary solid phase by mineral dissolution and authigenic mineral precipitation can take place in connection with sulfate reduction during non-steady-state diagenesis - i.e. phases of deposition during which sedimentary conditions are not constant over time and sediment geochemistry

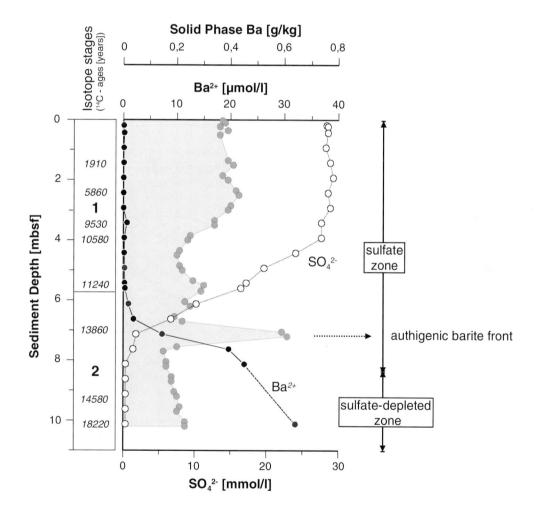

**Fig 8.6** Geochemical results for core GeoB 1023-4 recovered off north Angola (17°09.6'S, 10°59.9'E, 2047 m water depth). Modified from Gingele et al. (in press), after Kölling (1991). Barium and sulfate pore-water concentration profiles as well as the distribution of solid-phase barium indicate the precipitation of authigenic barite at a front slightly above the depth of complete sulfate consumption. At the base of the sulfate zone, sulfate reduction occurs at high rates and barite becomes undersaturated and is thus subject to dissolution due to the depletion of sulfate from pore water. Dissolved barium diffuses upwards into the sulfate zone where the mineral barite becomes supersaturated and barite precipitates at a front at the base of the sulfate zone.

consists of amorphous silica and/or barite and anhydrite instead of sulfides) is their central chalco-
pyrite-lined orifice. In the models of Haymon
(1983) and Goldfarb et al. (1983), anhydrite is precipitated around a black smoker vent at the leading edge of chimney growth, where hot hydrothermal fluids first encounter cold seawater. The
anhydrite is precipitated from $Ca^{2+}$ in the vent
fluids and $SO_4^{2-}$ in ambient seawater, although
some anhydrite in the outer walls of the chimney
or in the interior of hydrothermal mounds (cf.
Humphris et al. 1995) may also be formed simply
by the conductive heating of seawater above

150°C. The anhydrite which forms the initial wall
of the chimney is gradually displaced by high-
temperature Cu-Fe-sulfides as the structure grows
upward and outward. Because of its retrograde
solubility, anhydrite is not well preserved in older
chimney complexes and eventually dissolves at
ambient temperatures and seafloor pressures
(Haymon and Kastner 1981). As a result, many
black smokers that are cemented by anhydrite are
inherently unstable and ultimately collapse to become part of the sulfide mound (Fig. 13.3).

Most chimneys have growth rates that are
rapid in comparison to the half-life of $^{210}Pb$ (22.3

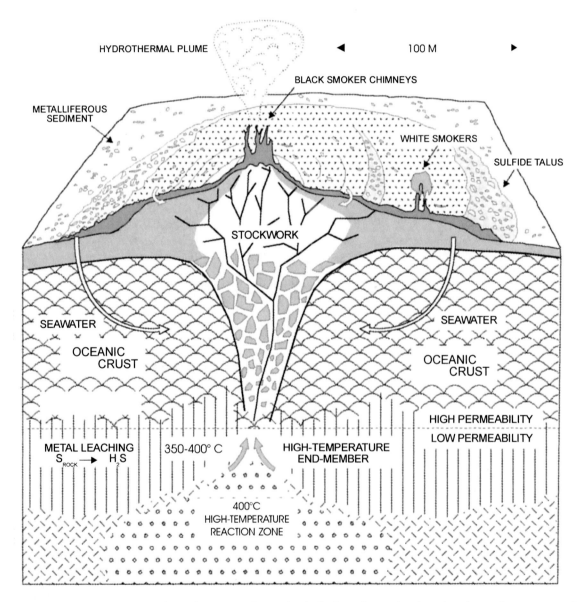

**Fig. 13.3** Surface features and internal structure of an active hydrothermal mound and stockwork complex at an oceanic
spreading center.

years) and radioisotope ages of several days or less for precipitates of some active vents are consistent with growth rates observed from submersibles (5-10 cm per day: Hekinian et al. 1983, Johnson and Tunnicliffe 1985). Larger vent complexes commonly have measured ages on the order of decades (Koski et al. 1994), but the data for entire vent fields may span several thousands of years (Lalou et al. 1993).

At typical black smoker vents, a very large proportion (at least 90%) of the metals and sulfur carried in solution are lost to a hydrothermal plume in the overlying seawater column rather than deposited as chimneys (Converse et al. 1984). The metals are precipitated as sulfide particles in the plume above the black smokers and are rapidly oxidized and dispersed over distances of several kilometers from the vent (Feely et al. 1987, 1994a,b, Mottl and McConachy 1990). Due to oxidation and dissolution, these particles also release some of the metals back into seawater (Feely et al. 1987, Metz and Trefry 1993). Particle settling models indicate that only a small fraction of the metals is likely to accumulate as plume fallout in the immediate vicinity of the vents (Feely et al. 1987) and observations of particulate Fe dispersal confirm that most of the metals produced at a vent site are carried away by buoyant plumes (Baker et al. 1985, Feely et al. 1994a,b).

Black smokers usually grow on hydrothermal mounds that are large enough to be thermally and chemically insulated from the surrounding seawater (Fig. 13.3). The sulfide mounds also serve to trap rising hydrothermal fluids and impede the loss of metals and sulfur by direct venting into the hydrothermal plume. The largest sulfide deposits are often composite bodies which appear to have evolved from several smaller hydrothermal mounds (Embley et al. 1988). Many of the original chimney structures are overgrown and eventually incorporated in the larger mound, destroying primary textural and mineralogical relationships by hydrothermal replacement. Large sulfide mounds are also constructed from the accumulation of sulfide debris produced by collapsing chimneys, and most large deposits are littered with the debris of older sulfide structures. In many places, new chimneys can be seen growing on top of the sulfide talus, and this debris is eventually overgrown, cemented, and incorporated within the mound (Rona et al. 1993, Hannington et al. 1995). At the same time, high-temperature fluids circulating or trapped beneath the deposit precipi-

tate new sulfide minerals in fractures and open spaces. Chimneys that are perched on the outer surface of an active mound apparently tap these high-temperature fluids but account for only a small part of the total mass of the deposit.

The common presence of high-temperature chimneys at the tops of the deposits and lower-temperature chimneys on their flanks suggests that most large mounds are zoned similar to many ancient deposits on land (cf. Franklin et al. 1981). The most common arrangement of mineral assemblages is a high-temperature Cu-rich core and a cooler, Zn-rich outer margin. Mineralogical zonation within a deposit is principally a result of hydrothermal reworking, whereby low-temperature phases such as sphalerite are dissolved by later, higher-temperature fluids and redistributed to the outer margins of the deposit. Sulfide samples that have been recovered from the interiors of hydrothermal mounds show signs of extensive hydrothermal recrystallization and annealing, especially when compared to the delicate, fine-grained sulfides found in surface precipitates (Humphris et al. 1995). As a result of the extensive hydrothermal reworking in large sulfide mounds, it must be considered that hydrothermal fluids arriving at the seafloor are also likely to have been substantially modified by interaction with pre-existing hydrothermal precipitates in their path (e.g. Janecky and Shanks 1988), and caution should be used when interpreting measurements made at the surface of a large mound to infer processes related solely to the direct venting of an end-product fluid.

## 13.4 Physical and Chemical Characteristics of Hydrothermal Vent Fluids

Most black smoker fluids are strongly buffered close to equilibrium with pyrite-pyrrhotite-magnetite, although the proximity of the fluids to this buffer assemblage is not necessarily a reflection of the state of saturation of the minerals in solution (Janecky and Seyfried 1984, Bowers et al. 1985, Tivey et al. 1995). In addition, because of the high concentrations of reduced components such as ferrous iron and $H_2S$, the fluids do not deviate significantly from the pyrite-pyrrhotite redox buffer, even after substantial mixing and cooling, and the common occurrence of both py-

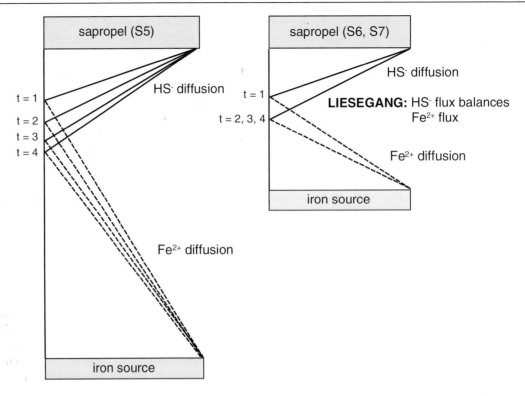

**Fig. 8.9** Schematic model for iron sulfide formation below sapropels. Two situations are distinguished, one in which the system remains HS⁻ dominated (left) and another one in which the Fe source is located closer to the sapropel whereby Liesegang phenomena (Berner 1969) occur (right). In the Liesegang situation, the front at which downward diffusing hydrogen sulfide and upward diffusing iron react to form iron sulfides is fixed at a particular level below the sapropel for a prolonged period of time. Modified from Passier et al. (1996).

et al. 1993, Schulz et al. 1994 and Niewöhner et al. 1998). The motivation for using this method is the fact that, in most cases, early diagenetic processes leave characteristic signatures in the chemistry of the pore water and the sedimentary solid phase. For example, the depletion of a pore-water constituent points to the consumption of this chemical compound in a specific reaction, e.g. the use of interstitial sulfate for the degradation of organic matter. The pore-water distributions of the different compounds consumed in the oxidation of organic matter as well as the mineralization products liberated into the pore water are therefore used to determine the reaction rates of the particular processes.

Other authors have used incubation experiments to determine the rate of sulfate reduction in sediments (e.g. Martens and Berner 1974, Goldhaber et al. 1977). In these experiments - which are typically conducted over periods of days to months - homogenized sediment from desired depths is enclosed in jars or bags and analyzed sequentially for consumption of oxidants,

or liberation of reaction end-products (Reeburgh 1983).

However, the specific reactions and reaction rates cannot always be deduced from the solid phase or the pore water of sediments, because the products of the specific chemical reactions may undergo further reactions and not accumulate in the sedimentary solid phase or pore water (Reeburgh 1983). This has also been demonstrated above by the description of the numerous biotic and abiotic reactions between sulfur compounds and other pore-water or solid-phase constituents within marine deposits. In order to accurately quantify the process of sulfate reduction for the degradation of organic matter and diagenesis in anoxic sediments, radiotracer techniques have been developed (Ivanov 1968, Sorokin 1962, Jørgensen 1978, Fossing and Jørgensen 1989, 1990). These methods use ³⁵S-labeled sulfate for either core injection (Jørgensen 1977) or sediment mixing (Ivanov 1968, Sorokin 1962). They allow a quantification of the gross sulfate reduction rates rather than the net sulfate reduction rates derived from pore-water concentration profiles or incubation experiments (Reeburgh 1983).

Radiotracer techniques for the determination of sulfate reduction rates were significantly improved in recent years (e.g. Devol and Ahmed 1981, Devol et al. 1984, Fossing and Jørgensen 1989, 1990). A description of the methodology of $^{35}S$ radiotracer techniques is given in Chapter 5.

A comparison of sulfate reduction rates derived from radiotracer techniques and those deduced from pore-water concentration profiles of sulfate often reveals large discrepancies. The sulfate reduction rates calculated from pore-water gradients are significantly lower than those obtained from direct measurements by the $^{35}SO_4^{2-}$ technique.

Studies of sulfate reduction in sediments of the Benguela upwelling region can serve as valuable examples to illustrate this observation. Based on the $^{35}SO_4^{2-}$-technique, Ferdelman et al. (in press) determined sulfate reduction rates in surface sediments at 17 stations in the Benguela upwelling area. Highest sulfate reduction rates were found in the upper 2 to 5 cm of the sediments. Fossing et al. (submitted) presented pore-water data and $^{35}SO_4^{2-}$ derived sulfate reduction rates for two stations in the Benguela upwelling area. At both stations, the highest sulfate reduction rates ($^{35}SO_4^{2-}$ technique) were measured at the sediment surface and then exponentially decreased with depth. In contrast, the pore-water concentration profile of sulfate determined by Niewöhner et al. (1998) at station GeoB 3714 (Fig. 8.4), which was also studied by Ferdelman et al. (in press) and Fossing et al. (submitted), showed linear depletion in the upper meters, indicating little or no net sulfate consumption around this depth (compare Chap. 3).

The comparison of sulfate reduction rates based on direct measurements ($^{35}SO_4^{2-}$ labelling) with rates obtained from pore-water profiles reveals that the areal rates of net sulfate reduction calculated from pore-water profiles underestimate the (gross) sulfate reduction rates as determined by the $^{35}SO_4^{2-}$ method. The largest discrepancies between the two methods occur close to the sediment surface where the highest sulfate reduction rates were measured by the radiotracer method. At the depth of the sulfate-methane transition the pore-water based sulfate reduction rates are in good agreement with the sulfate reduction rates obtained from the direct $^{35}SO_4^{2-}$ measurements. The total rates of organic matter mineralization

via sulfate reduction (i.e. gross sulfate reduction rates) may be seriously underestimated by only examining the net sulfate consumption rates. The sulfate reduction rates based on sulfate pore-water gradients may rather be a measure of deep methane flux than of organic carbon decomposition coupled directly to sulfate reduction as has been inferred by Borowski et al. (1996) and Niewöhner et al. (1998) (compare above).

In order to reconcile the modelled and measured sulfate reduction rates, Fossing et al. (submitted) employed a pore-water exchange function which considers irrigation using the fitting procedures of Berg et al. (1998). From the results of this modelling, they suggested irrigation of pore water in tubes or burrows by activities of tube-dwelling animals or by some other physical disturbance (e.g. bubble ebullition) as a possible transport mechanism to account for the observed discrepancy between measured and modelled (considering only diffusion) sulfate reduction rates.

As an alternative explanation for the discrepancy between sulfate reduction rates measured by radiotracer techniques and those derived from pore-water concentration gradients in the near-surface sediments, we propose the re-oxidation of reduced sulfur species formed during sulfate reduction. We assume that the mixing between Fe and Mn oxide particles and solid phase as well as dissolved sulfides taking place during bioturbation leads to a very effective oxidation of the reduced sulfur compounds. For this reason, we neither observed a build-up of an amount of authigenic iron sulfides corresponding to the determined gross sulfate reduction rates nor did we measure a substantial change in sulfate pore-water gradients in these surface sediments. As has been demonstrated by the studies of Ferdelman et al. (in press) and Fossing et al. (submitted), the largest discrepancies between measured and modelled sulfate reduction rates occur close to the sediment surface. This finding supports our assumption of the re-oxidation of sulfides, since bioturbation is restricted to the top few centimeters of the sediment.

Cycling of oxidized and reduced Fe and Mn phases as well as of organic matter in surface sediments due to bioturbation and mixing have been described e.g. by Aller and Rude (1988), Thamdrup et al. (1994b) and Haese (1997). According to studies of Haese (1997), Fe-oxides can be recycled up to 50 times in near-surface marine deposits (c.f. Chap. 7). The bioturbation-driven

the water depth is less than 1600 m, the fluid will begin to boil beneath the seafloor and may separate a vapor-rich phase. At these lower pressures, the process of boiling differs from that of phase separation in the deeper parts of the mid-ocean ridges, because the density difference between the vapor and liquid is much greater and therefore may facilitate the separation of a low-salinity, gas-rich phase. Such fluids have been encountered at vents in the caldera of Axial Seamount (1540 m water depth, fluid temperature 349°C) where low salinity and high gas contents have been measured (Massoth et al. 1989, Butterfield et al. 1990).

## 13.5 The Chemical Composition of Hydrothermal Vent Fluids and Precipitates

Since the first discovery of high-temperature hydrothermal vents at the East Pacific Rise 21°N in 1979, hydrothermal fluids have been sampled at numerous sites at mid-ocean ridges and back-arc spreading centers. Both Mg and sulfate show a negative correlation with temperature, and an extrapolation to the zero values of Mg and $SO_4$ intersects the temperature axis at a point referred to as the end-member temperature (Fig. 13.5). Published data on the chemical composition of these fluids are summarized in Von Damm (1990) and Von Damm (1995) and some examples are given in Table 13.3. Time series measurements on the order of a decade at a number of sites have indicated that the chemical composition of vent fluids at individual sites does not show significant temporal variability and is "steady-state" once the system has stabilized after a new volcanic event (Von Damm 1995). Variability in the chemical composition of the hydrothermal fluids does not appear to be related to the depth of the individual vent site to reflect changes as a function of temperature. Phase separation is responsible for most of the chemical variability observed (cf. Von Damm 1995, see below). In some areas, contributions from a degassing magma are also inferred (Lupton et al. 1991, de Ronde 1995, Herzig et al. 1998).

Despite the long-term stability of vent fluids at individual sites a wide range of salinities has been observed between different sites ranging from about 30% below (176 mM/kg; Von Damm

and Bischoff 1987) to 200% above (1,090 mM/kg; Massoth et al. 1989) seawater concentrations (546 mM/kg). Salinities are important for the chemical composition of vent fluids as Cl is the major complexing anion in hydrothermal systems. These extreme salinities cannot be accounted for by hydration of the oceanic crust (Cathles 1983) or precipitation and dissolution of Cl-bearing mineral phases (Edmond et al. 1979b, Seyfried et al. 1986) but are interpreted to be a result of supercritical phase separation at the top of the magma chamber followed by mixing of the brines and the vapor phases during ascent (cf. Nehlig 1991, Palmer 1992, James et al. 1995).

Most hydrothermal fluids contain considerable amounts of $CH_4$ and $^3He$. $^3He$ is significantly enriched in vent fluids over saturated seawater, with $^3He/^4He$ ratios between about 7-9. These ratios represent typical mantle values (Stuart et al. 1995), indicating that $^3He$ originates from MORB magma, which is degassing into a hydrothermal system (Lupton 1983, Baker and Lupton 1990). Lupton and Craig (1981) have demonstrated that $^3He$ behaves extremely conservatively in the water column and can be traced over distances up to 2000 km from the point of origin (Fig. 13.6). In unsedimented areas, methane is a product of inorganic chemical reactions within the hydrothermal system and also related to magmatic degassing. Methane in combination with TDM (total dissolvable Mn) have been found to be enriched about $10^6$-fold over ambient seawater in high-temperature vent fluids (e.g. Von Damm et al. 1985b). In diluted buoyant hydrothermal plumes, these values are still 100-fold enriched relative to seawater (Klinkhammer et al. 1986, Charlou et al. 1991) which makes $CH_4$ and Mn valuable tracers for prospecting hydrothermal vent areas (Herzig and Plüger 1988, Plüger et al. 1990; Fig. 13.7).

Concentrations of trace elements such as Ag, As, Cd, Co, Se, and Au have been measured in some vent fluids (Von Damm 1990, Campbell et al. 1988a,b, Fouquet et al. 1993a, Trefry et al. 1994, Edmond et al. 1995; Table. 13.3) and certain metals such as Cu, Co, Mo, and Se appear to show a strong positive temperature-concentration relationship which likely accounts for observed enrichments of these elements in the sulfides from high-temperature chimneys (Hannington et al. 1991). Other elements such as Zn, Ag, Cd, Pb, and Sb appear to be significantly enriched in lower-temperature fluids (Trefry et al. 1994). However, these data are scarce, and this may re-

**Table. 13.3** Chemical composition of hydrothermal fluids from selected hydrothermal fields at mid-ocean ridges and back-arc spreading centers in comparison to seawater.

| | $T_{max}$ (°C) | pH (25°C) | Cl (ppt) | $H_2S$ (ppm) | Na (ppm) | K (ppm) | Ca (ppm) | Ba (ppm) | Sr (ppm) | Fe (ppm) | Mn (ppm) | Zn (ppm) | Cu (ppm) | Si (ppm) | Reference |
|---|---|---|---|---|---|---|---|---|---|---|---|---|---|---|---|
| MARK (MAR 23°N) | 350 | 3.9 | 19.8 | 201 | 11,725 | 931 | 421 | - | 5 | 122 | 27 | 3 | 1.0 | 514 | Von Damm (19 |
| TAG (MAR 26°N) | 366 | 3.8 | 22.5 | 119 | 12,805 | 669 | 1,235 | - | 9 | 313 | 37 | 3 | 10.0 | 583 | Edmond et al. ( |
| Lucky Strike (MAR 37°N) | 332 | - | 19.3 | - | - | - | - | - | - | 35 | 21 | - | - | - | Von Damm (19 |
| EPR (11°N) | 347 | 3.7 | 24.3 | 416 | 13,265 | 1,287 | 1,411 | - | 12 | 361 | 51 | - | - | 579 | Von Damm (19 |
| EPR (13°N) | (380) | 3.3 | 26.9 | 279 | 13,702 | 1,165 | 2,204 | - | 16 | 603 | 160 | - | - | 618 | Von Damm (19 |
| EPR (21°N) | 355 | 3.8 | 20.5 | 286 | 11,725 | 1,009 | 834 | 2.2 | 9 | 136 | 55 | - | - | 548 | Von Damm (19 |
| Guyamas Basin | 315 | 5.9 | 22.6 | 204 | 11,794 | 1,924 | 1,663 | 7.4 | 22 | 10 | 13 | - | - | 388 | Von Damm (19 |
| Escanaba Trough | 217 | 5.4 | 23.7 | 51 | 12,874 | 1,580 | 1,339 | - | 18 | 1 | 1 | - | - | 194 | Von Damm (19 |
| Middle Valley (JFR) | 276 | 5.5 | 20.5 | 102 | 9,150 | 731 | 3,247 | 2.1 | 23 | 1 | 4 | 0.1 | 0.1 | 298 | Von Damm (19 |
| South Cleft (JFR) | 285 | 3.2 | 31.8 | (102) | 15,196 | 1,459 | 3,395 | - | 20 | 575 | 143 | 24 | 1.0 | 640 | Von Damm (19 |
| North Cleft (JFR) | 327 | 3.0 | 31.0 | 124 | 15,679 | 1,580 | 2,922 | - | 20 | 165 | 65 | 16 | 0.5 | 559 | Von Damm (19 |
| Axial Seamount (JFR) | 328 | 3.5 | 22.1 | 242 | 11,472 | 1,048 | 1,876 | 3.6 | 17 | 59,477 | 63,178 | 7 | 0.6 | 424 | Von Damm (19 |
| Endeavour (JFR) | 370 | 4.4 | 16.2 | 167 | 8,230 | 1,004 | 1,447 | - | 12 | 54,451 | 14,339 | 2 | 1.3 | 455 | Von Damm (19 |
| Lau Basin (SW-Pacific) | 334 | 2.0 | 28.0 | - | 13,564 | 3,089 | 1,655 | > 5.4 | 2 | 140 | 390 | 196 | 2.2 | 393 | Fouquet et al. ( |
| Seawater | 2 | 7.8 | 19.4 | - | 10,759 | 399 | 413 | 0.02 | 8 | $6 \cdot 10^{-5}$ | $5 \cdot 10^{-5}$ | $7 \cdot 10^{-4}$ | $5 \cdot 10^{-4}$ | 5 | |

Howarth, R.W., 1979. Pyrite: Its rapid formation in a salt marsh and its importance in ecosystem metabolism. Science 203: 49-50

Ivanov, M.V., 1968. Microbiological processes in the formation of sulfur deposits. Israel Program for Scientific Translations, Jerusalem.

Iversen, N. and Blachburn, T.H., 1981. Seasonal rates of methane oxidation in anoxid marine sediments. Appl. Environ. Microbiology, 41: 1295-1300.

Iversen, N. and Jørgensen, B.B., 1985. Anaerobic methane oxidation rates at the sulfate-methane transition in marine sediments from Kattegat and Skagerrak (Denmark). Limnology and Oceanography, 30: 944-955.

Jørgensen, B.B., 1977. The sulfur cycle of a coastal marine sediment (Limfjorden, Denmark). Limnology and Oceanography, 22: 814-832.

Jørgensen, B.B., 1978a. A comparison of methods for the quantification of bacterial sulfate reduction in coastal marine sediments. I. Measurement with radiotracer techniques. Geomicrobiol. Journal, 1: 11-27.

Jørgensen, B.B., 1978b. A comparison of methods for the quantification of bacterial sulfate reduction in coastal marine sediments. II. Calculation from mathematical models. Geomicrobiol. Journal, 1: 29-47.

Jørgensen, B.B., 1982. Mineralization of organic matter in the sea bed - the role of sulphate reduction. Nature, 296: 643-645.

Jørgensen, B.B., 1983. Processes at the sediment-water interface. In: Bolin, B. and Cook, R.C. (eds), The major biogeochemical cycles and their interactions. SCOPE , pp. 477-509.

Jørgensen, B.B., 1990. A thiosulfate shunt in the sulfur cycle of marine sediments. Science, 249: 152-154.

Karlin, R. and Levi, S., 1983. Diagnosis of magnetic minerals in recent hemipelagic sediments. Nature, 303: 327-330.

Karlin, R. and Levi, S., 1985. Geochemical and sedimentological control of the magnetic properties of hemipelagic sediments. Journal of Geophysical Research, 90: 10373-10392.

Kasten, S., Freudenthal, T., Gingele, F.X., von Dobeneck, T. and Schulz, H.D., 1998. Simultaneous formation of iron-rich layers at different redox boundaries in sediments of the Amazon Deep-Sea Fan. Geochimica et Cosmochimica Acta 62: 2253-2264.

Kelly, D.P., 1988. Oxidation of sulfur compounds. In: Cole, A.S. and Ferguson, S.J. (eds), The Nitrogen and Sulfur Cycles. Soc. Gen. Microbiol., 42, pp. 65-98.

Kölling, A., 1991. Frühdiagnetische Prozesse und Stoff-Flüsse in marinen und ästuarinen Sedimenten. Berichte, 15, Fachbereich Geowissenschaften, Universität Bremen, 140 pp.

Krein, E.B. and Aizenshtat, Z., 1995. Proposed thermal pathways for sulfur transformations in organic macromolecules: Laboratory simulation experiments. In: Vairavamurthy, M.A. and Schoonen, M.A.A. (eds), Geochemical Transformations of Sedimentary Sulfur, ACS symposium series, 612, Washington, DC, pp. 110-137.

Kremling, K., 1985. The distribution of cadmium, copper, nickel, manganese, and aluminium in surface waters of the open Atlantic and European shelf area. Deep-Sea Research, 32, 531-555.

Lovley, D.R. and Phillips, E.J.P., 1988. Manganese inhibition of microbial iron reduction in anaerobic sediments. Geomicrobiol. Journal, 6: 145-155.

Luther III, G.W., Giblin, A., Howarth, R.W. and Ryans, R.A., 1982. Pyrite and oxidized iron mineral phases formed from pyrite oxidation in salt marsh and estuarine sediments. Geochimica et Cosmochimica Acta, 46: 2665-2669.

Luther III, G.W. and Church, T.M., 1991. An overview of the environment chemistry of sulfur in wetland systems. In: Howarth, R.W. et al (eds), Sulfur cycling on the continents. John Wiley, pp: 125-144.

Martens, C.S. and Berner, R.A., 1974. Methane production in the interstitial waters of sulfate-depleted marine sediments. Science, 185: 1167-1169.

Middelburg, J.B.M., 1990. Early diagnesis and authigenic mineral formation in anoxic sediments of Kau Bay, Indonesia. PhD Thesis, Universitiy of Utrecht, Utrecht, 177 pp.

Murray, J.W., Grundmanis, V. and Smethie, W.M. Jr, 1978. Interstitial water chemistry in sediments os Saanich Inlet. Geochimica et Cosmochimica Acta, 42: 1011-1026.

Niewöhner, C., Hensen, C., Kasten, S., Zabel, M. and Schulz, H.D., 1998. Deep sulfate reduction completely mediated by anaerobic methane oxidation in sediments of the upwelling area off Namibia. Geochimica et Cosmochimica Acta, 62: 455-464.

Orr, W.L. and White, C.M. (eds), 1990. Geochemistry of sulfur in fossil fuels. ACS Symposium Series, 429, Washington, DC.

Passier, H.F., Middelburg, J.J., Os, B.J.H.v. and Lange, G.J.d., 1996. Diagenetic pyritisation under eastern Mediterranean sapropels caused by downward sulphide diffusion. Geochimica et Cosmochimica Acta, 60: 751-763.

Postma, D., 1985. Concentration of Mn and separation from Fe in sediments - I. Kinetics and stoichiometry of the reaction between birnessite and dissolved Fe(II) at 10°C. Geochimica et Cosmochimica Acta, 49: 1023-1033.

Pyzik, A.J. and Sommer, S.E., 1981. Sedimentary iron monosulfides: kinetics and mechanism of formation. Geochimica et Cosmochimica Acta, 45: 687-698.

Raiswell, R., 1982. Pyrite texture, isotopic composition and the availability of iron. American Journal of Science, 282: 1244-1265.

Raiswell, R., 1988. Chemical model for the origin of minor limestone-shale cycles by anaerobic methane oxidation. Geology, 16: 641-644.

Reeburgh, W.S., 1976. Methane consumption in Cariaco Trench waters and sediments. Earth and Planetary Science Letters, 28: 337-344.

Reeburgh, W.S. and Heggie, D.T., 1977. Microbial methane consumbtion reactions and their effect on methane distributions in freshwater and marine environments. Limnology and Oceanography, 22: 1-9.

Reeburgh, W.S., 1980. Anaerobic methan oxidation: rate depth disributions in Skan Bay sediments. Earth and Planetary Science Letters, 47: 345-352.

Reeburgh, W.S., 1982. A major sink and flux control for methane in sediments: Anaerobic consumption. In: Fanning, K.A. and Manheim, F.T. (eds), The dynamic environment. Heath, Lexington, MA, pp. 203-217.

Reeburgh, W.S., 1983. Rates of biogeochemical processes in anoxic sediments. Ann. Rev. Earth Planet. Sci., 11: 269-298.

Reeburgh, W.S. and Alperin, M.J., 1988. Studies on anaerobic methane oxidation. SCOPE/ UNEP, 66: 367-375.

Rickard, D.T., 1975. Kinetics and mechanisms of pyrite formation at low temperatures. American Journal of Science, 275: 636-652.

Rickard, D. and Luther III, G.W., 1997. Kinetics of pyrite formation by the H2S oxidation of iron(II) monosulfide in aqueous solutions between 25 and 125°C: The rate equation. Geochimica et Cosmochimica Acta, 61: 115-134.

Sagemann, J., Jørgensen, B.B. and Greeff, O., 1998. Temperature dependence and rates of sulfate reduction in cold sediments of Svalbard, Arctic Ocean. Geomicrobiol. Journal, 15: 85-100.

Schinzel, U., 1993. Laborversuche zu frühdiagnetischen Reaktionen von Eisen (III)-Oxidhydraten in marinen Sedimenten. Berichte, 36, Fachbereich Geowissenschaften, Universität Bremen, 189 pp

Schouten, S., Eglington, T.I., Sinninghe Damsté, J.S. and de Leeuw, J.W., 1995. Influence of sulfur cross-linking on the molecular size distribution of sulfur-rich macromolecules in bitumen. In: Vairavamurthy, M.A. and Schoonen, M.A.A. (eds), Geochemical transformations of sedimentary sulfur, ACS Symposium Series, 612, Washington, DC, pp. 80-92.

Schulz, H.D., Dahmke, A., Schinzel, U., Wallmann, K. and Zabel, M., 1994. Early diagenetic processes, fluxes and reaction rates in sediments of the South Atlantic. Geochimica et Cosmochimica Acta, 58: 2041-2060.

Schulz, H.D., cruise participants 1996. Report and preliminary result of METEOR cruise M 34/2 Walvis Bay-Walvis Bay, 29.01.1996-18.02.1996. Berichte, 78, Fachbereich Geo-wissenschaften, Universität Bremen, 133 pp.

Sorokin, Y.I., 1962. Experimental investigation of bacteriel sulfate reduction in the Black Sea using S35. Microbiology, 31: 329-335.

Sweeney, R.E. and Kaplan, I.R., 1973. Pyrite Framboid Formation: Laboratory Synthesis and Marine Sediments. Economic Geology, 68: 618-634.

Swider, K.T. and Mackin, J.E., 1989. Transformation of sulfur compounds in marsh-flat sediments. Geochimica et Cosmochimica Acta, 53: 2311-2323.

Thamdrup, F., Finster, K., Hansen, J.W. and Bak, F., 1993. Bacterial disproportionation of elemental sulfur coupled to chemical reduction of iron or manganese. Applied and Environmental Microbiology, 59: 101-108.

Thamdrup, B., Finster, K., Fossing, H., Hansen, J.W. and Jørgensen, B.B., 1994a. Thiosulfate and sulfite distributions in porewater of marine related to manganese, iron, and sulfur geochemistry. Geochimica et Cosmochimica Acta, 58: 67-73.

Thamdrup, B., Glud, R.N. and Hansen, J.W., 1994b. Manganese oxidation and in situ manganese fluxes from a coastal sediment. Geochimica et Cosmochimica Acta, 58: 2563-2570.

Thamdrup, B. and Canfield, D.E., 1996. Pathways of carbon oxidation in continental margin sediments off central Chile. Limnology and Oceanography, 41: 1629-1650.

Torres, M.E., Brumsack, H.J., Bohrmann, G. and Emeis, K.C., 1996. Barite fronts in continental margin sediments: A new look at barium remobilization in the zone of sulfate reduction and formation of heavy barites in diagenetic fronts. Chemical Geology, 127: 125-139.

Vairavamurthy, M.A., Orr, W.L. and Manowitz, B., 1995. Geochemical transformation of sedimentary sulfur: an introduction. In: Vairavamurthy, M.A. and Schoonen, M.A.A. (eds), Geochemical tranformation of sedimentary sulfur. ACS Symposium, 612, Washington, DC, pp. 1-17.

Widdel, F., 1988. Microbiology and ecology of sulfate-and sulfur-reduction bacteria. In: Zehnder, A.J.B. (ed), Biology of anaerobic microorganisms. Wiley & Sons, NY, pp. 469-585.

Widdel, F. and Hansen, T.A., 1991. The dissimilatory sulfate-and sulfur-reducing bacteria. In: Balows, H. et al. (eds), The procaryotes. Springer, pp. 583-624.

**Table 13.5** Bulk chemical composition of seafloor poly-metallic sulfides from mid-ocean ridges and back-arc spreading centers (after Herzig and Hannington 1995).

| Element | Mid-Ocean Ridges[1] | Back-Arc Ridges[2] |
|---------|:---------:|:---------:|
| n | 890 | 317 |
| Fe (wt%) | 23.6 | 13.3 |
| Cu | 4.3 | 5.1 |
| Zn | 11.7 | 15.1 |
| Pb | 0.2 | 1.2 |
| As | 0.03 | 0.1 |
| Sb | 0.01 | 0.01 |
| Ba | 1.7 | 13.0 |
| Ag (ppm) | 143 | 195 |
| Au | 1.2 | 2.9 |

[1] Explorer Ridge, Endeavour Ridge, Axial Seamount, Cleft Segment, East Pacific Rise, Galapagos Rift, TAG, Snake Pit, Mid-Atlantic Ridge 24.5°N

[2] Mariana Trough, Manus Basin, North Fiji Basin, Lau Basin

identify as they are usually masked by the large amount of seawater in the circulation system.

Deposits on sediment-covered mid-ocean ridges commonly have lower Cu and Zn contents and higher Pb contents than bare-ridge sulfides, and some deposits such as Guaymas Basin contain abundant carbonate. The low metal contents of deposits in the Guaymas Basin are a consequence of the higher pH values of the fluids that arise from chemical buffering by the carbonate in the sediments, and the high $CO_2$ in the fluids is derived directly from the sediments themselves (Bowers et al. 1985, Von Damm et al. 1985a). Furthermore, the fluids in this environment are strongly reduced (pyrrhotite stability field) as a consequence of interaction of the fluids with organic components in the sediments, and metal deficient relative to a volcanic-hosted hydrothermal system as a result of sulfide precipitation within the sedimentary sequence. A high ammonium content reflects the thermocatalytic cracking of immature planktonic carbon (cf. Von Damm et al. 1985a). Interaction of the hydrothermal fluids with sediments in the upflow zone also may result in significant enrichments in certain trace elements derived from the sediments (e.g., Pb, Sn, As, Sb, Bi, Se: Koski et al. 1988, Zierenberg et al. 1993). The higher Pb contents, in particular, reflect the destruction of feldspars from continentally-derived turbidites, a process supported by Pb isotope studies (LeHuray et al. 1988).

## 13.6    Characteristics of Cold Seep Fluids at Subduction Zones

In addition to hydrothermal activity at divergent plate boundaries, low-temperature fluid venting at convergent plate margins is an important global process. Cold seeps have now been documented at numerous sites along the circum-Pacific subduction zones and it is estimated that this type of oceanic venting is also of major significance for the chemical budget of seawater. The fluids are expelled from thick organic-rich marine sediments that are trapped in accretionary wedges along the subduction trenches (e.g., Nankai Trench, Oregon and Alaskan margins). Pore fluids in the sediment account for as much as 50-70% of their volume, and this fluid is literally squeezed from the sediment through diffuse flow at the toe of the accretionary wedge or by focussed flow along major fault structures in the accreted sediments.

In order to estimate the mass flux from cold seeps, flow rates have to be known, and these have been difficult to measure because of the large areas of diffuse flow involved. Furthermore, fluid flow rates determined from simple advection-diffusion modeling of temperature and chemical profiles have shown to be unreliable if applied to sites which are densely populated with bottom macrofauna. Calculations by Wallmann et al. (1997) which take into account the high pumping rates of bivalves have arrived at a mean value of 5.5 +/- 0.7 L m$^{-2}$ d$^{-1}$ based on a biogeochemical approach using oxygen flux and vent fluid analyses. Von Huene et al. (1997) used sediment porosity reduction to calculate a fluid flow of 0.02 L m$^{-2}$ d$^{-1}$. Suess et al. (1998) calculated an average rate of 0.006 L m$^{-2}$ d$^{-1}$ based on the occurrence of vent biota at the seafloor. Estimates of the fluid flux at subduction zones suggest that they recycle the volume of water in the oceans every 500 Ma, compared to 5-10 Ma along the mid-ocean ridges. Despite the rather slow circulation of fluids expelled from subduction zones, they are a significant player in the carbon cycle of the oceans as a result of the recycling of organic matter in marine sediments.

Methane plumes have been found to be a characteristic feature of cold seep areas. The concentrations of dissolved $CH_4$ in some areas are lower than in hydrothermal vent areas at mid-ocean ridges, but in other areas they exceed ridge values by more than an order of magnitude and have been successfully used as a tracer for detecting cold seep venting. The methane and carbon dioxide that are expelled with the pore waters are derived mainly from the breakdown of organic matter in the sediments. Methane together with $H_2S$ support abundant chemosynthetic bacteria, pogonophorans, vestimentifera, and bivalves which differ from those that inhabit high-temperature vent sites on the mid-ocean ridges (Suess et al. 1998). As the temperatures of venting are low compared to hydrothermal vents at the mid-ocean ridges, metals are not significantly mobilized.

Typical precipitates for areas of subduction venting are carbonate and barite crusts together with cemented sediment (Dia et al. 1993, Suess et al. 1998). The formation of carbonate is related to the anaerobic microbial oxidation of methane which in turn causes $CaCO_3$ to precipitate (Wallmann et al. 1997). The source for barium is thought to be the high concentration of biogenic barite buried in sediments at high productivity areas which is remobilized within the sediment column due to sulfate depletion. Barite in the cold seep areas forms as a result of mixing of Ba-rich fluids upwelling from the sediment with seawater sulfate (Ritger et al. 1987, Torres et al. 1996).

Ice-like gas hydrates (clathrates) have also been found at numerous sites along the convergent plate margins (e.g. Suess et al. 1997). Methane hydrates are stable in solid form only in a narrow temperature-pressure window (Dickens and Quinby-Hunt 1994; Fig. 13.10). In theory, $1 m^3$ of methane hydrate can contain up to 164 $m^3$ of methane gas at standard conditions (Kvenvolden 1993). It has been estimated that the amount of carbon in gas hydrates considerably exceeds the total of carbon occurring in all known oil, gas and coal deposits worldwide (Kvenvolden and McMenamin 1980, Kvenvolden 1988). This raises the possibility that gas hydrates may be a future energy source of global importance.

If disturbed by a thermal anomaly or pressure change, large volumes of gas hydrates can break down into water, methane and carbon-dioxide. The dissolution of metastable gas hydrates is a natural consequence of tectonic uplift of accretionary prisms at plate margins and is seen to be partly responsible for the extensive methane plumes reported from these areas (Suess et al. 1997).

As gas hydrates have a very limited stability (low temperature and high pressure), they are of major relevance for climate considerations and the budget of greenhouse gases (Suess et al. 1997). The destabilization and dissolution of gas hydrates due to environmental changes (increase of ocean bottom water temperature, change of sea level) could liberate enormous amounts of methane to the water column and eventually to the atmosphere where they could potentially accelerate greenhouse warming and have an effect on future global climate (Leggett 1990).

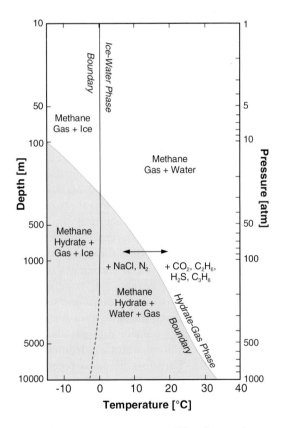

**Fig. 13.10** Pressure-temperature stability diagram for natural gas hydrates (after Katz et al. 1959).

ticles which have their origin in biotic and abiotic processes, or by massive reefs and platforms built up by skeleton-forming organisms. On a global scale, these shallow-water carbonates in the modern environment are mainly constituted by particles of skeletal origin. However, aside from the corals, the understanding of the physico-chemical and vital factors affecting the biomineral composition of shallow platform calcareous sediments in warm waters is still incomplete. Shallow environment precipitates form ooids and aragonitic needle muds, whereby the former involve primarily abiotic processes and the latter have both an abiotic and biotic source.

For long time, the classical picture of shallow water carbonates was suggesting that most of their formation was restricted to tropical and subtropical regions within the 22°C isotherm of annual mean surface water temperatures (e.g. Berger and Seibold 1993), but it now has become evident that a significant amount of carbonate can also be formed as so-called 'cool-water' carbonate banks and reefs in temperate and cold latitudes (review by James (1997)). In the temperate and cold-water zones, particularly shelf areas with only very low inputs of terrigenous sediments are covered by cool-water carbonate bioherms (Fig. 9.1a). Different from the warm-water environment, where the major portion of skeletal carbonate is predominantly formed by an association of hermatypic corals and green algae referred to as '*Photozoan Association*' (James 1997), the cool-water carbonates can be composed of molluscs, foraminifers, echinoderms, bryozoans, barnacles, ostracods, sponges, worms, ahermatypic corals and coralline algae. For differentiation from the warm-water *Photozoan Association*, the group of organisms forming cool-water carbonates in shallow waters that are colder than 20°C is defined by James 1997 as *Heterozoan Association* (see also Fig. 9.1). However, the portion of cool-water carbonate on the total amount of calcium carbonate accumulation in shallow-water environments is still questionable. Therefore the most recent estimates of Wollast (1994) and Milliman and Droxler (1996) for worldwide shelf areas are considered here. For budget considerations it seems reasonable to separate shallow-water carbonates according to Milliman (1993) into coral reefs, carbonate platforms which consist of non-reef habitats, such as banks and embayments dominated by the sedimentation of biogenic and abiogenic calcareous particles, and shelves which can

be further subdivided into carbonate-rich and carbonate-poor shelves (Table 9.1 adopted from Milliman (1993) and Milliman and Droxler (1996).

*Reefs*

Hermatypic coralalgal reefs occupy an area of about $0.6 \cdot 10^6$ km² and are considered as the most productive carbonate environments in modern times. Carbonate accretion is the result of corals and green algae and, to a minor extent also of benthic foramifera. Measured on a global scale, the mean calcium carbonate accumulation in coral reefs is in the order of 1500 g $CaCO_3$ m⁻²yr⁻¹. The total present-day global $CaCO_3$ production by coral reefs then is about $9 \cdot 10^{12}$ mol yr⁻¹ from which $7 \cdot 10^{12}$ mol yr⁻¹ accumulate, while the rest ($2 \cdot 10^{12}$ mol yr⁻¹) undergoes physical erosion and offshore transport, as well as biological destruction (Milliman 1993).

*Carbonate platforms*

These platforms are the second important tropical to subtropical environment where high amounts of carbonate are produced and accumulated at water depths shallower than 50 m. The areal extension is about $0.8 \cdot 10^6$ km². In contrast to reefs, on carbonate platforms production is mainly carried out by benthic red/green algae, mollusks and benthic foraminifera. Estimates of biotic and, to a much lesser extent, abiotic carbonate production on platforms range between 300-500 g $CaCO_3$ m⁻² yr⁻¹, which amounts to $4 \cdot 10^{12}$ mol yr⁻¹ on a global scale. Accumulation of platform carbonate is difficult to assess because a lot of it is dissolved or can be found as exported material in several 10 to 100 m thick sediment wedges of Holocene age at the fringes of the platforms. Milliman (1993) estimates that only one half of the carbonate accumulates on shallow platforms where it is produced ($2 \cdot 10^{12}$ mol yr⁻¹), while the other half leaves the shallow water environment in the form of sediment lobes or wedges downslope.

*Continental shelves*

Only very little knowledge exists concerning the quantitative carbonate production, export and its accumulation on continental shelves. Two shelf types, carbonate-rich and carbonate-poor shelves, are distinguished by Milliman and Droxler

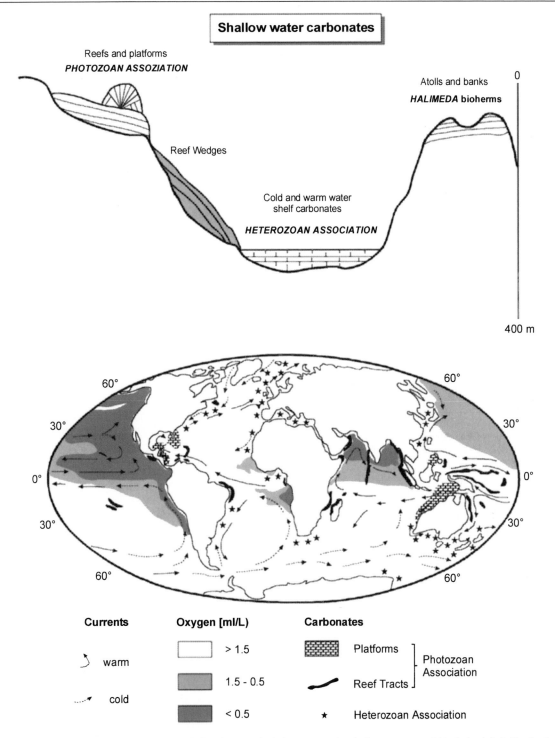

**Fig. 9.1** Marine calcium carbonate accumulation in a) typical depocenters in shallow waters and b) their global distribution (James and Clarke 1997).

(1996). Their areas amount to 15 and $10 \cdot 10^{12}$ km², respectively (Hay and Southam 1977). However for the two shelf types well-constrained estimates of how much carbonate is produced are missing. In the context of shelves it may be important to separate two other specific bioherms which could have a great potential in shallow-water carbonate production. These are sedimentary carbonates exclusively built up by the calcareous green algae *Halimeda* in tropical latitudes (e.g. Roberts and

deposits on the seafloor. Geological Survey of Canada Open File Report, 2915C: Map 1:35,000,000 and CD-ROM.

Hannington, M.D., Jonasson, I.R., Herzig, P.M. and Petersen, S., 1995. Physical, chemical processes of seafloor mineralization at mid - ocean ridges. In: Humphris, S.E. et al. (eds) Seafloor Hydrothermal Systems. Physical, Chemical, Biological and Geological Interactions, AGU Geophysical Monograph, 91: 115-157.

Haymon, R.M. and Kastner, M., 1981. Hot spring deposits on the East Pacific Rise at 21°N: Preliminary descripton of mineralogy and genesis. Earth and Planetary Science Letters, 53: 363-381.

Haymon, R.M., 1983. Growth history of hydrothermal black smoker chimneys. Nature, 301: 695-698.

Haymon, R.M.; Fornari, D.J., von Damm, K.L., Lilley, M.D., Perfit, M.R., Edmond, J.M., Shanks, W.C. III, Lutz, R.A., Grebmeier, J.M., Carbotte, S., Wright, D., McLaughlin, E., Smith, M., Beedle, N. and Olson, E., 1993. Volcanic eruption of the mid - ocean ridge along the East Pacific Rise crest at 9°45-52'N: Direct submersible observations of seafloor phenomena associated with an eruption event in April, 1991. Earth and Planetary Science Letters, 119: 85-101.

Hedenquist, J.W. and Lowenstern, J.B., 1994. The role of magmas in the formation of hydrothermal ore deposits. Nature, 370: 519-526.

Hekinian, R., Francheteau, J., Renard, V., Ballard, R.D., Choukroune, P., Cheminée, J.L., Albaréde, F., Minster, J.F., Charlou,J.L., Marty, J.C. and Boulégue, J., 1983. Intense hydrothermal activity at the rise axis of the axis of the East Pacific Rise near 13°N: Submersible withnesses the growth of sulfide chimney. Marine Geology Research, 6: 1-14.

Hemley, J.J., Cygan, G.L., Fein, J.B., Robinson, G.R. and D'Angelo, W.M., 1992. Hydrothermal ore - forminf processes in the light of studies in the rock-buffered systems: I. Iron-copper-zinc-lead sulfide solupility relations. Economic Geology, 87: 1-22.

Herzig, P.M. and Plüger, W.L., 1988. Exploration for hydrothermal mineralisation near the Rodriguez Triple Junction, Indian Ocean. Canadian Mineralogist, 26: 721-736.

Herzig, P.M. and Hannington, M.D., 1995. Polymetallic massive sulfides at the modern seafloor: A review. Ore Geology Reviews, 10: 95-115.

Herzig, P.M., Hannington, M.D. and Arribas, A.(jr.)., 1998. Sulfur isotopic composition of hydrothermal precipitates from the Lau back-arc: implications for magmatic contributions to seafloor hydrothermal systems. Mineralium Deposita, 33: 226-237.

Humphris, S.H., Herzig, P.M., Miller, D.J.,Alt, J.C., Becker, K., Brown, D., Brügmann, G., Chiba, H., Fouquet, Y., Gemmell, J.B., Guerin, G., Hannington, M.D., Holm, N.G., Honnorez, J.J., Itturino, G.J., Knott, R., Ludwig, R., Nakamura, K., Petersen, S., Reysenbach, A.L., Rona, P.A., Smith, S., Sturz, A.A., Tivey, M.K. and Zhao, X., 1995. The internal structure of an active sea-floor massive sulfide deposit. Nature, 377: 713-716.

James, R.H., Elderfield, H. and Palmer, M.R., 1995. The chemistry of hydrothermal fluids from Broken Spur site, 29°N Mid-Atlantic Ridge. Geochimica et Cosmochmica Acta, 59: 651-659.

Janecky, D.R. and Seyfried, W.E.(jr.), 1984. Formation of massiv sulfide deposits on ocean ridge crests: Incremental reaction models for mixing between hydrothermal solutions and seawater. Geochimica et Cosmochmica Acta, 48: 2723-2738.

Janecky, D.R. and Shanks, W.C.III, 1988. Computational modelling of chemical and isotopic reaction processes in seafloor hydrothermal systems: chimneys, massive sulfides, and subjacent alteration zones. Canadian Mineralogist, 26: 805-825.

Johnson, H.P. and Tunnicliffe, V., 1985. Time series measurements of hydrothermal activity on the northern Juan de Fuca Ridge. Geophysical Research Letters, 12: 685-688.

Kadko, D. and Moore, W., 1988. Radiochemical constraints on the crustal residence time of submarine hydrothermal fluids: Endeavour Ridge. Geochimica et Cosmochimica Acta, 52: 659-668.

Kastner, M., Craig, H. and Sturz, A., 1987. Hydrothermal deposition in the Mariana Trough: Preliminary mineralogical investigations. American Geophysical Union Transactions, 68: 1531.

Katz, D.L., Cornell, D., Kobayashi, R., Poettmann, F.H., Vary, J.A., Elenblass, J.R. and Weinaug, C.F., 1959. Handbook of Natural Gas Engineering, 802. McGraw-Hill, NY, 802 pp.

Keays, R.R., 1987. Principles of mobilization ( dissolution ) of metals in mafic and ultramafic rocks - The role of immiscible magmatic sulphides in the generation of hydrothermal gold and volcanogenic massiv sulphide deposits. Ore Geology Reviews, 2: 47-63.

Klinkhammer, G.P., Elderfield, H., Greaves, M., Rona, P.A. and Nelson, T., 1986. Manganese geochemistry near high-temperature vents in the Mid-Atlantic Ridge rift valley. Earth and Planetary Science Letters, 80: 230-240.

Klinkhammer, G.P., Chin, C.S., Wilson, C. and German, C.R., 1995. Venting from the Mid-Atlantic Ridge at 37°17'N: the Lucky Strike hydrothermal site. In. Parson, L.M., Walker, C.L. and Dixon, D.R. (eds) Hydrothermal vents and processes. Geological Society Special Publication, 87: 87-96.

Koski, R.A., Shanks, W.C.I., Bohrson, W.A. and Oscarson, R.L., 1988. The composition of massiv sulfide deposits from the sediment-covered floor of Escanaba Trough, Gorda Ridge: implications for depositional processes. Canadian Mineralogist, 26: 655-673.

Koski, R.A., Jonasson, I.R., Kadko, D.C., Smith, V.K. and Wong, F.L., 1994. Compositions, growth mechanisms, and temporal relations of hydrothermal sulfide-sulfate-silica chimneys at the northern Cleft segment, Juan de Fuca Ridge. Journal of Geophysical Research, 99: 4813-4832.

Kvenvolden, K.A. and McMenamin, 1980. Hydrate of natural gas. A review of their geological occurrence. U. S. Geological Survey Circular, 825: 11.

Kvenvolden, K.A., 1988. Methane hydrate - a major reservoir of carbon in the shallow geosphere. Chemical Geology, 71: 41-51.

Kvenvolden, K.A., 1993. Gas hydrates - Geological Perspective and global change. Reviews of Geophysics, 31: 173-187.

Lafitte, M., Maury, R., Perseil, E.A. and Boulegue, J., 1985. Morphological and analytical study of hydrothermal sulfides from 21° north East Pacific Rise. Earth and Planetary Science Letters, 73: 53-64.

Lalou, C., Reyss, J.L., Brichet, E., Arnold, M., Thompson, G., Fouquet, Y. and Rona, P.A., 1993. New age data for Mis - Atlantic Ridge hydrothermal sites: TAG and Snakepit chronology. Journal of Geophysical Research, 98: 9705-9713.

Leggett, J., 1990. The nature of the greenhouse threat. In: Leggett, J. (ed) Global Warming, The Greenpeace Report. Oxford University Press, NY: 14-43.

LeHuray, A.P., Church, S.E., Koski, R.A. and Bouse, R.M., 1988. Pb isotopes in sulfides from mid-ocean ridge hydrothermal sites. Geology, 16: 362-365.

Lowell, R.P., 1991. Modeling continental and submarine hydrothermal systems. Reviews in Geophysics, 29: 457-476.

Lowell, R.P., Rona, P.A. and von Herzen, R.P., 1995. Seafloor hydrothermal systems. Journal of Geophysical Research, 100: 327-352.

Lupton, J.E. and Craig, H., 1981. A major helium-3 source at 15°S on the East Pacific Rise. Science, 214: 13-18.

Lupton, J.E., 1983. Fluxes of helium-3 and heat from submarine hydrothermal systems; Guaymas Basin vesus 21°N EPR. American Geophysical Union Transactions, 64: 723.

Lupton, P., Lilley, M., Olson, E. and von Damm, K.L., 1991. Gas chemistry of vent fluids from 9°-10°N on the East Pacific Rise. American Geophysical Union Transactions, 72: 481.

Massoth, G.J., Butterfield, D., Lupton, J.E., McDuff, R.E., Lilley, M.D. and Jonasson, I.R., 1989. Submarine venting of phase-separated hydrothermal fluids at Axial Volcano, Juan de Fuca Ridge. Nature, 340: 702-7Я5.

Metz, S. and Trefry, J.H., 1993. Field and laboratory studies of metal uptake and release by hydrothermal precipitates. Journal of Geophysical Research, 98: 9661-9666.

Michard, A., Albarede, F., Michard, G., Minster, J.F. and Charlou, J.L., 1983. Rare-earth elements and uranium in high-temperature solutions from East Pacific Rise hydrothermal vent field (13°N). Nature, 303: 795-797.

Michard, A., 1989. Rare earth element systematics in hydrothermal fluids. Geochimica et Cosmochimica Acta, 53: 745-750.

Miller, A.R., Densmore, C.D., Degens, E.T., Hathaway, F.C., Manheim, F.T., McFarlen, P.F., Pocklington, H. and Jokela, A., 1966. Hot brines and recent iron deposits in deeps of the Red Sea. Geochimica et Cosmochimica Acta, 30: 341-359.

Morton, J.L. and Sleep, N.H., 1985. A mid - ocean ridge thermal model: Constraints on the volume of axial heat flux. Journal of Geophysical Research, 90: 11345-11353.

Mottl, M.J., 1983. Metabasalts, axial hot - springs, and the structure of hydrothermal systems at mid-ocean ridges. Geological Society of American Bulletin, 94: 161-180.

Mottl, M.J. and McConachy, T.F., 1990. Chemical processes in buoyant hydrothermal plumes on the East Pacific Rise near 21°N. Geochimica et Cosmochimica Acta, 54: 1911-1927.

Nehlig, P., 1991. Salinity of oceanic hydrothermal fluids: a fluid inclusion study. Earth and Planetary Science Letters, 102: 310-325.

Oudin, E., 1983. Hydrothermal sulfide deposits of the East Pacific Rise ( 21°N ) part I: descriptive mineralogy. Marine Mining, 4: 39-72.

Palmer, M.R., 1992. Controls over the chloride concentration of submarine hydrothermal vent fluids: evidence from Sr/Ca and $^{87}Sr/^{86}Sr$ rations. Earth and Planetary Science Letters, 109: 37-46.

Perfit, M.R., Fornari, D.J., Malahoff, A. and Embley, R.W., 1983. Geochemical studies of abyssal lavas recovered by DSRV Alvin from eastern Galapagos Rift, Inca Transform, and Equador Rift, 3. Trace Element abundances and pertogenesis. Journal of Geophysical Research, 88: 10551-10572.

Plüger, W.L., Herzig, P.M., Becker, K.P., Deissmann, G., Schöps, D., Lange, J., Jenisch, A., Ladage, S., Richnow, H.H., Schulze, T. and Michaelis, W., 1990. Discovery of hydrothermal fields at the Central Indian Ridge. Marine Mining, 9: 73-86.

Richardson, C.J., Cann, J.R., Richards, H.G. and Cowan, J.G., 1987. Metal - depleted zones of the Troodos ore - forming hydrothermal system, Cyprus. Earth and Planetary Science Letters, 84: 243-253.

Ritger, S., Carson, B. and Suess, E., 1987. Methane - derived authigenic carbonates formed by subduction - induced pore-water explution along the Oregon/Washington margin. Geological Society of American Bulletin, 98: 147-156.

Rona, P.A., 1988. Hydrothermal mineralization at ozean ridges. Canadian Mineralogist, 26: 431-465.

Rona, P.A. and Scott, S.D., 1993. A spezial issue on sea - floor hydothermal mineralization: new perspectives. Canadian Mineralogist, 88: 1935-1976.

Rona, P.A., Hannington, M.D., Raman, C.V., Thompson, G., Tivey, M.K., Humphris, S.E., Lalou, C. and Petersen, S., 1993. Active and relict sea-floor hydrothermal mineralization at the TAG Hydrothermal Field, Mid-Atlantic Ridge. Economic Geology, 88: 1989-2017.

Schöps, D. and Herzig, P.M., 1990. Sulfide composition and microthermometry of fluid inclusions in quartz - sulfide veins from the leg 111 dike section of ODP Hole 504B, Costa Rica Rift. Journal of Geophysical Research, 95: 8405-8418.

Seyfried, W.E.j. and Mottl, M.J., 1982. Hydrothermal alteration of basalt by seawater under seawater-dominated conditions. Geochimica et Cosmochimica Acta, 46: 985-1002.

Seyfried, W.E.j., Berndt, M.E. and Janecky, D.R., 1986. Chloride depletions and enrichments in seafloor hydrothermal fluids: Constraints from experimental basalt alteration studies. Geochimica et Cosmochimica Acta, 50: 469-475.

Seyfried, W.E.j., Berndt, M.E. and Seewald, J.S., 1988. Hydrothermal alteration processes at mid - ocean ridges: constraints from diabase alteration experiments, hot - spring fluids and composition of the oceanic crust. Canadian Mineralogist, 26: 787 - 804.

Spiess, F.N., Macdonald, K.C., Atwater, T., Ballard, R., Carranza, A., Cordoba, D., Cox, C., Diaz Gracia, V.M., Francheteau, J., Guerro, J., Hawkins, J.W., Haymon, R., Hessler, R., Juteau, T., Kastner, M., Larson, R., Luyendyk, B., Macdougall, J.D., Miller, S., Normark, W., Orcutt, J. and Rangin, C., 1980. East Pacific Rise. Hot springs and geophysical experiments. Science, 207: 1421-1433.

Stein, C.A. and Stein, S., 1994. Constrains on hydrothermal

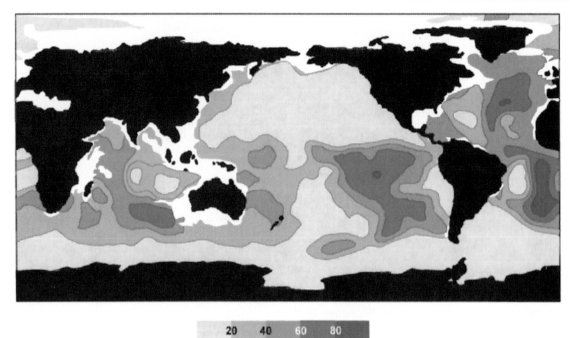

**CaCO₃ in surface sediments [%]**

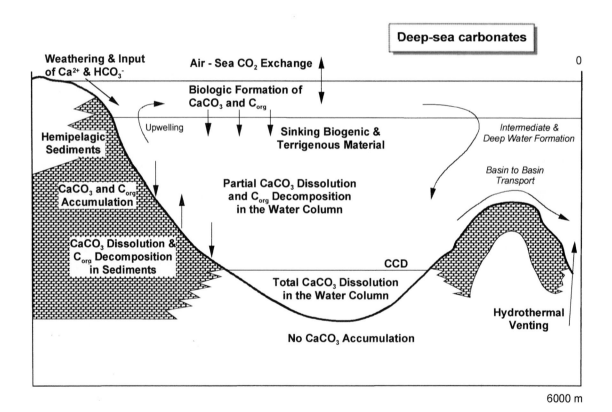

**Fig. 9.2** Environments of pelagic carbonate deposition: a) deep-sea distribution of carbonate-rich sediments (new compilation from Archer 1996a) and b) factors controlling pelagic carbonate production and dissolution (modified from Morse and Mackenzie 1990).

Essentially all the above mentioned factors have the potential to change the calcite-carbonate equilibrium in the ocean over time and thus exert major control on the distribution and amount of calcareous sediments in the deep sea. On the other hand, given the 60 times greater carbon reservoir of the ocean compared to that of the atmosphere, changes in the oceanic calcite-carbonate equilibrium can modify the oceanic $CO_2$ uptake/ release balance with respect to the atmosphere (Archer and Maier-Reimer 1994, Berger 1982, Broecker and Peng 1987, Maier-Reimer and Bacastow 1990, Opdyke and Walker 1992, Siegenthaler and Wenk 1984, Wolf-Gladrow 1994; see also review in Dittert et al. 1998). Therefore, the principles of the calcite-carbonate system will be re-examined in the following Section 9.3 and examples will be given for modeling this system under specific boundary conditions typical for the modern ocean. As we know from the geological record of calcareous sediment distribution, this system has changed dramatically in

the past and definitely will change in the future, leaving us with the questions of how much, how fast and in which direction the ocean carbonate system will respond to anthropogenic disturbances of the carbon cycle in the future.

## 9.3 The Calcite-Carbonate-Equilibrium in Marine Aquatic Systems

The calcite-carbonate-equilibrium is of particular importance in aquatic systems wherever carbon dioxide is released into water by various processes, and wherever a concurrent contact with calcite (or other carbonate minerals) buffers the system by processes of dissolution/precipitation. Carbon dioxide may either originate from the gaseous exchange with the atmosphere, or is formed during the oxidation of organic matter. In such a system this equilibrium controls the pH-value - an essential system parameter which, directly or indirectly, influences a number of secondary processes, e.g. on iron (cf. Chap. 7) and on manganese (cf. Chap. 11). In the following, the calcite-carbonate equilibrium will be described in more detail and with special attention given to its relevant primary reactions.

The description of Sections 9.3.1 and 9.3.2 will assume in principle a solution at infinite dilution in order to reduce the system to really important reactions. The Sections 9.3.3 and 9.3.4 provide examples calculated for seawater, including all constituents of quantitative importance. In such calculations, two different approaches are possible:

The use of measured concentrations (calcium, carbonate, pH) together with so-called 'apparent' equilibrium constants. These apparent constants (e.g. Goyet and Poisson 1989, Roy et al. 1993, Millero 1995) take the difference between activities and concentrations into account, as well as the fact, that for instance only part of the measured calcium is present in the form of $Ca^{2+}$-ions, whereas the rest forms complexes or ion pairs ($CaSO_{4\,aq}^{0}$, $CaHCO_{3\,aq}^{+}$, or others). This is true for seawater, since the activity coefficients of these complexes or ion pairs with sufficient accuracy exert a constant influence.

The other approach employs a geochemical computer model, such as PHREEQC (Parkhurst 1995; also Chap. 14) with an input of a complete

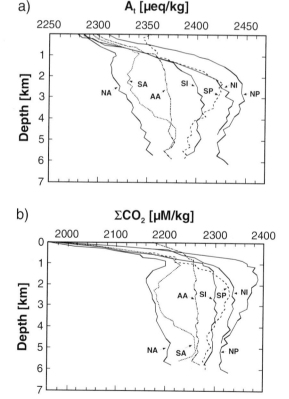

**Fig. 9.3** The mean vertical distribution of a) alkalinity and b) total $CO_2$ concentration normalized to the mean world ocean salinity value of 34.78 (after Takahashi et al. 1980, adapted from Morse and Mackenzie 1990)

time by Truesdell and Jones (1974). Unfortunately, this was done in the computer language PL1 which has become completely forgotten by now. Afterwards a FORTRAN version had followed (WATEQF) released by Plummer et al. (1976) as well as another revised version (WATEQ4F) by Ball and Nordstrom (1991).

All these programs share the common feature that they start out from water analyses which were designed as comprehensive as possible, along with conceiving the measured values as cumulative parameters (sums of all aquatic species), only to subsequently allocate them to the various aquatic species. This was done by applying iteration calculus that accounts for the equilibria related to the various aquatic species.

For instance, the measured value of calcium in an ocean water sample containing 10.6 mmol/l is

**Table 14.1** This ocean water analysis carried out by Nordstrom et al. (1979) has been frequently used to test and compare various geochemical model programs.

### Seawater from Nordstrom et al. (1979)

| | | |
|---|---|---|
| pH | = | 8.22 |
| pe | = | 8.451 |
| Ionic strength | = | 6.750E-1 |
| Temperature (°C) | = | 25.0 |

| Elements | Molality |
|---|---|
| Alkalinity | 2.406E-3 |
| Ca | 1.066E-2 |
| Cl | 5.657E-1 |
| Fe | 3.711E-8 |
| K | 1.058E-2 |
| Mg | 5.507E-2 |
| Mn | 3.773E-9 |
| N (-III) | 1.724E-6 |
| N (V) | 4.847E-6 |
| Na | 4.854E-1 |
| O (0) | 3.746E-4 |
| S (VI) | 2.926E-2 |
| Si | 7.382E-5 |

allocated to 9.51 mmol/l $Ca^{2+}$, 1.08 mmol/l $CaSO_4^0$, 0.04 mmol/l $CaHCO_3^+$, and to a series of low concentrated calcium species detectable in solution. The analysis of sulfate yielding 29.26 mmol/l is allocated accordingly to 14.67 mmol/l $SO_4^{2-}$, 7.30 mmol/l $NaSO_4^-$, 1.08 mmol/l $CaSO_4^0$, 0.16 mmol/l $KSO_4^-$, and, likewise, to a number of low concentrated sulfate species. The procedure is similarly applied to all analytical values. Table 14.2 summarizes the species allocations for the ocean water analysis previously shown in Table 14.1.

Based on the activities of the non-complex aquatic species (e.g. $[Ca^{2+}]$ or $[SO_4^{2-}]$), the next step consists in calculating the saturation indices (SI) of each compound mineral:

$$SI = \log\left(\frac{IAP}{K_{sp}}\right) \tag{14.1}$$

In this equation IAP denotes the Ion Activity Product (in the example of gypsum or anhydrite these would be ($[Ca^{2+}]\cdot[SO_4^{2-}]$). $K_{SP}$ is the solubility product constant of the respective mineral. A saturation index SI = 0 describes the condition in which the solution of the corresponding mineral is just saturated, SI > = 0 describes the condition of supersaturation of the solution, SI < 0 its undersaturation. The activity [A] of a substance is calculated according to the equation:

$$[A] = (A) \cdot \gamma_A \tag{14.2}$$

Here, (A) is the concentration of a substance measured in mol/l and $\gamma_A$ is the activity coefficient calculated as a function of the overall ionic strength in the solution. For infinitesimal diluted solutions, the activity coefficients assume the value of 1; hence, activity equals concentration. In ocean water, the various monovalent ions, or ion pairs, display activity coefficients somewhere around 0.75; whereas the various divalent ions , or ion pairs, display values around 0.2 (cf. Table 14.2, last column).

After transformation, the equation 14.1 to calculate the saturation index of the mineral gypsum reads:

$$SI_{gypsum} = \log\left(\frac{[Ca^{2+}]\cdot[SO_4^{2-}]}{K_{sp,gypsum}}\right) \tag{14.3}$$

If we now insert the solubility product constant for gypsum $K_{sp,gypsum} = 2.63\cdot10^{-5}$ in the equation

**Table 14.2** For an analysis of Table 14.1, this allocation of the aquatic species was calculated with the geochemical model program PHREEQC.

| Species | Molality | Activity | Molality log | Activity log | Gamma log |
|---|---|---|---|---|---|
| $OH^-$ | 2.18E-06 | 1.63E-06 | -5.661 | -5.788 | -0.127 |
| $H^+$ | 7.99E-09 | 6.03E-09 | -8.098 | -8.220 | -0.122 |
| $H_2O$ | 5.55E+01 | 9.81E-01 | -0.009 | -0.009 | 0.000 |
| **C (IV)** | 2.181E-03 | | | | |
| $HCO_3^-$ | 1.52E-03 | 1.03E-03 | -2.819 | -2.989 | -0.171 |
| $MgHCO_3^+$ | 2.20E-04 | 1.64E-04 | -3.659 | -3.785 | -0.127 |
| $NaHCO_3$ | 1.67E-04 | 1.95E-04 | -3.778 | -3.710 | 0.068 |
| $MgCO_3$ | 8.90E-05 | 1.04E-04 | -4.050 | -3.983 | 0.068 |
| $NaCO_3^-$ | 6.73E-05 | 5.03E-05 | -4.172 | -4.299 | -0.127 |
| $CaHCO_3^+$ | 4.16E-05 | 3.10E-05 | -4.381 | -4.508 | -0.127 |
| $CO_3^{-2}$ | 3.84E-05 | 7.98E-06 | -4.415 | -5.098 | -0.683 |
| $CaCO_3$ | 2.72E-05 | 3.17E-05 | -4.566 | -4.499 | 0.068 |
| $CO_2$ | 1.21E-05 | 1.42E-05 | -4.916 | -4.849 | 0.068 |
| $MnCO_3$ | 3.18E-10 | 3.71E-10 | -9.498 | -9.430 | 0.068 |
| $MnHCO_3^+$ | 7.17E-11 | 5.36E-11 | -10.144 | -10.271 | -0.127 |
| $FeCO_3$ | 1.96E-20 | 2.29E-20 | -19.709 | -19.641 | 0.068 |
| $FeHCO_3^+$ | 1.64E-20 | 1.22E-20 | -19.785 | -19.912 | -0.127 |
| **Ca** | 1.066E-02 | | | | |
| $Ca^{+2}$ | 9.51E-03 | 2.37E-03 | -2.022 | -2.625 | -0.603 |
| $CaSO_4$ | 1.08E-03 | 1.26E-03 | -2.967 | -2.900 | 0.068 |
| $CaHCO_3^+$ | 4.16E-05 | 3.10E-05 | -4.381 | -4.508 | -0.127 |
| $CaCO_3$ | 2.72E-05 | 3.17E-05 | -4.566 | -4.499 | 0.068 |
| $CaOH^+$ | 8.59E-08 | 6.41E-08 | -7.066 | -7.193 | -0.127 |
| $CaHSO_4^+$ | 5.96E-11 | 4.45E-11 | -10.225 | -10.352 | -0.127 |
| **Cl** | 5.656E-01 | | | | |
| $Cl^-$ | 5.66E-01 | 3.52E-01 | -0.247 | -0.453 | -0.206 |
| $MnCl^+$ | 1.13E-09 | 8.41E-10 | -8.948 | -9.075 | -0.127 |
| $MnCl_2$ | 1.11E-10 | 1.29E-10 | -9.956 | -9.888 | 0.068 |
| $MnCl_3^-$ | 1.68E-11 | 1.26E-11 | -10.774 | -10.901 | -0.127 |
| $FeCl^{+2}$ | 9.58E-19 | 2.97E-19 | -18.019 | -18.527 | -0.508 |
| $FeCl_2^+$ | 6.27E-19 | 4.68E-19 | -18.203 | -18.330 | -0.127 |
| $FeCl^+$ | 7.78E-20 | 5.81E-20 | -19.109 | -19.236 | -0.127 |
| $FeCl_3$ | 1.41E-20 | 1.65E-20 | -19.850 | -19.783 | 0.068 |
| **Fe (II)** | 5.550E-19 | | | | |
| $Fe^{+2}$ | 3.85E-19 | 1.19E-19 | -18.415 | -18.923 | -0.508 |
| $FeCl^+$ | 7.78E-20 | 5.81E-20 | -19.109 | -19.236 | -0.127 |
| $FeSO_4$ | 4.84E-20 | 5.65E-20 | -19.315 | -19.248 | 0.068 |
| $FeCO_3$ | 1.96E-20 | 2.29E-20 | -19.709 | -19.641 | 0.068 |
| $FeHCO_3^+$ | 1.64E-20 | 1.22E-20 | -19.785 | -19.912 | -0.127 |
| $FeOH^+$ | 8.23E-21 | 6.15E-21 | -20.084 | -20.211 | -0.127 |
| **Fe (III)** | 3.711E-08 | | | | |
| $Fe(OH)_3$ | 2.84E-08 | 3.32E-08 | -7.547 | -7.479 | 0.068 |
| $Fe(OH)_4^-$ | 6.60E-09 | 4.92E-09 | -8.181 | -8.308 | -0.127 |
| $Fe(OH)_2^+$ | 2.12E-09 | 1.58E-09 | -8.674 | -8.801 | -0.127 |
| $FeOH^{+2}$ | 9.46E-14 | 2.94E-14 | -13.024 | -13.532 | -0.508 |
| $FeSO_4^+$ | 1.09E-18 | 8.16E-19 | -17.962 | -18.089 | -0.127 |
| $FeCl^{+2}$ | 9.58E-19 | 2.97E-19 | -18.019 | -18.527 | -0.508 |
| $FeCl_2^+$ | 6.27E-19 | 4.68E-19 | -18.203 | -18.330 | -0.127 |
| $Fe^{+3}$ | 3.88E-19 | 2.80E-20 | -18.411 | -19.554 | -1.143 |
| $Fe(SO_4)_2^-$ | 6.36E-20 | 4.75E-20 | -19.196 | -19.323 | -0.127 |
| $FeCl_3$ | 1.41E-20 | 1.65E-20 | -19.850 | -19.783 | 0.068 |
| **K** | 1.058E-02 | | | | |
| $K^+$ | 1.04E-02 | 6.49E-03 | -1.982 | -2.188 | -0.206 |
| $KSO_4^-$ | 1.64E-04 | 1.22E-04 | -3.786 | -3.913 | -0.127 |

is referred to as open to $CO_2$. It is characteristic of such systems that the reactions can cause the release of $CO_2$ into the atmosphere at any time and that $CO_2$ can be drawn from the atmosphere without directly affecting total atmospheric $pCO_2$.

### 9.3.2    Primary Reactions of the Calcite-Carbonate-Equilibrium without Atmospheric Contact

A system which is closed to $CO_2$ exists wherever the final calcite-carbonate-equilibrium is reached without any concurrent uptake or release of atmospheric $CO_2$ in its process. This implies that the Equations 9.1 and 9.2 are no longer valid. In their stead, a balance of various C-species is related to calcium. This balance maintains that the sum of C-species must equal the calcium concentration in solution, since both can enter the solution only by dissolution of calcite or aragonite:

$$(Ca^{2+}) = (H_2CO_3{}^0) + (HCO_3{}^-) + (CO_3{}^{2-}) \quad (9.13)$$

Analogous to Equations 9.11 and 9.12, Equation 9.13 is only valid if the initial solution consists of pure water. In the normal case, in which the solution will already contain dissolved carbonate- and/or calcium ions, the difference between equilibrium concentrations $(...)_f$ and the initial concentrations $(...)_i$ needs to be regarded:

$$(Ca^{2+})_f - (Ca^{2+})_i =$$
$$(H_2CO_3{}^0)_f - (H_2CO_3{}^0)_i + (HCO_3{}^-)_f -$$
$$(HCO_3{}^-)_i + (CO_3{}^{2-})_f - (CO_3{}^{2-})_i \quad (9.14)$$

The six Equations 9.4, 9.6, 9.8, 9.10, 9.12, and 9.14 must now be solved for the six variables $(Ca^{2+})$, $(H^+)$, $(OH^-)$, $(H_2CO_3{}^0)$, $(HCO_3{}^-)$, $(CO_3{}^{2-})$. There is no analytical solution for this non-linear equation system. If there were one, it would be irrelevant because an iterative solution, similar to the one described in the previous section, would always be easier to handle and besides be more reliable. However, still independent of the method of mathematical solution is the fact that the equilibrium is exclusively determined by the carbonate concentration which the system previously contained and which is defined by the concentrations $(...)_i$. At this particular point, the amount of $CO_2$ would also have to be considered which is liberated into such a system, e.g. from the oxidation of organic matter. This amount would be simply added to the 'initial' concentrations and would influence the equilibrium in the same way as the pre-existing carbonate.

### 9.3.3    Secondary Reactions of the Calcite-Carbonate-Equilibrium in Seawater

In seawater, the differences between activities and concentrations must always be considered (cf. Sect. 14.1.1). The activity coefficients for monovalent ions in seawater assume a value around 0.75, for divalent ions this value usually lies around 0.2. In most cases of practical importance, the activity coefficients can be regarded with sufficient exactness as constants, since they are, over the whole range of ionic strengths in solution, predominately bound to the concentrations of sodium, chloride, and sulfate which are not directly involved in the calcite-carbonate-equilibrium. The proportion of ionic complexes in the overall calcium or carbonate content can mostly be considered with sufficient exactness as constant in the free water column of the ocean. Yet, this cannot be applied to pore water which frequently contains totally different concentrations and distributions of complex species due to diagenetic reactions.

Figure 9.4 shows the distribution of carbonate and calcium species in ocean water and in an anoxic pore water, calculated with the program PHREEQC (Parkhurst 1995). It is evident that about 10% of total calcium is prevalent in the form of ionic complexes and 25 – 30% of the total dissolved carbonate in different ionic complexes other than bicarbonate. These ionic complexes are not included in the equations of Sections 9.3.1 and 9.3.2. Accordingly, the omission of these complexes would lead to an erraneous calculation of the equilibrium. The inclusion of each complex shown in Figure 9.4 implies further additions to the system of equations, consisting in another concentration variable (the concentration of the complex) and a further equation (equilibrium of the complex concentration relative to the non-complexed ions). If only the ion complexes shown in Figure 9.4 are added, as there are $(MgHCO_3{}^+)$, $(NaHCO_3{}^0)$, $(MgCO_3{}^0)$, $(NaCO_3{}^-)$, $(CaHCO_3{}^+)$, $(CaCO_3{}^0)$, and $(CaSO_4{}^0)$, the system would consequently be extended to 6+7=13 variable concentrations and just as many non-linear equations. Such a system of equations is solvable with the aid of an appropriate spread-sheet calculation program, however, that would be certainly not reasonable and would also be more laborious

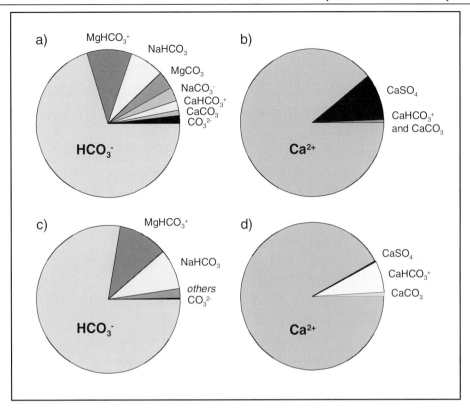

**Fig. 9.4** Distribution of carbonate species (a and c) and calcium species (b and d) in seawater after Nordstrom et al. 1979 (a and b) and in an anoxic pore water (c and d). The pore water sample was extracted from the core previously shown in Figure 3.1 and was taken from a depth of 14.8 m below the sediment surface. The calculation of species distributions was performed with the program PHREEQC (Parkhurst 1995)

than using an equilibrium model of the PHREEQC type (Parkhurst 1995). Function and application of PHREEQC are described in Section 14.1 in more detail. In the following section, some calculation examples with regard to carbonate will be presented.

### 9.3.4 Examples for Calculation of the Calcite-Carbonate-Equilibrium in Ocean Waters

Here, exemplary results will be introduced which were obtained from model calculations of the calcite-carbonate-equilibrium under applied boundary conditions close to reality. To this end, the model PHREEQC (Parkhurst 1995) has been used. The chemical composition of the solution studied in these examples is based on analytical data published by Nordstrom et al. 1979 (cf. Table 14.1).

In the first example shown in Table 9.2, a warm (25 °C) water from the ocean's surface (1 atm) has been modeled. In the zone near the equator, where this water sample had been

taken, carbon dioxide partial pressures of 400 μatm (equivalent to a $pCO_2$ of 0.0004 atm or a log $pCO_2$ of -3.40) have been measured which is somewhat higher than the corresponding atmospheric value. This sample of ocean water thus displays a $CO_2$-gradient directed towards the atmosphere and therefore continually releases $CO_2$ into the atmosphere. This situation is accounted for in the model by pre-setting $pCO_2$ to 0.0004 atm as an open boundary condition with regard to $CO_2$. Accordingly, a state of supersaturation ensues equivalent to a $SI_{calcite}$ value of 0.77 or an $\Omega_{calcite}$ value of 5.9[1]. Such a supersaturation state is, according to the analytical evidence, obviously permanent by its

---

[1] $\Omega_{calcite}$ is still in use in chemical oceanography for describing the state of saturation. It corresponds to the saturation index without the logarithm, hence: log $\Omega_{calcite} = SI_{calcite}$. However, the SI value is more useful, since the same amounts of undersaturation and supersaturation respectively display the same SI values, only distinguished by different signs (+) for supersaturation and (-) for undersaturation. A SI value of zero describes the state of saturation.

**Table 14.3**　According to the species distribution in Table 14.2, the PHREEQC model (Parkhurst 1995) was also applied to calculate these saturation indices. Here, 'CO$_2$ (g)', 'H$_2$(g)', 'NH$_3$(g)', 'O$_2$(g)' do not stand for mineral phases, but for gaseous phases in which the saturation index is obtained as the logarithm of the respective partial pressure.

| Phase | SI | log IAP | log KT | |
|---|---|---|---|---|
| Anhydrite | -0.84 | -5.20 | -4.36 | CaSO$_4$ |
| Aragonite | 0.61 | -7.72 | -8.34 | CaCO$_3$ |
| Artinite | -2.03 | 7.57 | 9.60 | MgCO$_3$:Mg(OH)$_2$:3H$_2$O |
| Birnessite | 4.81 | 5.37 | 0.56 | MnO$_2$ |
| Bixbyite | -2.68 | 47.73 | 50.41 | Mn$_2$O$_3$ |
| Brucite | -2.28 | 14.56 | 16.84 | Mg(OH)$_2$ |
| Calcite | 0.76 | -7.72 | -8.48 | CaCO$_3$ |
| Chalcedony | -0.51 | -4.06 | -3.55 | SiO$_2$ |
| Chrysotile | 3.36 | 35.56 | 32.20 | Mg$_3$Si$_2$O$_5$(OH)$_4$ |
| Clinoenstatite | -0.84 | 10.50 | 11.34 | MgSiO$_3$ |
| CO$_2$ (g) | -3.38 | -21.53 | -18.15 | CO$_2$ |
| Cristobalite | -0.48 | -4.06 | -3.59 | SiO$_2$ |
| Diopside | 0.35 | 20.25 | 19.89 | CaMgSi$_2$O$_6$ |
| Dolomite | 2.40 | -14.69 | -17.09 | CaMg(CO$_3$)$_2$ |
| Epsomite | -2.36 | -4.50 | -2.14 | MgSO$_4$:7H$_2$O |
| Fe(OH)$_{2.7}$Cl$_{0.3}$ | 5.52 | -6.02 | -11.54 | Fe(OH)$_{2.7}$Cl$_{0.3}$ |
| Fe(OH)$_3$ (a) | 0.19 | -3.42 | -3.61 | Fe(OH)$_3$ |
| Fe$_3$(OH)$_8$ | -12.56 | -9.34 | 3.22 | Fe$_3$(OH)$_8$ |
| Forsterite | -3.24 | 25.07 | 28.31 | Mg$_2$SiO$_4$ |
| Goethite | 6.09 | -3.41 | -9.50 | FeOOH |
| Greenalite | -36.43 | -15.62 | 20.81 | Fe$_3$Si$_2$O$_5$(OH)$_4$ |
| Gypsum | -0.64 | -5.22 | -4.58 | CaSO$_4$:2H$_2$O |
| H$_2$ (g) | -41.22 | 1.82 | 43.04 | H$_2$ |
| Halite | -2.51 | -0.92 | 1.58 | NaCl |
| Hausmannite | 1.78 | 19.77 | 17.99 | Mn$_3$O$_4$ |
| Hematite | 14.20 | -6.81 | -21.01 | Fe$_2$O$_3$ |
| Huntite | 1.36 | -28.61 | -29.97 | CaMg$_3$(CO$_3$)$_4$ |
| Hydromagnesite | -4.56 | -13.33 | -8.76 | Mg$_5$(CO$_3$)$_4$(OH)$_2$:4H$_2$O |
| Jarosite (ss) | -9.88 | -43.38 | -33.50 | (K$_{0.77}$Na$_{0.03}$H$_{0.2}$)Fe$_3$(SO$_4$)$_2$(OH)$_6$ |
| Magadiite | -6.44 | -20.74 | -14.30 | NaSi$_7$O$_{13}$(OH)$_3$:3H$_2$O |
| Maghemite | 3.80 | -6.81 | -10.61 | Fe$_2$O$_3$ |
| Magnesite | 1.07 | -6.96 | -8.03 | MgCO$_3$ |
| Magnetite | 3.96 | -9.30 | -13.26 | Fe$_3$O$_4$ |
| Manganite | 2.46 | 6.28 | 3.82 | MnOOH |
| Melanterite | -19.35 | -21.56 | -2.21 | FeSO$_4$:7H$_2$O |
| Mirabilite | -2.49 | -3.60 | -1.11 | Na$_2$SO$_4$:10H$_2$O |
| Mn$_2$(SO$_4$)$_3$ | -54.60 | -9.29 | 45.31 | Mn$_2$(SO$_4$)$_3$ |
| MnCl$_2$:4H$_2$O | -12.88 | -10.17 | 2.71 | MnCl$_2$:4H$_2$O |
| MnSO$_4$ | -14.48 | -11.81 | 2.67 | MnSO$_4$ |
| Nahcolite | -2.91 | -13.79 | -10.88 | NaHCO$_3$ |

**Table 14.3**   continued

| Phase | SI | log IAP | log KT | |
|-------|-----|---------|--------|--|
| Natron | -4.81 | -6.12 | -1.31 | $Na_2CO_3{:}10H_2O$ |
| Nesquehonite | -1.37 | -6.99 | -5.62 | $MgCO_3{:}3H_2O$ |
| $NH_3$ (g) | -8.73 | 2.29 | 11.02 | $NH_3$ |
| Nsutite | 5.85 | 5.37 | -0.48 | $MnO_2$ |
| $O_2$ (g) | -0.70 | -3.66 | -2.96 | $O_2$ |
| Portlandite | -9.00 | 13.80 | 22.80 | $Ca(OH)_2$ |
| Pyrochroite | -8.01 | 7.19 | 15.20 | $Mn(OH)_2$ |
| Pyrolusite | 7.03 | 5.37 | -1.66 | $MnO_2$ |
| Quartz | -0.08 | -4.06 | -3.98 | $SiO_2$ |
| Rhodochrosite | -3.20 | -14.33 | -11.13 | $MnCO_3$ |
| Sepiolite | 1.15 | 16.91 | 15.76 | $Mg_2Si_3O_{7.5}OH{:}3H_2O$ |
| Siderite | -13.13 | -24.02 | -10.89 | $FeCO_3$ |
| $SiO_2$ (a) | -1.35 | -4.06 | -2.71 | $SiO_2$ |
| Talc | 6.04 | 27.44 | 21.40 | $Mg_3Si_4O_{10}(OH)_2$ |
| Thenardite | -3.34 | -3.52 | -0.18 | $Na_2SO_4$ |
| Thermonatrite | -6.17 | -6.05 | 0.12 | $Na_2CO_3{:}H_2O$ |
| Tremolite | 11.36 | 67.93 | 56.57 | $Ca_2Mg_5Si_8O_{22}(OH)_2$ |

on the supersaturation of several manganese minerals in various depths, and thus in differing redox environments. Figure 14.1 shows the depth profiles for the saturation indices of the minerals manganite (MnOOH), birnessite ($MnO_2$), manganese sulfide (MnS), and rhodochrosite ($MnCO_3$).

The precipitation of manganese oxides, and manganese hydroxides, is very obvious at a depth of about 0.1m below the sediment surface. For the minerals manganite and birnessite, exemplifying a group of similar minerals, the saturation indices lie around zero. In deeper core regions these and all other oxides and hydroxides are in a state of undersaturation.

The slight but very constant supersaturation (saturation index about +0.5) found for the mineral rhodochrosite over the whole depth range between 0.2 and 12 m below the sediment surface is quite interesting as well. This can be understood as an indication for a new formation of the mineral in this particular range of depth. Below 13 m rhodochrosite and MnS are both *under*saturated, obviously absent in the solid phase, and therefore cannot have controlled the low manganese concentrations still prevalent in greater depths by means of mineral equilibria.

*Calculation of Predominance Field Diagrams*

The graphical representations of species distributions and/or mineral saturation as a function of $E_H$-value and pH-value, which are commonly typified as predominance field diagrams, or sometimes even not very correctly as stability field diagrams, can be done very accurately with the aid of geochemical model programs. The diagrams related to manganese, iron and copper shown in Chapter 11 (Figs. 11.2, 11.4, and 11.6) as well as the Figure 14.2 referring to arsenic, have all been made by applying the model program PHREEQC.

The procedure consisted in scanning the whole $E_H$/pH-range in narrow intervals with several thousand PHREEQC-calculations and a given specific configuration of concentrations (in this particular case an ocean water analysis at the ocean bottom, at the site of manganese nodule formation). Thus, each special case is individually calculated and adjusted in agreement to the appropriate temperature, the correct ionic strength (and hence the correct activity coefficient) and accounting for all essential aquatic species and all the minerals eventually present.

**Table 9.4** Model calculation applying the computer program PHREEQC (Parkhurst 1995) to a sample of cold surface water of the ocean from higher latitudes. The constant of the solubility product for calcite is accordingly corrected for temperature.

---

## Model of cold surface seawater

---

### input concentrations:

dissolved constituents from Nordstrom et al. 1979, cf. Tabel 14.1

### boundary conditions:

| | | |
|---|---|---|
| temperature | 2 °C | |
| $pCO_2$ | 280 µatm | (i.e. log $pCO_2$ = -3.55) |
| log k calcite | -8.34 | (at 2 °C and 1 atm pressure) |

### input situation without calcite-carbonate-equilibrium:

| | | |
|---|---|---|
| pH | 8.23 | |
| sum of carbonate species (TIC) | 2.28 mmol/l | |
| sum of calcium species | 10.63 mmol/l | |
| $SI_{calcite}$ | 0.43 | (i.e. $\Omega_{calcite}$ = 2.7) |

### reactions:

calcite supersaturation constant at SI = 0.50

### PHREEQC model results:

| | | |
|---|---|---|
| pH | 8.30 | |
| sum of carbonate species (TIC) | 2.30 mmol/l | |
| sum of calcium species | 10.65 mmol/l | |
| $SI_{calcite}$ | 0.50 | (i.e. $\Omega_{calcite}$ = 3.2) |

---

## 9.4 Carbonate Reservoir Sizes and Fluxes between Particulate and Dissolved Reservoirs

Present-day production of calcium carbonate in the pelagic ocean is calculated to be in the order of 6 to 9 billion tons (bt) per year (60-90·$10^{12}$ mol $yr^{-1}$), from which about 1.1 to 2 bt (11 to 20·$10^{12}$ mol $yr^{-1}$) accumulate in sediments (Tables 9.1 and 9.6). Together with the accumulation in shallow waters of 14.5·$10^{12}$ mol $yr^{-1}$ (Table 9.1, Fig. 9.5), the total carbonate accumulation in the world ocean amounts to 3.5 bt per year. This latter number is twice as much calcium as is brought into the ocean by rivers and hydrothermal activity (1.6 bt, Wollast 1994). Wollast (1994) and Milliman and Droxler (1996)

therefore consider the carbonate system of the modern ocean to be in non-steady state conditions, because production is not equal to the input of $Ca^{2+}$ and $HCO_3^-$ by rivers and hydrothermal vents (Fig. 9.5). Consequently, the marine carbonate system is at a stage of imbalance, or the output controlled by calcite sedimentation or input conveiled by dissolution has been overestimated or underestimated, respectively. On the other hand, one or more input sources may have been not detected so far. Published calcium carbonate budgets vary strongly, because the various authors have used different data sets and made different assumptions with respect to the production and accumulation of carbonates (e.g. Table 9.6), as well as for the sources of dissolved calcium and carbonate in marine waters (see summaries in Milliman 1993, Milliman and Droxler 1996, Wollast 1994).

**Table 9.5**  Model calculation using the computer program PHREEQC (Parkhurst 1995) on deep-sea waters of the ocean. The constant of the solubility product for calcite is accordingly corrected for temperature and pressure. A comparable decomposition of organic matter as in contained in Table 9.3 was excluded in this example.

---

## Model of seawater at 6000 m depth

---

### input concentrations:

model calculation of cold surface seawater

### boundary conditions:

| | | |
|---|---|---|
| temperature | 2 °C | |
| $pCO_2$ | 280 µatm | (i.e. log $pCO_2$ = -3.55) |
| log k calcite | -7.75 | (at 2 °C and 600 atm pressure) |

### input situation without calcite-carbonate-equilibrium:

| | | |
|---|---|---|
| pH | 8.23 | |
| sum of carbonate species (TIC) | 2.28 mmol/l | |
| sum of calcium species | 10.63 mmol/l | |
| $SI_{calcite}$ | 0.43 | (i.e. $\Omega_{calcite}$ = 2.7) |

### no reactions, but pressure changed to 600 atm

### PHREEQC model results:

| | | |
|---|---|---|
| pH | 8.23 | |
| sum of carbonate species (TIC) | 2.28 mmol/l | |
| sum of calcium species | 10.63 mmol/l | |
| $SI_{calcite}$ | -0.16 | (i.e. $\Omega_{calcite}$ = 0.7) |

---

### 9.4.1　Production *Versus* Dissolution of Pelagic Carbonates

This chapter summarizes the most recent compilations of carbonate reservoir size in the ocean and sediments, as well as the particulate and dissolved fluxes (Fig. 9.5) provided by the above mentioned authors. Coral reefs are probably the best documented shallow-water carbonate environment. Carbonate production on reef flats range as high as 10.000 g $CaCO_3$ m$^{-2}$yr$^{-1}$, with a global mean of about 1500 g $CaCO_3$ m$^{-2}$yr$^{-1}$. Totally this amounts to 24.5·10$^{12}$ mol yr$^{-1}$ (Table 9.1) from which 14.5·10$^{12}$ mol yr$^{-1}$ accumulate and 10·10$^{12}$ mol yr$^{-1}$ are transported to the deep-sea either by particulate or dissolved export. One of the most uncertain numbers in all these budget calculations are the estimates of the global carbonate production in the open ocean. Milliman's (1993) estimate was only about 24·10$^{12}$ mol yr$^{-1}$, based on carbonate flux rates at about 1000 m water depth, measured by long-term time series of sediment trap moorings, which is approximately 8 g $CaCO_3$ m$^{-2}$yr$^{-1}$ accounting for a global flux rate of particulate pelagic carbonate to be in the range of 24·10$^{12}$ mol yr$^{-1}$. However, as discussed in Wollast (1994), in order to produce the measured water column profiles of total inorganic carbon and carbonate alkalinity (Fig. 9.3) a much higher surface ocean carbonate production is required. Thus estimates reported more recently are in the order of 60 to 90·10$^{12}$ mol yr$^{-1}$ (Table 9.6). The discrepancy between the very high global production rates and measured fluxes obtained in sediment traps then can only be explained if one accepts that a substantial portion (30 to 50%) of carbonate pro-

## Seawater from Nordstrom et al. (1979)

| Phase | Si | log IAP | log KT |
|-------|-----|---------|--------|
| Calcite | 0.00 | -8.48 | -8.47 |
| $CO_2$ (g) | -3.47 | -21.62 | -18.15 |
| $O_2$ (g) | -0.68 ´ | 82.44 | 83.12 |

| | | | |
|------|---|--------|--------------------------|
| pH | = | 7.902 | Charge balance |
| pe | = | 12.713 | Adjusted to redox equilibrium |

| | | |
|-------------------|---|---------|
| Ionic strength | = | 6.734E-1 |
| Temperature (°C) | = | 25.0 |

| Elements | Molality | |
|----------|----------|---------------------------------|
| C | 8.139E-4 | Equilibrium with Calcite and $pCO_2$ |
| Ca | 9.876E-3 | Equilibrium with Calcite |
| Cl | 5.657E-1 | not changed |
| Fe | 3.711E-8 | not changed |
| K | 1.058E-2 | not changed |
| Mg | 5.507E-2 | not changed |
| Mn | 3.773E-9 | not changed |
| N | 6.571E-6 | Adjusted to redox equilibrium |
| Na | 4.854E-1 | not changed |
| O(0) | 3.951E-4 | Equilibrium with $pCO_2$ |
| S | 2.93E-02 | not changed |
| Si | 7.382E-5 | not changed |

**Table 14.4** In the seawater analyses shown in Tables 14.1 to 14.3, an equilibrium adaptation to the mineral calcite (SI = 0) and to the atmosphere with log $pCO_2$ = -3.47 (equivalent to $pCO_2$ = 0.00034) and log $pO_2$ = -0.68 (equivalent to $pO_2$ = 0.21) was calculated. This model calculation was performed with the program PHREEQC (Pankhurst 1995).

21% $O_2$ and 0.034% $CO_2$. The composition of this water sample is summarized in Table 14.4. Organic substance containing amounts of nitrogen and phosphorus according to the Redfield ratio of C:N:P = 106:16:1 (cf. Sect. 3.2.5) is then gradually added to this solution. Simplified, such a theoretical organic substance is composed of $(CH_2O)_{106}(NH_3)_{16}(H_3PO_4)_1$ or, divided by 106: $(CH_2O)_{1.0}(NH_3)_{0.15094}(H3PO4)_{0.009434}$. The horizontal axis in Figure 14.3 records the substance addition in terms of µmol per liter. If the situation in the sediment is to be described, an equilibrium of calcite should exist as well. The system under study should not be open to the gaseous phase any more, instead, these influences should now be determined by the consumption of oxygen within the system and by the concomitant release of $CO_2$.

In the model, the organic substance added to the reaction automatically triggers a sequence of redox reactions which becomes evident due to the concentrations involved which are demonstrated in Figure 14.3. First, the concentration of oxygen declines, and the concentration of nitrate and phosphorus increases according to their proportion in the original organic substance. As soon as the oxygen is consumed, nitrogen will be utilized for further oxidation of the continually added organic substance. Subsequently, the system shifts over to an anoxic state.

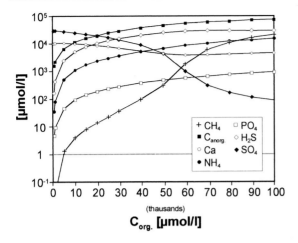

**Fig. 14.4** Geochemical model calculation using the program PHREEQC. In an anoxic system (state at the end of the model calculation from Fig. 14.3), the gradual addition of organic matter to the redox reaction is continued, whereby the system is kept open for calcite equilibrium and sealed from the gaseous phase. Initially, the dissolved sulfate will be consumed, in the course of which low amounts (logarithmic scale) of methane will emerge. Only after the sulfate concentration has become sufficiently low, will the generation of methane display its distinct increase.

*Decomposition of Organic Substance under Anoxic Conditions*

An anoxic system, as obtained from the previous reaction sequence, will be capable of further on-going reaction, if the supply of organic matter is continued and the reaction is held under the same boundary conditions. The result consists in the continuation of the processes shown in Figure 14.3. The outcome of this experiment is presented in Figure 14.4. Here as well, the horizontal axis is calibrated with regard to the added amounts of organic matter in terms of µmol per liter.

As expected, the sulfate concentration in solution initially decreases at a swift rate and, concomitantly, low amounts of methane already emerge. Only after the concentration of sulfate falls under a certain level, will methane be released in substantial amounts. The permanently maintained calcite equilibrium will result, due to precipitation, in the depression of the concentration of dissolved calcium ions (compare Schulz et al. 1994).

The sequence of reactions, as described and modeled above, has naturally been well established and frequently documented ever since the publications made by Froelich et al. (1979). However, it should not be overlooked that these processes were not pre-determined for the modeling procedure, but rather 'automatically' evolved from the fundamental geochemical data supporting the model as well as from the pre-set, although very generalized, starting and boundary conditions. During the entire model calculation procedure, all quantitatively important species, the complex aquatic species, as well as the activity coefficients related to these solutions, were invariably taken into account. As a realistic and also quite well-known reaction, calcite dissolution/precipitation was employed for an adjustment of the mineral phase equilibrium. Such a system can be made more complex, almost at will, by specifically operating with additional mineral phases (e.g. iron minerals or manganese minerals) that can be eventually dissolved, if they are present and if the condition of undersaturation prevails, or which are allowed to precipitate in the case of supersaturation. However, the simulation of such mineral equilibria requires a very profound knowledge of the specific reactions of mineral dissolution or precipitation in the sytem under study.

## 14.2 Analytical Solutions for Diffusion and Early Diagenetic Reactions

Analytical solutions for Fick's second law of diffusion were already discussed in Chapter 3, Section 3.2.4, without taking diagenetic reaction into account. With a diffusion coefficient $D_{sed}$, which describes the diffusion inside the pore cavity of sediments, Fick's second law of diffusion is stated as:

$$\frac{\partial C}{\partial t} = D_{sed} \cdot \frac{\partial^2 C}{\partial x^2} \qquad (14.5)$$

This partial differential equation has different solutions for particular configurations and boundary conditions. In Section 3.2.4 the following solution was presented:

$$C_{x,t} = C_0 + \left(C_{bw} - C_0\right) \cdot erfc\left(\frac{x}{2 \cdot \sqrt{\left(D_{sed} \cdot t\right)}}\right)$$

$$(14.6)$$

The solution of the equation yields the concentration $C_{x,t}$ at a specific point in time t and a specific

last glacial and the sedimentary calcium carbonate concentrations in deep-sea sediments. The calcite dissolution by oxic respiration of organic matter might therefore be able to mask effects of changes in carbonate productivity and deep-water chemistry in the sedimentary carbonate record (Martin and Sayles 1996).

For a long time, it was not possible to calculate the benthic total carbon dioxide or alkalinity flux by *ex-situ* methods because of artifacts introduced by decompression processes during core recovery. Moreover, there was no established method available to predict a true concentration profile or a benthic flux (Murray et al. 1980, Emerson and Bender 1981, Emerson et al. 1980, Emerson et al. 1982). What happens during the recovery of cores from several thousand meters of water depth is that the solubility of $CO_2$ in the pore water is increasingly reduced due to decompression and warming. Probably, dependent on the calcium carbonate content of the sediment providing nucleation sites, calcium carbonate is then precipitated from the pore water on its way through the water column. Figure 9.6 shows the differences between measurements and model results of pore water alkalinity concentrations at a location on the continental slope off Southwest Africa. The model alkalinity curve was calculated by the transport-reaction model CoTAM (Hamer and Sieger 1994), assuming the decay of organic matter through oxygen and nitrate and calcite at equilibrium. The equilibrium calculation was carried out by using the geochemical calculation program PHREEQE (Parkhurst et al. 1980; see also Sect. 9.3.4) as a subroutine of CoTAM (e.g. Hensen et al. 1997). Even a best fit application to the pore water results underestimates the diffusive alkalinity flux across the sediment-water interface by about 90% compared to what the model predicts.

In the last decade *in-situ* techniques have been developed to overcome these problems. Profiling lander systems were deployed to record the pore water microprofiles of oxygen, pH and $pCO_2$, whereas benthic chambers were deployed to measure solute fluxes across the sediment-water interface directly. In most cases, one-dimensional transport-reaction models allow to match measured data profiles now and to distinguish between different mechanisms of $CaCO_3$ dissolution. Generally, inorganic carbon fluxes show a dependence on the rate of calcium carbonate dissolution.

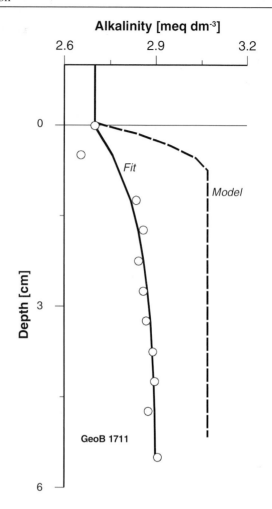

**Fig. 9.6** Measured, fitted and modeled alkalinity depth profiles of GeoB station 1711 off Southwest Africa from about 2000 m water depth. Model results indicate the underestimation of alkalinity concentration in the pore water due to decompression and warming upon recovery.

Dissolution rates of calcium carbonate in different environments and calcium carbonate kinetics have been studied intensively over the last decades. One of the pioneering work is that of Plummer et al. (1978) who defined a rate law for calcite dissolution dependent on the activities of $[H^+]$, $[H_2CO_3^*]$, $[Ca^{2+}]$, and $[HCO_3^-]$. This is widely known as the PWP model:

$$R_{PWP} = k_1 [H^+] + k_2 [H_2CO_3^*] + k_3 [H_2O] - k_4 [Ca^{2+}][HCO_3^-] \quad (9.16)$$

where $R_{PWP}$ is the rate of calcite dissolved (mol cm$^{-2}$s$^{-1}$), $k_1$, $k_2$, and $k_3$ are temperature dependent rate constants, $k_4$ is the rate constant for the reverse reaction which is dependent on temperature and $pCO_2$, and $H_2CO_3^*$ comprises $CO_{2(aq)} + H_2CO_3^0$. The PWP model therefore implies three general mechanistic reactions occurring at the crystal surface driving the dissolution of calcium carbonate:

$$CaCO_3 + H^+ \leftrightarrow Ca^{2+} + HCO_3^- \qquad (9.17a)$$

$$CaCO_3 + H_2CO_3^0 \leftrightarrow Ca^{2+} + 2\,HCO_3^- \qquad (9.17b)$$

$$CaCO_3 + H_2O \leftrightarrow Ca^{2+} + HCO_3^- + OH^- \qquad (9.17c)$$

This model has been extensively tested for its applicability in describing dissolution processes occurring on pure phases and natural calcium carbonates. It has been found that model results overestimate the dissolution rate in comparison to experimental data (Morse 1978, Plummer et al. 1979, Palmer 1991, Svensson and Dreybrodt 1992).This is true for conditions close to equilibrium and for natural calcium carbonates. The lower rate of dissolution is suggested to be related to the occupation of surface sites by adsorbed $Ca^{2+}$, heavy metal ions, or phosphate ions (Morse and Berner 1979, Svensson and Dreybrodt 1992). Therefore, in most recently published studies, the calcium carbonate dissolution in sea water and in the pore water of surface sediments is assumed to follow a kinetic process that can be described by the following equation:

$$R_d = k_d\,(1 - \Omega)^n \qquad (9.18)$$

and

$$\Omega = \frac{[Ca^{2+}][CO_3^{2-}]}{K}$$

or

$$SI = log\,\frac{[Ca^{2+}][CO_3^{2-}]}{K} \qquad (9.19)$$

(Keir 1980, Morse 1978), where $R_d$ is the calcite dissolution rate, $k_d$ is the calcite dissolution rate constant, and $\Omega$ or SI describes the degree of saturation (ion activity product divided by $K$[2], and $K$ the solubility constant of the calcium carbonate species in question. Often $K'$ is used instead of $K$ and is defined as the apparent equilib-

rium constant which is related to concentrations instead of activities. Compilations of equilibrium constants are available, e.g. by Mehrbach et al. (1973). In several studies, a further dependence of $R_d$ on the calcium carbonate content (respectively the surface area) in the sediment is considered.

In most recent studies that have applied the Equation 9.18, the superscript n is 4.5 for calcite and 4.2 for aragonite. The value of $k_d$ is, however, much less constrained than n. Laboratory experiments of Keir (1980) indicate a rate value of about 1000% $d^{-1}$ for $k_d$[3], but field observations require much lower values for $k_d$. In different model calculations of data derived from various locations in the Pacific and Atlantic Oceans, $k_d$ varies by several orders of magnitude 0.01-150% $d^{-1}$ (Berelson et al. 1994, Hales and Emerson 1997a). The reason for this discrepancy is not clearly known. Important and regionally variable factors, however, may be the grain size and thus the surface area of calcium carbonate crystals in the sediments, or adsorbed coatings like phosphate ions protecting calcium carbonate grains from the action of corrosive pore waters (see above; Jahnke et al. 1994, Hales and Emerson 1997a). In contrast, Hales and Emerson (1997b) found evidence that in-situ pH measurements in pore waters of calcite-rich deep-sea sediments are more consistent with a first-order rather than 4.5th-order dependence. Applying these data, they rewrote Equation 9.18 to

$$R_d = 38\,(1 - \Omega)^1 \qquad (9.20)$$

reducing the range of required $k_d$ values to less than one order of magnitude. However they state, that pore water measurements resulted in dissolution rate constants at least 2 orders of magnitude lower as compared to those determined in laboratory studies.

It is also important where in the sediment dissolution occurs. Metabolically produced $CO_2$ released at the sediment-water interface is probably much less effective in carbonate dissolution than

---

[2] Ion-activities might change dramatically in deeper sediment layers compared to sea-water conditions due to redox processes. The appropriate values of $\Omega$ or $SI$ can be determined by computer programs like PHREEQC (Parkhurst 1995) or as implemented in CoTReM (Adler et al. in press).

[3] The unit % $d^{-1}$ originates from experimental studies (e.g. Morse 1978, Keir 1980) and is used in most studies dealing with carbonate dissolution in marine sediments. The use is, however, not always consistent regarding the units of the dissolution rate and the parameters used in the equation and should, therefore, generally be evaluated with caution.

in deeper sediment strata, because the neutralization with bottom water $CO_3^{2-}$ might occur instead of dissolution. If the particulate organic matter is more rapidly mixed downwards, i.e. by bioturbation, and oxidized in deeper sediment strata, the $CO_2$ released into the pore waters can probably become more effective in carbonate dissolution (Martin and Sayles 1996). Generally, measurements of calcium carbonate dissolution in deep-sea surface sediments indicate a strong im-

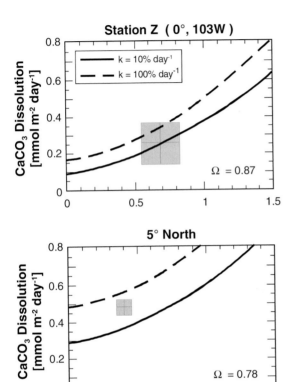

**Fig. 9.7** Model results of calcium carbonate dissolution rates as a function of oxygen uptake rates in Central Pacific Sediments and the degree of saturation with different values of $k_d$ from Berelson et al. (1994). The boxes represent averaged benthic lander fluxes for each station (see text for explanation).

pact of metabolic $CO_2$ so that generally both processes contribute to the release of dissolved carbon from the sediments. In summary, Figure 9.7 shows model calculations of calcium carbonate dissolution fluxes as a function of oxygen fluxes and bottom water calcite saturation state for different values of $k_d$ (Berelson et al. 1994). Several correlations can be observed:

1) It is obvious that, when oxygen fluxes are low, the relative importance of bottom water undersaturation ($\Omega$) is the driving force for $CaCO_3$ dissolution and that dissolution is exceptionally driven by $k_d$ and $\Omega$ at the y-axis intercept.

2) Equation 9.18 implores that a higher degree of undersaturation increases the differences of dissolution fluxes between the chosen values of $k_d$ (distance between hatched and solid lines).

3) Higher oxygen fluxes (higher amount of metabolically released $CO_2$) seem to be more efficient in carbonate dissolution when bottom water undersaturation is high (lower $\Omega$).

4) For high oxygen fluxes the model predicts a ratio of calcite dissolution and oxygen flux of 0.85 which is close to the stoichiometrical ratio in Equation 9.15.

All the factors mentioned above might explain why it is difficult to generalize the interdependences of the parameters and to develop a more general model to predict release rates of inorganic and organic carbon from the deep-sea sediments. Therefore, even estimates of total carbon release imply a great inaccuracy. This is the reason why only few compilations exist until today. Table 9.7 summarizes some recent data compilations that should help to constrain the carbon flux from deep-sea sediments. Three flux categories are given in Table 9.7: The $CO_2$ produced in oxidative respiration, the alkalinity as a collective parameter for calcium carbonate dissolution and $CO_2$ from oxic respiration, and, hitherto neglected in the discussion, the dissolved organic carbon (DOC).

As pointed out above, it is difficult to estimate the contribution of $CO_2$ from the oxidation of organic matter to the total carbon flux, since the bottom water saturation plays a major role in this context. The studies of Martin and Sayles (1996) and Hales and Emerson (1996), for example, indicate that more than 50% of the dissolution of calcium carbonate on the Ceará Rise (western Atlantic) and the Ontong Java Plateau (western Pacific) can be attributed to the neutralization of metabolic acids. Results reported by Hammond et al. (1996) from studies conducted in the equatorial Pacific, and by Reimers et al. (1992) conducted

**Table 9.7** Carbon fluxes from deep-sea sediments (below 1000 m water depth) in $10^{12}$ mol yr$^{-1}$ estimated by using different parameters. Global estimations of regional data compilations are made by multiplication with surface area factors.

| Parameter | Area Production | Flux | Source |
|---|---|---|---|
| **Respiratory CO$_2$**[1] | South Atlantic | 6 - (11) | after Hensen et al. (1998) |
| | Global | 60 - (110) | after Hensen et al. (1998) |
| | | 40 | after Jahnke (1996) |
| **Calcite Dissolution** | Global | 27 - 54 | Archer (1996b) |
| **Alkalinity / TCO$_2$** | Pacific | 55 | Berelson et al. (1994) |
| | Indo-Pacific | 91 | Berelson et al. (1994)[2] |
| | Global | 100 | after Berelson et al. (1994) |
| | | 120 | after Berelson et al. (1994)[2] |
| | | 120 | Mackenzie et al. (1993) |
| | | 5 | Morse and Mackenzie (1990) |
| **DOC** | Atlantic [3] | 4 | Otto (1996) |
| | Global [3] | 18 | Otto (1996) |

[1] Estimated as oxic respiration of organic matter.

[2] Using data of Broecker and Peng (1987).

[3] Including water depths above 1000m.

off central California revealed that both carbonate dissolution and oxidation of organic carbon contribute each about one half to the total carbon fluxes. Since the amount of carbon released by the oxidation of organic matter cannot be measured directly, it has to be estimated from the depletion of oxidants on the basis of stoichiometric relationships (Eq. 9.15). In deep-sea sediments, organic matter oxidation can be mainly related to benthic oxygen fluxes. In areas displaying higher organic carbon accumulation rates, i.e. upwelling areas and shelves, other electron acceptors like nitrate, metal-oxides and sulfate are of more importance. Oxygen is also consumed in the reoxidation of upward-diffusing, reduced species (see Section 6.3.1.2). In spite of the uncertainties we give rough estimations of the global benthic CO$_2$ release based on global benthic oxygen fluxes (Jahnke 1996) and benthic nitrate release (from oxic respiration) in the South Atlantic (Hensen et al. 1998; Table 9.7). Figure 9.8 shows

a distribution map of nitrate-equivalent carbon release rates for the South Atlantic below 1000 m water depth. The indicated flux ranges for each level refer to C:N ratios of decomposed organic matter between 4 and 7; nitrate fluxes are not corrected for denitrification. Generally, it is believed that decomposition occurs in the stoichiometric ratio for C and N of about 6.625 as given in Equation 9.15. However, there are several reasons why nitrate fluxes do not necessarily reflect the degradation of organic matter with that particular C:N ratio: (1) Preferential use of freshly deposited, low C/N material, (2) oxidation of other nitrogen sources (NH$_3$, dissolved organic nitrogen - DON), and (3) artificially increased subsurface nitrate concentrations due to *ex-situ* sampling of pore waters (see Chap. 6). A detailed discussion of these problems is beyond the scope of this summary, but for this reason the upper limit of carbon release fluxes after Hensen et al. (1998) in Figure 9.8 and in Table 9.7 is thought to be over-

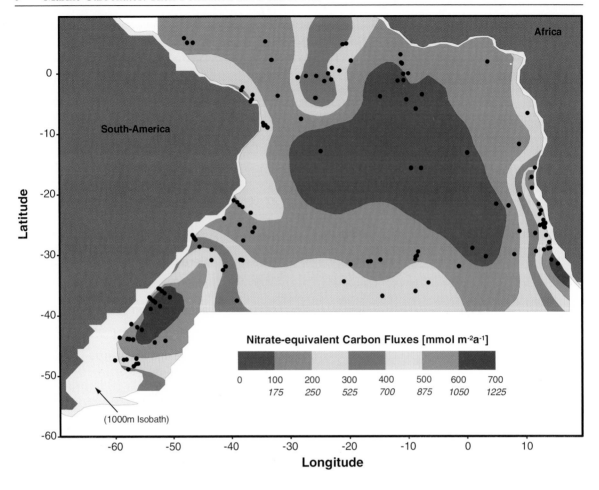

**Fig. 9.8** Estimation of equivalent carbon fluxes from sediments below 1000 m in the South Atlantic estimated from regional distribution of nitrate fluxes (from oxic respiration) after Hensen et al. (1998). The fluxes refer to C/N ratios of decomposed organic matter between 4-7. The lower end calculation was chosen to compensate low C/N oxic respiration, ammonium oxidation and possible core recovery artifacts and seems to be more realistic.

estimated, whereas the lower limit seems to be more realistic. The use of a range of different C:N ratios is, strictly seen, not correct regarding the above mentioned factors, but it provides an easy applicable tool to cope with the given uncertainties. Both estimates in Table 9.7 reveal that there is an uncertainty of nearly a factor of 2 for this parameter, similar to the error given by Milliman and Droxler (1996) for pelagic calcite production and dissolution in the water column (Table 9.1).

Berelson et al. (1994) made estimations for the benthic alkalinity input to the deep ocean for the Pacific and the Indo-Pacific (Table 9.7). They suggested that most of the carbonate dissolution in the deep ocean (Fig. 9.5) occurs within the sediments (85%). However, it has to be admitted that the proportions of $CaCO_3$ dissolution in the water column and in the sediment are still under

debate and that estimates for the dissolution in the water column range between 0-90% (Archer 1996b). A global projection for benthic alkalinity flux based on the calculations of Berelson et al. (1994) and using the surface areas for the global deep ocean (see Table 9.7) seems not to be appropriate. The application of their results from the Pacific and Indian Ocean to the Atlantic Ocean, leading to $120 \cdot 10^{12}$ mol $yr^{-1}$ of global dissolved carbon fluxes from sediments may be regarded critical because of the completely different deep-water conditions prevalent in the Indo-Pacific and the Atlantic. Deep-ocean waters in the Indian and Pacific Oceans are known to be much older and depleted in $CO_3^{2-}$ implying that a much higher proportion of calcite dissolution contributes to the total alkalinity input there. In this regard, Milliman (1993) estimated that the Atlantic

Ocean probably accounts for about 40% of the global deep-ocean calcium carbonate accumulation with about 20-25% of the global deep-sea area. However, despite this problem of different bottom-water saturation conditions, extending the Berelson et al. (1994) approach for a global estimate is in agreement to estimations made by Mackenzie et al. (1993). In contrast, reported by Morse and Mackenzie (1990) the global alkalinity flux seems to be drastically underestimated (Table 9.7). To compare the estimates of Berelson et al. with the results of others, the alkalinity fluxes are assumed to be equivalent to molar concentrations of carbon, which means that $HCO_3^-$ is assumed to be the dominant carbonate species.

Upon comparing the estimates of total carbon release based on benthic alkalinity fluxes with those from organic matter decay (Table 9.7; after Hensen et al. 1998), one can assume that metabolic $CO_2$ may contribute about 50% ($60 \cdot 10^{12}$ mol yr$^{-1}$) to the total carbon release of $120 \cdot 10^{12}$ mol yr$^{-1}$ from deep-sea sediments, meaning that the other 50% come from $CaCO_3$ dissolution. This is consistent with model results of Archer (1996b) who proposed an upper limit of global carbon fluxes measuring up to $54 \cdot 10^{12}$ mol yr$^{-1}$ due to calcite dissolution from deep-sea sediments. Thus, the total carbon release from deep-sea sediments of about $120 \cdot 10^{12}$ mol yr$^{-1}$ seems to be a relatively good approximation regarding all sources of uncertainty. The global estimate of $50 \cdot 10^{12}$ mol yr$^{-1}$ to $60 \cdot 10^{12}$ mol yr$^{-1}$ for sedimentary inorganic carbon release would mean that nearly all dissolved total inorganic carbon in the deep ocean originates from the calcite dissolution in the sediment and in the water column, which then questions the low carbonate flux at 1000 m water depth in comparison with the high estimates of global pelagic calcite production (Table 9.1 and Fig. 9.6). However, as pointed out above, the estimates for the pelagic realm, with the exception of the carbonate accumulation in deep-sea sediments, are not very well constrained on a global base and therefore more efforts have to be undertaken to verify such estimates in future. Moreover, an important factor in carbon budget calculations in the deep-sea which is not considered here is the carbon flux of dissolved organic carbon (DOC) which is not in the focus of carbon dioxide. For DOC, results reported by Otto (1996) indicate an additional release of extra global organic carbon of about $20 \cdot 10^{12}$ mol yr$^{-1}$ (Table 9.7) from the sediments into the ocean.

## Summary

The main subjects addressed in Chapter 9 are listed below:

- The major site of marine carbonate accumulation is the neritic environment, including coral reefs, banks and continental shelves, and pelagic calcite-rich sediments. In total, about $35 \cdot 10^{12}$ mol $CaCO_3$ accumulate per annum in the marine realm.

- Based on budget calculations of calcium carbonate, reservoir sizes in the world ocean and exchange fluxes between reservoirs the carbonate system is not in steady state.

- However, calcium carbonate budget calculations are strongly biased by inexact estimations of calcite production in the surface ocean and of the dissolution of pelagic biogenic calcite in the water column and in sediments above the calcite lysocline. In addition, the uncertainty is enhanced by the difficulty to estimate dissolved inorganic carbon release from sediments.

- The total carbon release from deep-sea sediments is estimated to be about $120 \cdot 10^{12}$ mol yr$^{-1}$, but is subject to great uncertainty due to the complexity of processes controlling carbon remobilization.

- Both bottom water undersaturation and organic matter decay are responsible for calcium carbonate dissolution in the sediments at more or less equal levels.

- The efficiency of calcium carbonate dissolution by metabolic $CO_2$ strongly depends on the organic carbon / calcium carbonate rain ratio at the sediment surface, the oxidation rate of organic matter (and the depth horizon, where oxidation occurs), as well as the saturation state of bottom water ($\Omega$) and the dissolution rate constant $k_d$.

This is contribution No 258 of the Special Research Program SFB 261 (*The South Atlantic in the Late Quaternary*) funded by the Deutsche Forschungsgemeinschaft (DFG).

# References

Adler, M., Hensen, C. and Schulz, H.D., in press. Computer simulation of deep-sulfate reduction in sediments off the Amazon Fan. Geol. Rdsch.

Andersen, N.R. and Malahoff, A., 1977. The fate of fossil fuel CO2 in the Oceans. Plenum Press, NY, 749 pp.

Archer, D.E., 1991. Modeling the calcite lysocline. Journal of Geological Research, 96: 17037-17050.

Archer, D. and Maier-Reimer, E., 1994. Effect of deep-sea sedimentary calcite preservation on atmospheric CO2 concentration. Nature, 367: 260-263.

Archer, D.E., 1996a. An atlas of the distribution of calcium carbonate in sediments of the deep sea. Global Biochemical Cycles, 10: 159-174.

Archer, D.E., 1996b. A data-driven model of the global calcite lysocline. Global Biochemical Cycles, 10: 511-526.

Berelson, W.M., Hammond, D.E. and Cutter, G.A., 1990. In situ measurements of calcium carbonate dissolution rates in deep-sea sediments. Geochimica et Cosmochimica Acta, 54: 3013-3020.

Berelson, W.M., Hammond, D.E., McManus, J. and Kilgore, T.E., 1994. Dissolution kinetics of calcium carbonate in equatorial Pacific sediments. Global Biogeochemical Cycles, 8: 219-235.

Berger, W.H., 1976. Biogenic deep-sea sediments: Production, preservation and interpretation. In: Riley, J.P. and Chester, R. (eds) Chemical Oceanography, 5, Academic Press, London, pp. 266-388.

Berger, W.H., 1982. Increase of carbon dioxide in the atmosphere during deglaciation: The coral reef hypothesis. Naturwissenschaften, 69, 87.

Broecker, W.S. and Peng, T.-H., 1982. Tracers in the Sea. Lamont-Doherty Geol. Observation, Eldigo Press, Palisades, NY, 690 pp.

Broecker, W.S. and Peng, T.-H., 1987. The Role of CaCO3 compensation in the glacial to interglacial atmospheric CO2 change. Global Biogeochemical Cycles, 1: 15-29.

DeBaar, H.J.W. and Suess, E., 1993. Ocean carbon cycle and climate change - An introduction to the interdisciplinary Union Symposium. Global and Planetery Change, 8: VII-XI.

Dittert, N., Baumann, K.H., Bickert, T., Henrich, R., Huber, R., Kinkel, H. and Meggers, H., in press. Carbonate dissolution in the deep ocean: Methods, quantification and paleoceanography application. In: Fischer, G. and Wefer, G. (eds) Use of proxies in paleoceanography: examples from the South Atlantic, Springer, Berlin, Heidelberg, NY.

Emerson, S.R., Jahnke, R., Bender, M., Froelich, P., Klinkhammer, G., Bowser, C. and Setlock, G., 1980. Early diagenesis in sediments from the eastern equatorial Pacific. 1. Pore water nutrient and carbonate results. Earth Planet Science Letters, 49: 57-80.

Emerson, S. and Bender, M., 1981. Carbon fluxes at the sediment-water interface of the deep-sea: calcium carbonate preservation. Journal of Marine Research, 39: 139-162.

Emerson, S., Grundmanis, V. and Graham, D., 1982. Carbonate chemistry in marine pore waters: MANOP sites C and

S. Earth and Planetary Science Letters, 61: 220-232.

Goyet, C. and Poisson, A., 1989. New determination of carbonic acid dissociation constants in seawater as a function of temperature and salinity. Deep-Sea-Research, 36: 1635-1654.

Hales, B. and Emerson, S., 1996. Calcite dissolution in sediments of the Ontong-Java Plateau: In situ measurements of pore water O2 and pH. Global Biogeochemical Cycles, 10: 527-541.

Hales, B. and Emerson, S., 1997a. Calcite dissolution in sediments of the Ceara Rise: In situ measurements of porewater O2, pH, and CO2(aq). Geochimica et Cosmochimica Acta, 61: 501-514.

Hales, B. and Emerson, S., 1997b. Evidence is support of first-order dissolution kinetics of calcite in seawater. Earth and Planetary Science Letters, 148: 317-327.

Hamer, K. and Sieger, R., 1994. Anwendung des Modells CoTAM zur Simulation von Stofftransport und geochemischen Reaktionen. Verlag Ernst und Sohn, Berlin, 186 pp.

Hammond, D.E., McManus, J., Berelson, W.M., Kilgore, T.E. and Pope, R.H., 1996. Early diagenesis of organic material in equatorial Pacific sediments: stoichiometry and kinetics. Deep-Sea Research, 43: 1365-1412.

Hay, W.W. and Southam, J.R., 1977. Modulation of marine sedimentation by continental shelves. In: Andersen, N.R. and Malahoff, A. (eds) The fate of fossil fuel CO2 in the Oceans. Plenum Press, NY, pp. 564-604.

Hensen, C., Landenberger, H., Zabel, M., Gundersen, J.K., Glud, R.N. and Schulz, H.D., 1997. Simulation of early diagenetic processes in continental slope sediments in Southwest Africa: The computer model CoTAM tested. .Marine Geology, 144: 191-210.

Hensen, C., Landenberger, H., Zabel, M. and Schulz, H.D., 1998. Quantification of diffusive benthic fluxes of nitrate, phosphate and silicate in the Southern Atlantic Ocean. Global Biogeochemical Cycles, 12: 193-210.

Jahnke, R.A., Craven, D.B. and Gaillard, J.-F., 1994. The influence of organic matter diagenesis on CaCO3 dissolution at the deep-sea floor. Geochimica et Cosmochimica Acta, 58: 2799-2809.

Jahnke, R.A., 1996. The global ocean flux of particulate organic carbon: Areal distribution and magnitude. Global Biogeochemical Cycles, 10: 71-88.

Jahnke, R.A., Craven, D.B., McCorkle, D.C. and Reimers, C.E., 1997. CaCO3 dissolution in California continental margin sediments: The influence of organic matter remineralization. Geochimica et Cosmochimica Acta, 61: 3587-3604.

James, N.P. and Clarke, J.A.D., 1997. Cool-water carbonates, SEPM Spec. Publ., 56, Tulsa, Oklahoma, 440 pp.

Keir, R.S., 1980. The dissolution kinetics of biogenic calcium carbonates in seawater. Geochimica et Cosmochimica Acta, 44: 241-252.

Kharaka, Y.K., Gunter, W.D., Aggarwal, P.K., Perkins, E.H. and DeBraal, J.D., 1988. SOLMINEQ88: a computer program for geochemical modeling of water-rock-interactions. Water-Recources Invest. Report, 88-4227, US Geol. Surv., 207 pp.

Lisitzin, A.P., 1996. Oceanic sedimentation: Lithology and Geochemistry (English Translation edited by Kennett, J.P.). Amer. Geophys. Union, Washington, D.C., 400 pp.

Mackenzie, F.T., Ver, L.M., Sabine, C., Lane, M. and Lerman, A., 1993. C, N, P, S global biogeochemical cycles and modeling of global change. In : Wollast, R., Mackenzie, F.T. and Chou, L. (eds), Interactions of C, N, P and S biogeochemical cycles and global change. NATO ASI Series, 14, Springer Verlag, pp 1-61.

Maier-Reimer, E. and Bacastow, R., 1990. Modelling of geochemical tracers in the ocean. Climate-Ocean Interaction. In: Schlesinger, M.E. (ed), Climate-ocean interactions, Kluwer, pp. 233-267.

Martin, W.R. and Sayles, F.L., 1996. CaCO3 dissolution in sediments of the Ceara Rise, western equatorial Atlantic. Geochimica et Cosmochimica Acta, 60: 243-263.

Mehrbach, C., Culberson, C., Hawley, J.E. and Pytkowicz, R.M., 1973. Measurement of the apparent dissociation constants of carbonic acid in seawater at atmospheric pressure. Limnology and Oceanography, 18: 897-907.

Millero, F.J., 1995. Thermodynamics of the carbon dioxide systems in the oceans. Geochimica et Cosmochimica Acta, 59: 661-677.

Milliman, J.D., 1993. Production and accumulation of calcium carbonate in the ocean: budget of a nonsteady state. Global Biogeochemical Cycles, 7: 927-957.

Milliman, J.D. and Droxler, A.W., 1996. Neritic and pelagic carbonate sedimentation in the marine environment: ignorance is not a bliss. Geologische Rundschau, 85: 496-504.

Morse, J.W., 1978. Dissolution kinetics of calcium carbonate in sea water: VI. The near-equilibrium dissolution kinetics of calcium carbonate-rich deep-sea sediments. American Journal of Science, 278: 344-353.

Morse, J.W. and Berner, R.A., 1979. Chemistry of calcium carbonate in the deep ocean. In: Jenne, E.A. (ed), Chemical modelling in aqueous systems. Am. Chem. Soc., Symp. Ser., 93, pp. 499-535.

Morse, J.W. and Mackenzie, F.T., 1990. Geochemistry of sedimentary carbonates. Elsevier, Amsterdam, 707 pp.

Murray, J.W., 1897. On the distribution of the pelagic foraminifera at the surface and on the sea floor of the ocean. Nat. Sci., 11: 17-27.

Murray, J.W., Emerson, S. and Jahnke, R.A., 1980. Carbonate saturation and the effect of pressure on the alkalinity of interstitial waters from the Guatemala Basin. Geochimica et Cosmochimica Acta, 44: 963-972.

Nordstrom, D.K., Plummer, L.N., Wigley, T.M.L., Wolery, T.J., Ball, J.W., Jenne, E.A., Basset, R.L., Crerar, D.A., Florence, T.M., Fritz, B., Hoffman, M., Holdren, G.R. Jr., Lafon, G.M., Mattigod, S.V. McDuff, R.E., Morel, F., Reddy, M.M., Sposito, G. and Thrailkill, J., 1979. A comparision of computerized chemical models for equilibrium calculations in aqueous systems. In: Jenne, E.A. (ed), Chemical modeling in aqueous systems, speciation, sorption, solubility, and kinetics, 93, American Chemical Society, pp. 857-892.

Opdyke, B.D. and Walker, J.C.G., 1992. Return of the coral reef hypothesis: basin to shelf partitioning of CaCO3 and its effects on atmospheric CO2. Geology, 20: 733-736.

Otto, S., 1996. Die Bedeutung von gelöstem organischen Kohlenstoff (DOC) für den Kohlenstofffluß im Ozean. Berichte, 87, Fachbereich Geowissenschaften, Universität Bremen, 150 pp.

Palmer, A.N., 1991. The origin and morphology of limestone caves. Geological Society American Bulletin, 103: 1-21.

Parkhurst, D.L., Thorstensen, D.C. and Plummer, L.N., 1980. PHREEQE - a computer program for geochemical calculations. Water-Recources Invest. Report, 80-96, US Geol. Surv., 219 pp.

Parkhurst, D.L., 1995. User's guide to PHREEQC: a computer model for speciation, reaction-path, advective-transport, and inverse geochemical calculation. Water-Resources Invest. Report, 95-4227, US Geol. Surv., 143 pp.

Plummer, L.N., Wigley, T.M.L. and Parkhurst, D.L., 1978. The kinetics of calcite dissolution in CO2-water systems at 5°C to 60°C and 0.0 to 1.0 atm CO2. Am. J. Sci., 278: 179-216.

Plummer, L.N., Wigley, T.M.L. and Parkhurst, D.L., 1979. Critical review of the kinetics of calcite dissolution and precipitation. In: Jenne, E.A. (ed), Chemical modelling in aqueous systems. Am. Chem. Soc., Symp. Ser., 93, pp. 537-572.

Ragueneau, O., Tréguer, P., Anderson, R.F., Brezinski, M.A., DeMaster, D.J., Dugdale, R.C., Dymond, J., Fischer, G., Francois, R., Heinze, C., Leynaert, A., Maier-Reimer, E., Martin-Jézéquel, V., Nelson, D.M. and Quéguiner, B., subm. Understanding the Si cycle in the modern ocean: A pre-requisite for the use of biogenic opal as a paleoproductivity proxy. Global and Planetary Change.

Redfield, A.C., 1958. The biological control of chemical factors in the environment. Am. Scientist, 46: 206-2226.

Reimers, C.E., Jahnke, R.H. and McCorkle, D.C., 1992. Carbon fluxes and burial rates over the continental slope and rise off central California with implications for the global carbon cycle. Global Biogeochemical Cycles, 6: 199-224.

Roberts, H.H. and Macintyre, I.G. (eds), 1988. Special issue: Halimeda. Coral Reefs, 6(3/4), 121-280.

Roy, R.N., Roy, L.N., Vogel, K.M., Moore, C.P., Pearson, T., Good, C.E., Millero, F.J. and Campbell, D.M., 1993. Determination of the ionization constance of cabonic acid in seawater. Marine Chemistry, 44: 249-268.

Siegenthaler, H.H. and Wenk, T., 1984. Rapid atmospheric CO2 variations and ocean circulation. Nature, 308: 624-626.

Sundquist, E.T. and Broeker, W.S., 1985. The carbon cycles and atmospheric CO2: natural variations archean to present. American Geophysical Union, Washington, D.C., 627 pp.

Svensson, U. and Dreybrodt, W., 1992. Dissolution kinetics of natural calcite minerals in CO2-water systems approaching calcite equilibrium. Chemical Geology, 100: 129-145.

Wolf-Gladrow, D., 1994. The ocean as part of the global carbon cycle. Environ. Sci. & Pollut. Res., 1: 99-106.

Wollast, R., 1994. The relativ importance of biomineralisation and dissolution of CaCO3 in the global carbon cycle. In: Doumenge, F., Allemand, D. and Toulemont, A. (eds), Past and present biomineralisation processes: Considerations about the carbonate cycle. Bull. de l'Institute océanographique, 13, Monaco, pp. 13-35.

# 10 Influence of Geochemical Processes on Stable Isotope Distribution in Marine Sediments

Torsten Bickert

## 10.1 Introduction

Stable isotope geochemistry has become an essential part of marine geochemistry and has contributed considerably to the understanding of the ocean's changing environment and the processes therein. In some fields, such as paleoceanography, the application of stable isotopes is still growing due to new microanalytical techniques, permitting a relatively precise analysis of very small samples or single compounds, which allows the investigation of a new generation of problems. Stable isotopes have become useful tracers for reconstructing past temperatures, salinities, productivity, $pCO_2$, nutrients, etc. However, it has become evident that some limitations exist on the application of these tracers. Diagenetic processes may considerably alter the primary signals, due to preferential preservation, decomposition or relocation of particular tracers. On the other hand, stable isotope geochemistry offers the chance to identify and better understand such diagenetic processes, which is essential for the interpretation of isotope variability in marine sediments of the past.

Within this chapter, we focus on four elements (C, O, N, S), which participate in most marine geochemical reactions and which are important elements in the biological system. We summarize the influence of geochemical processes on the stable isotope distribution of those elements in ocean water and marine sediments. After a short review on the fundamentals of stable isotope fractionation and mass spectrometry, the most important fractionation mechanisms for each element within the marine environment are described, followed by the identification of geochemical processes responsible for the syn- or postsedimentary changes in isotope signals.

## 10.2 Fundamentals

Excellent summaries of the use of stable isotopes in the study of sediments and the environment in which they are deposited are given by Arthur et al. (1983), by Faure (1986) and by Hoefs (1997). The fundamentals of isotope fractionation are also presented in the book on isotopes in hydrogeology by Clarke and Fritz (1997).

### 10.2.1 Principles of Isotopic Fractionation

The stable isotopic compositions of elements having low atomic numbers (e. g. H, C, N, O, S) vary considerably in nature as a consequence of the fact that certain thermodynamic properties of molecules depend on the masses of the atoms of which they are composed. The partitioning of isotopes between two substances or two phases of the same substance with different isotope ratios is called isotopic fractionation. In general, isotopic fractionation occurs during several kinds of physical processes and chemical reactions:

- Isotope exchange reactions involving the redistribution of isotopes of an element among different molecules containing that element.

- Kinetic effects, which are associated with unidirectional and incomplete processes such as condensation or evaporation, crystallization or melting, adsorption or desorption, biologically mediated reactions, and diffusion.

In general, light isotopes are more mobile and more affected by such processes than heavy isotopes. The isotope fractionation that occurs during these processes is indicated by the fractionation factor $\alpha$ which is defined as the ratio $R_A$ of the heavy to the light isotopes in one compound or phase A divided by the corresponding isotope ratio $R_B$ for the compound or phase B:

$$\alpha_{A-B} = R_A / R_B \qquad (10.1)$$

For example, the fractionation factor for the exchange of $^{18}O$ and $^{16}O$ between water and calcium carbonate is expressed as:

$$H_2{}^{18}O + 1/3\ CaC^{16}O_3 \Leftrightarrow$$
$$H_2{}^{16}O + 1/3\ CaC^{18}O_3 \qquad (10.2)$$

with the fractionation factor $\alpha_{CaCO3\ -\ H2O}$ defined as:

$$\alpha_{CaCO3\ -\ H2O} =$$
$$(^{18}O\ /\ ^{16}O)_{CaCO3}\ /\ (^{18}O\ /\ ^{16}O)_{H2O} =$$
$$1.031\ \text{at}\ 25°C \qquad (10.3)$$

Because isotopic fractionation factors are close to 1, they can be expressed in ‰ with the introduction of the $\varepsilon$-value defined as

$$\varepsilon_{A-B} = (\alpha_{A-B} - 1) \cdot 1000 \qquad (10.4)$$

For geochemical purposes, the dependence of isotope fractionation factors on temperature is the most important property. In principle, fractionation factors for isotope exchange reactions are also slightly pressure-dependent, but experimental studies have shown the pressure dependence to be of no importance within the outer earth environments (Hoefs 1997). Occasionally, the fractionation factors can be calculated by means of partition functions derivable from statistical mechanics. However, the interpretation of observed variations of the isotope distribution in nature is largely empirical and relies on observations in natural environments or experimental results obtained in laboratory studies. A brief summary of the theory of isotope exchange reactions is given by Hoefs (1997).

### 10.2.2  Analytical Procedures

Stable isotope measurements on light elements are made on gases, i.e. $H_2$ for hydrogen, $CO_2$ for carbon and oxygen, $N_2$ for nitrogen, and $SO_2$ or $SF_6$ for sulfur isotopes. A variety of techniques are used to convert samples to a compound suitable for analysis. The most important aspect of sample preparation is to avoid isotopic fractionation. Since molecules with different isotopic masses have different reaction rates, procedures with less than a quantitative (i.e. 100%) output may produce a reaction product that does not have the same isotopic composition as the original sample. Furthermore, a pure gas is necessary to avoid interference by contaminants in the mass spectrometer. Contamination may result from incomplete evacuation of the vacuum preparation system or degassing of the sample, as well as from unwanted side reactions in the preparation procedures. In general, the error attributable to sample preparation is greater than the instrumental analysis of the product gas.

Isotopic abundance measurements for geochemical research are determined using mass spectrometry. A mass spectrometer separates and detects ions based on their motions in magnetic or electrical fields. For detailed information on mass spectrometry and the according analytical techniques we refer the reader to the comprehensive review by Hoefs (1997). This volume also gives an introduction in new microanalytical techniques, such as laser-assisted ablation, which allows the on-line transfer of submilligram quantities of mineral into a standard gas-source spectrometer (Kyser 1995), and the gas chromatography combined with mass spectrometry, which allows the determination of the isotopic composition of single compounds previously separated by means of a gas chromatograph (Brand 1996).

In Earth sciences, the relative differences in isotopic ratios between a sample and a standard are mostly used for reporting stable isotope abundances and variations. The reason is that the absolute value of an isotopic ratio is difficult to determine with sufficient accuracy for geochemical applications. The reporting notation employed is the $\delta$-value, defined as

$$\delta\ \text{in}\ ‰ = (R_{sample} - R_{standard})\ /\ R_{standard} \cdot 1000$$

$$(10.5)$$

where $R_{sample}$ is the isotopic ratio of the sample ($^{13}C/^{12}C$, $^{18}O/^{16}O$, $^{15}N/^{14}N$, $^{34}S/^{32}S$, etc.) and $R_{standard}$ is the corresponding rate in a standard. Nevertheless, the determination of absolute isotope ratios (Table 10.1) is essential, since these

**Table 10.1**   Absolute isotope ratios of international standards and laboratory standards (after Hoefs 1997)

| Standard | Ratio | Accepted value $\cdot 10^6$ within 95% confid. interval | Lab standard | δ-value [$^0/_{00}$] |
|---|---|---|---|---|
| **SMOW** | D/H | 155.8 ± 0.1 | VSMOW | 0.00 |
| Atandard Mean Ocean Water | | | SLAP | -428.00 |
| | $^{18}O/^{16}O$ | 2005.2 ± 0.4 | VSMOW | 0.00 |
| | | | SLAP | -55.50 |
| **PDB** | $^{13}C/^{12}C$ | 11237.2 ± 2.9 | NBS 19 (calcite) | +1.95 |
| Pee Dee Belemnite | $^{18}O/^{16}O$ | 2067.1 ± 2.1 | NBS 19 (calcite) | -2.20 |
| | | | NBS 19 (carbonatite) | -23.01 |
| **N$_2$ (atm.)** | $^{15}N/^{14}N$ | 3676.5 ± 8.1 | Air nitrogen | 0.00 |
| Air nitrogen | | | | |
| **CDT** | $^{34}S/^{32}S$ | 45004.5 ± 9.3 | CDT (FeS) | ± 0.04 |
| Canyon Diablo Troilite | | | | |

numbers form the basis for the calculation of the relative differences.

Isotope laboratories use different reference gases or working standards for the measurement of relative isotope ratios by mass spectrometry. However, all results are reported relative to an internationally accepted standard (Table 10.1). The selection of standards is an important procedure in isotope geochemistry because their definition and availability controls the extent to which results from different laboratories can be compared. Since the supply of PDB, the working standard introduced by H. C. Urey's laboratory at the University of Chicago, as well as of SMOW, a water sample prepared by H. Craig for distribution by the IAEA, have been exhausted for years, some confusion and irregularities occurred in the past regarding standards, particularly oxygen isotope standards. These problems may be resolved following the recommendations of the Commission on Atomic weights and isotopic abundances of the International Union of Pure and Applied Chemistry, published in 1995 (see Appendix in Coplen 1996). Isotope reference materials may be obtained from the National Institute of Standards and Technology, Gaithersburg, MD, or the International Atomic Energy Agency, Vienna, Austria (addresses are given in Coplen 1996).

## 10.3   Geochemical Influences on $^{18}O / ^{16}O$ Ratios

### 10.3.1   $δ^{18}O$ of Seawater

*Principles of Fractionation*

The oxygen isotopic composition of seawater ($δ^{18}O_w$) is controlled by fractionation effects due to evaporation and precipitation at the sea surface, freezing of ice in polar regions, the admixing of water masses containing different $^{18}O/^{16}O$ ratios such as melt water, river runoff, etc., and the global isotope content of the oceans (Craig and Gordon 1965, Broecker 1974). Since the salinity of seawater is similarly affected by these processes, Craig and Gordon (1965) and later Fairbanks et al. (1992) defined a set of regression relationships between salinity and $δ^{18}O_w$ with different slopes for several modern water masses, varying between 0.1 for humid tropical and 1.0 for arid polar surface water masses with a global mean of 0.49 (Fig. 10.1). Higher slopes represent areas where evaporation exceeds precipitation, and vice versa. However, the extent of oxygen isotope enrichments due to evaporation is limited due to the recycling of atmospheric moisture by the different pathways of precipitation. The slope of the global trend extrapolates to a $\delta^{18}O$-value of −17‰ at zero salinity, reflecting the influx of high-latitude precipitation and glacial meltwater. For the Antarctic continental ice, even $\delta^{18}O$-values as low as −54‰ have been determined (Weiss et al. 1979, Jacobs et al. 1985). However, the slope of the $\delta^{18}O_w$-salinity relationship may have changed through geological time (see discussion below).

For water masses deeper than 2000m, Zahn and Mix (1991) obtained a slope as high as 1.53. This gradient is explained with the formation of sea-ice in the source areas especially for southern component water masses. Since the freezing of polar surface waters raises the salinity, but does not fractionate oxygen isotopes, southern source deep water masses, like the Antarctic Bottom Water, exhibit relatively low $\delta^{18}O_w$ values, and so do other water masses, which are derived from the admixture of south polar water masses.

*Modern Range of Values and Historical Variability*

The modern $\delta^{18}O_w$ values of seawater are close to 0‰ (V-SMOW) and vary only within narrow lim-

its. From the GEOSECS $\delta^{18}O$ sections for the to-day's world oceans, compiled by Birchfield (1987), a range of +1.0‰ in the mid-latitude surface Atlantic to −0.6‰ in the northern surface Pacific is present. Deep water mass $\delta^{18}O_w$ ranges from +0.3‰ in the core of North Atlantic Deep Water to −0.1‰ in the Circumpolar Deep Water. Since $\delta^{18}O_w$ exhibits only a narrow range within open ocean conditions, this proxy is an excellent tracer for indicating the influence of freshwater input to ocean water masses, since river discharge or meltwater release is always depleted in $^{18}O$ (Craig and Gordon 1965). For example, Mackensen et al. (1996) showed isotopically light meltwaters of the Antarctic Peninsula shelves to cascade down slopes the Wedell Sea Basin and thereby contributing to bottom water formation.

In geologic history, the $\delta^{18}O_w$ has been shown to vary considerably. For the sea-level low stand of −120m during the last glacial maximum, Fairbanks (1989) showed Barbados coral oxygen isotopic composition to be enriched by 1.2‰, which coincides with an isotopic change of 0.10 ‰ corresponding to a 10m sea-level change earlier estimated by Shackleton and Opdyke (1973). Therefore, in an ice-free world, like in the Creta-

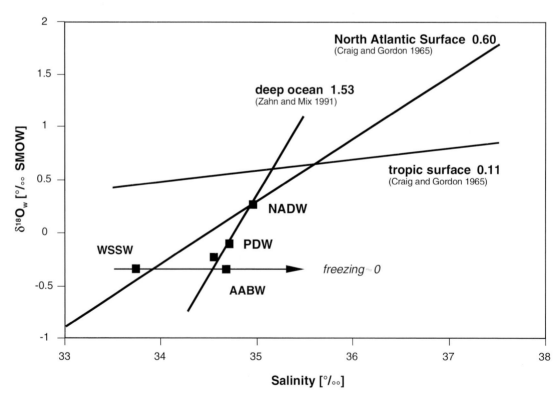

**Fig. 10.1** Relationship between salinity and $\delta^{18}O_w$ of major water masses (after Craig and Gordon 1965; deep water line according to Zahn and Mix 1991).

ceous, the global mean should have been depleted by another $-0.8‰$. On longer time scales, additional processes affecting the $\delta^{18}O_w$ have to be considered. Several authors have suggested a lower global isotopic composition for the Paleozoic oceans on the order of $-1‰$ to $-3.5‰$ (see for a recent summary Veizer et al. 1997). The depletion in $^{18}O$ is explained with enhanced interactions of seawater and fresh, silicate rocks at lower temperatures during the Paleozoic compared to younger epochs (Carpenter et al. 1991, Walker and Lohmann 1989). This explanation has been questioned because studies of the dynamics of mid-ocean-ridge / seawater interactions suggest that the oxygen isotopic composition of seawater should have been buffered at values close to the modern ocean composition (Hoffman et al. 1986, Muehlenbachs 1986). On the other hand, Barrett and Friedrichsen (1989) showed that low temperature exchange reactions may dominate in some ophiolite sequences, implying a possible imbalance between reactions controlling the $\delta^{18}O$ of seawater. Walker and Lohmann (1991) suggested that possible rates of change appear to be quite slow ($1‰$ in $10^8$ years) because of the large size of the oceanic reservoir. Railsback (1990) approximated the rising of $\delta^{18}O_w$ during the Paleozoic by a third-order relation to simulate the slow and non-reversible changing ocean. However, the large isotopic shifts observed for the entire Paleozoic, which include data from fossils (see review in Veizer et al. 1997) and abiotic marine calcites (Lohmann and Walker 1989, Carpenter et al. 1991), exceed by far the variability in isotopes observed in Neogene and Quaternary times and are too fast to be explained with ocean crust-seawater interactions. Those changes must be attributed to changes in earth climate systems, like ice volume and ocean circulation (e.g. Railsback 1990, Bickert et al. 1997).

### 10.3.2  $\delta^{18}O$ in Marine Carbonates

*Principles of Fractionation*

The oxygen isotope ratios in carbonates are a function of both temperature and the $\delta^{18}O_w$ of the surrounding seawater. Since the early equation given by Epstein et al. (1953), many equations have been published, which substantiated the potential of oxygen isotope paleothermometry for biogenically precipitated calcite. The first equation based on laboratory experiments with plank-

tonic foraminifera was generated by Erez and Luz (1983). Their measurements on the cultured symbiotic species *G. sacculifer* were approximated by the second order polynom

$$T = 17.0 - 4.52 \cdot (\delta^{18}O_c - \delta^{18}O_w)$$
$$+ 0.03 \cdot (\delta^{18}O_c - \delta^{18}O_w)^2 \qquad (10.6)$$

with $T$ standing for the *in-situ* temperature during calcite precipitation (°C), $\delta^{18}O_c$ representing the oxygen isotopic composition of the calcite (as ‰ PDB), and $\delta^{18}O_w$ representing the $\delta^{18}O$ value (‰ PDB) of the seawater from which the calcite has been precipitated. Since oxygen isotope analyses of waters are commonly reported relative to SMOW, the conversion of $\delta^{18}O_w$ to the PDB scale can be calculated according to Hoefs (1997)

$$\delta^{18}O_{PDB} = 0.97002 \cdot \delta^{18}O_{SMOW} - 29.98 \quad (10.7)$$

A $\delta^{18}O_w$ correction of $-0.27‰$ is therefore necessary to compare $\delta^{18}O$ values measured in $CO_2$ produced by the reaction of calcite with $H_3PO_4$ with those measured in $CO_2$ equilibrated with water.

Most of the paleotemperature equations appear to be similar to the one stated above (Eq. 10.6), but temperature reconstructions can differ as much as 2°C when ambient temperature varies between 5°C and 25°C (Fig.10.2; for a recent review see Bemis et al. 1998). The reason is that, in addition to *in-situ* temperature and water isotopic composition, the shell $\delta^{18}O$ may be affected by the photosynthetic activity of algal symbionts and by the carbonate-ion concentration in seawater.

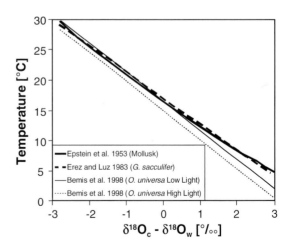

**Fig. 10.2** Comparison of the results of different paleotemperature equations (see Bemis et al. 1998 for review).

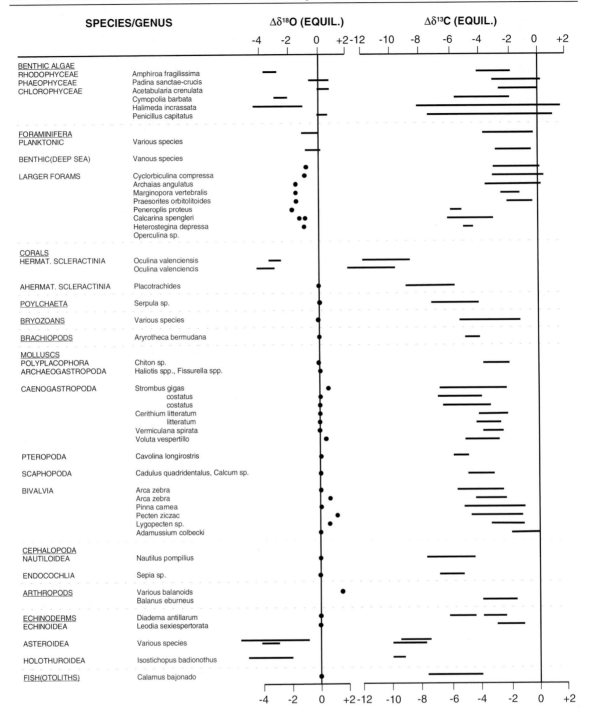

**Fig. 10.3** $\delta^{18}$O and $\delta^{13}$C deviations from equilibrium isotope composition of selected calcareous species (adopted from Wefer and Berger 1991).

Wefer and Berger (1991) summarized the importance of such effects, previously called 'vital effects', on a broad spectrum of organisms (Fig. 10.3). For oxygen isotopes, most organisms are shown to precipitate CaCO$_3$ close to equilibrium with the water in which they live. However, some organisms, such as planktonic foraminifera and hermatypic corals, exhibit significant differences from isotopic equilibrium. Laboratory experiments with live planktonic foraminifera demonstrated that an increase in the symbiont photosynthetic activity results in a decrease in shell $\delta^{18}O$ values (Spero 1992, Spero and Lea 1993). The mechanisms driving the effects of symbiont photosynthesis on shell $\delta^{18}O$ are not well understood, but appear to be linked to the carbonate ion concentration. During photosynthetic activity, $CO_2$ uptake by the symbionts increases the pH in the microenvironment around the shell (Jørgensen et al. 1985, Rink 1998). Consequently, more alkaline conditions correspond to locally elevated $[CO_3^{2-}]$. The $^{18}O$ depletion of shells due to higher symbiont photosynthetic activity is consistent with the effect of higher ambient $[CO_3^{2-}]$. Spero et al. (1997) obtained a $\delta^{18}O/[CO_3^{2-}]$ slope of $-0.002\%_o$ $\mu mol^{-1}kg^{-1}$ from experiments with symbiotic (*O. universa*) and non-symbiotic (*G. bulloides*) foraminifer species. This previously undocumented carbonate isotope effect may help to solve inconsistencies in temperature reconstructions by applying oxygen isotope paleothermometry relative to other marine or terrestrial temperature proxies (see recent discussions in Spero et al. 1997, Bemis et al. 1998).

*Diagenesis*

The isotopic composition of a carbonatic shell will remain unchanged until the shell material dissolves and recrystallizes during diagenesis. Diagenetic modification, however, could begin immediately after deposition or even in the water column due to corrosive deep ocean or pore waters. Such waters are generally enriched in $CO_2$ due to the respiration of organic matter mediated by specific bacteria. In shells of the planktonic foraminifera *P. obliquiloculata* sampled in a depth profile in the western equatorial Pacific Wu and Berger (1989) showed that below the depth of the modern lysocline the $\delta^{18}O$ of this species increased with water depth due to increasing calcite dissolution. Close to the modern depth of calcium carbonate compensation, the observed deviation reached a maximum value of $+0.9\%_o$. This effect of differential dissolution, i.e. the preferential removal of the light isotope $^{16}O$, is an important effect to be considered within sediment sequences affected by fluctuations in dissolution intensity.

The conversion of sediment into limestone within deep-sea sediments results from pressure and temperature rises which both increase with burial depth. Within carbonate sediments, diagenesis generally transforms the less stable arago-

nite and Mg-calcite into a low-Mg calcitic cement by means of a dissolution-reprecipitation process. Theoretically, the oxygen isotope composition of carbonates should not change significantly with burial, since the $^{18}O$ in pore waters originates from seawater. However, in many cores recovered during DSDP/ODP drilling, deep-sea carbonates and often pore waters as well, exhibit $^{18}O$ depletions by several permil (e.g. Lawrence 1989). Mass balance calculations by Matsumoto (1992) indicated that the $^{18}O$ shift in pore water towards lower values in sediments of the Japan Sea is controlled by a low-temperature alteration of basement basalts (see also discussion in Sect. 10.3.1), which is slightly compensated by the transformation of biogenic opal to quartz. Furthermore, detailed measurements on different generations of carbonate cements suggest that late cements exhibit lower $\delta^{18}O$ values compared to early precipitates. This $\delta^{18}O$ trend may be due to the increasing temperatures with increasing burial depth or to the isotopic evolution of pore waters during precipitation (Hoefs 1997).

## 10.4  Geochemical Influences on $^{13}C / ^{12}C$ Ratios

### 10.4.1  $\delta^{13}C_{\Sigma CO2}$ of Seawater

*Principles of Fractionation*

The carbon isotopic composition of $\Sigma CO_2$ in seawater is mainly controlled by two processes, the biochemical fractionation due to the formation and decay of organic matter, and the physical fractionation during gas exchange at the air-sea boundary (Broecker and Maier-Reimer 1992). Surface water is enriched in $^{13}C$, because photosynthesis preferentially removes $^{12}C$ from the $CO_2$. Deeper water masses have lower $\delta^{13}C$ values, since nearly all of the organic matter that is produced by photosynthesis is subsequently remineralized in the water column. Broecker and Maier-Reimer (1992) showed that if there were no air-sea gas exchange, the relationship between $\delta^{13}C$ and $PO_4$ in the ocean would be

$$\delta^{13}C - \delta^{13}C_{m.o.} =$$

$$\varepsilon_p / \Sigma CO_{2\,m.o.} \cdot C / P_{org} \cdot (PO_4 - PO_{4\,m.o.})$$

$$(10.8)$$

where $\varepsilon_p$ (‰) is the isotopic effect associated with the photosynthetic fixation of carbon, $C/P_{org}$ is the Redfield Ratio, and the subscript m.o. stands for mean ocean values. When reasonable values are substituted ($\delta^{13}C_{m.o.} = 0.3$‰, $\varepsilon_p = -19$‰, $\Sigma CO_{2\ m.o.} = 2200$ μmol kg$^{-1}$, $C/P_{org} = 128$, $PO_{4\ m.o.} = 2.2$ μmol kg$^{-1}$), the predicted relationship closely matches the relationship for waters in the deep Indian and Pacific Oceans ($\delta^{13}C = 2.7 - 1.1 \cdot PO_4$). This is to be expected, as the effect of air-sea exchange should be constant for these deep water masses due to the homogeneity of source waters for the deep Indian and Pacific Oceans.

On the other hand, carbon isotope fractionation during air-sea gas exchange is also an important factor in determining the isotopic composition of carbon in surface water (Charles and Fairbanks 1990, Broecker and Peng 1992, Lynch-Stieglitz et al. 1995). If the $CO_2$ in the atmosphere were in isotopic equilibrium with the dissolved inorganic carbon in the ocean, the dissolved inorganic carbon would be enriched in $^{13}C$ by about 8‰ at 20°C relative to the atmosphere $CO_2$ (Zhang et al. 1995). This thermodynamic fractionation depends on the temperature of equilibration, with $\Sigma CO_2$ becoming more enriched relative to the atmospheric value by about 1‰ per 10°C cooling (Fig. 10.4; Mook et al. 1974). If the surface ocean were in complete isotopic equilibrium with atmospheric $CO_2$, one expect a 3‰ range in oceanic $\delta^{13}C$ for the 30°C range in ocean temperatures, similar to the magnitude of $\delta^{13}C$ change induced by biological processes. In fact, $CO_2$ exchange rates between the ocean and the atmosphere are slow enough (relative to mass transport of $\Sigma CO_2$ within the ocean) that the range of $\delta^{13}C_{\Sigma CO2}$ in surface waters is less than 3‰. Moreover, Zahn and Keir (1994) showed in their ocean-atmosphere box-model that even in the absence of the temperature effect, the air-sea gas exchange modifies the $\delta^{13}C$ distribution of the upper ocean and the Atlantic deep water. The observed deviations occur because net biological production in the Southern Ocean is low relative to the upwelling fluxes of nutrients and dissolved carbon, allowing low $^{13}C/^{12}C$ ratios in the dissolved carbon to outcrop and exchange with the atmosphere. As a consequence, a dynamic balance is achieved where $CO_2$ evading from Antarctic waters to the atmosphere has a lower $\delta^{13}C$ than the invading $CO_2$, while the isotope ratio of the evading $CO_2$ from the warm ocean is greater than that of the return flux of $CO_2$.

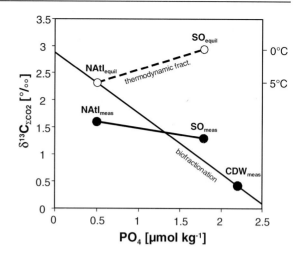

**Fig. 10.4** Schematic diagram showing the effect of biofractionation and thermodynamic effects on the carbon isotopic composition of total dissolved inorganic carbon (according to Zahn and Keir 1994).

### Modern Range of Values and Historical Variability

The modern $\delta^{13}C_{\Sigma CO2}$ values of seawater are close to 0‰ (PDB) and vary only within a small range. From the GEOSECS $\delta^{13}C_{\Sigma CO2}$ sections for today's world oceans, compiled by Kroopnick (1985), a range of +2.5‰ in the mid-latitude surface Atlantic to +0.7‰ in the northern surface Pacific has been determined. Deep water mass $\delta^{13}C_{\Sigma CO2}$ ranges from +1.2‰ in the core of North Atlantic Deep Water and +0.4‰ in the Circumpolar Deep Water to −1.0‰ in the northern Pacific Deep Water. As stated above, $\delta^{13}C_{\Sigma CO2}$ of a deep water mass behaves like a conservative tracer for phosphate, since the water mass left its sea surface source area. On the other hand, paired $\delta^{13}C_{\Sigma CO2}$ and $PO_4$ measurements on two sections in the Southern Ocean clearly support the thermodynamic imprint during Wedell Sea Bottom Water formation by a significant deviation from the Redfield $\delta^{13}C_{\Sigma CO2}/PO_4$ relationship (Mackensen et al. 1996).

The $\delta^{13}C_{\Sigma CO2}$ varied considerably in geologic history. Generally, three explanations are given for changes of $\delta^{13}C_{\Sigma CO2}$ distribution in the ocean: (1) changes in the surface-ocean productivity which cause variable fractionation between surface and deep water carbon isotopic composition, (2) changes in the gas exchange rates between ocean and atmosphere due to changes in surface temperatures and ocean circulation, and (3) changes in the marine carbon budget by variations in the reservoirs of the atmosphere, the ocean, or the lithosphere. Of course, these different processes act on different time scales. Variations on a scale of $10^1$ to $10^5$ years might be attributed to changes in the effectivity of the biologic pump (Berger and Vincent 1986), to rapid changes between land and ocean reservoirs (Shackleton 1977, Broecker 1982), or to changes in ocean circulation responding to climate variability (e.g. Sarnthein et al. 1994, Bickert and Wefer 1996, Raymo et al. 1996).

On scales of $10^5$ years and longer, changes in the influx and in the burial of sedimentary inorganic and organic carbon exert a primary control on the $\delta^{13}C$ of the ocean and the atmosphere (Holser 1997). Main sources for carbon comprise the erosional flux of sedimentary carbon (as $C_{org}$ and carbonate) and the degassing of volcanic $CO_2$; main sinks are the burial of organic matter and the deposition of carbonate. Changes in the oceanic $\delta^{13}C$ are assumed to be the result of variations in the ratio of inorganic to organic carbon contributed to sediments (e.g. Derry and France-Lanord 1996). An increase in the burial of or-

ganic carbon would preferentially remove $^{12}C$ from seawater, so that the ocean reservoir would become isotopically heavier, and vice versa.

### 10.4.2    $\delta^{13}C$ in Marine Organic Matter

*Principles of Fractionation*

The carbon isotopic composition of marine organic matter produced in the photic zone depends on the isotope ratio of the total dissolved inorganic carbon and the degree to which this inorganic pool is utilized (Degens et al. 1968). Although more than 90% of the inorganic carbon pool is in the form of $HCO_3^-$, phytoplankton utilizes carbon of the very small reservoir (1%) of dissolved $CO_{2(aq)}$. While the equilibrium exchange fractionation is only small between atmospheric $CO_2$ and dissolved $CO_{2(aq)}$ (about 1‰), and moderate between $CO_{2(aq)}$ and $HCO_3^-$ (with a range of 10‰ to 8‰ between 5°C and 25°C; Mook et al. 1974), a large kinetic fractionation accompanies the biological carbon fixation during photosynthesis. The main isotope-discriminating steps are (1) the uptake and intracellular diffusion

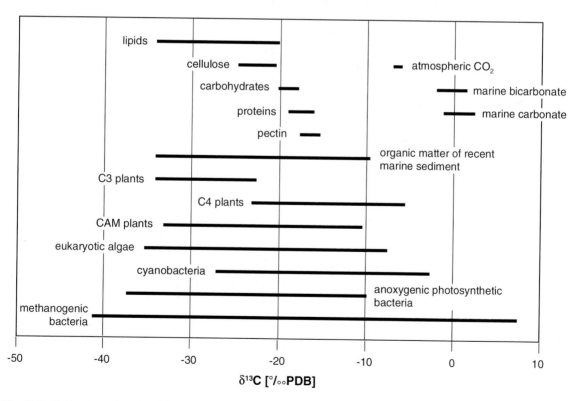

**Fig. 10.5** Carbon isotopic composition of autotrophic marine and terrestrial organisms in comparison to the range observed in anorganic compounds (after Schidlowski 1988) and organic constituents (after Degens 1969).

of $CO_2$, and (2) the enzymatic carbon fixation (Park and Epstein 1960). The first, reversible step is associated with a moderate fractionation of about $-4‰$, the second, irreversible step causes a kinetic fractionation up to $-40‰$ (O'Leary 1981).

The $\delta^{13}C_{org}$ values of marine phytoplankton range between $-10‰$ and $-31‰$, but most of the warm-water plankton exhibits values between $-17‰$ to $-22‰$ (Fig. 10.5; Degens et al. 1968, 1969). The two-step model of carbon fixation clearly suggests that isotope fractionation depends on the concentration of $CO_{2(aq)}$. The fractionation decreases with decreasing $CO_2$ availability, such that in warm marine surface waters one would expect higher $\delta^{13}C_{org}$ values than in colder waters, since the solubility of $CO_2$ increases with decreasing temperature. However, other factors, such as species composition, light intensity, growth rate and cell geometry may also influence organic $\delta^{13}C$ values of particulate organic matter (e.g. Hayes 1993, Popp et al. 1998).

An important application of carbon stable isotope ratios lies in the reconstruction of the carbon dioxide concentration in surface waters and subsequently in the atmosphere at the time the organic matter was produced. Experimental and field studies have shown that the fractionation of stable carbon isotopes during photosynthesis by plankton depends on the concentration of ambient dissolved molecular carbon dioxide ($CO_{2(aq)}$), which led to the suggestion that sedimentary $\delta^{13}C_{org}$ may serve as a proxy for surface water $CO_{2(aq)}$ (Popp et al. 1989, Rau et al. 1991). The applicability of this tool in paleoceanographic studies has been shown repeatedly (e.g. Jasper and Hayes 1990, 1994, Fontugne and Calvert 1992, Müller et al. 1994), but the reconstruction of paleo-$pCO_2$ has become more and more complex due to the growing knowledge of factors influencing carbon isotopic fractionation during organic matter construction. Recent studies suggest that estimations of ancient concentrations of

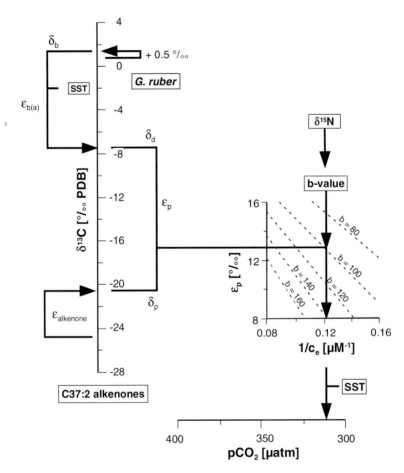

**Fig. 10.6**  Diagram outlining the estimation of surface ocean $CO_{2(aq)}$ and $pCO_2$, when $\delta^{13}C_{alkenones}$, $\delta^{13}C_{calcite}$, $\delta^{15}N$ as a proxy for carbon demand b, and surface ocean temperature are provided (Andersen et al. 1998; modified from Jasper and Hayes 1994).

$CO_{2(aq)}$ in oceanic surface waters will require knowledge of both the stable carbon dioxide fractionation associated with photosynthesis ($\varepsilon_p$) and phytoplankton growth rates ($\mu$) (Laws et al. 1995, Bidigare et al. 1997). Furthermore, a large effect of cell geometry on carbon isotopic fractionation of marine phytoplankton was demonstrated by Popp et al. (1998), consistent with results of physiological models introduced by Rau et al. (1997).

For algae where $CO_2$ is thought to reach the photosynthetic site only by passive diffusion, $\varepsilon_p$ can be calculated by the following equation (Bidigare et al. 1997):

$$\varepsilon_p = \varepsilon_f - b / [CO_{2(aq)}] \qquad (10.9)$$

where $\varepsilon_f$ (‰) is the maximum isotopic effect associated with the photosynthetic fixation of carbon mediated by the enzyms Rubisco and $\beta$-carboxylase ($\approx 25$‰; Bidigare et al. 1997), and the variable b reflects the intracellular carbon demand. Because the b-value combines a suite of factors, like growth rate, membrane permeability, cell geometry, and boundary layer thickness, significant variations in b ($-109$‰$\mu$M to $-164$‰$\mu$M) have been observed in empirical fits to field and experimental data (Laws et al. 1995). For example, higher growth rates lead to higher b-values (Bidigare et al. 1997, Rau et al. 1997). Based on chemostat experiments with the coccolithophorid *E. huxleyi*, Bidigare et al. (1997) found a close correlation between $\varepsilon_p$ and the growth rate $\mu$:

$$\varepsilon_p = 24.6 - 137.9 \cdot \mu / [CO_{2(aq)}] \qquad (10.10)$$

To obtain $CO_{2(aq)}$, Equation 10.9 has to be rearranged, and values for $\varepsilon_p$ are determined from the carbon isotopic compositions of the primary photosynthate $\delta^{13}C_p$ (‰) and of the ambient dissolved molecular carbon dioxide $\delta^{13}C_d$ (‰) (Fig. 10.6):

$$\varepsilon_p = ((\delta^{13}C_p + 1000) / (\delta^{13}C_d +1000) - 1) \cdot 1000$$

$$(10.11)$$

Assuming that diagenetic isotopic alterations can be neglected (see below), $\delta^{13}C_p$ may be directly substituted by the measured $\delta^{13}C_{org}$ value. Values for $\delta^{13}C_d$ are estimated on the basis of the $\delta^{13}C$ of calcite tests assuming that the calcite was precipitated in equilibrium with the $\Sigma CO_2$, cor-

recting for the temperature-dependent fractionation $\varepsilon_{b(a)}$ between $CO_{2(aq)}$ and dissolved bicarbonate according to Mook et al. (1974):

$$\delta^{13}C_d \quad = \delta^{13}C_{\Sigma CO2} - \varepsilon_{b(a)}$$
$$= \delta^{13}C_{\Sigma CO2} - 24.12 - 9866 / T$$

$$(10.12)$$

T is the temperature of surface waters measured in K. The $CO_{2(aq)}$ concentrations are then converted to $CO_2$ partial pressure values (pCO$_2$ in $\mu$atm) using Henry's Law:

$$pCO_2 = CO_{2(aq)} / \alpha \qquad (10.13)$$

where the solubility coefficient $\alpha$ is mainly a function of temperature and, to a minor extent, of salinity (Rau et al. 1991). $\alpha$ may be calculated according to Weiss (1974). The success of the equations in determining surface $CO_{2(aq)}$ is dependent on the estimation of the sea-surface temperatures and on the determination of the complex variable b for the past. While several proxies exist for reconstructing past SST, the determination of ancient b-values is difficult. A promising approach is the use of bulk sediment $\delta^{15}N$ as proxy for b (Andersen et al. 1998). However, the use of $\delta^{15}N$ as a proxy for carbon demand introduces a number of potential problems associated with the complex fractionation mechanisms of nitrogen isotopes (see Sect. 10.5).

A topic of much concern regarding the use of bulk $\delta^{13}C_{org}$ in paleoceanographic investigations is the masking of the marine $\delta^{13}C_{org}$ signal by terrestrially-derived organic matter. Since terrigenous material is usually substantially depleted in $^{13}C$ (estimated mean value of $-27$‰, since 90% of land plants are $C_3$ plants) compared to marine-derived matter (mean value of $-19$‰), the carbon isotopic composition of ocean sediments have been used to trace the origin of sedimentary organic carbon (Newman et al. 1973, Rühlemann et al. 1996). However, the underlying assumption regarding the mixing of a marine and a terrestrial $\delta^{13}C_{org}$ end member is not always supported because of the possible variability of $\delta^{13}C_{org}$ values produced in terrestrial systems (e.g. change in the relative contribution of $C_3$ and $C_4$ plants). Also, marine values may change by processes occurring in the water column (e.g. changing phytoplankton growth conditions, see above).

*Diagenesis*

Early diagenesis begins in the photic zone of the oceans, continues during the sinking of particles, and is intense in the bioturbated surface layer of sediments. Therefore, only a few percent of the initially produced organic matter becomes buried in the sediments (Berger et al. 1989). However, despite the extensive loss of organic matter due to remineralization, the carbon isotopic composition of particulate organic matter appears to undergo only little change (see review in Popp et al. 1997). On the other hand, large and nonsystematic differences have been observed between the isotopic composition of total organic carbon and single organic compounds, e.g. phytoplankton biomarkers, in the same deposit (Fig. 10.5; Degens et al. 1969). Therefore, the assumption that the carbon isotopic composition of marine sedimentary organic matter directly or indirectly reflects the carbon isotopic composition of phytoplanktonic matter must be taken cautiously (Popp et al. 1997). Furthermore, since marine organic matter is easily digestible, whereas terrestrial organic matter is rather resistant, such preferential decomposition of organic matter has great impact on the estimation of the average marine and terrestrial $\delta^{13}C_{org}$ component (de Lange et al. 1994).

The carbon isotopic composition of pore waters reflects both the decomposition of organic matter, which releases $^{13}C$-depleted $CO_2$ to the pore water, and the dissolution of $CaCO_3$, which adds $CO_2$ relatively enriched in $^{13}C$. Since the depletion of organic matter $\delta^{13}C$ exceeds by far the range of $\delta^{13}C$ exhibited in sedimentary carbonate, the net result of these two processes is to make dissolved $CO_2$ in pore water isotopically lighter compared to the overlying bottom water. Below the sediment/water interface, where pore water has a $\delta^{13}C$-value near that of seawater, a $\delta^{13}C$-gradient exists in the uppermost few centimeters. McCorkle et al. (1985) and McCorkle and Emerson (1988) showed that the gradient of $\delta^{13}C$-profiles observed in several box cores vary systematically with the rain of organic matter to the seafloor. In slowly deposited oxic sediments containing little amounts of organic carbon, only a small decrease in $\delta^{13}C$ relative to seawater $\delta^{13}C$ occurs. In organic-carbon rich sediments, the rates of organic matter decomposition and the production of reduced nitrogen compounds would rapidly decrease the pore water $\delta^{13}C$. In these anoxic sediments, the situation is even more complex due to bacterial methanogenesis, which follows the sulfate reduction during early diagenesis. Since methane-producing bacteria metabolize methane highly enriched in $^{12}C$ (–50‰ to –100‰; Deines 1981), the pore water becomes significantly enriched in $^{13}C$. Such trends in $\delta^{13}C$ of total dissolved $CO_2$ from pore waters within anoxic sediments have been recovered in various Deep Sea Drilling sites (Arthur et al. 1983). However, because pore waters are not truely closed systems, other factors like carbon loss due to upward diffusion of methane and other dissolved carbon species as well as precipitation of authigenic carbonates may also affect pore water $\delta^{13}C$. For example, if significant amounts of methane are utilized in sulfate reduction, the rate of $\delta^{13}C$ decrease associated with the $\Sigma CO_2$ increase would be anomalously high (Arthur et al. 1983).

### 10.4.3    $\delta^{13}C$ in Marine Carbonates

*Principles of Fractionation*

Precipitation of carbonate, largely from total dissolved carbon which is present primarily as $HCO_3^-$, involves a much smaller fractionation of carbon isotopes compared to the photosynthetic fixation of carbon. In fact, the $\delta^{13}C$ of calcite is relatively insensitive to changes in temperature (e.g. ≈0.035‰ °C$^{-1}$, Emrich et al. 1970) such that carbonate minerals can be used generally to monitor changes in the $\delta^{13}C_{\Sigma CO2}$ in the waters from which they precipitate. However, the possible effects of biologically mediated carbonate precipitation have been considered already for oxygen isotopes (see Sect. 10.3.2). Observations summarized by Wefer and Berger (1991) reveal that most organisms exhibit $\delta^{13}C$-values which are in disequilibrium with the $\delta^{13}C_{\Sigma CO2}$ of the waters, from which shell carbonate is precipitated (Fig. 10.3). As discussed for the influences on oxygen isotopic composition in carbonate, there may be several reasons for these deviations. McConnaughy et al. (1997) distinguish metabolic and kinetic effects on the carbon isotopic disequilibria. Kinetic isotope effects result from the discrimination against $^{13}C$ during hydration and hydroxylation of $CO_2$. Such effects appear to be associated with rapid calcification, as observed in corals. Metabolic effects apparently result from changes in the $\delta^{13}C_{\Sigma CO2}$ of the microenvironment due to photosynthesis and respiration. For example, infaunal benthic foraminifera precipitate their calcite in

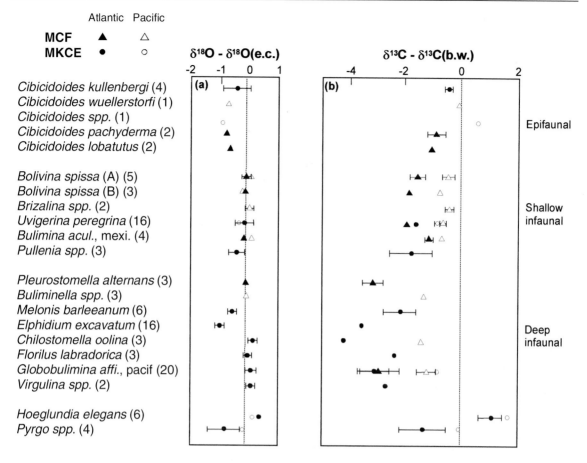

**Fig. 10.7**  $\delta^{18}O$ and $\delta^{13}C$ values of benthic foraminifera species in comparison to equilibrium calcite $\delta^{18}O$ values and to the pore water $\delta^{13}C_{\Sigma CO2}$ gradient, respectively (McCorkle et al. 1990).

equilibrium with ambient pore water $\delta^{13}C_{\Sigma CO2}$, and therefore monitor the vertical carbon isotope gradient due to organic matter remineralization rather than bottom water $\delta^{13}C_{\Sigma CO2}$ (Fig. 10.7; McCorkle et al. 1990, 1997). Mackensen et al. (1993) discuss several ways, how the $\delta^{13}C$ of even epibenthic living foraminifer species could be influenced by the decay of organic matter. An example for the consequences of such a 'Mackensen' effect is discussed by Bickert and Wefer (1996) for the late Quaternary reconstruction of South Atlantic deep water circulation. An extreme depletion in $\delta^{13}C$ is recorded in endobenthic foraminifers sampled from ODP Site 680. Wefer et al. 1994 attributed these excursions to the release of methane in the continental margin off Peru sediments, which lowers substantially the carbon isotopic composition of $CO_2$ in the porewater. Furthermore, the laboratory experiments with live planktonic foraminifera which demonstrated symbiont photosynthetic effect on shell $\delta^{18}O$ values

(Spero 1992, Spero and Lea 1993) revealed a similar effect on shell $\delta^{13}C$ values. Spero et al. (1997) obtained from experiments with symbiotic (*O. universa*) and non-symbiotic (*G. bulloides*) foraminifer species a $\delta^{13}C/[CO_3^{2-}]$ slope of $-0.007\%_o$ $\mu mol^{-1}kg^{-1}$. This effect might partly be responsible for shifts observed in foraminiferal calcite previously attributed to global changes in $\delta^{13}C_{\Sigma CO2}$ of the ocean (see recent discussion in Spero et al. 1997).

*Carbonate Dissolution and Precipitation*

Some aspects of the influence of diagenesis on the isotopic composition of carbonate have already been discussed for oxygen isotopes (see Sect. 10.3.2). The isotopic composition of a carbonatic shell will remain unchanged until the shell material dissolves and recrystallizes during diagenesis. As stated above, diagenetic modification could begin immediately after deposition or even in the water column due to corrosive deep ocean or pore waters. McCorkle et al. (1995) showed in shells of the benthic foraminifera *C. wuellerstorfi* sampled in a depth profile in the western equatorial Pacific that below the depth of the modern lysocline the $\delta^{13}C$ of this species decrease with water depth. Close to the modern depth of calcium carbonate compensation, the observed deviation reached a minimum value of –0.35‰. Since the effect of differential dissolution would cause a preferential removal of the light isotope $^{12}C$, an increase would be expected to occur with increasing dissolution. The opposite is observed and requires therefore other explanations, beside the effect of carbonate-ion concentration (Spero et al. 1997).

Under oxic porewater conditions, no further change in the carbon isotopic composition of shell material is expected, until dissolution occurs under deep burial. But while a temperature-effect is negligible (Emrich et al. 1970), an alteration of the primary $\delta^{13}C$ value may arise during mineral transformations (aragonite $\rightarrow$ low Mg-calcite), depending on the amount of carbon present in the post-depositional solutions. Under suboxic conditions in sediments with high organic carbon contents, the carbon isotope signal may be affected by early diagenesis (Nissenbaum et al. 1972, Irwin et al. 1977). The degradation of organic matter will cause an intense degree of carbonate dissolution and the reprecipitation of authigenic calcite standing in equilibrium with porewater $\delta^{13}C_{\Sigma CO2}$. Schneider et al. (1992) estimated for continental slope sediments off Angola that a precipitation of 5% authigenic calcite on foraminiferal tests would produce a decrease of –1‰ in the $\delta^{13}C$ of these tests. Again, the fractionation of $\delta^{13}C$ during precipitation has only a minor effect on the carbon isotope composition of the solid phase, the major effect is determined by the $\delta^{13}C_{\Sigma CO2}$ of the pore water which itself is controlled by the degradation of organic matter.

# 10.5   Geochemical Influences on $^{15}N / ^{14}N$ Ratios

The nitrogen isotope fractionation in the marine environment has been reviewed by Altabet and Francois (1994) and Montoya (1994). These studies mainly represent the basis of the following introduction.

## 10.5.1   $\delta^{15}N$ in Marine Ecosystems

*Principles of Fractionation*

The use of nitrogen isotopes in marine sediments is a relatively new tool in the field of paleoceanography. The usefulness of this proxy lies in its recording changes of nutrient dynamics in the water column. Major biological transformations of nitrogen in marine systems include the utilization of the dissolved forms of inorganic nitrogen ($NO_2^-$, $NO_3^-$, $NH_4^+$) by phytoplankton, the consumption of phytoplankton by grazers, and the remineralization of organic nitrogen by animals and bacteria. In addition, various marine prokaryotes are able to carry out a suite of reactions that move nitrogen in and out of the major inorganic pools. These include nitrogen fixation and dissimilatory transformations of nitrogen, such as nitrification and denitrification. Thus, the marine nitrogen cycle involves multiple inorganic and organic pools which are coupled by rapid biological transformations of nitrogen (cf. Chap. 6).

Since most marine autotrophs require combined nitrogen as a substrate for growth, the isotopic composition of dissolved inorganic nitrogen acts as a sort of master variable in setting the baseline isotopic composition of marine plankton. A number of biological processes may alter the isotopic composition of the marine pool of inorganic nitrogen. The most important are nitrogen fixation, denitrification and nitrification, all of which may lead to a net transfer between the oceanic pool of combined nitrogen and the atmosphere (for a recent summary see Montoya 1994). The basic biochemical reactions are the following:

Nitrogen fixation:

$$N_2 \text{ atm.} + 3\,H_2O \Rightarrow 2\,NH_3 + 3/2\,O_2$$

$$(10.14)$$

It is well known that symbiontic bacteria in the roots of plants, but also bacteria and algae in the ocean can fix nitrogen. The high energy needed to break the triple bond of the nitrogen gas molecule makes natural nitrogen fixation a very inefficient process.

Deamination:

$$N_{org} \Rightarrow NH_3 \qquad (10.15)$$

Nitrification:

$$NH_3 + 3/2\ O_2 \Rightarrow HNO_2 + H_2O \qquad (10.16)$$
$$HNO_2 + 1/2\ O_2 \Rightarrow HNO_3 \qquad (10.17)$$

Although ammonia is the direct product of decomposition of nitrogenous organic compounds (deamination), most inorganic nitrogen in the ocean occurs in the form of nitrate. Conversion of ammonia into nitrate is carried out by nitrifying organism.

Denitrification:

$$5\ CH_2O + 4\ NO_3^- + 4\ H^+ \Rightarrow$$
$$5\ CO_2 + 7\ H_2O + 2\ N_2 \qquad (10.18)$$

After oxygen is exhausted, nitrate can serve as electron acceptor in the degradation of organic matter. Denitrification results in the conversion of nitrate to nitrogen gas and therefore balances the biological fixation of nitrogen. If there were no denitrification, atmospheric nitrogen would be exhausted in less than 100 million years. Denitrification occurs in stratified or stagnant water masses of the ocean (e.g. within oxygen minimum layers).

The three main processes involved in the biogenic utilization of nitrogen are all associated with kinetic fractionation effects, nevertheless they exhibit different fractionation factors. The isotope effect of nitrogen fixation ($\alpha = 1.000$ to 1.004) is small relative to the effects of bacterial nitrification ($\alpha = 1.02$ to 1.04) and denitrification ($\alpha = 1.02$ to 1.04), respectively (see Montoya et al. 1994: Table 1). The results by Miyake and Wada (1971) indicate that little overall isotope fractionation occurs in the bacterial mineralization of organic nitrogen. Equilibrium isotope exchange reactions, which commonly occur in nature, are the ammonia volatilization and the solution of nitrogen gas. The former process has a significant isotopic effect ($\alpha = 1.034$), whereas the latter process induces only a small fractionation ($\alpha = 1.0008$).

*Modern Range of Values and Historical Variability*

Relatively few measurements of the isotopic composition of dissolved inorganic nitrogen have yet been published (Fig. 10.8). $NO_3^-$ dominates the oceanic pool of combined nitrogen, and the available data indicate that the $\delta^{15}N$ of $NO_3^-$ from oxygenated deep waters ranges between 3‰ and 7‰ with a mean value around 6‰ (Liu and Kaplan 1989). The $\delta^{15}N$ of $NO_3^-$ is significantly higher than this mean in and around anoxic water masses where denitrification occurs (Liu and Kaplan 1989). For the eastern tropical North Pacific Ocean values as high as 18.8‰, for the western Caribbean Sea values up to 12‰ have been reported to occur within the active denitrification zone (Cline and Kaplan 1975).

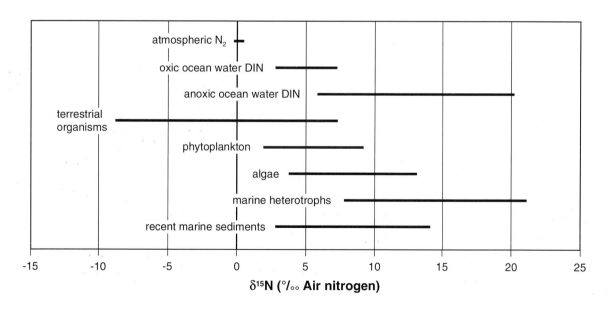

**Fig. 10.8**  $\delta^{15}N$ ranges of some important nitrogen compounds (according to Arthur et al. 1983).

On geological time-scales, the entire nitrogen cycle has to be considered to describe the factors controlling the nitrogen isotopic balance in the ocean. The major input of nitrogen compounds into the ocean results from precipitation, river discharge and the fixation of molecular nitrogen. Nitrogen is removed from the ocean mainly by burial in the sediment and denitrification. Although it is generally assumed that the nitrogen balance in the ocean is in a steady state, it is difficult to estimate the budget for mass balance as well as for isotopic balance (for a recent review of estimates see Gruber and Sarmiento 1997). However, denitrification in the water column and in the sediment is the dominant process maintaining mass balance. It is also the only process that can produce a large fractionation. All other processes add or remove nitrogen compounds with $\delta^{15}N$ values mostly in the range between $-2‰$ and $+8‰$. Changes in the global rate of denitrification could therefore lead to significant changes in the $\delta^{15}N$ of marine $NO_3^-$, which in turn would affect the $\delta^{15}N$ of marine plankton, since the isotopic composition of the phytoplankton at the base of the food web depends at least in part on the $\delta^{15}N$ of the dissolved inorganic nitrogen available (Altabet and Curry 1989). Today, it appears that denitrification is the principal mechanism that keeps the marine nitrogen compounds at a higher $\delta^{15}N$ value than atmospheric nitrogen.

### 10.5.2   $\delta^{15}N$ in Marine Organic Matter

*Principles of Fractionation*

The nitrogen isotopic composition of marine organic matter produced in the photic zone depends on the isotope ratio of nitrate and the degree to which this inorganic pool is utilized (Wada 1980, Altabet et al. 1991, Voss et al. 1996). The preferential uptake of $^{14}NO_3^-$ leads to a depletion of $^{15}N$ in organic matter relative to the dissolved inorganic nitrogen used as a substrate for growth (Montoya 1994). Subsequently, the remaining nitrogen pool becomes progressively enriched in $^{15}N$ according to Rayleigh fractionation kinetics (Cifuentes et al. 1988). The degree of isotopic fractionation associated with primary productivity varies between taxa and with growth conditions. Cultural experiments suggest that diatoms may discriminate more strongly than flagellates (Wada and Hattori 1978). The ensuing transfer of nitrogen through trophic levels is associated with a

systematic increase in $\delta^{15}N$, with each trophic step resulting in an enrichment of $3.5‰$ (Montoya 1994). This effect, however, should not affect the bulk sedimentary $\delta^{15}N$ values because of mass balance considerations.

An important use of nitrogen stable isotope ratios is the reconstruction of the degree of nitrate utilization in surface waters at the time the organic matter was produced. The applicability of this tool in paleoceanographic studies has been shown repeatedly (e.g. Altabet and Francois 1994, Holmes et al. 1997). To determine changes in the fraction of unutilized nitrate in surface waters, Altabet and Francois (1994) defined the following equations:

$$\delta^{15}NO_3^-{}_{(f)} = \delta^{15}NO_3^-{}_{(f=1)} - \varepsilon_u \cdot \ln (f)$$

$$(10.19)$$

$$\delta^{15}N\text{-}PN_{(f)} = \delta^{15}NO_3^-{}_{(f)} - \varepsilon_u$$

(instantaneous product)    (10.20)

$$\delta^{15}N\text{-}PN_{(f)} =$$
$$\delta^{15}NO_3^-{}_{(f=1)} + f/(1-f) \cdot \varepsilon_u \cdot \ln (f)$$

(accumulated product)    (10.21)

where f is the fraction of unutilized $NO_3^-$ remaining (i.e. $[NO_3^-]/[NO_3^-]_{initial}$), $\varepsilon_u$ is the fractionation factor associated with the $NO_3^-$ uptake, and $\delta^{15}N\text{-}PN$ and $\delta^{15}NO_3^-$ refers to the measured $\delta^{15}N$ of the particulate nitrogen and of $NO_3^-$, respectively. The success of these equations in determining surface nitrate utilization depends on the estimation of the fractionation factor $\varepsilon_u$ exhibited by phytoplankton during photosynthesis. The magnitude of $\varepsilon_u$ sets an upper limit to the amplitude in $\delta^{15}N$ observed. $\varepsilon_u$ may be obtained from the slope of the regression line between the photosynthate $\delta^{15}N$ (routinely, sedimentary $\delta^{15}N$ is used) and $\ln[NO_3^-]_{surface}$. A limited number of culture experiments with marine phytoplankton have shown significant variations in $\varepsilon_u$ between phylogenetic groups ($1‰$ to $9‰$) and growth conditions ($0‰$ to $16‰$; Montoya 1994). Field estimates of $\varepsilon_u$ in regions of moderate to high nitrate concentrations in the open ocean have fallen within a narrower range of $5‰$ to $9‰$ (Altabet and Francois 1994). At 90% utilization of $NO_3^-$, this latter range in $\varepsilon_u$ corresponds to an increase of $12‰$ to $21‰$ in $\delta^{15}N$.

$[NO_3^-]_{initial}$ is the nitrate concentration of newly upwelled water or surface waters before the onset of springtime productivity. Equation 10.19 underscores that changes in $\delta^{15}N$ are in reality a function of nitrate utilization and not simply concentration per se. Equation 10.20 refers to particulate nitrogen produced at any one point in the course of reaction (instantaneous product) (Fig.10.9). Observations would fit Equation 10.20, only if particulate nitrogen is rapidly removed from the system. On the other hand, if there is no removal of particulate nitrogen from the system, the change of $\delta^{15}N$ within this accumulating pool is described by Equation 10.21, the integral of Equations 10.19 and 10.20. However, this latter case almost never occurs (Altabet and Francois 1994).

A topic of much concern regarding the application of bulk $\delta^{15}N$ in paleoceanographic investigations is the effect of terrestrially derived organic matter on the marine $\delta^{15}N$ signal. Since terrigenous material is mostly substantially depleted in $^{15}N$ relative to marine-derived matter, the nitrogen isotopic composition of ocean sediments have been used to trace the origin of sedimentary nitrogen (Sweeny and Kaplan 1980, Wada et al. 1987). However, the underlying assumption regarding the mixing of a marine and terrestrial $\delta^{15}N$ end members is not always conclusive because of the wide range of $\delta^{15}N$ values produced in terrestrial systems (Sweeney et al. 1978). Furthermore, marine values may be subject to changes produced by processes in the water column (as discussed above).

*Diagenesis*

Reconstructing past variations of nutrient conditions in surface waters requires the preservation of the nitrogen isotope surface-water signal in the sedimentary organic matter. Diagenesis, however, may significantly alter the nitrogen isotopic composition of particles even before they are buried at the seafloor. Altabet and McCarthy (1985) suggested that bacterial remineralization of sinking particles resulted in a $^{15}N$ enrichment in the residual organic matter. Francois et al. (1992) reported a $^{15}N$ enrichment of 5‰ to 7‰ in core top values of Southern Ocean sediments compared to measurements of organic matter in the photic zone. However, Altabet and Francois (1994) demonstrated that there was only little offset between sinking particles at 150m water depth in the equatorial Pacific and the respective core tops. They suggested that the difference in preservation of

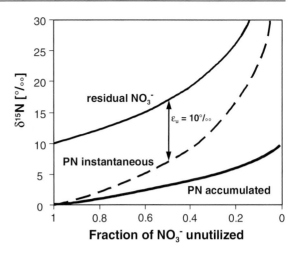

**Fig. 10.9** Effects of nitrate utilization (assuming an $\varepsilon_u$ of 10 ‰) on the $\delta^{15}N$ of the residual nitrate (solide line; Eq. 10.19), the instantaneous product (broken line; Eq. 10.20) and the accumulated product (heavy line; Eq. 10.21) of a reaction (according to Altabet and Francois 1994).

the isotopic signal existent in the two regions may be explained with the high opal content in the Southern Ocean sediments, in which organic matrices with high $\delta^{15}N$ may be captured. In any case, the findings of these authors suggest that where differences between the surface-generated signal and sediments exist, the offset appears to be relatively constant within geographic regions.

Further decomposition of organic matter occurs at the sediment/water interface and in the upper sections of the sediment. However, observations from equatorial Pacific sediments indicate that the loss of organic carbon as well as of nitrogen in the uppermost few millimeters is not accompanied by a corresponding shift in $\delta^{15}N$ over this depth interval (Altabet and Francois 1994). With further diagenesis of organic matter, nitrogen may occur as ammonium incorporated in the lattice of clay minerals, where it replaces potassium. But even this complex substitution is not associated with an isotopic effect, since this nitrogen exhibits an isotopic composition very similar to the organic matter from which it is derived (Williams 1995). Therefore, beside the possible fractionation mechanisms, which may alter the nitrogen isotopic composition of organic matter in the water column, there seems to be no significant diagenetic alteration of the isotope signal within the sediments. The nitrogen isotopic composition of sediments is primarily determined by the source organic matter.

## 10.6   Geochemical Influences on $^{34}S / ^{32}S$ Ratios

The discussion of the isotopic composition of sedimentary sulfur in the marine environment follows the review presented very recently by Strauss (1997).

### 10.6.1   $\delta^{34}S$ of Seawater and Pore Waters

*Principles of Fractionation*

In the marine environment, sulfur occurs most commonly in its oxidized form as dissolved sulfate in seawater or as precipitated sulfate in evaporites and in its reduced form as sedimentary pyrite. The ratio of $^{34}S/^{32}S$ is a sensitive indicator for the transfer of sulfur between the different reservoirs, which is often associated with a change in the oxidation state of the element. The main processes to be considered in the sulfur cycle are the riverine input of sulfate as a weathering product of sulfur-bearing rocks, the precipitation of evaporites from seawater and the biological reduction of seawater sulfate and subsequent formation of sedimentary pyrite (Fig. 10.10). The associated fractionation mechanisms are therefore

- exchange reactions between sulfate and sulfides

- kinetic isotope effects in the bacterial reduction of sulfate

Only a small fractionation is associated with the precipitation of sulfates in seawater. Theoreti-

cal considerations as well as sulfur-isotope measurements in natural environments revealed an isotopic difference between seawater sulfate and crystallized gypsum of 0 to +2.4‰ (Raab and Spiro 1991). Therefore, within the generally observed variability in an evaporitic deposit, calcium sulfate is assumed to record the isotopic composition of seawater sulfate at the time of deposition (Claypool et al. 1980, Strauss 1997).

By far the largest fractionation is associated with the bacterial sulfate reduction. Due to the activity of sulfate-reducing bacteria, such as *Desulfovibrio desulfuricans*, organic matter is oxidized according to the following equation:

$$2CH_2O + SO_4^{2-} \Rightarrow H_2S + 2HCO_3^- \quad (10.22)$$

The resulting hydrogen sulfide reacts with sedimentary iron, which is available in the reactive non-silicate bond form (oxy-hydroxides), and is fixed as iron sulfide (e.g. pyrite). In general, a substantial depletion of $^{34}S$ occurs due to preferential utilization of the $^{32}S$-isotope by the sulfate reducing bacteria. As a result, sedimentary sulfide is depleted in $^{34}S$ relative to ocean water sulfate (Kaplan and Rittenberg 1964). The depletion is usually in the order of 20‰ to 70‰ (Hartmann and Nielsen 1969; Ohmoto 1990), although cultural experiments yielded only depletions between 10‰ and 30‰, with a maximum reported value of 46‰ (Kaplan and Rittenberg 1964). The much larger fractionation occurring under natural conditions is explained with a disproportionation of thiosulfate and/or elemental sulfur, which are both intermediate reaction products in the biological sulfur cycle (Jørgensen 1990). Such a disproportionation might cause an additional fractionation in the oxidative part of the sulfur cycle (for a recent discussion see Habicht and Canfield 1997).

However, the extent of depletion during bacterial sulfate reduction is affected by the magnitude of reduction rates and by the availability of sulfate. Generally, an increasing rate of sulfate reduction leads to a decrease in fractionation. Furthermore, if the availability of sulfate is reduced, e.g. through limited diffusion in non-bioturbated or impermeable sediments, the isotopic composition of both reactant and product shifts towards higher values. Such a Rayleigh fractionation process has been shown by Hartmann and Nielsen (1969) for dissolved sulfate and sulfide in the pore water of a core retrieved from the western

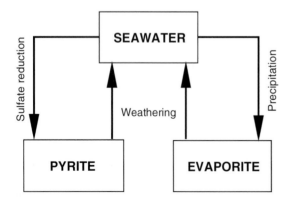

**Fig. 10.10** Main reservoirs of the sedimentary sulfur cycle (adopted from Strauss 1997).

Baltic Sea (Fig. 10.11). The $\delta^{34}S$-values of both reactant and product increase with depth, but the difference between the isotopic values remains constant at 58.9‰ below a depth of 10cm, which indicates that the sediment becomes closed to sulfate after burial under a layer of about that thickness.

### Modern Range of Values and Historical Variability

The isotopic composition of sulfur in modern ocean sulfate is constant within very narrow limits and is represented by a $\delta^{34}S$-value of +20‰ with a standard deviation of ±0.12‰ (Longinelli 1989). The small isotopic range over a great variety of localities and at various depths is explained with the residence time of sulfate in the ocean which is well in excess of the ocean mixing time (Holland 1978). This sulfate homogeneity should be considered for the evaluation of isotope data analyzed in ancient deposits.

The most recent comprehensive review of the sulfur isotopic composition in oceanic sulfate through Phanerozoic time has been published by Strauss (1997), who substantially improved the older reference curve of Claypool (1980) by incorporating new data sets and refining the age model. The main features of the new curve consist in a Cambrian maximum of about +30‰, a decrease to a Permian minimum of about +10‰ and an increase towards the modern ocean value of +20‰. Superimposed over this long-term secular variation is a pronounced short-term variability exhibiting about the same isotopic range as documented for the entire Phanerozoic. The observed variations are assumed to reflect changes in the sulfur redox cycle within the ocean. Since bacterial sulfate reduction and the subsequent formation of sedimentary pyrite is associated with a substantial fractionation, any changes in the isotopic mass balance are generally interpreted as changes in the burial of reduced sulfur. Furthermore, both the rate of sulfur input and its isotopic

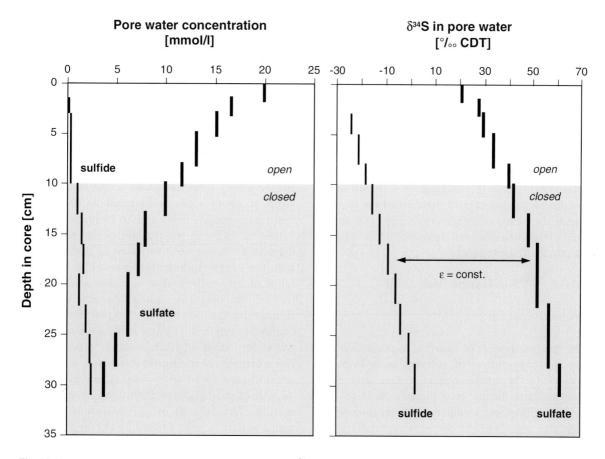

**Fig. 10.11** Variation of sulfate and sulfide concentration and $\delta^{34}S$ values of pore water in core 2092 from the western Baltic Sea (Hartmann and Nielsen 1969).

composition may vary as a function of time depending on the weathering rates of different sedimentary rocks, or on the intensity of volcanic activity. For example, increased erosion of black shales containing $^{34}$S-depleted sulfides may lower the $\delta^{34}$S value of oceanic sulfate. On the other hand, the increased reduction of sulfate by bacteria and removal of $^{34}$S-depleted sulfide from the ocean may increase its $\delta^{34}$S value. However, the fact that the deposition of evaporites in the Phanerozoic history occurred in rather short episodes raises the question about the validity of the sulfate isotope curve with respect to representativeness and homogeneity. The specific causes for the variation of $\delta^{34}$S of marine sulfate at a particular geologic time interval are still largely speculative (for further discussion see Strauss 1997).

A special case in the modern ocean sulfur isotope distribution is the isotopic composition of hydrogen sulfide in anoxic basins. Fry et al. (1991) compared depth profiles of the Black Sea and the Cariaco Basin. Sulfide isotopic compositions in deep waters of the Black Sea were roughly constant near –41.5‰, while those of the Cariaco Trench averaged –31‰. In contrast, in the uppermost 50m of the Black Sea's sulfidic waters (that is in water depths between 100 and 150m) the sulfide isotopic compositions change significantly in a region of sulfide consumption, increasing up to 5‰ vs. the deep-water background. These increases may be due to sulfide oxidation mediated by $MnO_2$ or oxygen, but are not consistent with sulfide oxidation by photosynthetic bacteria (Fry et al. 1991). Growth experiments with sulfate-reducing bacteria suggested that part of the increase in sulfide isotopic compositions could be attributed to rapid rates of sulfate reduction in the oxic/anoxic interface region.

## 10.6.2   $\delta^{34}$S in Marine Sediments

*Evaporitic Sulfate*

The general aspects of isotopic fractionation during the evaporation of sulfates have been addressed in the previous section. Due to a minor fractionation during precipitation, marine evaporitic sulfates are assumed to reflect the original seawater signature. However, exceptions might occur in sulfates precipitated during late stages of evaporation. Sulfur isotope measurements from the Permian Zechstein in Germany (e.g. Nielsen and

Ricke 1964) and experiments by Raab and Spiro (1991) indicate that sulfate deposited within the halite or even the potash-magnesia facies is depleted in $^{34}$S in comparison to sulfate precipitated within the calcium sulfate facies. Progressive crystallization of $^{34}$S-enriched sulfate minerals results in a residual brine depleted in $^{34}$S, by possibly as much as 4‰. Therefore, to obtain a true record of seawater sulfate isotopic composition in the past no late-stage sulfates should be analyzed.

*Sedimentary Pyrite*

Modern sedimentary pyrites are generally depleted in $^{34}$S relative to ocean water sulfate with average $\delta^{34}$S-values around –20‰ and a range between +20‰ and –50‰. The majority of isotopic compositions of pyrites in modern sediments indicate the sedimentary sulfides having formed through bacterial sulfate reduction under open system conditions with respect to sulfate availability. These measurements agree well with experimental data as discussed above. However, higher values have been reported for pyrites formed under anoxic bottom-water conditions like in the Black Sea (Vinograd et al. 1962, Lyons and Berner 1992). They are interpreted to result from enhanced rates of bacterial sulfate reduction in the water column and from a Rayleigh fractionation due to the reduced availability of sulfate within the sediment.

The isotope record for sedimentary sulfides of Phanerozoic age shows a much higher variability than observed for marine sulfates (Strauss 1997). Beside a long-term trend with a Cambrium maximum of about +2‰, a Permian to Cretaceous minimum around –30‰ and an increase to about –10‰ to –20‰ in the modern ocean sediments, there is a considerable variation as high as 80‰ within a single sedimentary unit. The observed range clearly reflects that in addition to the factors which control the sulfate isotopic composition in ancient seawater, changes in the geochemical conditions during sulfate reduction and pyrite formation may affect the $\delta^{34}$S of sulfides. For example, changing bottom-water conditions (oxic to anoxic and vice versa) during the time of deposition may alter the conditions for pyrite formation. The amount of pyrite formed in sediments is known to be limited by (1) the amount of metabolizable organic matter (2) the amount of sulfate and (3) the amount of reactive iron. Under normal marine conditions, the amount of organic matter limits the reduction process (Raiswell and Berner 1986). This is documented in a positive correlation of organic carbon and pyrite sulfur through geologic history (Berner and Raiswell 1983). In anoxic environments, reactive iron appears to be the limiting factor for pyrite formation, since the amount of $H_2S$ exceeds its removal by the reaction with iron (Raiswell and Berner 1985). In that case, some $H_2S$ diffuses back to the sediment surface, where it is reoxidized, whereas some $H_2S$ undergoes further reactions. All these processes are associated with different fractionation effects on the sulfur isotopic composition of sulfides.

In addition to these primary, syngenetic effects, the isotope variability may reflect the diagenetic history of the sediments with a progressive evolution from early to late diagenetic sulfides (i.e. framboids to concretions or overgrowth). In summary, due to the large number of processes that might affect the isotopic composition of sulfides, their potential for interpreting features of the Phanerozoic sulfur cycle appears to be rather limited (Strauss 1997). However, sulfur isotopes can provide a valuable tool for understanding the mechanism of parts of the sedimentary pyrite genesis, such as fossil pyritization (Raiswell 1997).

This is contribution No 259 of the Special Research Program SFB 261 (*The South Atlantic in the Late Quaternary*) funded by the Deutsche Forschungsgemeinschaft (DFG).

# References

Altabet, M. and McCathy, J., 1985. Temporal and spatial variation in the natural abundance of $^{15}N$ in PON from a warm core ring. Deep-Sea Research, 32: 755-722.

Altabet, M., 1988. Variations in nitrogen isotopic composition between sinking and suspended particles: implication for nitrogen cycling and particle transformation in the open ocean. Deep-Sea Research, 35: 535-554.

Altabet, M.A. and Curry, W.B., 1989. Testing models of past ocean chemistry using foraminifera $^{15}N/^{14}N$. Global Biogeochemical Cycles, 3: 107-119.

Altabet, M.A., Deuser, W.G., Honjo, S. and Stienen, C., 1991. Seasonal and depth-related changes in the source of sinking particles in the North Atlantic. Nature, 354: 136-139.

Altabet, M. and Francois, R., 1994. Sedimentary nitrogen isotopic ratio as recorder for surface ocean nitrate utilization. Global Biogeochemical Cycles, 8: 103-116.

Altenbach, A.V. and Sarnthein, M., 1989. Producitivity record in benthic foraminifera. In Berger W.H., Smetacek V.S., Wefer, G. (eds), Producitivity of the ocean: Present and Past. John Wiley & Sons, pp. 255-269.

Andersen, N., Müller, P.J., Kirst, G. and Schneider, R.R., in press. Late Quantenary $pCO_2$ variations in the Angola Current interred from alkenone $d^{13}C$ and carbon demand estimated by $d^{15}N$. In: Fischer, G. and Wefer, G. (eds), Use of proxies in paleoceanography: examples from the South Atlantic. Springer Verlag, Berlin, Heidelberg, NY.

Arthur, M.A., F., A.T., Kaplan, I.R., Veizer, J. and Land, L.S., 1983. Stable isotops in sedimentary geology. SEPM Short Course, 10, SEPM, Tulsa, OK, 432 pp.

Barret, T.J. and Friedrichsen, H., 1989. Stable isotopic composition of atypical ophiolitic rocks from east Liguria, Italy. Chemical Geology, 80: 71-84.

Bemis, B.E., Spero, H.J., Bijma, J. and Lea, D.W., 1998. Reevaluation of the oxygen isotopic composition of planktonic ferominifera: Experimantal results and revised paleotemperature equations. Paleoceanogaphy, 13: 150-160.

Berger, W.H. and Vincent, E., 1986. Deep-sea carbonates: Reading the carbon-isotope signal. Geologische Rundschau, 75: 249-269.

Berger, W.H., Smetacek, V.S. and Wefer, G., (eds) 1989. Productivity of the ocean: Present and Past. Wiley & Sons, NY, 471 pp.

Berner, R.A. and Raiswell, R., 1983. Burial of organic carbon and pyrit sulfur in sediments over Phanerozoic time: a new theory. Geochimica et Cosmochimica Acta, 47: 885-892.

Bickert, T. and Wefer, G., 1996. Late Quaternary deep water circulation in the South Atlantic: Reconstruction from carbonate dissolution and benethic stable isotopes. In: Wefer, G., Berger, W.H., Siedler, G. and Webb, D. (eds), The South Atlantic: present and past circulation. Springer, Berlin, pp. 599-620.

Bickert, T., Pätzold, J., Samtleben, C. and Munnecke, A., 1997. Paleoenvironmental changes in the Silurian indicated by stable isotopes in brachiopod shells from Gotland, Sweden. Geochimica et Cosmochimica Acta, 61: 2717-2739.

Bidigare, R.R., Fluegge, A, Freeman, K.H., Hanson, K.L., Hayes, J.M., Hollander, D., Jasper, J., King, L.L., Laws, E.A., Milder, J., Millero, F.J., Pancost, R., Popp, B.N., Steinberg, P.A. and Wakeham, S.G., 1997. Consistance fractionation of $^{13}C$ in nature and in the laboratory: Growth rate effects in some haptophyte algae. Global Biogeochemical Cycles, 11: 279-292.

Birchfield, G.E., 1987. Changes in deep-ocean water $d^{18}O$ and temperature from last glacial maximum to present. Paleoceanography, 2: 431-442.

Brand, W., 1996. High precision isotope ratio monitoring techniques in mass spectrometry. Journal of Mass Spectrometry, 31: 225-235.

Broecker, W.S., 1982. Ocean chemistry during glacial time. Geochimica et Cosmochimica Acta, 46: 1689-1705.

Broecker, W.S. and Maier-Reimer, E., 1992. The influence of

air and sea exchange on the carbon isotope disribution in the sea. Global Biogeochemical Cycles, 6: 315-320.

Carpenter, S.J., Lohmann, KC, Holden, P, Walter, LM, Huston, TJ, Halliday, AN, 1991. d18O values, 87Sr/86Sr and Sr/Mg rations of Late Devonian abiotic arine calcite: Implications for the composition of ancient seawater. Geochimica et Cosmochimica Acta, 55: 1991-2010.

Charles, C.D. and Fairbanks, R.G., 1990. Glacial to interglacial changes in the isotopic gradients of the Southern Ocean surface water. In Bleil, U. and Thiede, J. (eds), Geological history of the Polar Oceans: Artic vesus Antarctic. Kluwer, Dordrecht, pp. 519-538.

Cifuentes, L.A., Sharp, J.H. and Fogel, M.L., 1988. Stable natron and nitrogen isotope biogeochemistry in the Delaware estuary. Limnology and Oceanography, 33: 1102-1115.

Clark, I.D. and Fritz, P., 1997. Environmental isotopes in hydrogeology. Press/Lewis Puplishers, Boca Raton, 328 pp.

Claypool, G.E., Holser, W.T., Kaplan, I.R., Sakai, H. and Zak, I., 1980. The age curves of sulfur and oxygen isotopes in marine sulfate and their mutual interpretation. Chemical Geology, 28: 190-260.

Cline, J.D. and Kaplan, I.R., 1975. Isotopic fractionation of dissolved nitrate during denitrification in the eastern tropical North Pacific Ocean. Marine Chemistry, 3: 271-299.

Coplen, T.B., 1996. More uncertainty than necessary. Paleoceanography, 11: 369-370.

Craig, H. and Gordon, L.I., 1965. Deuterium and oxygen-18 variations in the ocean and marine atmosphere. In: Tongiori, E. (ed) Stable isotopes in oceanic studies and paleotemperatures. Consiglio Nazionale Delle Ricerche, Laboratorio di Geologia Nucleare, Pisa, pp. 9-130.

De Lange, G.J., van Os, B., Pruysers, P.A., Middelburg, J.J., Castradori, D., van Santvoort, P., Müller, P.J., Eggenkamp, H. and Prahl, F.G., 1994. Possible early diagenetic alteration of paleo proxies. In: Zahn, R., Pederson, T.F., Kaminski, M.A. and Labeyrie, L. (eds), Carbon cycling in the glacial ocean: Constraints on the ocean's role in global climate. NATO ASI Series, Springer Verlag, Berlin, pp. 225-258.

Degens, E.T., Guillard, R.R.L., Sackett, W.M. and Hellebust, J.A., 1968. Metabolic fractionation of carbon isotopes in marine plankton. I. temperature and respiration experiments. Deep-Sea Research, 15: 1-9.

Degens, E.T., 1969. Biogeochemistry of stable carbon isotopes. In: Eglington, G. and Murphy, M.T.J. (eds); Organic geochemistry. Methods and results. Springer Verlag, Berlin, pp. 304-329.

Deines, P., 1981. The isotopic composition of reduced organic carbon. In: Fritz, P. and Fontes, J.C., (eds), Handbook of environmental geochemistry, 1. Elsevier, NY, pp. 239-406.

Derry, L.A. and France-Lanord, C., 1996. Neogene growth of the sedimentary organic carbon. Paleoceanography, 11: 267-276.

Emrich, K., Ehhalt, D.H. and Vogel, J.C., 1970. Carbon isotope fractionation during the precipitation of Calcium carbonate. Earth and Planetary Science Letters, 8: 363-371.

Epstein, S., Buchsbaum, R., Lowenstam, H.A. and Urey, H.C., 1953. Revised carbonate-water isotopic temperature scale. Bull. Geol. Soc. Am., 64: 1315-1325.

Erez, J. and Luz, B., 1983. Experimantal paleotemperature equation for planktonic foraminifera. Geochimicet et Cosmochimica Acta, 47: 1025-1031.

Fairbanks, R.G., 1989. A 17,000-year glacio-eustatic sea level record: Influence of glacial melting rates on the Younger Dryas event and deep-ocean circulation. Nature, 342: 637-642.

Fairbanks, R.G., Charles, C.D. and Wright, J.D., 1992. Origin of global meltwater pulses. In: Tayler, R.E. (ed), Radiocarbon after four decades. Springer, NY, pp. 473-500.

Faure, G., 1986. Principles of isotope geology. Wiley & Sons, NY, 589 pp.

Fontugne, M.R. and Calvert, S.E., 1992. Late Pleistocene variability of the carbon isotopic composition of organic and atmospheric matter in the eastern Mediterranean: Monitor of changes in carbon sources and atmospheric CO2 concentrations. Paleoceanography, 7: 1-20.

Francois, R., Altabet, M.A. and Burckle, L.H., 1992. Glacial to interglacial changes in surface nitrate utilization in the Indian sector of the Southern Ocean as recorded by sediment d15N. Paleoceanography, 7: 589-606.

Fry, B., Jannasch, H.W., Molyneaux, S.J., Wirsen, C.O., Muramato, J.A. and King, S., 1991. Stable isotopes of the carbon, nitrogen and sulfur cycles in the Black Sea and the Cariaco Trench. Deep-Sea Research, 38: 1003-1019.

Gruber, N. and Sarmiento, J.L., 1997. Global patterns of marine nitrogen fixation and denitrification. Global Biogeochemical Cycles, 11: 235-266.

Habicht, K.S. and Canfield, D.E., 1997. Sulfur isotope fractionation during bacterial sulfate reduction in organic-rich sediments. Geochimica et Cosmochimica Acta, 61: 5351-5361.

Hartmann, M. and Nielson, H., 1969. 34S Werte in rezenten Meeressedimenten und ihre Deutung am Beispiel einiger Sedimentprofile aus der weslichen Ostsee. Geologische Rundschau, 58: 621-655.

Hayes, J.M., 1993. Factors controlling 13C contents of sedimentary organic compounds: Principles and evidence. Marine Geology, 113: 111-125.

Hoefs, J., 1997. Stable Isotop Geochemistry. Springer, Berlin, Heidelberg, NY, 201 pp.

Hoffmann, S.E., Wilson, M. and Stakes, D.S., 1986. Inferred oxygen isotope profile of Archean crust, Onverwacht Group, South Africa. Nature, 321: 55-58.

Holland, H.D., 1978. The chemistry of the atmosphere and oceans. Wiley, NY, 351 pp.

Holmes, M.E., Müller, P.J., Schneider, R.R., Segl, M. and Wefer, G., 1997. Reconsruction of past nutrient utilization in the eastern Angola Basin based on sedimentary 15N/14N rations. Paleoceanography, 12: 604-614.

Holser, W.T., 1997. Geochemical events documented in inorganic carbon isotopes. Paleogeography, Paleoclimatology, Paleoecology, 132: 173-182.

Irwin, H., Curtis, C. and Coleman, M., 1977. Isotopic evidence for source of diagenetic carbonates formed during burial of organic-rich sediments. Nature, 269: 209-213.

Jacobs, S.S., Fairbanks, R.G. and Horibe, Y., 1985. Origin and evolution of water masses near the antarctic continental margin: Evidence from H2 18O/H2 16O ratios in seawater. Antarctic Research Series, 43: 59-85.

Jasper, J.P. and Hayes, J.M., 1990. A carbon isotope record of $CO_2$ levels during the late Quantenary. Nature, 347: 462-464.

Jasper, J.P. and Hayes, J.M., 1994. Reconstraction of Paleoceanic $pCO_2$ levels from carbon isotopic compositions of sedimentary biogenic components. In: Zahn, R., Pederson, T.F., Kaminski, M.A. and Labeyrie, L. (eds), Carbon cycling in the glacial ocean: Constraints on the ocean´s role in global climate. NATO ASI Series, Springer Verlag, Berlin, pp. 323-342.

Jørgensen, B.B., Erez, J., Revsbech, N.P. and Cohen, Y., 1985. Symbiontic photosynthesis in a planktonic foraminifera, Globigerinoides sacculifer (Brady), studied with microelectrodes. Limnol. Oceanogr., 30: 1253-1267.

Jørgensen, B.B., 1990. A thiosulfate shunt in the sulfur cycle of marine sediments. Science, 249: 152-154.

Kaplan, I.R. and Rittenberg, S.C., 1964. Microbiological fractionation of sulfur isotopes. J. Gen. Microbiol., 34: 195-212.

Kroopnick, P., 1985. The distribution of $^{13}C$ of $SCO_2$ in the world oceans. Deep-Sea Research, 32: 57-84.

Kyser, T.K., 1995. Micro-analytical techniques in stable isotope geochemistry. Can. Mineral, 33: 261-278.

Lawrence, J.R., 1989. The stable isotope Geochemistry of deep-sea pore water. In: Fritz, P., and Fontes, J.C., (eds), Handbook of environmental isotope geochemistry, 3, Elsevier, Amsterdam, pp. 317-354.

Laws, E.A., Popp, B.N., Bidigare, R.R., Kennicutt, M.C. and Macko, S.A., 1995. Dependence of phytoplankton carbon isotopic composition on growth rate and $(CO_2)aq$: Theoretical considerations and experimental results. Geochimica et Cosmochimica Acta, 59: 1131-1138.

Liu, K.K. and Kaplan, I.R., 1989. The eastern tropical Pacific as a source of $^{15}N$-enriched nitrate in seawater of southern California. Limnol. Oceanography, 34: 820-830.

Longinelli, A., 1989. Oxygen-18 and sulphur-34 in dissolved oceanic sulphate and phosphate. In: Fritz, P., Fontes, J.C. (eds), Handbook of environmental isotope geochemistry, 3. Elsevier, Amsterdam, pp. 219-255.

Lynch-Stieglitz, J., Stocker, T.F., Broecker, W.S. and Fairbanks, R.G., 1995. The influence of air-sea exchange on the isotopic composition of oceanic carbon: observations and modeling. Global Biogeochemical Cycles, 9: 653-665.

Lyons, T.W. and Berner, R.A., 1992. Carbon-sulfur-iron systematics of the uppermost deep-water sediments of the Black-Sea. Chemical Geology, 99: 1-27.

Mackensen, A., Hubberten, H.W., Bickert, T., Fischer, G. and Fütterer, D.K., 1993. $d^{13}C$ in benthic foraminiferal tests of Fontbotia wuellerstorfi ( SCHWAGER ) relative to $d^{13}C$ of dissolved inorganic carbon in Southern Ocean deep water: implications for glacial ocean circulation models. Paleoceanography, 8: 587-610.

Mackensen, A., Hubberten, H.W., Scheele, N. and Schlitzer, R., 1996. Decoupling of $d^{13}CSCO_2$ and phosphate in recent Weddell Sea Deep and Bottom Water: Implcations for glacial Southern Ocean paleoceanography. Paleoceanography, 11: 587-610.

Matsumoto, R., 1992. Causes of the oxygen isotopic depletion of interstitial waters from sites 798 and 799, Japan Sea, Leg. 128, Proc. ODP, Sci. Res., 127/128: 697-703.

McConnaughey, T.A., Burdett, J., Whelan, J.F. and Paull, C.K., 1997. Carbon isotopes in biological carbonates: Respiration and photosynthesis. Geochimica et Cosmochimica Acta, 61: 611-622.

McCorkle, D.C., Emerson, S.R. and Quay, P.D., 1985. Stable carbon isotopes in marine porewaters. Earth Planetary Science Letters, 74: 13-26.

McCorkle, D.C. and Emerson, S.R., 1988. The relationship between pore water carbon isotopic composition and bottom water oxygen concentration. Geochimica et Cosmochimica Acta, 52: 1196-1178.

McCorkle, D.C., Keigwin, L.D., Corliss, B.H. and Emerson, S.R., 1990. The influence of microhabitats on the carbon isotopic composition of deep-sea benthic foraminifera. Paleoceanography, 5: 161-185.

McCorkle, D.C., Martin, P.A., Lea, D.W. and Klinkhammer, G.P., 1995. Evidence of a dissolution effect on benethic foraminiferal shell chemistry: $d^{13}C$, Cd/Ca, Ba/Ca, and Sr/Ca results from the Ontong Java Plateau. Paleoceanography, 10: 699-714.

McCorkle, D.C., Corliss, B.H. and Farnham, C.A., 1997. Vertical distributions and stable isotopic compositions of live (stained) benthic foraminifera from the North Carolina and California continental margins. Deep-Sea Research, 44: 983-1024.

Miyake, Y. and Wada, E., 1971. The isotope effect on the nitrogen in biochemical oxidation-reduction reactions. Rec. Oceanogr. Works Japan, 11: 1-6.

Montoya, J.P., 1994. Nitrogen fractionation in the modern ocean: Implications for the sedimentary record. In Zahn, R., Pedersen, T.F., Kaminski, M.A. and Labeyrie, L. (eds), Carbon cycling in the glacial ocean: Constraints on the ocean's role in global change. NATO ASI Series, Springer, Berlin, pp. 259-279

Mook, W.G., Bommerson, J.C. and Staverman, W.H., 1974. Carbon isotope fractionation between dissolved bicarbonate and gaseous carbon dioxide. Earth and Planetary Science Letters, 22: 169-176.

Muehlenbach, K., 1986. Alteration of the oceanic crust and the $^{18}O$ history of seawater. Mineral Soc. Amer. Rev. in Mineral, 16: 425-444.

Müller, P.J., Schneider, R. and Ruhland, G., 1994. Late Quaternary $pCO_2$ variations in the Angola Current: Evidence from organic carbon $d^{13}C$ and alkenone temperatures. In: Zahn,.R., Pedersen, T.F., Kaminski, M.A. and Lebeyrie, L. (eds), Carbon cycling in the glacial ocean: constraints on the ocean´s role in global change. NATO ASI Series, Springer Verlag, Berlin, Heidelberg, pp. 343-361.

Newman, J.W., Parker, P.L. and Behrens, E.W., 1973. Organic carbon isotope ratios in Quaternary cores from the Gulf of Mexico. Geochimica et Cosmochimica Acta, 37: 225-238.

Nielsen, H. and Ricke, W., 1964. Schwefel-Isotopen-Verhältnisse von Evaporiten aus Deutschland: ein Beitrag zur Kenntniss von $^{34}S$ im Meerwasser-Sulfat. Geochimica et Cosmochimica Acta, 28: 577-591.

Nissenbaum, A., Presley, B.J. and Kaplan, I.R., 1972. Early diagnosis in a reducing fjord, Saanich Inlet, British Columbia - I. Chemical and isotopic changes in major components of interstitial water. Geochimica et Cosmochimica Acta, 36: 1007-1027.

O'Leary, M.H., 1981. Carbon isotope fractionation in plants. Phytochemistry, 20: 553-567.

Ohmoto, H., Kaiser, C.J. and Geer, K.A., 1990. Systematics of sulphur isotopes in recent marine sediments and ancient sediment-hosted basement deposits. University Western Australia Publ., 23: 70-120.

Park, R. and Epstein, S., 1960. Carbon isotope fractionationduring photosynthesis. Geochimica et Cosmochimica Acta, 21: 110-126.

Popp, B.N., Anderson, T.F. and Sandberg, P.A., 1986. Brachiopods as indicators of original isotopic compositions in some Paleozoic limestones. Geological Society America Bulletin, 97: 1262-1269.

Popp, B.N., Tagiku, R., Hayes, J.M., Louda, J.W. and Baker, E.W., 1989. The post-paleozoic chronology and mechanism of d$^{13}$C depletion in primary marine organic matter. American Journal of Science, 289: 436-454.

Popp, B.N., Parekh, P., Tilbrook, B., Bidigare, R.R. and Laws, E.A., 1997. Organic carbon d$^{13}$C variations in sedimentary rocks as chemostratigraphic and paleoenvironmental tools. Paleogeography, Paleoclimatology, Paleoecology, 132: 119-132.

Popp, B.N., Laws, EA, Bidigare, RR, Dore, JE, Hanson, KL and Wakeham, SG, 1998. Effect of phytoplankton cell geometry on carbon isotopic fractionation. Geochimica et Cosmochimica Acta, 62: 69-77.

Raab, M. and Spiro, B., 1991. Sulfur isotopic variations during seawater evaporation with fractional crystallization. Chemical Geology, 86: 323-333.

Railsback, L.B., 1990. Influenze of changing deep ocean circulation on the Phanerozoic oxygen isotopic record. Geochimica et Cosmochimica Acta, 54: 1501-1509.

Raiswell, R. and Berner, R.A., 1985. Pyrit formation in euxinic and semi-euxenic sediments. American Journal of Science, 285: 710-724.

Raiswell, R. and Berner, R.A., 1986. Pyrit and organic matter in Phanerozoic normal marine shales. Geochimica et Cosmochimica Acta, 50: 1967-1976.

Raiswell, R., 1997. A geochimical framework for the application of stable sulphur isotopes to fossil pyritization. Journal of the Geological Society, 154: 343-345.

Rau, G.H., Froehlich, P.N., Takahashi, T. and Des Marais, D.J., 1991. Does sedimentary organic d$^{13}$C record variations in Quaternary ocean [$CO_{2(aq)}$]? Paleoceanography, 6: 335-347.

Rau, G.H., Riebesell, U. and Wolf-Gladrow, D., 1997. $CO_{2aq}$-dependet photosynthetic $^{13}$C franctionation in the ocean: A model versus measurements. Global Biogeochemical Cycles, 11: 267-278.

Raymo, M.E., Grant, B., Horowitz, M. and Rau, G.H., 1996. Mid-Pliocene warmth: stronger greenhouse and stronger conveyor. Marine Micropaleontology, 27: 313-326.

Rink, S., Kühl, M., Bijma, J. and Spero, H.J., 1998. Microsensor studies of photosynthesis and respiration in the symbiotic foraminifera O. universa. Marine Geology, 131: 583-596.

Rühlemann, C. , Frank, M., Hale, W., Mangini, A., Mulitza, S., Müller, PJ. and Wefer, G., 1996. Late Quaternary productivity changes in the western equatorial Atlantic: Evidence from $^{230}$Th-normalized carbonate and organic carbon accumulation rates. Marine Geology, 135: 127-152.

Sarnthein, M., Winn, K., Jung, SJA., Duplessy, JC., Labeyrie, L., Erlenkeuser, H. and Ganssen, G., 1994. Changes in

east Atlantic deepwater circulation over the last 30,000 years: Eigth time slice reconstuctions. Paleoceanography, 9: 209-268.

Schneider, R., Dahmke, A., Kölling, A., Müller, PJ., Schulz, HD. and Wefer, G., 1992. Strong deglacial minimum in the d$^{13}$C record from planktonic foraminifera in the Benguala upwelling region: palaeoceanographic signal or early diagenetic imprint. In: Summerhayes, CP., Prell, WL. and Emeis, KC (eds), Upwelling systems: Evolution since the early Miocene. Geological Society Special Publication, 63, pp. 285-297.

Shackleton, N.J. and Opdyke, N.D., 1973. Oxygen isotope and paleomagnetic stratigraphy of equatorial Pacific core V 28-238: Oxygen isotope temperatures and ice volumes on a 10^5 year scale. Quantenary Research, 3: 39-55.

Shackleton, N.J., 1977. Tropical rainforest history and the equatorial Pacific carbonate dissolotion cycles. In: Anderson, NR. and Malahoff, A. (eds), Fate in fossil foel $CO_2$ in the oceans. Plenum, NY, pp. 401-427.

Spero, H.J., 1992. Do planktonic foraminifera accurately record shifts in th ecarbon isotopic composition of $CO_2$? Marine Micropaleontology, 19: 275-285.

Spero, H.J. and Lea, D.W., 1993. Intraspecific stable isotope variability in the planktic foraminifera Globigerinoides sacculifer: Results from laboratory experiments. Marine Micropaleontology, 22: 221-234.

Spero, H.J., Bijma, J., Lea, D.W. and Bemis, B.E., 1997. Effect of seawater carbonate chemistry on planktonic foraminiferal carbon and oxygen isotope values. Nature, 390: 497-500.

Strauss, H., 1997. The isotopic composition of sedimentary sulfur through time. Paleogeography, Paleoclimatology, Paleoecology, 132: 97-118.

Sweeney, R.E., Liu, K.K. and Kaplan, I.R., 1978. Oceanic nitrogen isotopes and their uses in determinig the source of sedimentary nitrogen. In: Robinson, B.W. (ed), Stable isotopes in earth sciences. Dept. Scientific and Industrial Research, Wellington, pp. 9-26.

Sweeney, R.E. and Kaplan, I.R., 1980. Natural abundances of $^{15}$N as a source indicator for near-shore marine sedimentary and dissolved nitrogen. Marine Chemistry, 9: 81-94.

Veizer, J., Brukschen, P., Pawellek, F., Diener, A., Podlaha, O.G., Carden, G.A.F., Jasper, T., Korte, C., Strauss, H., Azmy, K. and Ala, D., 1997. Oxygen isotope evolution of Phanerozoic seawater. Palaeogeography, Palaeoclimatology, Palaeoecology, 132: 159-172.

Vinogradov, A.P., Grinenko, V.A. and Ustinov, V.I., 1962. Isotopic composition of sulfur compounds in the Black Sea. Geokhimiya, 10: 973-997.

Voss, M., Altabet, M.A. and von Bodungen, B., 1996. d$^{15}$N in sedimenting particles as indicator of euphotic - zone processes. Deep-Sea Research, 43: 33-47.

Wada, E. and Hattori, A., 1978. Nitrogen assimilation affects in the assimilation of inorganic nitrogenous compounds by marine diatoms. Geomicrobiology J, 1: 85-101.

Wada, E., Minagawa, M., Mizutani, H., Tsuji, T., Imaizumi, R. and Karaswa, K., 1987. Biogeochemical studies on the transport of organic matter along the Otsuchi River watershed, Japan. Estuarine, Coastal and Shelf Sciences, 25: 321-336.

Walker, J.G.C. and Lohmann, K.C., 1989. Why the oxygen

isotopic composition of seawater changes through time. Geophysical Research Letters, 16: 323-326.

Wefer, G. and Berger, W.H., 1991. Isotope paleontology: growth and composition of extnant calacareous species. Marine Geology, 100: 207-248.

Wefer, G., Heinze, P.M. and Berger, W.H., 1994. Clues to ancient methane release. Nature, 369: 282.

Weiss, R.F., 1974. Carbon dioxide in water and seawater: The solubility of a non-ideal gas. Marine Chemistry, 2: 203-215.

Weiss, R.F., Östlund, H.G. and Craig, H., 1979. Geochemical studies of the Wedell Sea. Deep-Sea Research, 26: 1093-1120.

Williams, L.B. Ferell, R.E., Hutcheon, I., Bakel, A.J., Walsh, M.M. and Krouse, H.R., 1995. Nitrogen isotope geochemistry of organic matter and minerals during diagneses and hydrocarbon migration. Geochimica et Cosmochimica Acta, 59: 765-779.

Wu, G. and Berger, W.H., 1989. Planktonic foraminifera: differential dissolution and the Quanternary stable isotope record in the west Equatorial Pacific. Paleoceanography, 4: 181-198.

Zahn, R. and Mix, A.C., 1991. Benethic foraminifera $d^{18}O$ in the ocean's temperature - salinty - density field: Constraints on ice age thermohaline circualtion. Paleoceanography, 6: 1-20.

Zahn, R. and Keir, R., 1994. Tracer-nutrient correlations in the upper ocean: observatorial and box model constraints on the use of benethic foraminiferal $d^{13}C$ and Cd/Ca as paleo-proxies for the intermadiate-depth ocean. In: Zahn, R., Pedersen, T.F., Kaminski, M.A. and Labeyrie, L. (eds), Carbon cycling in the glacial ocean: Constraints on the ocean's role in global change. NATO ASI Series, I 17, Springer, Berlin, pp. 195-223.

# 11 Manganese: Predominant Role of Nodules and Crusts

Geoffrey P. Glasby

## 11.1 Introduction

The importance of manganese in the marine environment can be deduced from the fact that it is the tenth most abundant element in the earth's crust (av. conc. 0.093%) and is available in two valency states whose stability boundary lies within the range of the natural environment (Glasby 1984). Manganese oxides also have a high adsorption capacity. $\delta MnO_2$, for example, has a surface area of about 260 $m^2$ $g^{-1}$ and a $pH_{zpc}$ of 2.25 and can therefore adsorb cations such as $Ni^{2+}$, $Cu^{2+}$ and $Zn^{2+}$ from natural waters. By comparison, iron is the fourth most abundant element in the earth's crust (av. conc. 5.17%) giving an average Mn/Fe ratio of 0.02. It also occurs in two valency states whose stability boundary lies within the range of the natural environment. Fe oxyhydroxides have a high adsorption capacity and large surface area. Goethite has a $pH_{zpc}$ of 7.1 and can adsorb both cations and anions (e.g. REE, P, Mo, W etc.).

Both elements can therefore migrate under the influence of redox gradients. They can also fractionate from each other, particularly under acid or reducing conditions such as in lakes and shallow seas or in marine sediments. $FeS_2$ plays a major role in separating Mn and Fe in anoxic marine environments such as in Baltic Sea sediments where $SO_4^{2-}$ ions are present. The reduced form of manganese, Ca-rhodochrosite ($Ca$-$MnCO_3$), is much rarer than $FeS_2$ but it is found in environments such as Gotland Deep sediments. Mn is more mobile than Fe. Mn therefore migrates more readily than Fe but Fe deposits more readily than Mn. This leads to the fractionation of Mn from Fe as, for example, in submarine hydrothermal systems or in anoxic environments such as Baltic Sea deeps.

The formation of huge quantities of manganese nodules and crusts on the deep-sea floor is a function of the fact that manganese and iron are relatively abundant in the earth's crust and migrate from less oxidizing to more oxidizing environments. The increased glaciation of Antarctica about 12 Ma led to the initiation of Antarctic Bottom Water (AABW) flow and the increased ventilation of the deep ocean. The modern, well-oxygenated deep-sea has therefore become the ultimate repository for manganese. Deep-sea Mn nodules formed since the lower Miocene unconformity (12 Ma) hold about $10^{11}$ tons of Mn (about 16 times the total Mn in terrestrial deposits) reflecting the importance of manganese nodules in the global cycle of manganese (Glasby 1988a). Strictly speaking, the last 12 Ma of the earth's history can be designated the manganese nodule era. The high contents of Co, Ni, Cu and Zn in manganese nodules and crusts which make these deposits an important potential economic resource for these elements are a function of the sorption characteristics of the manganese and iron oxyhydroxides.

## 11.2 Manganese, Iron and Trace Elements in Seawater

Manganese occurs in seawater mainly as $Mn^{2+}$ or $MnCl^+$ (Bruland 1983). The dissolved Mn concentration in the open ocean is in range 0.2-3 nmol $kg^{-1}$ which is above the equilibrium concentration with respect to $MnO_2$ or $MnOOH$. This situation reflects the slow rate of oxidation of $Mn^{2+}$ in solution.

Figure 11.1 shows the vertical distribution of Mn in seawater at the VERTEX-IV site situated in the center of the central North Pacific subtropi-

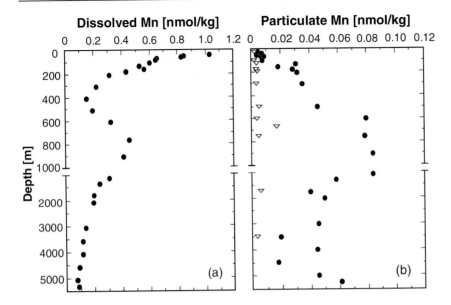

**Dissolved Mn [nmol/kg]**

**Particulate Mn [nmol/kg]**

Fig. 11.1   Vertical profiles of (a) dissolved manganese (nmol kg⁻¹) and (b) particulate manganese (nmol kg⁻¹) at the VERTEX-IV site (after Bruland et al. 1994). For particulate manganese, filled circles represent the acetic acid leachable fraction and open triangles represent refractory manganese.

cal gyre, a region of low biological productivity (Bruland et al. 1994). The high content of Mn in the surface waters reflects the input of eolian material into this region from Asia. Photoreduction of particulate $MnO_2$ to Mn(II) takes place in the surface waters resulting in 99% of the Mn in the surface waters (0-100 m) being in the dissolved form. By contrast, only 80% of the Mn in the deep water (500-4000 m) is in the dissolved form. Mn therefore behaves in seawater as a scavenged-type metal. Bacterial mediation plays a key role in the scavenging and oxidation of dissolved Mn in intermediate and deep water. The higher concentration of particulate Mn in North Atlantic deep water (0.15 nmol kg⁻¹) than in central North Pacific deep water (0.05 nmol kg⁻¹) reflects the higher input of eolian material into the Atlantic compared to the Pacific Ocean.

Mn is also influenced by redox processes in the water column. Both dissolved and particulate Mn display maxima at the oxygen minimum zone which occurs at a depth of 500-1000 m in the central North Pacific (Fig. 11.1). In this case, the maximum dissolved Mn concentration in the oxygen minimum zone is about 0.4 nmol kg⁻¹. This high concentration is a result of the lateral transport of Mn from the continental margins to the open ocean along the oxygen minimum zone (Martin et al. 1985). This process is particularly important in the highly oxygen-deficient waters of the oxygen minimum zone in parts of the eastern North Pacific Ocean (Burton and Statham 1988). A model for the formation of this manga-

nese maximum zone has recently been presented by Johnson et al. (1996).

About 90% of the Mn introduced to the oceans has a hydrothermal origin (Glasby 1988b). Hydrothermal Mn anomalies in seawater can be detected over 1000 km from the source in the Pacific Ocean (Burton and Statham 1988). When hydrothermal fluids are discharged at the sea floor, a buoyant hydrothermal plume is formed on mixing of the hydrothermal fluid with seawater (Lilley et al. 1995). The plume can rise tens to hundreds of meters above the sea floor to a level of neutral buoyancy where it forms a distinct hydrographic layer with a distribution extending tens to thousands of kilometers from the vent. The dilution factor of the vent fluid with respect to seawater is of the order of $10^4$-$10^5$. In the first centimeters to meters above the vent, upto 50% of the iron is precipitated as sulfides. The chalcophile elements (Cu, Zn, Cd and Pb) tend to be incorporated in the sulfide minerals at this stage. The remaining Fe is precipitated over a longer time period as fine-grained iron oxyhydroxide particles. The half life for Fe (II) precipitation is 2-3 minutes. The iron oxyhydroxide particles scavenge anionic species such as $HPO_4^{2-}$, $CrO_4^{2-}$, $VO_4^{2-}$ and $HAsO_4^{2-}$ as well as the rare earth elements (REE). Precipitation of particulate Mn oxides takes place much more slowly, mainly in the neutrally buoyant plume where the oxidation is bacterially mediated. Because of the slow precipitation rate, particulate Mn concentrations increase in the plume to a maximum 80-150 km from the vent. 80% of the hydrothermal Mn is deposited on the sea floor within several hundred km of the vent field but the remaining Mn still raises the background concentration of Mn in seawater severalfold (Lavelle et al. 1992). The residence time of hydrothermal Mn in seawater is several years. German and Angel (1995) have estimated that the total hydrothermal Mn flux to the oceans is $6.85 \cdot 10^9$ kg yr$^{-1}$. This compares with the flux of Mn from the rivers of $0.27 \cdot 10^9$ kg yr$^{-1}$ (Elderfield and Schulz 1996) confirming the original calculation of Glasby (1988b).

The predominance field of Mn in seawater (Glasby and Schulz in press) is best illustrated by the use of an $E_H$, pH diagram (Fig. 11.2). At the conditions prevalent in seawater ($E_H$ +0.4 V, pH 8), the stable form of Mn is seen to be the aqueous species, $Mn^{2+}$, and not any of the solid phases of Mn. The fact that Mn oxides are abundant on the seafloor can be explained on the basis that the Mn oxyhydroxides initially formed in seawater (β-manganite) are not pure mineral phases but have significant concentrations of transition elements and are fine grained, both of which help to stabilize them (Glasby 1974).

Nonetheless, this situation probably explains the slow rate of oxidation of Mn in seawater.

The kinetics of oxidation of $Mn^{2+}$ in seawater have been discussed by Murray and Brewer (1977) and the following reaction sequence proposed.

$$Mn(II) + O_2 \rightarrow MnO_{2\,s} \qquad\qquad \textit{slow}$$

$$Mn(II) + MnO_{2\,s} \rightarrow (Mn(II).MnO_2)_s \quad \textit{fast}$$

$$(Mn(II).MnO_2)_s + O_2 \rightarrow 2MnO_{2\,s} \qquad \textit{slow}$$

$$-d(Mn(II))/\,dt \rightarrow k_o(Mn(II)) + \\ k_1(Mn(II))(MnO_{2\,s})(P_{O2})(OH^-)^2$$

This rate law demonstrates that the oxidation of Mn (II) in seawater is autocatalytic. From this equation, it was calculated that it would take about 1000 years to oxidize 90% of the Mn present in seawater. Surface catalysis on $MnO_2$ or FeOOH or bacterial oxidation are therefore required to remove $Mn^{2+}$ from solution (Cowen and Bruland 1985, Mandernack et al. 1985, Ehrlich 1996, Hastings and Emerson 1986, Tebo et al. 1997). Giovanoli and Arrhenius (1988) also proposed that the surface catalyzed oxidation of

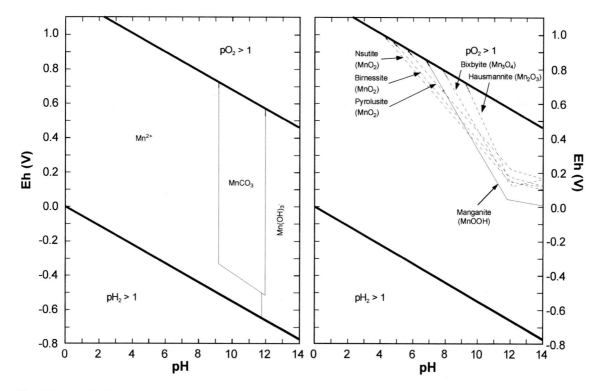

**Fig. 11.2**  $E_H$, pH diagram for Mn calculated for the chemical conditions prevailing in the deep sea (after Glasby and Schulz in press). Note that, under seawater conditions and at an $E_H$, of +0.4 V and a pH of 8, the stable form of Mn is the aqueous species, $Mn^{2+}$, and not any of the solid phases of Mn.

$Mn^{2+}$ by FeOOH is a rate-controlling step in the formation of marine manganese deposits. Rate equations and rate constants for the oxidation of manganese in seawater based on field data have been determined by Yeats and Strain (1990) and von Langen et al. (1997).

By contrast, the form and concentration of iron in seawater remain poorly known because of the pronounced tendency of Fe(III) species to hydrolyze in aqueous solution (Bruland 1983) (see Chap. 7). Iron occurs in seawater mainly in the hydrolyzed forms $Fe(OH)_3^0$ and $Fe(OH)_2^+$ and as $FeCl^+$. The concentration range of Fe in seawater is 0.1-2.5 nmol $kg^{-1}$ giving a Mn/Fe ratio of about unity which is much greater than that in the earth's crust.

Figure 11.3 shows the vertical distribution of Fe in seawater at the VERTEX-IV site (Bruland et al. 1994). Dissolved Fe shows a max. concentration in the surface mixed layer (0.35 nmol $kg^{-1}$) but this declines to a minimum in the subsurface stratified layer (70-100 m) (0.02 nmol $kg^{-1}$). The high concentration of Fe in the surface waters reflects the strong eolian input into this region. The Fe in the surface layer is rapidly consumed by plankton but is recycled within days. Below 100 m, dissolved Fe displays a nutrient-type distribution. In deep water, the dissolved Fe concentration reaches a value of 0.38 nmol $kg^{-1}$ Fe. Iron is a limiting nutrient in open-ocean surface waters characterized by high nitrate and low chlorophyll contents. Proposals have been made to seed the oceans with Fe in order to stimulate phytoplankton growth which would hopefully reduce the levels of atmospheric $CO_2$ (de Baar et al. 1995, Fitzwater et al. 1996). A model describing the factors controlling the dissolved Fe content in seawater has recently been presented by Johnson et al. (1997).

Particulate Fe does not show an eolian influence in the surface layer and its distribution is more constant with depth. 48% of Fe in surface water (0-100 m) is in dissolved form and 55% in deep water (500-4000 m). Most of the particulate Fe in the intermediate and deep water is in the form of refractory aluminosilicate minerals of eolian origin. The higher concentration of particulate Fe in North Atlantic deep water (1.2 nmol $kg^{-1}$) than in central North Pacific deep water (0.3 nmol $kg^{-1}$) again reflects the higher input of eolian material into the Atlantic compared to the Pacific Ocean.

Fe behaves in seawater in part as a scavenged-type element and in part as a nutrient-type element as shown by the strong correlation of dissolved Fe with nitrate and phosphate at depths below 100 m.

The stability field of Fe in seawater can be best illustrated by the use of an $E_H$, pH diagram (Fig. 11.4). $Fe(OH)_{3\,s}$ is shown as the metastable form of iron at the conditions prevalent in seawater ($E_H$ +0.4 V, pH 8). Actually, akagenéite (β-FeOOH) is the more stable form found in deep-sea manganese nodules (see Sect. 11.4.8) but thermodynamic data are not available for such very fine-grained minerals.

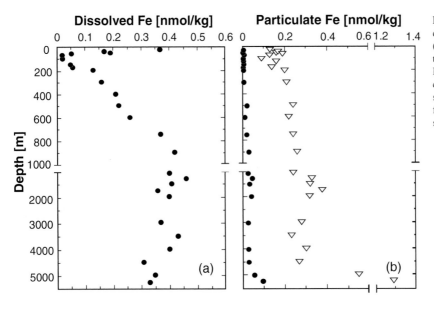

Fig. 11.3 Vertical profiles of (a) dissolved iron (nmol $kg^{-1}$) and (b) particulate iron (nmol $kg^{-1}$) at the VERTEX-IV site (after Bruland et al. 1994). For particulate iron, filled circles represent the acetic acid leachable fraction and open triangles represent refractory iron.

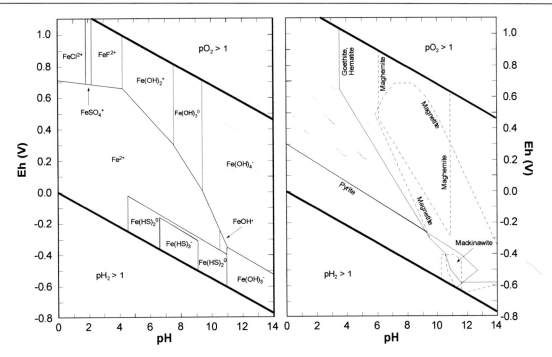

**Fig. 11.4** $E_H$, pH diagram for Fe calculated for the chemical conditions prevailing in the deep sea (after Glasby and Schulz in press). Note that, under seawater conditions and at an $E_H$, of +0.4 V and a pH of 8, the metastable form of Fe is $Fe(OH)_3$ and the stable forms magnetite $((Fe, Mg) Fe_2O_4)$ and maghemite $(Fe_2O_3)$.

Of the trace elements that are of most interest in the formation of manganese nodules, Co is present in deep ocean water at concentrations of <0.1 nmol kg$^{-1}$, mainly as Co$^{2+}$, CoCO$_3^0$ and CoCl$^+$ (Bruland 1983, Burton and Statham 1988, Nozaki 1997). Its extremely low concentration suggests that it is rapidly removed from seawater, probably scavenged by manganese oxides. Ni is present in deep ocean water at concentrations of about 10 nmol kg$^{-1}$, mainly as Ni$^{2+}$, NiCO$_3^0$ and NiCl$^+$, and displays a nutrient-type behavior in seawater. Cu is present in deep ocean water at concentrations of about 6 nmol kg$^{-1}$, mainly as CuCO$_3^0$, CuOH$^+$ and Cu$^{2+}$, and has a distribution intermediate between that of nutrient-type elements and Mn. Zn is present in deep ocean water at concentrations of about 8 nmol kg$^{-1}$, mainly as Zn$^{2+}$, ZnOH$^+$, ZnCO$_3^0$, and ZnCl$^{2+}$, and displays a nutrient-type behavior in seawater.

Trace metals in seawater can be characterized as scavenged-type or nutrients-type elements (Bruland et al. 1994). On this basis, Co is mainly a scavenged-type element, Ni and Zn are nutrient-type elements, Fe displays an intermediate behavior and Mn is a scavenged-type element strongly influenced by redox processes. Nutrient-type elements have much longer deep-sea resi-

dence times than scavenged-type elements. For Zn, this has been estimated to be 22,000-45,000 years, for Fe 70-140 years and for Mn 20-40 years. The residence times for Fe and Mn are short compared to the general oceanic turnover time of about 1500 years (Bender et al. 1977). The rapid removal of Mn from seawater accounts for its significant fractionation between oceanic basins as well as the widespread occurrence of manganese deposits on the deep-sea floor.

## 11.3    Sediments

### 11.3.1    Manganese, Iron and Trace Elements in Deep-Sea Sediments

The distribution of elements in deep-sea sediments has been discussed by Bischoff et al. (1979), Stoffers et al. (1981, 1985), Meylan et al. (1982), Aplin and Cronan (1985), Chester (1990), Glasby (1991) and Miller and Cronan (1994).

Deep-sea sediments cover more than 50% of the earth's surface and consist of carbonates, red clay and siliceous ooze (cf. Chap. 1). On average, red clay covers about 31% of the world's ocean basins but its abundance is much higher in the Pacific (49%) than in the Atlantic (26%) and Indian (25%) Oceans (Glasby 1991). Carbonates act as a diluent for the transition elements in deep-sea sediments because of the low contents of these elements in them and the composition of deep-sea sediments is therefore often presented on a carbonate-free basis.

Red clays are mainly allogenic in origin (Glasby 1991). In the Pacific, this allogenic component is dominantly eolian dust. The high input of dust from the deserts of central Asia explains the higher sedimentation rates of the noncarbonate fraction of sediments from the North Pacific ($0.5\text{-}6$ mm $10^3\text{yr}^{-1}$) than those of sediments from the South Pacific ($0.4$ mm $10^3\text{yr}^{-1}$). Because of its composite origin, red clay has a similar composition to that of average shale. However, a number of elements are enriched in red clay relative to average shale. These include the transition elements (Mn, Co, Ni, Cu) and Ba (Table 11.1). The transition elements constitute the authigenic fraction of the sediments whereas Ba is biogenically introduced into the sediments as barite. The Fe contents of red clay and average shale, on the other hand, are similar. Red clays are enriched in Mn relative to average shale by a factor of 7, in Co by a factor of 4, in Ni by a factor of 3, in Cu by a factor of 5 and in Fe by a factor of 1.4. By contrast, mildly reducing or seasonally oxidizing near-shore sediments do not incorporate an authigenic fraction and the composition of these sediments is similar to that of average shale. Nonetheless, elevated concentrations of manganese in surficial coastal sediments are well documented (Overnell et al. 1996).

The occurrence of an authigenic fraction in red clays reflects the fact that Mn occurs in two valencies and can migrate to regions where more oxidizing conditions prevail. Within red clays, Mn is dominantly in the tetravalent state with an average O:Mn ratio of $1.89 \pm 0.05$. The presence of Mn oxides in the sediments leads to the scavenging of transition elements such as Co, Ni and Cu.

The hemipelagic/pelagic transition is important in defining the nature of red clays. This boundary is a redox transition zone. Red clays form in regions of low sedimentation rate where the rate of organic matter deposition is very low. In these sediments, diffusion of oxygen into the sediments exceeds its rate of consumption in oxi-

dizing organic matter. The sediments therefore remain brown throughout their length and Mn, Co, Ni and Cu are enriched in the authigenic fraction of the sediments. Oxic sediments are characterized by sedimentation rates of <40 mm $10^3\text{yr}^{-1}$ in regions of moderate productivity. In these sediments, iron is present dominantly in the trivalent state and reduction of nitrate in the pore waters by organic carbon does not occur.

An inverse relationship has been observed between the transition metal contents (Mn, Co, Ni and Cu) of red clays and the sedimentation rate. This relationship suggests that these sediments are characterized by a uniform rate of deposition of the authigenic elements superimposed on a variable deposition rate of detrital elements. This model explains the high concentration of authigenic elements in Pacific pelagic clays where sedimentation rates are low. The higher Mn content of red clays from the South Pacific compared to those from the North Pacific reflects the lower rates of sedimentation there. Chemical leaching techniques have been used to identify the forms of elements in red clays. About 90% Mn, 80% Co and Ni and 50% Cu are considered to be of authigenic origin whereas >90% of Fe is thought to be of detrital origin. Fe therefore occurs dominantly in the allogenic phase of these sediments. Table 11.2 suggests that most of the transition elements are delivered to the sediment/water interface associated with large organic aggregates.

Chemical analyses of red clays taken on a series of transects across the Southwestern Pacific Basin have shown that there is a systematic change in the composition of the red clays across the basin. On a transect from the base of the New

**Table 11.1** Comparison of the transition metal and Ba contents of Pacific Pelagic Clay and average shale. Mn and Fe in per cent; Co, Ni, Cu and Ba in ppm (after Glasby 1991).

|     | Pacific Pelagic Clay | Average shale |
| --- | --- | --- |
| Mn | 0.43 | 0.05 |
| Fe | 5.4 | 5.2 |
| Co | 113 | 8 |
| Ni | 210 | 29 |
| Cu | 230 | 45 |
| Ba | 3900 | 250 |

**Table 11.2** Rates of deposition ($\mu$g cm$^{-2}$ 10$^3$yr$^{-1}$) of transition elements from eolian dust and organic aggregates and into red clays. (after Glasby 1991).

|    | Dust | Organic aggregates | Red clay |
|----|------|--------------------|----------|
| Mn | 20-70 | 1565 | 1020-1230 |
| Fe | 3000-5800 | 16175 | 1560-3500 |
| Co | 2.7-4 | 12 | 4-14 |
| Ni | 14-20 | 15 | 20 |
| Cu | 20-30 | 490 | 28-34 |

Zealand continental slope to Rarotonga, Mn was shown to increase from 0.3-1.4%, Fe from 3-8 %, Co from 25-250 ppm, Ni from 50-250 ppm and Cu from 75-325 ppm (Meylan et al. 1982). However, relative to each other, the elements show an enrichment sequence along the transect of Co>Ni>Mn=Cu>Fe. Mössbauer studies also showed a marked increase in the Fe$^{3+}$:Fe$^{2+}$ ratio in the sediments with increasing distance from New Zealand which was attributed to the incorporation of Fe oxyhydroxides, probably ferrihydrite, into the sediments (Johnston and Glasby 1982). By contrast, Fe$^{2+}$ is thought to occur in the sediments mainly in montmorillonite and chlorite. A decrease in sedimentation rate from 32 to 2 mm 10$^3$yr$^{-1}$ was also observed along this transect (Schmitz et al. 1986). The sediments on this transect therefore show a decrease in grain size, an increased darkening of the sediments from pale yellowish brown to dusky brown, increasing transition metal contents, increasing oxidation state of iron, decreasing sedimentation rate and increased abundance of deep-sea manganese nodules with increasing distance from New Zealand. The dusky brown sediments are rich in phillipsite and manganese micronodules (Glasby et al. 1980). Several of the changes along the transect were thought to reflect an increase in the degree of oxidation of the sediments with decreasing sediment accumulation rate caused by a longer contact time of the sediment surface with well-oxygenated ocean bottom water. On a transect from the crest of the East Pacific Rise to New Zealand at 42°S, Stoffers et al. (1985) found similar increases in the contents of Mn, Fe, Co, Ni and Cu in sediments with increasing distance from New Zealand.

In a detailed comparison of sediments from the equatorial North Pacific high productivity zone (Area C) and the low productivity SW Pacific subtropical anticyclonic gyre (Area K), Stoffers et al. (1981) showed that the siliceous oozes from the equatorial N. Pacific have much higher contents of Mn, Ni, Cu and Ba but lower contents of Fe and Co than the red clays from the SW Pacific (Table 11.3). Sedimentation rates on a carbonate-free basis for the two areas are of the same order (1-3 mm 10$^3$yr$^{-1}$ for the equatorial North Pacific and 0.5-1 mm 10$^3$yr$^{-1}$ SW Pacific). Differences in the transition metal contents of these sediments were therefore considered to be controlled by sediment type rather than sedimentation rate. Calculations based on the equations of Bischoff et al. (1979) confirmed that the hydrogenous (authigenic) component was much higher in Area C (7.9%) than in Area K (3.2%) sediments.

In addition to the authigenic component, there may also be a hydrothermal component in deep-sea sediments. During DSDP cruise 92, the variation in composition of sediments from three drill cores and several piston cores taken on a transect away from the crest of the East Pacific Rise was determined (Lyle et al. 1986, Marchig and Erzinger 1986). It was shown that the Mn accumulation rate in sediments falls off rapidly with increasing distance becoming relatively small 1000 km from the ridge crest. At 19-20°S, for example, the Mn accumulation rate in the surface sediments declined from 36 mg cm$^{-2}$ 10$^3$yr$^{-1}$ at the ridge crest to 0.2 mg cm$^{-2}$ 10$^3$yr$^{-1}$ 1130 km away. The corresponding decrease for Fe was from 120 mg cm$^{-2}$ 10$^3$yr$^{-1}$ to 0.68 mg cm$^{-2}$ 10$^3$yr$^{-1}$. In a similar study at 42°S, the hydrothermal component was shown to decline from about

**Table 11.3** Comparison of the transition metal and Ba contents of sediments (on a carbonate-free basis) from Areas C and K. Mn and Fe in per cent; Co, Ni, Cu and Ba in ppm (after Stoffers et al. 1981).

|    | Area C | Area K |
|----|--------|--------|
| Mn | 1.97 | 0.8 |
| Fe | 4.8 | 8.39 |
| Co | 129 | 155 |
| Ni | 677 | 235 |
| Cu | 1044 | 275 |
| Ba | 3921 | 730 |

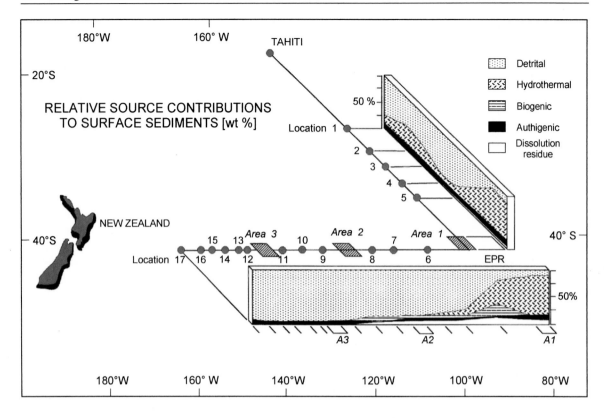

**Fig. 11.5** Distribution of the normative sediment components; weight percent of the five individual components present in the sediments along the Tahiti - EPR - New Zealand transect (after Stoffers et al. 1985)

75% at the ridge crest to zero about 1000 km away (Stoffers et al. 1985) (Fig. 11.5).

### 11.3.2  Diagenetic Processes in Deep-sea Sediments

The nature of the diagenetic changes occurring in pelagic sediments depends on the influx of decomposable organic matter to the sediment and the metabolic rate of oxidation (Müller et al. 1988). Three types of diagenetic processes can be distinguished: oxic diagenesis, suboxic diagenesis and anoxic diagenesis. Oxic diagenesis takes place when oxygen remains in the pore waters as in red clays. Mn concentrations in the pore waters remain extremely low (of the order of 2 µg $l^{-1}$) compared to a concentration of about 0.2 µg $l^{-1}$ in ocean bottom water. Suboxic diagenesis takes place when nitrate reduction occurs in the core and the oxygen content in the pore waters becomes very low. Dissolved Mn concentrations in the pore water can then increase by several orders of magnitude (>1000 µg $l^{-1}$) compared to ocean bottom water. Anoxic diagenesis takes place in

stratified anoxic basins characterized by a well-developed halocline which prevents mixing of the anoxic basinal waters with the overlying seawater as well as in the deeper layers of sediments from well-oxygenated coastal and upwelling environments. This leads to extensive diffusion of Mn and Fe into the water column from the underlying sediments. As an example, the dissolved Mn and Fe contents attain concentrations of <700 and 120 µg $l^{-1}$ respectively in the anoxic waters of the Gotland Deep, Baltic Sea (Glasby et al. 1997). The diagenetic pathway taken depends principally on the rate of organic carbon accumulation in the sediment.

Within red clays, the diffusive flux of Mn is small (23 µg $cm^{-2}$ $10^3yr^{-1}$) and corresponds to about 7% of the total sedimentation flux of Mn (Glasby 1991). Mn is therefore largely immobilized in red clays. By contrast, 96% of the Cu in red clays is regenerated from the sediment into the bottom water as a result of the diagenetic flux across the sediment/water interface. Red clays therefore provide a relatively low flux of Mn and transition elements to the sediment surface and this is not an important source of metals for manganese nodule formation in red clay areas.

Müller et al. (1988) have distinguished between 'deep diagenesis' and 'surficial diagenesis'. In Pacific red clays, deep diagenesis results in no significant net upward flux of Mn, Fe, Ni or Cu. Surficial diagenesis is more significant. In siliceous ooze sediments, the regeneration rate of Mn in the surface sediments is of the same order as the accretion rate of Mn in the associated Mn nodules. Surficial diagenesis is therefore a significant source of metals to manganese nodules in siliceous ooze areas. 96% of the metals in the associated manganese nodules come from this source.

In general, the thickness of the oxidized layer in the sediment increases from near-shore and hemipelagic to pelagic environments. This is illustrated in Figure 11.6 which shows the trends for the eastern equatorial Pacific. This diagram confirms the inverse relationship between the thickness of the oxidized layer in the sediment and the biological productivity in the overlying surface waters. In general, there is a transition from tan to green within these sediments resulting from the *in-situ* reduction of Fe (III) to Fe (II) in smectites at the iron redox transition-zone (Lyle 1983, Köning et al. 1997).

In a comparison of pore water profiles of red clays and hemipelagic sediments, Sawlan and Murray (1983) showed that Mn and Fe are below the detection limit in the pore waters of red clays, Ni is present in the same concentration as in ocean bottom water and Cu shows a pronounced maximum at the sediment-water interface. In hemipelagic clays, denitrification becomes important and remobilization of Mn and Fe takes place. Ni correlates with Mn in the pore waters suggesting that it is associated with the Mn oxides in the solid phase. Cu is regenerated very rapidly at the sediment/water interface. The diffusive flux of Mn in hemipelagic sediments was determined to be in the range 2200-33,000 $\mu g$ $cm^{-2}$ $10^3 yr^{-1}$. More detailed studies of pore water profiles in five different areas of the Californian Borderland confirmed the importance of Mn recycling in the surface sediments when the oxygen content of the bottom waters exceeded 0.1 ml $l^{-1}$ (Shaw et al. 1990). Co and Ni appeared to be scavenged by Mn oxides and trapped in the surface sediments whereas the accumulation of Cu appeared to be more closely related to the flux of biogenic material to the sediment.

Based on a detailed study of pore water profiles in sediment cores from the eastern equatorial Atlantic, Froelich et al. (1979) were able to show that oxidants are consumed in the order of decreasing energy production per mole of organic carbon oxidized ($O_2$>Mn oxides=nitrate>Fe oxides>sulfate). A schematic representation of the profiles is shown in Figure 11.7. From this diagram, it is seen that the reduction and remobilization of Mn takes place in zone 4. This is followed by the upward diffusion and reoxidation of Mn in zone 3. This process enables Mn to be stripped from the sediments as they accumulate and to be redeposited as a discrete layer within the sediment column. An example of this process is given in Figure 3.8. Of course, the depth at which these processes take place is controlled by the influx of organic matter to the sediments which governs the nature of the diagenetic process occurring there. In regions of extremely high productivity characterized by organic-rich hemipelagic sediments (such as found in the Panama Basin), burrowing by macrofauna can markedly increase the rate of recycling of Mn in the bioturbated zone of these

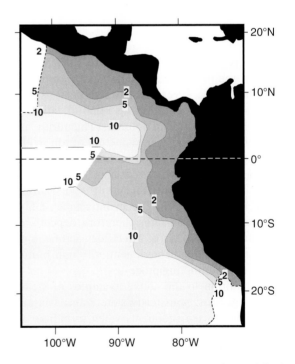

**Fig. 11.6** Variation of the thickness (in cm) of the oxidized surface layer of sediments from the eastern equatorial Pacific (after Müller et al. 1988). In conjunction with the regional distribution of the biological productivity of the surface waters (Fig. 11.12), this pattern indicates that early diagenesis in the sediments is controlled on a regional scale by the input of biological detritus into the sediments.

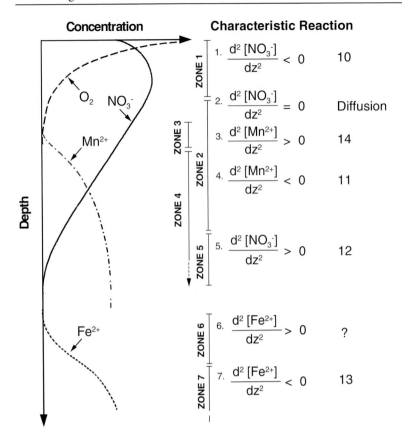

**Fig. 11.7**   Schematic representation of trends in pore water profiles for the principal oxidants in marine sediments. Depths and concentrations in arbitrary units (after Froelich et al. 1979). This pattern reflects the sequence of reduction of the principal oxidants in the sediment column ($O_2$>Mn oxides=nitrate>Fe oxides>sulfate).

sediments (Aller 1990). In this situation, Mn-oxide rich sediments and organic matter are mixed into the anoxic layers of the sediment by bioturbation, thereby permitting remobilization of $Mn^{2+}$ in the sediment column (Thamdrup and Canfield 1996). However, reduction of Mn oxides was shown to play only a minor role in the oxidation of organic carbon in continental margin sediments taken off Chile (Thamdrup and Canfield 1996) (see Chapter 7). The maintenance of high Mn oxide reduction rates therefore depends on the continuous mixing of Mn oxides and fresh organic matter into the sediments through bioturbation or other mixing processes.

In addition to the above examples of steady state diagenesis, non-steady state diagenesis may occur in deep-sea sediments when turbidites are deposited in abyssal plains (Thomson et al. 1987). A color difference is often seen in the upper layers of the turbidite. This is a consequence of the 'oxidation front' in which oxygen from the overlying pelagic sediment, often a carbonate ooze, diffuses down into the underlying turbidite sequence and oxidizes organic carbon there. As a result, oxygen and nitrate are reduced to almost

zero below the front but Mn and Fe are mobilized in the sediment. Mn migrate upwards and is immobilized at the oxidation front. Ultimately, it may be fixed in the sediment as a manganese carbonate. Within a long sediment core, a number of turbidite sequences, and therefore fossil oxidation fronts, may be seen. Other examples of the non-steady state deposition of Mn in deep-sea sediments have been recorded at the glacial/interglacial boundary (Wallace et al. 1988, Gingele and Kasten 1994).

## 11.4    Manganese Nodules and Crusts

There are three principal types of manganese deposits in the marine environment. *Manganese nodules* generally accumulate in deep water (>4000 m) in oceanic basins where the sedimentation rates are low. They usually grow concentrically around a discrete nucleus. Growth occurs mainly at sediment-water interface. *Manganese crusts* accumulate on submarine seamounts and plateaus at depths >1000 m where bottom currents prevent sediment accumulation. They generally form on submarine outcrops. *Ferromanganese concretions* occur in shallow marine environments (e.g. the Baltic and Black Seas) and in temperate-zone lakes. They grow much faster than deep-sea nodules and are quite different in shape, mineralogy and composition from them.

There are also three principal modes of formation of these deposits. *Hydrogenous deposits* form directly from seawater in an oxidizing environment. They are characterized by slow growth (about 2 mm $10^6 yr^{-1}$). Manganese nodules tend to form on red clays and Co-rich manganese crusts on rock substrates. The high Mn/Fe ratio in deep sea water compared to the earth's crust is mainly responsible for the formation of hydrogenous manganese deposits with Mn/Fe ratios of about unity. *Diagenetic deposits* result from diagenetic processes within the underlying sediments leading to upward supply of elements from the sediment column. These deposits are characterized by faster growth rates (10-100 mm $10^6 yr^{-1}$) and are often found on siliceous oozes. *Hydrothermal deposits* precipitate directly from hydrothermal solutions in areas with high heat flow such as mid-ocean ridges, back-arc basins and hot spot volcanoes. They are characterized by high to extremely high growth rates (>1000 mm $10^6 yr^{-1}$) and low to very low trace element contents. They tend to be associated with hydrothermal sulfide deposits and iron oxihydroxide crusts.

### 11.4.1 Deep-Sea Manganese Nodules

Deep-sea manganese nodules occur mainly in deep-ocean basins characterized by low sedimentation rates (i.e. <5 mm $10^3 yr^{-1}$) where inputs of calcareous ooze, turbidity flows and volcanic ash are low. They therefore occur in highest abundances on red clay and siliceous ooze far from land. They occur worldwide and are found in most major oceanic basins. Their distribution is also related to the patterns of oceanic bottom water flow and, to a lesser extent, to availability of potential nuclei on which they grow such as weathered volcanic rock, pumice, whales earbone, sharks teeth, fragments of older nodules and indurated sediment. Antarctic Bottom Water (AABW) is the major oceanic bottom current in the Pacific. Its influence is seen in the lowered sedimentation rates, and therefore increased nodule abundance, along its flow path. A huge litera-

ture on deep-sea nodules has developed over the last 30 years, including a number of standard texts on the subject (Mero 1965, Horn 1972, Glasby 1977, Anon 1979, Bischoff and Piper 1979, Sorem and Fewkes 1979, Cronan 1980, Varentsov and Grasselly 1980, Roy 1981, Teleki et al. 1987, Baturin 1988, Halbach et al. 1988).

In the Pacific Ocean, the distribution of manganese nodules has been mapped as part of the Circum-Pacific Map Project using data from 2500 bottom camera stations and from sediment cores (Piper et al. 1987) (Fig. 11.8). Although considerable variability in nodule abundance within individual stations was observed, high coverage of nodules was recorded in five main regions: between the Clarion and Clipperton Fracture Zones (C-C F.Z.) in the equatorial North Pacific and extending westwards into the northern sector of the Central Pacific Basin, in the abyssal plain area around the Musicians Seamounts in the Northeast Pacific Basin, in the central sector of the Southwestern Pacific Basin, in an E-W trending belt in the Southern Ocean coincident with the Antarctic Convergence, and in the northern sector of the Peru Basin.

In discussing the composition of deep-sea manganese nodules, it should be born in mind that the composition of nodules varies within individual nodules as seen in the discrete microbanding in the nodules, locally (on the scale of hundreds of meters) and regionally (over thousands of km) (e.g von Stackelberg and Marchig 1987). In spite of these limitations, regional patterns in the composition of nodules are commonly observed such that we can reasonably compare and contrast the characteristics of nodules from different physio-graphic provinces of the world ocean as attempted here (cf. Cronan 1977, Piper and Williamson 1977, Sorem and Fewkes 1979).

In the following section, the distribution, mineralogy and composition of manganese nodules from three of these regions is considered in order to illustrate the different modes of formation of nodules in different settings. A detailed comparison of the characteristics of the nodules from these three regions has already been presented by Glasby et al. (1982).

*Southwestern Pacific Basin*

The Southwestern Pacific Basin has an area of $10 \cdot 10^6$ km². It is bounded by New Zealand-Tonga-Kermadec Arc, the East Pacific Rise and

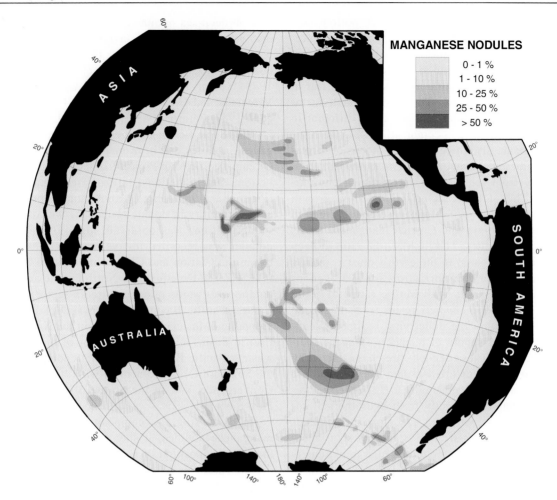

**Fig. 11.8**   Schematic map showing the distribution of manganese nodules in the Pacific Ocean compiled as part of the Circum-Pacific Map Project (Piper et al. 1985). The contours represent the percentage cover of the ocean floor by manganese nodules. Diagram prepared with the permission of D.Z. Piper, USGS.

the Polynesian island chain and has a maximum depth 5800 m. It lies beneath subtropical anticyclonic gyre which is a low productivity area. Two cruises of R.V. Tangaroa were undertaken in 1974 and 1976 to study the distribution and mode of formation of nodules in the Southwestern Pacific and Samoan Basins (Glasby et al. 1980). On a transect from New Zealand to Rarotonga, Cook Islands, it was shown that the maximum abundance of nodules (>20 kg m$^{-2}$) occurs on the dusky brown clays in the region 220-745 km S.W. of Rarotonga. The western sector of the basin, particularly N.E. of New Zealand, was largely devoid of nodules. This was attributed to the influx of terrigenous sediments from the New Zealand landmass which raised the sedimentation rate above the threshold for nodule formation. The morphology of the nodules in the Southwestern

Pacific Basin is somewhat variable but those taken S.W. of Rarotonga are 40% s[S]m, 17% m[S]m and 16% s[E]m (see box for explanation). 72% of these nodules are small (<30 mm), 26% medium (30-60 mm) and 2% large (>60 mm). The nodules are dominantly spheroidal with smooth surface texture when small but become more ellipsoidal and develop equatorial rims with increasing size. The larger nodules also tend to exhibit differences in the surface texture between the upper and lower surfaces. This reflects the fact that the larger nodules have been static at the seafloor for longer than is necessary to form the external layer of the nodule (i.e. their rate of rolling is slower than the rate of growth). Mineralogically, the nodules consist of $\delta MnO_2$, quartz and feldspar. Table 11.4 lists the average composition of nodules from S.W. Pacific Basin and the

346

adjacent Samoan Basin. The average Ni+Cu+Co content of the nodules on the transect from New Zealand to Rarotonga is 1.00%. This is well below the level considered necessary for economic exploitation even though the Southwestern Pacific Basin is thought to contain $10 \cdot 10^9$ tons of nodules. A subsequent E-W transect across the Southwestern Pacific Basin at 42°S undertaken during cruise SO-14 of R.V. Sonne showed that the highest abundances of nodules occur in the far eastern sector of the basin furthest from land (Plüger et al. 1985).

Descriptive classification of manganese nodule types (after Glasby et al. 1980).

Prefix:      (Nodule size based on maximum diameter)

s =   small (<30 mm)
m =   medium (30-60 mm)
l =   large (>60 mm)

Bracketed:   (Primary nodule shape)

S =   speroidal
E =   ellipsoidal
D =   discoidal
P =   polynucleate
T =   tabular
F =   faceted (polygonal)
V =   scoriaceous (volcanic)
B =   biological (shape determined by shark's tooth)

Suffix:      (Nodule surface texture)

s =   smooth
m =   microbotryoidal
b =   botryoidal
r =   rough

Examples:    m[S]m = medium-sized spheroidal nodule with microbotryoidal surface texture.

$l[D]^b_r$ = large discoidal nodule with botryoidal upper surface texture and rough lower surface texture.

s[S-F]s = small spheroidal nodule with significant facing, smooth surface texture.

Southwestern Pacific Basin nodules are considered to be hydrogenous in origin based on their average Mn/Fe ratio of about unity, Ni+Cu contents of <1%, $\delta MnO_2$ as the principal Mn oxide phase and growth rates of 1-2 mm $10^6 yr^{-1}$ (Dymond et al. 1984).

### Clarion-Clipperton Fracture Zone belt

The C-C F.Z. belt is the area bounded by the Clarion F.Z. to north and the Clipperton F.Z. to the south. It is a high productivity area characterized by the occurrence of siliceous sediments. The sedimentation is controlled largely by bottom current activity. The area is dominated by an abyssal hill topography. There is a great variability in nodule distribution and type which depends on the sediment type and accumulation rate, bottom current activity, benthic biological activity (bioturbation), bottom topography, availability of potential nodule nuclei and productivity of the surface waters (Friedrich et al. 1983, Halbach et al. 1988, von Stackelberg and Beiersdorf 1991, Skornyakova and Murdmaa 1992, Jeong et al. 1994, 1996, Knoop et al. 1998).

Within the C-C F.Z., there appear to be two discrete nodule types; larger 'mature' nodules and smaller 'immature' nodules. The larger 'mature' nodules make up 26% of the nodules by number but 92% by weight. The larger nodules are dominantly $m-l[E,D]_b^s$, the so-called hamburger-shaped nodules which have an equatorial rim corresponding to the sediment-water interface. The nodules become flatter with increasing size and have a maximum diameter of 140 mm. The nodules are characterized by hydrogenous growth on the upper surface and diagenetic growth on the lower surface. Figures 11.9 and 11.10 show the principal features of these hamburger nodules. The morphology of the nodules from the slopes of seamounts is s-m[S,P]s-m. The nodules form at the sediment surface and are characterized by hydrogenous growth. The principal minerals present are todorokite and $\delta MnO_2$. Table 11.4 lists the average composition of nodules from the Clarion-Clipperton F.Z. belt. There is an inverse relation between nodule grade and abundance which vary from 3.4% Ni+Cu and 3 kg/m² to 0.3% Ni+Cu and 16 kg m⁻². The C-C F.Z. was the prime target for economic exploitation of manganese nodules (cf. Morgan 1999).

C-C F.Z. nodules may be considered to be diagenetic in origin based on their average Mn/Fe

**Fig. 11.9**   Hamburger-shaped nodule from the Clarion-Clipperton F.Z. region displaying a smooth surface texture on the upper surface and a rough surface texture on the underside (1 [D] $^s_r$). Sample collected at Stn 321 GBH, R.V Sonne cruise SO-25 (7°57.30'N, 143°28.92'W, 5153 m). Photograph courtesy of U. von Stackelberg, BGR.

ratio of about 2.5, Ni+Cu contents of <2.5%, 10 Å manganate as the principal Mn oxide phase and growth rates of 10-50 mm 10$^6$yr$^{-1}$ (Dymond et al. 1984).The influence of oxic diagenesis on nodule formation is the result of low sedimentation rates and low inputs of organic carbon to sediments. The sediments are oxidizing with an $E_H$ of +0.4 V at 8m depth in the sediment column (Müller et al. 1988). Ni and Cu on the undersides of the nodules are ultimately derived from the dissolution of siliceous tests in the sediment (Glasby and Thijssen 1982). Dissolved silicate concentrations in the sediment pore waters may attain values of upto 900 µM. $Ni^{2+}$ and $Cu^{2+}$ substitute in the interlayer spacings of 10 Å manganate on the undersides of the nodules resulting in differences in the compositions of the upper and lower surfaces of nodules (Dymond et al. 1984). In general, the tops of these nodules are enriched in Fe, Co and Pb and the bottoms in Mn, Cu, Zn and Mo (Raab and Meylan 1977).

*Peru Basin*

The Peru Basin is a relatively shallow oceanic basin with depths typically in the range 3900-4300 m

**Table 11. 4**   Comparison of the average compositions of Mn nodules from S.W. Pacific Basin and Samoan Basin (after Glasby et al. 1980), the Clarion-Clipperton Fracture Zone region (after Friedrich et al. 1983) and the Peru Basin (after Thijssen et al. 1985).

|  | S.W. Pacific Basin | Samoan Basin | C-C F.Z. | Peru Basin |
|---|---|---|---|---|
| Mn (%) | 16.6 | 17.3 | 29.1 | 33.1 |
| Fe (%) | 22.8 | 19.6 | 5.4 | 7.1 |
| Co (%) | 0.44 | 0.23 | 0.23 | 0.09 |
| Ni (%) | 0.35 | 0.23 | 1.29 | 1.4 |
| Cu (%) | 0.21 | 0.17 | 1.19 | 0.69 |
| Mn/Fe | 0.73 | 0.88 | 5.4 | 4.7 |

3 cm

**Fig. 11.10** Vertical section of a hamburger-shaped nodule from the Clarion-Clipperton F.Z. region displaying a layered growth structure (l D $^s_r$). The nodule nucleus is a nodule fragment. Sample collected at Stn 178 GBH, R.V Valdivia cruise VA 13/2 (9°22.5'N, 145°53.1'W, 5231 m). Photograph courtesy of U. von Stackelberg, BGR.

which is close to the depth of the Carbonate Compensation Depth (CCD) (4250 m) and lies in a region of high biological productivity. Manganese nodules were initially recovered during cruises SO-04 and SO-11 of R.V. Sonne and investigated in part because of their economic potential (Thijssen et al. 1985). Subsequently, the nodule field was investigated during cruises SO-61, SO-79 and SO-106 as part of the DISCOL program, a long-term, large-scale disturbance-recolonization experiment to assess the potential environmental impact of nodule mining (von Stackelberg 1997).

Peru Basin nodules are quite different from those in other regions of the ocean floor. They tend to be large and mamillated with an average Mn/Fe ratio of 4.7 and an average Ni+Cu content of about 2.1% (Thijssen et al. 1985). Table 11.4 lists the average composition of nodules from the Peru Basin. They are friable and contain fragments of fish bones or broken nodules as nodule nuclei. Nodule density averages >10 kg m$^{-2}$ but

achieves a maximum value of 53 kg m$^{-2}$. The highest abundance of nodules is found near the CCD. Although only 3.8% of the nodules by number are in the >80 mm size class, these nodules make up 56% of the total weight of the nodules. The largest nodule recovered was 190 mm in diameter. The morphology of the nodules is a function of the size class. Nodules <20 mm in diameter are dominantly fragments of larger nodules and less commonly discrete ellipsoidal nodules with botryoidal surface texture. Nodules in the size class 20-60 mm are dominantly ellipsoidal to discoidal with the position of the sediment-water interface clearly defined. The upper surface of these nodules is mamillated with microbotryoidal surface texture whereas the lower surface is less mamillated and smoother. Nodules in the size class >60 mm are dominantly highly mamillated and spheroidal with botryoidal surface texture on the upper surface and smooth to microbotryoidal on the lower surface. Figures. 11.11 and 11.12 show the principal features of these large nodules. For every nodule encountered at the sediment surface, 1.25 nodules occur to a depth of 2.5 m in the sediment column. Buried nodules have the same morphology as surface nodules but different surface texture indicating that material has been removed from the surface of the nodule by dissolution after burial. von Stackelberg (1997) has shown that the redox boundary separating soft oxic surface sediments from stiffer suboxic sediments occurs at a depth of about 100 mm in the sediment column. Highest growth rates occur on the underside of large nodules immediately above the redox boundary. Strong bioturbation occurs throughout the region and helps maintain the nodules at the sediment surface.

Peru Basin nodules are considered to be diagenetic in origin with maximum Mn/Fe ratios of >50, corresponding to Ni+Cu in these samples of <1.4%. 10 Å manganate as the principal Mn oxide phase and growth rates of 100-200 mm $10^6yr^{-1}$ (Dymond et al. 1984). Nodule compositional data for this assessment are taken from Halbach et al. (1980) and Thijssen et al. (1985) and nodule growth rates from Reyss et al. (1985) and Bollhöffer et al. (1996). The influence of suboxic diagenesis on nodule growth is the result of higher sedimentation rates and higher inputs of organic carbon into the sediments. Mn is strongly

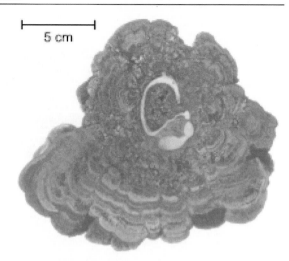

**Figs 11.12**  Vertical section of an asymmetric nodule from the Peru Basin with a cetacean earbone nucleus (l [S] $^r_s$). Sample collected at Stn 63 KG, R.V Sonne cruise SO-79 (6°45.58'N, 90°41.48'W, 4257 m). Photograph courtesy of U. von Stackelberg, BGR.

remobilized by dissolution of Mn micronodules within the sediment column. This leads to the high Mn/Fe ratios and low Ni+Cu contents in the nodules. The nodules are characterized by increasing substitution of $Mn^{2+}$ in the interlayer spacings of 10 Å manganate.

### 11.4.2  Influence of Diagenesis on Nodule Growth

Perhaps the most comprehensive explanation of the role of hydrogenetic and diagenetic processes on manganese nodule accretion has been presented by Dymond et al (1984). On their classification, *hydrogenous deposition* involves the direct precipitation or accumulation of colloidal metals oxides from seawater. Strictly, this involves deposition of manganese oxides on surfaces in contact with seawater such as involved in the formation of manganese crusts. In practice, manganese nodules formed on red clays have characteristics very similar to those of manganese crusts and are therefore considered to be hydrogenous in origin. However, Aplin and Cronan (1985) have suggested that no nodules resting on or in marine sediments can be considered entirely free of diagenetic influences. *Oxic diagenesis* refers to processes occurring within oxic sediments. Decomposition and oxidation of labile organic matter and the dissolution of labile biogenic components such as siliceous tests may release bio-

**Figs 11.11**  Sperical nodule with a cauliflower-shaped surface from the Peru Basin displaying a rough surface texture on the upper surface and a smooth surface texture on the underside (l [S] $^r_s$). Sample collected at Stn 25 KG, R.V Sonne cruise SO-79 (6°51.99'N, 90°26.52'W, 4170 m). Photograph courtesy of U. von Stackelberg, BGR.

logically-bound metals into the sediment pore waters which may ultimately be incorporated into the nodules. Dissolution of siliceous tests may also introduce silica into the sediment pore waters which may react with amorphous ferromanganese oxides to form nontronite (Dymond and Eklund 1978). This process fixes silica and Fe in the sediment column and releases transition elements such as Mn, Co, Ni, Cu and Zn into the sediment pore waters for incorporation into the nodules. *Suboxic diagenesis* involves the reduction of Mn (IV) to Mn (II) within the sediment column and then reoxidation of Mn (II) to Mn (IV) during the formation of the nodules. These diagenetic processes are controlled by redox processes within the sediment column and are therefore dependent on the depth in the sediment column. In most cases (as in the Peru Basin), these nodules are formed in oxic sediments overlying reducing sediments. The diagenetic supply of metals to the nodules normally takes place at or near the sediment surface. Because the metals are released within the sediment column and migrate upwards, they tend to be incorporated on the underside of nodules. This frequently results in differences in composition between the upper and lower surfaces of diagenetic nodules. Manganese micronodules may play a key role in retaining transition elements in the sediment column until they are released on further burial of the micronodules. Diagenetic processes may therefore involve recycling of elements within the sediment column prior to their incorporation in nodules. The characteristics of each of these nodule types has been given in the

previous sections. In addition to influencing the mineralogy and composition of the nodules, diagenetic processes also influence their surface texture; hydrogenous nodules tend to have smooth surface texture and diagenetic nodules botryoidal to rough surface texture.

From the above comments, it will be seen that the principal factor driving the diagenetic milieu in the sediments is the biological productivity of the oceanic surface waters. A map of biological productivity of Pacific ocean surface waters shows that the productivity of the northern sector of the Southwestern Pacific Basin is 50-100 gC m$^{-2}$ yr$^{-1}$, in the C-C F.Z. is in the range 100-150 gC m$^{-2}$ yr$^{-1}$ and in the Peru Basin is 100-150 gC m$^{-2}$ yr$^{-1}$ (Fig. 12.5) (cf. Cronan 1987, 1997). The importance of the productivity of the surface waters to the diagenetic component of manganese nodules in the C-C F.Z. is well illustrated in Figure 11.13 which shows that the diagenetic component of the nodules increases as the equatorial zone of high productivity is approached (cf. Morgan 1999).

The influence of productivity on nodule composition is perhaps best illustrated by the hyperbolic regression curve of Cu and Ni against Mn/Fe for nodules from C-C F.Z. and Peru Basin (Fig. 11.14). This diagram can be divided into three parts. Hydrogenous nodules deposited entirely from seawater have Mn/Fe ratios of about unity and low Ni+Cu contents. Nodules influenced by oxic diagenesis have Mn/Fe ratios of upto 5 (but more typically about 2.5) with correspondingly high Ni+Cu contents. Nodules influenced by suboxic diagenesis have Mn/Fe ratios of

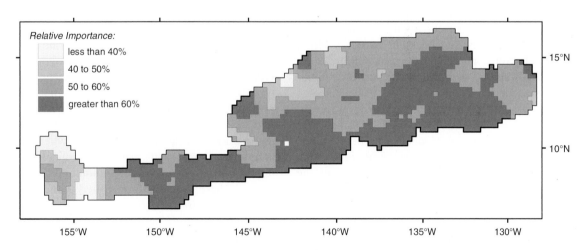

**Fig. 11.13** Contour plot of the diagenetic accretion compositional end member of manganese nodules which shows that this end member increases in relative proportion as the equatorial zone of high productivity is approached as a consequence of the increase in both the primary productivity and sedimentation rate (after Knoop et al. 1998).

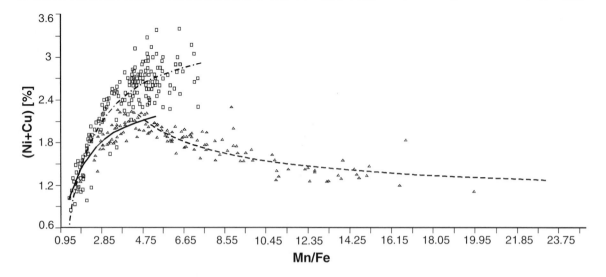

**Fig. 11.14** Hyperbolic regression curves of Ni+Cu against Mn/Fe for nodules from the Clarion-Clipperton F.Z. region (upper curve) and Peru Basin (lower curve) (after Halbach et al. 1981).

upto 50 but lower Ni+Cu contents. The maximum Ni+Cu contents of nodules corresponds to a Mn/Fe ratio of about 5 and is found at the so-called point of reversal.

For this purpose, we may therefore consider hydrogenous deep-sea nodules to be a baseline. When oxic diagenesis takes place, Mn, Ni and Cu are released into the sediment pore waters and are ultimately incorporated into the nodules. $Ni^{2+}$ and $Cu^{2+}$ substitute in the phyllomanganate lattice (see below). This explains the high Mn/Fe ratios and Ni+Cu contents of oxic nodules. When suboxic diagenesis takes place, $Mn^{2+}$ is remobilized into the reducing sediments and migrates upwards towards the sediment surface where it is either oxidized to Mn (IV) at the nodule surface or substitutes in the phyllomanganate lattice as $Mn^{2+}$. This behavior explains both the very high Mn/Fe ratios of suboxic nodules and their correspondingly low Ni+Cu contents. The boundary between oxic and suboxic diagenesis is marked by a Mn/Fe ratio in the nodules of about 5. The nature of the diagenetic processes occurring in the sediments depends directly on the input of organic carbon and therefore on the productivity of the overlying surface waters.

Based on a study of manganese nodules taken on the Aitutaki-Jarvis Transect in the S.W. Pacific, Cronan and Hodkinson (1994) also demonstrated the influence of the CCD on nodule composition. Above the CCD, accumulation of carbonate tests was thought to dilute the organic carbon in the sediments, thus inhibiting diagenesis, whereas, below the CCD, the decay of organic material in the water column was thought to reduce its effectiveness in driving diagenetic reactions. Cronan and Hodkinson (1994) indeed found that Mn, Ni, Cu and Zn were most concentrated in nodules taken near the CCD at the north of the transect in an region of high productivity and least concentrated in nodules taken away the CCD at the south of the transect in a region of low productivity. They therefore inferred that diagenetic cycling of Mn, Ni, Cu and Zn to the nodules is enhanced as a result of the decay of organic material near the CCD.

On a more local scale, the influence of diagenetic processes on nodule morphology and composition has been related to variations in sedimentation rate in areas of hilly topography within the C-C F.Z. (von Stackelberg and Marchig 1987, von Stackelberg and Beiersdorf 1991). On the flanks of hills and in parts of basins, sedimentation rates are lower and dominantly diagenetic nodules are formed whereas, in areas of sediment drift, sedimentation rates are higher and dominantly hydrogenous nodules are formed. These processes are related to the degree of decomposition of organic matter within the sediment column. Where sedimentation rates are low, organic matter is rapidly consumed within the sediment releasing $Mn^{2+}$ into the pore water and resulting in the formation of dominantly diagenetic nodules. In areas of sediment drift, on the other hand, the organic carbon is buried within the sediment stimulating bioturbation and the resultant biogenic lifting of the nodules to the sediment surface.

In addition to the above, Calvert and Piper (1984) have proposed a diagenetic source of metals derived from sediments far away for nodules occurring in an erosional area with thin sediment cover in the C-C F.Z. The metals were thought to be transported to the site of deposition by oceanic bottom water, probably AABW. Although this process is well known in shallow-water continental margin areas such as the Baltic Sea, only limited evidence to support this hypothesis has been presented for the deep-sea environment.

### 11.4.3  Rare Earth Elements (REE) as Redox Indicators

A key parameter in understanding nodules formation is the redox milieu of the environment at the time of deposition. This can influence the mineralogy and therefore composition of the nodules (see below). In this regard, Glasby (1973) used the Ce/La ratio of nodules as a redox indicator. He showed that deep-sea nodules from the NW Indian Ocean had much higher Ce/La ratios (4.4) and $\Sigma$REE contents (490 ppm) than those from shallow-water continental margin environments such as Loch Fyne, Scotland (1.9 and 29 ppm, respectively). The higher REE contents of the deep-sea nodules were taken to reflect the more oxidizing conditions in the deep sea which facilitated the oxidation of the trivalent $Ce^{3+}$ in seawater to $CeO_2$ on the surface of the nodules. The lower $\Sigma$REE contents of the shallow-water continental margin concretions was taken to reflect the remobilization of Mn in the sediment column which resulted in the fractionation of Mn from Fe, Co, Ni, Cu and REE in the concretions. It was suggested that the REE were adsorbed from seawater onto colloidal FeOOH in the nodules. The high rate of manganese remobilization in the shallow-water continental margin sediments also resulted in growth rates of the associated concretions two orders of magnitude higher than those of the deep-sea nodules (Ku and Glasby 1972).

Based on these ideas, Glasby et al. (1987) attempted to use the Ce/La ratios of deep-sea manganese nodules as a means of tracing the flow path of the AABW in the S.W. and Central Pacific Ocean. On this basis, the Ce/La ratios in the deep-sea nodules should decrease along the flow path of the AABW as the oxygen in the bottom waters is consumed. The data confirmed that the $\Sigma$REE contents of the nodules are strongly correlated with Fe resulting in higher REE contents in nodules from the Southwestern Pacific Basin compared to those from the C-C F.Z. (La 169 v 93 ppm). They also showed a systematic decrease in the Ce/La ratios of the nodules from the Southwestern Pacific Basin to the C-C F.Z. and on to the Peru Basin as follows: Area III (9.6) > Area K (8.0) > Area C (3.7) > Peru Basin (2.0) > Area G (1.5) > Area F (1.4). The Peru Basin and Areas G and F are situated in the equatorial South Pacific in regions which do not lie directly under the flow path of the AABW and which are characterized by sluggish bottom water flow (Nemoto and Kroenke 1981, Davies 1985). This factor explains the low Ce/La ratios of these samples. The sequence of these ratios confirms that the Ce/La ratio of deep-sea manganese nodules can be used to trace the flow path of the AABW and therefore that the Ce/La ratios of manganese nodules can be used as a redox indicator in the deep-sea environment. Subsequent work on the REE contents of manganese nodules taken on an E-W transect across the Southwestern Pacific Basin at 42°S showed that the highest Ce/La ratios and $\Sigma$REE contents of the nodules occur in the central part of the basin which is exposed to the strongest flow of AABW (and therefore the most oxygenated conditions) (Kunzendorf et al. 1993).

More recently, Kasten et al. (1998) have determined the REE contents of a suite of manganese nodules from the South Atlantic to see if a similar relationship between the Ce/La ratios of the nodules and AABW flow could be observed there. In fact, the Ce/La ratios of the nodules decrease in the sequence Lazarev Sea, Weddell Sea (10.4 and 9.7) > East Georgia Basin (6.5-7.1) > Argentine Basin (5.0) but then increase in the Brazil Basin (6.2) and Angola Basin (9.8 and 15.1). A further decrease was observed in the Cape Basin (7.6). In addition, an extremely high Ce/La ratio of 24.4 had already been determined for nodules sampled north of the Nares Abyssal Plain in the North Atlantic. These data reflect the more complicated pattern of bottom water flow in the Atlantic compared to the Pacific Ocean. It is believed that the penetration of more oxygenated North Atlantic Deep Water (NADW) into the South Atlantic accounts for the higher Ce/La ratios in the nodules from the Angola and Brazil Basins. The influence of AABW could therefore be traced only as far north as the Argentine Basin. These data confirm the potential importance of REE studies in deep-sea manganese nodules in tracing oceanic bottom water flow.

### 11.4.4   Co-Rich Mn Crusts

Co-rich Mn crusts may be defined as hydrogenous manganese crusts having a Co content >1% (Manheim 1986, Mangini et al. 1987). These crusts are typically 5-100 mm thick and occur on older seamounts (100-60 Ma) in many of the seamount chains in the equatorial Pacific such as the Mid-Pacific Mountains and Line Islands. The crusts are commonly found on exposed rock on seamount slopes or on the summits of oceanic plateaus at water depths of 3000-1100 m. Principal substrates include basalts, hyaloclastites, indurated phosphorite and claystone. A vertical section of a Co-rich Mn crust is shown in Fig. 11.15. On plateaus and flat terraces where fragments of rock or manganese crusts have accumulated, manganese nodules may be seen lying on the surface of calcareous ooze. Ripple marks indicating the presence of strong bottom currents are sometimes observed on this ooze. The average composition of crusts from the 1500-1100 m depth zone in the Mid-Pacific Mountains has been reported as Mn 28.4%, Fe 14.3%, Co 1.18%, Ni 0.50%, Cu 0.03%, Pt 0.5 ppm, Mn/Fe 2.0. $\delta MnO_2$ is the principal manganese mineral present and they have a growth rate of 1-2 mm $10^6 yr^{-1}$. The crusts have attracted economic interest as a potential source of Co and, to a lesser extent, Pt. The areas in which these crusts generally form lies well above the CCD. Crusts therefore tend to form in regions of strong bottom current activity, perhaps

influenced by AABW flow, which can prevent the deposition of calcareous ooze by erosion.

The variation in the composition of manganese crusts from the Mid-Pacific Mountains with water depth is presented in Table 11.5. These data show that these crusts are hydrogenous in origin (based on their Mn/Fe ratios) but they tend to have much higher Co and lower Cu contents than deep-sea nodules from the same region. The Mn/Fe ratios, Co and Ni contents are highest but the Cu contents lowest in crusts in the depth range 1900-1100 m. The positive correlation of Mn, Co and Ni reflects that association of these elements in $\delta MnO_2$. Overall, the Pt content of the crusts is very high, in the range 0.2-1.2 ppm with an average of 0.5 ppm.

Halbach and Puteanus (1984) showed that the dissolution of calcareous tests in the water column plays a key role in the incorporation of Fe into these crusts. The calcareous tests contain about 500 ppm Fe. The flux of Fe to the surface of the crusts derived from the release of colloidal Fe oxyhydroxide particles on dissolution of the calcareous tests was estimated to be about 15 $\mu g\ cm^{-2}yr^{-1}$ which is almost equivalent to the flux of Fe in the concretions of 22.4-44.8 $\mu g\ cm^{-2}yr^{-1}$. The rate of incorporation of Fe into the crusts is therefore related to the position of the lysocline.

The highest Co contents are found in crusts occurring in the depth range 1900-1100 m which corresponds to the depth of the oxygen minimum zone. In samples taken from summits and the upper parts of slopes at depths less than 1500 m, the Co contents of the crusts sometimes exceed 2%. At this depth, Mn tends to remain in solution because the deposition of $MnO_2$ is not favored at these low oxygen contents. The deposition rate

**Table 11.5**   Variation of the composition of manganese crusts from the Mid-Pacific Mountains with water depth (after Mangini et al. 1987). Analyses expressed as wt. % of dried material.

| water depth (m) | Mn | Fe | Co | Ni | Cu | Mn/Fe |
|---|---|---|---|---|---|---|
| 1100 - 1500 | 28.4 | 14.3 | 1.18 | 0.50 | 0.03 | 1.99 |
| 1500 - 1900 | 24.7 | 15.3 | 0.90 | 0.42 | 0.06 | 1.61 |
| 1900 - 2400 | 25.5 | 16.1 | 0.88 | 0.41 | 0.07 | 1.58 |
| 2400 - 3000 | 20.5 | 19.5 | 0.69 | 0.18 | 0.09 | 1.05 |
| 3000 - 4000 | 20.5 | 18.0 | 0.63 | 0.35 | 0.13 | 1.41 |
| 4000 - 4400 | 19.7 | 16.7 | 0.67 | 0.24 | 0.10 | 1.17 |

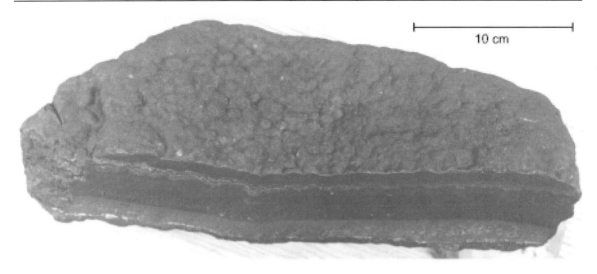

10 cm

**Fig. 11.15** Vertical section of a Co-rich Mn crust from the flanks of a guyot on the Ogasawara Plateau; N.W. Pacific (25°18.9'N, 143°54.8'E; 1515 m) collected by dredge during a cruise of the GSJ with R.V. Hakurei-maru in 1986. The substrate (not clearly seen) is phosphatized limestone. The crust is more than 20 cm across, about 10 cm thick and has a knobby surface texture. Element contents are: Mn 21.7%, Fe 18.9%, Co 0.81%, Ni 0.31%, Cu 0.04%, Pb 0.26%Pb and Pt 0.29 ppm. The upper layer of the crust displays the highest Co content and the bottom layer the highest Pt (0.78 ppm). $\delta MnO_2$ is the principal mineral present with minor quartz and plagioclase. Photograph courtesy of A. Usui, GSJ.

of $MnO_2$ is therefore at a minimum and the Mn content in seawater at a maximum of about 2 nmol kg$^{-1}$. The flux of Co to the surface of the crusts remains constant with water depth at about 2.9 µg cm$^{-2}$ 10$^{-3}$yr$^s$. The Co content of crusts is therefore a maximum just below the oxygen minimum zone and no special source of Co is required (Puteanus and Halbach 1988).

Pt also shows a significant enrichment in Co-rich crusts (Hein et al. 1988, 1997, Halbach et al. 1989). However, the mechanism of enrichment in the crusts is not well understood. Hodge et al. (1985) argued that Pt is oxidized from the $PtCl_4^{2-}$ state in seawater to the tetravalent state in manganese nodules. This process was thought to be responsible for the anomalously high Pt/Pd ratios in nodules (50-1000) compared to seawater (4.5). However, Halbach et al. (1989) considered that this process would not be possible because the first formed tetravalent species, $PtCl_6^{2-}$, would be very stable in oxygenated seawater. Instead, they proposed that the $PtCl_4^{2-}$ would be reduced to Pt metal in the crusts with some minor amounts of Pt being introduced from cosmic spherules. $Mn^{2+}$ was assumed to be the reducing agent for the reduction of the $PtCl_4^{2-}$. This uncertainty argues for a closer study of the distribution of the platinum group elements (PGE) in marine manganese deposits.

For Cu, the $E_H$-pH diagram shows that the boundary between $Cu^{2+}$ and $CuCl_3^{2-}$ as the dominant species in seawater lies slightly above the normal range of redox conditions in seawater ($E_H$ of +0.48 V) (Fig. 11.16). At an $E_H$ of +0.4 V, the concentration of $Cu^{2+}$ in seawater would still be sufficient for it to be incorporated into manganese nodules by sorption on the surface of negatively charged $MnO_2$ (Glasby 1974, Glasby and Thijssen 1982). At somewhat lower $E_H$ values, however, the concentration of $Cu^{2+}$ in seawater would decline drastically and the anionic species $CuCl_3^{2-}$ would be the dominant species. Its sorption on $MnO_2$ would be inhibited by charge considerations. This may well explain the high Ni/Cu ratios observed in cobalt-rich manganese crusts (max. 15) formed adjacent to the oxygen minimum zone where less oxidizing conditions prevail (Mangini et al., 1987, Meylan et al., 1990).

In many crusts, two generations of crustal growth can be observed, an older crust (18-12 m.yrs B.P.) and a younger crust (<12 m.yrs B.P.) which are separated by a thin phosphorite horizon (e.g. Mangini et al. 1987, Halbach et al. 1989). Differences in the composition of these two layers depend on variations in paleoceanographic conditions at the seamount with time. These include changes in oceanic productivity, depth of the lysocline, rate of $CaCO_3$ dissolution and

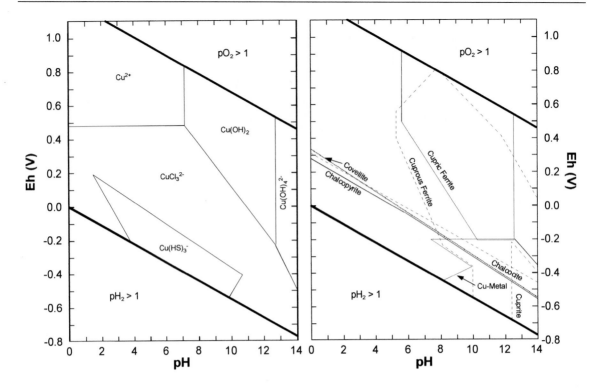

**Fig. 11.16** $E_H$, pH diagram for Cu calculated for the chemical conditions prevailing in the deep sea (after Glasby and Schulz in press). Note that, at pH 8, the metastable form of Cu is $Cu(OH)_2$ at an $E_H$, > +0.48 V. However, if the $E_H$, drops to +0.4 V corresponding to the $E_H$ of seawater, the anionic species, $CuCl_3^{3-}$, becomes the more stable species. It is believed that the dominance of $CuCl_3^{3-}$ in seawater in the oxygen minimum zone accounts for the low Cu contents in Co-rich Mn crusts.

strength of AABW flow. However, McMurtry et al. (1994) have disputed the Miocene maximum ages for seamount Mn crusts obtained by these authors and proposed a Cretaceous of almost 80 m.yrs B.P. for the same crust.

In order to be of potential economic interest, crusts should have Co contents >0.8%, average crustal thicknesses >40 mm and be situated in an area of subdued small-scale topography. Based on their extensive experience of studying Co-rich Mn crusts, Hein et al. (1987) listed a number of other criteria for locating interesting crusts. The crusts should be situated on large volcanic edifices shallower than 1500-1000 m and older than 20 m.yrs B.P. occurring in areas of strong oceanic bottom currents. The volcanic structure should not be capped by large modern atolls or reefs and the seamount slopes should be stable. The area should not be influenced by the input of abundant fluvial or eolian debris and there should be no local active volcanism. Most importantly, the area should be characterized by a shallow and stable oxygen minimum zone. In particular, the importance of mass wasting of coral debris from atolls

and guyots in tropical environments is stressed. In a detailed sampling program around islands and seamounts in the area of the Manihiki Plateau in the equatorial South Pacific, Meylan et al. (1990) recovered only thin Mn crusts with a maximum thickness of 20 mm because of the extensive mass wasting of limestone debris in the region which gave insufficient time for thick crusts to develop.

### 11.4.5   Shallow-Marine Ferromanganese Concretions

Shallow-water continental margin ferromanganese concretions have been reported in a number of areas such as the Baltic Sea, Black Sea, Kara Sea, Loch Fyne, Scotland, and Jervis Inlet, British Columbia (Calvert and Price 1977). In fact, these were the first marine Mn deposits to be discovered, during the 1868 Sofia expedition to the Kara Sea led by A.E. Nordenskiøld (Earney 1990).

Ferromanganese concretions from the Baltic Sea have been described in detail by Glasby et al. (1997). Three main types of concretions occur there (spheroidal, discoidal and crusts). The concretions frequently, although not always, form around a glacial erratic seed or nucleus and display alternate banding of Fe- and Mn-rich layers tens to hundreds of µm thick. Mineralogically, the concretions consist of 10 Å manganate with abundant quartz and lesser amounts of feldspar and montmorillonite. The composition of the concretions is highly variable reflecting, in part, the variable amount of erratic material in the concretion. The composition is controlled by the redox characteristics of the environment. In particular, the higher Mn/Fe ratios of concretions from Kiel Bay compared to those from other areas of the Baltic reflects the diagenetic remobilization of Mn from the adjacent muds into the concretions during the summer anoxia (Table 11.6). Cu, Zn and Pb may be trapped as sulfides in the associated sediments and this may explain their low contents in the concretions compared to deep-sea nodules. The growth rates of the concretions have been estimated to be 3-4 orders of magnitude higher than those of deep-sea nodules. A 20 mm diameter concretion is therefore about 500-800 years old (Glasby et al. 1996). This means that these concretions are a transient feature on the sea floor. In active areas of the sea floor, concretions may be buried by sediment during storms and new concretions then begin to form on erratic material exposed at the sediment surface. Certain anthropogenic elements, notably Zn, are enriched in the outer layers of the concretions as a result of the pollution of Baltic seawater over the last 160 years or so. Zn profiles in the outer layers of concretions can be used to monitor heavy-metal pollution in Baltic (Hlawatsch 1998). This technique will be useful if the planned clean up of the Baltic over the next century is to be achieved.

Baltic Sea concretions can be classified into three main types based on their abundance, morphology, composition and mode of formation: those from the Gulfs of Bothnia, Finland and Riga, from the Baltic Proper and from the western Belt Sea.

Concretions from the Gulf of Bothnia are most abundant in Bothnian Bay where the abundance reaches 15-40 kg m$^{-2}$ in an area of about 200 km$^2$. This is equivalent to about 3 million tons of concretions and has led to these deposits being evaluated as a possible economic source of Mn. These concretions are mainly spheroidal upto 25-30 mm in diameter and are formed in the uppermost water-rich sediment layers at well-oxidized sites. They are most abundant where sedimentation rates are <0.4 mm yr$^{-1}$.

Concretions from the Baltic Proper are found mainly around the margins of the deep basins in a depth range 48-103 m. The concretions are mainly discoidal 20-150 mm in diameter and crusts. Their abundance is mainly sporadic and more rarely common to abundant. Locally, abundances of 10-16 kg m$^{-2}$ are attained. Their formation is the result of the build up of Mn and Fe in the anoxic waters of the deep basins of the Baltic Proper. During major inflows of North Sea water (>100 km$^3$) into the Baltic which occur on average once every 11 years, the anoxic waters are flushed out of the basins. Mn and Fe precipitate out as an unstable gel but are ultimately incorporated into the concretions as oxyhydroxides. The

Table 11.6   Average composition of ferromanganese concretions from various basins of the Baltic Sea. Mn and Fe in per cent; Co, Ni, Cu and Zn in ppm (after Glasby et al. 1997).

|       | Gulf of Bothnia | Gulf of Finland | Gulf of Riga | Gotland Region | Gdansk Bay | Kiel Bay |
|-------|-----------------|-----------------|--------------|----------------|------------|----------|
| Mn    | 14.6            | 13.3            | 9.7          | 14             | 8.7        | 29.3     |
| Fe    | 16.6            | 19.7            | 22.8         | 22.5           | 18.5       | 10.1     |
| Co    | 140             | 96              | 64           | 160            | 91         | 77       |
| Ni    | 260             | 35              | 47           | 750            | 148        | 97       |
| Cu    | 80              | 9               | 17           | 48             | 42         | 21       |
| Zn    | 200             | 113             | 135          | 80             | 137        | 340      |
| Mn/Fe | 0.88            | 0.68            | 0.43         | 0.62           | 0.47       | 2.9      |

concretions occur mainly on lag deposits in the vicinity of the halocline where strong bottom currents occur.

Concretions from Kiel Bay in the western Belt Sea occur in a narrow depth range of 20-28 m at the boundary between sands and mud in zones of active bottom currents. They occur as coatings on molluscs and as spheroidal and discoidal concretions. The formation of the concretions is influenced by the development of summer anoxia which leads to the diagenetic remobilization and lateral transport of Mn. This accounts for the high Mn/Fe ratios of these concretions.

### 11.4.6   Hydrothermal Manganese Crusts

At mid-ocean ridges, three types of submarine hydrothermal minerals are found, sulfide minerals associated with silicates and oxides, sharply fractionated oxides and silicates of localized extent and widely dispersed ferromanganese oxides. The ferromanganese oxides are generally considered to have precipitated last in this sequence and are thought to represent a late-stage, low-temperature hydrothermal phase with temperatures of deposition estimated to be in the range 20-5°C (Burgath and von Stackelberg 1995). The hydrothermal Mn deposits are characterized by high Mn/Fe ratios and low contents of Cu, Ni, Zn, Co, Pb and detrital silicate minerals. They have growth rates exceeding 1000 mm $10^6$yr$^{-1}$ in some cases, more than three orders of magnitude faster than that of hydrogenous deep-sea nodules and crusts.

Compared to hydrogenous Mn nodules and crusts, hydrothermal Mn crusts are relatively restricted in the marine environment and make up less than 1% of the total Mn deposits in the world ocean. These crusts occur in all types of active oceanic environments such as at active mid-ocean spreading centers in the depth range 250-5440 m, in back-arc basins in the depth range 50-3900 m, in island arcs in the depth range 200-2800 m (Eckhardt et al. 1997), in mid-plate submarine rift zones in the depth range 1500-2200 m (Hein et al. 1996) and at hot spot volcanoes in the depth range 638-1260 m (Eckhardt et al. 1997).

Submarine hydrothermal Mn crusts have been reported from Enareta and Palinuro seamounts in the Tyrrhenian Sea. (Eckhardt et al. 1977), along the Izu-Bonin-Mariana Arc (Usui and Terashima

**Fig. 11.17**   A photograph showing the upper surface of a hydrothermal Mn crust from the Tyrrhenian Sea (39°31.40'N, 14°43.67'E, 996 m). Part of the surface displays the characteristic black metallic sheen of hydrothermal crusts (after Eckardt et al. 1997).

1997, Usui and Glasby 1998) and at the Pitcairn hotspot (Glasby et al. 1997). The Tyrrhenian Sea crusts consist of porous, black, layered Mn oxides up to 45 mm thick. In some cases, the surface has a black metallic sheen. The crusts overlie substrates such as calcareous sediment, siltstone and oyster shells. A photograph of a hydrothermal Mn crust from the Tyrrhenian Sea is shown in Figure 11.17. They consist dominantly of 10Å manganate and 7Å manganate with minor quartz, illite, montmorillonite, plagioclase and goethite. The sample having the highest Mn content contained 54.2% Mn, 0.07% Fe, 33 ppm Ni, 200 ppm Cu, 20 ppm Zn, 11 ppm Pb and 910 ppm Ba with a Mn/Fe ratio of 774. It also had a low REE abundance and a negative Ce anomaly.

Submarine hydrothermal Mn crusts are also relatively common along the Izu-Bonin-Mariana Arc. Recent hydrothermal manganese crusts associated with active hydrothermal systems tend to occur on seamounts or rifts located about 5-40 km behind the volcanic front on the Shichito-Iwojima Ridge. Fossil hydrothermal Mn crusts associated with inactive hydrothermal systems occur on seamounts located on older ridges running parallel to the volcanic front in both forearc and backarc settings. Fossil hydrothermal Mn crusts are generally overlain by hydrogenetic Mn oxides. The

thickness of the overlying hydrogenetic Mn crust depends on the length of time since hydrothermal activity ceased. Figure 11.18 shows the distribution of recent and fossil submarine hydrothermal Mn crusts across the Izu-Bonin Arc. Mineralogically, the hydrothermal Mn deposits consist of 10 Å manganate and/or 7 Å manganate. The Mn/Fe ratios of these deposits range from 10 to 4670 and the contents of Cu, Ni, Zn, Co and Pb from 20 to 1000, 1 to 1403, 1 to 1233, 6 to 209 and 0 to 93 ppm, respectively.

At the Pitcairn hotspot, massive hydrothermal Mn crusts display the highest Mn/Fe ratios (2440), the lowest contents of Ni (18ppm) and Zn (21) as well as the lowest aluminosilicate fraction (<1%) in individual horizons (sample 69-3 DS at a depth of 7-8 mm). This type of crust may therefore be considered to represent the most extreme hydrothermal endmember. The other types of hydrothermal Mn crust are probably formed as a result of the interpenetration and replacement of volcanoclastic sands or biogenic carbonates by hydrothermal Mn oxides. The low contents of Fe, Ni and Zn in the massive crusts were thought to reflect the rapid incorporation of these elements into sulfide minerals within the interior of the hotspot volcano such that only a small proportion of these elements are available for incorporation

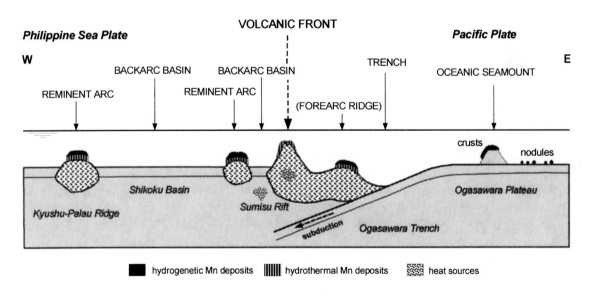

**Fig. 11.18**  A schematic diagram showing the distribution of hydrogenous and hydrothermal Mn deposits across an E-W section of the northwest Pacific island arc, south of Japan (after Usui and Terashima 1997). Four types of deposit are observed: i. hydrogenous Mn crusts and nodules occurring on Pacific seamounts and in deep-sea basins, ii. hydrogenous crusts on marginal seamounts of the remnant arcs, iii. modern hydrothermal Mn deposits associated with submarine volcanoes and backarc rifts in the active volcanic ridge, and iv. fossil hydrothermal Mn deposits usually overlain by younger hydrogenous Mn crusts from seamounts of the remnant arc. Each type of deposit has its own characteristic mineralogy, composition and growth rate. The growth of these deposits is closely related the evolution of the island arc system.

in the crusts. The small positive Eu anomaly in the crusts on a NASC-normalized basis indicated lower temperatures of the hydrothermal fluids within the hotspot volcano (<250°C) compared to those at mid-ocean ridges (c.350°C). A laser-ablation ICP-MS profile in one of the crusts revealed varying REE concentrations and patterns in the different layers of the crust. These data showed that the hydrothermal component was variable during the formation of the crust and was almost 100% in the upper layers of this crust but about 80% in the lower layers.

## 11.4.7  Micronodules

Manganese micronodules are an important reservoir for Mn and associated transition elements in oxic deep-sea sediments (Stoffers et al. 1984). They are generally <1 mm in diameter. By contrast, macronodules are generally >3-6 mm. There is therefore a size gap in which no Mn deposits are found. This is a consequence of the fact that macronodules require a discrete nucleus several mm in diameter on which to form (Heath 1981). The abundance of micronodules appears to be inversely related to the sedimentation rate with highest abundances occurring in dark brown clays where sedimentation rates are low. The shape and internal structure of the micronodules is dependent on the nature of the seed material around which they form, of which calcareous tests, siliceous tests and volcaniclastic material are the most important. Micronodules tend to predominate in the coarse fraction of the sediment. The mineralogy and composition of the micronodules tend to be similar to that of the associated nodules, although differences do exist. Micronodules tend to have higher Mn/Fe ratios than nodules from the same sediment core. These differences are related to the depth of occurrence of the micronodules within the sediment column and reflect the importance of diagenetic processes in remobilizing elements within the sediment. In this sense, micronodules may be considered a transient feature which serve as a medium for recycling elements within the sediment and ultimately into manganese nodules. The study of micronodules has received relatively little attention because of the difficulty of hand picking them from the sediment.

In a detailed study of micronodules from various areas of the Pacific, Stoffers et al. (1984) showed that micronodules are characterized by regional variations in shape, texture, mineralogy and composition. Micronodules from the Southwestern Pacific Basin are most abundant in the dark brown clays of the central Southwestern Pacific Basin. Mineralogically, these micronodules consist of 10 Å manganate and $\delta MnO_2$ with traces of quartz, phillipsite, quartz and feldspar. They have Mn/Fe ratios of 2.4 and Ni+Cu contents of 1.3%. These Mn/Fe ratios are somewhat higher than for the associated nodules but the Ni+Cu contents somewhat lower. The micronodules are botryoidal with smooth surface texture. Under the SEM, the surfaces of the micronodules display a honeycomb texture.

Micronodules from the C-C F.Z. occur mainly within siliceous ooze. They consist of 10 Å manganate with traces of quartz and sometimes phillipsite. They have Mn/Fe ratios of 4.7 and Ni+Cu contents of 1.7%. Again, the Mn/Fe ratios are somewhat higher than for the associated nodules but the Ni+Cu contents somewhat lower. The micronodules are dominantly spheroids or have rod-like structures. Under the SEM, the surfaces of the micronodules consist of plates.

Micronodules from the Peru Basin occur in three stratigraphic horizons. The upper 50-100 mm of the core consists of a homogeneous brown calcareous-siliceous mud containing a high abundance of micronodules. The micronodules are larger (125-500 μm) than those taken deeper in the core (40-125μm) and have a mamillated botryoidal surface texture similar to that of the associated macronodules. They consist of 10 Å manganate with traces of quartz and feldspar. They have Mn/Fe ratios of 8.7 and Ni+Cu contents of 2.1%. These sediments are underlain by a highly bioturbated light yellowish brown mud containing varying amounts of calcareous and siliceous organisms. These micronodules have a smooth surface texture. The replacement of radiolaria and foraminifera can be clearly seen under the SEM. The micronodules are coarsely crystalline with well-defined rod-like structures. They consist of 10 Å manganate with traces of quartz and feldspar and have Mn/Fe ratios of 1.9 and Ni+Cu contents of 0.9%. The sharp drop in the Mn/Fe ratios of these micronodules compared with those from the sediment surface suggests that remobilization of Mn and associated transition elements has taken place within this sediment horizon. At the base lies a dark overconsolidated clay with a low content of siliceous tests but containing volcanic glass and fish teeth. These micronodules have Mn/Fe ratios of 5.0 and Ni+Cu contents of 1.3%. The high Mn/Fe ratios of these micronodules suggests a well-oxidized sedimentary environment from which no diagenetic remobilization has taken place. The importance of element remobilization in the yellowish brown muds is confirmed by the fact that micronodule abundance is high in the surface sediments, low in the yellowish brown muds and high again in the dark brown clays.

A plot of Ni+Cu against Mn/Fe for micronodules from various areas in the Pacific is presented in Figure 11.19. It is seen that, although Mn/Fe ratios in micronodules >20 can occur, particularly in the Peru Basin, there is no well-defined point of reversal such as observed for Mn nodules (Fig. 11.14), although there is a flattening of the curve at a Mn/Fe ratio of about 10. This is a consequence of the fact that the micronodules occur within the sediment column where conditions are less oxidizing than at the sediment surface. Pore waters are not as enriched in Mn, Ni and Cu within the sediment as at the sediment surface and the Mn/Fe ratios and Ni+Cu contents of micronodules are never as high as in the asso-

ciated macronodules. Diagenetic fractionation of elements in micronodules is therefore not as pronounced as in the associated macronodules.

### 11.4.8 Mineralogy

There are three principal phases in manganese nodules, Mn oxides which tend to incorporate cationic transition metal species such as $Ni^{2+}$, $Cu^{2+}$ and $Zn^{2+}$, Fe oxihydroxides which tend to incorporate anionic species such as $HPO_4^{2-}$, $HAsO_4^{2-}$, $HVO_4^{2-}$, $MoO_4^{2-}$ and $WO_4^{2-}$ as well as the REE and $Co^{3+}$ and detrital aluminosilicates which consist of elements such as $SiO_2$, $Al_2O_3$, $TiO_2$ and $Cr_2O_3$. In addition, carbonate fluorapatite (francolite) also occurs in the older crust generation and as a discrete layer between the older and younger crust generations in Co-rich manganeses crusts. Concentrations of individual elements in deep-sea Mn nodules and crusts therefore tend to covary with the relative amounts of these four phases.

Three principal Mn oxide minerals are found in Mn nodules and crusts. Their principal X-ray diffraction peaks are given below together with their alternative mineral names (in parenthesis):

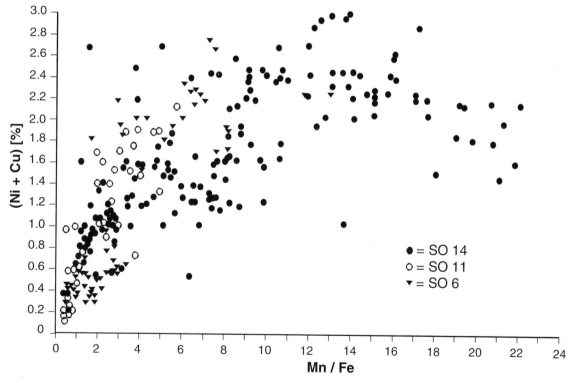

**Fig. 11.19** A plot of Ni+Cu against Mn/Fe for micronodules from various areas in the Pacific (after Stoffers et al. 1984). SO-06 samples taken from the Clarion-Clipperton F.Z. region, SO-11 from the Peru Basin and SO-14 from the Southwestern Pacific Basin.

10 Å manganate   9.7 Å   4.8 Å   2.4 Å 1.4Å
        (todorokite, buserite, 10 Å manganite)

7 Å manganate        7.3 Å 3.6 Å 2.4 Å 1.4Å
        (birnessite, 7 Å manganite)

$\delta MnO_2$                     2.4 Å 1.4Å
        (vernadite)

Mn oxide minerals are characterized by fine grain size in the range 100-1000 Å. These minerals are poorly crystalline and give diffuse X-ray diffraction patterns. FeOOH minerals also tend to be fine grained with sizes of the order of 100 Å and are generally X-ray amorphous (Johnson and Glasby 1969). In chloride-bearing solutions such as seawater, akagenéite ($\beta$-FeOOH) has been identified as the principal iron oxihydroxide mineral present by Mössbauer spectroscopy (Johnston and Glasby 1982). In fact, there are no independent Fe-bearing minerals in Mn nodules. Rather, $\delta MnO_2$ may be considered to be a randomly stacked Fe-Mn mineral. Some authors have suggested that there is usually an intimate association of $\delta MnO_2$ and $FeOOH \cdot xH_2O$ in deep-sea hydrogenous nodules due to the transport of minerals as colloidal particles and subsequent coagulation of these particles (Halbach et al. 1981). Detrital minerals such as quartz and feldspar normally give much stronger X-ray diffraction peaks than the Mn and Fe oxides. Within various types of nodules and crusts, the following Mn and Fe oxide minerals tend to be found.

| | |
|---|---|
| Shallow-water concretions (Baltic Sea, Loch Fyne, Scotland) | 10 Å manganate |
| Hydrothermal Mn crusts | 10 Å manganate + 7 Å manganate |
| Deep-sea hydrogenous Mn nodules and crusts | $\delta MnO_2$ |
| Deep-sea diagenetic Mn nodules | 10 Å manganate |

Within the Mn oxide minerals, there is a well-known dehydration sequence, 10 Å manganate → 7 Å manganate → $\delta MnO_2$. Samples collected on board ship must therefore be preserved moist in seawater to prevent mineralogical change on drying. A similar redox sequence 10 Å manganate → 7 Å manganate → $\delta MnO_2$ is also observed in nodules (Glasby 1972). This means that nodules formed in less-oxidizing environments tend to

contain 10 Å manganate whereas those formed in more-oxidizing environments tend to contain $\delta MnO_2$. This trend is confirmed by the O/Mn ratios in nodules which vary from 1.60 in shallow-water concretions to 1.95 in deep-sea nodules. 98% of the Mn in deep-sea nodules is therefore in the Mn (IV) form (Murray et al. 1984, Piper et al. 1984).

Modern ideas on the structure of Mn oxides are based largely on the work of R. Giovanoli of the University of Berne. Much of this work was carried out on synthetic 10 Å ($Na^+$)-manganate prepared by the rapid oxidation of $Mn(OH)_2$ with $O_2$ for 5 hrs. Two types of 10 Å manganate were considered (Giovanoli and Arrhenius 1988, Mellin and Lei 1993, Kuma et al. 1994, Usui and Mita in 1995, Lei 1996). Todorokite is a large tunnel-structure mineral based on $MnO_6$ octahedral which can not expand or contract on heating to 100°C (or even 400°C) (Turner and Buseck 1981, Waychunas 1991). Buserite, on the other hand, is a phyllomanganate mineral with an expandable or contractible sheet-like structure (Waychunas 1991, Manceau et al. 1997). It contains exchangeable interlayer cations ($Ca^{2+}$, $Mg^{2+}$, $Cu^{2+}$, $Ni^{2+}$, $Co^{2+}$, $2Na^+$) which occupy specific lattice sites. The ratio of these metals to Mn is 1:6 or 1:7. The uptake sequence of these metals into the buserite structure is $Cu^{2+} > Co^{2+} > Ni^{2+} > Zn^{2+} > Mn^{2+} > Ca^{2+} > Mg^{2+} > Na^+$. Fe can not enter the buserite lattice because any $Fe^{2+}$ would be oxidized to insoluble FeOOH by $Mn^{4+}$. Instead, any excess FeOOH in the sediments reacts with the dissolving siliceous tests to form nontronite, and therefore fix Fe, in siliceous sediments. This explains the high Mn/Fe ratios in diagenetic nodules from oxic environments. Buserite may be considered to be an expandable or contractible sheet which is able to accommodate hydrated stabilizing interlayer cations. It can therefore be distinguished from todorokite by expanding the interlayer spacing to 25 Å on treatment with dodecylammonium hydrochloride or contracting it to 7 Å on heating in air to 100°C.

Mellin and Lei (1993) have explained the mineralogy of the principal marine Mn deposits in the following terms. Low-temperature hydrothermal 10 Å manganate has a less stable todorokite-like structure with tunnel walls composed of $Mn^{2+}O_{2x}^{2-}(OH)_{6-2x}$ octahedral. High-temperature hydrothermal 10 Å manganate has a more stable todorokite-like structure as a result of the oxidation of interlayer $Mn^{2+}$. In both cases, divalent cations such as $Cu^{2+}$, $Ni^{2+}$ and $Zn^{2+}$ are deposited as sulfides prior to the deposition of Mn minerals which explains the low contents of these metals in these deposits. Diagenetic 10 Å manganate, on the other hand, has an unstable buserite-like structure. Divalent cations such as $Ni^{2+}$, $Cu^{2+}$, $Zn^{2+}$, $Mg^{2+}$ and $Ca^{2+}$ can substitute for $2Na^+$ in the interlayer spacing. Diagenetic deep-sea nodules formed in oxic environments have high $Cu^{2+}$ and $Ni^{2+}$ contents (upto 2%) as a result of the release of these elements from siliceous tests. The resulting high contents of Ni+Cu in these nodules explains their commercial interest. Diagenetic shallow-water concretions have low $Cu^{2+}$ and $Ni^{2+}$ contents (<0.1%) as a result of trapping these elements in the sediment column as sulfides. The composition of marine manganese deposits therefore reflects both the mineralogy of the sample and the availability of the transition metal ions for nodule formation (Glasby and Thijssen 1982).

The anomalous position of Co in the geochemistry of nodules has been interpreted by Glasby and Thijssen (1982) in terms of its crystal field characteristics. $Co^{3+}$ ($d^6$) is stable in nodules in the low spin state with octahedral coordination and has an ionic radius of 0.53 Å which is almost identical to that of $Fe^{3+}$ and $Mn^{4+}$. As such, Co will preferentially substitute in either FeOOH or $MnO_2$ as the trivalent ion but not in the interlayer spacing of 10 Å manganate as the divalent ion as is the case for $Ni^{2+}$, $Cu^{2+}$, $Zn^{2+}$ and $Mn^{2+}$. The ability of Co to substitute in both Mn and Fe oxyhydroxides leads to its correlation with either Mn or Fe in nodules or crusts depending on the environment of formation (Burns and Burns 1977, Halbach et al. 1983, Giovanoli and Arrhenius 1988). Experimental evidence has confirmed that $Co^{3+}$ is the dominant form of Co in Mn nodules (Dillard et al. 1984, Hem et al. 1985, Manceau et al. 1997).

## 11.4.9  Dating

Deep-sea Mn nodules and crusts are amongst the slowest growing minerals on earth. They have a minimum growth rate of 0.8 mm $10^6yr^{-1}$ (Puteanus and Halbach 1988) which is equivalent to the formation of one unit cell every 100 years or so. Shallow-water concretions grow about $10^4$ times faster. Several methods have been used to date deep-sea manganese nodules and crusts (Ku 1977, Mangini 1988).

One of the earliest methods used involved K-Ar dating of the volcanic nucleus of the nodule. This method is of limited value because submarine weathering of the core often invalidates the results. In addition, the method makes no allowance for any time gap between the formation of the nucleus and subsequent Mn accretion or for hiatuses in growth of the Mn deposit. Only a minimum average growth rate is therefore obtained. Although this method was used by early workers (e.g. Barnes and Dymond 1967), it is too crude to be of much value now.

Some attempts have been made to date Mn nodules and crusts by paleontological methods. Cowen et al. (1993) determined the ages of different layers of a Mn crust from the Schumann Seamount north of Hawaii based on identification of coccolith imprints. The ages of the individual layers were not well constrained (from 1-4 m.yrs to >10 m.yrs B.P.) and the minimum age for the upper 27 mm of the crust was determined to be Eocene giving an average growth rate of about 0.5 mm $10^6yr^{-1}$. Since the crust was originally 95 mm thick, this would result in a much older age for the crust than is usually accepted. Additional studies of this crust supported the older age but revealed some disparities with other methods of dating casting some doubt on their validity (McMurtry et al. 1994).

Most modern methods of dating Mn nodules and crusts therefore rely on radiometric determinations. The $^{238}U$ and $^{235}U$ decay-series method is based on the fact that $^{230}Th$ and $^{231}Pa$ are generated in seawater by the decay of U isotopes. The isotopes are then carried to the sea floor on particles and incorporated into the nodules and crusts. The residence times for Th and Pa in seawater are short (less than 40 and 160 yrs respectively). The half life for $^{230}Th$ 75,000 yrs and for $^{231}Pa$ 35,000 yrs. The distribution of $^{230}Th$ and $^{231}Pa$ with depth in nodule or crust can be measured by counting α particles after separation from the nodule material and plating on a planchet or more rapidly by α-track counting on nuclear emulsion plates. The growth rate of the nodule or crust can then be calculated from the formula

$$dC/dt = S\ dC/dx - \lambda C$$

$$\text{or} \quad C_{(x)} = C_0 \exp(-x\lambda/S)$$

where $C_{(x)}$ is the concentration at depth x, $\lambda$ is the decay constant of the nucleus (=ln $2/\tau_{1/2}$ ), $\tau_{1/2}$ the half life of the nuclide and S the nodule growth rate.

The major limitation of the $^{230}$Th method is that it can only date sample to an age of 300,000 yrs (4 half lives) which is equivalent to the outer few mm of deep-sea nodule or crust. It was not therefore possible to determine discontinuities in the growth pattern of the nodule or crust. However, the recent development of the thermal-ionization mass spectrometric (TIMS) method for determining U and Th isotopes has increased the precision and extended the possibilities of this method (Chabaux et al. 1995). As an example, Böllhofer et al. (1996) determined the distribution of excess $^{230}$Th with depth in a diagenetic nodule with a diameter of about 100 mm from the Peru Basin. The growth rate of the outer 25 mm of the nodule was determined to be 110 mm $10^6$yr$^{-1}$ and from a depth of 33 to 50 mm to be 60 mm $10^6$yr$^{-1}$. Unfortunately, the position of the nodules in the sediment was not noted so it is not known if these results refer to the upper or lower surface of the nodule. Within the Peru Basin, the factors controlling nodule growth rates are complex. Von Stackelberg (1997) has shown that the highest growth rates in these nodules are found on the underside of large nodules which repeatedly sink to a level immediately above the redox boundary where diagenetic recycling of Mn is at a maximum. The layering of these nodules is therefore believed to result from the lifting of these nodules by benthic organisms within the oxic surface sediments from a diagenetic to a hydrogenous environment. This explanation seems preferable to that of Böllhofer et al. (1996) that the layering of these nodules is the result of climatic change.

$^{10}$Be-dating involves a measurement of the depth profile of $^{10}$Be in deep-sea nodule or crust. The half life of $^{10}$Be is 1.5 m.yrs. $^{10}$Be is a cosmogenic nuclide which is deposited at the sea surface before being mixed into the oceans. Its residence time in the ocean is about 1000 yrs. Because of its greater half life, $^{10}$Be can be measured to much greater depths within the nodule or crust (> 40 mm) than the U decay-series isotopes (1-2 mm). Post-depositional diffusion of $^{10}$Be is not considered a problem in determining the growth rates of the nodule or crust (cf. Kusakabe and Ku 1984). The development of accelerator mass spectrometry (AMS) led to a reduction in sample sizes and counting times by several orders of magnitude compared to measurements based on radioactive decay. Much better resolution can therefore be obtained and dating to 15 m.yrs B.P. is now possible. This age precedes that of the

lower Miocene Antarctic glaciation. Discontinuities in the growth rates of nodules and crusts can also be determined from breaks in the $^{10}$Be depth profile. As an example, Koschinsky et al. (1996) determined the growth rates of the surface layers of two NE Atlantic manganese crusts to be 3 and 4.5 mm $10^6$yr$^{-1}$. Mangini et al. (1990) and Usui et al. (1993) determined the growth rates of S.W. Pacific nodules.

Sr isotope stratigraphy has been developed as a high resolution stratigraphic method for determining the growth rates of hydrogenous and hydrothermal Mn crusts based on a comparison of the Sr isotope ratios of individual layers in the crust with Sr isotope curves for seawater (Futa et al. 1988). In principal, this method can be used to identify hiatuses in the growth of the crusts to ages in excess of 20 m.yrs B.P. However, von der Haar et al. (1995) have recently demonstrated that Sr in the crusts is extracted from both the phosphatic and detrital phases of the crust during leaching and that the Sr in the Mn oxide phase exchanges with seawater throughout the history of the crust. The method therefore appears to be of little value in dating Mn crusts.

In addition to the above, indirect methods of dating such as the one based on the inverse relationship between the Co content in Mn crusts and the rate of accumulation of the crusts have been developed (Manheim and Lane-Bostwick 1988, Puteanus and Halbach 1988).

Growth rates of shallow-water ferromanganese concretions from the Baltic Sea have been determined using $^{210}$Pb dating, but the few results so far obtained are not directly comparable with these obtained by other methods (Hlawatsch 1998).

### 11.4.10   Mn Crusts as Paleoceanographic Indicators

Mn crusts may be considered to be condensed stratigraphic sections that record paleoceanographic conditions. High resolution dating of Mn crusts can therefore be used to record major paleoceanographic events. A number of such studies has been undertaken (Segl et al. 1989, McMurtry et al. 1994, Koschinsky et al. 1996).

One of the most thoroughly studied Mn crusts was collected on top of an abyssal seamount at 4830 m water depth in the equatorial N. Pacific in 1976 during cruise VA 13/2 of R.V. Valdivia (sample 237 KD). It is a large hemispherical Fe-

Mn crust 500 mm in diameter and upto 250 mm thick partly covering a basalt substrate. It is quite different in character from the Co-rich Mn crusts described in Section 11.4.4. The volcanic seamount on which it occurs was formed about 65 m.yrs B.P. near crest of the East Pacific Rise. Submarine weathering of glass on the surface of the submarine basalt led to formation of a 8 mm thick nontronite layer. The seamount migrated north across the equatorial high productivity belt during which time it was above the Carbonate Compensation Depth (CCD).

The mode of formation of this crust has been outlined by von Stackelberg et al. (1984). Below a depth of 40 mm, the crust was characterized by higher Fe contents which could be attributed to the dissolution of calcareous tests resulting in an increased supply of iron to the crust. Goethite was the main mineral formed and could be seen as yellowish-brown flecks. Calcareous tests were also observed within the crust. At 40 mm, there is an abrupt change in composition of the crust with higher Mn/Fe (1.6) and Ce/La (3.5) ratios and higher contents of Ni (0.45%) and Cu (0.25%) above that boundary. These changes were associated with the development of a strong AABW at 12 m.yrs B.P. From 10-0 mm, the chemical composition was again similar to that below 40 mm reflecting the weak influence of AABW at present.

[10]Be dating of the crust was undertaken by Segl (1984, 1989). Figure 11.20 shows the variation in growth rates within the crust. An age of 12-13 m.yrs B.P. was determined at a depth of 36 mm in the crust. This was taken to correspond to the build up of east Antarctic ice sheet, the onset of high AABW velocities, increased ventilation of the deep ocean and a major erosional event in Pacific. At 6.2-6.7 m.yrs B.P. at a depth of 16 mm in the crust, there was a change of growth rate and a visible change in crustal structure. This was taken to correspond to a decrease in global $\delta C^{13}$, a lowering of sea level, isolation and drying of Mediterranean, shoaling of the Panama Isthmus and an increase in bottom water circulation rates and oceanic fertility. At 3 m.yrs B.P. at a depth of 8 mm in the crust, there was a visible change in crustal structure. This was taken to correspond to closure of the Panama Isthmus and the onset of northern hemisphere glaciation. These events are recorded in many manganese crusts in the World Ocean. Subsequent determination of the distribution of Nd isotopes within this crust showed a decrease in the $\varepsilon_{Nd}$ from 3-5 m.yrs B.P. to the present. These data suggest that the closure of the Panama Isthmus about 3-4 m.yrs B.P. may have played a role in reducing the inflow of Atlantic-derived Nd into the Pacific Ocean (Burton et al. 1997, Ling et al. 1997).

High resolution [230]Th profiles were also obtained by Eisenhauer et al. (1992) for this crust as part of a detailed study of growth rates during last 300,000 yrs. 69 samples were taken in the upper 1.4 mm of crust at intervals of 0.02 mm. Fig. 11.21 shows the high resolution growth rates for the crust. From 0-0.88 mm, the growth rate was determined to be 6.6 ± 1.1 mm $10^6$yr$^{-1}$, from 0.9-1.28 mm 6.1 ± 1.7 mm $10^6$yr$^{-1}$ and from 1.3-1.4 mm 5.8 ± 1.4 mm $10^6$yr$^{-1}$. Two breaks in the growth rate were observed. These correspond to standstills in growth at 284-244 ka and 138-128 ka which are equivalent to glacial stages 8 and 6. The growth rates of the crust were seen to be higher during interglacial (stages 1 and 5) and lower during glacial stages. These data imply a correlation between the growth rate of nodules and crusts and climate in the late Quaternary.

These data demonstrate that radiometric dating of deep-sea manganese crusts can be used to determine major changes in paleoceanographic conditions since the Miocene including changes in the Pleistocene.

### 11.4.11 Economic Prospects

In 1965, J.L. Mero proposed that manganese nodules could serve as an economic resource for Ni,

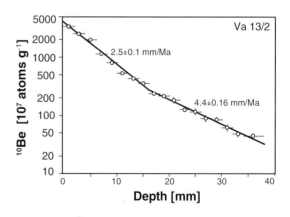

**Fig. 11.20** Profile of [10]Be with depth in the Mn crust VA 13/2 from the Clarion-Clipperton F.Z. region showing the change in a growth rate in the crust at a depth of 16 mm (after Segl et al. 1989). Other time markers in the crust were marked by changes in the structure of the crust.

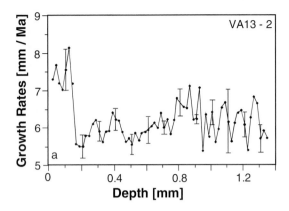

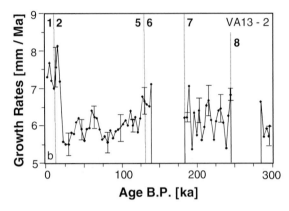

**Fig. 11.21** High resolution growth rates for Mn crust VA 13/2 plotted as a function of depth in the crust (a) and of time (b) (after Eisenhauer et al. 1992). The major climatic changes are marked by dotted lines. High growth rates are mainly associated with interglacial stages (especially during the Holocene and stage 5) whereas low growth rates are associated with glacial stages. Periods of growth standstills are associated with glacial stages 6 and 8.

Cu and Co (Mero 1965). Although he grossly overestimated their economic potential, his work set in train a number of major national programs to explore the resource potential of manganese nodules. As such, it was extremely influential. In 1972-1982, the U.S. International Decade of Ocean Exploration (IDOE) program was implemented to evaluate deep-sea Mn nodules in the C-C F.Z. Strong German, French and Soviet programs were also undertaken at this time to evaluate nodules as an economic resource. From the 1960s onwards, there were 100 Soviet cruises to all parts of World Ocean to evaluate manganese nodules. In 1981, the emphasis shifted and German Midpac and USGS cruises to Mid-Pacific Mountains began to evaluate Co-rich Mn crusts in order to assess their potential as a source of Co which was then considered to be a strategic metal. Since 1989, limited German interest has been di-

rected to the environmental aspects of nodule mining through the DISCOL program (Disturbance and Recolonization of a Mn Nodule Area in the Peru Basin of the South Pacific). This exploration work indicated that manganese nodules and crusts are not an economic resource in the short term. Nonetheless, Mn nodules are still considered to have economic potential in the medium- to long-term in some quarters. In the 1980s, Japan, South Korea, China and India became involved in evaluating nodule resources and these countries appear to be the leading candidates to begin nodule mining. Summaries of some of the latest developments in nodule exploration have been presented in Glasby (1986), Halbach et al. (1988), Anon (1995) and Lenoble (1996). Recent estimates of manganese nodule resources in the world ocean have been made by Andreev and Gramburg (1998) and Morgan (1999).

For deep-sea nodules, only those nodules with Cu+Ni+Co >2.5% and abundance >10 kg m$^{-2}$ can be considered a potential economic resource. This represents only a small percentage of total amount of nodules (i.e. diagenetic nodules from the C-C F.Z. and Central Indian Basin). It is technically feasible to mine deep-sea nodules. To be economic, this would require a 20 year-mine-site producing 3 million tons/yr of nodules. This would cover an area >6000 km$^2$.

For Co-rich manganese crusts, the crusts should contain >0.8% Co and be >40 mm thick. They occur in shallower water depths (1000-2500 m) than deep-sea nodules (>4000 m) and could be mined within national Exclusive Economic Zones (EEZs) within 200 n.m. of the shore and therefore under national jurisdiction. Technically, they are more difficult to mine than nodules because they occur as slabs in steep areas of the seafloor and it would be necessary to remove the substrate at the seafloor prior to recovering the Mn crust.

At present, world metal prices are depressed and land-based mines working at less than full capacity. It is not clear that deep-sea nodules can be recovered more cheaply than mineral deposits from land-based mines. The UN Law of the Sea convention does not favor commercial mining of the deep sea by industrialized countries but rather the role of the International Seabed Authority and the setting up of a 'New Economic Order'. Nonetheless, a number of registered pioneer investors have been allocated areas for future mining in the C-C F.Z. by the International Seabed Authority (Fig. 11.22). Those countries and organizations that have accepted pioneer investor status are obliged to fulfill a work program within their mine site area.

**Future Prospects**

The preceding discussion has demonstrated that the marine geochemistry of manganese is a large field in which much remains to be done. Although the boom period of research into deep-sea Mn nodules and Co-rich Mn crusts which accompanied the drive to establish these deposits as an economic resource in the 1970s and 1980s is over, there remain many interesting topics which have been somewhat neglected over the last two decades. These include the study of shallow-marine concretions, lake concretions, hydrothermal Mn crusts, ancient Mn deposits and the use of crusts as paleoceanographic indicators and as tracers of oceanic water masses. Time will tell if deep-sea manganese nodules and crusts are to become a major source of metals for the world as once thought.

**Acknowledgements**

The author would like to thank G.M. McMurtry and C.L. Morgan (University of Hawaii), A. Usui (GSJ) and U. von Stackelberg (BGR) for their helpful reviews of the manuscript. This is contribution No 260 of the Special Research Program SFB 261 (*The South Atlantic in the Late Quaternary*) funded by the Deutsche Forschungsgemeinschaft (DFG).

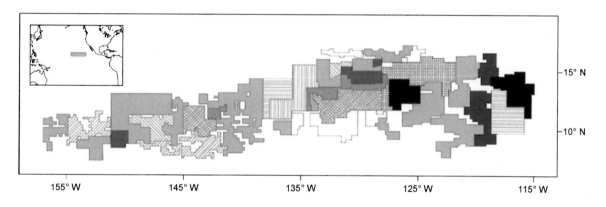

**Fig. 11.22** Schematic map showing the disposition of sectors of registered pioneer areas in the C-C F.Z. (after Kotlinski 1995). This map demonstrates that most of the consortia interested in deep-sea mining have already made extensive studies of their future 'mine sites' in the C-C F.Z. and staked their claims there. The economic potential of this region is demonstrated by the fact that a large proportion of the area is already under claim. Areas not subject to claims are often topographically unsuitable (too mountainous).

# References

Aller, R.C., 1990. Bioturbation and manganese cycling in hemipelagic sediments. Phil. Trans. R. Soc. Lond., A331: 51-68.

Andreev, S.I. and Gramburg, I.S., 1998. The explanatory note to the metallogenic map of the world ocean, VNIIOkeanologia (St Petersburg) and InterOceanMetall (Szczecin), 212 pp. + illustrations (in English and Russian).

Anon, 1979. La Genèse des nodules de manganese. Colloques Internationaux du Centre National de la Recherche Scientifique (CNRS), 287: 410 pp.

Anon, 1995. Proceedings of the ISOPE (The International Society of Offshore and Polar Engineers) - Ocean Mining Symposium, November 21-22, Tsukuba, Japan.

Aplin, A.C. and Cronan, D.S., 1985. Ferromanganese deposits from the central Pacific Ocean, II. Nodules and associated sediments. Geochimica et Cosmochmica Acta, 49: 437-451.

Barnes, S.S. and Dymond, J., 1967. Rates of accumulation of ferro-manganese nodules. Nature, 213: 1218-1219.

Baturin, G.N., 1988. The geochemistry of manganese and manganese nodules in the ocean. Reidel, D., Dordrecht, 342 pp.

Bender, M.L., Klinkhammer, G.P. and Spencer, D.W., 1977. Manganese in seawater and the marine manganese balance. Deep-Sea Research, 24: 799-812.

Berger, W.H., Fischer, K., Lai, C. and Wu, G., 1987. Ocean

productivity and organic carbon flux. Scripps Inst. Oceanogr. Ref. Ser. 87-30.

Bischoff, J.L. and Piper, D.Z., 1979. Marine geology and oceanography of the Pacific manganese nodule province. Plenum Press, NY, 842 pp.

Bischoff, J.L., Heath, G.R. and Leinen, M.L., 1979. Geochemistry of deep-sea sediments from the Pacific manganese nodule province: DOMES Sites A, B, and C. In: Bischoff, J.L. and Piper, D.Z. (eds), Marine geology and oceanography of the Pacific manganese nodule province. Plenum Press, NY, pp. 397-436.

Böllhofer, A., Eisenhauer, A., Frank, N., Pech, D. and Mangini, A., 1996. Thorium and uranium isotopes in a manganese nodule from the Peru basin determined by alpha spectrometry and thermal ionization mass spectrometry (TIMS): Are manganese supply and growth related to climate? Geol. Rdsch., 85: 577-585.

Bruland, K.W., 1983. Trace elements in sea-water. In: Riley, J.P. and Chester, R. (eds), Chemical oceanography. Academic Press, London, pp. 157-220.

Bruland, K.W., Orians, K.J. and Cowen, J.P., 1994. Reactive trace metals in the stratified central North Pacific. Geochimica et Cosmochimica Acta, 58: 3171-3182.

Burgarth, K.P. and von Stackelberg, U., 1995. Sulfide - impregnated volcanics and ferromanganese incrustations from the southern Lau Basin (Southwest Pacific). Mar. Georesourc. Geotechnol., 13: 263-308.

Burns, R.G. and Burns, V.M., 1977. Mineralogy. In: Glasby, G.P. (ed), Marine manganese deposits. Elsevier, Amsterdam, pp. 185-248.

Burton, J.D. and Statham, P.J., 1988. Trace metals as tracers in the ocean. Phil Trans R. Soc., 325: 127-145.

Burton, K.W., Ling, H.F. and O'Nions, R.K., 1997. Closure of the central American Isthmus and its effect on deep-water formation in the North Atlantic. Nature, 386: 382-385.

Calvert, S.E. and Price, N.B., 1977. Shallow water and continental margin and lacustrine nodules: Distribution and geochemistry. In: Glasby, G.P. (ed), Marine manganese deposits. Elsevier, Amsterdam, pp. 45-86.

Calvert, S.E. and Piper, D.Z., 1984. Geochemistry of ferromanganese nodules from DOMES Site A, Northern Equatorial Pacific: Multiple diagentic metal sources in the deep sea. Geochim. et Cosmochim. Acta, 48: 1913-1928.

Chabaux, F., Cohen, A.S., O'Nions, R.K. and Hein, J.R., 1995. 238U - 235U - 230Th chronometry of Fe-Mn crusts: Growth processes and recovery of thorium isotopic ratios of seawater. Geochimica et Cosmochimica Acta, 59: 633-638.

Chester, R., 1990. Marine geochemistry. Chapman & Hall, London, 698 pp.

Cowen, J.P. and Bruland, K.W., 1985. Metal deposits associated with bacteria: implications for Fe and Mn marine geochemistry. Deep-Sea Research, 32: 253-272.

Cowen, J.P., DeCarlo, E.H. and McGee, D.L., 1993. Calcareous nannofossil biostratigraphic dating of a ferromanganese crust from Schumann Seamount. Marine Geology, 115: 289-306.

Cronan, D.S., 1977. Deep-sea nodule: distribution and chemistry. In: Glasby, G.P. (ed), Marine manganese deposits. Elsevier, Amsterdam, pp. 11-44.

Cronan, D.S., 1980. Underwater minerals. Academic Press, London, 362 pp.

Cronan, D.S., 1987. Controls on the nature and distribution of manganese nodules in the western equatorial Pacific Ocean. In: Teleki, P.G., Dobson, M.R., Moore, J.R. and von Stackelberg, U. (eds), Marine minerals advances in research and resouce assessment. D. Reidl, Dordrecht, pp. 177-188.

Cronan, D.S. and Hodkinson, R.A., 1994. Element supply to surface manganese nodule along the Aitutaki-Jarvis Transect, South Pacific. J. Geol. Soc., 151: 392-401.

Cronan, D.S., 1997. Some controls on the geochemical variability of manganese nodules with particular reference to the tropical South Pacific. In: Nicholson, K., Hein, J.R., Bühn, B. and Dasgupta, S. (eds), Manganese mineralization: Geochemistry and mineralogy of terrestrial and marine deposits. Geol. Soc. Spec. Publ., pp. 139-151.

Davies, T.A., 1985. Mesozoic and Cenozoic sedimentation in the Pacific Ocean Basin. In: Nairn, A.E.M., Stehli, F.G. and Uyeda, S. (eds), The ocean basins and margins 7A The Pacific Ocean. Plenum Press, NY, pp. 65-88.

DeBaar, H.J.W. et al., 1995. Importance of iron for pankton blooms and carbon dioxide drawdown in the Southern Ocean. Nature, 373: 412-415.

Dillard, J.G., Crowther, D.L. and Calvert, S.E., 1984. X - ray photoelectron spectroscopi study of ferromanganese nodule: Chemical speciation for selected transition metals. Geochimica et Cosmochimica Acta, 48: 1565-1569.

Dymond, J. and Eklund, W., 1978. A microprobe study of metalliferous sediment component. Earth Planetary Science Letters, 40: 243-251.

Dymond, J., Lyle, M., Finney, B., Piper, D.Z., Murphy, K., Conard, R. and Pisias, N., 1984. Ferromanganese nodules from MANOP Sites H, S, and R - Control of mineralogical and chemical composition by multiple accretionary processes. Geochimica et Cosmochimica Acta, 48: 931-949.

Earny, F.C.F., 1990. Marine mineral recources. Routledge, London, 387 pp.

Eckhardt, J.D., Glasby, G.P., Puchelt, H. and Berner, Z., 1997. Hydrothermal manganese crusts from Enareta and Palinuro seamounts in the Tyrrhenian Sea. Marine Georesourc. Geotechnol., 15: 175-209.

Ehrlich, H.L., 1996. Geomicrobiology. Marcel Dekker, NY, 719 pp.

Eisenhauer, A., Gögen, K., Pernicka, E. and Mangini, A., 1992. Climatic influences on the growth rates of Mn crusts during the Late Quaternary. Earth and Planetary Science Letters, 109: 25-36.

Elderfield, H. and Schulz, A., 1996. Mid-ocean ridge hydrothermal fluxes and the chemical composition of the ocean. Ann. Review Earth and Planetary Science, 24: 191-224.

Fitzwater, S.E., Coale, K.H., Gordon, M., Johnson, K.S. and Ondrusek, M.E., 1996. Iron deficiancy and plankton growth in the equatorial Pacific. Deep-Sea Research, 43: 995-1015.

Friedrich, G., Glasby, G.P., Thijssen, T. and Plüger, W.L., 1983. Morphological and geochemical characteristics of manganese nodules collected from three areas on an equatorial Pacific transect by R.V. Sonne. Marine Mining, 4: 167-253.

Froelich, P.N., Klinkhammer, G.P., Bender, M.L., Luedtke,

N.A., Heath, G.R., Cullen, D., Dauphin, P., Hammond, D., Hartmann, B. and Maynard, V., 1979. Early oxidation of organic matter in pelagic sediments of the eastern equatorial Atlantic: suboxic diagenesis. Geochimica et Cosmochimica Acta, 43: 1075-1090.

Futa, K., Peteman, Z.E. and Hein, J.R., 1988. Sr and Nd isotopic variations in ferromanganese crusts from the Central Pacific: Implications for age and source provenance. Geochimica et Cosmochimica Acta, 52: 2229-2233.

German, C.R. and Angel, M.V., 1995. Hydrothermal fluxes of metal to the oceans: a comparison with anthropogenic discharges. In: Parson, L.M., Walker, C.L. and Dixon, D.R. (eds), Hydrothermal vents and processes. Geol. Soc. Spec. Publ., 87: 365-372.

Gingele, F.X. and Kasten, S., 1994. Solid-phase manganese in Southeast Atlantic sediments: Implications for the paleoenvironment. Mar. Geol., 121: 317-332.

Giovanoli, R. and Arrhenius, G., 1988. Structural chemistry of marine manganese and iron minerals and synthetic model compounds. In: Halbach, P., Friedrich, G. and von Stackelberg, U. (eds), The Manganese Nodule Belt of the Pacific Ocean geological environment, Nodule formation and mining aspects. Enke Verlag, Stuttgart, pp. 20-37.

Glasby, G.P., 1972. the mineralogy of manganese nodules from a range of marine environments. Marine Geology, 13: 57-72.

Glasby, G.P., 1973. Mechanism of enrichment of the rarer elements in marine manganese nodules. Marine Chemistry, 1: 105-125.

Glasby, G.P., 1974. Mechanism of incorporation of manganese and associated trace elements in marine manganese nodules. Oceanogr. Mar. Ann. Rev., 12: 11-40.

Glasby, G.P. (ed), 1977. Marine manganese deposits. Elsevier, Amsterdam, 523 pp.

Glasby, G.P., Meylan, M.A., Margolis, S.V. and Bäcker, H., 1980. Manganese deposits of the Southwestern Pacific Basin. In: Varentsov, I.M. and Grasselly, G.Y. (eds), Geology and geochemistry of manganese. Hungarian Academy of Sciences, Budapest, pp. 137-183.

Glasby, G.P. and Thijssen, T., 1982. Control of the mineralogy and composition of the manganese nodules by the supply of divalent transition metal ions. Neues Jb Mineral. Abh., 145: 291-307.

Glasby, G.P., Stoffers, P., Sioulas, A., Thijssen, T. and Friedrich, G., 1983. Manganese nodule formation in the Pacific Ocean: a general theory. Geological Marine Letters, 2: 47-53.

Glasby, G.P., 1984. Manganese in the marine environment. Oceanography Marine Biology, 22: 169-194.

Glasby, G.P., 1986. Marine Minerals in the Pacific. Oceanography and Marine Biology An Annual Review, 24: 11-64.

Glasby, G.P., Gwozdz, R., Kunzendorf, H., Friedrich, G. and Thijssen, T., 1987. The distribution of rare earth and minor elements in manganese nodules and sediments from the equatorial and S.W. Pacific. Lithos, 20: 97-113.

Glasby, G.P., 1988a. Manganese deposition through geological time: Dominance of the Post-Eocene environment. Ore Geology reviews, 4: 135-144.

Glasby, G.P., 1988b. Hydrothermal manganese deposits in island arcs and related to subduction processes: A possible model for genesis. Ore Geology reviews, 4: 145-153.

Glasby, G.P., 1991. Mineralogy and geochemistry of Pacific red clays. Geological Geophysical, 34: 167-176.

Glasby, G.P., Uscinowicz, S.Z. and Sochan, J.A., 1996. Marine ferromanganese concretions from the Polish exclusive economic zone: Influence of Major Inflows of North Sea Water. Marine Georesourc. and Geotechnol., 14: 335-352.

Glasby, G.P., Stüben, D., Jeschke, G., Stoffers, P. and Garbe-Schönberg, C.-D., 1997. A model for the formation of hydrothermal manganese crusts from the Pitcairn Island hotspot. Geochim. et Cosmochim. Acta, 61: 4583-4597.

Glasby, G.P., Emelyanov, E.M., Zhamoida, V.A., Baturin, G.N., Leipe, T., Bahlo, R. and Bonacker, P., 1997. Environments of formation of ferromanganese concretions in the Baltic Sea: a critical review. In: Nicholson, K., Hein, J.R., Bühn B. and Dasgupta, S. (eds), Manganese mineralization: Geochemistry and mineralogy of terrestrial and marine deposits. Geol. Soc. Spec. Publ., 119: 213-237.

Glasby, G.P. and Schulz, H.D., 1999. $E_H$, pH diagrams for Mn, Fe, Co, Ni, Cu and As under seawater conditions: Application of two new types of $E_H$, pH diagrams to the study of specific problems in marine geochemistry. Aquatic Geochemistry, 5: 227-248.

Halbach, P., Marchig, V. and Scherhag, C., 1980. Regional variations in Mn, Cu, and Co of ferromanganese nodules from a basin in the Southeast Pacific. Marine Geology, 38: M1-M9.

Halbach, P., Scherhag, C., Hebisch, U. and Marchig, V., 1981. Geochemical and mineralogical control of different genetic types of deep-sea nodules from Pacific Ocean. Mineral Deposita, 16: 59-84.

Halbach, P. and Puteanus, D., 1984. The influence of the carbonate dissolution rate on the growth and composition of Co-rich ferromanganese crusts from the central Pacific seamount areas. Earth and Planetary Science Letters, 68: 73-87.

Halbach, P., Friedrich, G. and von Stackelberg, U. (eds), 1988. The Manganese Nodule Belt of the Pacific Ocean geological environment, nodule formation, and mining aspects. Enke Verlag, Stuttgart, 254 pp.

Halbach, P., Kriete, C., Prause, B. and Puteanus, D., 1989. Mechanism to explain the platinum concentration in ferromanganese seamount crusts. Chemical Geology, 76: 95-106.

Hastings, D. and Emerson, M., 1986. Oxidation of manganese by spores of a marine bacillius: Kinetics and thermodynamic considerations. Geochimica et Cosmochimica Acta, 50: 1819-1824.

Heath, G.R., 1981. Ferromanganese nodules of the deep sea. Econ. Geol., 75: 736-765.

Hein, J.R., Morgenson, L.A., Clague, D.A. and Koski, R.A., 1987. Cobalt-rich ferromanganese crusts from the exclusive economic zone of the United States and nodules from the oceanic Pacific. In: Scholl, D.W., Grantz, A. and Vedder, J.G. (eds), Geology and resource potential of the continental margin of western North America and the adjacent oceans-Beaufort Sea to Baja California. Circum-Pacific Council for Energy and Mineral Recources, Earth Science Series, Houston, Texas, pp. 753-771.

Hein, J.R., Schwab, W.C. and Davis, A.S., 1988. Cobalt- and platinum-rich ferromanganese crusts and associated substrate rocks from the Marshall Islands. Marine Geology, 78: 255-283.

Hein, J.R., Gibbs, A.E., Clague, D.A. and Torresan, M., 1996. Hydrothermal mineralization along submarine rift zones, Hawaii. Mar. Georesourc. Geotechnol., 14: 177-203.

Hem, J.D., Roberson, C.E. and Lind, C.J., 1985. Thermodynamic stability of CoOOH and its coprecipitation with manganese. Geochim. et Cosmochim. Acta, 49: 801-810.

Hlawatsch, S., 1998. Mn-Fe-Akkumulate als Indikator für Schad- und Nährstoffflüsse in der westlichen Ostsee, unpubl. Dr.rer.nat. Diss., Universität Kiel, 113 pp + Appendix.

Hodge, V.F., Stallard, M., Koide, M. and Goldberg, E.D., 1985. Platinum and the platinum anomaly in the marine environment. Earth Planetary Science Letters, 72: 158-162.

Horn, D.R., 1972. Ferromanganese deposits of the ocean floor. National Science Foundation, Washington, D. C., 293 pp.

Jeong, K.S., Kang, J.K. and Chough, S.K., 1994. Sedimentary processes and manganese nodule formation in the Korea Deep Ocean Study (KODOS) area, western part of Clarion-Clipperton fracture zones, northeast equatorial Pacific. Marine Geology, 122: 125-150.

Jeong, K.S., Kang, J.K., Lee, K.Y. Jung, H.S., Chi, S.B. and Ahn, S.J., 1996. Formation and distribution of manganese nodule deposits in the western margin of Clarion-Clipperton fracture zones, northeast equatorial Pacific. Geo-Mar. Letts., 16: 123-131.

Johnson, C.E. and Glasby, G.P., 1969. Mössbauer Effect determination of particle size in microcrystalline iron - manganese nodule. Nature, 222: 376-377.

Johnson, K.S., Coale, K.H., Berelson, W.M. and Gordon, R.M., 1996. On the formation of the manganese maximum in the oxygen minimum. Geochimica et Cosmochimica Acta, 60: 1291-1299.

Johnson, C.E., Gordon, R.M. and Coale, K.H., 1997. What controls dissolved iron concentrations in the world ocean. Marine Chemistry, 57: 137-161.

Johnston, J.H. and Glasby, G.P., 1982. A Mössbauer spectroscopic and X-ray diffraction study of the iron mineralogy of some sediments from the Southwest Pacific Basin. Marine Chemistry, 11: 437-448.

Kasten, S., Glasby, G.P., Schulz, H., Friedrich, G. and Andreev, S.I., 1998. Rare earth elements in manganese nodules from the South Atlantic Ocean as indicators of oceanic bottom water flow. Marine Geology, 146: 33-52.

Knoop, P.A., Owen, R.M. and Morgan, C.L., 1998. Regional variability in ferromanganese nodule composition: northeastern tropical Pacific Ocean. Marine Geol., 147: 1-12.

Koschinsky, A., Halbach, P., Hein, J.R. and Mangini, A., 1996. Ferromanganese crusts as indicators for paleoceanographic events in the NE Atlantic. Geol. Rdsch., 85: 567-576.

Kotlinski, R., 1995. InterOceanMetal Joint Organization: Archievements and Challenges. Proceedings of the ISOPE (The International Society of Offshore and Polar Engineers)-Ocean Mining Symposium, Tsukuba, Japan, November 21-22, 5-7.

Köning, I., Drodt, M., Suess, E. and Trautwein, A.X., 1997. Iron reduction through the tan-green color transition in deep-sea sediments. Geochimica et Cosmochimmica Acta, 61: 1679-1683.

Ku, T.L. and Glasby, G.P., 1972. Radiometric evidence for the rapid growth rates of shallow - water, continental margin manganese nodules. Geochimica et Cosmochimica Acta, 36: 699-703.

Ku, T.L., 1977. Rates of accretion. In: Glasby, G.P. (ed) Marine manganese deposits. Elsevier, Amsterdam, pp. 249-267.

Kuma, K., Usui, A., Paplawsky, W., Gedulin, B. and Arrhenius, G., 1994. Crystal structures of synteteic 7 Å and 10 Å manganates substituted by mono - and divalent cations. Mineral. Mag., 58: 425-447.

Kunzendorf, H., Glasby, G.P., Stoffers, P. and Plüger, W.L., 1993. The distribution of rare earth and minor elements in manganese nodules, micronodules and sediments along an east-west transect in the southern Pacific. Lithos, 30: 45-56.

Kusukabe, M. and Ku, T.L., 1984. Incorporation of Be isotopes and other trace metals into marine ferromanganese deposits. Geochim. et Cosmochim. Acta, 48: 2187-2193.

Lavelle, J.W., Cowen, J.P. and Massoth, G.J., 1992. A model for the deposition of hydrothermal manganese near mid-ocean ridge crests. Journal of Geophysical Research, 97: 7413-7427.

Lei, G., 1996. Crystal structure and metal uptake capacity of 10 Å-manganates: An overview. Marine Geology, 133: 103-112.

Lenoble, 1996. Polymetallic modules of the deep sea: 30 years of activities around the world. Chronique de la Recherche Minière, 524: 15-39.

Lilley, M.D., Feely, R.A. and Trefry, J.H., 1995. Chemical and biochemical transformations in hydrothermal plumes. In: Humphris, S.E. Zierenberg, R.A., Mullineaux, L.S. and Thompson, R.E. (eds), Seafloor Hydrothermal systems: physical, chemical, biological, and geological interactions. Am. Geophys. Un. Geophys. Monogr., 91: 369-391.

Ling, H.F., Burton, K.W., O'Nions, R.K., Kamber, B.S., von Blanckenberg, F., Gibb, A.J. and Hein, J.R., 1997. Evolution of Nd and Pb isotopes in central Pacific seawater from ferromanganese crusts. Earth and Planetary Science Letters, 146: 1-12.

Lyle, M., 1983. The brown-green color transition in marine sediments: A marker of the Fe(III)-Fe(II) redox boundary. Limnol. Oceanogr., 28: 1026-1033.

Lyle, M., Owen, R.M. and Leinen, M., 1986. History of hydrothermal sedimentation at the East Pacific Rise, 19°S. In: Initial Reports of the Deep Sea Drilling Project. U.S. Government Printing Office, Washington, D.C., 92: 585-596.

Maceau, A., Drits, V.A., Silvester, E., Bartoli, C. and Lanson, B., 1997. Structural mechanism of Co2+ oxidation by the phyllomanganate buserite. Am. Mineral., 82: 1150-1175.

Mandernack, K.W., Post, J. and Tebo, B.M., 1995. Manganese mineral formation by bacterial spores of the marine Bacillus, strain SG-1: Evidence for the direct oxidation of Mn(II) to Mn(IV). Geochimica et Cosmochimica Acta, 59: 4393-4408.

Mangini, A., Halbach, P., Puteanus, D. and Segl, M., 1987. Chemistry and growth history of Central Pacific Mn - crusts and their economic importance. In: Teleki, P.G., Dobson, M.R., Moore, J.R. and von Stackelberg, U. (eds), Marine minerals advances in research and recource assessment. D. Reidl, Dordrecht, pp. 205-220.

Mangini, A., 1988. Growth rates of manganese nodules and crusts. In: Halbach, P., Friedrich, G. and von Stackelberg, U. (eds), The Manganese Nodule Belt of the Pacific Ocean geological environment, nodule formation, and mining aspects. Enke Verlag, Stuttgart, pp. 142-151.

Mangini, A., Segl, M., Glasby, G.P, Stoffers, P. and Plüger, W.L., 1990. Element accumulation rates in and growth histories of manganese nodules from the Southwestern Pacific Basin. Marine Geology, 94: 97-107.

Manheim, F.T., 1986. Marine cobalt recources. Science, 232: 600-608.

Manheim, F.T. and Lane-Bostwick, C.M., 1988. Cobalt in ferromanganese crusts as a monitor of hydrothermal discharge on the Pacific sea floor. Nature, 335: 59-62.

Marchig, V. and Erzinger, J., 1986. Chemical composition of Pacific sediments near 20°S: changes with increasing distance from the East Pacific Rise. In: Initial Reports of the Deep Sea Drilling Project. U.S. Government Printing Office, Washington, D.C., pp. 371-381.

Martin, J.H., Knauer, G.A. and Broenkow, W.W., 1985. VERTEX: the lateral transport of manganese in the northeast Pacific. Deep-Sea Research, 32: 1405-1427.

McMurtry, G.M., von der Haar, D.L., Eisenhauer, A., Mahoney, J.J. and Yeh, H.W., 1994. Cenozoic accumulation history of a Pacific ferromanganese crust. Earth and Planetary Science Letters, 125: 105-118.

Mellin, T.A. and Lei, G., 1993. Stabilization of 10 Å-manganates by interlayer cations and hydrothermal treatment: Implications for the mineralogy of marine manganese concretions. Marine Geology, 115: 67-83.

Mero, J.L., 1965. The Mineral Resource of the Sea. Elsevier, Amsterdam, 312 pp.

Meylan, M.A., Glasby, G.P., McDougall, J.C. and Kumbalek, S.C., 1982. Lithology, colour, mineralogy, and geochemistry of marine sediments from the Southwestern Pacific and Samoan Basin. N.Z. Jl. Geol. Geophys., 25: 437-458.

Meylan, M.A., Glasby, G.P., Hill, P.J., McKelvey, B.C., Walter, P. and Stoffers, P., 1990. Manganese crusts and nodules from the Manihiki Plateau and adjacent areas: Results of HMNZS Tui cruises. Mar. Mining, 9: 43-72.

Miller, S. and Cronan, D.S., 1994. Element supply to surface sediments and interrelationships with nodules along the Aitutaki-Jarvis Transect, South Pacific. J. Geol. Soc. Lond., 151: 403-412.

Morgan, C.L., 1999. Resource estimates of the Clarion - Clipperton manganese nodule deposits. In: Cronan, D.S. (ed), Handbook of marine minerals. CRC Press, Boca Raton, Florida.

Murray, J.W. and Brewer, P.G., 1977. Mechanism of removal of manganese, iron and other trace metals from seawater. In: Glasby, G.P. (ed), Marine manganese deposits. Elsevier, Amsterdam, pp. 291-325.

Murray, J.W., Balistrieri, L.S. and Paul, B., 1984. The oxidation state of manganese in marine sediments and ferromanganese nodules. Geochimica et Cosmochimica Acta, 48: 1237-1247.

Müller, P.J., Hartmann, M. and Suess, E., 1988. The chemical environment of pelagic sediments. In: Halbach, P., Friedrich, G. and von Stackelberg, U. (eds), The Manganese Nodule Belt of the Pacific Ocean geological environ-ment, nodule formation, and mining aspects. Enke Verlag, Stuttgart, pp. 70-99.

Nemoto, K. and Kroenke, L.W., 1981. Marine Geology of the Hess Rise 1. Bathymetry, surface sediment distribution and environment of deposition. J. Geophys. Res., 86: 10734-10752.

Nozaki, Y., 1997. A fresh look at element distribution in the North Pacific. EOS Trans. Am Geophysics, 78: 221.

Overnell, J., Harvey, S.M. and Parkes, R.J., 1996. A biogeochemical comparison of sea loch sediments. Manganese and iron contents, sulphate reduction rates and oxygen uptake rates. Oceanol. Acta, 19: 41-55.

Piper, D.Z. and Williamson, M.E., 1977. Composition of Pacific Ocean ferromanganese nodules. Marine Geology., 23: 285-303.

Piper, D.Z., Basler, J.R. and Bischoff, J.L., 1984. Oxidation state of marine manganese nodules. Geochimica et Cosmochimica Acta, 48: 2347-2355.

Piper, D.Z., Swint, T.R., Sullivan, L.G. and McCoy, F.W., 1985. Manganese nodules, seafloor sediment, and sedimentation rates in the Circum-Pacific region. Circum-Pacific Council for Energy and Mineral Recources Circum-Pacific Map Project. American Association of Petroleum Geologists, Tulsa, Oklahoma.

Piper, D.Z., Swint-Iki, T.R. and McCoy, F.W., 1987. Distribution of ferromanganese nodules in the Pacific Ocean. Chem. Erde, 46: 171-184.

Plüger, W.L., Friedrich, G. and Stoffers, P., 1985. Environmental controls of the formation of deep-sea ferromanganese concretions. Monogr. Ser. Mineral Deposits, 25: 31-52.

Puteanus, D. and Halbach, P., 1988. Correlation of Co concentration and growth rate-a method for age determination of ferromanganese crusts. Chemical Geology, 69: 71-85.

Raab, W.J. and Meylan, M.A., 1977. Morphology. In: Glasby, G.P. (ed), Marine manganese deposits. Elsevier, Amsterdam, pp. 109-146.

Reyss, J.L., Lemaitre, N., Ku, T.L., Marchig, V., Southon, J.R., Nelson, D.E. and Vogel, J.S., 1985. Growth of manganese nodule from the Peru Basin: A radiochemical anomaly. Geochimica et Cosmochimica Acta, 49: 2401-2408.

Roy, S., 1981. Manganese deposits. Academic Press, London, 458 pp.

Sawlan, J.J. and Murray, J.W., 1983. Trace metal remobilization in the interstitial waters of red clay and hemipelagic marine sediments. Earth and Planetary Science Letters, 64: 213-230.

Schmitz, W., Mangini, A., Stoffers, P., Glasby, G.P. and Plüger, W.L., 1986. Sediment accumulation rates in the Southwestern Pacific Basin and Aitutaki Passage. Marine Geology, 73: 181-190.

Segl, M., Mangini, A., Bonani, G., Hofmann, H.J., Nessi, M., Suter, M., Wölfli, W., Friedrich, G., Plüger, W.L., Wiechowski, A. and Beer, J., 1984. 10Be-dating of a manganese crust from the central North Pacific Ocean and implications for ocean palaeocirculation. Nature, 309: 540-543.

Segl, M., Mangini, A., Beer, J., Bonani, G., Suter, M. and Wölfli, W., 1989. Growth rate variations of manganese nodules and crusts induces by paleoceanographic events. Paleoceanography, 4: 511-530.

Shaw, T.J., Gieskes, J.M. and Jahnke, R.A., 1990. Early diagenesis in differing depositional environments: The response of transition metals in pore waters. Geochimica et Cosmochimica Acta, 54: 1233-1246.

Skornyakova, N.S. and Murdmaa, I.O., 1992. Local variation in distribution and composition of ferromanganese nodules in the Clarion - Clipperton Nodule Province. Marine Geology, 103: 381-405.

Sorem, R.K. and Fewkes, R.H., 1979. Manganese nodule research data and methods of investigation. IFI/Plenum Press, NY, 723 pp.

Stoffers, P., Glasby, G.P., Thijssen, T., Shrivastava, P.C. and Melguen, M., 1981. The geochemistry of co - existing manganese nodules, micronodules, sediments and pore waters from five areas in the equatorial and South - West Pacific. Chem. Erde, 40: 273-297.

Stoffers, P., Glasby, G.P. and Frenzel, G., 1984. Comparison of the characteristics of manganese micronodules from the equatorial and south - west Pacific. TMPM Tschermaks Min. Petr. Mitt., 33: 1-23.

Stoffers, P., Schmitz, W., Glasby, G.P., Plüger, W.L. and Walter, P., 1985. Mineralogy and geochemistry of sediments in the Southwestern Pacific Basin: Tahiti - east Pacific Rise - New Zealand. N.Z. Jl. Geol. Geophys., 28: 513-530.

Tebo, B.M., Ghiorse, W.C., van Waasbergen, L.G., Siering, P.L. and Caspi, R., 1997. Bacterially mediated mineral formation: insights into manganese (II) oxidation from molecular genetic and biochemical studies. Revs. Mineral., 35: 225-266.

Teleki, P.G., Dopson, M.R., Moore, J.R. and von Stackelberg, U. (eds), 1987. Marine minerals advances in research and recource assessment. D. Reidel, Dordrecht, 588 pp.

Thamdrup, B. and Canfield, D.E., 1996. Pathways of carbon oxidation in continental margin sediments off central Chile. Limnol. Oceanogr., 41: 1629-1650.

Thijssen, T., Glasby, G.P., Friedrich, G., Stoffers, P. and Sioulas, A., 1985. Manganese nodules in the central Peru Basin. Chem. Erde, 44: 1-46.

Thomson, J., Higgs, N.C., Hydes, D.J., Wilson, T.R.S. and Sorensen, J., 1987. Geochemical oxidation fronts in NE Atlantic distal turbidites and their effects in the sedimentary record. In: Weaver, P.P.E. and Thompson, J. (eds) Geology and geochemistry of abyssal plains. Geol. Soc. Spec. Publ., 31: 167-177.

Turner, S. and Buseck, P.R., 1981. Todorokites: A new family of naturally occuring manganese oxides. Science, 212: 1024-1027.

Usui, A., Nishimura, A. and Mita, N., 1993. Composition and growth history of surficial and buried manganese nodules in the Penrhyn Basin, Southwestern Pacific. Marine Geology, 114: 133-153.

Usui, A. and Mita, N., 1995. Geochemistry and mineralogy of a modern buserite deposit from a hot spring in Hokkaido, Japan. Clays Clay Mins., 43: 116-127.

Usui, A. and Terashima, S., 1997. Deposition of hydrogenetic and hydrothermal manganese minerals in the Ogasawara (Bonin) arc area, northwest Pacific. Mar. Georecourc. Geotechnol., 15: 127-154.

Usui, A. and Glasby, G.P., 1998. Submarine hydrothermal manganese deposits in the Izu-Bonin-Mariana arc: An overview. The Island Arc, 7: 422-431.

Varentsov, I.M. and Grasselly, G.Y. (eds), 1980. Geology and geochemistry of manganese. Hungarian Academy of Sciences, Budapest, 3 volumes.

von der Haar, D.L., Mahoney, J.J. and McMurtry, G.M., 1995. An evaluation of strontium isotopic dating of ferromanganese oxides in a marine hydrogenous crust. Geochimica et Cosmochimica Acta, 59: 4267-4277.

von Langen, P., Johnson, K.S., Coale, K.H. and Elrod, V.A., 1997. Oxidation kinetics of manganese (II) in seawater in nanomolar concentrations. Geochimica et Cosmochimica Acta, 61: 4945-4954.

von Stackelberg, U., Kunzendorf, H., Marchig, V. and Gwozdz, R., 1984. Growth history of a large manganese crust from the Equatorial North Pacific Nodule Belt. Geol. Jb., 75: 213-235.

von Stackelberg, U. and Marchig, V., 1987. Manganese nodule from the equatorial North Pacific Ocean. Geol. Jb., 87: 123-227.

von Stackelberg, U. and Beiersdorf, H., 1991. The formation of manganese nodules between the Clarion and Clipperton fracture zones southeast of Hawaii. Marine Geology, 98: 411-423.

von Stackelberg, U., 1997. Growth history of manganese nodules and crusts of the Peru Basin. In: Nicholson, K., Hein, J.R., Bühn, B. and Dasgupta S. (eds), Manganese mineralization: Geochemistry and mineralogy of terrestrial and marine deposits. Geol. Soc. Spec. Publ., 119: 153-176.

Wallace, H.E. et al., 1988. Active diagenetic formation of metal-rich layers in N.E. Atlantic sediments. Geochimica et Cosmochimica Acta, 52: 1557-1569.

Waychunas, G.A., 1991. Crystal chemistry of oxides and oxyhydroxides. Revs. Minerol., 25: 11-68.

Yeats, P.A. and Strain, P.M., 1990. The oxidation rate of manganese in seawater: Rate constants based on field data. Estuarine Coastal Shelf Sci., 31: 11-24.

# 12 Back to the Ocean Cycles: Benthic Fluxes and Their Distribution Patterns

MATTHIAS ZABEL, CHRISTIAN HENSEN
AND MICHAEL SCHLÜTER

## 12.1 Introduction

The observations made in the preceding chapters have demonstrated that the upper sediment layers usually contain the highest biochemical reaction rates within marine deposits.

The decomposition of organic matter represents the major motive force of the benthic system and its processes which are mostly catalyzed by microbial activity. What importance does the boundary between free water mass and the pore water system have? In this chapter, we will be concerned with finding answers to this question.

It is well established by a large number of long-term studies on the fluxes of particulate matter through the water column, that most part of the primary organic substance is remineralized in the upper water zones after the organisms died (cf. Chap. 4 and Sect. 5.5). Estimates on the marine carbon cycle, which were made for pelagic and hemipelagic domains of the ocean, demonstrate this process (Fig. 12.1). According to Berger et al. (1989), only about 1% of the organic matter on averages reaches sediment surface in the open ocean. Approximately 97% of this amount is decomposed in early diagenetic reactions so that only 0.03% of the organic carbon produced in the photic zone will ultimately become buried. Later we will more thoroughly discuss the differences between pelagic and coastal regions of the ocean where the burial rates are estimated to be 27 times higher. Although the information contained in Figure 12.1 might vary depending on the applied method (e.g. de Baar and Suess 1993), in most cases relationships and dimensions usually remain quite comparable. As will be seen later, this approximate balance may also apply to most nutrient cycles in the ocean.

Regarding the water body simplistically as an indivisible entity, without giving thought to its numerous internal cycles, only the boundary conditions of the system will ultimately determine its potential changes in the course of time. Under such an assumption, the following boundary systems will have to be considered: 1) the atmosphere, 2) the continents and 3) the ocean floor (Fig. 12.2).

Of all ocean boundaries the contact made with the atmosphere is surely of greatest significance

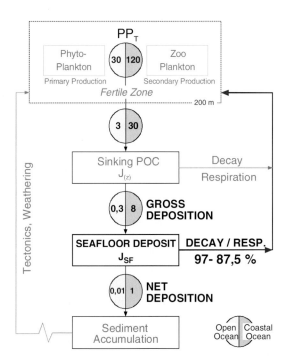

**Fig. 12.1** Schematic representation of the particulate, organic carbon cycle in the ocean according to Berger et al. (1989). The numerical values have the dimension of g C m$^{-2}$y$^{-1}$. Differences between the open ocean and coastal areas are shown. Of the 1% primary production (PP$_T$) that reaches the ocean floor, only 3% are embedded in the open ocean for a period of geological time. In contrast, 97% are decomposed by microbial activity and returned to the water column in the form of dissolved constituents (cf. Fig. 6.6).

373

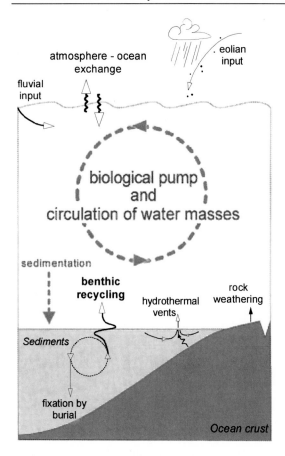

**Fig. 12.2** Simplified representation of the imports and exports pertaining to oceanic cycles.

entry into marine systems for the most elements (e.g. Chester (1990) gives an extensive overview especially of processes with terrestrial origin). In comparison, the local inputs originating from an erosion/weathering of the oceanic crust, or from hydrothermal activity (cf. Chap. 13), are quantitatively far less significant on a global scale.

Apart from the two boundary systems mentioned before, the role of the benthic interface is not to be defined exclusively by one transport direction. On the one hand, the deposit of marine sediments is quite the reverse from the routes of input already mentioned. Here, burial processes withdraw the elements from their marine cycles over geological time periods. Yet, as the example of the carbon cycle shown above already demonstrated, an immense proportion of accumulated particles are subject to dissolution or microbial decomposition in the course of early diagenesis. Marine sediments therefore also act as a 'secondary' source of remineralized dissolved components. The coexistence of these two fundamental opposite directed classes of processes constitutes one of the most essential phenomena on the seafloor. Next to studies on particle fluxes through the water column and element-specific accumulation rates, the quantification of benthic flux rates across the sediment/water boundary represents the third pre-condition for obtaining a complete balance of the marine material cycles.

The object of this chapter is to clarify the approaches to interpret benthic material fluxes, whereby emphasis is put on the regional and global interactions with other parameters. In this regard, three different concepts for the investigation of spatial distribution patterns will be introduced. The comparison of results from geostatistical methods and the opportunities given by modern geographic information systems (GIS) will be placed in the center of the observations. A critical evaluation of the existing database is fundamental for all procedures. A brief summary of this aspect of analytical and mathematical basics will therefore be given.

with regard to medium-term alterations in global cycles. Accordingly, one of the most fundamental issues in the discussion of climate change is concerned with the exchange processes of dissolved and free gases across this boundary (cf. Chap. 10). Considering the numerous regions with carbonate-rich sediments (cf. Chap. 1), the marine deposits are often explained in terms of a sink that, in the long run, works in opposition to the increase of green-house gases in the atmosphere (cf. Chap. 9). However, under the assumption of a continuous equilibration with the ocean, the source or sink function of the atmosphere may be neglected for the moment in the context presented here.

The fluvial input of dissolved and suspended material to the ocean presents a boundary condition which can be far better estimated. Apart from eolian particle transport, which may be of great importance in areas adjacent to the great desert regions, this mode represents the main route of

## 12.2 Fundamental Considerations and Assessment Criteria for Benthic Flux Rates

Locally conducted investigations are the basis of all territorial descriptions. Depending on the subject under study, the quality of regional and global observation, in principle, improves in accordance with the quantity of single studies and the type of their spatial distribution. However, it lies in the nature of scientific progress that the criterion for comparison is often only included to limited extent into the multitude of available data. This may have various causes. A first indication is given in Chapter 3 by a comprehensive survey of technical equipment and analytical methods currently employed in the geochemical studies of benthic boundaries. As the knowledge of detailed results obtained from applying these methods is an imperative criterion for selecting literature references

on benthic flux rates, a brief compilation of the essential aspects related to this subject will follow.

Irrespective of the differences between *in situ* and *ex situ* measurements (see below), there are two ways to determine the transport rates of dissolved substances across the water/sediment boundary: determination of the concentration differences along the depth profiles (gradient approach) and measurement of the time-dependent concentration changes of dissolved species within a closed reservoir (chamber experiments).

The difference between both approaches lies in the participation of different mechanisms of transport. Whereas only a diffusion-controlled transport of material can be measured by applying concentration profiles in the diffusion boundary layer (DBL) and in the pore water (a) (cf. Sect. 5.2.3), enrichment and/or depletion in the overlying bottom water deliver rates of total exchange (b). The particular transport processes are demonstrated schematically in Figure 12.3.

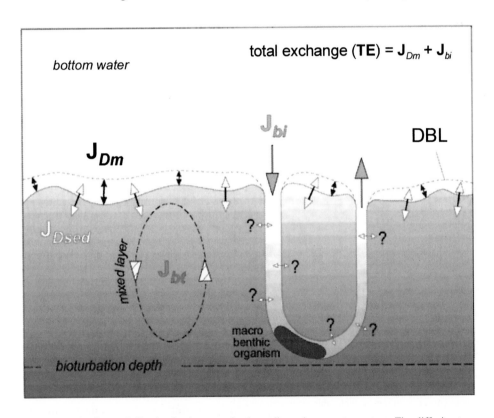

**Fig. 12.3** Transport mechanisms of dissolved substances in the sediment/pore water system. The diffusive transport is substance-specific and is solely based on the distribution of material concentrations. Its calculation is performed on the basis of concentration gradients in pore water ($J_{Dsed}$) and/or within the diffusive boundary layer ($J_{Dm}$). The morphology of the sediment surface is usually not considered. To determine total exchanges (TE), transport processes induced by macrobenthic activity must be additionally included. Whilst the unspecific transport by means of bioturbation ($J_{bt}$) can mostly be neglected, bioirrigation ($J_{bi}$) assumes great importance especially in densely populated and nutrient-rich sediments. A rarely noticed boundary exchange takes place across the walls of housing or feeding habitats which are densely populated by microbes.

### 12.2.1  Depth Resolution
### of Concentration Profiles

Let us first take a look at depth profiles of concentrations and how they are interpreted. Since the DBL represents the transition zone between the sediment/pore water system and the turbulent movements within the free water body, all diffusive exchange must pass across this layer of laminar flow. Since, biogeochemical reactions within the DBL are negligible relative to the surface sediments, measurements of concentration gradients across this zone provide an accurate estimate

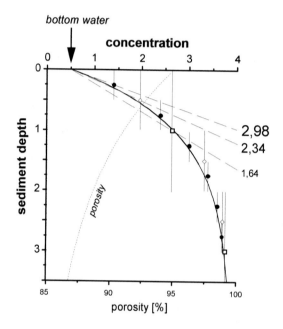

**Fig. 12.4**  Effects of the depth resolution in pore water concentration profiles on calculating the rates of diffusive transport. Three samples drawn from surface sediments are shown to possess different resolutions (intervals: 0.5 cm - dots, 1.0 cm diamonds, 2.0 cm - squares). All values are sufficient to plot the idealized concentration profile within the bounds of analytical error, yet very different flux rates are calculated in dependence on the depth resolution values. In the demonstrated example, the smallest sample distance indicates the highest diffusion (2.98 mmol cm$^{-2}$yr$^{-1}$). As soon as the vertical distance between single values increases, or, when the sediment segments under study grows in thickness, the calculated export across the sediment-water boundary diminishes (2.34 – 1.64 mmol cm$^{-2}$yr$^{-1}$). In our example, this error which is due to the coarse depth resolution can be reduced by applying a mathematical Fit-function. A truncation of 0.05 cm yields a flux rate of 2.84 mmol cm$^{-2}$yr$^{-1}$.(The indicated values were calculated under the assumption of the presented porosity profile according to Fick's first law of diffusion - see Chap. 3. A diffusion coefficient of 1 cm$^{-2}$yr$^{-1}$ was assumed. Adaptation to the resolution interval of 2.0 cm was accomplished by using a simple exponential equation).

of diffusive benthic exchange. As has already been explained in the Chapters 3 and 6, the necessary application of microsensor technology in deep-sea environments is only feasible for analyzing very few parameters ($O_2$, $CO_2$, Ca and pH). All other profiling measurements are consequently based on the conventional methods of pore water extraction, or are conducted by applying ion exchange resins in so-called gelpeepers. The sampling methods mentioned differ in their underlying principles of measurement and in their power of depth resolution. This is of enormous importance when the rates of diffusive exchange are to be determined. As concentration changes in the pore water fraction situated immediately below the DBL display linearity only in exceptional cases, the distance between each single reported concentration value markedly influences the determination of the exchange rate (cf. Chap. 3). An increased resolution of sampling sites could therefore easily lead to changes of a factor 2. Usually this leads to more pronounced gradients, as for example shown in Figure 12.4, i.e. higher rates of transport. The problem of having to decide upon which result should be used as the basis for regionalizing is likely to occur, especially when older and recent studies are compared.

A very careful examination of the database will become necessary should the flux rates, calculated on the basis of pore water profiles of variant resolution, also show some degree of discrepancy. A commonly used method to minimize such discrepancies consists in the application of mathematical Fit-functions (e.g. Sayles et al. 1996). As long as the sampling procedure includes at least all essential features of an anticipated concentration curve (zones of release and fixation), the application of non-linear functions will permit a 'theoretical' depth resolution of any desired precision and thus a theoretical flux calculation in close proximity to the boundary.

### 12.2.2  Diffusive *Versus*
### Total Solute Exchange

Figure 12.3 demonstrates that the diffusion-controlled exchange of substance represents only one more or less large proportion of the total transport activity between the sediment and the ocean bottom water. Studies on the influences of the macrobenthic population density and the role of bioirrigation in the overall fluxes revealed that negligence of biological transport processes in regions of high productivity, as well as in marginal - hence shallower areas of the ocean, might lead to considerable underestimates (e.g. Glud et al. 1994). However, in most oligotrophic deep-sea areas, these processes may be ignored in comparison to molecular diffusion (Sayles and Martin 1995). Frequently, a 1000m isobath is described as a 'boundary' of water depth below which the influence of macroorganisms strongly diminishes and diffusive fluxes are approximately identical to the total transport rates (cf. Sect. 6.3.2). However, strictly speaking, this assumption is merely an arbitrary simplification.

### 12.2.3  *In-Situ Versus Ex-Situ*

In order to assess the comparability of literature values, it is furthermore relevant to distinguish measurements and sample withdrawals which are conducted directly at the ocean floor (*in-situ*) from those conducted on board of research vessels or in laboratories (*ex-situ*). In Section 6.3.2 it was shown in detail that identical conditions of measuring the highly reactive elements oxygen and nitrogen (in the form of $NO_2$, $NO_3$, and $NH_4$) are of decisive importance for examining stoichiometric balances during early diagenesis. Indeed, the same is true for regional differences of single parameters. Depending on the sensitivity to the influences of extreme pressure and temperature changes, the analyzed concentrations might differ strongly (e.g. Jahnke et al. 1982, Glud et al. 1994, cf. Fig. 6.10). This applies to both profile determinations and to incubation experiments. From this, an important, yet hardly quantifiable, effect on the measurement of benthic transport rates will result.

### 12.2.4  Time-Dependent Variances and Spatial Variations in the Micro-Environment

Estimates of the significance of seasonal and micro-environmental variations of the benthic system are very difficult to perform. They are important for assessing the comparability of results obtained from single case studies. However, both factors could be essential for reliable data analysis. For instance, the reproduction of measurements carried out at 'identical' locations, but performed in successive years and in different seasons, in fact exhibit a significant variation of concentrations within the surface-near zone (e.g.

Zabel et al. 1998). Two reasons for this phenomenon are presently being discussed: a) seasonal variations of the particle flux down to the ocean floor and b) differences in relatively small natural environments.

Interannual variations of export production have long been known from long-term sediment trap experiments (e.g. Wefer and Fischer 1993). However, important effects of sediment pulses on biogeochemical reactions at the sediment surface are primarily confined to eutrophic areas, predominately along continental margins (Sayles et al. 1994). Such brief deposition events of organic-rich aggregates known as 'marine snow' induce a very strong increase of benthic reaction rates, especially as far as the oxygen consumption is concerned (cf. Fig. 6.17). The interannual variation of benthic flux rates in the pelagic zone of the deep-sea has been estimated to be rather low (Smith et al. 1992).

Apart from the variances in particulate input, there are also micro-environmental variations at the ocean floor which are time-independent (see Fig. 3.20). They primarily depend on the morphology of the sediment surface and the intensity of the bottom water current (compare with Chaps. 3 and 6). A connection between the sediment surface relief and marked sedimentation events was documented with the aid of time-lapse recordings of the ocean floor (e.g. Lampitt 1996). According to these, the so-called 'fluffy layer' is not a homogenous layer on top of the sediment, instead, deposits of phytodetritus accumulate in little hollows in the sediment surface. It is evident that, in such a case, significantly different rates are determined in the course of profiling measurements, depending on where the multicorer core has been removed, or where the micro-electrode measurement has been conducted.

Additionally to this heterogeneities on a scale of some centimetres, pore water profiles can also be influenced significantly by chemical variations on a much smaller space. The reasons are suboxic microenvironments, which are formed in small niches within the sediment or in single large particles like fragile fecal pellets, respectively. The existence of these phenomena presuppose conditions of rapid rates of consuming electron acceptors, and slow internal diffusivity (e.g. Jahnke 1985). Locally restricted deficits of oxidants, especially of oxygen and nitrate, are the consequence. In this context Jørgensen (1977) reported reducing microniches of about 50-200µm diameter, were sulfate reduction takes place in a basically oxidized surrounding. The retardation of diffusion may dependent on the presence of organic membranes or due to biotic and chemical processes constricting the porespace. Developing a simple mathematical model, Jahnke (1985) could already show that processes in microenvironments can effect pore water profiles in a different way. Applying a three dimensional model to results of high resolution DGT profiles (see Sect. 3.5) Harper et al. (sub.) recently deduce that the common one dimensional view of the system may underestimate relevant process rates by a factor of at least 3.

Just to be complete, advective effusions also need to be mentioned here (see Chap. 13). Although their significance for global balances of materials (e.g. $CO_2$, see Chap. 5) has been hardly investigated yet, the hot vents and cold seeps, usually locally confined, are rather unimportant for the interaction described here.

In summary, it remains that, on the one hand, knowledge of the applied methods and the local conditions is indispensable for the compilation of a homogenous database which should be as homogenous as possible. On the other hand, brief indications have also shown that a clearly defined objective is required for characterizing and balancing patterns of regional and global distribution, if an illicit comparison should be successfully avoided. As we will see, to merely resign to data produced under identical conditions is by no means a realistic conclusion. Good judgment is always of ultimate importance. Whether it helps to exclude certain results published in the literature depends on each question at hand. Table 12.1 gives an overview on the important aspects of this section.

## 12.3    The Interpretation of Patterns of Regionally Distributed Data

One important objective in characterizing regional differences of early diagenetic processes is to obtain information on the rate-determining control parameters and the benthic material fluxes. Previous comments have stated that, for various reasons, there can be no database which covers all of the depository zones and biogeochemical and sedimentological environments globally. It is therefore only possible to design global balances of geochemical material cycles by means of deterministic approaches. It seems reasonable, for the necessary interpolations, to focus

**Table 12.1**  Factors for assessing the comparability of differently indicated benthic flux rates.

| methods* | ex situ | | | | | in situ | | | | |
|---|---|---|---|---|---|---|---|---|---|---|
| | conventional pore water sampling | whole core squeezer | gelpeeper DET / DGT | microelectrodes / -optodes | incubation experiments | harpoon sampler | whole core squeezer | gelpeeper DET / DGT | microelectrodes / -optodes | incubation experiments |
| **main characteristics in this context** | | | | | | | | | | |
| flux calculation via gradient | yes | yes | yes | yes | **direct (!)** | yes | yes | yes | yes | **direct (!)** |
| depth resolution | some mm | mm | µm | µm | / | cm | mm | µm | µm | / |
| concentration | average | average | relative | "real" | "average" | average | average | relative | "real" | "average" |
| parameter set | most | most | many | some | "all" | most | most | many | some | "all" |
| transport mechan. | diffusive | diffusive | diffusive | diffusive | (total) | diffusive | diffusive | diffusive | diffusive | total !! |
| **effects on flux calculations** | | | | | | | | | | |
| reliability** | + | ++ | ++ | +++ | +++ | -- | ++ | ++ | +++ | +++ |
| bioturbation is neglected | yes | yes | yes | yes | (yes) | yes | yes | yes | yes | no |
| artefacts during sed. recovery | yes | yes | yes | yes | (yes) | no | no | no | no | no |

\*   for the detailed descriptions see Chapter 3
\*\*  this critical examination excludes mathematical data fits or subsequent computer modeling

on those control parameters of which we have attained the most information, for instance, primary production (see Sect. 12.5.2) or the frequently documented spatial distribution pattern of the benthic $C_{org}$ content (see Sect. 12.5.1).

The connections between the individual areas of the oceans have already been outlined at the beginning. Owing to an intensive research work during the last decades, many principles of interaction and modes of possible interference among single processes have been demonstrated (e.g. a high rate of primary production yields large amounts of organic substance in the sediment, which subsequently leads to a high density benthic population). According to the principle of actuality, any statement on environmental changes in Earth's history, as well as any future directed prognoses, are based on presently understanding of interrelated factors. Mostly, empirically raised and mathematically described conjunctions of one or more parameters are only valid for a limited range of locations. In the following, the most important control parameters will be presented, along with a brief comment on the potential relevance they have for the benthic interface.

As most subjects and factors have already been comprehensively discussed in the preceding chapters and sections, the subsequent observations will be exclusively limited to environmental aspects which are relevant to the subject.

### 12.3.1   Input and Accumulation of Organic Substance

The amount of dissolved nutrients returned to the hydrological cycles of the ocean is closely related to the availability of organic substance which is biodegradable by microbes upon and within the near-surface sediments (cf. Chaps. 3 - 10). There are two fundamental strategies to exploit this correlation as a control parameter for the benthic system and as a tool for the estimation of benthic material fluxes.

The direct way is to determine the organic carbon concentration in surface sediments or to determine the organic carbon burial rate. Although thousands of such measurements already exist worldwide (e.g. Premuzic et al. 1982, Romankevich 1984), problems related to the internal comparability of the data may arise here due to sampling artifacts or the different analytical methods applied (see Sect. 12.5.1). After all, organic substances are not identically prone to the

same degree of microbial decomposition (see Chap. 5). Since the analytical effort is enormous to classify the organic carbon either according to its origin (either as terrigenous or marine, e.g. by means of $\delta^{13}C$ -measurements, see Chap. 10), or its labile and refractory proportions, most $C_{org}$ amounts are reported as unspecified total values. Furthermore, despite the huge amount of data available, the measurements focus on selected oceanic regions whereas in numerous others substantial gaps exist. Yet, this approach to describe the dependence between the $C_{org}$ - content and benthic flux rates has proven successful in specific regions where the oceanographic conditions are comparable. (Figs 6.16b and 6.18b).

The second, indirect way to estimate the amount of organic substance in the benthic system relies on the correlation of the biological phytoplankton reproduction in the photic zone with the $C_{org}$ rain rate down to the sea floor. The advantage of this approach consists in the spatially and temporally high resolution obtained upon determining primary production. Although extensive territorial representations of primary production distribution patterns depended on correlations among relatively easy determinable parameters a few years ago, such as the well estimated amount of disposable nutrients or the latitude-dependent irradiance or light intensity (e.g. Berger et al. 1987), nowadays the determinations are conducted by evaluating satellite-aided measurements of thermal or optical features (e.g. Antoine et al. 1996, Behrenfeld and Falkowski 1997). One characteristic concept in this context is *remote sensing*. This collective term encompasses all aircraft and satellite survey techniques designed to measure parameters indirectly. Among these are the territorial observations, such as records of surface-water temperatures or pigment concentrations. Figure 12.5 demonstrates the global distribution of the mean annual primary production of phytoplankton as was accomplished on modeling various remote sensing parameters by Antoine et al. (1996).

The regional subdivision of the ocean resulted from a combination of the classic methods with the new options provided by the techniques of satellite survey. This classification into *biogeochemical provinces* was done on the basis of the knowledge obtained in surface oceanography and of the effects which the atmospheric circulation takes on the production of phytoplankton. According to Sathyendranat et al. (1995) and Longhurst

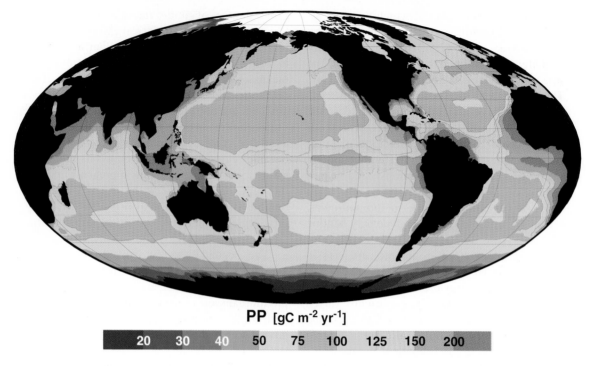

**PP** [gC m⁻² yr⁻¹]

$$PP\ [\text{gC m}^{-2}\text{yr}^{-1}]$$

| 20 | 30 | 40 | 50 | 75 | 100 | 125 | 150 | 200 |

**Fig. 12.5** Map representing the mean annual primary production [gCm⁻²yr⁻¹]. Among other parameters, the distribution of pigment concentration, surface-water temperature and surface irradiance were used for its construction. The representation is the result of model calculations (after Antoine et al. 1996).

et al. (1995) four zones are distinguished: 1) the polar zones 2) the west wind zone 3) the trade wind zone, and 4) the coastal zone. As a result of subdividing further, Longhurst et al. (1995) finally characterized 57 biogeochemical provinces worldwide that display different dynamics regarding the processes of primary production (Fig. 12.6). We will recognize later that this approach

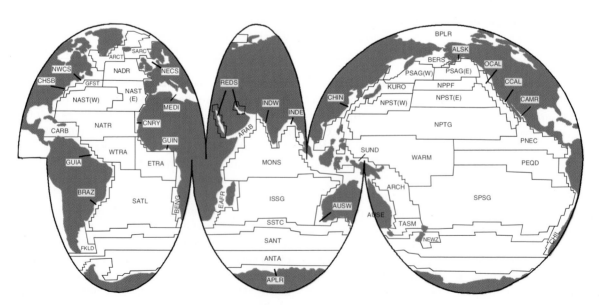

**Fig. 12.6** Subdivision of the world's oceans into 57 biogeochemical provinces. The geographical boundaries of the single regions have been set by inclusion of the subsurface distribution of chlorophyll, parameters of the photosynthesis-light relationship, the sun angle and cloudiness (after Longhurst et al. 1995).

can also be used for interpreting regional differences in benthic flux rates. For now, let us return to the estimation of accumulated biogenic material on the ocean floor.

Yet, the application of primary production as a control parameter for benthic flux rates does have disadvantages. These consist in the sophisticated processes in the course of sedimentation. On the one hand, the export production cannot be estimated directly on the basis of remote sensing data, because further empirical data will be required. On the other hand, results obtained from sediment trap experiments have revealed that the transport of particulate matter down to the ocean floor occurs only very seldom in a vertical fashion. Thus the pre-condition of a locally bound and vertical projection of biochemical processes in surface waters to the accumulation of biogenic particles on the ocean floor is not absolutely fulfilled (see Sect. 12.3.2). Another unknown factor in this context is represented by the sometimes extremely variable sedimentation velocities of single particles and aggregates. The time particles spend in the water column effectively influences the rates of dissolution and microbial decomposition reactions in the course of the sedimentation process.

Despite these limitations, empirically determined transfer functions are frequently applied which predominately describe the relation between water depth and particle flux. Thus determined fluxes of particulate organic carbon down to ocean floor is then applied as the rate limiting control parameter (see Sect. 12.5.2). Some of these mathematical equations are described in Chapter 6 (see Fig. 6.3). Bishop (1989) compares the most commonly used functions with results from sediment trap studies.

### 12.3.2    Vertical *Versus* Lateral Input of Particulate Matter

As already mentioned, currents in the water column effect a variable and strong lateral drift as the particles descend from the upper water zones vertically down to the ocean floor. This effect is particularly strong in the region of continental slopes, yet results from sediment traps have shown its occurrence in the open ocean as well. In spite of microbial decomposition reactions in the water column, deeper suspended sediment traps do not often display higher yields and flux rates than collections made several hundred or thousand meters

vertically above. Similar discrepancies are obtained when particle flux rates from sediment trap experiments are compared with benthic reaction rates.

Chapter 3 contains some examples which demonstrate how theoretically decomposed amounts of organic substance can be very easily calculated from pore water profiles or the depletion of the corresponding electron acceptors ($O_2$, $NO_3^-$, etc.) by making few definite basic assumptions (such as the negligence of oxidation of reduced species). Should the $C_{org}$ burial rate be additionally known from the depth profile of the $C_{org}$ content, then the sum of decomposition and preservation should be identical to the primarily accumulated amount of organic substance. A number of studies have proven, however, that a comparison of the particle flux either determined in the water column or calculated on the basis of the primary production (results derived from benthic studies) do not produce a conclusive picture. This is especially the case at continental margins (e.g. Walsh 1991, Rowe et al. 1994, Hensen et al., in rev.). Lateral drifts in the vertical-gravitative sedimentation process frequently do not suffice to produce an explanatory model. If the velocities of the currents increase, then resuspension and, in extreme cases, even the erosion of older sediments will turn into additional transport components. Whilst erosive processes are essentially limited to deep-ocean channels and passages with morphologically narrowed flow diameters, the so-called *winnowing effect* is of significance for the balances made for most shelf regions. Depending on the velocity of the bottom currents, the accumulation of the smaller or lighter components (like organic material) can be reduced or totally prevented. The effect of resuspension processes result in a rise in the number of particles in the water layer near the bottom, the *nepheloid boundary layer*. Its thickness might range from several tens to hundreds of meters. If the currents subside the particles previously held in suspension will become deposited. This accumulation process is referred to as *focusing* and occurs at the site of continental slopes, especially in the medium and lower slope regions. The experimental results shown in Figure 12.7 were obtained from a study on the central Californian continental margin. They demonstrate the lateral shift of mass from the shelf and the upper slope region down to the deep-sea.

Although the flux relationships and distribution briefly outlined so far are intended for specific regions, it is evident that reliable interpretations of the distribution patterns of biogeochemical reactions are otherwise impossible to make.

### 12.3.3    Composition of the Sediment

Whereas Chapter 1 presented a detailed introduction into the distribution and diversity of marine sediments, a brief reference will be made here to aspects of the sedimentary composition which might effect the exchange of dissolved components moving across the benthic interface (see Section 12.3.1 for $C_{org}$-content).

The problems that occur from applying biogenic sediment components as control parameters of benthic flux rates is shown in the example of the marine silicon cycle. The dispersal and the proportion of biogenic silicates in sediments depends on three factors: a) the rain rate of biogenic silicates, b) the accumulation of other particles, and c) the degree of preservation in the sediment. At first sight the significance of these three factor appears to be trivial. Yet, upon closer inspection, very sophisticated interactions with other environmental conditions may be quickly conceived. The rain rate of biogenic silicate as well as its degree of preservation in the sediment are both a function of spe-

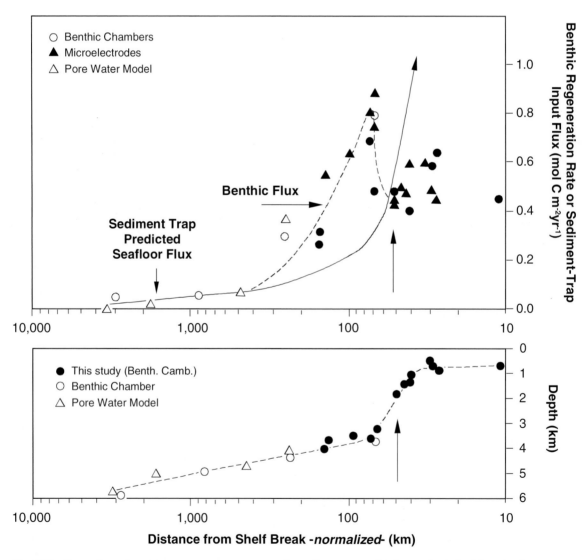

**Fig. 12.7**  Comparison of particle transport measurements using sediment traps with the results derived from *in situ*- measurements at benthic boundaries (after Jahnke et al. 1990). The lateral set-off of maximal values of organic material import, as predicted by trap studies, and the highest $O_2$-depletion rates (converted in terms of organic substance decomposition) indicates a $C_{org}$ - transport close to the ocean bottom, which is directed from the shelf and upper slope region to the deep-sea margins.

cific dissolution rates. It often occurs that the amount of biogenic opal in the sediment is below the limit of analytical detection, despite a significant amount of siliceous plankton within the upper water layers. From a thermodynamic point of view, the dissolution of 'pure' opal is a function of pressure, temperature and dissolved substances. However, since pure phases almost never occur in reality, the description of the dissolution processes proves insufficient as an approach to interpret the measured benthic material fluxes. It is known from numerous studies that the dissolution kinetics of biogenic opal is determined essentially by the concentration of absorbed trace elements and by organic coatings (e.g. van Bennekom et al. 1991). Whether trace elements are already incorporated during microbial excretion, or whether they are absorbed and bound after the organism ceases to exist, is not essential for our observation of the processes occurring at benthic interfaces. At any rate, the presence and amount of trace metals depends on the availability and the concentration of adequate lithogenic elements and is therefore regionally rather diversified. On the basis of these connections, van Cappellen and Qui (1997) were able to describe the relation between the primarily asymptotical increase of silicon concentrations in pore water, the silicon-specific ratio of lithogenic components, and the content of opal (Fig. 12.8). Since no correlation between the contents of pore water and the concentration of opal could be ascertained, it does not suffice to know the sediment's opal content in order to estimate the extent of the benthic reflux.

Another example showing the significance of the sediment's composition in benthic material fluxes lies in the effect of dilution. This process is equivalent to the effect of the accumulation rate of all biogeochemical non-reactive particles (NRP). More simply, a high accumulation rate, especially of carbonate (foraminifera, coccolithophorids etc.) and or lithogenic particles (clay minerals, quartz) produces an increase in burial efficiency of all other components ($C_{org}$, opal etc.). In case of constant inputs, for instance of organic substance, the time these substances spend in the especially reactive surface layer diminishes with an increase in the rain rate of inorganic compounds. Organic substance is withdrawn from the efficient microbial decomposition reactions which take place in the sediment-water transition zone. A reversal of the effect, the reduction of the burial rate due to decreasing di-

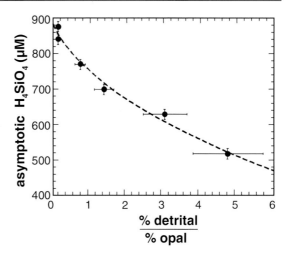

**Fig. 12.8** Correlation between the asymptotical concentration of $H_4SiO_4$ in pore water and the ratio of lithogenic particles and opal content (after van Cappellen and Qui 1997).

luted phases, may be induced by physico-chemical dissolution of carbonate below the CCD (see Chap. 9). Figure 12.9 demonstrates both effects schematically as simplified balances.

The intention of the precedent observations is to show that profound knowledge of potential control processes and/or control parameters is essential for the interpretation of distributive patterns of benthic transports.

## 12.4 Conceptual Approaches and Methods for Regional Balancing

Should the database be limited in its size, basin-wide or global balancing of geochemical material cycles will require extrapolation procedures over the area. Various conceptual approaches are available to this end. In this section, fundamental aspects to the most frequently used methods will be briefly presented. For more advanced insights into the matter described, we will have to refer to the specialized literature (e.g. Journel and Huijbregts 1978, Davis 1986, Akin and Siemes 1988, Schlüter 1996, Wackernagel 1996).

### 12.4.1 Statistical Key Parameters and Regression Analysis

The probably easiest method of regional balancing is based on the calculation of statistical key parameters, such as the mean value or the median value. In this context, the key parameters are always related to geographically defined regions. Therefore, the regions must be defined prior to statistically evaluating the presumed representative data. The definition of the region is not the result of the evaluation process. The usual calculatory operations are applied, statistical tests such as the Mann-Whitney U-test (Davis 1986) in order to calculate the key parameters from the single measured values. Having knowledge of the regional distribution in each district is consequently an important pre-condition to the application of the procedure. The sufficient fulfillment of these conditions permits balancing performance on a global level (e.g. Nelson et al. 1996).

The options and conclusions to be drawn from regression analysis are essentially more diverse.

In this case, the measured values are related to one or more variables of another sort. The interrelation between the parameters under study is described by functions of mathematical regression and correlation (Fig. 12.8). Such mathematical descriptions of relations between the single measured values and potential control parameters enable the statistical extrapolation of isolated measured values across the area (see Sect. 12.5.1). Furthermore, they can be employed for the characterization and demarcation of provinces, i.e. region-dependent validity assessment for the correlations to be investigated (see Sect. 12.5.2).

### 12.4.2  Variograms and Kriging

Basic publications related to this complex geostatistical procedure originate from Krige (1951)

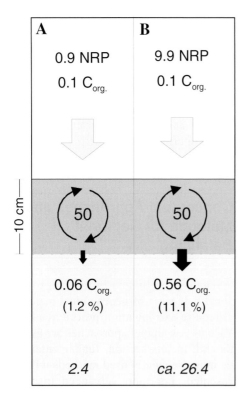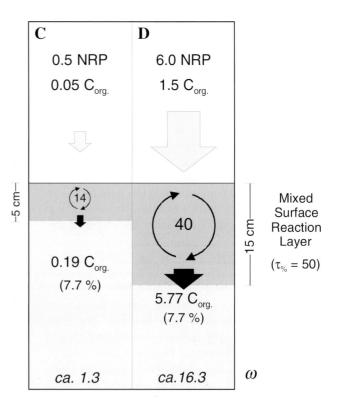

**Fig. 12.9**  Simplified representation of the burial efficiency of reactive substances (e.g. $C_{org}$) dependent on the accumulation rate of non-reactive components (NRP, esp. carbonate and lithogenic particles), after Jahnke (1996), modified. The following assumptions are fundamental to the simulations shown: 1) The sedimenting particles have a density of 2.5 g cm$^{-3}$, 2) the surface-near reaction layer is subject to constant bioturbation (homogenization); 3) the porosity is constantly 85%. Flux rates are reported in the unit mg cm$^{-2}$yr$^{-1}$, whereas the resultant sedimentation rates ($\omega$) are indicated as cm kyr$^{-1}$. The decomposition of organic carbon compounds ($\tau$) is reported as % / 50 yrs. The juxtaposition of scenarios A1 and A2 demonstrates the dependence of the $C_{org}$ burial rate on the rate of NRP accumulation. In case of identical decomposition rates, an increase of the NRP/$C_{org}$ ratio (10/1-100/1) will result in a more efficient burial of $C_{org}$ (1.2%-11.1%). The scenarios B and C show the relevant principle differences between the open ocean (B) and the coastal ocean (C). Compared to Figure 1, the selected Corg values are equivalent to rain rates of 50, or respectively, 225 gC m$^{-2}$yr$^{-1}$ of primary production. The realistic assumptions demonstrate that identical burial rates can prevail in both systems due to dissimilar marginal conditions (decomposition efficiency, thickness of the mixed layer); cf. Fig. 6.6).

and Matheron (1963). In the broadest sense, geostatistics deals with the spatial variability of location-dependent variables. Beginning with investigations on the storability of raw materials, a huge number of specialized procedures were developed in order to derive reliable estimations as to territorial structures of greater dimension on the basis of a size-limited database containing only local measurements. The problem can be formulated more simply: are single and spatially dispersed values capable of giving a representation of coherent structures, i.e. do results of local measurements stand in a spatial context? Usually, the solution to this fundamental problem can be found with the aid of *variogram analysis*. In contrast to regression analysis, only the relationships within one and the same parameter are examined (e.g. values of benthic $O_2$ - depletion). Information on eventual control parameters, such as the biodegradable amount of organic substance, as mentioned in the example above, are not required. As long as the single values are interrelated, their mathematical description by the various kriging methods will serve to provide the desired territorial interpolation. This strongly simplified presentation of the problem makes it evident that the application of kriging methods demands a precedent variogram analysis. Although the theoretical basics essential to this subject will be briefly mentioned in the following, we can merely make reference to the multitude of aspects concerned. Therefore we have to point out to publications of the geostatistics reference library listed in Section 12.4.

The construction of iso-linear maps implicates that the individual values obtained from local measurements are related to each other. These interrelations are usually subject to directional changes, i.e. closely neighbored values are more similar than distant ones. Such 'trend structures' showing decreasing similarity with an increase of distance can be studied by means of variogram analysis, since not only the values themselves are included but their geographical location as well. This is quite different from merely calculating characteristic values of statistical parameters. This connection is made by adding vectors to the measured values. In particular, variograms include vector functions designed to determine the one-half, medium, and squared differences between measurements conducted at two discrete locations (variances). The calculation of these variances will consequently consider the distance

between the points. Thus, distance clusters need to be defined to permit an allocation (Eq. 12.1) of each variogram value (as a sum function - $\gamma_{(h)}$) to a number of point pairs ($n_{(h)}$) defined by the chosen distance criterion (h). In the example shown in Figure 12.10, a cluster size of 0.5 (e.g. ° or km) was chosen, i.e. differences between measured values with distances of up to 0.5 are plotted in the first variogram, should the distance lie between 2 and 2.5, the difference value will be entered in the fifth variance calculation.

$$\gamma_{(h)} = \frac{1}{2n_{(h)}} \cdot \sum_{i=1}^{n} \left( z_{(x_i)} - z_{(x_i + h)} \right)^2 \qquad (12.1)$$

$z_{(x_i)}$ and $z_{(x_i + h)}$ : known values at the locations $x_i$ and $x_i + h$

Mostly, data are related up to a certain distance. The variogram values usually fluctuate at greater distances around one sill value which is equivalent to the total variance of the parameter studied (transitive variogram type). The distance required to reach the sill indicates the range of structural dependence. Frequently the ranges of a variogram

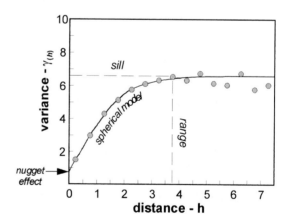

**Fig. 12.10** Schematic representation of a variogram. The mean distances of point-pair clusters are plotted (h) and the corresponding variances ($\gamma$(h)). The connection between individual values diminishes with an increasing distance, i.e. the $\gamma$(h)-values grow. However, the structural connections between the locally bound point pairs can only be demonstrated up to a specific range or span. After the variances have increase steadily, the variogram switches into a more or less level section that reveals fluctuations of mean variance (sill). The nugget effect, or the apparent failure of the variogram to go through the origin, indicative of a regionalized variable that is highly variable over distances less than the sampling/cluster interval. A spherical model was adapted to the idealized data.

vary depending on the chosen direction (geo-metrical anisotropy).

Mathematical model functions are adaptable to the variances (linear, spherical, exponential, etc.) which describe the relationships found between the single local measurements or their specific spatial orientation. These functions build the basis of the subsequently carried out regionalization procedure. There are various geostatistical software packages designed for performing variogram analyses (share ware: e.g. GEO-EAS (Englund and Sparks 1988); commercial: e.g. VARIOWIN (Pannatier 1996)).

If the data points originate from dispersed locations and the shape of the variogram is also known, kriging procedures may help to estimate values for any areal position which was left unexamined.

Single data selected from variogram analysis are subjected to a weighting process for estimations of defined locations (grid of interpolation) in such a manner that the estimated variances become minimal. Equation 12.2 demonstrates the formulation of ordinary kriging.

$$z^*_{(x)} = \sum_{i=1}^{n} \lambda_i \cdot z_{(x_i)} \qquad (12.2)$$

$z^*_{(x)}$ : estimated value for a real, but unknown value $z_{(x)}$

$z_{(x_i)}$ : known values at the locations $x_i$

$\lambda_i$ :   weighting factor dependent on distance and eventually on direction

A comprehensive system of equations needs to be solved in order to solve the optimization problem, making the sum of weighting factors turn to zero, i.e. not allowing distortion/strain occurs in the course of the estimation procedure. The limitation on applying this powerful regionalization procedure consists in the geographical distribution, or density, of the existing data. Apart from the ordinary kriging method briefly outlined, there are a number of complex extensions (universal kriging, co-kriging, external-drift-kriging) with which, for example, additional information on parameter interactions can be used for interpolation purposes. Unfortunately, the more detailed consideration of the methods of multivariate statistics will require more space than can be afforded within the limited scope of this textbook.

## 12.4.3   Geographical Information Systems (GIS)

Apart from mapping, the spatial distributions of benthic material fluxes (e.g. in the form of contour plots), global balancing of geochemical material cycles frequently requires the linkage of different levels of information. For instance, to study the amounts of organically bound carbon, or biogenic opal, which are mineralized in one sub-region of the ocean (e.g. the continental slope) bathymetric information on the geographical location of the continental slope, the deep-sea, or selected oceanic regions, are just as necessary as the contour plots of benthic fluxes. Geographical information systems (Bonham-Carter 1996) offer efficient methods for plotting contour maps and enable the conjunction of the different information levels.

One partial aspect of all territorial geochemical balancing methods consists in the representation and the calculation of areas of irregularly separated work fields. The surface exactness of specific oceanic regions, or the bathymetric depth zones, account for the fact that flux rate reports are always related to surface areas (e.g. mol m$^{-2}$yr$^{-1}$). To survey the Earth's surface and to plot it in a two-dimensional diagram various methods of projection can be applied (Krause and Tomczak 1995). In the conception of GIS all available parameters represent informational levels which dimension depends on the expansion of the area to be studied.

GIS programs are capable of connecting the various informational levels to each other, i.e. they put them in a relationship of mutual dependence (Bonham-Carter 1996). The systems are usually made of a database and a cartographic module which permits the construction and evaluation of maps as well as isoline plots. Apart from locally confined data, such as the benthic fluxes at definite locations, most GIS programs process grid data or vector data. One example for grid data are the already mentioned remote sensing data of pigment distribution in surface water, or the ice cover of the ocean which is currently recorded by the *Coastal Zone Color Scanner* of NASA[1] or the earth-orbiting NOAA-satelites (compare with Sect.12.3.1). These data are available as a grid of discrete image points and cover

---

[1]   An internet address list with references is to be found in the appendix.

extensive surface areas of the ocean. Another example is the ETOPO5 data set which contains global information on topographic height or bathymetric depth. This data set has a resolution of 5.5 minutes in circular measure.

Contrary to grid data, only such local coordinates are stored in a vector approach that describe a line (e.g. the facies boundary between deep-sea clay and globigerina ooze, Fig. 12.11). Bathymetric information is also available in the form of vector data (e.g. GEBCO data set). In this particular case, the coordinates are stored in clusters allocated to specific water depths. By connecting single points, the depth-line can be constructed directly (Fig. 12.11). This is not directly possible with a grid data set, which requires an interpolation algorithm to determine the local coordinates of a line, for instance the 2000m depth-line.

Contrary to computer-aided programs used in map construction, GIS enables a linkage between various mapping levels and levels of information. For instance, set operations are performed to blend and calculate intersections, or enable sectional map coupling procedures. A bathymetric map can be, for instance, combined with the distributive pattern of benthic particle fluxes. As a result, the transport across the sediment/water interface is balanced for specific regions, such as the deep-sea, the continental slope, and the shelf area.

## 12.5    Applications

Following the theoretical explanations of the preceding sections, the different conceptual approaches of regionalization and balancing will now be demonstrated on several practical examples.

### 12.5.1    Global Distribution of Benthic Oxygen Depletion Rates - An Example of Applied Regression Analysis

The immense importance of dissolved free oxygen for the microbial decomposition of organic substance has already been discussed at length in the Chapters 3, 5 and 6. Due to its chemical properties which make it one of the most effective energy sources and a reactive oxidizing agent, dissolved oxygen is usually consumed in the uppermost layers of the sediment. Since the benthic concentrations of oxygen are additionally affected by nearly all processes and factors mentioned in Section 12.2, the determination of benthic flux rates proves to be difficult. The number of reliable and comparable measurements is consequently very low. Moreover, most field studies are limited to selected oceanic areas. The low

**a)**

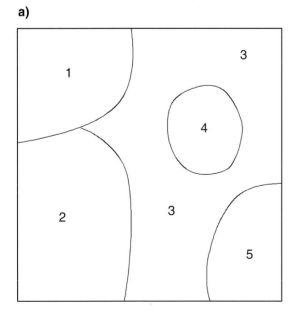

**b)**

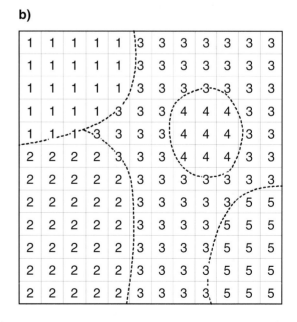

**Fig. 12.11**    Vector data set (a) used for representations of contour lines or sediment facies boundaries. The coordinates of each line stored. In a raster data set (b), the informations are stored for each pixel. Remote sensing data as primary production maps or sea ice distribution are available in raster format (after Bonham-Carter 1996, modified).

number of analyses and the extreme heterogeneity of the geographical distribution exclude an application of kriging procedures for regionalization as well as the construction of contour plots with the aid of GIS methods. Jahnke (1996) therefore chose regression analysis for the interpolation and extrapolation for a set of data consisting of 68 single measurements.

The close relation of the prevalent unstable organic matter was used as the rate-limiting control parameter in benthic oxygen consumption. Jahnke (1996) determined an empirical function for estimating flux rates across the sediment/water boundary, based on the correlation between the oxygen flux rate, the concentrations of $C_{org}$ and carbonate, and the accumulation rate. In this procedure, oxygen depletion is related to the burial rate of organic carbon compounds (Fig. 12.12). By including the concentration of carbonate and the rate of accumulation, the variations in evaluating the dilution effect will be compensated for (see Sect. 12.3.3). The latter is obtained on account of the different determinations obtained from carbonate-rich and carbonate-poor regions. The verification of this easy method merely produced a maximum deviation from real values of factor 2.

The result obtained from the regression between single measurement points and the control parameters was then applied to extrapolating the

area. Global oxygen consumption was estimated with the aid of a 2° grid, even in regions that had not revealed any results from field studies. Figure 12.13 demonstrates the thus estimated global pattern of oxygen fluxes in the deep-sea (> 1000m wd, see Sect. 12.2.2). The structures roughly follow the pattern of primary production. This does not take wonder as primary production is closely related to the available amount of benthic organic matter (compare Sect. 12.3.1) and this parameter was included into the function of estimation in the form of the concentration of $C_{org}$. Despite all inaccuracies adhering to details, some of which can be traced back to the applied 2° grid, the regionalization process did at last make the process of global balancing possible. Jahnke (1996) calculated a rate of global consumption for dissolved oxygen as $1.2 \cdot 10^{14}$ mol yr$^{-1}$. Assuming that microbes have consumed the oxygen completely in the biodegradation of the organic substance, which is done stoichiometrically (compare with Fig. 3.11), and by applying the estimated burial rates of $C_{org}$, a flux rate of particulate organic carbon (POC) was estimated to be $3.3 \cdot 10^{13}$ mol yr$^{-1}$ in the deep-sea. This value is equivalent to 45% of the POC-flux over the 1000m depth horizon ($7.2 \cdot 10^{13}$ mol C yr$^{-1}$), estimated on the basis of production.

### 12.5.2   Balancing the Diffusion Controlled Flux of Benthic Silicate in the South Atlantic - Applications of Kriging

Compared to the previously described conditions for investigating the global distribution of benthic oxygen flux rates, most of the other parameters display two differences. Firstly, no globally valid functions of correlation can be described on account of the multitude of different factors involved. Ultimately, all processes of early diagenesis are in fact influenced by the import of organic substance, still other control parameters are also of relevance for the reflux rates of most nutrients (e.g. C:N:P - ratio of the remineralized material, compare with Sect. 6.2). The complexity of the benthic system therefore demands to be treated with spatial differentiation. Like the biogeochemical provinces applied in calculating primary production (compare with Sect. 12.3.1), correlational dependencies between benthic flux rates and rate-limiting control parameters only possess regionally restricted validity. The second difference as compared to oxygen depletion consists in the greater number of reliable analytical

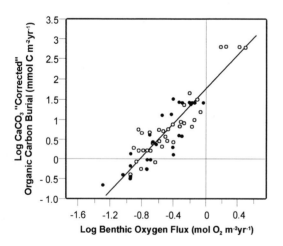

**Fig. 12.12**  Correlation between $C_{org}$-burial rates and benthic oxygen consumption. To compensate for deviant determinations made in regions with high and low $C_{org}/CaCO_3$ ratios, $CaCO_3$-burial rates are subtracted from the rates of total accumulation, thus producing *new* $C_{org}$ burial rates (COB) by multiplying the concentrations of $C_{org}$ with the accumulation rate corrected for $CaCO_3$. Open symbols represent Atlantic sites; solid symbols represent Pacific sites (after Jahnke 1996; cf. Fig. 6.6).

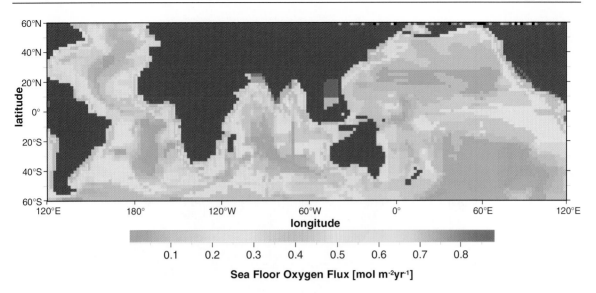

**Fig. 12.13**   Distribution pattern of benthic oxygen consumption (after Jahnke 1996).

results. In some oceanic regions, the available database, as well as its internal geographical resolution, already permit regionalizations based on kriging. The following example of the regional distribution of benthic silicon fluxes in the South Atlantic will make both aspects evident.

On the basis of comparative pore water measurements carried out at 76 locations in the eastern part of the South Atlantic, Zabel et al. (1998) demonstrated the distribution pattern of rates of benthic silicate release. As explained in Section 12.4.2 this extrapolation method always averages the measured values. From this it follows that the estimated result of regionalization may strongly differ from the real values in regions showing great differences between neighboring locations (especially in regions of the intensively studied continental slope). This effect depends on the resolution of the chosen grid (clustering) and can require a subsequent manual adaptation of the kriging results to the database. Figure 12.14a shows the corrected result of kriging-regionalization.

To regionalize the entire South Atlantic, Hensen et al. (1998) used an extended database which contained 180 single measurements. Contrary to the more detailed map of the eastern parts, the distribution pattern obtained was not corrected for, which was partially due to the rougher subdivisions made by the isolines (Fig. 12.14b). As for many regions, the results obtained from both procedures reflect the distributive pattern of surface-water activity (see Fig. 12.5).

There are, however, differences to be observed in two areas: a) in the eastern equatorial Atlantic and b) the western part of the Argentine Basin. Despite a high level of production, the benthic silicon release is quite low in Guinea and the northern part of the Angola Basin. There are no obviously high production rates to match the high solubility rates in the western part of the Argentine Basin (see Fig. 12.13). What is the reason for this peculiar type of distribution?

As already described in Section 12.3.3, the solubility rate of biogenic opal depends on a large number of processes and environmental conditions. The quality of the various opal phases has a crucial influence in this regard (Archer et al. 1993). The Niger River and the Zaire River, two of the greatest streams worldwide, both flow into the ocean at the northern part of the Angola Basin where they release tremendous amounts of dissolved and particulate substance freight into the ocean (compare with Fig. 1.2). Several studies have shown that the solubility of opal decreases when the proportion of aluminiferous lithogenic components increases (e.g. van Cappellen and Qui 1997). The high opal content in the sediments of river estuaries should therefore result from an effective preservation of amorphous silicates. Distant from equatorial upwelling, another explanation should be considered, one which is also discussed for similar anomalies observed in benthic distribution patterns of organic decay. Remineralization processes in the water column,

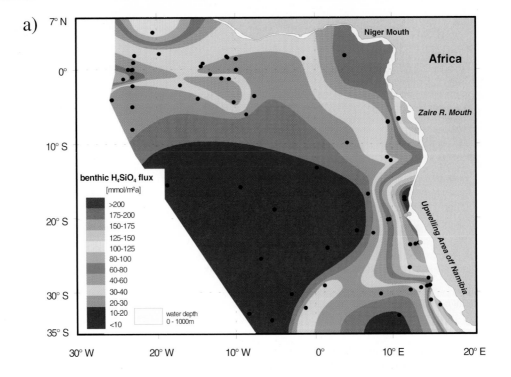

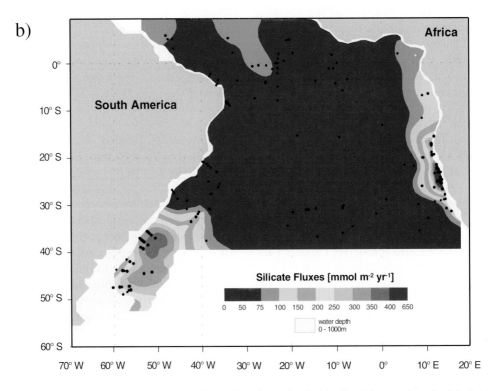

**Fig. 12.14**   Distribution pattern of rates of benthic silicon release in the South Atlantic. a) Corrected result of the kriging procedure on the basis of 76 measurement points in the eastern part of the South Atlantic (redrawn after Zabel et al. 1998). b) results of regionalization not corrected for, valid for the entire Atlantic ocean on the basis of 180 single measurements (redrawn after Hensen et al. 1998). The results of both procedures in general reflect the distribution pattern of surface-water productivity. Single significant deviations (western Argentinean Basin, eastern equatorial region) can be explained as due to various influences exerted on the particle flux down to the seafloor.

more intensive than described by generalizing transfer functions (cf. Fig. 6.3), lead to a reduction of specific accumulation rates. Thus, an increased solubility rate prior to deposition, and a reduced solubility process afterwards, would give cause to the low benthic silicon releases.

The sedimentary structures of the Argentine Basin are distinguished by powerful currents (Ledbetter and Klaus 1987). Sinking particles, or particles already deposited, are subjected to a lateral drift over wide passages, or become re-suspended (compare with Sect. 12.3.2). As a result for the benthic interface, relatively low amounts of biogenic material accumulates on the extensive shelf. Especially the organic compounds are translocated (*winnowing*) and then accumulate in the area of the continental rise (*focusing*). To a certain extent, the high production rates of shallow regions find their equivalent in the deep-sea, however, more pronounced. This effect is enhanced by the additional import of readily soluble opal from the Southern Ocean.

What are the conclusion we can draw from this type of regionalization? Figure 12.13b shows that approximately 80% of the total release occurs in the predominately oligotrophic open ocean. The balance for the South Atlantic (>1000m water depth) revealed a silicon recycling rate of $2.1 \cdot 10^{12}$ mol Si yr$^{-1}$. Assuming that the area under study specifies a representative part of the world's oceans, the global release rate may be extrapolated up to a value of $2.0 \cdot 10^{13}$ mol Si yr$^{-1}$ (Hensen et al., 1998). These rough approximations probably underestimate the real flux rate, since opal-rich sediments from the Southern Ocean, and from the equatorial Pacific, are not realistically taken into account (cf. Sect. 12.5.3). In all cases, these rough balances demonstrate that the benthic reflux of silicon outweighs the import by rivers and streams by a factor of 3 ($7.3 \cdot 10^{12}$ mol Si yr$^{-1}$, Wollast and Mackenzie 1983). The balances confirm the results of other studies on the marine silicon cycle (e.g. Tréguer et al. 1995).

Figure 6.12 shows an additional application of the Kriging method.

### 12.5.3 Benthic Si-Release in the Southern Ocean (Atlantic Sector) - A GIS-Supported Balance

One example for an application of the geographical information system is shown in Figure 12.15. Here, bathymetric information, the zonal subdivision into the oceanographic regions, for which a balance calculation is to be performed and finally the benthic fluxes of dissolved silicate across the sediment/water interface are depicted.

The bathymetrics of the area were derived from the GEBCO data set. Furthermore, a vector data set was entered which specifies the sub-regions for which the surface-area balance is to be calculated. For map construction, the surface-exact Lamberth-Azimuthal projection was chosen.

A method of neighborhood analysis (Bonham-Carter 1996) was chosen to translate the silicon flux rates, which were determined at the measuring stations, to the surface areas. The Thiessen-polygons shown in Figure 12.15c are calculated by using an algorithm which determines the surface area representative of one particular measurement site on the basis of the regional distribution of all sites included. Due to this unequivocal procedure, a first estimate on the flux of dissolved silicon is obtained with respect to the area through which silicon is returned to the substance cycle in the bottom water. The Thiessen polygons provide a conservative and reproducible estimate when data is scant and the construction of the isoline map of the material fluxes consequently depends strongly on the expected regional distribution of the studied parameter. As compared to the estimates made on the basis of statistical features, like the mean number or the median number, this method has the advantage that it includes the spatial distribution of measurement points (Fig. 12.16). However, size and shape of Thiessen polygons are not determined by the properties of the measured values as such, but exclusively by the spatial distribution of the measurement sites.

The suitability of balancing procedures on the basis of Thiessen polygons should therefore always be reconsidered depending on the indivual case at hand. Should the reflux study, like the variogram analysis (see Sect. 12.4.2), reveal that the isoline map is valid and reliable, then those balances (e.g. performed by Kriging methods or by combination with GIS) must certainly be preferred which are based on contour plots. Unfortunately, this is not common in supraregional balances of marine material cycles. Although a rather comprehensive data set of more than 140 single measurements was available in the example shown, the construction of an isoline map had not been unequivocally possible.

The GIS option to process various levels of information permits the combination of the bathymetric map with the oceanographic subregions and the Thiessen polygons which were determined on the basis of distributed measurement sites. This becomes evident in the example of the central part of the Weddell Sea (Fig. 12.16). A flux rate of dissolved silicate was calculated for the polygon area, or the partial areas that are encompassed by this irregularly defined region possessing a water depth of more than 3500m. The procedure allows to estimate the flux of dissolved silicate down to the bottom water zone of the central part of the Weddell Sea. The result is also not affected by faulty surface area calculations. By conducting an analogous procedure in other sub-regions shown in Figure 12.14, the release of benthic silicate can be balanced for the entire region. Schlüter et al.

(1998) calculated a diffusive Si reflux rate of 0.84 $10^{12}$ mol yr$^{-1}$ (compare with Sect. 12.5.2) for a benthic interface measuring $9.5 \cdot 10^6$ km$^2$.

Owing to the efficient processing options of working with GIS, the evaluation can always include additional measurements or additional informational levels, like the regional distribution of primary production, at a later time. Balances can therefore be updated quickly and without problem.

## 12.6    Concluding Remarks

The regionalization and balancing examples have emphasized the value of such investigations for solving a variety of problems. They have also shown that 1) many aspects related to the selec-

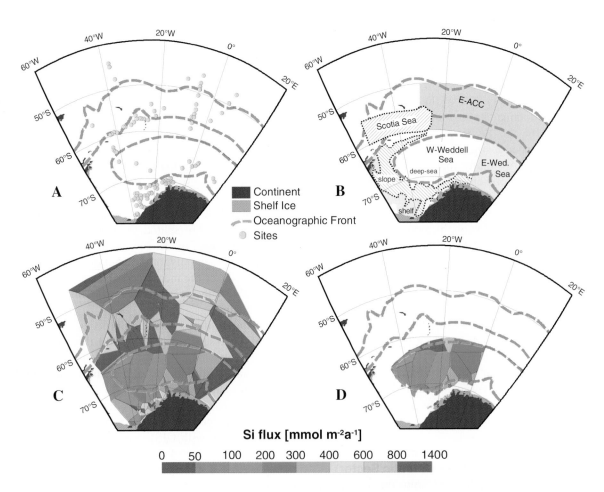

**Fig. 12.15** GIS application for the calculation of benthic Si-release in the Southern Ocean (Atlantic sector). Bathymetric data (a) and information on oceanographic regions (b) were used. The GIS-controlled construction of single areas (Thiessen polygons; Bonham-Carter 1996) is based on the spatial distribution of the measuring stations. The surface-exact Lamberth-Azimuthal projection was chosen for making the balances (c). By combining the informational levels, balances are made for definite regions, for example for the western part of the Weddell Sea (d). Redrawn after Schlüter et al. (1998).

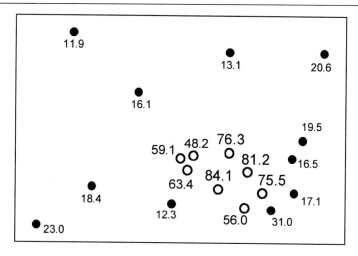

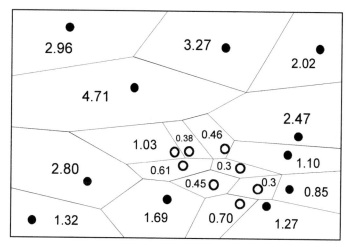

**Fig. 12.16** Example showing the influence of the spatial distribution of measuring stations in surface area balances of material cycles. If the fluxes are balanced exclusively on the basis of the mean value obtained from the single measured values (A), a value of 1122 mol yr$^{-1}$ will result. A more differentiated assessment of the regional distribution of the measured values by applying Thiessen polygons (B, areas are shown) leads to the value of 695 mol yr$^{-1}$ (= S value$_i$ · area$_i$). This notable difference results from the fact that relatively high values are concentrated in one confined area. The calculation of the mean value, including the entire data, consequently leads to an overestimation. The construction of Voroni or Thiessen polygons accounts for the distribution of the values and their limitation to a relatively small area.

tion and compilation of the database need to be noticed prior to the performance territorial studies, 2) that the interpretation of distributive patterns requires the knowledge of regional oceanographic (e.g. amount of export production or oceanic current patterns) and chemical conditions (e.g. sediment composition). Beyond this, regionalizations can be used as tools to prove the plausibility of prevailing concepts on processes. Thus, hypotheses on supraregional/global biogeochemical interactions can be verified. Therefore, more and more attention is given to database applications and the attempts of balancing studies to join as many results as possible obtained from many

successful, national and international research programs. Apart from the balance of essential single components, and as follows from the current publications on the distribution of primary production, it is the characterization of the benthic biogeochemical provinces that can be now formulated as the main objective.

This is contribution No 261 of the Special Research Program SFB 261 (*The South Atlantic in the Late Quaternary*) funded by the Deutsche Forschungsgemeinschaft (DFG).

## Appendix

Some useful internet addresses for the search of data and remote sensing projects.

*British Oceanographic Data Centre (BODC)*
www.nbi.ac.uk/bodc/bodcmain.html

*Climate and Environmental Data Retrieval and Archive (CERA)*
www.dkrz.de/forschung/project/cera.html

*Geochemical Ocean Sections Study (GEOSECS)*
ingrid.idgo.columbia.edu/SOURSES/
GEOSECS

*Joint Global Ocean Flux Study – Data inventory*
ads.smr.uib.no/jgofs/inventory/
index.html*NASA's Earth Science Enterprise*
www.earth.nasa.gov/missions/index.html

*National Geophysical Data Center*
www.ngdc.noaa.gov

*SeaWIFS Project, NASA – Gridded Space Flight Center*
seawifs.gsfc.nasa.gov/SEAWIFS.html

*World Data Centers*
cdiac.esd.ornl.gov/wdca/wdcinfo.html

*World Data Center-A Marine Geology & Geophysics*
www.ngdc.noaa.gov/mgg/aboutmgg/
wdcamgg.html

*5-Minute Gridded Elevation Data Selection*
www.ngdc.noaa.gov/mgg/global/settopo.html

*PANGAEA – Network for Geological and Environmental Data*
www.pangaea.de

# References

Akin, H. and Siemes, H., 1988. Praktische Geostatistik - Eine Einführung für den Bergbau und die Geowissenschaften. Springer Verlag, Berlin, Heidelberg, NY, 304 pp.

Antoine, D., André, J.-M. and Morel, A., 1996. Oceanic primary production; 2. Estimation at global scale from satellite (coastal zone color scanner) chlorophyll. Global Biogechemical Cycles, 10(1): 57-69.

Archer, D., Lyle, M., Rodgers, K. and Froelich, P., 1993. What controls opal preservation in tropical deep-sea sediments. Paleoceanography, 8: 7-21.

Behrenfeld, M.J. and Falkowski, P.G., 1997. Photosynthetic rates derived from satellite-based chlorophyll concentration. Limnology and Oceanography, 42(1): 1-20.

Berger, W.H., Fischer, K., Lai, C. and Wu, G., 1987. Ocean producitvity and organic carbon flux. I. Overview and maps of primary production and export production. University California, San Diego, SIO Reference, 8: 7-30.

Berger, W.H., Smetacek, V.S. and Wefer, G., 1989. Ocean productivity and paleoproductivity - an overview. In: Berger, W.H., Smetacek, V. and Wefer, G. (eds) Productivity of the ocean: present and past. Wiley & Sons, pp 1-34.

Bishop, J.K.B., 1989. Regional extremes in particulate matter composition and flux: effects on the chemistry of the ocean interior. In: Berger, W.H., Smetacek, V. and Wefer, G. (eds) Productivity of the ocean: present and past. Wiley & Sons, pp 117-137.

Bonham-Carter, G.F., 1996. Geographic information systems for geosciences: Modeling with GIS. Pergamon Press, NY, 435 pp.

Chester, R., 1990. Marine Geochemistry. Chapman & Hall, London, 698 pp.

Davis , J.D., 1986. Statistics and Data Analysis in Geology. Wiley & Sons, NY, 646 pp.

DeBaar, H.J.W. and Suess, E., 1993. Ocean carbon cycle and climate change - An introduction to the interdisciplinary Union Symposium. Global and Planetery Change, 8: VII - XI.

Englund, E. and Sparks, A., 1991. Geostatistical environment assessment software - User's guide -. US - EPA Report #600/8 - 91/008, EPA - EMSL, Las Vegas, Nevada.

Glud, R.N., Gundersen, J.K., Jorgensen, B.B., Revsbech, N.P. and Schulz, H.D., 1994. Diffusive and total oxygen uptake of deep-sea sediments in the eastern South Atlantic Ocean: in situ and laboratory measurements. Deep-Sea Research, 41: 1767-1788.

Harper, M.P., Davison, W. and Tych, W., subm. One dimensional views of three dimensional sediments. Environmental Science and Technology.

Hensen, C., Landenberger, H., Zabel, M. and Schulz, H.D., 1998. Quantification of diffusive benthic fluxes of nitrate, phosphate and silicate in the Southern Atlantic Ocean. Global Biogeochemical Cycles, 12(1): 193-210.

Hensen, C., Zabel, M. and Schulz, H.D., in press. A comparision of benthic nutrient fluxes from deep - dea sediments of Namibia and Argentinia. In: Gansen, G.M. and Wefer, G. (eds) Particle flux and its preservation in deep-sea sediments.Deep-Sea Research II.

Jahnke, R.A., Heggie, D., Emerson, S. and Grundmanis, V.,

1982. Pore waters of the central Pacific Ocean: nutrient results. Earth Planetary Science Letters, 61: 233-256.

Jahnke, R.A., 1985. A Model of Microenvironments in Deep-Sea Sediments: Formation and Effects on Porewater Profiles. Limnology Oceanography, 30(5): 956-965.

Jahnke, R.A., Reimers, C.E. and Craven, D.B., 1990. Intensification of recycling of organic matter at the sea floor near ocean margins. Nature, 348: 50-54.

Jahnke, R.A., 1996. The global ocean flux of particulate organic carbon: Areal distribution and magnitude. Global Biogeochemical Cycles, 10: 71-88.

Jørgensen, B.B., 1977. Bacterial sulfate reduction within reduced microniches of oxidized marine sediments. Marine Biology, 41: 7-17.

Journel, A.G. and Huijbregts, C., 1978. Mining Geostatistics. Academic Press, London, 600 pp.

Kraus, G. and Tomaczak, M., 1995. Do marine scientists have a scientific view of the earth? Oceanography, 8: 11-16.

Krige, D.G., 1951. A statistical approach to some basic mine valuation problems an the Witwatersrand. Journal Chem. Metall. Min. Soc. South Africa, 52: 119-139.

Lampitt, R.S., 1996. Snow falls in the open ocean. In: Summerhayes, C.P. and Thorpe, S.A. (eds) Oceanography - An illustrated guide. Manson Publ., Southampton Oceanogr. Centre, pp 96-112.

Ledbetter, M.T. and Klaus, A. (eds), 1987. Influence of bottom currents on sediment texture and sea-floor morphology in the Argentine Basin. Geology and Geochemistry of Abyssal Plains, Geological Society Special Publication, 31: 23-31.

Longhurst, A., Sathyendranath, S., Platt, T. and Caverhill, C., 1995. An estimate of global primary producion in the ocean from satellite radiometer data. Journal of Plankton Research, 17: 1245-1271.

Matheron, G., 1963. Principles of geostatistics. Economic Geology, 58: 1246-1266.

Nelson, D.M., Tréguer, P., Brzezinski, M.A., Leynaert, A. and Quéguiner, B., 1995. Production and dissolution of biogenic silica in the ocean: Revised global estimates, comparison with regional data and relationship to biogenic sedimentation. Global Biogeochemical Cycle, 9: 359-372.

Pannatier, Y., 1996. Variowin: Software for spatial data analysis in 2D. Springer Verlag, Berlin, Heidelberg, NY, 91 pp.

Premuzic, E.T., Benkovitz, C.M., Gaffney, J.S. and Walsh, J.J., 1982. The nature and distribution of organic matter in the surface sediments of world oceans and seas. Organic Geochemistry, 4: 63-77.

Romankevich, E.A., 1984. Geochemistry of organic matter in the ocean. Springer Verlag, Berlin, Heidelberg, NY, 334 pp.

Rowe, G.T., Boland, G.S., Phoel, W.C., Anderson, R.F. and Biscaye, P.E., 1994. Deep - sea floor respiration as an indication of lateral input of biogenic detritus from continental margins. Deep-Sea Research, 41: 657-668.

Sathyendranath, S., Longhurst, A., Caverhill, C.M. and Platt, T., 1995. Regionally and seasonally differentiated primary production in the North Atlantic. Deep-Sea Research, 42(10): 1773-1802.

Sayles, F.L., Martin, W.R. and Deuser, W.G., 1994. Response of benthic oxygen demand to particulate organic carbon supply in the deep sea near Bermuda. Nature, 371: 686-689.

Sayles, F.L. and Martin, W.R., 1995. In Situ tracer studies of solute transport across the sediment - water interface at the Bermuda Time Series site. Deep-Sea Research, 42(1): 31-52.

Sayles, F.L., Deuser, W.G., Goudreau, J.E., Dickinson, W.H., Jickells, T.D. and King, P., 1996. The benthic cycle of biogenic opal at the Bermuda Atlantic Time Series site. Deep-Sea Research, 43(4): 383-409.

Schlüter, M., 1996. Einführung in geostatische Verfahren und deren Programmierung. Enke Verlag, Stuttgart, 326 pp.

Schlüter, M., Rutgers van der Loeff, M.M., Holby, M. and Kuhn, G., 1998. Silica cycle in surface sediment of the South Atlantic. Deep-Sea Research, 45: 1085-1109.

Smith, K.L.j., Baldwin, R.J. and Williams, P.M., 1992. Reconciling particulate organic carbon flux and sediment community oxygen consumption in the deep North Pacific. Nature, 359: 313-316.

Tréguer, P., Nelson, D., Van Bennekom, A.J., DeMaster, D.J., Leynaert, A. and Quéguiner, B., 1995. The Silica Balance in the World Ocean: A Reestimate. Science, 268: 375-379.

Van Bennekom, A.J., Buma, A.G.J. and Nolting, R.F., 1991. Dissolved aluninium in the Weddel-Scotia Confluence effect of Al on the dissolution kinetics of biogenic silica. Marine Chemistry, 35: 423-434.

Van Cappellen, P. and Qui, L., 1997. Biogenic silica dissolution in sediments of the Southern Ocean. I. Solubility. Deep-Sea Research, 44: 1109-1128.

Wackernagel, H., 1996. Multivariate Geostatistics. Springer Verlag, Berlin, Heidelberg, NY, 256 pp.

Walsh, J.J., 1991. Importance of continental margins in the marine biogeochemical cycling of carbon and nitrogen. Nature, 350: 53-55.

Wefer, G. and Fischer, G., 1993. Seasonal Patterns of vertical Particle Flux in equatorial and coastal Upwelling Areas of the Eastern Atlantic. Deep-Sea Research, 40: 1613-1645.

Wollast, R. and Mackenzie, F.T., 1983. The global cycle of silica. In: Aston, S.R. (ed) Silicon Geochemistry and Biogeochemistry. Academic Press, London, pp 39-76.

Zabel, M., Dahmke, A. and Schulz, H.D., 1998. Regional Distribution of Phosphate and Silicon Fluxes across the Sediment Water Interface in the Eastern South Atlantic. Deep-Sea Research, 45: 277-300.

# 13 Input from the Deep: Hot Vents and Cold Seeps

PETER M. HERZIG AND MARK D. HANNINGTON

The discovery of black smokers, massive sulfides and vent biota at the crest of the East Pacific Rise at 21°N in 1979 (Francheteau et al. 1979, Spiess et al. 1980) confirmed that the formation of new oceanic crust through seafloor spreading is intimately associated with the formation of metallic mineral deposits at the seafloor. It was documented that the 350°C hydrothermal fluids discharging from the black smoker chimneys at this site at a water depth of about 2600 m continuously precipitate metal sulfides in response to mixing of the high-temperature hydrothermal fluids with ambient seawater. The metal sulfides including pyrite, sphalerite, and chalcopyrite eventually accummulate at and just below the seafloor and have the potential to form a massive sulfide deposit. It has also been documented that the circulation of seawater through the oceanic crust is the principal process responsible for leaching and transport of metals in this environment. Seawater which deeply penetrates into the oceanic crust at seafloor spreading centers is being converted to a hydrothermal fluid with low pH, low $E_H$, and high temperature during water-rock interaction above a high-level magma chamber. This fluid is than capable of leaching and transporting metals and other elements which eventually precipitate as massive sulfides at the seafloor or as stockwork and replacement sulfides in the subseafloor. The resulting massive sulfide deposits can reach considerable size. The TAG hydrothermal mound at the Mid-Atlantic Ridge at 26°N for example has a diameter of about 200 m, a height above seafloor of about 50 m and is characterized by a black smoker complex on the top which consists of several black smokers venting hydrothermal fluids at a temperature of more than 360°C.

Seafloor hydrothermal activity at mid-ocean ridges and back-arc spreading centers has a major impact on the chemistry of the oceans (Edmond et al. 1979a, 1982) and has been responsible for extensive alteration of modern and ancient oceanic crust (Alt 1995). It has been estimated that 25-30% of the Earth´s total heat flux is transferred from the lithosphere to the hydrosphere by the circulation of seawater through oceanic spreading centers (Stein and Stein 1994, Lowell 1991). Early calculations of the total discharge of hydrothermal vents at oceanic ridges arrived at values of about $5 \cdot 10^6$ L/s, which requires that the total amount of water in the oceans is circulated through thermally active seafloor rift zones every 5-11 Ma (Wolery and Sleep 1976, Morton and Sleep 1985). Another assessment of the available data has led Lowell et al. (1995) to suggest that this time span is as short as one million years. Neglecting any component of diffuse flow, the estimated flux of high-temperature fluids would require at least one black smoker with a mass flux of approximately 1 kg/s and an estimated power of 1.5 megawatts (Converse et al. 1984) for every 50 meters of ridge crest (55,000 km in total). The rate of circulation of hydrothermal fluids through ridge-crest hydrothermal systems is small by comparison with the fluxes from rivers, but nevertheless significantly affects the geochemical budget of certain elements that are highly concentrated in vent fluids.

In this chapter, we have choosen to focus on ridge crest hydrothermal systems which are characterized by high-temperature black- and white smoker discharge (250-400°C), the formation of polymetallic massive sulfide deposits, and the occurrence of typical vent biota and bacterial communities. These systems are currently far better known than the widespread diffuse low-temperature discharge on the ridge flanks. However, off-axial diffuse flow accounts for as much as 70-80% of the total heat loss at oceanic ridges (Wheat and Mottl 1994, Stein and Stein 1994) and thus represents an important component of seafloor hydrothermal activity, although the actual chemical fluxes through the ridge flanks have not yet been documented. According to Wheat and Mottl (1994) and Ginster et al. (1994), up to 90% of the heat in the neovolcanic axial rift areas may also be removed as a result of diffuse discharge which originates from subseafloor mixing of high-temperature hydrothermal fluids with cold seawater. The significance of axial and off-axial diffuse discharge indicates that the chemical fluxes to the ocean cannot be simply calculated from the composition and venting rates of high-temperature hydrothermal fluids at the ridge crests (cf. Alt 1995).

Following the initial discovery of ore-forming hydrothermal systems in the Red Sea (Miller et al. 1966), black smokers, polymetallic massive sulfide deposits, and vent biota were located at the Galapagos Spreading Center (Corliss et al. 1979) and the East Pacific Rise at 21°N (Francheteau et al. 1979, Spiess et al. 1980). This initiated an intensive investigation of the mid-ocean ridge systems in the Pacific, Atlantic and Indian Ocean, which resulted in the delineation of numerous new sites of hydrothermal activity. In 1986, the first inactive hydrothermal sites were found at the

active back-arc spreading center of the Manus Basin in the Southwest Pacific (Both et al. 1986). Subsequently, active hydrothermal systems and associated sulfide deposits were reported from the Marianas back-arc (Craig et al. 1987, Kastner et al. 1987), the North Fiji back-arc (Auzende et al. 1989), the Okinawa Trough (Halbach et al. 1989), and the Lau back-arc (Fouquet et al. 1991). Today, more than 100 sites of hydrothermal mineralization are known on the modern seafloor (Rona 1988, Rona and Scott 1993, Hannington et al. 1994) including at least 25 sites with high-temperature (350-400°C) black smoker venting.

## 13.1  Hydrothermal Convection and Generation of Hydrothermal Fluids at Mid-Ocean Ridges

At mid-ocean ridges, seawater penetrates deeply into layers 2 and 3 of the newly formed oceanic crust along cracks and fissures, which are a response to thermal contraction and seismic events typical for zones of active seafloor spreading (Fig. 13.1). The seawater circulating through the oceanic crust at seafloor spreading centers is con-

**Table 13. 1**  Conversion of seawater to a hydropthermal fluid through water/rock interaction above a high-level magma chamber.

| temperature: | 2°C | $\rightarrow$ | > 400°C | magma chamber (1200°C) |
|---|---|---|---|---|
| pH (acidity): | 7.8 | $\rightarrow$ | < 4 | $H_2O \rightarrow OH^- + H^+$ |
| | | | | $2OH^- + Mg^{2+} \rightarrow Mg(OH)_2$ (> 350°C: Ca instead of Mg) |
| | | | | $Mg(OH)_2$ fixed in   smectite (<200°C)   chlorite (>200°C) |
| | | | | excess $H^+$ = pH decrease |
| $E_H$ (redox state): | + | $\rightarrow$ | - | $Fe^{2+} \rightarrow Fe^{3+}$ (in basalt) |
| | | | | seawater $SO_4^{2-} \rightarrow S^{2-}$ ($H_2S$) at > 250°C |
| | | | | note: a significant amount of the reduced $S^{2-}$ in the hydrothermal fluid results from leaching of sulfide inclusions in the basalt |

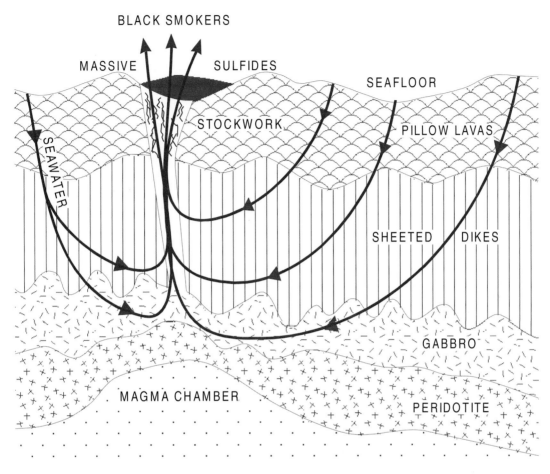

**Fig. 13.1** Model showing a seawater hydrothermal convection system above a subaxial magma chamber at an oceanic spreading center. Radius of a typical convection cell is about 3-5 km. Depth of the magma chamber usually varies between 1.5 and 3.5 km (see text for details).

verted into an ore-forming hydrothermal fluid in a reaction zone which is situated close to the top of a subaxial magma chamber. Major physical and chemical changes in the circulating seawater include (i) increasing temperature, (ii) decreasing pH, and (iii) decreasing $E_H$ (Table. 13.1).

The increase in temperature from about 2°C to values >400°C (Richardson et al. 1987, Schöps and Herzig 1990) is a result of conductive heating of a small percentage of seawater close to the frozen top of a high-level magma chamber (Cann and Strens 1982). This drives the hydrothermal convection system and gives rise to black smokers at the seafloor. High-resolution seismic reflection studies have indicated that some of these magma reservoirs may occur only 1.5-3.5 km below the seafloor (Detrick et al. 1987, Collier and Sinha 1990; Fig. 13.2). The crustal residence time of seawater in the convection system has been constrained to be 3 years or less (Kadko and Moore 1988). Data from water/rock interaction experiments indicate that, with increasing temperatures, the $Mg^{2+}$ dissolved in seawater (about 1,280 ppm) combines with OH-groups (which originate from the dissociation of seawater at higher temperatures) to form $Mg(OH)_2$. The $Mg^{2+}$ is incorporated in secondary minerals such as smectite (<200°C) and chlorite (>200°C) (Hajash 1975, Seyfried and Mottl 1982, Seyfried et al. 1988, Alt 1995). The removal of OH-groups creates an excess of $H^+$ ions, which is the principal acid-generating reaction responsible for the drop

399

in pH from seawater values (pH 7.8 at 2°C) to values as low as pH 2 (cf. Fouquet et al. 1993a). Further exchange of $H^+$ for $Ca^{2+}$ and $K^+$ in the rock, releases these elements into the hydrothermal fluid. The leaching of $Ca^{2+}$ balances the removal of $Mg^{2+}$ from seawater which is quantitative. End-member hydrothermal fluids are defined as presumed deep-seated high-temperature fluids computed by extrapolating compositions and physical parameters back to Mg=0 on the assumption of quantitative removal of Mg. At high temperatures, however, the formation of epidote (Ca fixation) also results in an excess of $H^+$ which further contributes to the acidity of the hydrothermal fluid. These reactions take place at water/rock ratios of less than five and commonly close to one (Von Damm 1995). The oxygen which is present in the circulating seawater in the form of sulfate is removed partly by precipitation of anhydrite and partly through reaction of igneous pyrrhotite to secondary pyrite and oxidation of $Fe^{2+}$ to $Fe^{3+}$ in the basalt (Alt 1995). The reduction of seawater $SO_4^{2-}$ contributes to the formation of $H_2S$, but most of the reduced S in the fluid is derived from the rock.

This highly corrosive fluid is now capable of leaching elements such as Li, K, Rb, Ca, Ba, the

**Table 13.2** Temperature dependence of water/rock interaction in the oceanic crust.

| Temperature | Basalt | Seawater / Hydrothermal Fluid |
|---|---|---|
| < 150°C | K, Rb, Li, B, $^{18}O$ | ← |
| > 70°C | Mg | ← |
| > 350°C | Ca | ← |
| >> 150°C | → | Fe, Mn, Zn, Cu, Ba, Si, Ca, Rb, Li, K, B, $H^+$, $^{18}O$, S |
| all T | $H_2O$, $CO_2$, Na | ← |

transition metals Fe, Mn, Cu, Zn, together with Au, Ag and some Si from the oceanic basement (Mottl 1983; Table 13.2). Sulfide droplets in the basalt are considered to be the major source for metals and S (Keays 1987). The metals are mainly transported as chloride complexes at high tem-

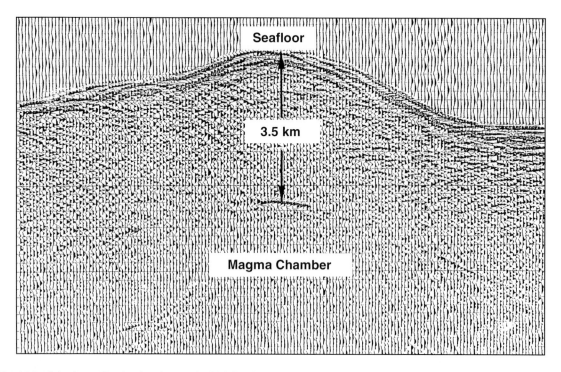

**Fig. 13.2** Seismic profile showing the top of a high-level magma chamber at a depth of 3.5 km below the seafloor at a spreading center in the southwest Pacific (after M. Shinha, unpublished).

peratures and, in some cases, as bisulfide complexes (in particular Au) at lower temperatures.

Due to its increased buoyancy at high temperatures, the hydrothermal fluid rises rapidly from the deep-seated reaction zone to the surface along major faults and fractures within the rift valley or close to the flanks of the rift. In particular the intersections of faults running parallel and perpendicular to the ridge axis are the loci of high-velocity discharge black smokers and massive sulfide mounds. The sulfide precipitation within the upflow zone (stockwork) and at the seafloor (massive sulfides; Fig. 13.1) is a consequence of changing physical and chemical conditions during mixing of high-temperature (250-400°C), metal-rich hydrothermal fluids with cold (about 2°C), oxygen-bearing seawater.

## 13.2   Onset of Hydrothermal Activity

Regardless of how a hydrothermal system evolves at the seafloor, it must start out by displacing a large volume of cold seawater that occupies pore spaces within the crust. The initial flushing of the system may be very rapid, especially on medium- and fast-spreading ridges that are undergoing extension at tens of centimeters per year. Where there are frequent intrusions of magma close to the seafloor and fissure-fed eruptions along the axial rift, seafloor hydrothermal activity may begin with the rapid release of a large volume of hydrothermal fluid forming a "megaplume" in the overlying water column. Observations at new hydrothermal fields on the Juan de Fuca Ridge (Embley et al. 1993, Baker 1995) and the East Pacific Rise (Haymon et al. 1993) suggest that hydrothermal activity is directly linked to discrete volcanic eruptions and that megaplumes are triggered by dike emplacement (Baker 1995, Embley and Chadwick 1994). The intrusion of the dikes close to the seafloor and major eruptions of lava are coincident with the displacement of large volumes of hydrothermal fluid. Shortly after these eruptions, widespread diffuse flow of low-temperature (≤100°C) fluids begins through fractures in the fresh lavas and between new pillows (Butterfield and Massoth 1994). Within a period of about 5-10 years, a low-temperature vent field may be sealed by hydrothermal precipitates, allowing sub-seafloor temperatures to rise and fluid discharge to become focussed into deeper frac-

tures. On a fast-spreading ridge such as the East Pacific Rise, the cycle of dike injection, eruption, and hydrothermal discharge may repeat itself with each new eruption, perhaps as often as every 3-5 years (Haymon et al. 1993).

Diffuse venting typically occurs throughout the life of a hydrothermal system. It may be the earliest form of discharge in a new hydrothermal field (see above) but commonly also occurs at the margins of existing high-temperature upflow where rising hydrothermal fluids mix with cold seawater. Diffuse venting also typically dominates the last stages of activity in a waning hydrothermal system as high-temperature upflow collapses around a cooling subvolcanic intrusion. Periods of diffuse flow can sustain a vibrant biological community, but are not generally associated with extensive sulfide mineralization because the low temperatures of the fluids (<10°C to 50°C) do not allow transport of significant concentrations of dissolved metals. The mineral precipitates associated with diffuse venting typically consist of amorphous Fe-oxyhydroxides, Mn-oxides, and silica.

## 13.3   Growth of Black Smokers and Massive Sulfide Mounds

Black smoker activity begins when the hydrothermal fluids contain enough metals and sulfur to cause precipitation of sulfide particles during mixing at the vent orifice. In order to carry these metals in solution, fluids arriving at the seafloor are usually hotter than 300°C. At many black smoker chimneys measurements of vent temperatures in the range of 350°-400°C are common. The sequence of mineral precipitation which attends mixing of these high-temperature fluids with cold seawater has been modeled extensively (Janecky and Seyfried 1984, Bowers et al. 1985, Tivey 1995, Tivey et al. 1995). Most descriptive models of the growth of black smoker chimneys involve an early assemblage of chalcopyrite, pyrrhotite, and anhydrite at high temperatures, followed by pyrite and sphalerite at lower temperatures (Haymon and Kastner 1981, Haymon 1983, Goldfarb et al. 1983, Oudin 1983, Graham et al. 1988, Zierenberg et al. 1984, Lafitte et al. 1985, Fouquet et al. 1993b).

The characteristic feature that distinguishes black smoker vents from white smokers (smoke

consists of amorphous silica and/or barite and anhydrite instead of sulfides) is their central chalco-
pyrite-lined orifice. In the models of Haymon
(1983) and Goldfarb et al. (1983), anhydrite is precipitated around a black smoker vent at the leading edge of chimney growth, where hot hydrothermal fluids first encounter cold seawater. The
anhydrite is precipitated from $Ca^{2+}$ in the vent
fluids and $SO_4^{2-}$ in ambient seawater, although
some anhydrite in the outer walls of the chimney
or in the interior of hydrothermal mounds (cf.
Humphris et al. 1995) may also be formed simply
by the conductive heating of seawater above

150°C. The anhydrite which forms the initial wall
of the chimney is gradually displaced by high-
temperature Cu-Fe-sulfides as the structure grows
upward and outward. Because of its retrograde
solubility, anhydrite is not well preserved in older
chimney complexes and eventually dissolves at
ambient temperatures and seafloor pressures
(Haymon and Kastner 1981). As a result, many
black smokers that are cemented by anhydrite are
inherently unstable and ultimately collapse to become part of the sulfide mound (Fig. 13.3).

Most chimneys have growth rates that are
rapid in comparison to the half-life of [210]Pb (22.3

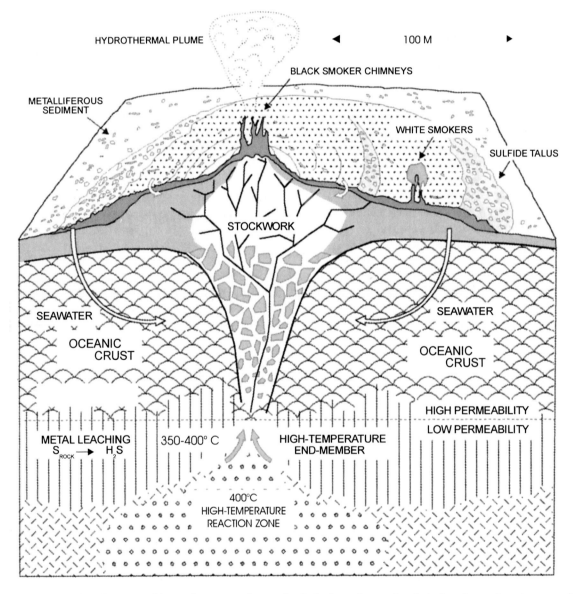

**Fig. 13.3** Surface features and internal structure of an active hydrothermal mound and stockwork complex at an oceanic
spreading center.

years) and radioisotope ages of several days or less for precipitates of some active vents are consistent with growth rates observed from submersibles (5-10 cm per day: Hekinian et al. 1983, Johnson and Tunnicliffe 1985). Larger vent complexes commonly have measured ages on the order of decades (Koski et al. 1994), but the data for entire vent fields may span several thousands of years (Lalou et al. 1993).

At typical black smoker vents, a very large proportion (at least 90%) of the metals and sulfur carried in solution are lost to a hydrothermal plume in the overlying seawater column rather than deposited as chimneys (Converse et al. 1984). The metals are precipitated as sulfide particles in the plume above the black smokers and are rapidly oxidized and dispersed over distances of several kilometers from the vent (Feely et al. 1987, 1994a,b, Mottl and McConachy 1990). Due to oxidation and dissolution, these particles also release some of the metals back into seawater (Feely et al. 1987, Metz and Trefry 1993). Particle settling models indicate that only a small fraction of the metals is likely to accumulate as plume fallout in the immediate vicinity of the vents (Feely et al. 1987) and observations of particulate Fe dispersal confirm that most of the metals produced at a vent site are carried away by buoyant plumes (Baker et al. 1985, Feely et al. 1994a,b).

Black smokers usually grow on hydrothermal mounds that are large enough to be thermally and chemically insulated from the surrounding seawater (Fig. 13.3). The sulfide mounds also serve to trap rising hydrothermal fluids and impede the loss of metals and sulfur by direct venting into the hydrothermal plume. The largest sulfide deposits are often composite bodies which appear to have evolved from several smaller hydrothermal mounds (Embley et al. 1988). Many of the original chimney structures are overgrown and eventually incorporated in the larger mound, destroying primary textural and mineralogical relationships by hydrothermal replacement. Large sulfide mounds are also constructed from the accumulation of sulfide debris produced by collapsing chimneys, and most large deposits are littered with the debris of older sulfide structures. In many places, new chimneys can be seen growing on top of the sulfide talus, and this debris is eventually overgrown, cemented, and incorporated within the mound (Rona et al. 1993, Hannington et al. 1995). At the same time, high-temperature fluids circulating or trapped beneath the deposit precipi-

tate new sulfide minerals in fractures and open spaces. Chimneys that are perched on the outer surface of an active mound apparently tap these high-temperature fluids but account for only a small part of the total mass of the deposit.

The common presence of high-temperature chimneys at the tops of the deposits and lower-temperature chimneys on their flanks suggests that most large mounds are zoned similar to many ancient deposits on land (cf. Franklin et al. 1981). The most common arrangement of mineral assemblages is a high-temperature Cu-rich core and a cooler, Zn-rich outer margin. Mineralogical zonation within a deposit is principally a result of hydrothermal reworking, whereby low-temperature phases such as sphalerite are dissolved by later, higher-temperature fluids and redistributed to the outer margins of the deposit. Sulfide samples that have been recovered from the interiors of hydrothermal mounds show signs of extensive hydrothermal recrystallization and annealing, especially when compared to the delicate, fine-grained sulfides found in surface precipitates (Humphris et al. 1995). As a result of the extensive hydrothermal reworking in large sulfide mounds, it must be considered that hydrothermal fluids arriving at the seafloor are also likely to have been substantially modified by interaction with pre-existing hydrothermal precipitates in their path (e.g. Janecky and Shanks 1988), and caution should be used when interpreting measurements made at the surface of a large mound to infer processes related solely to the direct venting of an end-product fluid.

## 13.4 Physical and Chemical Characteristics of Hydrothermal Vent Fluids

Most black smoker fluids are strongly buffered close to equilibrium with pyrite-pyrrhotite-magnetite, although the proximity of the fluids to this buffer assemblage is not necessarily a reflection of the state of saturation of the minerals in solution (Janecky and Seyfried 1984, Bowers et al. 1985, Tivey et al. 1995). In addition, because of the high concentrations of reduced components such as ferrous iron and $H_2S$, the fluids do not deviate significantly from the pyrite-pyrrhotite redox buffer, even after substantial mixing and cooling, and the common occurrence of both pyrite and pyrrhotite in many sulfide chimneys reflects conditions close to pyrite-pyrrhotite equilibrium throughout mineralization. A common scenario during mixing of a 21°N-type fluid with seawater is the precipitation of pyrrhotite at high-temperatures, followed by pyrite or marcasite which may replace the pyrrhotite at lower-temperatures. Janecky and Seyfried (1984) note that, until relatively large amounts of seawater mix with the end-member fluids, the calculated $f_{O_2}$ ($f_x$ = fugacity of gaseous species) of the mixture does not approach ambient seawater values, a characteristic which reflects the abundance of reducing agents in the vent fluids and the ineffective buffer capacity of seawater.

Because vent fluid compositions are determined largely by fluid-rock interactions that take place in the source region, the equilibrium mineral assemblage that is precipitated during mixing will depend to a large extent on the chemistry of the source rocks. Fluids that are buffered to lower $f_{O_2}$ and $f_{S2}$ values should produce a pyrite-pyrrhotite-magnetite assemblage; fluids at higher $f_{O_2}$ and $f_{S2}$ values will precipitate only pyrite. Whereas the buffer assemblage in most volcanic rocks is close to pyrite-pyrrhotite-magnetite, fluids that have reacted extensively with organic-rich sediments may have significantly more reduced compositions. As a result, seafloor spreading in areas of high sedimentation near the continental margins (e.g., Guaymas Basin) has given rise to a class of deposits that are mineralogically quite different from those of bare-ridge massive sulfides. In these deposits, reaction of heated seawater with basaltic rocks controls the initial composition of the hydrothermal fluid, but interaction of the fluid with sediments in the upflow zone has modified the fluid chemistry (Zierenberg et al. 1993, see below). Chemical buffering of the fluids by sediments results in generally higher pH, and reactions with organic matter in the sediments result in significantly lower $f_{O_2}$ than in bare-ridge systems (Von Damm et al. 1985a, Bowers et al. 1985). Pyrrhotite tends to be the dominant Fe-sulfide phase, and other minerals typically formed at low $f_{O_2}$ may also be common (e.g. Koski et al. 1988, Zierenberg et al. 1993).

The metal zonation within black smoker chimneys and hydrothermal mounds is a consequence of the precipitation of sulfide minerals, in open spaces or as replacements of pre-existing minerals, according to their respective solubilities at different temperatures and pH. For the most part,

metals such as Cu and Zn are carried in solution as aqueous chloride complexes with stabilities that are enhanced at high temperatures and low pH (Bourcier and Barnes 1987, Crerar and Barnes 1976, Hemley et al. 1992), and during mixing with and cooling by seawater the deposition of chalcopyrite and sphalerite is mainly a function of decreasing temperature and increasing pH. For example, the Cu-rich cores of black smokers reflect saturation with respect to chalcopyrite at high temperatures, and marginal Zn-rich zones reflect saturation with sphalerite at lower temperatures caused by mixing through the chimney walls. During mixing, the pH of a hydrothermal fluid can also increase dramatically from its starting value (Janecky and Seyfried 1984). However, during conductive cooling, pH changes in the fluid are moderated by the production of acid during sulfide precipitation, and the lower pH of the fluids may inhibit the precipitation of certain minerals.

An important constraint on the P-T path for some seafloor hydrothermal fluids is the two-phase curve for seawater (Bischoff and Rosenbauer 1984, Bischoff and Pitzer 1985). Large variations in the salinities of some vent fluids indicate that phase separation is occurring in some seafloor hydrothermal systems (Von Damm 1988, 1990; see below). At the depth range of the most mid-ocean ridge vent sites (2500-3000 m), the two-phase boundary of seawater is between 385°C and 405°C (Fig.13.4). For the most part, measured tempera-

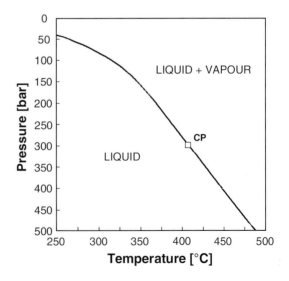

**Fig. 13.4** Pressure-temperature curve for seawater (CP = critical point for seawater). Note that 100 bar equal 1,000 m water depth.

tures of vent fluids at these sites are well below the two-phase curve for seawater and therefore they are unlikely to undergo phase separation. However, some vent fluids have been observed that are clearly within the two-phase region (e.g., 420°C and 220 bars at Endeavour Ridge: Delaney et al. 1984), and similar conditions may be common in the high-temperature reaction zones or upflow conduits of some systems. At pressures and temperature below the critical point for seawater (407°C and 298 bars), fluids which intersect the two-phase curve will separate a small amount of low-salinity, vapor-rich fluid. At temperatures and pressures above the critical point, phase separation involves the condensation of a small amount of high salinity brine (i.e., super-

critical phase separation). The existence of such high-salinity fluids at depth has implications for the development of metal-rich brines. Bischoff and Rosenbauer (1987) noted that during supercritical phase separation, both the acidity and the concentration of heavy metals increase in the chloride-rich phase, and the solubilities of metals as aqueous chloride complexes in these fluids may be several orders of magnitude greater than in fluids of ordinary seawater composition.

At shallow water depths, subcritical boiling may also have a major impact on seafloor mineralization. At 350°C, a 21°N-type fluid will intersect the two-phase curve for seawater when the hydrostatic pressure is equivalent to about 1600 m of water depth (160 bars or 16 Mpa; Fig. 13.4). If

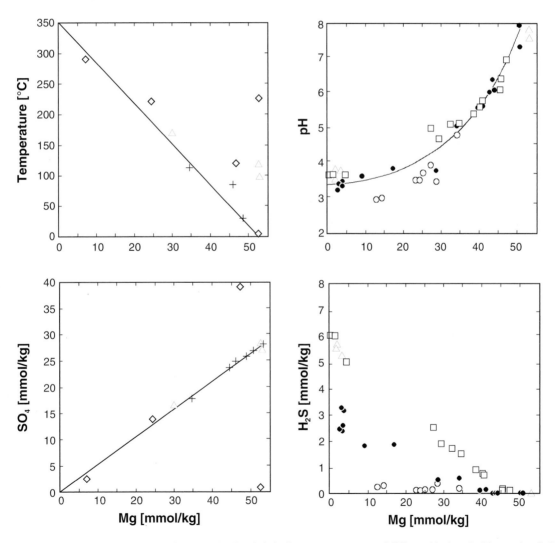

**Fig. 13.5** Temperature, pH, SO$_4$ and H$_2$S versus Mg. Symbols indicate measurements of different black and white smoker fluids from the East Pacific Rise 21°N, the TAG site at 26°N Mid-Atlantic Ridge, and the Snakepit hydrothermal field at 23°N Mid-Atlantic Ridge (after Von Damm et al. 1983 and Edmond et al. 1995).

the water depth is less than 1600 m, the fluid will begin to boil beneath the seafloor and may separate a vapor-rich phase. At these lower pressures, the process of boiling differs from that of phase separation in the deeper parts of the mid-ocean ridges, because the density difference between the vapor and liquid is much greater and therefore may facilitate the separation of a low-salinity, gas-rich phase. Such fluids have been encountered at vents in the caldera of Axial Seamount (1540 m water depth, fluid temperature 349°C) where low salinity and high gas contents have been measured (Massoth et al. 1989, Butterfield et al. 1990).

## 13.5  The Chemical Composition of Hydrothermal Vent Fluids and Precipitates

Since the first discovery of high-temperature hydrothermal vents at the East Pacific Rise 21°N in 1979, hydrothermal fluids have been sampled at numerous sites at mid-ocean ridges and back-arc spreading centers. Both Mg and sulfate show a negative correlation with temperature, and an extrapolation to the zero values of Mg and $SO_4$ intersects the temperature axis at a point referred to as the end-member temperature (Fig. 13.5). Published data on the chemical composition of these fluids are summarized in Von Damm (1990) and Von Damm (1995) and some examples are given in Table 13.3. Time series measurements on the order of a decade at a number of sites have indicated that the chemical composition of vent fluids at individual sites does not show significant temporal variability and is "steady-state" once the system has stabilized after a new volcanic event (Von Damm 1995). Variability in the chemical composition of the hydrothermal fluids does not appear to be related to the depth of the individual vent site to reflect changes as a function of temperature. Phase separation is responsible for most of the chemical variability observed (cf. Von Damm 1995, see below). In some areas, contributions from a degassing magma are also inferred (Lupton et al. 1991, de Ronde 1995, Herzig et al. 1998).

Despite the long-term stability of vent fluids at individual sites a wide range of salinities has been observed between different sites ranging from about 30% below (176 mM/kg; Von Damm

and Bischoff 1987) to 200% above (1,090 mM/kg; Massoth et al. 1989) seawater concentrations (546 mM/kg). Salinities are important for the chemical composition of vent fluids as Cl is the major complexing anion in hydrothermal systems. These extreme salinities cannot be accounted for by hydration of the oceanic crust (Cathles 1983) or precipitation and dissolution of Cl-bearing mineral phases (Edmond et al. 1979b, Seyfried et al. 1986) but are interpreted to be a result of supercritical phase separation at the top of the magma chamber followed by mixing of the brines and the vapor phases during ascent (cf. Nehlig 1991, Palmer 1992, James et al. 1995).

Most hydrothermal fluids contain considerable amounts of $CH_4$ and $^3He$. $^3He$ is significantly enriched in vent fluids over saturated seawater, with $^3He/^4He$ ratios between about 7-9. These ratios represent typical mantle values (Stuart et al. 1995), indicating that $^3He$ originates from MORB magma, which is degassing into a hydrothermal system (Lupton 1983, Baker and Lupton 1990). Lupton and Craig (1981) have demonstrated that $^3He$ behaves extremely conservatively in the water column and can be traced over distances up to 2000 km from the point of origin (Fig. 13.6). In unsedimented areas, methane is a product of inorganic chemical reactions within the hydrothermal system and also related to magmatic degassing. Methane in combination with TDM (total dissolvable Mn) have been found to be enriched about $10^6$-fold over ambient seawater in high-temperature vent fluids (e.g. Von Damm et al. 1985b). In diluted buoyant hydrothermal plumes, these values are still 100-fold enriched relative to seawater (Klinkhammer et al. 1986, Charlou et al. 1991) which makes $CH_4$ and Mn valuable tracers for prospecting hydrothermal vent areas (Herzig and Plüger 1988, Plüger et al. 1990; Fig. 13.7).

Concentrations of trace elements such as Ag, As, Cd, Co, Se, and Au have been measured in some vent fluids (Von Damm 1990, Campbell et al. 1988a,b, Fouquet et al. 1993a, Trefry et al. 1994, Edmond et al. 1995; Table. 13.3) and certain metals such as Cu, Co, Mo, and Se appear to show a strong positive temperature-concentration relationship which likely accounts for observed enrichments of these elements in the sulfides from high-temperature chimneys (Hannington et al. 1991). Other elements such as Zn, Ag, Cd, Pb, and Sb appear to be significantly enriched in lower-temperature fluids (Trefry et al. 1994). However, these data are scarce, and this may re-

**Table. 13.3**   Chemical composition of hydrothermal fluids from selected hydrothermal fields at mid-ocean ridges and back-arc spreading centers in comparison to seawater.

| | $T_{max}$ (°C) | pH (25°C) | Cl (ppt) | $H_2S$ (ppm) | Na (ppm) | K (ppm) | Ca (ppm) | Ba (ppm) | Sr (ppm) | Fe (ppm) | Mn (ppm) | Zn (ppm) | Cu (ppm) | Si (ppm) | Reference |
|---|---|---|---|---|---|---|---|---|---|---|---|---|---|---|---|
| MARK (MAR 23°N) | 350 | 3.9 | 19.8 | 201 | 11,725 | 931 | 421 | - | 5 | 122 | 27 | 3 | 1.0 | 514 | Von Damm (19 |
| TAG (MAR 26°N) | 366 | 3.8 | 22.5 | 119 | 12,805 | 669 | 1,235 | - | 9 | 313 | 37 | 3 | 10.0 | 583 | Edmond et al. ( |
| Lucky Strike (MAR 37°N) | 332 | - | 19.3 | - | - | - | - | - | - | 35 | 21 | - | - | - | Von Damm (19 |
| EPR (11°N) | 347 | 3.7 | 24.3 | 416 | 13,265 | 1,287 | 1,411 | - | 12 | 361 | 51 | - | - | 579 | Von Damm (19 |
| EPR (13°N) | (380) | 3.3 | 26.9 | 279 | 13,702 | 1,165 | 2,204 | - | 16 | 603 | 160 | - | - | 618 | Von Damm (19 |
| EPR (21°N) | 355 | 3.8 | 20.5 | 286 | 11,725 | 1,009 | 834 | 2.2 | 9 | 136 | 55 | - | - | 548 | Von Damm (19 |
| Guyamas Basin | 315 | 5.9 | 22.6 | 204 | 11,794 | 1,924 | 1,663 | 7.4 | 22 | 10 | 13 | - | - | 388 | Von Damm (19 |
| Escanaba Trough | 217 | 5.4 | 23.7 | 51 | 12,874 | 1,580 | 1,339 | - | 18 | 1 | 1 | - | - | 194 | Von Damm (19 |
| Middle Valley (JFR) | 276 | 5.5 | 20.5 | 102 | 9,150 | 731 | 3,247 | 2.1 | 23 | 1 | 4 | 0.1 | 0.1 | 298 | Von Damm (19 |
| South Cleft (JFR) | 285 | 3.2 | 31.8 | (102) | 15,196 | 1,459 | 3,395 | - | 20 | 575 | 143 | 24 | 1.0 | 640 | Von Damm (19 |
| North Cleft (JFR) | 327 | 3.0 | 31.0 | 124 | 15,679 | 1,580 | 2,922 | - | 20 | 165 | 65 | 16 | 0.5 | 559 | Von Damm (19 |
| Axial Seamount (JFR) | 328 | 3.5 | 22.1 | 242 | 11,472 | 1,048 | 1,876 | 3.6 | 17 | 59,477 | 63,178 | 7 | 0.6 | 424 | Von Damm (19 |
| Endeavour (JFR) | 370 | 4.4 | 16.2 | 167 | 8,230 | 1,004 | 1,447 | - | 12 | 54,451 | 14,339 | 2 | 1.3 | 455 | Von Damm (19 |
| Lau Basin (SW-Pacific) | 334 | 2.0 | 28.0 | - | 13,564 | 3,089 | 1,655 | > 5.4 | 2 | 140 | 390 | 196 | 2.2 | 393 | Fouquet et al. ( |
| Seawater | 2 | 7.8 | 19.4 | - | 10,759 | 399 | 413 | 0.02 | 8 | $6 \cdot 10^{-5}$ | $5 \cdot 10^{-5}$ | $7 \cdot 10^{-4}$ | $5 \cdot 10^{-4}$ | 5 | |

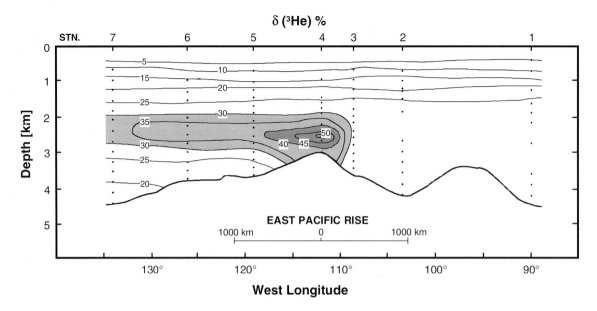

**Fig 13.6**    Helium plume at the East Pacific Rise (after Lupton and Craig 1981).

flect difficulties in reliably measuring trace metal concentrations in the vent fluids.

The typical REE pattern of end-member hydrothermal fluids exhibits a strong enrichment of LREE and a pronounced positive Eu anomaly (Michard et al. 1983, Campbell et al. 1988c, Michard 1989, Fig. 13.8), in contrast to MORB which have a nearly flat, LREE depleted spectrum, and to seawater which is characterized by a strong negative Ce anomaly. The overall pattern is likely controlled by leaching of plagioclase at hydrothermal conditions (Campbell et al. 1988c).

The chemical compositions of volcanic and sedimentary rocks with which the end-member hydrothermal fluids react have a major impact on

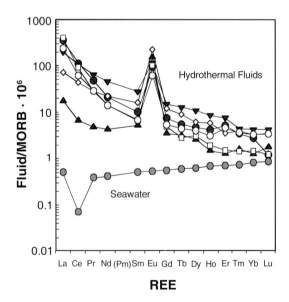

**Fig. 13.8**    Rare earth element (REE) concentrations in hydrothermal fluids versus Atlantic seawater at 2500 m (Elderfield and Greaves 1982) normalized to MORB (Sun 1980). Symbols indicate measurements of different vent fluids from the Lucky Strike hydrothermal field at 37°17'N Mid-Atlantic Ridge (after Klinkhammer et al. 1995).

**Fig. 13.7**    Positive correlation of Mn and $CH_4$ in a hydrothermal plume (after Herzig and Plüger 1988).

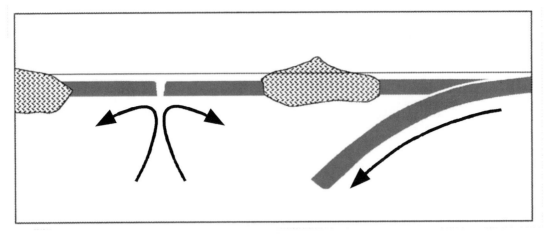

**Fig. 13.9** Schematic diagram showing the geotectonic setting of a seafloor back-arc spreading center.

the metal supply and nature of subseafloor alteration. In particular, the bulk compositions of the
deposits generally reflect the rock types from
which the metals were leached (Doe 1994, Herzig
and Hannington 1995). In mid-ocean ridge environments, the dissolution of sulfides and the destruction of ferromagnesian minerals in basalt are
the major sources of Cu, Fe, Zn, Au, and S in the
hydrothermal system. In contrast, elements such
as Pb and Ba are derived mainly from the destruction of feldspars. Na-K feldspars are particularly
abundant in felsic volcanic rocks but less so in
mid-ocean ridge basalts (MORB). Even in mid-
ocean ridge settings, the bulk compositions of the
deposits appear to be quite sensitive to the chemistry of the source rocks. For example, the high Ba
contents of some mid-ocean ridge deposits may
be related to the enrichment of incompatible elements in some lava suites due to an enriched melt
source or greater degrees of fractionation (Perfit et
al. 1983, Delaney et al. 1981, Clague et al. 1981).

Hydrothermal fluids at immature back-arc rifts,
which form behind island-arcs in areas where oceanic crust is subducted beneath oceanic crust (Fig.
13.9), differ from mid-ocean ridge fluids as they
clearly reflect the compositional differences of the
source rocks (felsic, calc-alkaline lavas versus
MORB). The fluids at immature back-arcs are
typically enriched in Zn, Pb, As and Ba (and depleted in Fe) over mid-ocean ridge fluids, which is
related to the higher concentrations of these elements in the calc-alkaline lavas (Herzig and
Hannington 1995; Table 13.4). These differences
in fluid chemistry are also reflected in the chemical composition of the hydrothermal sulfides (Table 13.5).

Recently, evidence for direct magmatic fluid
and gas contributions to seafloor hydrothermal
systems were also detected in back-arc vent fluids
and precipitates (Gamo et al. 1997, Herzig et al.
1998). This is of major importance, as magmatic
fluids and gases can be responsible for a significant input of metals into the hydrothermal system, as they are known to be highly concentrated
(Hedenquist and Lowenstern 1994). Magmatic
components, however, are extremely difficult to

**Table. 13.4** Chemical composition of hydrothermal fluids at
mid-ocean ridges (EPR, East Pacific Rise 21°N) and back-arc
rifts (Valu Fa Ridge, Lau Basin) relative to seawater (sw).

| | EPR 21°N [1] | Lau Basin [2] | seawater |
|---|---|---|---|
| **T (°C)** | 350 | 334 | 2 |
| **pH** | 3.6 | 2 | 7.8 |
| **salinity** | sw | 1.5 · sw | |
| **Fe (ppm)** | 80 | 140 | $6 \cdot 10^{-5}$ |
| **Mn (ppm)** | 49 | 390 | $5 \cdot 10^{-5}$ |
| **Cu (ppm)** | 1.4 | 2.2 | $5 \cdot 10^{-4}$ |
| **Zn (ppm)** | 5.5 | 196 | $7 \cdot 10^{-4}$ |
| **Ba (ppm)** | 1.4 | 5.4 | $2 \cdot 10^{-2}$ |
| **Pb (ppb)** | 54 | 808 | $2 \cdot 10^{-3}$ |
| **As (ppb)** | 17 | 450 | 1.7 |

[1] Von Damm et al. (1985a); [2] Fouquet et al. (1993a)

**Table 13.5** Bulk chemical composition of seafloor poly-metallic sulfides from mid-ocean ridges and back-arc spreading centers (after Herzig and Hannington 1995).

| Element | Mid-Ocean Ridges [1] | Back-Arc Ridges [2] |
|---|---|---|
| n | 890 | 317 |
| Fe (wt%) | 23.6 | 13.3 |
| Cu | 4.3 | 5.1 |
| Zn | 11.7 | 15.1 |
| Pb | 0.2 | 1.2 |
| As | 0.03 | 0.1 |
| Sb | 0.01 | 0.01 |
| Ba | 1.7 | 13.0 |
| Ag (ppm) | 143 | 195 |
| Au | 1.2 | 2.9 |

[1] Explorer Ridge, Endeavour Ridge, Axial Seamount, Cleft Segment, East Pacific Rise, Galapagos Rift, TAG, Snake Pit, Mid-Atlantic Ridge 24.5°N

[2] Mariana Trough, Manus Basin, North Fiji Basin, Lau Basin

identify as they are usually masked by the large amount of seawater in the circulation system.

Deposits on sediment-covered mid-ocean ridges commonly have lower Cu and Zn contents and higher Pb contents than bare-ridge sulfides, and some deposits such as Guaymas Basin contain abundant carbonate. The low metal contents of deposits in the Guaymas Basin are a consequence of the higher pH values of the fluids that arise from chemical buffering by the carbonate in the sediments, and the high $CO_2$ in the fluids is derived directly from the sediments themselves (Bowers et al. 1985, Von Damm et al. 1985a). Furthermore, the fluids in this environment are strongly reduced (pyrrhotite stability field) as a consequence of interaction of the fluids with organic components in the sediments, and metal deficient relative to a volcanic-hosted hydrothermal system as a result of sulfide precipitation within the sedimentary sequence. A high ammonium content reflects the thermocatalytic cracking of

immature planktonic carbon (cf. Von Damm et al. 1985a). Interaction of the hydrothermal fluids with sediments in the upflow zone also may result in significant enrichments in certain trace elements derived from the sediments (e.g., Pb, Sn, As, Sb, Bi, Se: Koski et al. 1988, Zierenberg et al. 1993). The higher Pb contents, in particular, reflect the destruction of feldspars from continentally-derived turbidites, a process supported by Pb isotope studies (LeHuray et al. 1988).

## 13.6 Characteristics of Cold Seep Fluids at Subduction Zones

In addition to hydrothermal activity at divergent plate boundaries, low-temperature fluid venting at convergent plate margins is an important global process. Cold seeps have now been documented at numerous sites along the circum-Pacific subduction zones and it is estimated that this type of oceanic venting is also of major significance for the chemical budget of seawater. The fluids are expelled from thick organic-rich marine sediments that are trapped in accretionary wedges along the subduction trenches (e.g., Nankai Trench, Oregon and Alaskan margins). Pore fluids in the sediment account for as much as 50-70% of their volume, and this fluid is literally squeezed from the sediment through diffuse flow at the toe of the accretionary wedge or by focussed flow along major fault structures in the accreted sediments.

In order to estimate the mass flux from cold seeps, flow rates have to be known, and these have been difficult to measure because of the large areas of diffuse flow involved. Furthermore, fluid flow rates determined from simple advection-diffusion modeling of temperature and chemical profiles have shown to be unreliable if applied to sites which are densely populated with bottom macrofauna. Calculations by Wallmann et al. (1997) which take into account the high pumping rates of bivalves have arrived at a mean value of 5.5 +/- 0.7 L $m^{-2}$ $d^{-1}$ based on a biogeochemical approach using oxygen flux and vent fluid analyses. Von Huene et al. (1997) used sediment porosity reduction to calculate a fluid flow of 0.02 L $m^{-2}$ $d^{-1}$. Suess et al. (1998) calculated an average rate of 0.006 L $m^{-2}$ $d^{-1}$ based on the occurrence of vent biota at the seafloor. Estimates of the fluid flux at subduction zones suggest that they recycle the volume of water in the oceans every 500 Ma, compared to 5-10 Ma along the mid-ocean ridges. Despite the rather slow circulation of fluids expelled from subduction zones, they are a significant player in the carbon cycle of the oceans as a result of the recycling of organic matter in marine sediments.

Methane plumes have been found to be a characteristic feature of cold seep areas. The concentrations of dissolved $CH_4$ in some areas are lower than in hydrothermal vent areas at mid-ocean ridges, but in other areas they exceed ridge values by more than an order of magnitude and have been successfully used as a tracer for detecting cold seep venting. The methane and carbon dioxide that are expelled with the pore waters are derived mainly from the breakdown of organic matter in the sediments. Methane together with $H_2S$ support abundant chemosynthetic bacteria, pogonophorans, vestimentifera, and bivalves which differ from those that inhabit high-temperature vent sites on the mid-ocean ridges (Suess et al. 1998). As the temperatures of venting are low compared to hydrothermal vents at the mid-ocean ridges, metals are not significantly mobilized.

Typical precipitates for areas of subduction venting are carbonate and barite crusts together with cemented sediment (Dia et al. 1993, Suess et al. 1998). The formation of carbonate is related to the anaerobic microbial oxidation of methane which in turn causes $CaCO_3$ to precipitate (Wallmann et al. 1997). The source for barium is thought to be the high concentration of biogenic barite buried in sediments at high productivity areas which is remobilized within the sediment column due to sulfate depletion. Barite in the cold seep areas forms as a result of mixing of Ba-rich fluids upwelling from the sediment with seawater sulfate (Ritger et al. 1987, Torres et al. 1996).

Ice-like gas hydrates (clathrates) have also been found at numerous sites along the convergent plate margins (e.g. Suess et al. 1997). Methane hydrates are stable in solid form only in a narrow temperature-pressure window (Dickens and Quinby-Hunt 1994; Fig. 13.10). In theory, $1 m^3$ of methane hydrate can contain up to $164 m^3$ of methane gas at standard conditions (Kvenvolden 1993). It has been estimated that the amount of carbon in gas hydrates considerably exceeds the total of carbon occurring in all known oil, gas and coal deposits worldwide (Kvenvolden and McMenamin 1980, Kvenvolden 1988). This raises the possibility that gas hydrates may be a future energy source of global importance.

If disturbed by a thermal anomaly or pressure change, large volumes of gas hydrates can break down into water, methane and carbon-dioxide. The dissolution of metastable gas hydrates is a natural consequence of tectonic uplift of accretionary prisms at plate margins and is seen to be partly responsible for the extensive methane plumes reported from these areas (Suess et al. 1997).

As gas hydrates have a very limited stability (low temperature and high pressure), they are of major relevance for climate considerations and the budget of greenhouse gases (Suess et al. 1997). The destabilization and dissolution of gas hydrates due to environmental changes (increase of ocean bottom water temperature, change of sea level) could liberate enormous amounts of methane to the water column and eventually to the atmosphere where they could potentially accelerate greenhouse warming and have an effect on future global climate (Leggett 1990).

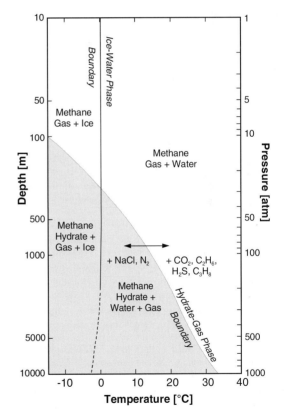

**Fig. 13.10** Pressure-temperature stability diagram for natural gas hydrates (after Katz et al. 1959).

# References

Alt, J.C., 1995. Subseafloor processes in mid - ocean ridge hydrothermal systems. Seafloor Hydrothermal Systems: Physical, Chemical, Biological and Geological Interacions, AGU Geophysical Monograph, 91: 85-114 pp.

Auzende, J.M., Urabe, T., Deplus, C., Eissen, J.P., Grimaud, D., Huchon, P., Ishibashi, J., Joshima, M., Lagabrielle, Y., Mevel, C., Naka, J., Ruellan, E., Tanaka, T. and Tanahashi, M., 1989. Le cadre geologique d'un site hydrothermal actif. la campagne Starmer 1 du submersible nautile dans le bassin Nord Fidjien. C. R. Acad. Sci. Paris, 309: 1787-1795.

Baker, E.T., Lavelle, J.W. and Massoth, G.J., 1985. Hydrothermal particel plumes over the southern Juan de Fuca Ridge. Nature, 316: 342-344.

Baker, E.T. and Lupton, J.E., 1990. Changes in submarine hydrothermal 3He/ heat ratios as an indicator of magmatic/ tectonic activity. Nature, 346: 556-558.

Baker, E.T., 1995. Characteristics of hydrothermal discharge following a magmatic intrusion. Hydrothermal Vents and Processes, In: Parson, L.M., Walker, C.L. and Dixon, D.R. (eds) Geological Society Special Puplication, 87: 65-76 pp.

Bischoff, J.L. and Rosenbauer, R.J., 1984. The critical point and two-phase boundery of seawater, 200°C - 500°C. Earth and Planetary Science Letters, 68: 172-180.

Bischoff, J.L. and Pitzer, K.S., 1985. Phase relation and adiabats in boiling seafloor geothermal systems. Earth and Planetary Science Letters, 75: 327-338.

Bischoff, J.L. and Rosenbauer, R.J., 1987. Phase seperation in seafloor geothermal systems: An experimental studie of the effects on metal transport. American Journal of Science, 287: 953-978.

Both, R.A., Crook, K., Taylor, B., Brogan, S., Chapell, B., Frankel, E., Liu, L., Sinton, J. and Tiffin, D., 1986. Hydrothermal chimneys and associated fauna in the Manus back - arc basin, Papua New Guinea. American Geophysical Union Translations, 67: 489-490.

Bourcier, W.L. and Barnes, H.L., 1987. Ore solution chemistry Vll. Stabilities of chloride and bisulfide complexes of zinc to 350°C. Economic Geology, 82: 1839-1863.

Bowers, T.S., von Damm, K.L. and Edmond, J.M., 1985. Chemical Evolution of mid - ocean ridge hot springs. Geochimica et Cosmochimica Acta, 49: 2239-2252.

Butterfield, D.A., Massoth, G.J., McDuff, R.E., Lupton, J.E. and Lilley, M.D., 1990. Geochemistry of hydrothermal fluids from Axial seamount hydrothermal emmissions study vent field, Juan de Fuca Ridge: Subseafloor boiling and subsequent fluid rock and interaction. Journal of Geophysical Research, 95: 12895-12921.

Butterfield, D.A. and Massoth, G.J., 1994. Geochemistry of north Cleft segment vent fluids: Temporal changes in chlorinity and their possible relationto recent volcanism. Journal of Geophysical Research, 99: 4951-4968.

Campbell, A.C., Bowers, T.S., Measures, C.I., Falkner, K.K., Khadem, M. and Edmond, J.M., 1988a. A time - series of vent fluid composition from 21°N, EPR (1979, 1981, 1985), and the Guaymas Basin, Gulf of California (1982, 1985). Journal of Geophysical Research, 93: 4537-4549.

Campbell, A.C., German, C., Palmer, M.R. and Edmond, J.M., 1988b. Preliminary report on the chemistry of hydrothermal fluids from the Escanaba Trough. American Geophysical Union Transactions, 69: 1271.

Campbell, A.C., Palmer, M.R.,Klinkhammer, G.P., Bowers, T.S., Edmond, J.M., Lawrence, J.R., Casey, J.F., Thompson, G., Humphris, S., Rona, P. and Karson, J.A. 1988c. Chemistry of hot springs of the Mid–Atlantic Ridge. Nature, 335: 514-519.

Cann, J.R. and Strens, M.R., 1982. Black smokers fuelled by freezing magma. Nature, 298: 147-149.

Cathless, L.M., 1983. An analysis of the hydrothermal system responsible for massive sulfide deposition in the Hokuroku Basin of Japan. Economic Geology, Monograph, 5: 439-487.

Charlou, J.L., Bougalt, H., Appriu, P., Nelsen, T. and Rona, P., 1991. Different TDM/CH4 hydrothermal plume signatures: TAG site at 26°N and serpentinized ultrabasic diapier at 15°05'N on the Mid-Atlantic Ridge. Geochimica et Cosmochimica Acta, 55: 3209-3222.

Clague, D.A., Frey, F.A., Thompson, G. and Rindge, S., 1981. Minor and trace element geochemistry of volcanic rocks dredged from the Galapagos spreading center: Role of crystal fractionation and mantle heterogeneity. Journal of Geophysical Research, 86: 9469-9482.

Collier, J. and Sinha, N., 1990. Seismic images of a magma chamber beneath the Lau Basin back - arc spreading center. Nature, 346: 646-648.

Converse, D.R., Holland, H.D. and Edmond, J.M., 1984. Flow rates in the axial hot springs of the East Pacific Rise (21°N) : implications for the heat budget and the formation of massiv sulfide deposits. Earth and Planetary Sience Letters, 69: 159-175.

Corliss, J.B., Dymond, J., Gordon, L.I., Edmond, J.M., von Herzen, R.P., Ballard, R.D. Green, K., Williams, D., Bainbridge, A., Crane, K. and van Andel, T.H., 1997. Submarine thermal springs on the Galapagos Rift. Science, 203: 1073-1083.

Craig, H., Horibe, Y., Farley, K.A., Welhan, J.A., Kim, K.R. and Hey, R.N., 1987. Hydrothermal vents in the Mariana Trough: Results of the first Alvin dives. American Geophysical Union Transactions, 68: 1531.

Crerar, D. and Barnes, H.L., 1976. Ore solution chemistry V. Solubilities of chalacopyrite and chalcocite assemblages in hydrothermal solutions at 200°C to 350°C. Economic Geology, 71: 772-794.

de Ronde, C.E.J., 1995. Fluid chemistry and isotopic characteristics of seafloor hydrothermal systems and associated VMS deposits: potential for magmatic contributions. In: Thompson, JFH (ed) Magmas, Fluids, and Ore Deposits. Mineralogical Association of Canada Short Course Notes, 23: 479-509.

Delaney, J.R., Johnson, H.P. and Karsten, J.L., 1981. The Juan de Fuca Ridge-hot spot-propagating rift system. Journal of geophysical Research, 86: 11747-11750.

Delaney, J.R., McDuff, R.E. and Lupton, J.E., 1984. Hydrothermal fluid temperatures of 400°C of the Endeacour Segment, northern Juan de Fuca Ridge. American Geophysical Union Transactions, 65: 973.

Detrick, R.S.P., Buhl, E., Vera, J., Mutter, J., Orcutt, J., Madsen, J. and Brocher, T., 1987. Multi-channel seismic imaging of a crustal magma chamber along the East Pacific Rise. Nature, 326: 35-41.

Dia, A.N., Aquilina, L., Boulègue, J., Bourgois, J., Suess, E. and Torres, M., 1993. Origin of fluids and related barite deposits and vent sites along the Peru convergent margin. Geology, 21: 1099-1102.

Dickens, G.R. and Quinby - Hunt, M.S., 1994. Methane hydrate stability in seawater. Geophysical Research Letters, 21: 2115-2118.

Doe, B.R., 1994. Zinc, copper, and lead in mid - ocean ridge basalts and the source rock control on Zn/Pb in ocean-ridge hydrothermal deposits. Geochimica et Cosmochimica Acta, 58: 2215-2223.

Edmond, J.M., Measures, C.I., McDuff, R.E., Chan, L.H., Collier, R., Grant, B., Gordon, L.I. and Corliss, J.B., 1979a. Ridge crest hydrithermal activity and the balance of the major and minor elements in the ocean: The Galapagos data. Earth and Planetary Science Letters, 46: 1-18.

Edmond, J.M., Measures, C.I., Mangum, B., Grant, B., Sclater, F.R., Collier, R., Hudson, A., Gordon, L.I. and Corliss, J.B., 1979b. On the formation of metal-rich deposits at ridge crests. Earth and Planetary Science Letters, 46: 19-30.

Edmond, J.M., von Damm, K.L., McDuff, R.E. and Measures, C.I., 1982. Chemistry of hot springs on the East Pacific Rise and their effluent dispersal. Nature, 297: 187 - 191.

Edmond, J.M., Campbell, A.C., Palmer, M.R., Klinkhammer, G.P., German, C.R., Edmonds, H.N., Elderfield, H., Thompson, G. and Rona, P., 1995. Time series studies of vent fluids from the TAG and MARK sites (1986, 1990) Mid - Atlantic Ridge: a new solution schemistry model and a mechanism for Cu/Zn zonation in massiv sulphide orebodies. In: Parson, L.M., Walker, C.L. and Dixon, D.R. (eds). Hydrothermal Vents and Processes, Geological Society Special Publication, 87: 77-86.

Elderfield, H. and Greaves, M.J., 1982. The rare earth elements in seawater. Nature, 296: 214-219.

Embley, R.W., Jonasson, I.R., Perfit, M.R., Franklin, J.M., Tivey, M.A., Malahoff, A., Smith, M.F. and Francis, T.J.G., 1988. Submersible investigation of an extinct hydrothermal system on the Galapagos Ridge: Sulfide mounds, stockwork zone, and differentiated lavas. Canadian Mineralogist, 26: 517-539.

Embley, R.W., Chadwick, W.W., Jonasson, I.R., Petersen, S., Butterfield, D., Tunnicliffe, V. and Juniper, K., 1993. Geologic inference from a response to the first remotely detectederuption on the mid - ocean ridge: Coaxial Segment, Juan de Fuca Ridge. American Geophysical Union Transactions, 74: 619.

Embley, R.W. and Chadwick, W.W.(jr.), 1994. Volcanic and hydrothermal processes associated with a recent phase of seafloor spreading at the northern Cleft segment: Juan de Fuca Ridge. Journal of Geophysical Research, 99: 4735-4740.

Feely, R.A., Lewison, M., Massoth, G.J., Robert-Galdo, G., Lavelle, J.W., Byrne, R.H., von Damm, K.L. and Curl, H.C.(jr.), 1987. Composition and dissolution of black smoker particulates from active vents on the Juan de Fuca Ridge. Journal of Geophysical Research, 92: 11347-11363.

Feely, R.A., Massoth, G.J., Trefry, J.H., Baker, E.T., Paulson, A.J. and Lebon, G.T., 1994. Composition and sedimenta-tion of hydrothermal plume particles from North Cleft segment, Juan de Fuca Ridge. Journal of Geophysical Research, 99: 4985-5006.

Feely, R.A., Gedron, J.F., Baker, E.T. and Lebon, G.T., 1994. Hydrothermal plumes along the East Pacific Rise, 8°40' to 11°50'N: Particle distribution and composition. Earth and Planetary Science Letters, 128: 19-36.

Forquet, Y., von Stackelberg, U., Charlou, J.L., Donval, J.L., Erzinger, J., Foucher, J.P., Herzig, P.M., Mühe, R., Soakai, S., Wiedicke, M. and Whitechurch, H., 1991. Hydrothermal activity and metallogenesis in the Lau back - arc basin. Nature, 349: 778-781.

Forquet, Y., von Stackelberg, U., Charlou, J.L., Erzinger, J., Herzig, P.M., Mühe, R. and Wiedicke, M., 1993a. Metallogenesis in back-arc enviroment: the Lau Basin example. Economic Geology, 88: 2154-2181.

Forquet, Y., Auclair, P., Cambon, P. and Etoubleau, J., 1993b. Geological setting, mineralogical, and geochemical investigation on sulfide deposits near 13°N on the East Pacific Rise. Marine Geology, 84: 145-178.

Francheteau, J., Needham, H.D., Choukroune, P., Juteau, T., Seguret, M., Ballard, R.D., Fox, P.J., Normark, W., Carranza, A., Cordoba, D., Guerrero, J., Rangin, C., Bougault, H., Cambon, P. and Hekinian, R., 1979. Massiv deep-sea sulphide ore deposits discovered on the East Pacific Rise. Nature, 277: 145-178.

Franklin, J.M., Lydon, J.W. and Sangster, D.F., 1981. Volcanic - associated massiv sulfide deposits. Economic Geology, 75: 485-627.

Gamo, T., Okamura, K., Charlou, J.L., Urabe, T., Auzende, J.M., Ishibashi, J., Shitashima, K., Chiba, H., ManusFlux Shipboard Scientific Party, 1997. Acidic and sulfate - rich hydrothermal fluids from the Manus back-arc basin, Papua New Guinea. Geology, 25: 139-142.

Ginster, U., Mottl, M.J. and von Herzen, R.P., 1994. Heat flux from black smokers on the Endeavour anf Cleft segments, Juan de Fuca Ridge. Journal of Geophysical Research, 99: 4937-4950.

Goldfarb, M.S., Converse, D.R., Holland, H.D. and Edmond, J.M., 1983. The genesis of hot spring deposits on the East Pacific Rise, 21°N. Economic Geology, Monograph, 5: 184-197.

Graham, U.M., Bluth, G.J. and Ohmoto, H., 1988. Sulfide-sulfate chimneys on the East Pacific Rise, 11° and 13°N latitudes. Part 1: Mineralogy and paragenesis. Canadian Mineralogist, 26: 487-504.

Hajash, A., 1975. Hydrothermal processes along Mid-Ocean Ridges: an experimental inverstigation. Contributions in Mineralogy and Petrology, 53: 205-226.

Halbach, P., Nakamura, K., Wahsner, M., Lange, J., Sakai, H., Käselitz, L., Hansen, R.D., Yamano, M., Post, J., Prause, B., Seifert, R., Michaelis, W., Teichmann, F., Kinoshita, M., Märten, A., Ishibashi, J., Czerwinski, S. and Blum, N., 1989. Probable modern analoque of Kuroko - type massiv sulphide deposits in the Okinawa Trough back-arc basin. Nature, 338: 496-499.

Hannington, M.D., Herzig, P.M., Scott, S.D., Thompson, G. and Rona, P.A., 1991. Comparative mineralogy and geochemistry of gold-bearing sulfide deposits on the mid-ocean ridges. Marine Geology, 101: 217-248.

Hannington, M.D., Petersen, S., Jonasson, I.R. and Franklin, J.M., 1994. Hydrothermal activity and associated mineral deposits on the seafloor. Geological Survey of Canada Open File Report, 2915C: Map 1:35,000,000 and CD-ROM.

Hannington, M.D., Jonasson, I.R., Herzig, P.M. and Petersen, S., 1995. Physical, chemical processes of seafloor mineralization at mid - ocean ridges. In: Humphris, S.E. et al. (eds) Seafloor Hydrothermal Systems. Physical, Chemical, Biological and Geological Interactions, AGU Geophysical Monograph, 91: 115-157.

Haymon, R.M. and Kastner, M., 1981. Hot spring deposits on the East Pacific Rise at 21°N: Preliminary descripton of mineralogy and genesis. Earth and Planetary Science Letters, 53: 363-381.

Haymon, R.M., 1983. Growth history of hydrothermal black smoker chimneys. Nature, 301: 695-698.

Haymon, R.M.; Fornari, D.J., von Damm, K.L., Lilley, M.D., Perfit, M.R., Edmond, J.M., Shanks, W.C. III, Lutz, R.A., Grebmeier, J.M., Carbotte, S., Wright, D., McLaughlin, E., Smith, M., Beedle, N. and Olson, E., 1993. Volcanic eruption of the mid - ocean ridge along the East Pacific Rise crest at 9°45-52'N: Direct submersible observations of seafloor phenomena associated with an eruption event in April, 1991. Earth and Planetary Science Letters, 119: 85-101.

Hedenquist, J.W. and Lowenstern, J.B., 1994. The role of magmas in the formation of hydrothermal ore deposits. Nature, 370: 519-526.

Hekinian, R., Francheteau, J., Renard, V., Ballard, R.D., Choukroune, P., Cheminée, J.L., Albaréde, F., Minster, J.F., Charlou,J.L., Marty, J.C. and Boulégue, J., 1983. Intense hydrothermal activity at the rise axis of the axis of the East Pacific Rise near 13°N: Submersible withnesses the growth of sulfide chimney. Marine Geology Research, 6: 1-14.

Hemley, J.J., Cygan, G.L., Fein, J.B., Robinson, G.R. and D'Angelo, W.M., 1992. Hydrothermal ore - forminf processes in the light of studies in the rock-buffered systems: I. Iron-copper-zinc-lead sulfide solupility relations. Economic Geology, 87: 1-22.

Herzig, P.M. and Plüger, W.L., 1988. Exploration for hydrothermal mineralisation near the Rodriguez Triple Junction, Indian Ocean. Canadian Mineralogist, 26: 721-736.

Herzig, P.M. and Hannington, M.D., 1995. Polymetallic massive sulfides at the modern seafloor: A review. Ore Geology Reviews, 10: 95-115.

Herzig, P.M., Hannington, M.D. and Arribas, A.(jr.)., 1998. Sulfur isotopic composition of hydrothermal precipitates from the Lau back-arc: implications for magmatic contributions to seafloor hydrothermal systems. Mineralium Deposita, 33: 226-237.

Humphris, S.H., Herzig, P.M., Miller, D.J.,Alt, J.C., Becker, K., Brown, D., Brügmann, G., Chiba, H., Fouquet, Y., Gemmell, J.B., Guerin, G., Hannington, M.D., Holm, N.G., Honnorez, J.J., Itturino, G.J., Knott, R., Ludwig, R., Nakamura, K., Petersen, S., Reysenbach, A.L., Rona, P.A., Smith, S., Sturz, A.A., Tivey, M.K. and Zhao, X., 1995. The internal structure of an active sea-floor massive sulfide deposit. Nature, 377: 713-716.

James, R.H., Elderfield, H. and Palmer, M.R., 1995. The chemistry of hydrothermal fluids from Broken Spur site, 29°N Mid-Atlantic Ridge. Geochimica et Cosmochmica Acta, 59: 651-659.

Janecky, D.R. and Seyfried, W.E.(jr.), 1984. Formation of massiv sulfide deposits on ocean ridge crests: Incremental reaction models for mixing between hydrothermal solutions and seawater. Geochimica et Cosmochmica Acta, 48: 2723-2738.

Janecky, D.R. and Shanks, W.C.III, 1988. Computational modelling of chemical and isotopic reaction processes in seafloor hydrothermal systems: chimneys, massive sulfides, and subjacent alteration zones. Canadian Mineralogist, 26: 805-825.

Johnson, H.P. and Tunnicliffe, V., 1985. Time series measurements of hydrothermal activity on the northern Juan de Fuca Ridge. Geophysical Research Letters, 12: 685-688.

Kadko, D. and Moore, W., 1988. Radiochemical constraints on the crustal residence time of submarine hydrothermal fluids: Endeavour Ridge. Geochimica et Cosmochimica Acta, 52: 659-668.

Kastner, M., Craig, H. and Sturz, A., 1987. Hydrothermal deposition in the Mariana Trough: Preliminary mineralogical investigations. American Geophysical Union Transactions, 68: 1531.

Katz, D.L., Cornell, D., Kobayashi, R., Poettmann, F.H., Vary, J.A., Elenblass, J.R. and Weinaug, C.F., 1959. Handbook of Natural Gas Engineering, 802. McGraw-Hill, NY, 802 pp.

Keays, R.R., 1987. Principles of mobilization ( dissolution ) of metals in mafic and ultramafic rocks - The role of immiscible magmatic sulphides in the generation of hydrothermal gold and volcanogenic massiv sulphide deposits. Ore Geology Reviews, 2: 47-63.

Klinkhammer, G.P., Elderfield, H., Greaves, M., Rona, P.A. and Nelson, T., 1986. Manganese geochemistry near high-temperature vents in the Mid-Atlantic Ridge rift valley. Earth and Planetary Science Letters, 80: 230-240.

Klinkhammer, G.P., Chin, C.S., Wilson, C. and German, C.R., 1995. Venting from the Mid-Atlantic Ridge at 37°17'N: the Lucky Strike hydrothermal site. In. Parson, L.M., Walker, C.L. and Dixon, D.R. (eds) Hydrothermal vents and processes. Geological Society Special Publication, 87: 87-96.

Koski, R.A., Shanks, W.C.I., Bohrson, W.A. and Oscarson, R.L., 1988. The composition of massiv sulfide deposits from the sediment-covered floor of Escanaba Trough, Gorda Ridge: implications for depositional processes. Canadian Mineralogist, 26: 655-673.

Koski, R.A., Jonasson, I.R., Kadko, D.C., Smith, V.K. and Wong, F.L., 1994. Compositions, growth mechanisms, and temporal relations of hydrothermal sulfide-sulfate-silica chimneys at the northern Cleft segment, Juan de Fuca Ridge. Journal of Geophysical Research, 99: 4813-4832.

Kvenvolden, K.A. and McMenamin, 1980. Hydrate of natural gas. A review of their geological occurrence. U. S. Geological Survey Circular, 825: 11.

Kvenvolden, K.A., 1988. Methane hydrate - a major reservoir of carbon in the shallow geosphere. Chemical Geology, 71: 41-51.

Kvenvolden, K.A., 1993. Gas hydrates - Geological Perspective and global change. Reviews of Geophysics, 31: 173-187.

Lafitte, M., Maury, R., Perseil, E.A. and Boulegue, J., 1985. Morphological and analytical study of hydrothermal sulfides from 21° north East Pacific Rise. Earth and Planetary Science Letters, 73: 53-64.

Lalou, C., Reyss, J.L., Brichet, E., Arnold, M., Thompson, G., Fouquet, Y. and Rona, P.A., 1993. New age data for Mis - Atlantic Ridge hydrothermal sites: TAG and Snakepit chronology. Journal of Geophysical Research, 98: 9705-9713.

Leggett, J., 1990. The nature of the greenhouse threat. In: Leggett, J. (ed) Global Warming, The Greenpeace Report. Oxford University Press, NY: 14-43.

LeHuray, A.P., Church, S.E., Koski, R.A. and Bouse, R.M., 1988. Pb isotopes in sulfides from mid-ocean ridge hydrothermal sites. Geology, 16: 362-365.

Lowell, R.P., 1991. Modeling continental and submarine hydrothermal systems. Reviews in Geophysics, 29: 457-476.

Lowell, R.P., Rona, P.A. and von Herzen, R.P., 1995. Seafloor hydrothermal systems. Journal of Geophysical Research, 100: 327-352.

Lupton, J.E. and Craig, H., 1981. A major helium-3 source at 15°S on the East Pacific Rise. Science, 214: 13-18.

Lupton, J.E., 1983. Fluxes of helium-3 and heat from submarine hydrothermal systems; Guaymas Basin vesus 21°N EPR. American Geophysical Union Transactions, 64: 723.

Lupton, P., Lilley, M., Olson, E. and von Damm, K.L., 1991. Gas chemistry of vent fluids from 9°-10°N on the East Pacific Rise. American Geophysical Union Transactions, 72: 481.

Massoth, G.J., Butterfield, D., Lupton, J.E., McDuff, R.E., Lilley, M.D. and Jonasson, I.R., 1989. Submarine venting of phase-separated hydrothermal fluids at Axial Volcano, Juan de Fuca Ridge. Nature, 340: 702-7ß5.

Metz, S. and Trefry, J.H., 1993. Field and laboratory studies of metal uptake and release by hydrothermal precipitates. Journal of Geophysical Research, 98: 9661-9666.

Michard, A., Albarede, F., Michard, G., Minster, J.F. and Charlou, J.L., 1983. Rare-earth elements and uranium in high-temperature solutions from East Pacific Rise hydrothermal vent field (13°N). Nature, 303: 795-797.

Michard, A., 1989. Rare earth element systematics in hydrothermal fluids. Geochimica et Cosmochimica Acta, 53: 745-750.

Miller, A.R., Densmore, C.D., Degens, E.T., Hathaway, F.C., Manheim, F.T., McFarlen, P.F., Pocklington, H. and Jokela, A., 1966. Hot brines and recent iron deposits in deeps of the Red Sea. Geochimica et Cosmochimica Acta, 30: 341-359.

Morton, J.L. and Sleep, N.H., 1985. A mid - ocean ridge thermal model: Constraints on the volume of axial heat flux. Journal of Geophysical Research, 90: 11345-11353.

Mottl, M.J., 1983. Metabasalts, axial hot - springs, and the structure of hydrothermal systems at mid-ocean ridges. Geological Society of American Bulletin, 94: 161-180.

Mottl, M.J. and McConachy, T.F., 1990. Chemical processes in buoyant hydrothermal plumes on the East Pacific Rise near 21°N. Geochimica et Cosmochimica Acta, 54: 1911-1927.

Nehlig, P., 1991. Salinity of oceanic hydrothermal fluids: a fluid inclusion study. Earth and Planetary Science Letters, 102: 310-325.

Oudin, E., 1983. Hydrothermal sulfide deposits of the East Pacific Rise ( 21°N ) part I: descriptive mineralogy. Marine Mining, 4: 39-72.

Palmer, M.R., 1992. Controls over the chloride concentration of submarine hydrothermal vent fluids: evidence from Sr/Ca and $^{87}Sr/^{86}Sr$ rations. Earth and Planetary Science Letters, 109: 37-46.

Perfit, M.R., Fornari, D.J., Malahoff, A. and Embley, R.W., 1983. Geochemical studies of abyssal lavas recovered by DSRV Alvin from eastern Galapagos Rift, Inca Transform, and Equador Rift, 3. Trace Element abundances and pertogenesis. Journal of Geophysical Research, 88: 10551-10572.

Plüger, W.L., Herzig, P.M., Becker, K.P., Deissmann, G., Schöps, D., Lange, J., Jenisch, A., Ladage, S., Richnow, H.H., Schulze, T. and Michaelis, W., 1990. Discovery of hydrothermal fields at the Central Indian Ridge. Marine Mining, 9: 73-86.

Richardson, C.J., Cann, J.R., Richards, H.G. and Cowan, J.G., 1987. Metal - depleted zones of the Troodos ore - forming hydrothermal system, Cyprus. Earth and Planetary Science Letters, 84: 243-253.

Ritger, S., Carson, B. and Suess, E., 1987. Methane - derived authigenic carbonates formed by subduction - induced pore-water explution along the Oregon/Washington margin. Geological Society of American Bulletin, 98: 147-156.

Rona, P.A., 1988. Hydrothermal mineralization at ozean ridges. Canadian Mineralogist, 26: 431-465.

Rona, P.A. and Scott, S.D., 1993. A spezial issue on sea - floor hydothermal mineralization: new perspectives. Canadian Mineralogist, 88: 1935-1976.

Rona, P.A., Hannington, M.D., Raman, C.V., Thompson, G., Tivey, M.K., Humphris, S.E., Lalou, C. and Petersen, S., 1993. Active and relict sea-floor hydrothermal mineralization at the TAG Hydrothermal Field, Mid-Atlantic Ridge. Economic Geology, 88: 1989-2017.

Schöps, D. and Herzig, P.M., 1990. Sulfide composition and microthermometry of fluid inclusions in quartz - sulfide veins from the leg 111 dike section of ODP Hole 504B, Costa Rica Rift. Journal of Geophysical Research, 95: 8405-8418.

Seyfried, W.E.j. and Mottl, M.J., 1982. Hydrothermal alteration of basalt by seawater under seawater-dominated conditions. Geochimica et Cosmochimica Acta, 46: 985-1002.

Seyfried, W.E.j., Berndt, M.E. and Janecky, D.R., 1986. Chloride depletions and enrichments in seafloor hydrothermal fluids: Constraints from experimental basalt alteration studies. Geochimica et Cosmochimica Acta, 50: 469-475.

Seyfried, W.E.j., Berndt, M.E. and Seewald, J.S., 1988. Hydrothermal alteration processes at mid - ocean ridges: constraints from diabase alteration experiments, hot - spring fluids and composition of the oceanic crust. Canadian Mineralogist, 26: 787 - 804.

Spiess, F.N., Macdonald, K.C., Atwater, T., Ballard, R., Carranza, A., Cordoba, D., Cox, C., Diaz Gracia, V.M., Francheteau, J., Guerro, J., Hawkins, J.W., Haymon, R., Hessler, R., Juteau, T., Kastner, M., Larson, R., Luyendyk, B., Macdougall, J.D., Miller, S., Normark, W., Orcutt, J. and Rangin, C., 1980. East Pacific Rise. Hot springs and geophysical experiments. Science, 207: 1421-1433.

Stein, C.A. and Stein, S., 1994. Constrains on hydrothermal

heat flux through the ocean lithosphere from global heat flow. Journal of Geophysical Research, 99: 3081-3095.

Stein, C.A., Stein, S. and Pelayo, A.M., 1995. Heat flow and hydrothermal circulation. In: Humphris, S.E. et al. (eds) Seafloor Hydrothermal Systems: Physical, Chemical, Biological and Geological Interactions, AGU Geophysical Monograph, 91: 425-445.

Stuart, F.M., Harrop, P.J., Knott, R., Fallick, A.E., Turner, G., Fouquet, Y. and Richard, D., 1995. Noble gase isotopes in 25,000 years of hydrothermal fluids from 13°N on the East Pacific Rise. In: Pason, L.M., Walker, C.L. and Dixon, D.R. (eds) Hydrothermal Vents and Processes, Geological Society Special Puplication, 87: 133-143.

Suess, E. Bohrmann, G., Greinert, J., Linke, P., Lammers, S., Zuleger, E., Wallmann, K., Sahling, H., Dählmann, A., Rickert, D. and von Mirbach, N., 1997. Methanhydratfund von der FS Sonne vor der Westküste Nordamerikas. Geowissenschaften, 15: 194-199.

Suess, E., Bohrmann, G., von Huene, R., Linke, P., Wallmann, K., Lammers, S. and Sahling, H., 1998. Fluid venting in the eastern Aleutian subduction zone. Journal of Geophysical Research, 103: 2597-2614.

Sun, S.S., 1980. Lead isotopic study of young volcanic rocks from mid-ocean ridges, ocean islands arcs. Philosophical Transactions of the Royal Society of London, 297: 409-445.

Tivey, M.K., 1995. Modeling chimney growth and associated fluid flow at seafloor hydrothermal vent sites. In: Humhris, S.E. et al. (eds) Seafloor Hydrothermal Systems: Physical, Chemical, Biological and Geological Interactions, AGU Geological Monograph, 91: 158-177.

Tivey, M.K., Humphris, S.E., Thompson, G., Hannington, M.D. and Rona, P.A., 1995. Deducing patterns of fluid flow and mixing within the active TAG mound using mineralogical and chemical data. Journal of Geophysical Research, 100: 2527-2556.

Torres, M.E., Bohrmann, G. and Suess, E., 1996. Authigenic barites and fluxes of barium associated with fluid seeps in the Peru subduction zone. Earth and Planetary Science Letters, 144: 469-481.

Trefry, J.H., Butterfield, D.B.,Metz, S., Massoth, G.J., Trocine, R.P. and Feely, R.A., 1994. Trace metals in hydrothermal solutions from cleft segment on the southern Juan de Fuca Ridge. Journal of Geophysical Research, 99: 4925-4935.

Von Damm, K.L., Grant, B. and Edmond, J.M., 1983. Preliminary report on the chemistry of hydrothermal solutions at 21° North, East Pacific Rise. In: Rona, P.A., Bostrom, K., Laubier, L. and Smith, L. (eds) Hydrothermal Processes at Seafloor Spreading Centers. Plenum Press: 369-390 pp.

Von Damm, K.L., Edmond, J.M., Grant, B. and Measures, C.I., 1985a. Chemistry of submarine hydrothermal solutions at 21°N, East Pacific Rise. Geochimica et Cosmochimica Acta, 49: 2197-2220.

Von Damm, K.L., Edmond, J.M., Measures, C.I. and Grant, B., 1985b. Chemistry of submarine hydrothermal solutions at Guayamas Basin, Gulf of California. Geochimica et Cosmochimica Acta, 49: 2221-2237.

Von Damm, K.L. and Bischoff, J.L., 1987. Chemistry of hydrothermal solutions from the southern Juan de Fuca Ridge. Journal of Geophysical Research, 92: 11334-11346.

Von Damm, K.L., 1988. Systematics of and postulated con-

trols on submarine hydrothermal solution chemistry. Journal of Geophysical Research, 93: 4551-4935.

Von Damm, K.L., 1990. Seafloor hydrothermal activity: Black smoker chemistry and chimneys. Annual Reviews in Earth and Planetary Science, 18: 173-204.

Von Damm, K.L., 1995. Controls on the chemistry and temporal variability of seafloor hydrothermal fluids. In: Humhris, S.E. et al. (eds) Seafloor Hydrothermal Systems: Physical, Chemical, Biological and Geological Interactions, AGU Geological Monograph, 91: 222-247.

Von Huene, R., Klaeschen, D., Gutscher, M. and Frühn, J., in press. Mass and fluid flux during accretion at the Alaska margin. Geological Socienty of American Bulletin.

Wallmann, K., Linke, P., Suess, E., Bohrmann, G., Sahling, H., Schlüter, M., Dählmann, A., Lammers, S., Greinert, J. and von Mirbach, N., 1997. Quantifying fluid flow, solute mixing, and biogeochemical turnover at cold vents of the eastern Aleutian subduction zone. Geochimica et Cosmochimica Acta, 61: 5209-5219.

Wheat, C.G. and Mottl, M.J., 1994. Hydrothermal circulation, Juan de Fuca Ridge eastern flank: Factors controllong basement water composition. Journal of Geophysical Research, 99: 3067-3080.

Wolery, T.J. and Sleep, N.H., 1976. Hydrothermal circulation and geochemical flux at mid-ocean ridges. Journal of Geology, 84: 249-275.

Zierenberg, R.A., Shanks, W.C.I. and Bischoff, J.L., 1984. Massiv sulfide deposits at 21°, East Pacific Rise: Chemical composition, stable isotopes, and phase equilibria. Geological Society of Americal Bulletin, 95: 922-929.

Zierenberg, R.A., Koski, R.A., Morton, J.L., Bouse, R.M. and Shanks, W.C.III., 1993. Genesis of massive sulfide deposits on a sediment-covered spreading center, Escanaba Trough, southern Gorda Ridge. Economic Geology, 88: 2069-2098.

# 14    Conceptual Models and Computer Models

Horst D. Schulz

Upon recording *processes* of nature quantitatively, the term model is closely related to the term system. A *system* is a segment derived from nature with either real or, at least, imagined boundaries. Within these boundaries, there are processes which are to be analyzed. Outside, there is the *environment* exerting an influence on the course of the procedural events which are internal to the system by means of the *boundary conditions*. A *conceptual model* contains principle statements, mostly translatable quantitatively, with regard to the processes in a system and the influence of prevalent boundary conditions. If the systems to be reproduced are especially complex, any significant realization of the conceptual model is often only possible by applying computer models.

Depending on the field of interest and the nature of the task, the processes in a model will be either somewhat more physical/hydraulic, geochemical, biogeochemical, or biological. The systems and processes studied might then expand in size over the great oceans and their extensive stream patterns down to small segments of deep ocean floor measuring only few cubic centimeters, including the communities of microorganisms living therein. Consequently, there are no pre-defined dimensions or structures which are valid for the systems and their models, instead, these are exclusively determined by the understanding of the respective processes of interest and their quantitative translatability. Since geochemical and biogeochemical reactions occur relatively fast, and are thus bound to spaces of smaller dimensions, the models dealt with in this chapter will be preferentially concerned with dimensions encompassing few centimeters and meters of sediment, to some hundred meters of water column.

In the field of applied geological sciences, models frequently assume a primarily prognostic

character. For instance, after accomplishing an adequate calibration, an hydraulic model applied to groundwater, or a model of solute transport in groundwater, is primarily supposed to predict the system's reaction as precisely as possible under the given boundary conditions. Such models will find very practical and often well paid applications. In marine geochemistry such a prognostic function of models occurs rather seldom; here, the objective of obtaining a quantitative concept of a system predominates in most cases, which is 'function' of a particular segment of nature. Computer models can then check the plausibility of model concepts of complex systems, can draw attention to the not sufficiently understood areas, and they might help to discover parameters with which the system can be described reasonably well and effectively.

Here, the zero-dimensional geochemical reaction models will be distinguished from one-dimensional models, in which, by various means, transport processes are coupled to geochemical and biogeochemical reactions. In the following subsections, approaches to very differing models will be introduced, as well as models that have advanced to very different stages of development.

## 14.1    Geochemical Models

### 14.1.1    Structure of Geochemical Models

Zero-dimensional geochemical models are concerned with the contents of solutions of an aquatic subsystem (e.g. pore water, ocean water, precipitation, groundwater), the equilibria between the various dissolved species as well as their adjoining gaseous and solid phases. A computer model by the name of WATEQ (derived from water-equilibria) was published for the first time by Truesdell and Jones (1974). Unfortunately, this was done in the computer language PL1 which has become completely forgotten by now. Afterwards a FORTRAN version had followed (WATEQF) released by Plummer et al. (1976) as well as another revised version (WATEQ4F) by Ball and Nordstrom (1991).

All these programs share the common feature that they start out from water analyses which were designed as comprehensive as possible, along with conceiving the measured values as cumulative parameters (sums of all aquatic species), only to subsequently allocate them to the various aquatic species. This was done by applying iteration calculus that accounts for the equilibria related to the various aquatic species.

For instance, the measured value of calcium in an ocean water sample containing 10.6 mmol/l is

**Table 14.1** This ocean water analysis carried out by Nordstrom et al. (1979) has been frequently used to test and compare various geochemical model programs.

### Seawater from Nordstrom et al. (1979)

| | | |
|---|---|---|
| pH | = | 8.22 |
| pe | = | 8.451 |
| Ionic strength | = | 6.750E-1 |
| Temperature (°C) | = | 25.0 |

| Elements | Molality |
|---|---|
| Alkalinity | 2.406E-3 |
| Ca | 1.066E-2 |
| Cl | 5.657E-1 |
| Fe | 3.711E-8 |
| K | 1.058E-2 |
| Mg | 5.507E-2 |
| Mn | 3.773E-9 |
| N (-III) | 1.724E-6 |
| N (V) | 4.847E-6 |
| Na | 4.854E-1 |
| O (0) | 3.746E-4 |
| S (VI) | 2.926E-2 |
| Si | 7.382E-5 |

allocated to 9.51 mmol/l $Ca^{2+}$, 1.08 mmol/l $CaSO_4^0$, 0.04 mmol/l $CaHCO_3^+$, and to a series of low concentrated calcium species detectable in solution. The analysis of sulfate yielding 29.26 mmol/l is allocated accordingly to 14.67 mmol/l $SO_4^{2-}$, 7.30 mmol/l $NaSO_4^-$, 1.08 mmol/l $CaSO_4^0$, 0.16 mmol/l $KSO_4^-$, and, likewise, to a number of low concentrated sulfate species. The procedure is similarly applied to all analytical values. Table 14.2 summerizes the species allocations for the ocean water analysis previously shown in Table 14.1.

Based on the activities of the non-complex aquatic species (e.g. $[Ca^{2+}]$ or $[SO_4^{2-}]$, the next step consists in calculating the saturation indices (SI) of each compound mineral:

$$SI = \log\left(\frac{IAP}{K_{sp}}\right) \quad (14.1)$$

In this equation IAP denotes the Ion Activity Product (in the example of gypsum or anhydrite these would be $([Ca^{2+}] \cdot [SO_4^{2-}])$. $K_{SP}$ is the solubility product constant of the respective mineral. A saturation index SI = 0 describes the condition in which the solution of the corresponding mineral is just saturated, SI > = 0 describes the condition of supersaturation of the solution, SI < 0 its undersaturation. The activity [A] of a substance is calculated according to the equation:

$$[A] = (A) \cdot \gamma_A \quad (14.2)$$

Here, (A) is the concentration of a substance measured in mol/l and $\gamma_A$ is the activity coefficient calculated as a function of the overall ionic strength in the solution. For infinitesimal diluted solutions, the activity coefficients assume the value of 1; hence, activity equals concentration. In ocean water, the various monovalent ions, or ion pairs, display activity coefficients somewhere around 0.75; whereas the various divalent ions , or ion pairs, display values around 0.2 (cf. Table 14.2, last column).

After transformation, the equation 14.1 to calculate the saturation index of the mineral gypsum reads:

$$SI_{gypsum} = \log\left(\frac{[Ca^{2+}] \cdot [SO_4^{2-}]}{K_{sp,gypsum}}\right) \quad (14.3)$$

If we now insert the solubility product constant for gypsum $K_{sp,gypsum} = 2.63 \cdot 10^{-5}$ in the equation

**Table 14.2**  For an analysis of Table 14.1, this allocation of the aquatic species was calculated with the geochemical model program PHREEQC.

| Species | Molality | Activity | Molality log | Activity log | Gamma log |
|---|---|---|---|---|---|
| $OH^-$ | 2.18E-06 | 1.63E-06 | -5.661 | -5.788 | -0.127 |
| $H^+$ | 7.99E-09 | 6.03E-09 | -8.098 | -8.220 | -0.122 |
| $H_2O$ | 5.55E+01 | 9.81E-01 | -0.009 | -0.009 | 0.000 |
| **C (IV)** | 2.181E-03 | | | | |
| $HCO_3^-$ | 1.52E-03 | 1.03E-03 | -2.819 | -2.989 | -0.171 |
| $MgHCO_3^+$ | 2.20E-04 | 1.64E-04 | -3.659 | -3.785 | -0.127 |
| $NaHCO_3$ | 1.67E-04 | 1.95E-04 | -3.778 | -3.710 | 0.068 |
| $MgCO_3$ | 8.90E-05 | 1.04E-04 | -4.050 | -3.983 | 0.068 |
| $NaCO_3^-$ | 6.73E-05 | 5.03E-05 | -4.172 | -4.299 | -0.127 |
| $CaHCO_3^+$ | 4.16E-05 | 3.10E-05 | -4.381 | -4.508 | -0.127 |
| $CO_3^{-2}$ | 3.84E-05 | 7.98E-06 | -4.415 | -5.098 | -0.683 |
| $CaCO_3$ | 2.72E-05 | 3.17E-05 | -4.566 | -4.499 | 0.068 |
| $CO_2$ | 1.21E-05 | 1.42E-05 | -4.916 | -4.849 | 0.068 |
| $MnCO_3$ | 3.18E-10 | 3.71E-10 | -9.498 | -9.430 | 0.068 |
| $MnHCO_3^+$ | 7.17E-11 | 5.36E-11 | -10.144 | -10.271 | -0.127 |
| $FeCO_3$ | 1.96E-20 | 2.29E-20 | -19.709 | -19.641 | 0.068 |
| $FeHCO_3^+$ | 1.64E-20 | 1.22E-20 | -19.785 | -19.912 | -0.127 |
| **Ca** | 1.066E-02 | | | | |
| $Ca^{+2}$ | 9.51E-03 | 2.37E-03 | -2.022 | -2.625 | -0.603 |
| $CaSO_4$ | 1.08E-03 | 1.26E-03 | -2.967 | -2.900 | 0.068 |
| $CaHCO_3^+$ | 4.16E-05 | 3.10E-05 | -4.381 | -4.508 | -0.127 |
| $CaCO_3$ | 2.72E-05 | 3.17E-05 | -4.566 | -4.499 | 0.068 |
| $CaOH^+$ | 8.59E-08 | 6.41E-08 | -7.066 | -7.193 | -0.127 |
| $CaHSO_4^+$ | 5.96E-11 | 4.45E-11 | -10.225 | -10.352 | -0.127 |
| **Cl** | 5.656E-01 | | | | |
| $Cl^-$ | 5.66E-01 | 3.52E-01 | -0.247 | -0.453 | -0.206 |
| $MnCl^+$ | 1.13E-09 | 8.41E-10 | -8.948 | -9.075 | -0.127 |
| $MnCl_2$ | 1.11E-10 | 1.29E-10 | -9.956 | -9.888 | 0.068 |
| $MnCl_3^-$ | 1.68E-11 | 1.26E-11 | -10.774 | -10.901 | -0.127 |
| $FeCl^{+2}$ | 9.58E-19 | 2.97E-19 | -18.019 | -18.527 | -0.508 |
| $FeCl_2^+$ | 6.27E-19 | 4.68E-19 | -18.203 | -18.330 | -0.127 |
| $FeCl^+$ | 7.78E-20 | 5.81E-20 | -19.109 | -19.236 | -0.127 |
| $FeCl_3$ | 1.41E-20 | 1.65E-20 | -19.850 | -19.783 | 0.068 |
| **Fe (II)** | 5.550E-19 | | | | |
| $Fe^{+2}$ | 3.85E-19 | 1.19E-19 | -18.415 | -18.923 | -0.508 |
| $FeCl^+$ | 7.78E-20 | 5.81E-20 | -19.109 | -19.236 | -0.127 |
| $FeSO_4$ | 4.84E-20 | 5.65E-20 | -19.315 | -19.248 | 0.068 |
| $FeCO_3$ | 1.96E-20 | 2.29E-20 | -19.709 | -19.641 | 0.068 |
| $FeHCO_3^+$ | 1.64E-20 | 1.22E-20 | -19.785 | -19.912 | -0.127 |
| $FeOH^+$ | 8.23E-21 | 6.15E-21 | -20.084 | -20.211 | -0.127 |
| **Fe (III)** | 3.711E-08 | | | | |
| $Fe(OH)_3$ | 2.84E-08 | 3.32E-08 | -7.547 | -7.479 | 0.068 |
| $Fe(OH)_4^-$ | 6.60E-09 | 4.92E-09 | -8.181 | -8.308 | -0.127 |
| $Fe(OH)_2^+$ | 2.12E-09 | 1.58E-09 | -8.674 | -8.801 | -0.127 |
| $FeOH^{+2}$ | 9.46E-14 | 2.94E-14 | -13.024 | -13.532 | -0.508 |
| $FeSO_4^+$ | 1.09E-18 | 8.16E-19 | -17.962 | -18.089 | -0.127 |
| $FeCl^{+2}$ | 9.58E-19 | 2.97E-19 | -18.019 | -18.527 | -0.508 |
| $FeCl_2^+$ | 6.27E-19 | 4.68E-19 | -18.203 | -18.330 | -0.127 |
| $Fe^{+3}$ | 3.88E-19 | 2.80E-20 | -18.411 | -19.554 | -1.143 |
| $Fe(SO_4)_2^-$ | 6.36E-20 | 4.75E-20 | -19.196 | -19.323 | -0.127 |
| $FeCl_3$ | 1.41E-20 | 1.65E-20 | -19.850 | -19.783 | 0.068 |
| **K** | 1.058E-02 | | | | |
| $K^+$ | 1.04E-02 | 6.49E-03 | -1.982 | -2.188 | -0.206 |
| $KSO_4^-$ | 1.64E-04 | 1.22E-04 | -3.786 | -3.913 | -0.127 |

Table 14.2    continued.

| Species | Molality | Activity | Molality log | Activity log | Gamma log |
|---|---|---|---|---|---|
| **Mg** | 5.507E-02 | | | | |
| $Mg^{+2}$ | 4.75E-02 | 1.37E-02 | -1.324 | -1.864 | -0.541 |
| $MgSO_4$ | 7.30E-03 | 8.53E-03 | -2.137 | -2.069 | 0.068 |
| $MgHCO_3^+$ | 2.20E-04 | 1.64E-04 | -3.659 | -3.785 | -0.127 |
| $MgCO_3$ | 8.90E-05 | 1.04E-04 | -4.050 | -3.983 | 0.068 |
| $MgOH^+$ | 1.08E-05 | 8.07E-06 | -4.966 | -5.093 | -0.127 |
| **Mn (II)** | 3.773E-09 | | | | |
| $Mn^{+2}$ | 1.89E-09 | 5.86E-10 | -8.724 | -9.232 | -0.508 |
| $MnCl^+$ | 1.13E-09 | 8.41E-10 | -8.948 | -9.075 | -0.127 |
| $MnCO_3$ | 3.18E-10 | 3.71E-10 | -9.498 | -9.430 | 0.068 |
| $MnSO_4$ | 2.38E-10 | 2.77E-10 | -9.624 | -9.557 | 0.068 |
| $MnCl_2$ | 1.11E-10 | 1.29E-10 | -9.956 | -9.888 | 0.068 |
| $MnHCO_3^+$ | 7.17E-11 | 5.36E-11 | -10.144 | -10.271 | -0.127 |
| $MnCl_3^-$ | 1.68E-11 | 1.26E-11 | -10.774 | -10.901 | -0.127 |
| $MnOH^+$ | 3.29E-12 | 2.45E-12 | -11.483 | -11.610 | -0.127 |
| $Mn(OH)_3^-$ | 5.36E-20 | 4.00E-20 | -19.271 | -19.398 | -0.127 |
| $Mn(NO_3)_2$ | 2.62E-20 | 3.06E-20 | -19.582 | -19.515 | 0.068 |
| **Mn (III)** | 7.108E-26 | | | | |
| $Mn^{+3}$ | 7.11E-26 | 5.12E-27 | -25.148 | -26.291 | -1.143 |
| **Mn (VI)** | 2.322E-28 | | | | |
| $MnO_4^{-2}$ | 2.32E-28 | 7.21E-29 | -27.634 | -28.142 | -0.508 |
| **Mn (VII)** | 1.127E-29 | | | | |
| $MnO_4^-$ | 1.13E-29 | 8.41E-30 | -28.948 | -29.075 | -0.127 |
| **N (-III)** | 1.724E-06 | | | | |
| $NH_4^+$ | 1.58E-06 | 1.18E-06 | -5.802 | -5.929 | -0.127 |
| $NH_3$ | 9.36E-08 | 1.09E-07 | -7.029 | -6.961 | 0.068 |
| $NH_4SO_4^-$ | 5.40E-08 | 4.03E-08 | -7.267 | -7.394 | -0.127 |
| **N (V)** | 4.847E-06 | | | | |
| $NO_3^-$ | 4.85E-06 | 3.62E-06 | -5.315 | -5.441 | -0.127 |
| $Mn(NO_3)_2$ | 2.62E-20 | 3.06E-20 | -19.582 | -19.515 | 0.068 |
| **Na** | 4,854E-01 | | | | |
| $Na^+$ | 4.79E-01 | 3.38E-01 | -0.320 | -0.471 | -0.151 |
| $NaSO_4^-$ | 6.05E-03 | 4.51E-03 | -2.219 | -2.346 | -0.127 |
| $NaHCO_3$ | 1.67E-04 | 1.95E-04 | -3.778 | -3.710 | 0.068 |
| $NaCO_3^-$ | 6.73E-05 | 5.03E-05 | -4.172 | -4.299 | -0.127 |
| **O (0)** | 3.746E-04 | | | | |
| $O_2$ | 1.87E-04 | 2.19E-04 | -3.728 | -3.660 | 0.068 |
| **S (VI)** | 2.926E-02 | | | | |
| $SO_4^{-2}$ | 1.47E-02 | 2.66E-03 | -1.833 | -2.575 | -0.741 |
| $MgSO_4$ | 7.30E-03 | 8.53E-03 | -2.137 | -2.069 | 0.068 |
| $NaSO_4^-$ | 6.05E-03 | 4.51E-03 | -2.219 | -2.346 | -0.127 |
| $CaSO_4$ | 1.08E-03 | 1.26E-03 | -2.967 | -2.900 | 0.068 |
| $KSO_4^-$ | 1.64E-04 | 1.22E-04 | -3.786 | -3.913 | -0.127 |
| $NH_4SO_4^-$ | 5.40E-08 | 4.03E-08 | -7.267 | -7.394 | -0.127 |
| $HSO_4^-$ | 2.09E-09 | 1.56E-09 | -8.680 | -8.807 | -0.127 |
| $MnSO_4$ | 2.38E-10 | 2.77E-10 | -9.624 | -9.557 | 0.068 |
| $CaHSO_4^+$ | 5.96E-11 | 4.45E-11 | -10.225 | -10.352 | -0.127 |
| $FeSO_4^+$ | 1.09E-18 | 8.16E-19 | -17.962 | -18.089 | -0.127 |
| $Fe(SO_4)_2^-$ | 6.36E-20 | 4.75E-20 | -19.196 | -19.323 | -0.127 |
| $FeSO_4$ | 4.84E-20 | 5.65E-20 | -19.315 | -19.248 | 0.068 |
| $FeHSO_4^{+2}$ | 4.24E-26 | 1.32E-26 | -25.373 | -25.881 | -0.508 |
| $FeHSO_4^+$ | 3.00E-27 | 2.24E-27 | -26.523 | -26.650 | -0.127 |
| **Si** | 7.382E-05 | | | | |
| $H_4SiO_4$ | 7.11E-05 | 8.31E-05 | -4.148 | -4.081 | 0.068 |
| $H_3SiO_4^-$ | 2.72E-06 | 2.03E-06 | -5.565 | -5.692 | -0.127 |
| $H_2SiO_4^{-2}$ | 7.39E-11 | 2.29E-11 | -10.131 | -10.639 | -0.508 |

and the exemplary values mentioned above for ocean water (activity coefficients of divalent ions $\approx 0.2$; concentration of $Ca^{2+} = 9.15/1000$ mol/l; concentration of $SO_4^{2-} = 14.67/1000$ mol/l), yields the following equation:

$$SI_{gypsum,\ seawater} =$$

$$\log\left\{\left(0.2\cdot\frac{9.15}{1000}\right)\cdot\left(0.2\cdot\frac{14.67}{1000}\right)\middle/2.63\cdot10^{-5}\right\} = -0.69$$

$$(14.4)$$

Matching to Tables 14.1 and 14.2 the saturation indices are demonstrated in Table 14.3, likewise calculated by applying the PHREEQC model (Parkhurst 1995). Here, the majority of the saturation indices have been calculated on the basis of mineral phases, however, there are also gaseous phases denoted as '$CO_2(g)$', '$H_2(g)$', '$NH_3(g)$', '$O_2(g)$'. Here, the saturation index stands for the common logarithm of the respective partial pressure. The number of minerals and gases listed in Table 14.3 is determined by the fact that calculations can only be performed after all the ions involved have been analyzed, and furthermore, if the database file of the program contains the appropriate thermodynamic data.

It needs to be stated clearly that the conditions of saturation thus calculated in the process of water analyses are determined *exclusively* on the basis of the prevalent analytical data and the employed data available on equilibria. A mineral shown to be supersaturated must not immediately, or at a later instance, be precipitated from this solution, but *can* be formed, for example, when other conditions are fulfilled, e.g. such standing in relation to the reaction kinetics. At the same time, a mineral found to be undersaturated does not have to become dissolved immediately or at a later time – after all, it is possible that this mineral will never come in contact with the solution. The result of such a model calculation should just be understood as the statement that certain minerals can be either dissolved or precipitated. It goes without saying that mostly such minerals are of particular interest that have a calculated saturation index close to zero, because this circumstance often refers to set equilibria and hence to corresponding reactions.

By now, there is a huge number of such geochemical models. Apart from the models belonging to the WATEQ-family which are by now

only of interest to science history, the model SOLMINEQ (Kharaka et al. 1988) is essential because, although representing a model for waters in crude oil fields, it contains a special pressure and temperature corrective for the applied constants which can also be important in marine environments as well. Other models make use of a temperature corrective only. Unfortunately, the SOLMINEQ model has not been developed any further nor improved since its first release, so that only a first, somewhat faulty version is now available.

Beside the model EQ 3/6 (Wolery 1993), the model PHREEQE (Parkhurst et al. 1980) and its recent successor PHREEQC (Parkhurst et al. 1995) are geochemical models in the true sense of the word. With these, one cannot just calculate any kind of saturation indices on the basis of a previously conducted analysis, which are to be interpreted afterwards more or less well by giving some meaning to them, instead, one can simulate almost any process, almost without any limitation, after pre-selecting boundary conditions (e.g. exchanges with the gaseous phase and precipitation of minerals), or specific processes (e.g. decomposition of organic matter with corresponding amounts of nitrogen and phosphorous, mixing of waters containing different solutes).

The only limitation is one's own knowledge as to the process to be modeled. The program PHREEQC can be obtained from the internet as public domain software, along with an elaborate, very informative description and many examples, under:

*brrcrftp.cr.usgs.GOV (136.177.112.5)*
The files reside in directories */geochem/pc/ phreeqc* and */geochem/unix/phreeqc*.

The examples presented in the following and in Section 9.2 have been calculated exclusively with the program PHREEQC.

### 14.1.2  Application Examples of Geochemical Modeling

*Determination of Saturation Indices in Pore water, Precipitation of Minerals*

Saturation indices were determined with the PHREEQC model for the core which pore water concentration profiles were previously shown in Chapter 3, Figure 3.1. Here, interest had focused

**Table 14.3** According to the species distribution in Table 14.2, the PHREEQC model (Parkhurst 1995) was also applied to calculate these saturation indices. Here, 'CO$_2$ (g)', 'H$_2$(g)', 'NH$_3$(g)', 'O$_2$(g)' do not stand for mineral phases, but for gaseous phases in which the saturation index is obtained as the logarithm of the respective partial pressure.

| Phase | SI | log IAP | log KT | |
|---|---|---|---|---|
| Anhydrite | -0.84 | -5.20 | -4.36 | $CaSO_4$ |
| Aragonite | 0.61 | -7.72 | -8.34 | $CaCO_3$ |
| Artinite | -2.03 | 7.57 | 9.60 | $MgCO_3{:}Mg(OH)_2{:}3H_2O$ |
| Birnessite | 4.81 | 5.37 | 0.56 | $MnO_2$ |
| Bixbyite | -2.68 | 47.73 | 50.41 | $Mn_2O_3$ |
| Brucite | -2.28 | 14.56 | 16.84 | $Mg(OH)_2$ |
| Calcite | 0.76 | -7.72 | -8.48 | $CaCO_3$ |
| Chalcedony | -0.51 | -4.06 | -3.55 | $SiO_2$ |
| Chrysotile | 3.36 | 35.56 | 32.20 | $Mg_3Si_2O_5(OH)_4$ |
| Clinoenstatite | -0.84 | 10.50 | 11.34 | $MgSiO_3$ |
| CO$_2$ (g) | -3.38 | -21.53 | -18.15 | $CO_2$ |
| Cristobalite | -0.48 | -4.06 | -3.59 | $SiO_2$ |
| Diopside | 0.35 | 20.25 | 19.89 | $CaMgSi_2O_6$ |
| Dolomite | 2.40 | -14.69 | -17.09 | $CaMg(CO_3)_2$ |
| Epsomite | -2.36 | -4.50 | -2.14 | $MgSO_4{:}7H_2O$ |
| Fe(OH)$_{2.7}$Cl$_{0.3}$ | 5.52 | -6.02 | -11.54 | $Fe(OH)_{2.7}Cl_{0.3}$ |
| Fe(OH)$_3$ (a) | 0.19 | -3.42 | -3.61 | $Fe(OH)_3$ |
| Fe$_3$(OH)$_8$ | -12.56 | -9.34 | 3.22 | $Fe_3(OH)_8$ |
| Forsterite | -3.24 | 25.07 | 28.31 | $Mg_2SiO_4$ |
| Goethite | 6.09 | -3.41 | -9.50 | $FeOOH$ |
| Greenalite | -36.43 | -15.62 | 20.81 | $Fe_3Si_2O_5(OH)_4$ |
| Gypsum | -0.64 | -5.22 | -4.58 | $CaSO_4{:}2H_2O$ |
| H$_2$ (g) | -41.22 | 1.82 | 43.04 | $H_2$ |
| Halite | -2.51 | -0.92 | 1.58 | $NaCl$ |
| Hausmannite | 1.78 | 19.77 | 17.99 | $Mn_3O_4$ |
| Hematite | 14.20 | -6.81 | -21.01 | $Fe_2O_3$ |
| Huntite | 1.36 | -28.61 | -29.97 | $CaMg_3(CO_3)_4$ |
| Hydromagnesite | -4.56 | -13.33 | -8.76 | $Mg_5(CO_3)_4(OH)_2{:}4H_2O$ |
| Jarosite (ss) | -9.88 | -43.38 | -33.50 | $(K_{0.77}Na_{0.03}H_{0.2})Fe_3(SO_4)_2(OH)_6$ |
| Magadiite | -6.44 | -20.74 | -14.30 | $NaSi_7O_{13}(OH)_3{:}3H_2O$ |
| Maghemite | 3.80 | -6.81 | -10.61 | $Fe_2O_3$ |
| Magnesite | 1.07 | -6.96 | -8.03 | $MgCO_3$ |
| Magnetite | 3.96 | -9.30 | -13.26 | $Fe_3O_4$ |
| Manganite | 2.46 | 6.28 | 3.82 | $MnOOH$ |
| Melanterite | -19.35 | -21.56 | -2.21 | $FeSO_4{:}7H_2O$ |
| Mirabilite | -2.49 | -3.60 | -1.11 | $Na_2SO_4{:}10H_2O$ |
| Mn$_2$(SO$_4$)$_3$ | -54.60 | -9.29 | 45.31 | $Mn_2(SO_4)_3$ |
| MnCl$_2$:4H$_2$O | -12.88 | -10.17 | 2.71 | $MnCl_2{:}4H_2O$ |
| MnSO$_4$ | -14.48 | -11.81 | 2.67 | $MnSO_4$ |
| Nahcolite | -2.91 | -13.79 | -10.88 | $NaHCO_3$ |

**Table 14.3**  continued

| Phase | SI | log IAP | log KT | |
|---|---|---|---|---|
| Natron | -4.81 | -6.12 | -1.31 | $Na_2CO_3{:}10H_2O$ |
| Nesquehonite | -1.37 | -6.99 | -5.62 | $MgCO_3{:}3H_2O$ |
| $NH_3$ (g) | -8.73 | 2.29 | 11.02 | $NH_3$ |
| Nsutite | 5.85 | 5.37 | -0.48 | $MnO_2$ |
| $O_2$ (g) | -0.70 | -3.66 | -2.96 | $O_2$ |
| Portlandite | -9.00 | 13.80 | 22.80 | $Ca(OH)_2$ |
| Pyrochroite | -8.01 | 7.19 | 15.20 | $Mn(OH)_2$ |
| Pyrolusite | 7.03 | 5.37 | -1.66 | $MnO_2$ |
| Quartz | -0.08 | -4.06 | -3.98 | $SiO_2$ |
| Rhodochrosite | -3.20 | -14.33 | -11.13 | $MnCO_3$ |
| Sepiolite | 1.15 | 16.91 | 15.76 | $Mg_2Si_3O_{7.5}OH{:}3H_2O$ |
| Siderite | -13.13 | -24.02 | -10.89 | $FeCO_3$ |
| $SiO_2$ (a) | -1.35 | -4.06 | -2.71 | $SiO_2$ |
| Talc | 6.04 | 27.44 | 21.40 | $Mg_3Si_4O_{10}(OH)_2$ |
| Thenardite | -3.34 | -3.52 | -0.18 | $Na_2SO_4$ |
| Thermonatrite | -6.17 | -6.05 | 0.12 | $Na_2CO_3{:}H_2O$ |
| Tremolite | 11.36 | 67.93 | 56.57 | $Ca_2Mg_5Si_8O_{22}(OH)_2$ |

on the supersaturation of several manganese minerals in various depths, and thus in differing redox environments. Figure 14.1 shows the depth profiles for the saturation indices of the minerals manganite (MnOOH), birnessite ($MnO_2$), manganese sulfide (MnS), and rhodochrosite ($MnCO_3$).

The precipitation of manganese oxides, and manganese hydroxides, is very obvious at a depth of about 0.1m below the sediment surface. For the minerals manganite and birnessite, exemplifying a group of similar minerals, the saturation indices lie around zero. In deeper core regions these and all other oxides and hydroxides are in a state of undersaturation.

The slight but very constant supersaturation (saturation index about +0.5) found for the mineral rhodochrosite over the whole depth range between 0.2 and 12 m below the sediment surface is quite interesting as well. This can be understood as an indication for a new formation of the mineral in this particular range of depth. Below 13 m rhodochrosite and MnS are both *under*saturated, obviously absent in the solid phase, and therefore cannot have controlled the low manganese concentrations still prevalent in greater depths by means of mineral equilibria.

*Calculation of Predominance Field Diagrams*

The graphical representations of species distributions and/or mineral saturation as a function of $E_H$-value and pH-value, which are commonly typified as predominance field diagrams, or sometimes even not very correctly as stability field diagrams, can be done very accurately with the aid of geochemical model programs. The diagrams related to manganese, iron and copper shown in Chapter 11 (Figs. 11.2, 11.4, and 11.6) as well as the Figure 14.2 referring to arsenic, have all been made by applying the model program PHREEQC.

The procedure consisted in scanning the whole $E_H$/pH-range in narrow intervals with several thousand PHREEQC-calculations and a given specific configuration of concentrations (in this particular case an ocean water analysis at the ocean bottom, at the site of manganese nodule formation). Thus, each special case is individually calculated and adjusted in agreement to the appropriate temperature, the correct ionic strength (and hence the correct activity coefficient) and accounting for all essential aquatic species and all the minerals eventually present.

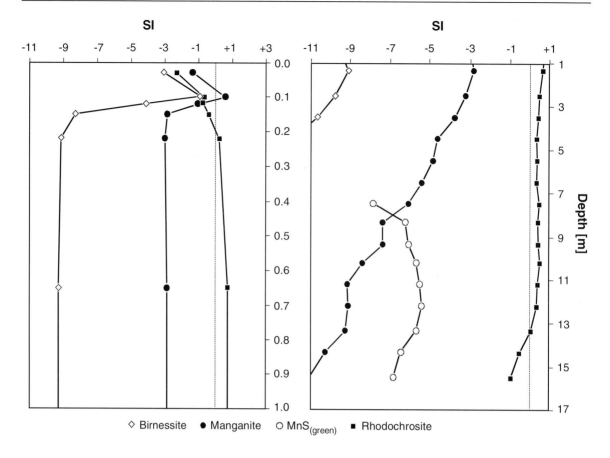

**Fig. 14.1**  Depth profiles of the saturation indices related to several manganese minerals in sediments approximately 4000 m below the water surface, off the Congo Fan. The diagram depicted on the left-hand side covers the region close to the sediment surface; whereas the diagram on the right reflects the range down to the core's ultimate depth in another depth scale. Concentration profiles of the pore water pertaining to this core are shown in Figure 3.1; the concentration profile of manganese is shown in more detail in Figure 3.8, its diffusive fluxes are discussed more thoroughly in Section 3.2.3.

Such diagrams as these are not only much more closely adapted to the respective geochemical environment, but also contain essentially more information than the otherwise customary diagrams in which boundaries mostly result in a simplified fashion from the given specific conditions. On the other hand, this information cannot any longer be included reasonably into one diagram only. Here, one diagram is used exclusively for the predominance field of aquatic species (left diagram), whereas a second is used for visualizing the saturation ranges of the various minerals (right diagram). This makes sense inasmuch as, wherever minerals become supersaturated, there will nevertheless be an according distribution of the aquatic species that cannot simply be omitted in preference of the minerals. The diagrams shown on the right-hand side reflect the saturation ranges. Here, it is reasonable to depict over-

lapping ranges of various minerals, because which mineral might precipitate – and which one will not – is not determined on account of these statements or the degree of supersaturation.

*Adjustment of Equilibria to Minerals and to the Gaseous Phase*

Calcite is supersaturated in seawater which is described with regard to its solutes in the Tables 14.1 to 14.3. For calcite, the degree of supersaturation is expressed by the saturation index of 0.76, as shown in Table 14.3. Moreover, the partial pressure values of the gaseous phases $pCO_2$ and $pO_2$ (log $pCO_2$ = -3.38; log $pO_2$ = -0.70) are indicated which differ only slightly from the respective atmospheric values. Now it would make sense to ask what composition of this seawater is to be expected, if the equilibria were adjusted to

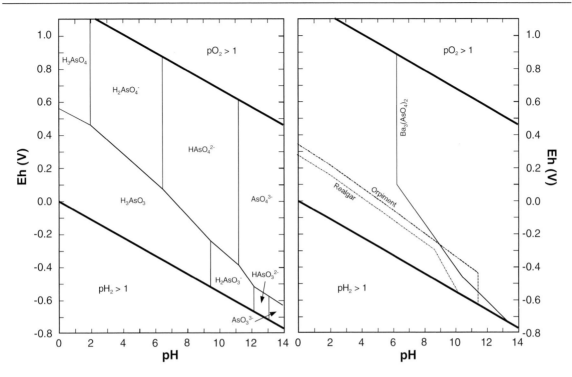

**Fig. 14.2** Predominance Field Diagram for arsenic under the condition of manganese nodule formation at the bottom of the deep sea (after Glasby and Schulz, in press). The calculations in this diagram were performed with the model PHREEQC. The partial diagram on the left only represents the predominant ranges of the aquatic species. The diagram on the right covers the ranges of several minerals which are, under the given conditions, supersaturated (SI > 0).

the atmospheric $CO_2$ and $O_2$-partial pressure, and, for example, to the mineral calcite that eventually might need to be precipitated. Table 14.4 shows the result of such a calculation with regard to the essential properties of the solution. Naturally, the program PHREEQC will then provide further detailed information on the distribution of the aquatic species and all the saturation indices according to the Tables 14.2 and 14.3, which presentation, however, was omitted in this context.

*Decomposition of Organic Matter by Oxygen and Nitrate*

If certain boundary conditions are maintained, the application of the model program PHREEQC leads to options of particular interest, whenever simulations of one, or several simultaneously running reactions are carried out. The oxidation of organic matter by dissolved oxygen – and after it is consumed, by nitrate – represents a very important reaction in seawater and especially in the sediment. In the model calculation presented in Figure 14.3 the following boundary conditions were selected: The seawater under study, or ma-

rine pore water, should be in a state of equilibrium with the mineral calcite at the beginning of the reaction and with a gaseous phase containing

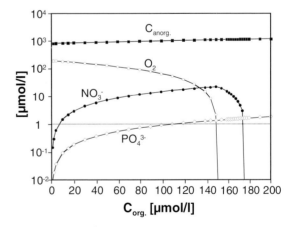

**Fig. 14.3** Geochemical model calculation using the program PHREEQC. To an oxic seawater with calcite in equilibrium (cf. Table 14.4) an organic substance is gradually added automatically leading to a redox reaction. The system should continue to be open to calcite equilibrium, but sealed from the gaseous phase environment. First, free oxygen, and subsequently nitrate, will be consumed. Afterwards, the system shifts to the anoxic state (cf. continuance of reactions in Fig. 14.4).

## Seawater from Nordstrom et al. (1979)

| Phase | Si | log IAP | log KT |
|-------|-----|---------|--------|
| Calcite | 0.00 | -8.48 | -8.47 |
| $CO_2$ (g) | -3.47 | -21.62 | -18.15 |
| $O_2$ (g) | -0.68 | 82.44 | 83.12 |

| | | | |
|-------|-----|---------|--------|
| pH | = | 7.902 | Charge balance |
| pe | = | 12.713 | Adjusted to redox equilibrium |
| Ionic strength | = | 6.734E-1 | |
| Temperature (°C) | = | 25.0 | |

| Elements | Molality | |
|----------|----------|---|
| C | 8.139E-4 | Equilibrium with Calcite and $pCO_2$ |
| Ca | 9.876E-3 | Equilibrium with Calcite |
| Cl | 5.657E-1 | not changed |
| Fe | 3.711E-8 | not changed |
| K | 1.058E-2 | not changed |
| Mg | 5.507E-2 | not changed |
| Mn | 3.773E-9 | not changed |
| N | 6.571E-6 | Adjusted to redox equilibrium |
| Na | 4.854E-1 | not changed |
| O(0) | 3.951E-4 | Equilibrium with $pCO_2$ |
| S | 2.93E-02 | not changed |
| Si | 7.382E-5 | not changed |

**Table 14.4** In the seawater analyses shown in Tables 14.1 to 14.3, an equilibrium adaptation to the mineral calcite (SI = 0) and to the atmosphere with log $pCO_2$ = -3.47 (equivalent to $pCO_2$ = 0.00034) and log $pO_2$ = -0.68 (equivalent to $pO_2$ = 0.21) was calculated. This model calculation was performed with the program PHREEQC (Pankhurst 1995).

21% $O_2$ and 0.034% $CO_2$. The composition of this water sample is summarized in Table 14.4. Organic substance containing amounts of nitrogen and phosphorus according to the Redfield ratio of C:N:P = 106:16:1 (cf. Sect. 3.2.5) is then gradually added to this solution. Simplified, such a theoretical organic substance is composed of $(CH_2O)_{106}(NH_3)_{16}(H_3PO_4)_1$ or, divided by 106: $(CH_2O)_{1.0}(NH_3)_{0.15094}(H3PO4)_{0.009434}$. The horizontal axis in Figure 14.3 records the substance addition in terms of μmol per liter. If the situation in the sediment is to be described, an equilibrium of calcite should exist as well. The system under study should not be open to the gaseous phase any more, instead, these influences should now be determined by the consumption of oxygen within the system and by the concomitant release of $CO_2$.

In the model, the organic substance added to the reaction automatically triggers a sequence of redox reactions which becomes evident due to the concentrations involved which are demonstrated in Figure 14.3. First, the concentration of oxygen declines, and the concentration of nitrate and phosphorus increases according to their proportion in the original organic substance. As soon as the oxygen is consumed, nitrogen will be utilized for further oxidation of the continually added organic substance. Subsequently, the system shifts over to an anoxic state.

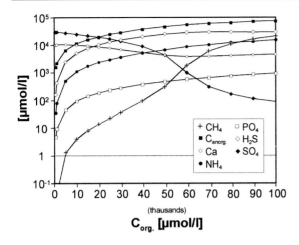

**Fig. 14.4** Geochemical model calculation using the program PHREEQC. In an anoxic system (state at the end of the model calculation from Fig. 14.3), the gradual addition of organic matter to the redox reaction is continued, whereby the system is kept open for calcite equilibrium and sealed from the gaseous phase. Initially, the dissolved sulfate will be consumed, in the course of which low amounts (logarithmic scale) of methane will emerge. Only after the sulfate concentration has become sufficiently low, will the generation of methane display its distinct increase.

*Decomposition of Organic Substance under Anoxic Conditions*

An anoxic system, as obtained from the previous reaction sequence, will be capable of further ongoing reaction, if the supply of organic matter is continued and the reaction is held under the same boundary conditions. The result consists in the continuation of the processes shown in Figure 14.3. The outcome of this experiment is presented in Figure 14.4. Here as well, the horizontal axis is calibrated with regard to the added amounts of organic matter in terms of μmol per liter.

As expected, the sulfate concentration in solution initially decreases at a swift rate and, concomitantly, low amounts of methane already emerge. Only after the concentration of sulfate falls under a certain level, will methane be released in substantial amounts. The permanently maintained calcite equilibrium will result, due to precipitation, in the depression of the concentration of dissolved calcium ions (compare Schulz et al. 1994).

The sequence of reactions, as described and modeled above, has naturally been well established and frequently documented ever since the publications made by Froelich et al. (1979). However, it should not be overlooked that these processes were not pre-determined for the modeling procedure, but rather 'automatically' evolved from the fundamental geochemical data supporting the model as well as from the pre-set, although very generalized, starting and boundary conditions. During the entire model calculation procedure, all quantitatively important species, the complex aquatic species, as well as the activity coefficients related to these solutions, were invariably taken into account. As a realistic and also quite well-known reaction, calcite dissolution/precipitation was employed for an adjustment of the mineral phase equilibrium. Such a system can be made more complex, almost at will, by specifically operating with additional mineral phases (e.g. iron minerals or manganese minerals) that can be eventually dissolved, if they are present and if the condition of undersaturation prevails, or which are allowed to precipitate in the case of supersaturation. However, the simulation of such mineral equilibria requires a very profound knowledge of the specific reactions of mineral dissolution or precipitation in the sytem under study.

## 14.2    Analytical Solutions for Diffusion and Early Diagenetic Reactions

Analytical solutions for Fick's second law of diffusion were already discussed in Chapter 3, Section 3.2.4, without taking diagenetic reaction into account. With a diffusion coefficient $D_{sed}$, which describes the diffusion inside the pore cavity of sediments, Fick's second law of diffusion is stated as:

$$\frac{\partial C}{\partial t} = D_{sed} \cdot \frac{\partial^2 C}{\partial x^2} \qquad (14.5)$$

This partial differential equation has different solutions for particular configurations and boundary conditions. In Section 3.2.4 the following solution was presented:

$$C_{x,t} = C_0 + \left(C_{bw} - C_0\right) \cdot erfc\left(\frac{x}{2 \cdot \sqrt{\left(D_{sed} \cdot t\right)}}\right)$$

$$(14.6)$$

The solution of the equation yields the concentration $C_{x,t}$ at a specific point in time t and a specific

427

depth x below the sediment surface. Here, t=0 is a point in time at which the concentration $C_0$ prevails in the profile's entire pore water compartment. At all later points in time the bottom water displayed a constant value $C_{bw}$. An approximation for the complementary error function (erfc) according to Kinzelbach (1986) can be found in Section 3.2.4, where examples are given that employ this analytical solution of Fick's second law.

Advection and very simple reactions can also be included in Fick's second law of diffusion (e.g. Boudreau 1997):

$$\frac{\partial C}{\partial t} = D_{sed} \cdot \frac{\partial^2 C}{\partial x^2} - v \cdot \frac{\partial C}{\partial x} - k \cdot C + R_{(x)} \quad (14.7)$$

Here, $v$ denotes the mean velocity of advection, and $k$ is a rate constant of a reaction with first order kinetics. The last term in the equation $R_{(x)}$ is an unspecified source or sink related term which is determined by its dependence on the depth coordinate x. Instead of $R_{(x)}$, one might occasionally find the expression $(\Sigma R_i)$ which emphasizes that actually the sum of different rates originating from various diagenetic processes should be considered (e.g. Berner 1980). Such reactions, still rather easy to cope with in mathematics, frequently consist of adsorption and desorption, as well as radioactive decay (first order reaction kinetics). Sometimes even solubility and precipitation reactions, albeit the illicit simplification, are concealed among these processes of sorption, and sometimes even reactions of microbial decomposition are treated as first order kinetics.

These, or smaller variations of the equation, are occasionally referred to as 'general diagenetic equation', thus stating a rather simplistic comprehension of (bio)geochemistry in favor of a more mathematically minded approach. A comprehensive and elaborate treatment of the analytical solutions of the equation is presented by Boudreau (1997) who, in particular, considers its multiple variations and boundary conditions. Currently, this important book represents the 'state of the art' for the diagenesis researcher who is more interested in mathematics. For the more practical diagenesis scientist thinking in terms of (bio)-geochemistry, this presentation remains somewhat unsatisfactory, despite the fact that it gives various inspiration of interest and contributes to a vast number of important compilations (e.g. as to the state of knowledge on diffusion coefficients in the sediment). Ultimately, the (bio)geochemical

applications of such a solution are in fact rather limited and illicitly restrict the complexity obtained by experimental measurement in the natural system.

In simpler stationary cases the author would always prefer solutions obtained by applying the 'Press-F9-method' (compare Sect. 14.3.1) and numerical solutions, particularly for non-stationary cases or in cases with complex reactions, by using the models CoTAM or CoTReM (compare Sect. 14.3.2).

Van Cappellen and Gaillard (1996) gave indications that the activity coefficients in marine pore water, and the various complex species as well, need to be included as a very important part of the solution (also compare Sects. 14.1.1 and 14.1.2). It has to be assumed that, for instance, the decomposition of sulfate via respective complex equilibria also affects the activity of other dissolved ions. Furthermore, it has to be considered that charged complexes and ions, transported by means of diffusion at different rates (differing diffusion coefficients!), must lead to a separation of the distributed charges in the course of transportation. The importance this has for the processes particularly taking place in the sediment is not yet sufficiently established. However, van Cappellen and Gaillard (1996) assume that such processes may probably be neglected in marine sediments as the charges there are essentially determined by sodium and chloride ions.

## 14.3   Numerical Solutions for Diagenetic Models

The previous section demonstrated that the analytical solutions of Fick's second law of diffusion can only be applied to a very limited number of cases. Frequently, highly simplified boundary conditions have to be assumed which actually cannot be found in a natural environment. Additionally, the effort is almost always limited to merely one component dissolved in pore water, or to its behavior in space and/or time. Any complex relations between the numerous dissolved components, or even concentrations of complexes and ion pairs which become variable due to reaction sequences, or accordingly taking into account the difference between concentrations and activities, all are simply not regarded in the mathematical solutions of diagenetic processes. Nevertheless, these multiple component processes do not make the rare exception but represent the normal case, and must be considered, the more one becomes concerned with measuring natural systems.

The solution to this problem consists in the application of numerical solutions when diagenetic processes are modeled. Such numerical solutions always divide the continuum of reaction space and reaction time into discrete cells and discrete time intervals. If one divides up the continuum of space and time to a sufficient degree into discrete cells and time steps (which is not the decisive problem with the possibilities given by today's computers), one will be able to apply much simpler and more manageable conditions within the corresponding cells, and with regard to the expansion of a time interval, so that, in their entirety, they still will describe a complex system. Thus, it is possible, for example, to apply the two-step-procedure (Schulz and Reardon 1983) in which the individual observation of physical transport (advection, dispersion, diffusion) or any geochemical multiple component reaction is made feasible within one interval of time.

### 14.3.1  Simple Models with Spreadsheet Software ('Press-F9-Method')

*Structure of a Worksheet and Oxygen/Nitrate Modeling*

Normal spreadsheet software (e.g. Microsoft - Excel® or Lotus 123) present especially eligible tools for modeling stationary and not too complex systems. The well-understood decomposition of organic substance by dissolved oxygen, and the concomitant release of nitrate, will serve as an example. Nitrate is also utilized for the oxidation of organic substance in deeper sections of the diffusion controlled profile. Figure 14.5 demonstrates the result of such a model procedure, the details of which will be further discussed in the following.

Table 14.5 shows excerpts from the worksheet used for modeling the decomposition of organic substance mediated by oxygen and nitrate. In the upper frame of the worksheet there are – apart from the headline – single parameters and constants that will be used in the lower part of the worksheet; they will be discussed later in that particular context.

In the lower three frames of Table 14.5 the actual worksheet is depicted in rows 11 to 61 that

divide up the one-dimensional profile into 50 lines, which resent 50 discrete cells. The first column (A) contains the depth below the sediment surface. In the uppermost cell (11) this depth is set to zero and represents the bottom water zone immediately on top of the sediment. Line 61 is already beyond the boundary of the modeled area and will be discussed later in the context of setting the boundary conditions. In cell A12 to A61 the depth is increased by the value dx = 0.001 m per cell, as prechosen in cell A6. The cell A12 contains the Excel® notation: =A11+$A$6. This field may be copied all the way down to field A61. In the next two columns (B) and (C) we find default porosity values (n) or diffusion coefficients ($D_{sed}$) valid for all cells. However, should the porosity constants or diffusion coefficients be known for the particular depth profile, then they may be entered individually into the respective cells. The columns (D), (E) and (F) are connected to each other by loops so that an error message

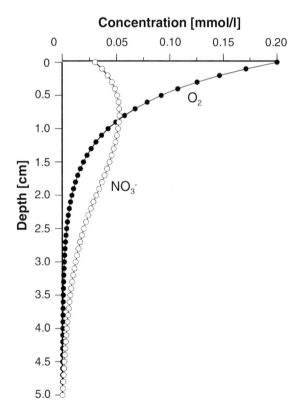

**Fig. 14.5**  Model of the decomposition of organic substance by dissolved oxygen and dissolved nitrate in a diffusion controlled pore water profile. Modeling was performed according to the 'Press-F9-method' with the spreadsheet software Excel®. Details pertaining to this model are explained in Table 14.5 and in the text.

**Table 14.5** Worksheet (excerpts) from modeling the decomposition of organic substance by dissolved oxygen and dissolved nitrate in a diffusion controlled pore water profile. The model procedure was performed according to the 'Press-F9-method' with the spreadsheet software Excel®. The columns are marked in the uppermost row with alphabetical characters, whereas the rows are numbered in the last column. Details pertaining to the model are further explained in the text.

| | A | B | C | D | E | F | G | H | I |
|---|---|---|---|---|---|---|---|---|---|
| 1 | | | | | | | | | |
| 2 | **Model oxidation $C_{org}$ with oxygen and nitrate** | | | | | | switch | C:N | $k(O_2)$ |
| 3 | | | | | | | 0 | 3.0 | 7.9E-09 |
| 4 | | | | -------------------- mmol $l^{-1}$ = mol m$^{-3}$ -------------------- | | | | | |
| 5 | dx(m) | | dt(sec) | $C_0(O_2)$ | $C_0(NO_3^-)$ | $C_b(O_2)$ | $C_b(NO_3^-)$ | | $k(NO_3^-)$ |
| 6 | 0.001 | | 1.0E+06 | 0.2 | 0.03 | 0.0 | 0.0 | | 3.2E-09 |
| 7 | | | | | | | | | |
| 8 | | | | | | | | | |
| 9 | Depth | $\phi$ | Ds | $O_2$-Flux | $O_2$-Red. | $O_2$ | $NO_3^-$-Flux | $NO_3^-$-Red. | $NO_3^-$ |
| 10 | m | | m$^2$ sec$^{-1}$ | -------- mol m$^{-2}$ s$^{-1}$ ----- | | mmol $l^{-1}$ | -------- mol m$^{-2}$ s$^{-1}$ ----- | | mmol $l^{-1}$ |
| 11 | 0.000 | 0.60 | 5.4E-10 | 9.36E-09 | | 2.00E-01 | -2.11E-09 | | 3.00E-02 |
| 12 | 0.001 | 0.60 | 5.4E-10 | 8.01E-09 | 1.35E-09 | 1.71E-01 | -1.66E-09 | 0.00E+00 | 3.65E-02 |
| 13 | 0.002 | 0.60 | 5.4E-10 | 6.85E-09 | 1.16E-09 | 1.46E-01 | -1.27E-09 | 0.00E+00 | 4.16E-02 |
| 14 | 0.003 | 0.60 | 5.4E-10 | 5.86E-09 | 9.90E-10 | 1.25E-01 | -9.40E-10 | 0.00E+00 | 4.55E-02 |

*continued*

| | A | B | C | D | E | F | G | H | I |
|---|---|---|---|---|---|---|---|---|---|
| 29 | 0.018 | 0.60 | 5.4E-10 | 5.65E-10 | 9.53E-11 | 1.21E-02 | 8.27E-10 | 0.00E+00 | 3.63E-02 |
| 30 | 0.019 | 0.60 | 5.4E-10 | 4.83E-10 | 8.16E-11 | 1.03E-02 | 8.55E-10 | 0.00E+00 | 3.37E-02 |
| 31 | 0.020 | 0.60 | 5.4E-10 | 4.13E-10 | 6.98E-11 | 8.83E-03 | 8.78E-10 | 0.00E+00 | 3.11E-02 |
| 32 | 0.021 | 0.60 | 5.4E-10 | 3.54E-10 | 5.97E-11 | 7.56E-03 | 8.07E-10 | 9.09E-11 | 2.84E-02 |
| 33 | 0.022 | 0.60 | 5.4E-10 | 3.03E-10 | 5.11E-11 | 6.46E-03 | 7.42E-10 | 8.29E-11 | 2.59E-02 |
| 34 | 0.023 | 0.60 | 5.4E-10 | 2.59E-10 | 4.37E-11 | 5.53E-03 | 6.81E-10 | 7.56E-11 | 2.36E-02 |

*continued*

| | A | B | C | D | E | F | G | H | I |
|---|---|---|---|---|---|---|---|---|---|
| 55 | 0.044 | 0.60 | 5.4E-10 | 1.15E-11 | 1.40E-12 | 1.77E-04 | 1.30E-10 | 7.13E-12 | 2.23E-03 |
| 56 | 0.045 | 0.60 | 5.4E-10 | 1.04E-11 | 1.12E-12 | 1.41E-04 | 1.25E-10 | 5.84E-12 | 1.83E-03 |
| 57 | 0.046 | 0.60 | 5.4E-10 | 9.56E-12 | 8.62E-13 | 1.09E-04 | 1.20E-10 | 4.61E-12 | 1.44E-03 |
| 58 | 0.047 | 0.60 | 5.4E-10 | 8.93E-12 | 6.29E-13 | 7.96E-05 | 1.17E-10 | 3.42E-12 | 1.07E-03 |
| 59 | 0.048 | 0.60 | 5.4E-10 | 8.52E-12 | 4.11E-13 | 5.20E-05 | 1.15E-10 | 2.26E-12 | 7.07E-04 |
| 60 | 0.049 | 0.60 | 5.4E-10 | 8.32E-12 | 2.03E-13 | 2.57E-05 | 1.14E-10 | 1.13E-12 | 3.52E-04 |
| 61 | 0.050 | | | | | 0.00E+00 | | | 0.00E+00 |

will be announced, if the appropriate Excel® setting ('Iterative Calculation') should not be selected. The diffusive $O_2$ flux is calculated by employing Fick's first law of diffusion (Eq. 3.2 in Chap. 3), whereby the gradient obtained from the concentrations in column (F) is used. The concentrations in column (F) are calculated stepwise for each time interval from the preceding value, from the material flux in and out of the cell, and from the decomposition activity listed in column (D). Hence, the result in cell D11 reads in Excel® notation:

$$D11=B11*C11*((F11-F12)/\$A\$6)$$

D11 consequently contains the flux rate of material directed from the bottom water into the sediment. The cells D12 to D60 will be accordingly copied from this cell. Cell D61 remains empty.

The decomposition of organic substance in the uppermost sediment zone is written in the cell E12; cell E11 (bottom water) remains empty. In the example shown the decomposition of organic substance mediated by oxygen in a reaction following first order kinetics was still bound to the yet prevalent oxygen concentration. This does not mean to say that this must always be so. As for the model, any other kinetic, or, in general terms, any other value could be assigned to the respective cells. As for the first order kinetics example, the line E12 reads in Excel® notation as follows:

E12=$I$3*F12

Cells E13 to E60 are copied from this cell accordingly, cell E61 remains empty. Column (F) contains O₂-concentrations, whereby cell F11 contains the upper boundary condition via a constant bottom water concentration. Cell F61 contains the concentration at the lower boundary - which in this particular case is equivalent to 0.0. The cell F12 is calculated from cell F12 (hence from its own previous value) as well as from the cells D11, D12, and E12, each multiplied with the length of the time interval. Written in Excel® notation this reads:

F12=IF($G$3;$D$6;F12+(D11*$C$6)-(D12*$C$6)-(E12*$C$6))

(The function 'IF' refers to the English version of Excel®, e.g. for the German version this function would be 'WENN'.)

Cells F13 to F60 are copied from this cell accordingly. Cell F61 contains the lower boundary conditions (in this case 0.0) which will be further dealt with later. Working with the model, it appears of practical importance at this point of calculation, if one has the option of stopping the calculation procedure by means of a switch function. If the switch cell G3 assumes zero value the calculation will be performed. If the cell contains the value 1, then all concentrations will be adjusted to the value present in cell D6 (concentration in bottom water). Designing the structure of the worksheet, the switch should be set to a value of 1, in order to avoid unnecessary error messages due to references to not yet properly filled cells. Additionally, it is easy to rebuild the worksheet whenever ill-chosen conditions pertaining to the iteration have led to extreme values and/or oscillations.

Column (G) contains nitrate concentrations analogous to column (D) used for oxygen. In this example, the same diffusion coefficients were used for calculating the fluxes. Here, different individually adjusted coefficients can certainly also be used. Again, cell G11 contains the flux rate di-

rected from the bottom water to the sediment and reads written in Excel® notation:

G11=B11*C11*((I11-I12)/$A$6)

Cells G12 to G60 are copied from this cell accordingly, whereas cell G61 remains empty. Column (H) contains the nitrate mediated decomposition which can be voluntarily set, as it has been the case in oxygen mediated decomposition. In this example, a decomposition of 0.0 has been pre-determined for the cells H12 to H31, as well as the assumption of first order kinetics for the cells H32 to H60. Here, too, it is not intended to claim that this allocation must necessarily be as shown in the example. As for the model, any other type of kinetics would be possible as well as any otherwise set values. Hence, cell H32 reads in Excel® notation:

H32=$I$6*I32

Cells H33 to H60 are copied from this cell accordingly. Cell H61 remains empty.

Analogous to column (F), column (I) this time contains the calculated nitrate concentrations which loops again connect it to the columns (G) and (H). However, in the case of nitrate, there is also some release into the pore water fraction due to the oxygen mediated decomposition of organic substances. Thus, each cell belonging to column (I) is calculated from its previous value, from the diffusive import from above, from the diffusive export downwards, from the import from the oxidation of organic material mediated by oxygen (here, selective C:N ratio), as well as from the decomposition mediated by nitrate. For cell I12 this writes in Excel® notation as follows:

I12=IF($G$3;$E$6;I12+(G11*$C$6)-(G12*$C$6)+((1/$H$3)*E12*$C$6)-(H12*$C$6))

(The function 'IF' refers to the English version of Excel®, e.g. for the German version this function would be 'WENN'.)

Cells I13 to I60 are copied from this cell accordingly. Cell I11 contains the bottom water concentration as a fixed boundary condition, whereas cell I61 assumes the value of 0.0 as the concentration at the lower boundary. In this column it is also recommended to link the calculation performance to the operational 'switch' of cell G3.

The model is now finished and the essential margin values compiled in the upper frame of Table 14.5 are added. The switch in G3 is then set to zero. Now the key F9 (re-calculation of the worksheet) is repeatedly pressed until the figures in the worksheet do not (significantly) change, i.e., until a (practically) steady state is reached. Depending on the speed of the computer and the spreadsheet software used, this can take some few minutes, so that one can hold the F9-key down during that time.

At this point it needs to be clearly stated that only the final iterative steady state of the model provides a correct result, because the diffusive flux is calculated according to Fick's first law of diffusion in the columns (D) or (G) which is only valid when stationary conditions are provided. Even if time intervals of specific length are employed in the process of iteration, it is not permitted, by no means at all, to draw conclusions on the time necessary for reaching the steady state, since the columns (D) and (G), and consequently (F) and (I) as well, contain false information relative to each of the respective non-steady states.

The length of the time intervals (cell C6) is only a mechanism used for iteration, whereby too short intervals will result in an unnecessarily long pressing of the key F9, whereas too large intervals would lead to escalating oscillations within the model (in our example after about dt=2.3E6). Only after the steady state is reached will this 'false way' of the iteration become 'consigned to oblivion', and the result will be correct in all columns. Non-steady states and intervals prior to the development of particular concentration patterns can only be managed by applying Fick's second law of diffusion, either in form of analytical solutions (cf. Sect. 14.2) or in form of numerical solutions (cf. Sect. 14.3.2).

The concentration dependent boundary conditions assigned to the model's upper-boundary limits (unspecified but constant concentrations in the cells F11 and I11) are relatively unproblematic since they represent a constant concentration in the bottom water zone. This is an imperative prerequisite for assuming a steady state in pore waters from superficial sediments. The condition for the lower boundary of the profile is somewhat more problematical. In the above example, the concentrations of both oxygen and nitrate were set to zero. In the model, this is tolerable as long as the value 0.0, constituting the lower boundary within the confines of the model, already results

from the reactions, and as long as the set value 0.0 will not exert any practical influence on the lower parts of the profile. This 'zero concentration boundary condition' hence constitutes a 'zero flux boundary condition'. This also implies that the confines of the model must be extended downward to such a degree that this state will be sufficiently well assumed. This choice of zero concentration and, consequently 'zero flux boundary condition', is well suited in this example, but it must not always be an adequate boundary condition. In the following example, we will consider the case in which a value, clearly deviating from 0.0, is chosen, and thus, indirectly provides a steady diffusive flux across the lower boundary.

*Modeling the Reaction Sulfate/Methane*

Niewöhner et al. (1998) describe profiles of pore water which were obtained from upwelling areas off the shores of Namibia. In these pore water samples, a 1:1 reaction between sulfate and methane is indicated. Such an example has already been introduced in Chapter 3 (Fig. 3.31, also cf. Chaps. 5 and 8), in context with problems occurring in methane analytics. In this example, the measured values will now be presented in Figure 14.6 together with the result of a model designed according to the 'Press-F9- method'.

In principle, the worksheet for the modeling of Figure 14.6 does not differ very much from the one described previously, so that it does not demand special description. The essential features of this model are:

The decomposition of methane and sulfate (cf. Chap. 8) occur for both reactants in specific depths at identical rates which were entered into the spreadsheet. The adjustment to the measured profiles was committed only by these (microbial) decomposition rates. The decomposition parameter is set to 0.0 in all other cells, so that a diffusion controlled transport occurs with a constant concentration gradient.

The concentration of the upper boundary condition for the model is again provided by the appropriate bottom water concentrations: sulfate = 28 mmol/l; methane = 0.0 mmol/l.

The concentration of the lower boundary condition for sulfate is 0.0 mmol/l, since it was decomposed to 0.0 mmol/l long before the lower boundary of the model had been reached. However, the concentration of the lower boundary condition for methane has been concluded on the basis of pre-determining a steady decomposition in the same amount as for sulfate, and that this decomposition should continue in a specific depth until the value 0.0 is reached.

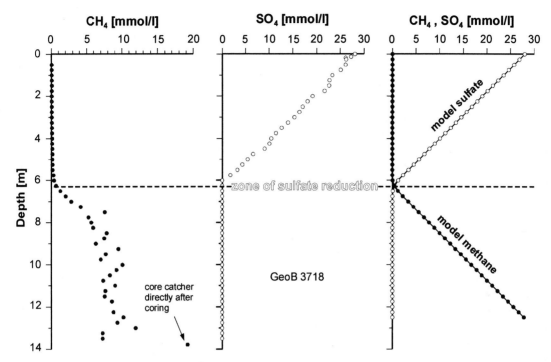

**Fig. 14.6**   Model results of the mutual decomposition of sulfate and methane as a 1:1-reaction in a diffusion controlled pore water profile. Modeling was performed according to the 'Press-F9-method' using the spreadsheet software Excel®. Details pertaining to the model and the calibration with data from a measured pore water profile obtained from an upwelling area off Namibia (Niewöhner et al. 1998) are discussed in the text.

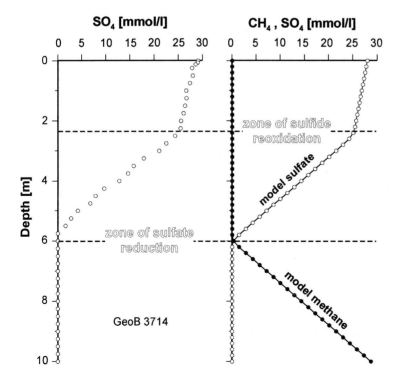

**Fig. 14.7**   Model results of the mutual decomposition of sulfate and methane as a 1:1-reaction in a diffusion controlled pore water profile. The design of the model was done in accordance with the 'Press-F9-method' using the spreadsheet software Excel®. Details pertaining to the model and the calibration of data from a measured pore water profile obtained from an upwelling area off Namibia (Niewöhner et al. 1998) are discussed in the text.

Figure 14.7 demonstrates the measured and appropriately modeled sulfate concentrations in pore waters sampled at a depth of 1300 m below sea level, in a sediment core also originating from the upwelling area off Namibia. The situation is, in principle, similar to the one described in the previous example of Figure 14.6. Here, however, there is an additional alteration of the sulfate concentration gradient which becomes evident at a depth of approximately 2.3 m below the sediment surface. Concentration profiles with this characteristic kink in different depths below the sediment surface (1.5 m up to 10 m) are in fact known from many locations of the ocean's organic-rich sediments. Currently, the processes inducing these kinks in the concentration profile are not yet conclusively identified and various possibilities are being discussed. One is the model concept applied in Figure 14.7, which shows the re-oxidation of sulfide (e.g. on iron oxides) to yield sulfate. Upon making adjustments to the entire measured profile shown in this example, the design of the model demands a specific reaction rate which plausibility needs to be investigated further with regard to the sedimentation rate, the iron content of the sediment, and by including the question of steady or non-steady state conditions in the sediment.

## 14.3.2  Two-Step Models for Combined Transport/Reaction Processes

The main problem encountered in modeling combinations of advective/dispersive and/or diffusive transport, on the one hand, and any (bio)geochemical reaction, on the other hand, consists in that at least two totally different processes interact simultaneously on the same object:

Physical transport including diffusion, advection and dispersion, when reactions are absent, is one part of the problem. In their entirety, these processes are quantitatively well understood in model concepts. They are also applicable, without raising principle problems, in analytical and numerical solutions to the general partial differential equations of material transport.

Geochemical and (bio)geochemical processes which are independent of transport processes present the other part of the problem. Here as well, there is a number of far advanced model concepts. The models belonging to the PHREEQE or PHREEQC - type may be referred to in this respect (cf. Sects 14.1.1 and 14.1.2). Other fields are object of the various chapters in this book.

First, attempts were made to develop models that could accomplish the coupling of both process groups in one single, either analytical or numerical procedure, although these often bore strong limitations for the application to geochemical reactions. Recent attempts now more often foresee solutions in two steps. This group of methods is referred to either as 'operator splitting', or as the 'two-step method'. Boudreau (1997) mentions this procedure only briefly at the close of his book and describes it as 'an apparently crazy idea that works rather well in practice'. The first versions of such models, which couple physical transport to geochemical reactions in groundwater are already nearly 20 years old (e.g. Schulz and Reardon 1983). Application examples for the 'two-step method' in modeling diagenetic processes in marine sediments over the recent years were published by Hamer and Sieger (1994), Hensen et al. (1997), and Wang and Van Cappellen (1996).

The model CoTAM (Hamer and Sieger 1994) is primarily designed for its application to groundwater and for trace metal transport and reaction with the solid phase of the aquifer. In the one-dimensional transport part of the model (as a continuance of the model DISPER by Flühler and Jury 1983) the model's territory is divided up into a variable number of REVs (representative elementary volumes). These REVs may differ in length, porosity, and in their dispersion and diffusion properties.

The time to be modeled is subdivided into a number of time intervals. First, an analytical solution is found for the partial differential equation of transport (including adsorption and desorption) with respect to all substances under study and every single time interval. Afterwards, the geochemical processes in each REV are modeled independent of transport. For this step, CoTAM runs a subroutine method called REDOX, as well as the geochemical equilibrium and reaction model PHREEQE. Thereafter, the next time interval is processed starting again with the physical transport, followed by the geochemical reactions etc., until the pre-determined operating time is reached. The structural principle of the model CoTAM is outlined in Figure 14.8.

Hensen et al. (1997) applied the model CoTAM to the processes of diagenesis in oxic marine sediments, using oxygen and nitrate as electron acceptors (cf. Sect. 6.3.1.1). Carbonate, calcium, and the pH-value were controlled by the

**Fig. 14.8** Fundamental
structure of the 'two-step
method' by Hamer and
Sieger (1994). The figure
originates from a model
application in a marine
environment carried out
by Hensen et al. (1997).
The upper box contains a
description of the trans-
port step for each time
interval; the two lower
boxes 'REDOX' and
'PHREEQE' describe for
each time interval the
geochemical reactions.

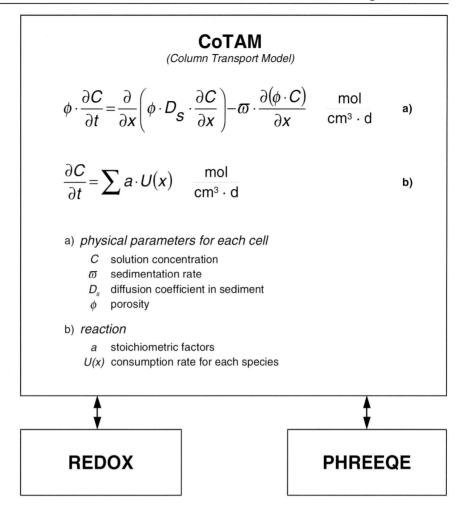

## CoTAM
*(Column Transport Model)*

$$\phi \cdot \frac{\partial C}{\partial t} = \frac{\partial}{\partial x}\left(\phi \cdot D_s \cdot \frac{\partial C}{\partial x}\right) - \varpi \cdot \frac{\partial(\phi \cdot C)}{\partial x} \qquad \frac{mol}{cm^3 \cdot d} \qquad \text{a)}$$

$$\frac{\partial C}{\partial t} = \sum a \cdot U(x) \qquad \frac{mol}{cm^3 \cdot d} \qquad \text{b)}$$

a) *physical parameters for each cell*

  $C$   solution concentration
  $\varpi$   sedimentation rate
  $D_s$   diffusion coefficient in sediment
  $\phi$   porosity

b) *reaction*

  $a$   stoichiometric factors
  $U(x)$   consumption rate for each species

**REDOX**                                             **PHREEQE**

model PHREEQE by means of the calcite-carbonate-equilibrium, accounting for all major components in pore water as well as the dissolved complexes and their respective activity coefficients.
Since the thermodynamic constants of equilibrium cannot be used for fitting the measured data,
this can only be carried out by applying the biogeochemical reaction rates and the C:N-ratio of
the decomposed organic substance.

As the result of a model calculation, Figure
14.9 shows the adjustment to pore water profiles
from cores of the upwelling area off the shores of
Angola (GeoB 1702, water depth 3094 m) and
Namibia (GeoB 1711, water depth 1965 m). Naturally, the model complies well to the measured
data with regard to oxygen and nitrate, as this is
made possible without limitation on account of
the respective reaction rates and the C:N ratio of
the decomposed organic material. The measured
profiles of pH, alkalinity, and calcium, however,
cannot be fitted since they are already determined

by the concomitant $CO_2$ release, and by means of
the well established constants of the calcite-carbonate-equilibrium. The good agreement nevertheless obtained suggests that neither the model
nor the measured data are fundamentally wrong.

The latest version of the model CoTAM and
its successor CoTReM are available, together
with a program description, at the following
internet web page:

http://www.geochemie.uni-bremen.de

The currently most versatile program designed
for modeling the essential processes in early diagenesis originates from Wang and Van Capellen
(1996), or Van Cappellen and Wang (1995). This
model STEADYSED1 can be obtained from the
authors as 'public domain software' and enables
the operator to change and adjust all essential sets
of data to a particular problem. Yet, one can also
select and apply far more unreasonable combina-

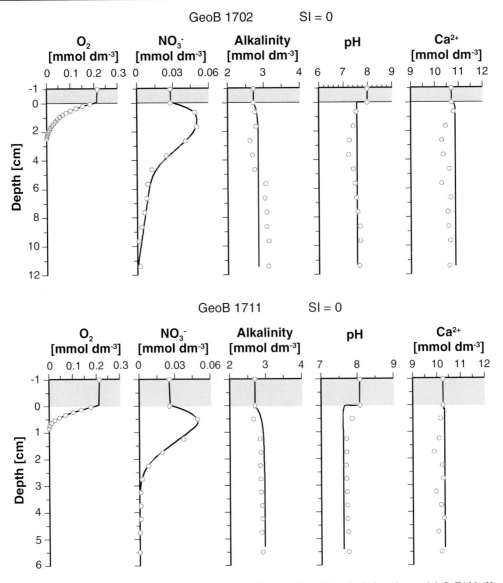

**Fig. 14.9** Modeling of diagenetic processes in oxic marine sediments. For this calculation the model CoTAM (Hamer and Sieger 1994) was used. The decomposition of organic matter with oxygen and nitrate as electron acceptors is demonstrated by the concentration profiles of these constituents. With PHREEQE as a subroutine of the model CoTAM, calcite was always kept in equilibrium (modified after Hensen et al. 1997).

tions than reasonable ones – which is in the nature of the model considering the complexity the model offers. The agreeable 'learning by playing' which is typical for so many simple models, turns out to be not so easy after all.

The model uses the following physical data: the sedimentation rate for solid phase advection, temperature for correcting diffusion, porosity and the derived formation factor. A description of bioturbation is obtained by utilizing a mixture-related coefficient, comparable to the diffusion coefficient. An indication of depth informs how far

below the sediment surface bioturbation will be effective.

Constant concentrations are assumed for the following components in bottom water: $O_2$, $NO_3^-$, $SO_4^{2-}$, $Mn^{2+}$, $Fe^{2+}$, $NH_4^+$, salinity, alkalinity and pH-value. As the model STEADYSED1 exclusively tolerates steady state situations, this assumption naturally requires constant concentrations in bottom water as a prerequisite.

As for the solid phase of sediment, the import of iron and manganese oxides is studied, the amount of organic substance, and the C:N:P ratio of the organic substance. The diagenetic decomposition of the organic substance mediated by the electron acceptors oxygen, nitrate, manganese and iron oxides, sulfate, all the way to methane fermentation, is measured according to the reactions published by Froelich et al. (1979). However, the user may also select the C:N:P ratio of the organic substance. The re-oxidation of iron sulfides, and manganese sulfides, or of methane is preferably performed as 'secondary redox reactions'. Moreover, the adsorption of ammonium, and the precipitation/dissolution of iron carbonate, manganese carbonate, and iron sulfide can be included. For all these reactions, the analyst is able to and *must* select the adequate rates.

This is where the general problems in the applying of such models lie: with all the parameters and boundary conditions that need to be set, there are so many 'buttons' and control options so that any kind of modeling of measured data will actually be nearly always feasible. Only if so many examples are available as Wang and Van Cappellen (1996) found in the data of Canfield et al. (1993a, b) based on measurements, will it be possible to come to substantial and reasonable results. Figure 14.10 shows concentration profiles for $O_2$, $NO_3^-$, $NH_4^+$, $Mn^{2+}$, $Fe^{2+}$ measured in the pore water of three cores. Figure 14.11 demonstrates the solid phase concentration profiles of Fe- and Mn participating in the reactions.

If one compares the two model programs CoTAM/CoTReM by Hamer and Sieger (1994), Hensen et al. (1997), Landenberger et al. (1997), Landenberger (1998), on the one hand, and the model STEADYSED1 published by Wang and Van Cappellen (1996) on the other, the following aspects need to be emphasized:

- Both models have been made freely available by the respective authors as public domain software and hence can be used by everyone.

- Both models are not easy to operate and afford at least some background knowledge in modeling procedures and early diagenetic reactions.

- Both models simulate the complex situation of combining diffusive, advective and bioturbational transports in pore water and in the solid phase. They contain the various reactions of early diagenesis and require a large number of reliable measurements. Otherwise, too many degrees of freedom would remain.

- Both models contain the possibility to simulate bioirrigation; STEADYSED1 generally treats bioirrigation and bioturbation with much more thoroughness and reliability.

STEADYSED1 is quite limited as far as its chemical processes are concerned (absence of real activities in the pore water, and hence only apparent equilibrium constants, complex species are not included). However, it contains all the essential processes currently known. CoTReM is much more flexible owing to the utilization of PHREEQC (Parkhurst 1995) as a subroutine, yet it requires proportionally more information.

The model CoTAM/CoTReM works with Fick's second law of diffusion and thus permits the calculation of any possible, especially non-steady state situations. STEADYSED1 can only be applied to calculate steady state situations which accordingly demand the existence of steady state boundary conditions.

## 14.4    Bioturbation and Bioirrigation in Combined Models

Bioturbation is normally modeled by applying a bio-diffusion coefficient to the pore water fraction *and* the solid phase. This means that a bio-diffusion in the solid phase will be simulated with the same model concept which is applied in the form of Fick's laws to the dissolved constituents in pore water. Under the assumption of a given steady state (e.g. STEADYSED1), this requires that a sufficient length of time is studied which allows the macroorganisms contributing to bioturbation to display activity all over and as many times as appears necessary. Only then will each part of the solid phase be turned over often enough in the statistical balance, hence providing a workable model concept. As for cases of non-steady state and short observation periods the model concept actually cannot be considered as valid.

Figure 14.12 shows, in one exemplary model calculation carried out with STEADYSED1, the conversion rates of iron and manganese in the upper sediment zone and their exchange with the supernatant bottom water layer. It is evident that the amounts of iron which ultimately become deposited in the sediment (burial flux) as FeS, $FeCO_3$, or in an adsorbed state, are essentially smaller

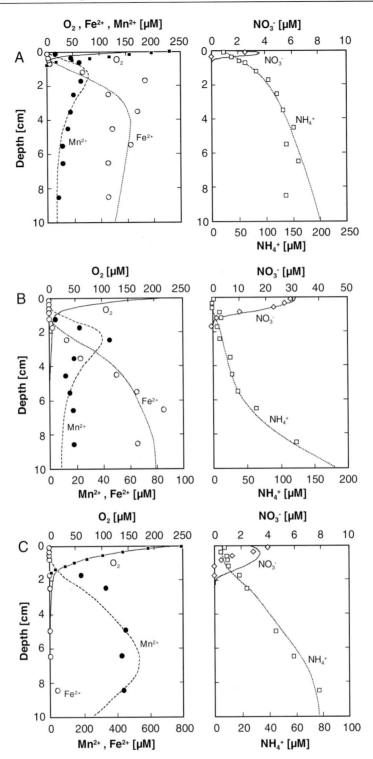

**Fig. 14.10**  Concentration profiles in pore water of three examples from marine sediments. Data represented by symbols were taken from Canfield et al. (1993 a,b), whereas the curves demonstrate the respective simulations carried out with the model STEADYSED1 by Wang and Van Cappellen (1996).

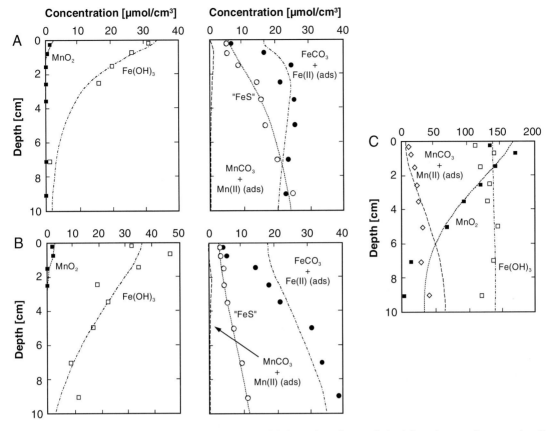

**Fig. 14.11** Concentration profiles of iron and manganese solid phases in sediments derived from three marine examples. Data represented by symbols were taken from Canfield et al. (1993 a,b), whereas the curves demonstrate the respective simulations carried out with the model STEADYSED1 by Wang and Van Cappellen (1996).

than the amounts undergoing conversions in the upper parts of the sediment and which have therefore undergone redox processes several times before. The lower part of Figure 14.12 demonstrates the corresponding turnovers of manganese with again quite specific ratios between the burial flux, conversion rates in the upper sediment zone, and exchange with the bottom water.

The simulation of bioirrigation is much more difficult, a circumstance which pertains to its nature, since the process, endowed with great inhomogeneity and strong dependence on the organism under study, combines bottom water with pore water at the particular depths concerned. How this combination becomes effective in any one particular case, and to which degree it also influences the pore water not involved directly, naturally depends on the macroorganisms presently responsible for the effect. Thus, the process actually eludes a generalized solution and should be treated individually for each special case. To this end, workable models are not available in

sufficient number, neither are there any adequately applicable modeling techniques at hand.

In Figure 14.13, and by using the 'Press-F9-method' described in Section 14.3.1, two steady states displaying bioirrigation have been calculated. The case shown on the left-hand side anticipates an upper sediment zone continually influenced by bioirrigation, whereas the case shown on the right merely assumes surface coupling in a specific depth – e.g., by the effect of a particular worm-hole. The result may be compared with the measurements conducted by Glud et al. (1994) and shown in Figure 3.32. Just as Wang and Van Cappellen (1994) have described it in one application of the model STEADYSED1, the situation here is that bioirrigation conveys nitrate into the sediment, whereas, at the same time, molecular diffusion transports nitrate from the sediment back into the bottom water. This results from the fact that both processes are coupled to the bottom water layer in different depth stages of the sediment. In the case of diffusion, these are the super-

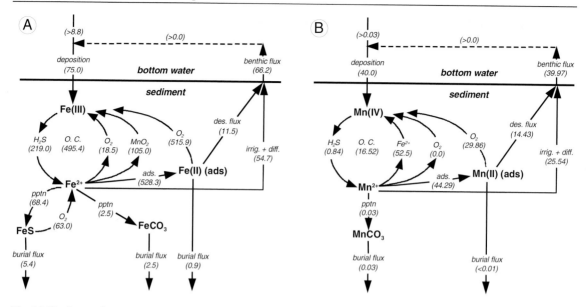

**Fig. 14.12**   Iron and manganese fluxes in superficial marine sediments and bottom water, as obtained from calculations using the model STEADYSED1 by Wang and Van Cappellen (1996). Here, it is significant that bioturbation and bioirrigation were included in the task.

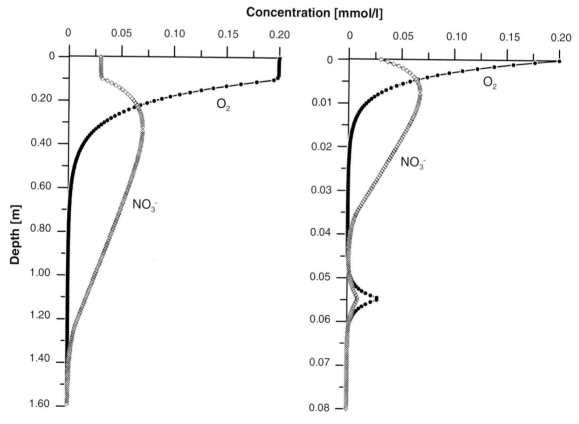

**Fig. 14.13**

Left:   Oxygen and nitrate profiles influenced by bioirrigation within the upper 10 cm as modeled with an Excel spreadsheet according to the 'Press F9 method'. The Simulation was performed on the basis of anticipating a partial coupling of pore water to the concentrations prevalent in bottom water in the upper 10 cm of the sediment.

Right:   Here, a quantity of bottom water was continually added in a specific depth. The result is related to the measurement shown in Figure 3.24 after Glud et al. (1994).

ficial nitrate-rich zones, whereas in the case of bioirrigation, the deeper situated zones are effective, which contain only small amounts of nitrate.

Ultimately, all these agreements between the measured profile and the resultant model are not really significant since they merely represent a direct application of the model concept of bioirrigation, i.e., the pore water input conducted from the bottom water at a certain depth in the sediment. Yet, these models are only interesting inasmuch as the non-diffusive transport between sediment and bottom water results from them. These are exemplified for iron and manganese in Figure 14.12, calculated with a similar model concept using the program STEADYSED1. It should be kept in mind that these calculations are only workable to the point as they comply to a rather simplified model concept and to a stationary situation anticipated not only in the 'Press F9 method' but in STEADYSED1 as well. Only by conducting measurements of the material fluxes *in-situ* at very dissimilar locations, and after repeated comparisons with the according models have been made, the previous model concepts and the calculated fluxes will ever become verified in the future.

This is contribution No 262 of the Special Research Program SFB 261 (*The South Atlantic in the Late Quaternary*) funded by the Deutsche Forschungsgemeinschaft (DFG).

# References

Ball, J.W. and Nordstrom, D.K., 1991. WATEQ4F - User's manual with revised thermodynamic data base and test cases for calculating speciation of major, trace and redox elments in nature waters. U.S. Geologie Surv., Open - File Report 90-129, 185 pp.

Berner, R.A., 1980. Early diagenesis: A theoretical approach. Princton Univ. Press, Princton, NY, 241 pp.

Boudreau, B.P., 1997. Diagenetic models and their impletation: modelling transport and reactions in aquatic sediments. Springer Verlag, Berlin, Heidelberg, NY, 414 pp.

Canfield, D.E., Thamdrup, B. and Hansen, J.W., 1993a. The anaerobic degradation of organic matter in Danish coastal sediments: Iron reduction, manganese reduction, and sulfate reduction. Geochimica et Cosmochimica Acta, 57: 3867-3883.

Canfield, D.E., Jørgensen, B.B., Fossing, H., Glud, R., Gundersen, J., Ramsing, N.B., Thamdrup, B., Hansen, J.W., Nielsen, L.P. and Hall, P.O.J., 1993b. Pathways of organic carbon oxidation in three continental margin sediments. Marine Geology, 113: 27-40.

Flühler, H. and Jury, W.A., 1983. Estimating solute transport using nonlinear, rate dependent, two - site - adsorption models. Microfiche, Eidg. Anstalt forstlicher Versuchswesen, 245, Zürich, 48 pp.

Froelich, P.N., Klinkhammer, G.P., Bender, M.L., Luetke, N.A., Heath, G.R., Cullen, D., Dauphin, P., Hammond, D. and Hartman, B., 1979. Early oxidation of organic matter in pelagic sediments of the eastern equatorial Atlantic: suboxic diagenesis. Geochimica et Cosmochimica Acta, 43: 1075-1090.

Garrels, R.M., 1960. Mineral Equilibria at Low Temperature and Pressure. Harper, NY, 254 pp.

Garrels, R.M. and Christ, C.L., 1965. Solutions, Minerals and Equilibria. Harper & Row, NY, Evanston, London. Weatherhill, Tokyo, 450 pp.

Glasby, G.P. and Schulz, H.D., 1999. $E_H$, pH diagrams for Mn, Fe, Co, Ni, Cu and As under seawater conditions: Application of two new types of $E_H$, pH diagrams to the study of specific problems in marine geochemistry. Aquatic Geochemistry, 5: 227-248.

Glud, R.N., Gundersen, J.K., Jørgensen, B.B., Revsbech, N.P. and Schulz, H.D., 1994. Diffusive and total oxygen uptake of deep-sea sediments in the eastern South Atlantic Ocean: in situ and laboratory measurements. Deep-Sea Research, 41: 1767-1788.

Hamer, K. and Sieger, R., 1994. Anwendung des Modells CoTAM zur Simulation von Stofftransport und geochemischen Reaktionen. Verlag Ernst & Sohn, Berlin, 186 pp.

Hensen, C., Landenberger, H., Zabel, M., Gundersen, J.K., Glud, R.N. and Schulz, H.D., 1997. Simulation of early diagenetic processes in continental slope sediments in Southwest Africa: The computer model CoTAM tested. Marine Geology, 144: 191-210.

Kharaka, Y.K., Gunter, W.D., Aggarwal, P.K., Perkins, E.H. and DeBraal, J.D., 1988. SOLMINEQ88: a computer program for geochemical Modeling of water - rock - interactions. US Geological Survey, Water - Recources Investigations Report, 88-4227, 207 pp.

Kinzelbach, W., 1986. Groundwater Modeling - An Introduction with Sample Programs in BASIC. Elsevier, Amsterdam, Oxford, NY, Tokyo: 333 pp.

Landenberger, H., Hensen, C., Zabel, M. and Schulz, H.D., 1997. Softwareentwicklung zur computergestützten Simulation frühdiagenetischer Prozesse in marinen Sedimenten. Zeitschrift der deutschen geologischen Gesellschaft, 148: 447-455.

Landenberger, H., 1998. CoTReM. ein Multi - Komponenten Transport- und Reaktions - Modell., Berichte, Fachbereich Geowissenschaften, Universität, Bremen, No 110, 142 pp.

Niewöhner, C., Hensen, C., Kasten, S., Zabel, M. and Schulz, H.D., 1998. Deep sulfate reduction completely mediated by anaerobic methane oxidation in sediments of the upwelling area off Namibia. Geochimica et Cosmochimica Acta, 62(3): 455-464.

Nordstrom, D.K., Plummer, L.N., Wigley, T.M.L., Woley, T.J., Ball, J.W., Jenne, E.A., Basset, R.L., Crerar, D.A., Florence, T.M., Fritz, B., Hoffman, M., Holdren, G.R.(jr.), Lafon, G.M., Mattigod, S.V., McDuff, R.E., Morel, F., Reddy, M.M., Sposito, G. and Thraikill, J., 1979. A comparision of computerized chemical models for equilibrium calculations in aqueous systems: in Chemical Modeling in aqueous systems, speciation, sorption, solubility, and kinetics. Chemical Modeling in aqueous systems speciation, sorption, solubility, and kinetics. In: Jenne, E.A. (ed) Series, American Chemical Society, 93: 857-892.

Parkhurst, D.L., Thorstensen, D.C. and Plummer, L.N., 1980. PHREEQE - a computer program for geochemical calculations. US Geological Survey, Water - Recources Investigations Report., 80-96, 219 pp.

Parkhurst, D.L., 1995. User's guide to PHREEQC: a computer model for speciation, reaction - path, advective - transport, and inverse geochemical calculation. US Geological Survey, Water - Resources Investigations Report, 95-4227, 143 pp.

Plummer, L.N., Jones, B.F. and Truesdell, A.H., 1976. WATEQF - a fortran 4 version of WATEQ, a computer program fpr calculating chemical equilibrium of natural waters. US Geological Survey , Water - Recources Investigations Report, 76-13, 614 pp.

Schulz, H.D. and Reardon, E.J., 1983. A combined mixing cell/analytical model to discribe two - dimensional reactiv solute transport for unidirectional groundwater flow. Water Recources Research, 19: 493-502.

Schulz, H.D., Dahmke, A., Schinzel, U., Wallmann, K. and Zabel, M., 1994. Early diagenetic processes, fluxes and reaction rates in sediments of the South Atlantic. Geochimica et Cosmochimica Acta, 58(9): 2041-2060.

Truesdell, A.H. and Jones, B.F., 1974. WATEQ - a computer program for calculating chemical equilibria on natural waters. Jour. Research US Geological Survey, Washington DC, 2: 233-248.

Van Cappellen, P. and Yifeng Wang, Y., 1995. STEADYSED1: A Steady - State Reaction - Transport Model for C, N, S, O, Fe and Mn in Surface Sediments. Version 1.0 User's Manual, Georgia Inst. Technol. 40 pp.

Van Cappellen, P. and Wang, Y., 1996. Cycling of iron and manganese in surface sediments: a general theory for the coupled transport and reaction of carbon, oxygen, nitrogen, sulfur, iron, and manganese. American Journal of Science, 296: 197-243.

Van Cappellen, P. and Gaillard, J.-F., 1996. Biogeochemical Dynamics in Aquatic Sediments. In: Lichtner, P.C., Steefel, C.I. and Oelkers, E.H. (eds) Reactive Transport in Porous Media. Reviews in Mineralogy, The Mineralogical Society of America, Washington, DC, 34: 335-376.

Wang, Y. and Cappellen, P.v., 1996. A multicomponent reactive transport model of early diagenesis: Application to redox cycling in coastal marine sediments. Geochimica et Cosmochimica Acta, 60(16): 2993-3014.

Wolery, T.J., 1993. EQ 3/6, A Software Package for Geochemical Modeling of Aqueous Systems. Lawrence Livermore National Laboratory, California, 247 pp.

# Index